Lecture Notes in Mechanical Engineering

Lecture Notes in Mechanical Engineering (LNME) publishes the latest developments in Mechanical Engineering—quickly, informally and with high quality. Original research reported in proceedings and post-proceedings represents the core of LNME. Volumes published in LNME embrace all aspects, subfields and new challenges of mechanical engineering. Topics in the series include:

- Engineering Design
- Machinery and Machine Elements
- Mechanical Structures and Stress Analysis
- Automotive Engineering
- Engine Technology
- Aerospace Technology and Astronautics
- Nanotechnology and Microengineering
- Control, Robotics, Mechatronics
- MEMS
- Theoretical and Applied Mechanics
- Dynamical Systems, Control
- Fluid Mechanics
- Engineering Thermodynamics, Heat and Mass Transfer
- Manufacturing
- Precision Engineering, Instrumentation, Measurement
- Materials Engineering
- Tribology and Surface Technology

To submit a proposal or request further information, please contact the Springer Editor of your location:

China: Dr. Mengchu Huang at mengchu.huang@springer.com
India: Priya Vyas at priya.vyas@springer.com
Rest of Asia, Australia, New Zealand: Swati Meherishi at swati.meherishi@springer.com
All other countries: Dr. Leontina Di Cecco at Leontina.dicecco@springer.com

To submit a proposal for a monograph, please check our Springer Tracts in Mechanical Engineering at http://www.springer.com/series/11693 or contact Leontina.dicecco@springer.com

Indexed by SCOPUS.

All books published in the series are submitted for consideration in Web of Science.

More information about this series at http://www.springer.com/series/11236

Mohammad Muzammil · Arunesh Chandra ·
Pavan Kumar Kankar · Harish Kumar
Editors

Recent Advances in Mechanical Engineering

Select Proceedings of ITME 2019

Springer

Editors
Mohammad Muzammil
Aligarh Muslim University
Aligarh, India

Pavan Kumar Kankar
Department of Mechanical Engineering
Indian Institute of Technology Indore
Indore, India

Arunesh Chandra
Department of Mechanical Engineering
KIET Group of Institutions
Ghaziabad, India

Harish Kumar
Department of Mechanical Engineering
National Institute of Technology Delhi
New Delhi, India

ISSN 2195-4356 ISSN 2195-4364 (electronic)
Lecture Notes in Mechanical Engineering
ISBN 978-981-15-8703-0 ISBN 978-981-15-8704-7 (eBook)
https://doi.org/10.1007/978-981-15-8704-7

This Springer imprint is published by the registered company Springer Nature Singapore Pte Ltd.
The registered company address is: 152 Beach Road, #21-01/04 Gateway East, Singapore 189721, Singapore

Preface

The First International Conference on Innovative Technologies in Mechanical Engineering (ITME-2019) has been the first conference of its series organized by Department of Mechanical Engineering of KIET Group of Institutions, Delhi-NCR, Ghaziabad, India. It was considered as a forum to bring together scientists, academicians, research scholars, graduate students, and mechanical engineers, presenting new science, technology, and engineering ideas and achievements.

The motive of this conference is of great importance. The time and location of the conference were quite excellent. Month of October in Ghaziabad is very pleasant with advantage of Delhi as the capital of the country with excellent connectivity by road, train, and air. The conference attracted many participants working in various fields of engineering: design, mechanics, materials, etc. The success of the conference inspired the organizers to turn the conference into series of event.

More than 275 papers were presented at the conference ITME-2019. They covered topics ranging from the mechanics of machines, material engineering, structural strength, and tribological behavior to transport technologies, machinery quality, and innovations, in addition to dynamics of machines, walking mechanisms, ergonomics, and computational methods. However, for this book, only 90 peer-reviewed papers, authored by research groups representing various universities and institutes, were selected for inclusion.

Furthermore, I thank the staff and management of the Institute for their cooperation and support, and especially Dr. Amik Garg, Hon. Director, KIET Group of Institutions, Ghaziabad, and all members of the Program Committee and the Organizing Committee for their work in preparing and organizing the conference. Last but not least, I thank Springer for its professional assistance and particularly Dr. Akash Chakraborty, Priya Vyas, Lokeshwaran Manickavasagam and Daniel Joseph Glarance who supported this publication.

Aligarh, India
Ghaziabad, India
Indore, India
New Delhi, India

Mohammad Muzammil
Arunesh Chandra
Pavan Kumar Kankar
Harish Kumar

Contents

About the Editors

Dr. Mohammad Muzammil is a Senior Professor in the Department of Mechanical Engineering of the Z H College of Engineering and Technology at Aligarh Muslim University Aligarh, India. He is currently in charge of industrial and production engineering and heads the ergonomics research division. He works in the field of ergonomics and human factors engineering. Much of his work has focused on improving the design and performance of industrial systems. His research interests include green ergonomics, active noise and vibration control, tool design and productivity improvement. He is also working on various projects in the field of biomedical engineering. Some of them are Parkinson's disease mitigation device, exoskeleton systems for stroke inflicted and old aged patients, bionic kidneys, etc. He has supervised more than 30 doctoral and postgraduate theses and has completed several research projects. He has published more than 100 research articles in leading journals and international and national conferences. Among the journals in which his articles have been published are Quality Engineering, International Review of Industrial Ergonomics, Theoretical Issues in Ergonomics Science, Occupational Ergonomics, Journal of Occupational Safety and Ergonomics, Work: A Journal of Prevention, Assessment and Rehabilitation and Rehabilitation and Journal of Low Frequency Noise, Vibration and Active Control, etc. He is a co-author of three books. He has provided consultancy to various industries. He is also a member of the editorial board of prestigious international journals. He has been a coordinator and member of the advisory committee/organizer of various international/national workshops and conferences. He has given numerous lectures at various renowned universities.

Dr. Arunesh Chandra is currently working as a Professor in the Department of Mechanical Engineering at KIET Group of Institutions, Ghaziabad, Uttar Pradesh, India. He obtained his B.Tech. degree in Mechanical Engineering from Govt. College, Satna, Madhya Pradesh, M.Tech. degree from National Institute of Foundry and Forge Technology, Ranchi, Jharkhand and Ph.D. from the Department of Mechanical Engineering of National Institute of Technology, Kurukshetra, Haryana. He has more than 16 years of teaching/research experience. He has more than 40 research papers in various international and national journals, conferences and a patent granted in the design of helmet by Government of India in the year 2019. His current research areas include ergonomics, quality management, virtual reality. He has reviewed various research papers and is an editorial board member of Journal of Mechanical and Mechanics Engineering, International Journal of Engineering Research in Mechanical and Civil Engineering and MAT Journals. He received a Certificate of Appreciation & Award from KIET Group of Institutions for superior performance in academics. He has also authored two books for undergraduate students Laboratory Manual Engineering Mechanics Lab and Elements of Mechanical Engineering.

Dr. Pavan Kumar Kankar is currently working as an Associate Professor in Discipline of Mechanical Engineering, Indian Institute of Technology Indore. He is having more than 14 years of teaching and research experience. He had obtained his Ph.D. from Indian Institute of Technology Roorkee, India. His research interests include vibration, design, condition monitoring of mechanical components, non-linear dynamics, soft computing, etc. He has published more than 100 papers in refereed journals and conferences. He is a reviewer of many journals including Journal of Sound and Vibration, Applied Acoustics, European Journal of Mechanics - A/Solids, Journal of Vibration and Control, IEEE Transactions on Industrial Electronics, Journal of Mechanical Science and Technology. He has been awarded certificate of outstanding contribution in reviewing for Applied Acoustics Journal in June 2017 and awarded outstanding reviewer status for Journal of Sound and Vibration in November 2015. He also served as a guest editor of the special issue of various journals including Advances in Mechanical Engineering, International Journal of Rotating Machinery and Shock and Vibration Journal. He is a member of professional bodies like the American Society of Mechanical Engineers, Society for Reliability and Safety (SRESA), Tribology Society of India and International Institute of Acoustics and Vibration (IIAV).

Dr. Harish Kumar is currently working as Assistant Professor in National Institute of Technology Delhi. He has more than 15 years of research and academic experience and served as Scientist at different grades in CSIR - National Physical Laboratory, India. He has been an active researcher in the areas of mechanical measurement and metrology. He has worked as a guest researcher at National Institute of Standards and Technology, USA in 2016. He has been instrumental in the ongoing redefinition of the kilogram in India. He has authored more than 70 publications in peer-reviewed journals and conferences. He is an active reviewer of many reputed journals related to measurement, metrology, and related areas.

A Review on Different Performance Enhancement Techniques for Solar Air Heaters

Ankur Dwivedi, Harshit Mishra, and Vishwas Nagrath

1 Introduction

The increasing energy demands due to exponential industrialization and population growth propelled the society toward renewable energy usage. Solar energy is considered as the highest potential renewable energy source with maximum contribution of about 65% to useful power generation [1]. Solar heaters are utilized for conversion of solar energy to heat energy of working medium and have been turned out to be a dominant area of interest of researchers for low-temperature engineering applications ranging from textile industries to space heating. Depending upon type of working fluids used, the solar heaters may be classified as either solar air heater (SAHs) (air as working fluid) or solar water heaters (SWHs) (water as working fluid), respectively. \query{Please check the edits made in the sentence 'Depending upon type of working fluids used...'.}SAHs have found relatively wide and advantageous domain over SWHs as air is freezing-free and non-corrosive in nature [2].

Although a major problem associated with SAH is their poor heat transfer characteristics due to low heat transfer coefficient values between working fluid (air) and heat exchanging surface, hence appropriate heat transfer enhancement methods are used for system performance improvement [3]. Different thermal enhancement methods are categorized as—active methods, passive methods and hybrid ones.

Active methods utilize external power to maintain heat transfer augmentation while passive methods deal with effective surface area variations through some relevant modifications to provide similar result. The passive methods are found more advantageous than active ones due to less complexity and cost associated. Also, it is not always possible to incorporate external power means to the system; hence, the use of active methods is practically limited. When active and passive methods are employed together in combination, such technique is kept under hybrid methods.

A. Dwivedi (✉) · H. Mishra · V. Nagrath
Allenhouse Institute of Technology, Kanpur 208008, India
e-mail: ank.nith@gmail.com

M. Muzammil et al. (eds.), *Recent Advances in Mechanical Engineering*, Lecture Notes in Mechanical Engineering, https://doi.org/10.1007/978-981-15-8704-7_1

There are numbers of literatures available regarding experimental and numerical works dealing with performance of solar air heater.

The present paper takes an opportunity to present an up-to-date collection of different literatures, as an exhaustive review regarding different heat transfer enhancement techniques employed.

2 Performance Parameters

A solar air heater arrangement comprises a glass cover, an absorber surface and a back plate, properly insulated to ensure minimum energy losses to the environment. Generally, three different performance characteristics are employed to measure the SAH's effectiveness, as discussed in the coming context.

2.1 *Heat Transfer Characteristics*

The heat transfer performance of solar air heater is calculated as the ratio of actual heat transfer between the absorber plate and air (Q_{act}) with respect to incident solar irradiation (G). Mathematically, it may be written as

$$\text{TP} = \frac{Q_{\text{act}}}{G} \tag{1}$$

2.2 *Hydraulic Characteristics*

The optimum power requirements for SAH are measured in terms of pressure drop (ΔP) within the channel as calculated by the Darcy-Wiesbach relation as below

$$\Delta P = \frac{4\rho f L V^2}{2D_h}. \tag{2}$$

2.3 *Thermo-Hydraulic Characteristics*

The performance parameter determining the cumulative thermal and hydraulic performances of roughened SAH was coined in work [4] as expressed in Eq. 3.

$$\mathrm{EEF} = \left(\frac{\mathrm{Nu}}{\mathrm{Nu}_s}\right)\left(\frac{f}{f_s}\right)^{-1/3} \tag{3}$$

3 Performance Enhancement Techniques

Various performance augmentation methodologies adopted have a prime objective to provide maximum heat transfer for minimum pressure loss penalty. Some relevant heat transfer augmentation techniques are discussed as below.

3.1 Extended Surfaces

Use of extended surfaces as heat transfer augmentation method comes under the category of passive methods that use effective surface variation to affect system performance without application of external power. The extended surfaces are generally employed in form of fins, turbulators, baffles, etc. A detailed review regarding applications of various cross-sectional fins affecting rectangular SAHs performance was presented by Sahu et al. [5]. Some relevant studies are given in Table 1.

3.2 Surface Roughness

The main idea to employ surface roughness for improving heat transfer characteristics is to disrupt the laminar sub layer over absorber plate; and hence, creating turbulence associated with some pressure loss. The selection of appropriate surface roughness is very important aspect to consider with an objective of optimum system performance. The roughness is generated using wire ribs, dimples and protrusions etc. The collective summary of some relevant literatures is presented in Table 2.

3.3 Porous Bed

Based on the position of the absorber surface, Gupta and Garg [16] classified the solar air heaters under the categories of porous and non-porous solar air heaters. The performance of porous SAH is reported better of two due to high heat transfer surface area density, minimum radiation losses and lower pressure drop values. The porous bed solar collectors are classified as: honeycomb-based [17]; overlapped glass plate [18] and matrix type [19–21]. The matrix porosity and thickness are some crucial parameters subjected to precise selection [22]. Naphon [23] suggested the porous

Table 1 Summary of various literatures on heat transfer enhancement using extended surfaces

Authors	Analysis type	Extended surfaces	Empirical correlations
Garg et al. [6]	Analytical approach	Rectangular fins	$\mathrm{Nu} = 0.0293\mathrm{Re}^{0.98}$ $f = 0.059\mathrm{Re}^{-0.2}$
Bopche and Tandale [7]	Experimental work	Inverted U-shaped turbulators	$\mathrm{Nu} = 0.5429\mathrm{Re}^{0.7054}\left(\frac{P}{e}\right)^{-0.1592}\left(\frac{e}{D_h}\right)^{0.3619}$ $f = 1.2134\mathrm{Re}^{-0.2076}\left(\frac{P}{e}\right)^{-0.4259}\left(\frac{e}{D_h}\right)^{0.3285}$
Bahrehmand et al. [8]	Mathematical model	Rectangular; triangular	$\mathrm{Nu} = 0.015\mathrm{Re}^{0.8}$ $f = \frac{64}{\mathrm{Re}}$(laminar flow) $f = 0.079\mathrm{Re}^{-0.25}$ (turbulent flow)
Chamoli et al. [9]	Numerical investigation	Winglet vortex generators	$\mathrm{Nu} = 0.2365\mathrm{Re}^{0.6689}\left(1+\frac{c}{a}\right)^{0.1866}\left(\frac{\alpha}{60}\right)^{-0.3227}$ $\exp\left(-0.076\ln\left(1+\frac{c}{a}\right)^{2}\right)\exp\left(-0.9576\ln\left(\frac{\alpha}{60}\right)^{2}\right)$ $f = 0.661\mathrm{Re}^{-0.2164}\left(1+\frac{c}{a}\right)^{0.982}\left(\frac{\alpha}{60}\right)^{0.1128}$ $\exp\left(-1.6592\ln\left(\frac{\alpha}{60}\right)^{2}\right)$

Table 2 Summary of various literatures on heat transfer enhancement using surface roughness

Authors	Roughness used	Parametric details	Empirical correlations
Karmare and Tikekar [10]	Metal grit ribs	$0.035 \le e/D_h \le 0.044$ $12.5 \le P/e \le 36$ $1 \le l/s \le 1.72$ $4000 \le Re \le 17000$	$Nu = 2.4 \times 10^{-3} Re^{1.3} (e/D_h)^{0.42} (l/s)^{-0.146} (P/e)^{-0.27}$ $f = 15.55 Re^{-0.26} (e/D_h)^{0.91} (l/s)^{-0.27} (P/e)^{-0.51}$
Saini and Saini [11]	Arc-shaped wire	$0.0213 \le e/D_h \le 0.04220.3333 \le \varphi/90 \le 0.5666$ $2000 \le Re \le 17000$	$Nu = 0.001047 Re^{1.3186} (e/D_h)^{0.3772} (\varphi/90)^{-0.1198}$ $f = 0.14408 Re^{-0.17103} (e/D_h)^{0.1765} (\varphi/90)^{0.1185}$
Alam and Kim [12]	Conical protrusion ribs	$0.020 \le e/D_h \le 0.044$ $6 \le P/e \le 12$ $4000 \le Re \le 16000$	$Nu = 2.29 \times 10^{-4} Re^{0.984} \left(\frac{e}{D_h}\right)^{0.280} \left(\frac{P}{e}\right)^{4.085} \exp\left(-0.922\left(\ln\left(\frac{P}{e}\right)\right)^2\right)$ $f = 2.19 \times 10^{4} Re^{-0.352} \left(\frac{e}{D_h}\right)^{5.839} \left(\frac{P}{e}\right)^{1.860} \exp\left(-0.523\left(\ln\left(\frac{P}{e}\right)\right)^2\right) \exp\left(0.739\left(\ln\left(\frac{e}{D_h}\right)\right)^2\right)$
Rajneesh et al. [13]	Forward facing chamfered rectangular ribs	$0 \le e'/e \le 1$ $0.24 \le e/w \le 1.5$ $0.018 \le e/D_h \le 0.043$ $4000 \le Re \le 17000$ $P/e = 12$	$Nu = 0.066 Re^{0.858} (e'/e)^{0.022} (e/w)^{0.112} (e/D_h)^{0.241} \exp\left(0.039(\ln(e'/e))^2\right)$ $f = 11.79 Re^{-0.58} (e'/e)^{0.169} (e/w)^{0.131} (e/D_h)^{0.318} \exp\left(0.203(\ln(e'/e))^2\right)$
Soi et al. [14]	Protruded geometry (rectangular, circular, triangular, semicircular)	$0.32 \le \psi \le 1$ $0.083 \le w/W \le 0.166$ $0.022 \le e/D_h \le 0.046$ $4000 \le Re \le 20000$	$Nu = 1.01 \times 10^{-10} Re^{1.116} \left(\frac{w}{W}\right)^{0.544} \left(\frac{e}{D_h}\right)^{-10.54} \psi^{0.387} \exp\left(-7.99\left\{\log\left(\frac{e}{D_h}\right)\right\}^2\right)$ $f = 0.312 Re^{-0.211} (w/W)^{0.291} (e/D_h)^{0.160} \psi^{0.147}$

(continued)

Table 2 (continued)

Authors	Roughness used	Parametric details	Empirical correlations
Rajneesh et al. [15]	Square ribs	$0.013 \leq e/D_h \leq 0.05$ $5 \leq P/e \leq 13$ $3900 \leq \text{Re} \leq 17900$	$\text{Nu} = 0.00415\text{Re}^{0.95}(P/e)^{1.217}(e/D_h)^{0.243}\exp\left(-0.265(\ln(P/e))^2\right)$ $f = 6.436\text{Re}^{-0.4}(P/e)^{-1.033}(e/D_h)^{0.278}\exp\left(-0.224(\ln(P/e))^2\right)$

Table 3 Summary of different literatures on heat transfer enhancement using porous bed

Authors	Absorber type	Empirical correlations	Key results
Varshney and Saini [24]	Wire mesh screen matrix	$J_n = 0.647\left[\left(\frac{1}{nP}\right)\left(\frac{p_t}{d_w}\right)\right]^{2.104} \mathrm{Re}^{-0.55}$ $f = 2.484\left[\left(\frac{1}{nP}\right)\left(\frac{p_t}{d_w}\right)\right]^{0.699} \mathrm{Re}^{-0.44}$	The system performance was reported to be very strong function of mesh geometry
Thakur et al. [25]	Low porosity packed bed	$J_n = 0.4\left[\left(\frac{1}{nP}\right)^{0.50}\left(\frac{p_t}{d_w}\right)^{0.25}\right]^{1.4} \mathrm{Re}^{-0.61}$ $f = 3.0\left[\left(\frac{1}{nP}\right)^{0.50}\left(\frac{p_t}{d_w}\right)^{0.25}\right]^{0.90} \mathrm{Re}^{-0.41}$	The performance index was found in between 1.14 and 1.58 confirming enhanced performance
Prasad et al. [26]	Wire screen matrix	$J_n = 0.2563\left(\frac{1}{nP}\right)^{0.609}\left(\frac{p_t}{d_w}\right)^{0.7954} \mathrm{Re}^{-0.63}$ $f = 3.5722\left(\frac{1}{nP}\right)^{1.0431}\left(\frac{p_t}{d_w}\right)^{1.1507} \mathrm{Re}^{-0.43}$	Maximum enhancement of 89.5% can be obtained
Lalji et al. [27]	Hexagonal-shaped matrix	$J_n = 0.1765\left[\left(\frac{1}{nP}\right)^{0.6156}\left(\frac{p_t}{d_w}\right)^{0.1229}\right]^{0.6651} \mathrm{Re}^{-0.4767}$ $f = 2.0291\left[\left(\frac{1}{nP}\right)^{0.3521}\left(\frac{p_t}{d_w}\right)^{0.0686}\right]^{0.6759} \mathrm{Re}^{-0.3897}$	The packed bed SAH with lower porosity was reported to perform better than highly porous SAH

mesh to be about 26% more efficient than non-porous mesh when analyzed for double-pass counter flow collectors. Table 3 summarizes various relevant literatures dealing with utilization of porous materials in different heat transfer applications.

3.4 Selective Coatings

The absorber plates are sometimes coated selectively with substances having maximum absorptivity (to absorb maximum incident energy) and minimum emissivity (to ensure minimum radiation losses). In this way, application of selective coating reduces radiation losses, reducing overall heat transfer coefficient vice-versa [28], and hence improves thermal efficiency. An experimental work to analyze the effects of nickel-plated and chromium-colored sheet as selective coating was performed by Togrul and Pehlivan [29] at Elazing, Turkey. The maximum increase in efficiencies for three different cases considered (selective coating only, Aluminum wire mesh packing and Combined packing and selective coating) was reported about 6%, 60% and 68%, respectively. El-Sebaii and Al-Snani [30] suggested Ni–Sn as best selective coating material with maximum efficiency 46% through their mathematical work considering the climatic conditions of Jeddah in Saudi Arabia.

4 Concluding Remarks

Different performance enhancement techniques utilized for solar air heaters are reviewed as investigated experimentally and analytically under various studies. Based on the studies reviewed, the following conclusions may be drawn.

The fin height, fin density and orientation are found to be important design factors in case of extended surface applications, while roughness geometry, height and pitch are reported important parameters in studies dealing with artificial roughness. Performance of porous bed depends strongly on its porosity. Application of appropriate coatings is reported as an effective way to improve system performance.

References

1. Indian Energy Statistics Report, vol 25, pp. 4 (2018)
2. Duffie, J.A., Beckman, W.A.: Solar Engineering Thermal Processes, 2nd edn. John Wiley, New York (1991)
3. Sukhatme, S.P., Nayak, J.P.: Solar Energy, 3rd edn. Tata McGraw Hill, New Delhi (2011)
4. Lewis, M.J.: Optimizing the thermohydraulic performance of rough surfaces. Int. J. Heat Mass Transf. **18**, 1943–1948 (1975)
5. Sahu, M.K., Sharma, M., Matheswaran, M.M., Maitra, K.: On the use of various configurations of fins to enhance the performance in rectangular duct of solar air heaters—a review. J. Solar Energy Eng. (2019)
6. Garg, H.P., Jha, R., Choudhury, C., Datta, G.: Theoretical analysis on a new finned type solar air heater. Energy **16**, 1231–1238 (1991)
7. Bopche, S.B., Tandale, M.S.: Experimental investigations on heat transfer and frictional characteristics of a turbulator roughened solar air heater duct. Int. J. Heat Mass Transf. **52**(11–12), 2834–2848 (2009)
8. Bahrehmand, D., Ameri, M., Gholampour, M.: Energy and exergy analysis of different solar air collector systems with forced convection. Renew. Energy **83**, 1119–1130 (2015)
9. Chamoli, S., Lu, R., Xu, D., Yu, P.: Thermal performance improvement of a solar air heater fitted with winglet vortex generators. Sol. Energy **159**, 966–983 (2018)
10. Karmare, S.V., Tikekar, A.N.: Heat transfer and friction factor correlation for artificially roughened duct with metal grit ribs. Int. J. Heat Mass Transf. **50**, 4342–4351 (2007)
11. Saini, S.K., Saini, R.P.: Development of correlations for Nusselt number and friction factor for solar air heater with roughened duct having arc-shaped wire as artificial roughness. Sol. Energy **82**(12), 1118–1130 (2008)
12. Alam, T.: Kim, Man-Hoe: Heat transfer enhancement in solar air heater duct with conical protrusion roughness ribs. Appl. Therm. Eng. **26**, 458–469 (2017)

13. Investigation of heat transfer augmentation and friction factor in triangular duct solar air heater due to forward facing chamfered rectangular ribs: A CFD based analysis. Renew. Energy **115**, 824–835 (2018)
14. Heat transfer and friction characteristics of solar air heater duct having protruded roughness geometry on absorber plate. Exp. Heat Transf. **31**(6), 571–585 (2018)
15. Kumar, R., Kumar, A., Goel, V.: Performance improvement and development of correlation for friction factor and heat transfer using computational fluid dynamics for ribbed triangular duct solar air heater. Renew. Energy **131**, 788–799 (2019)
16. Gupta, C.L., Garg, H.P.: Performance studies on solar air heaters. Sol. Energy **11**(1), 25–31 (1967)
17. Lalude, O., Buchberg, H.: Design and application of honeycomb porous-bed solar-air heaters. Sol. Energy **13**(2), 223–242 (1971)
18. Selçuk, K.: Thermal and economic analysis of the overlapped-glass plate solar-air heater. Sol. Energy **13**(2), 165–191 (1971)
19. Chiou, J.P., El-Wakil, M.M., Duffie, J.A.: A slit-and -expanded aluminum-foil matrix solar collector. Sol. Energy **9**, 73–80 (1965)
20. Sharma, S.P., Saini, J.S., Varma, H.K.: Thermal performance of packed bed solar air heaters. Sol. Energy **47**, 59–67 (1991)
21. Prasad, R.K., Saini, J.S.: Thermal performance of matrix bed solar air heaters. In: Proceedings of International Conference on Recent Advances in Solar Energy Conversion Systems, pp. 55–60. MANIT, Bhopal, India (2002)
22. Saini, P., Patil, D.V., Powar, S.: Review on integration of solar air heaters with thermal energy storage. In: Tyagi, H., Agarwal, A., Chakraborty, P., Powar, S. (eds.) Applications of Solar Energy, EES, Springer, Singapore (2018)
23. Naphon, P.: Effect of porous media on the performance of the double-pass flat plate solar air heater. Int. Commun. Heat Mass Transf. **32**, 140–150 (2005)
24. Varshney, L., Saini, J.S.: Heat transfer and friction factor correlations for rectangular solar air heater duct packed with wire mesh screen matrices. Sol. Energy **62**(4), 255–262 (1998)
25. Thakur, N.S., Saini, J.S., Solanki, S.C.: Heat transfer and friction factor correlations for packed bed solar air heater for a low porosity system. Sol. Energy **74**, 319–329 (2003)
26. Prasad, S.B., Saini, J.S., Singh, K.M.: Investigation of heat transfer and friction characteristics of packed bed solar air heater using wire mesh as packing material. Sol. Energy **83**, 773–783 (2009)
27. Lalji, M.K., Sarviya, R.M., Bhagoria, J.L.: Exergy evaluation of packed bed solar air heater. Renew. Sustain. Energy Rev. **16**, 6262–6267 (2012)
28. Duffie, J.A., Beckman, W.A.: Solar Engineering of Thermal Processes, 3rd edn. John Wiley & Sons, New Jersey (2006)
29. Togrul, I.T., Pehlivan, D.: Effect of packing in the airflow passage on the performance of a solar air-heater with conical concentrator. Appl. Therm. Eng. **25**, 1349–1362 (2005)
30. El-Sebaii, A.A., Al-Snani, H.: Effect of selective coating on thermal performance of flat plate solar air heaters. Energy **35**, 1820–1828 (2010)

Worker Fatigue Model and Its Ergonomic Application for a Dishwashing Task

P. Lasin and Vinay V. Panicker

1 Introduction

Fatigue is a workplace hazard that is related to the health and safety of the worker. Fatigue refers to the problem that appears from poorly designed shift patterns or excessive working time. It is generally considered to be a decrease in physical and/or mental performance that results from lengthy exertion, sleep loss and/or disruption of the internal clock. It is also related to workload, and in that, workers are more quickly fatigued if their work is machine-paced, monotonous, or complicated. Fatigue can cause a reduction in productivity, absent-mindedness, less coordination, and low concentration. It is often the root cause of major accidents.

This paper at hand contributes to a study on worker fatigue evaluation for a dishwashing task. In particular, it investigates the dishwashing task of female workers where the plates are washed after food in one of the hostel mess in an educational institute in the southern part of India. In the scenario considered, the worker will wash a specific number of plates per day.

The objective of this study is to evaluate the worker fatigue for a cycle of work by using digital human modeling software and compare the actual rest time and recommended rest time.

A questionnaire discomfort survey has been conducted for female dishwashing workers as an initial study. In the questionnaire, ten body parts are analyzed which are neck, elbows, forearms, wrist, thighs, knees, hips, lower back, upper back, and shoulders. From the survey, it has been found that 77.78% of workers have shoulder pain, and 66.78% of workers have back pain. Among the total health problems, 20% is an account for shoulder pain and 17% is an account for back pain. In the long run, these health problems will lead to severe musculoskeletal disorders (MSDs). These

P. Lasin · V. V. Panicker (✉)
Industrial Engineering Laboratory, Department of Mechanical Engineering, National Institute of Technology Calicut, Kozhikode, Kerala 673601, India
e-mail: vinay@nitc.ac.in

M. Muzammil et al. (eds.), *Recent Advances in Mechanical Engineering*, Lecture Notes in Mechanical Engineering, https://doi.org/10.1007/978-981-15-8704-7_2

health problems may arise due to excess fatigue generated during work, improper rest breaks, and unscientific design of the workplace or working posture. In this study, the dishwashing process is broken down into sub-activities and the analysis of work activities and cumulative loads on body joints is performed on the biomechanical digital human model on the basis of posture captured by using Xbox Kinect and hand forces by using Siemens Jack. The outcomes of this model are used to estimate different fatigue parameters for the dishwashing activity. The results show that the recommended rest time is very much higher than the actual rest time which is taking now.

The remaining of this paper is organized as follows. The related literatures are reviewed in the next section. Detail study of the problem, development of a digital human model for evaluating the core shoulder loads associated with the dishwashing task for female workers, and detailed explanation of the solution methodology adopted are discussed in Sect. 3. Section 4 describes the results of biomechanical analysis from the digital human model and numerical results of fatigue generated for the unit work cycle. Section 5 concludes the mini-project report and discusses directions for future work.

2 Review of Literature

A comprehensive review of the literature is provided in this section on different areas of fatigue evaluation and rest allowance. The literatures give a clear understanding of fatigue generation for workers under different conditions. Many previous epidemiological studies have conducted in the area of worker fatigue evaluation. So the relevance of assessment of worker fatigue is vital in the workplace or any industry.

2.1 Worker Fatigue Evaluation Model

Worker fatigue evaluation for the packaging process with an integrated cost concept is investigated by Glock et al. [1]. The objective of the paper is to minimize the overall packaging cost. The overall cost involves the cost of workers and the cost of packaging material like a box. There are three types of boxes: small, medium, and large.

The packaging process involves various steps such as place an item in box, tape and label the box, place the box into the trolley, and carry the box to the storage area.. The exponential function will calculate the fatigue generated by the worker after completing a specified time ta. A fatigue growth parameter is developed by combining the equations for maximum endurance time (MET) and maximum voluntary contraction (MVC). The results show that the smaller box will have low total cost and it is less sensitive to changes in Fmax for lower-wage cost. But the larger box is suitable for high-wage costs. The expression for % maximum voluntary contraction is taken from Norman et al. [2]. A fatigue recovery function developed which will give the residual fatigue retained in the body of the worker after taking some rest for time *t*.

Maximum endurance time (MET) is the maximum time in which the worker can able to work in the same posture. The expression for maximum endurance time is taken from the paper done by Rose et al. [3] and El-Ahrache and Imbeau [4]. It is an exponential function that will give the MET in seconds.

2.2 Rest Allowance for Static Muscular Work

The rest periods for static muscular works in the manufacturing industry and compares different rest allowance models have investigated. Rest allowances are the scheduled break times to be given to the workers for recovery from fatigue. Different rest allowance models are compared with the estimation of rest allowance using EN1005-3 standards. The rest allowance of Bystro and Fransson-Hall [5] is used in this paper.

A comprehensive review of literature on different topics related to the evaluation of worker fatigue, rest allowance helped in making a clear understanding of the topic and helped to identify the direction of future research. Generally, fatigue is expressed as the percentage of maximum permissible fatigue.

3 Problem Description and Solution Methodology

3.1 Problem Description

Fatigue is a more prevalent and serious workplace issue with respect to the productivity of the workforce. With a focus on an employee's health and well-being, it is essential for any employer to assess the fatigue at the workplace. In this paper, the assessment of the fatigue of dishwashing workers at a mess is proposed. The observational study will help to develop a digital human model to analyze the forces acting on the shoulder joints of the worker.

The dishwashing task from a hostel mess is carefully observed and the steps involved in a dishwashing activity have been summarized as follows:

Step 1 The plates to be washed are kept on left side of the worker. In this step, worker will reach to take a plate.
Step 2 The worker will grasp the plate.
Step 3 Take the plate and position it in front of the sink.
Step 4 Apply scrubber and soap to plate.
Step 5 Dip the scrubbed plate in the water which is stored in tub.
Step 6 Stand erect from bending posture.

Step 7 Cleaned plates are to be kept on the right side of the sink. In this step, the worker will reach to the place where the plates are to be placed.

Step 8 Collect all the cleaned plates and keep it in a specified place.

Sum of all these process steps is defined as one job cycle. The postures of different steps in the dishwashing task are captured using Xbox Kinect and the digital human models are developed by using Siemens Jack software (Figs. 1, 2, 3, 4, 5, 6, 7 and 8). The 50th percentile of the female Asian Indian population is considered as the subject anthropometric dimensions in the digital human modeling software 'Siemens jack'. Twelve female subjects with an average age of 42 is the sample size and the period of the study is March–April 2019. The assumptions in this study are:

- Only one worker is considered at a time.
- The activity carried out by the worker is dishwashing only. All other jobs are neglected.
- One worker will wash 150 plates in one period.
- The worker can withstand fatigue to a maximum of 99%.

For numerical analysis and evaluation, the biomechanical parameters of the female population are focused. Because comparatively female workers have less biomechanical tolerance level than male workers. So, it will induce a high risk for female workers when the processes and workplace are designed according to the characteristics of male workers. So the work setup would not ergonomically fit for female workers. In such a situation, the safety and health of female workers would not be treated appropriately. Likewise, any body part of interest can be studied, and here, the main focus is on the impact of fatigue generated on the shoulders.

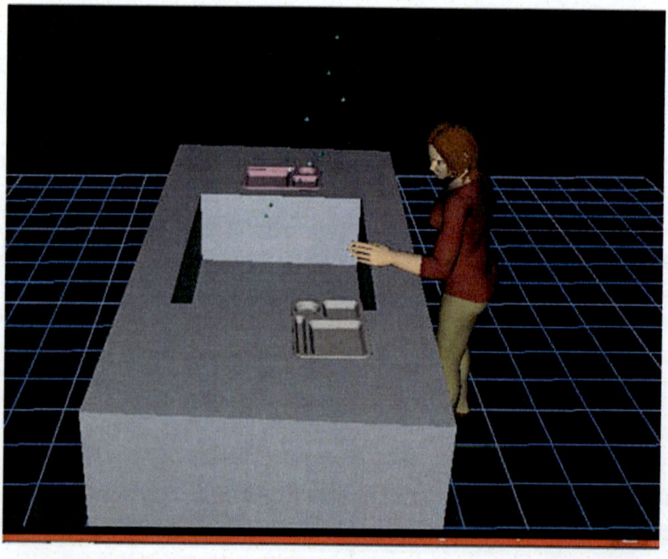

Fig.1 Digital human model of dishwashing task, Posture 1

Fig.2 Digital human model of dishwashing task, Posture 2

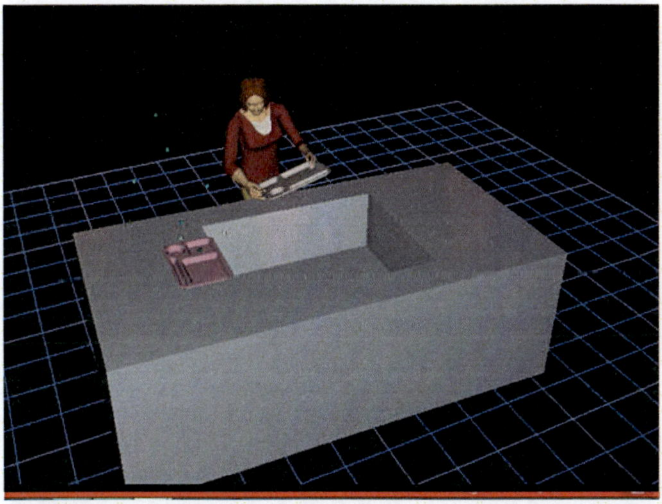

Fig.3 Digital human model of dishwashing task, Posture 3

3.2 Estimation of Expected Fatigue Parameters

Analysis of work activities and cumulative loads on body joints is performed on the biomechanical digital human model on the basis of posture and hand forces by using Siemens Jack. The outcomes of this model are used to estimate different fatigue parameters for the dishwashing activity. The dishwashing process is broken down into a variety of static actions like reach, grasp, scrubbing, washing, etc. The inputs

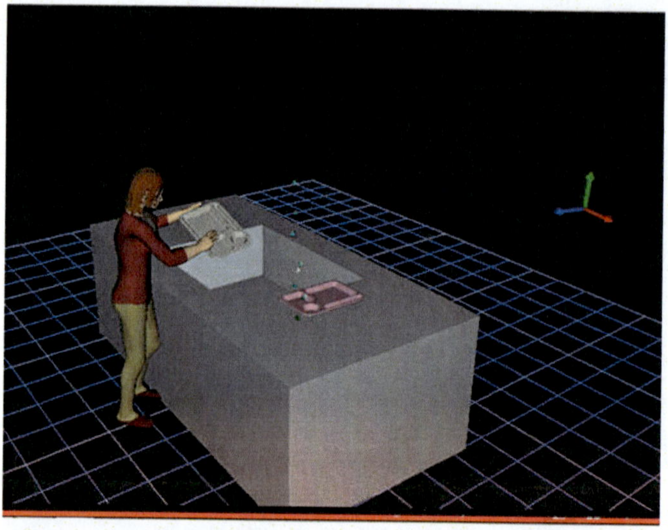

Fig.4 Digital human model of dishwashing task, Posture 4

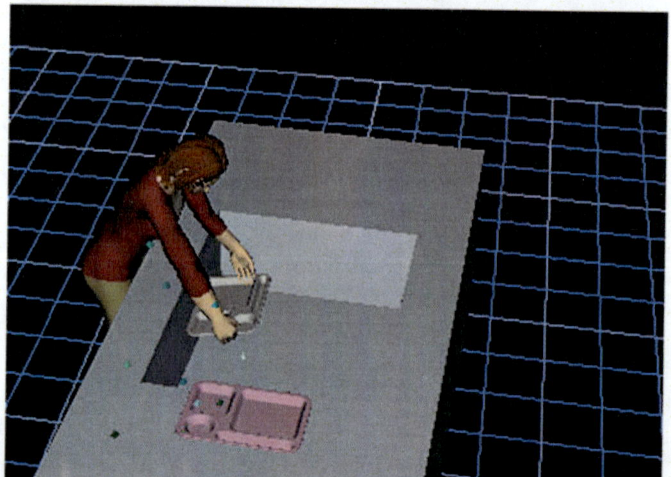

Fig.5 Digital human model of dishwashing task, Posture 5

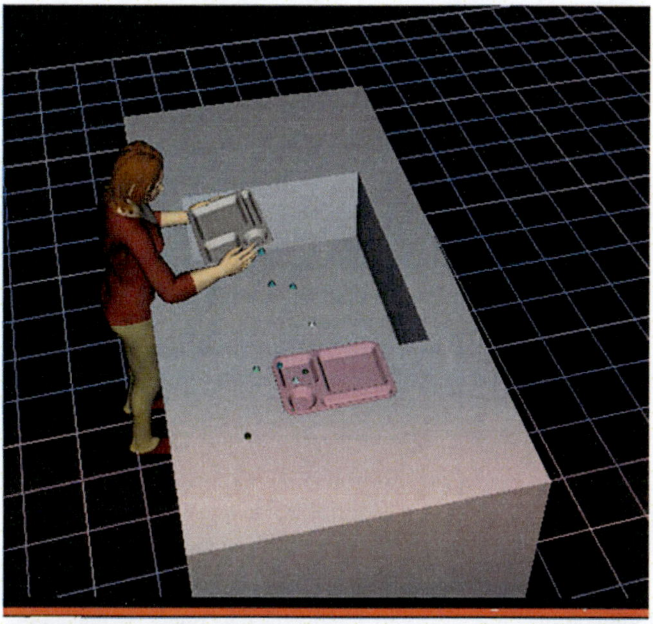

Fig.6 Digital human model of dishwashing task, Posture 6

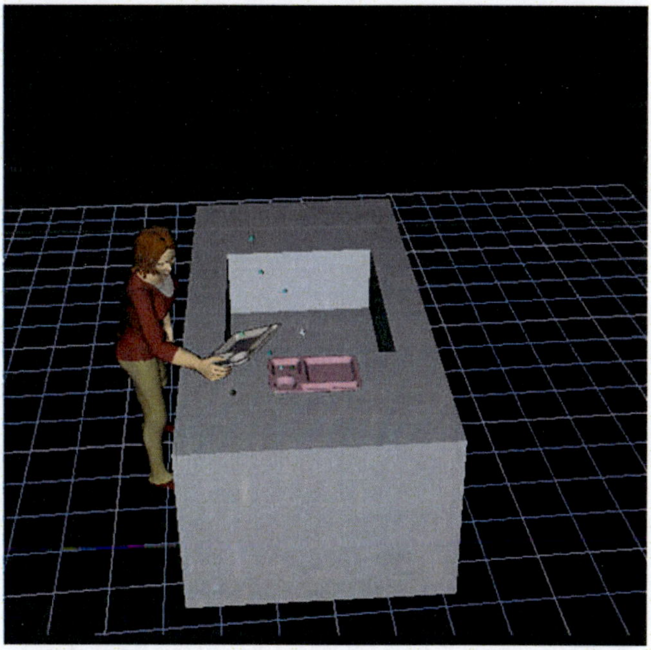

Fig.7 Digital human model of dishwashing task, Posture 7

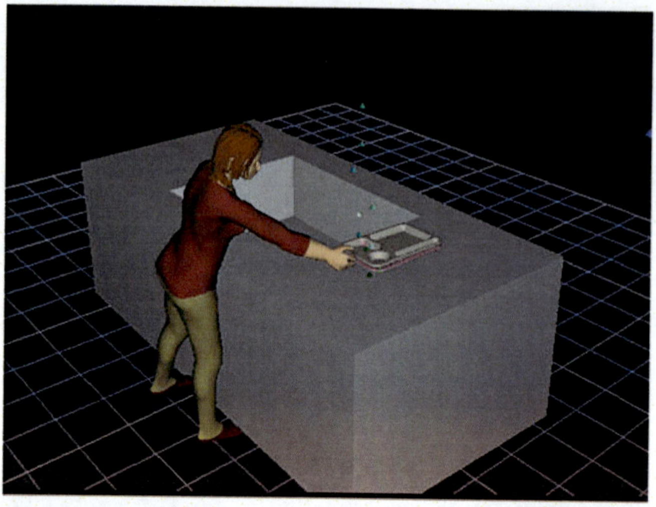

Fig.8 Digital human model of dishwashing task, Posture 8

chosen for undergoing the analysis are mainly the data related to the working posture, hand forces, and body dimension of the worker (stature, weight, etc.).

The analysis is performed based on the inputs by estimating peak loads on the shoulders, L4/L5, and other body joint and also help us to calculate the cumulative load on these parts. The results of the Siemens Jack software help us to estimate the fatigue parameters of dishwashing task performed by the worker.

3.3 *Fatigue Model*

The exponential model is used to track fatigue generation of workers [1].

$$\mathrm{FS(ta)} = 1 - \mathrm{e}^{-\theta \mathrm{ata}} \tag{1}$$

This function will calculate the accumulated fatigue. In (1), the generated fatigue at time ta is FS (ta). θ a is the fatigue growth parameter.θ a can also be called as the rate of fatigue accumulation, a low value of this parameter indicates slow fatigue accumulation, and a high value indicates fast fatigue accumulation. The concepts of maximum endurance time (MET), maximum voluntary contraction (MVC) are used to estimate the fatigue parameters. Earlier researches have shown that the muscle force capacity will reduce to a threshold value over time when a worker continuously doing an action.

Fatigue reaches its maximum level (Fmax ~100%) if this threshold value reached at the MET. After reaching this point, the worker will unable to retain the force required for performing action [1].

$$\text{METa} = 7.96\text{e}^{(-4.16*\%\text{MVCa})} \tag{2}$$

The maximum force exerted by a muscle when fully activated by a worker is defined as MVCa. Here, %MVCa for an action a is calculated as [1]:

$$\%\text{MVC} = \frac{\text{The moment of force for a given action a}}{\text{The population maximum strength in the required posture}} \tag{3}$$

The fatigue growth parameter of an action, a, is approximated by combining (1) and (2) due to the absence of empirical data [1]:

$$\theta = \frac{1}{7.96\text{e}^{(-4.16\%\ \text{MVCa})}} \ln\left(\frac{1}{1 - F\max}\right) \tag{4}$$

By accumulating all fatigue levels associated with the dishwashing actions, the total static fatigue (Fs(tw)) of the worker for the dishwashing process during the job cycle tw is estimated [1]:

$$\text{Fs (tw)} = \sum_{a} \text{Fs(ta)na} \tag{5}$$

Now, it is considered that the total fatigue is generated due to the static actions of the dishwashing task. But in reality, the worker is performing the jobs under dynamic conditions. The dynamic (real) fatigue generated by the dishwashing process is calculated by using the below-mentioned equation [1]:

$$\text{Ft (tw)} = \beta.\text{Fs (tw)} \tag{6}$$

where β is the dynamic coefficient of value 0.17 for manual handling operations.

Rest allowance (RA) indicates the time needed for sufficient rest followed by a static exertion. It is generally expressed as a percentage of working time. This rest allowance model takes only %MVC as an input parameter which is prior calculated for estimating the fatigue developed for the dishwashing task. The recommended rest allowance is for manual work is [4]:

$$\text{RA} = \left[\frac{\%\ \text{MVC}}{15} - 1\right]*100 \tag{7}$$

In this model, it is assumed that no fatigue when %MVC < 15%.

4 Results and Discussion

This work studied a manual dishwashing task. In particular, it studied the dishwashing task of female workers where the plates are washed after the food in one of the hostel mess in an educational institute. The results of the study are tabulated in Table 1.

Total static fatigue for one work cycle,

$\mathrm{Fs(tw)} = \sum_a \mathrm{Fs(ta)na} = 0.273.$

Total dynamic fatigue for one work cycle,

$$\begin{aligned}\mathrm{Ft(tw)} = \beta.\mathrm{FS\ (tw)} &= 0.17 \times 0.273 \\ &= 0.0465 \\ &= 4.65\%\end{aligned}$$

Actual rest time between cycles = 1.8 s
Total task time = 19 s
Cycle timer for one job cycle = 19 + 1.8 = 20.8 s
Total recommended rest allowance for one cycle = 0.426
Total recommended rest time for one cycle = 8.094 s
Recommended cycle time = 27.094 s

In one period, a worker will wash 150 plates.

- Actual total rest time taken for 150 cycles = 4.5 min
- Recommended rest time for 150 cycles = 20.23 min
- Actual time taken for washing 150 plates = 52 min
- Recommended time for washing 150 plates = 67.73 min.

This paper has studied the dishwashing task for a female worker. A biomechanical digital human model is developed in Siemens Jack software after capturing the different postures by using Xbox Kinect. The core shoulder moment is estimated from the force solver module in Siemens Jack software. The developed model was

Table 1 Results of the ergonomic assessment of different postures in dishwashing task

Posture	%MVC	MET (s)	θ	Ta (s)	na	Fatigue	mvc > 15	RA
1	4.65	393.5	0.01170	1	1	0.0116	0	0
2	7.77	346.6	0.01328	2	1	0.0262	0	0
3	15.37	262.6	0.01753	2	1	0.0344	0.0344	0.02466
4	16.78	237.6	0.01937	3	1	0.0564	0.0564	0.11866
5	4.43	397.2	0.01159	6	1	0.0671	0	0
6	9.48	321.9	0.01430	2	1	0.0282	0	0
7	9.43	322.6	0.01427	2	1	0.0281	0	0
8	19.25	214.4	0.02147	1	1	0.0212	0.0212	0.28333

analyzed in a digital human model simulation inspired by a case observed in daily life. The fatigue parameters and the recommended rest time are also estimated for the dishwashing task. From the numerical analysis, we obtained several managerial insights that can be summarized as follows:

- Total fatigue developed for one work cycle is 4.65%.
- Recommended rest time is very much higher than the actual rest time which is taking now.
- In the long run, this will cause serious health problems like musculoskeletal disorders
- To avoid this problem, give sufficient rest time for the workers.
- The rest break should be given intermittently.

The workers are unaware of their proper rest time required. In one period, each worker will wash 150 plates. In the actual condition, it will take 52 min for completing the task. But scientifically, by considering recommended rest times, the total time for completing the task requires 67.73 min. This indicates that the workers doing their works by ignoring the recommended rest time. Workers are unaware of the recommended rest times. This will lead to workers' health problems like shoulder pain and back pain. The fatigue generation can also be reduced by implementing an optimal workspace design.

5 Conclusion

This work deals with a common workplace issue which is often observed in manual handling workplaces. This work estimated the fatigue generation for the unit work cycle and the allowable rest times for the dishwashing task. Then, it compares the allotted rest time in the current scenario and the scientifically developed rest time for the dishwashing task. It is showing a significant difference. This means, the current rest time is very much less than the required rest time. This may result due to the lack of awareness about the recommended rest time. The improper rest time and work schedule will lead to serious musculoskeletal disorders (MSDs).

Although the results obtained from the developed model clearly illustrate the difference between actual rest times and recommended rest times. This paper has some limitations. One limiting factor is that only a preselected number of plates were considered. The activity would probably not be an issue in many cases. Only dishwashing task is considered for the analysis and all other tasks are neglected. But in actual situation, workers are doing various tasks in the kitchen.

References

1. Glock, C.H., Grosse, E.H., Kim, T., Neumann, W.P., Sobhani, A.: An integrated cost and worker fatigue evaluation model of a packaging process. Int. J. Prod. Econ. **207**, 107–124 (2019)
2. Norman, R., Wells, R., Neumann, W.P., Frank, J., Shannon, H., Kerr, M.: A comparison of peak vs cumulative physical work exposure risk factors for the reporting of low backpain in the automotive industry. Clin. BioMech. **13**(8), 561–573 (1998)
3. Rose, L., Ericsson, M., Glimskar, B., Nordgren, B., Ortengren, R.: Ergoindex:developmentof a model to determine pause needs after fatigue and pain reactions duringwork. In: Mattila, M. , Karwowski, W. (eds.) Computer Applications in Ergonomics, Occupational Safety and Health, pp. 461–468 (1992)
4. El-Ahrache, K., Imbeau, D.: Comparison of rest allowance models for static muscular work. Int. J. Ind. Ergon. **39**(1), 73–80 (2009)
5. Byström, S., Fransson-Hall, C.: Acceptability of intermittent handgrip contractions based on physiological response. Human Factors **36**(1), 158–171 (1994); Chaffin, D.B., Andersson, G.B.J., Martin, B.: Occupational Biomechanics, 3rd edn. Wiley, New York (1999)

Influence of Backpack Load on Energy Expenditure during Walking

A. Tamizhinian and Vinay V. Panicker

1 Introduction

Backpack has been commonly used by athletes, adults and students, since it is a better choice than carrying the load in hands, over the head, or using front packs. As the users feel that this may reduce physiological and biomechanical demand compared to other load carrying methods, they habitually carry backpacks. However, these backpack loads can cause potential problems like back pain and have adverse effects as mentioned by Gupta et al. [3]. The backpack weight can be considered in terms of body weight (BW) percentage. It is expected that the backpack load adds proportionately to the work needed to redirect the Center of Mass.

It is seen that most of the youngsters and students form a significant portion of backpack users. Therefore, it is essential to assess the influence of carrying a backpack on the energy expenditure of the students. This work focuses on the assessment of energy expenditure at different load carrying conditions at the back of adults under two different walking conditions.

The following are the objectives to be accomplished to assess the energy expended by the students under different backpack load carriages which are discussed in the later part of the article,

1. Define the different load conditions and terrain conditions taken for study.
2. Measure the energy expenditure of the students for each of the defined conditions.
3. Perform statistical comparison among the cases and analyze the results.

Rest of the article is organized as follows: Sect. 2 presents a review of relevant literature. Section 3 describes the problem considered. Section 4 explains the methodology adopted. Section 5 discusses the data analysis carried out on the collected data and their results. This article concludes in Sect. 6.

A. Tamizhinian · V. V. Panicker (✉)
Industrial Engineering Laboratory, Department of Mechanical Engineering, National Institute of Technology Calicut, Kozhikode, Kerala 673601, India
e-mail: vinay@nitc.ac.in

M. Muzammil et al. (eds.), *Recent Advances in Mechanical Engineering*, Lecture Notes in Mechanical Engineering, https://doi.org/10.1007/978-981-15-8704-7_3

2 Review of Related Literature

This section reports the review of the literature published related to energy expenditure helps in knowing the right choice of prediction equations for various activities, factors (gait parameters) influencing the energy expenditure, and whether to use trekking pole or not for field track walking while carrying backpack load.

2.1 Energy Expenditure

The total number of calories burnt by a person for each day is the total daily energy expenditure of the person. The energy consumption can change broadly from individual to individual and for an individual, it can fluctuate by activities involved. For example, an individual sitting discreetly around uses 10% lesser energy than individual who is dozing. Light force exercises, for example, work area work or moderate walking may use thrice the energy required for sitting discreetly. Moderate force exercises, for example, relaxed cycling or sexual activity may exhaust 3–6 times the energy required for sitting discreetly, while overwhelming power exercises, for example, running or doing pushups use in excess of multiple times the energy used for sitting unobtrusively.

2.2 Validation of Prediction Equations for Energy Expenditure and Motivation

The prediction equations for the energy expenditure of walking and running are compared with the measured values obtained from the actual experiment conducted by Hall et al. [1]. It is reported that there is no significant difference in energy expenditure between track and treadmill for both walking and running.

2.3 Effect of Trekking Pole on Energy Expenditure While Field Track Walking with Backpack

The impacts of the utilization of trekking shafts with or without burden carriage on energy expenditure and pace of apparent effort while walking on field tracks is researched by Brito et al. [2]. The outcomes propose that noteworthy contrasts in pulse and walking velocity are found in the correlations between walking with and without shafts, no distinctions were seen in pulse, pace of apparent effort and walking speed between walking with or without posts when carrying a knapsack load.

2.4 Evaluation of Modified Backpack Design for School Children

Carrying of knapsacks by younger students prompts postural and walk adjustments, prompting low back pain and musculoskeletal issue. The research carried out by Gupta et al. [3] has inspected step parameters, posture, and energy use differences between modified backpack (MBP) and the existing backpacks (EBPs) in school students. In this work, around 26 healthy participants were included utilizing three stacking states of 10, 20, and 30% of their body weight (BW) with the two different designs of backpacks. By maintaining the posture during the loaded condition around the plumb line/vertical axis similar to the neutral posture (unloaded condition), a modified design of the school backpack is found to reduce, higher bio-mechanical strains, gait variation, and metabolic cost seen in the existing designs of backpacks carried by students.

3 Problem Description

It is understood that the metabolic energy expenditure increases proportionately with the carrying load, and can easily become twice for a moderate load. While walking, there is a need for equilibrium and movement of arms and legs relative to the body and the load carriage places increasing requirements on equilibrium, as stated by increasing variability in step width. This variation contributes to energy expenditure. The amount of increasing energy expenditure with respect to the load carried by an individual gets modified in two different road scenarios. The scenarios are shown in Table 1.

Under both the scenarios, at the higher load carrying conditions, it is anticipated that students are suffering since they tend to expend more energy. The following work has been carried out to validate the energy expended by the students under above-mentioned scenarios at three different load conditions as follows:

1. No load condition (NL)
2. 5% of the subjects' BW (body weight) and
3. 10% of the subjects' BW.

Table 1 Scenarios considered for the study

Scenario 1		Scenario 2
Flat road—zero gradient		Steep road—comparatively more gradient
Sub-scenario 1	Sub-scenario 2	
Flat road	Climbing the steps	

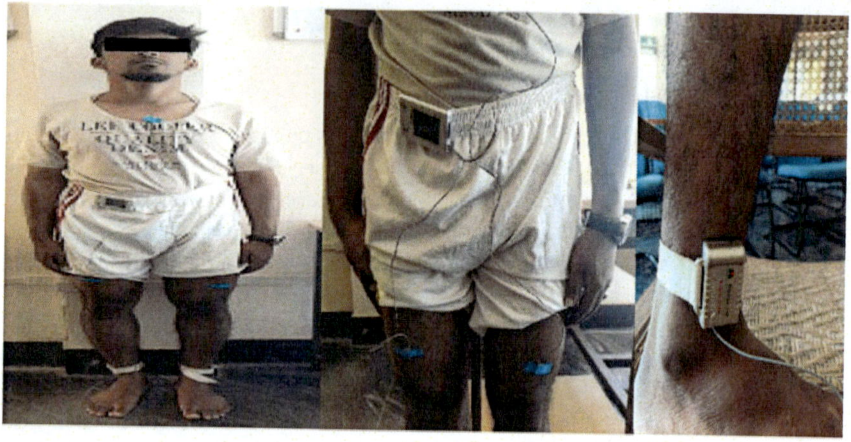

Fig. 1 Subject equipped with set of sensors, sub-recorders, and a main recorder

4 Methodology Adopted

The energy expended by the individuals under above two scenarios is measured using a device, IDEEA minisun. The obtained values are analyzed using ANOVA at a significance level of 5% and the results are interpreted to deduce certain conclusions.

4.1 IDEEA (Intelligent Device for Energy Expenditure)

MiniSun's new IDEEA®3 System is a breakthrough in recording and analysis of physical activities, gait, functional capacity, and energy expenditure using the latest technology. IDEEA3 hardware is comprised of IDEEA main recorder, sub-recorders, and accessories. The sensors are attached to the subjects' body at the proper positions as per the manual and the recorders are strapped to collect the data. The subject equipped with the IDEEA minisun device is shown in Fig. 1.

4.2 Data Collection

Data has been collected from eleven male subjects with the age group ranging between 22 and 25. The subjects are instructed to go for a walk at their self-selected pace with backpack at different load under two different road conditions. This work estimates the variation in energy expenditure during the walking under different load conditions and terrain conditions. The summary of the physical parameters of the samples required for the study is displayed in Table 2.

Table 2 Summary of physical parameters

Groups	Mean age (SD) (in years)	Mean weight (SD) (in kg)	Mean height (SD) (in m)	Mean BMI (SD) (in kg/m^2)
S	23.4 (1.04)	72.7 (9.8)	1.72 (0.07)	24.3 (2.65)

SD Standard deviation is given brackets

Table 3 Energy expenditure for two scenarios

Subject	Scenario 1—EE (kcal)			Scenario 2—EE (kcal)		
	NL	5% BW	10% BW	NL	5% BW	10% BW
1	106	125	109	92	99	99.2
2	90	93	99	94	97	97.9
3	107	119	118.3	99	100.7	100
4	92	95	99	93	96.8	98
5	88.2	96	100	90	94	98.5
6	89.3	98.4	100.3	93.6	93.4	95.9
7	98	101.7	109	98	98	104
8	82	89.7	91	86	88	89
9	90.4	95	102.5	93	99	98
10	93	100	107	97	99	99.5
11	92.6	98.9	104	94.7	96	100.2
Mean	93.5	101	103.5	93.6	96.5	98.2
S.D.	7.5	11	7.1	3.6	3.5	3.6

The energy expenditure measured for individual subjects under two scenarios such as flat road and steep road conditions at different load conditions using IDEEA Minisun is shown in Table 3.

In Scenario 1, the sample has to walk through a flat road as well as climb the steps. In order to compare the energy expenditure under these two sub-scenarios, same step count (about 321 steps) is taken. The obtained values for energy expenditure under flat road condition and climbing steps' condition are shown in Table 4.

5 Results and Discussion

Analysis of variance (ANOVA) is a statistical technique, used to check the significant difference in the means of two or more groups. ANOVA checks the impact of one or more factors by comparing the means of different samples. So, ANOVA is used here to compare the energy expenditure between the undertaken scenarios.

Table 4 Energy expenditure for flat road and climbing steps

Subject	Scenario 1—EE (kcal)					
	Flat road			Climbing steps		
	NL	5% BW	10% BW	NL	5% BW	10% BW
1	13.46	16.66	14.82	39.50	41.70	34.90
2	11.44	12.40	13.46	33.50	31.00	31.70
3	13.60	15.86	16.08	39.80	39.70	37.90
4	11.68	12.66	13.46	34.30	31.70	31.70
5	11.20	12.80	13.60	32.80	32.00	32.00
6	11.34	13.12	13.64	33.30	32.80	32.10
7	12.46	13.56	14.82	36.50	33.90	34.90
8	10.42	11.96	12.38	30.50	29.90	29.10
9	11.48	12.66	13.94	33.70	31.70	32.80
10	11.82	13.34	14.54	34.60	33.00	33.30
11	11.76	13.18	14.14	34.50	33.00	33.30
Mean	11.88	13.47	14.08	34.82	33.70	33.15
S.D.	0.95	1.46	0.97	2.79	3.66	2.31

5.1 *Energy Expenditure Between Flat Road and Steep Road*

Here, the two scenarios have been compared statistically by conducting one-way ANOVA under 5% level of significance and the results between two scenarios under no load condition, 5% BW condition, and 10% BW condition have been provided in Table 5.

It is clearly evident from the above table that there is no significant difference in energy expenditure between flat road and steep road at no load condition and at 5% BW load carriage ($p > 0.05$). An individual carrying a backpack load equivalent to 10% body weight is not recommended as it has more influence on the energy expenditure.

Table 5 ANOVA results—flat road vs steep road at three different load conditions

Comparative parameters	Flat road versus steep road		
	NL	5% BW	10% BW
Sum of squares between groups	0.15	117.30	157.69
Sum of squares within groups	697.31	1328.61	646.007
F-statistic	0.004	1.76	4.88
F-critical	4.35	4.35	4.35
p-value	0.94	0.19	0.039

Table 6 Statistical parameter among different load conditions under each scenario

Comparative parameters	NL versus 5% BW versus 10% BW	
	Flat road	Steep road
Sum of squares between groups	603.192	115.117
Sum of squares within groups	2275.453	396.472
F-statistic	3.976	4.355
F-critical	3.315	3.315
p-value	0.029	0.021

5.2 *Energy Expenditure at Different Load Conditions in Different Scenarios*

Three different load conditions have been compared statistically by conducting one-way ANOVA under 5% level of significance and are provided in Table 6.

There is a significant difference in energy expenditure among different load conditions under both flat road and steep road walking condition. Further to know among three different load conditions, between which cases significance exist, multiple comparison test (also known as Tukey's Honestly Significant Difference test) is conducted and the results obtained state that there is a significant difference in energy expenditure between no load and 10% BW under flat road walking (HSD ~ 9.16) as well as under steep road walking condition (HSD ~ 3.82).

5.3 *Energy Expenditure Between Flat Road and Climbing the Steps*

In scenario 1, the two sub-scenarios are compared statistically by conducting one-way ANOVA under 5% level of significance. The results are provided in Table 7.

Table 7 Statistical parameters for two sub-scenarios at different load conditions

Comparative parameters	Flat road versus climbing steps		
	NL	5% BW	10% BW
Sum of squares between groups	2894.34	2250.28	2001.11
Sum of squares within groups	87.084	155.42	62.604
F-statistic	664.71	289.56	639.331
F-critical	4.3512	4.3512	4.351
p-value	8.07E-17	2.3E-13	1.18E-16

Table 8 Statistical parameters among different load conditions under two sub-scenarios

Comparative parameters	NL versus 5% BW versus 10% BW	
	Flat road	Climbing steps
Sum of squares between groups	28.45	15.82
Sum of squares within groups	39.80	265.30
F-statistic	10.72	0.89
F-critical	3.31	3.31
p-value	0.0003	0.4193

It is clearly evident from the above table that there is no significant difference in energy expenditure between flat road and climbing steps at all the three different load carrying conditions ($p < 0.05$).

5.4 *Energy Expenditure at Different Load Conditions in Flat Road and Climbing Steps*

The three different load conditions have been compared statistically by conducting one-way ANOVA under 5% level of significance and are provided in Table 8.

From the above table, it is clear that there is no significant difference in energy expenditure among different load conditions while individuals are climbing the steps. Whereas there is a significant difference in energy expenditure among different load conditions under while walking on a flat road. Multiple comparison test is conducted and the results obtained state that there is a significant difference in energy expenditure between no load and 10% BW under flat road walking condition (HSD ~ 1.21).

6 Conclusion

From this work, it is concluded that while carrying 10% body weight of load in backpack, steeper road (road with gradient) is having more impact in expended energy in comparison with energy expenditure under 10% BW load carriage at flat road walking condition. There is no significant impact on the energy expenditure while climbing the steps at varied load carriage. On the flat road, the energy expenditure is proportionately increasing with the increase in load. Under both the road conditions, it is found that the 10% BW load to backpack is more influential than any other load conditions. In both the steeper road and flat road walking condition, its normal to carry 5% BW load in backpack but 10% is having greater influence on energy expenditure. At the same time, there is no significance increase in energy expenditure if a person shifts from 5 to 10% body weight load carriage. Further investigation can be carried out by increasing backpack load at the rate of 1% over 5% body weight to know exactly at which percentage, the energy expenditure is becoming influential between 5 and 10% of body weight load carriage in the backpack while walking. This work has mainly focused on the assessment of energy expenditure. This study can serve as one of the parameters in carrying out a research work on bio-mechanical strains due to backpack load and even the conditions of the road as a future work.

References

1. Hall, C., Figueroa, A., Fernhall, B.O., Kanaley, J.A.: Energy expenditure of walking and running: comparison with prediction equations. Med. Sci. Sports Exerc. **36**, 2128–2134 (2004)
2. Brito, J.P., Garrido, N., Romero, F., de Araújo Junior, A.T., Reis, V.M.: Effects of backpack load and trekking poles on energy expenditure during field track walking. Sports Med. Int. Open **2**(04), E117–E122 (2018)
3. Gupta, I., Kalra, P., Iqbal, R.: Evaluation of school backpack prototype-based on gait parameters, energy expenditure and posture of students. Curr. Sci. **115**(5), 930–936 (2018)

Low-Carbon Supply Chain Management: A Fuzzy-DEMATEL Analysis of Some Practical Issues of Indian Manufacturing Industries

Vivek Gupta and Arvind Jayant

1 Introduction

This paper includes some basic models and hypothesis developed by using questionnaires getting from many industries. By using these hypotheses, we can differentiate between some on paper applied phenomenon and the techniques actually applied. I have prepared some barriers by using questionnaires apply fuzzy DEMATEL tool to develop fuzzy triangular function and to develop total relation matrix.

This paper broadly examines different components that may add to constrained reception of this training. This paper at that point finishes a quantitative investigation using a fuzzy-oriented DEMATEL way to deal with distinguish the relative relationship and significance of different hindrances, particularly at the city level. At that point, a fuzzy-oriented DEMATEL way to deal with evaluates the connections of obstructions is presented. To assess the hindrances, this article at first gets contributed from a genuine government official accountable for acquisition.

Some Main Objectives of the Paper Are:

i. To represent and describe some related hypothesis which is to develop some common issues of 'low-carbon supply chain' and also some strategies developed by managers for the effectiveness of their supply chain [1].
ii. To check the effectiveness and efficiency of this hypothesis and to compare the importance of these selected issues to influencing these supply chain attributes.
iii. To apply fuzzy-DEMATEL approach for the analysis of the data.
iv. To check and relate applied research with managers to find out selected issues.

V. Gupta (✉) · A. Jayant
Sant Longowal Institute of Engineering and Technology, Sangrur, Punjab, India
e-mail: vivekgups@gmail.com

A. Jayant
e-mail: arvindjayant@gmail.com

M. Muzammil et al. (eds.), *Recent Advances in Mechanical Engineering*, Lecture Notes in Mechanical Engineering, https://doi.org/10.1007/978-981-15-8704-7_4

2 Background

The decision-making trial and evaluation laboratory (DEMATEL) method was first developed by the Geneva Research Center of the Bottle Memorial Institute to visualize complex causal relationships through matrix or diagrams. As a kind of structural modeling approach, it is particularly useful in analyzing cause-and-effect relationships between parts of a system. DEMATEL can determine the interdependence between factors and map development to reflect the relative relationships among them and can be used to investigate and solve complex and interrelated problems [2].

2.1 *Evaluating Low-Carbon Authorities' Practice Implementation*

A fuzzy triangular number:

$$K = (p,\ q,\ r)$$

$$0,\ x\ \leq\ p$$

$$y - p$$

$$-------,\ p < x \leq s$$

$$s - p$$

$$A_2(x) = t - x$$

$$-------,\ s < x \leq t$$

$$t - s$$

$$0, x > u$$

Let u_1 and u_2 are two triangular fuzzy numbers [3]. The essential tasks of those numbers are as per the following:

$$u_1 + u_2 = (p_1 + p_2, q_1 + q_2, r_1 + r_2) \tag{1}$$

$$u_1 - u_2 = (p_1 - p_2, q_1 - q_2, r_1 - r_2) \tag{2}$$

$$u_1 * u_2 = (p_1 * p_2, q_1 * q_2, r_1 * r_2) \tag{3}$$

$$u_1/u_2 = (p_1/p_2, q_1/q_2, r_1/r_2) \tag{4}$$

$$u_1 * n = (p_1 * n, q_1 * n, r_1 * n) \tag{5}$$

On this paper, the changing fuzzy actualities into fresh scores (CFCS) defuzzification procedure is pursued.

The CFCS strategy incorporates a five-step calculation permit [4].

$\xi_{ij}^{k} = \left(r_{ij}^{k}, s_{ij}^{k}, t_{ij}^{k}\right)$ demonstrate the fuzzy assessment of evaluator n ($n = 1, 2, 3, \ldots, n$) roughly the recognition to which the item I influences the article j. The CFCS strategy finishes the ensuing advances:

(1) Normalization

$$yp_{ij}^{n} = \frac{p_{ij}^{n} -- \min p_{ij}^{n})}{N_{min}^{max}} \tag{6}$$

$$yq_{ij}^{n} = \frac{q_{ij}^{n} -- \min q_{ij}^{n})}{N_{min}^{max}} \tag{7}$$

$$yr_{ij}^{k} = \frac{r_{ij}^{n} -- \min r_{ij}^{n})}{N_{min}^{max}} \tag{8}$$

where $N_{min}^{max} = \max q_{ij}^{n} - r_{ij}^{n}$

2.2 *DEMATEL*

The DEMATEL approach transformed into cutting edge by method for Geneva explores focus of Battelle commemoration institute. The connections between gadget-added substances with the quantitative depicted qualities of the connections are depicted by the frameworks or diagrams [5]. The DEMATEL procedure expects a contraption which contains a fixed of parts $C = C1, C2, \ldots, Cn$, with pair savvy relations that might be assessed.

DEMATEL might be deteriorated into four ordinary dimensions:

Step 1 Produce a pair-wise direct-relation matrix among machine components thru decision maker enter.

Step 2 Decide the preliminary have an impact on matrix by way of normalizing the direct-relation matrix [6].

Based on the general fresh direct-connection framework Z, the standardized direct-connection network m can be acquired through articulations (10):

$$k = q.t \tag{9}$$

$$T = \frac{1}{\text{Max} \sum_{J=1}^{k} K_{ij}}, O, \quad j = 1, 2, n, k$$

$$Q = \frac{1}{\text{Max} \sum_{J=1}^{k} K_{ij}}, O, \quad j = 1, 2, n, k \tag{10}$$

Step 3 Determination of total relation (influence) matrix.
The total relation matrix (T) is determined by expression (11) where I represent an $n \times n$ identity matrix.
Compilation of complete connection (impact) matrix.
The complete connection framework (T) is controlled by articulation (11) where I speak to a character grid.

$$T = L_1 + L_2 + L_3 + \cdots = \sum_{I=1}^{\infty} L_i = L(I - L)^{-1} \tag{11}$$

Step 4 Decide the thought process connections (noticeable causal graph) among the segments and relative qualities [7].
Right off the bat, push (R_i) and section (D_j) entireties for each line i and segment j from the entire connection network T are obtained, through the consequent detailing (12) and (13):

$$\mathrm{R_i} = \sum_{J=1}^{n} \mathrm{R_{ij}} \quad ¥\mathrm{O} \tag{12}$$

$$\mathrm{C_j} = \sum_{J=1}^{n} \mathrm{R_{ij}} \quad ¥\mathrm{U} \tag{13}$$

The column esteems R_i represents the general immediate and slanted impact of a hindrance I on various deterrents, even as the column values D_j speaks to the general, immediate and diagonal results of the considerable number of impediments on obstruction j.
Additionally, the general noteworthiness or unmistakable quality (P_i) of an obstruction 'I' and net impact (E_i) of hindrance 'I' is chosen, through articulations (14) and (15) [8].

$$P_i = \{R_i = D_j | I = j\} \tag{14}$$

$$E_i = \{R_i = D_j | I = j\} \tag{15}$$

P_i illustrates general procedure or significance of barrier '*I*' regarding the overall relationships with other obstacles. The bigger is the price of p_i, then the extra overall prominence of barrier '*I*'. If E_i > zero then barrier '*I*' is a net purpose for different inexperienced dealer development programs. If E_i < 0, barrier '*I*' is predicated on (internet impact of) operation of different obstacles [13].

The entire procedure of the fuzzy principally based DEMATEL method with consideration of FCM could be done as pursues.

3 Case Study

3.1 Some Applications of the Fuzzy-Applied DEMATEL

Underneath the weight of reasonable improvement, numerous close-by Indian governments have attempted to incite GLCP[1]. In any case, numerous boundaries obstruct the ground-breaking usage of GLCP. The data are taken from the northern part of India.

At that point, the twelve constraints are given to the authorities to finish the pair shrewd similar examination for DEMATEL. The accompanying case represents how one respondent completed our proposed fuzzy DEMATEL procedure to assess the hindrances of GLCP and their interrelationship[2] [9].

Stage one: Draw a fuzzy pair-wise influence appraisal scale for the evaluations. A five-phase scale with the things of *N* (no impact), VL (exceptionally low impact), *L* (low impact), *H* (high impact), and VH (extremely high impact) is utilized inside the paper. The fuzzy semantic scale is demonstrated in Table 1.

Table 1 Some barriers are used in study

No.	Barriers
*q*1	Short of buying employees global knowledge
*q*2	Short of expertise in the way to put into effect GLCP
*q*3	Problem of measuring total price on environmental impact
*q*4	Problem of measuring environmental effect of a few big impacts
*q*5	Notion of poor first-rate unpracticed merchandise
*q*6	Notion of very redundant operations for GLCP
*q*7	Lack of pinnacle management dedication
*q*8	Loss of middle management assists
*q*9	Awareness on assisting local price
*q*10	Company's subculture of resisting advancements
*q*11	Lack of mixing GLCP within different larger corporation initiatives
*q*12	Short of internal conversation [13]

Table 2 Fuzzy-based operating scale

Linguistic terms	Triangular fuzzy numbers
Zero change (*N*)	[0.001,0.001,0.25]
Sharp little change (VL)	[0.001,0.25,0.5]
Little change (*L*)	[0.25,0.51,0.75]
High change (*H*)	[0.5,0.75,1.0]
Very high change (VH)	[0.75,1.0,1.0]

Stage two: Increase the fuzzy direct-connection network X through presenting the fuzzy pair-wise sway connections between the restrictions in a lattice. Note that everyone has essential corner of components, to begin with set to a fresh cost of zero ("N" = no impact). Table 4demonstrates the etymological scale direct-connection grid completed by means of one respondent. At that point, the etymological scales in Table 2 might be converted into triangular numbers by utilizing the utilization of Table 2 [10].

Stage four: Setting up and perusing the basic model. In view of on the underlying direct connection fresh lattice Z, the standardized direct connection framework M is acquired. Further, the full-connection network T is resultant. At that point, the reason/sway connections (noticeable causal graph) among the parts and relative qualities can be chosen. Table 3 exhibits the unmistakable quality and web reason/sway estimations of the twelve limits. Figure 1 is the noticeable quality causal DEMATEL diagram [11].

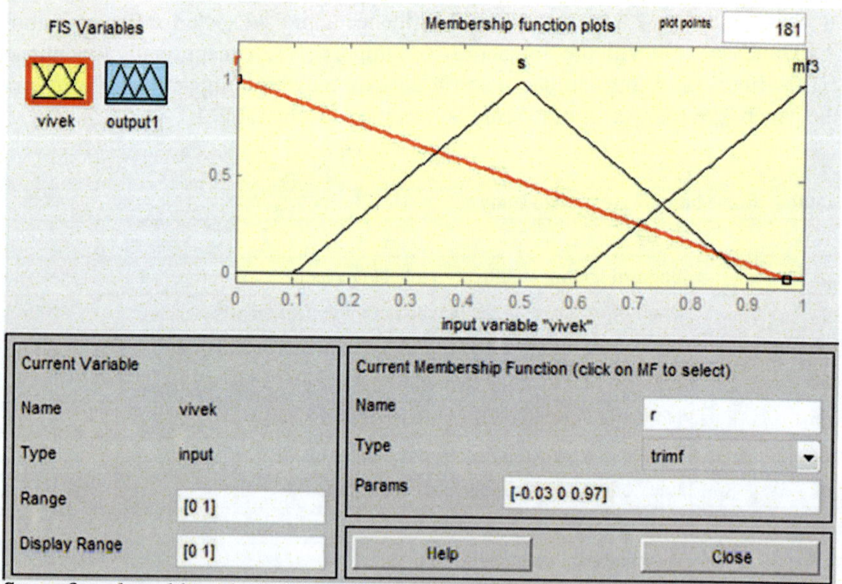

Stage five: bunching similar snags into different organizations. FCM is utilized to characterize the different boundaries into associations. Association 1 incorporates seven restrictions: $q1$, $q2$, $q3$, this fall, $q5$, $q9$, $q10$. Obstructions in foundation 1 are

Table 3 Direct relation matrix *Z*

GLCP	*q*1	*q*2	*q*3	*q*4	*q*5	*q*6	*q*7	*q*8	*q*9	*q*10	*q*11	*q*12
*q*1	0.03	0.50	0.03	0.04	0.03	0.04	0.04	0.04	0.73	0.73	0.97	0.03
*q*2	0.50	0.03	0.03	0.04	0.03	0.04	0.04	0.04	0.50	0.03	0.27	0.03
*q*3	0.50	0.73	0.03	0.04	0.73	0.04	0.04	0.04	0.50	0.03	0.03	0.97
*q*4	0.03	0.97	0.97	0.04	0.73	0.04	0.04	0.04	0.97	0.73	0.73	0.97
*q*5	0.73	0.73	0.73	0.04	0.03	0.04	0.04	0.04	0.73	0.03	0.73	0.97
*q*6	0.03	0.03	0.03	0.04	0.03	0.04	0.49	0.04	0.03	0.03	0.50	0.03
*q*7	0.03	0.03	0.27	0.04	0.03	0.25	0.04	0.04	0.03	0.03	0.03	0.03
*q*8	0.03	0.03	0.03	0.04	0.03	0.04	0.04	0.04	0.03	0.03	0.30	0.73
*q*9	0.03	0.03	0.03	0.04	0.27	0.04	0.04	0.04	0.03	0.97	0.73	0.97
*q*10	0.03	0.03	0.03	0.04	0.50	0.04	0.04	0.04	0.03	0.03	0.73	0.73
*q*11	0.03	0.03	0.03	0.04	0.50	0.04	0.04	0.04	0.03	0.97	0.03	0.73
*q*12	0.03	0.03	0.03	0.04	0.73	0.04	0.04	0.04	0.03	0.03	0.97	0.03

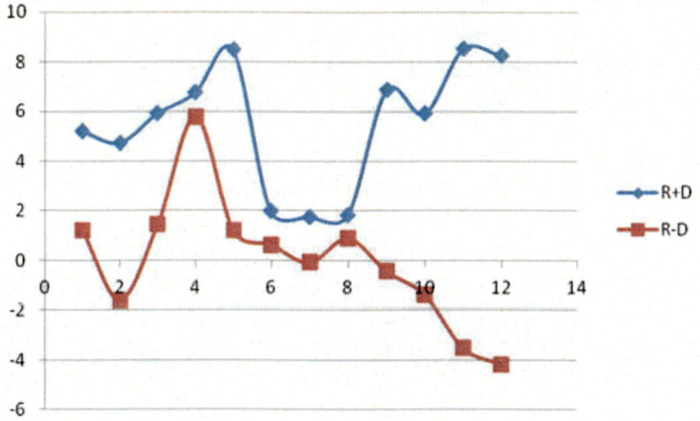

Fig. 1 The grouping after effect of noticeable causal DEMATEL chart

described by utilizing having remarkably high *r* and *d* esteems, even as they play increasingly pivotal position of affecting distinctive hindrances [12].

4 Discussion

Inside the exemplified case, the area government endeavored to comprehend the limits to GLCP through the utilization of the proposed philosophy. Right off the bat, *q*5 (degenerate practices penance potential chances of a couple of green providers) and *q*11 (conviction of larger top of the line of green items) are analyzed as the two most basic obstructions that require fundamental and exceptional attention. In venture with Table 4 and Fig. 1, $q5$ (2.53) and $q11$ (2.52), with over the top $R + D$ scores, are the two hindrances have the most dominant penchant of associations with various constraints. This implies defilement may understand by obstructs a few open doors for green providers and along these lines confinement the viable usage of GLCP. Full-measure intrigue wants to be given to tending to debasement as an approach to advance the viability of GLCP. Table 4 distinguishes factor this fall is the thought process boundary with the most noteworthy $r - d$ score. Indian neighborhood governments may likewise take in preparing from the specific reasonable GLCP rules starting from advanced worldwide areas/zones. Be that as it may, considering the excellent monetary and business skills existing among various Indian nearby governments, the outside encounters of test GLCP will expect changes to sound with Indian exact circumstances. Likewise, $q12$, with the least $r - d$ rating, is the effect hindrance which can be empowered by method of various obstructions. The deterrents of this fall and $q5$ were examined. With the second one most astounding $R - D$ score, component $q3$ (loss of top control responsibility) should likewise be assessed and considered. A few ecological bundles blur because of poor top administration help. Indian better authorities for the most part give more noteworthy consideration to their outstanding execution necessities. The cutting edge of general execution assessment framework in India stresses budgetary markers which incorporate GDP increment extent, bringing down accentuation on greening signs and practices.

Table 4 Level of noticeable quality and net reason/impact esteems

S. No.	*R* Addition	*D* Addition	$R + D$	$R - D$
$q1$	3.21	2	5.21	1.21
$q2$	1.58	3.17	4.75	−1.59
$q3$	3.68	2.24	5.92	1.44
$q4$	6.26	0.48	6.74	5.78
$q5$	4.84	3.64	8.48	1.2
$q6$	1.32	0.69	2.01	0.63
$q7$	0.85	0.93	1.78	−0.08
$q8$	1.37	0.48	1.85	0.89
$q9$	3.22	3.64	6.86	−0.42
$q10$	2.27	3.64	5.91	−1.37
$q11$	2.51	6.02	8.53	−3.51
$q12$	2.04	6.22	8.26	−4.18

5 Conclusions

In developing worldwide areas, specialists of low carbon obtainment may likewise go about as a fundamental riding power to increase feasible improvement. This truth is not constantly lost in India. However, government low carbon obtains in India keeps on being confronting limited usage. To address the issues experiencing GLCP reception in India, diverse variables that may add to restrict appropriation of this training are examined. Along these lines, this paper completes a similar assessment utilizing a fuzzy DEMATEL-based technique to find the relative relationship and significance of many boundaries, explicitly of the metropolitan stage. By methods for consolidating FCM into technique, hindrances are additionally named into one of kind groups. This hypothesis is beneficial for leather, metal, textile, iron, production industries, etc.

Our proposed methodology effectively expands the DEMATEL approach for incorporating each of a fuzzy set idea and fuzzy grouping approach. Accordingly, it can proficiently manage the inconveniences of ill defined and fragmented records. Furthermore, the grouping examination lets in an additional efficient comment of the interrelationships among GLCP hindrances.

With the guide of the utility of the proposed method in an Indian city, this paper gets contribution from genuine government officers. The evaluation results show the reason-impact pursuing among the limits and a lot of valuable experiences into successfully forcing GLCP had been given.

References

1. Jayant, A., Dhillon, M.S.: Use of analytic hierarchy process (AHP) to select welding process in high pressure vessel manufacturing environment. Int. J. Appl. Eng. Res. **10**(8), 586–595 (2014)
2. Arvind, J., Gupta, P., Garg, S.K.: Design and simulation of reverse logistics network: a case study. In: The Proceedings of World Congress on Engineering (WCE-2011), London, U.K., during July 6–8, (2011)
3. Jayant, A., Gupta, P., Garg, S.K.: Reverse logistics network design for spent batteries: a simulation study. Int. J. Logist. Syst. Manage. **18**(3), 343–365 (2014)
4. Jayant, A., Paul, V., Kumar, U.: Application of analytic network process (ANP) in business environment: a comprehensive literature review. Int. J. Res. Mech. Eng. Technol. **4**(3), 29–43
5. Jayant, A.: Evaluation of 3PL service provider in supply chain management: an analytic network process approach. Int. J. Bus. Insights Transf. IJBIT **6**(2), 78–82 (2013)
6. Damert, M., Feng, Y., Zhu, Q., Baumgartner, R.J.: Motivating low-carbon initiatives among suppliers: the role of risk and opportunity perception. Resour. Conserv. Recycl. **136**, 276–286 (2018)
7. Brammer, S., Walker, H.: Sustainable procurement practice in the public sector: an international comparative study. Int. J. Oper. Prod. Manage. **31**(4), 452–476 (2011)
8. Tseng, M.L., Zhu, Q., Sarkis, J., Chiu, A.S.: Responsible consumption and production in corporate decisionmaking models. Ind. Manage. Data Syst. (2018)
9. Carter, C.R., Ellram, L.M., Ready, K.J.: Environmental purchasing: benchmarking our German counterparts. J. Supply Chain Manage. **34**(4), 28–38 (1998)

10. Coggburn, J.D.: Achieving managerial values through green procurement? Public Perform. Manage. Rev. **28**(2), 236–258 (2004)
11. Day, C.: Buying green: the crucial role of public authorities. Local Environ. **10**(2), 201–209 (2005)
12. Fan, Z.P., Suo, W.L., Feng, B.: Identifying risk factors of IT outsourcing using interdependent information: an extended DEMATEL method. Expert Syst. Appl. **39**, 3832–3840 (2012)
13. Tzeng, G., Chiang, C., Li, C.: Evaluating intertwined effects in e-learning programs: anovel hybrid MCDM model based on factor analysis and DEMATEL. Expert Syst. Appl. **32**(4), 1028–1044 (2007). https://doi.org/10.1016/j.eswa.2006.02.004

Air Jet Erosion Behavior of AA 6082 T6 Aluminum Alloy

Gyanesh Mangal, Vinod Kumar, and Siddhartha

1 Introduction

Solid particle erosion wear is one of the modes of wear; it is the progressive loss of material due to striking of solid or liquid particles or mixture of both on the solid surface of material . This wear phenomenon occurs in many industrial equipments and machinery for example automobile components, hydraulic turbines, aeronautical equipments, wind turbines, marine parts, automobile parts, etc.[1]. In erosive wear environment, aluminum alloys are extensively used because of their ease of manufacturing, high strength to mass ratio, relatively low manufacturing cost and high corrosion resistance. These alloys replaced conventional materials due to their superior specific properties [2–4]. Many researchers have examined [5–10] that erosion rate of materials is dependent on different parameters. Impact velocity, impingement angle, erodent size and standoff distance are acknowledged as some factors that noticeably affect the erosion rate of various materials. In general, impact velocity influences the erosion rate more than any other parameters. Erosion rate is dependent on impact velocity of erodent and is given by power law [11] (Er = kVn). The value of 'n' varies between '2.0 and 3.5' for metallic materials. This velocity exponent (*n*) could go up to 6.5 for brittle materials [12]. Influence of impingement angle on erosive wear has been examined by many researchers. Neilson et al. [13] developed the relationship between erosion wear and impingement angle. Desale et al. [14] found that in ductile materials, maximum level of erosion occurs between impingement angles of 150–300 which decrease continuously with further increment of the impingement angle up to 900. According to different researchers, most extreme wear rate of ductile materials was found between 150 and 450 impingement angle, while for brittle material like glass, maximum wear rate occurred at 900. Bitter [15] examined that deformation wear occurred when the erodent particles impinge normally

G. Mangal · V. Kumar (✉) · Siddhartha
Department of Mechanical Engineering, NIT Hamirpur, Hamirpur, H.P. 177005, India
e-mail: vthakur19@gmail.com

M. Muzammil et al. (eds.), *Recent Advances in Mechanical Engineering*, Lecture Notes in Mechanical Engineering, https://doi.org/10.1007/978-981-15-8704-7_5

Table 1 Chemical composition of alloy sample (weight %)

Si	Mg	Mn	Fe	Cr
0.928	0.909	0.552	0.240	0.087
Cu	Ti	Zn	Ni	Al
0.022	0.029	0.016	0.003	Balance

to material surface, while cutting wear phenomenon was obtained on the parallel impingement of particles. Clark and Wong [16] observed that with reduction in the erodent size, the kinetic energy of particle also reduces, and the striking of smaller erodent may not be adequate to start material removal because of their lower kinetic energy. Rao et al. [17] examined that the solid particle erosive mass loss decreases with increment in standoff distance at a particular impingement angle. Chowdhary et al. [18] examined that erosion rate of aluminum alloys continuously decreases with increase in standoff distance.

2 Experimental Details

2.1 *Specimen and Properties of Material*

The materials used for the experimental purpose was AA 6082-T6 aluminum alloy which has excellent erosion corrosion resistance with medium fatigue strength and highest strength among 6xxx series. It is a heat-treatable alloy with excellent formability and simple to complex profiles can be produced by extrusion. AA 6082-T6 aluminum alloy used in this study was supplied by 'Bharat Aerospace Metals, Mumbai-400004' in the form of 300 mm × 300 mm × 5 mm sheets. Rectangular type samples of size 30 mm * 30 mm were prepared by using simple hacksaw. Acetone was used to clean the samples before and after erosion wear tests. The chemical composition of aluminum alloy was measured and listed in Table 1. The erodent particles used in experiments are dry quartz type silica sand which is separated in three groups of grain sizes (150–300, 300–450, 450–600 μm). The weight of the specimens was measured up to four decimal place by using electronic weighing machine. Erosion rate was calculated from weight loss per kilogram of erodent particles.

2.2 *Erosion Wear Testing Apparatus*

In present research, solid particle erosion wear tests were performed on DUCOM—Modal TR 517, air jet erosion tester as shown in Fig. 1. The tests were conducted in accordance with ASTM G 76 standard. Erodent particles were fed at the rate of 5 gm/min through the nozzle to perform the experiments. Sand particles were filled

Fig. 1 Air jet erosion wear testing apparatus

in the hopper which is placed at upper side of the apparatus. The erodent impinges at the surface of specimen with definite velocity through the 3 mm internal diameter and 50 mm long nozzle which is made up of tungsten carbide. The specimen of size 30 mm × 30 mm × 5 mm has been used for erosion tests.

2.3 Taguchi Experimental Analysis

According to Taguchi experiments, 27 experiments were performed as per L_{27} orthogonal array and erosion rate of performed experiment with S/N ratio are presented in Table 2. The investigation of the trial results was done with the help of software 'MINITAB18'.

The mean for S/N ratios is calculated to be −36.6821 dB for Al alloy. After analyzing the results, it can be recognized that set of factors impact velocity with 30 m/s, angle of impingement 900, erodent size of 600 μm and standoff distance 25 mm, i.e., A1B3C3D3 give minimum wear rate for aluminum alloy AA6082 T6 as shown in Fig. 2. From Fig. 3, it can be examined that interaction between impact velocity and impingement angle is strong because the lines interact with each other, while the moderate interaction occurs between the parameters velocity of impact and particle size, and between the parameters impingement angle and erodent size because lines are not parallel. Thus, it can be recognized from the present analysis that impact velocity is most influencing parameter of erosion characteristics of AA6082 T6 aluminum alloy.

Table 2 Experimental design using L_{27} orthogonal array for AA 6082 Al alloy

S. No.	Velocity of impact (m/s)	Impingement angle (deg)	Erodent size (micron)	Standoff distance (mm)	Erosion rate (mg/kg)	S/N ratio (dB)
1	30	30	300	15	57.6	35.2084
2	30	30	450	20	42.8	−32.6289
3	30	30	600	25	27.3	−28.7233
4	30	60	300	20	62.4	−35.9037
5	30	60	450	25	44.4	−32.9477
6	30	60	600	15	60.1	−35.5775
7	30	90	300	25	26.4	−28.4321
8	30	90	450	15	35.2	−30.9309
9	30	90	600	20	22.6	−27.0822
10	50	30	300	20	89.4	−39.0268
11	50	30	450	25	69.0	−36.7770
12	50	30	600	15	109.6	−40.7962
13	50	60	300	25	90.0	−39.0849
14	50	60	450	15	138.4	−42.8227
15	50	60	600	20	99.4	−39.9477
16	50	90	300	15	54.2	−34.6800
17	50	90	450	20	47.8	−33.5886
18	50	90	600	25	29.1	−29.2779
19	70	30	300	25	98.4	−39.8599
20	70	30	450	15	157.8	−43.9621
21	70	30	600	20	135.8	42.6580
22	70	60	300	15	186.4	45.4089
23	70	60	450	20	165.1	−44.3549
24	70	60	600	25	104.6	−40.3906
25	70	90	300	20	69.2	−36.8021
26	70	90	450	25	60.2	−35.5919
27	70	90	600	15	79.0	−37.9525

2.4 ANOVA and Influence of Factors on Erosion Rate

ANOVA is applied to determine statistical significance of various parameters on erosion wear rate. Table 3 shows outcome of ANOVA for erosion wear. ANOVA table consists of certain terminology and their value. It can be concluded from Table 3, that % contribution of velocity of impact and impingement angle are equal to 52.05% and 33.41%, respectively, and have high effect on the erosion wear in comparison with standoff distance (10.72%) and erodent size (1.46%) which are less significant. While

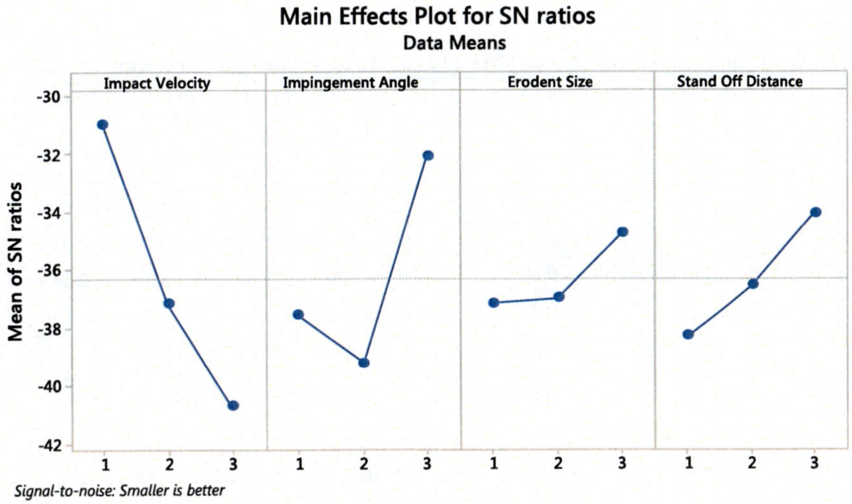

Fig. 2 Effect of control factor on erosion

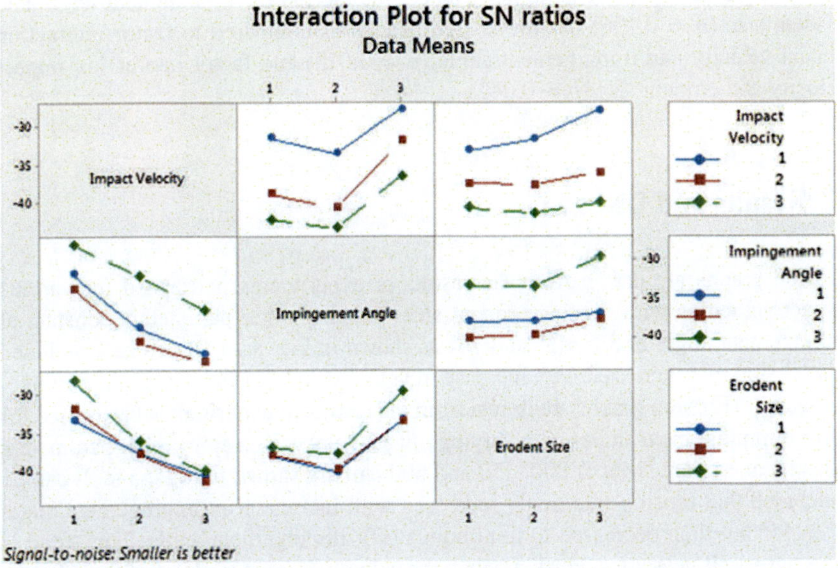

Fig. 3 Interaction graph for erosion

the interaction point of impact velocity and impingement angle (1.16%) has more critical effect on erosion rate as compared to interactions between impact velocity and erodent size (0.69%), impingement angle and erodent size (0.30%). *P*-values given in the table state significance level of main effects. From ANOVA table, it is seen that impact velocity has *p* value of 0.000, impingement angle has *p* value

Table 3 ANOVA

Source	DOF	Seq. SS	Contribution (%)	Adj. SS	Adj. MS	*F*-value	*P*-value
A	2	357.265	52.05	357.265	178.633	737.69	0.000
B	2	229.295	33.41	229.295	114.647	473.45	0.000
C	2	10.003	1.46	10.003	5.001	20.65	0.002
D	2	73.593	10.72	73.593	36.796	151.96	0.000
A * *B*	4	7.968	1.16	7.968	1.992	8.23	0.013
A * *C*	4	4.766	0.69	4.766	1.192	4.92	0.042
B * *C*	4	2.058	0.30	2.058	0.514	2.12	0.196
Error	6	1.453	0.21	1.453	0.242		
Total	26	686.401	100.00				

of 0.000, erodent size has *p* value of 0.002, standoff distance has p value of 0.000, and hence, significance level of all these four parameters is higher. However, in the analysis of interaction of impact velocity and angle of impingement has *p* value of 0.013 lesser than the other two sets of interactions. If the *p* value is lesser, it shows that its significance level is higher. Hence, interaction of impingement angle and erodent size ($p = 0.196$) has lower significance as compared to factor interaction impact velocity and impingement angle ($p = 0.013$) and factor interaction impact velocity and erodent size ($p = 0.042$).

3 Results and Discussion

Erosion graphs are drawn from the results of experiments performed for various impact velocities with varying erodent sizes keeping other parameters constant at impingement angle of 30°, 45° and 75° as shown in Fig. 4a–c. It can be concluded that erosion rate increased with increase in velocity and erodent size at all impingement angle. Erosion graphs are drawn from the results of experiments performed for various impingement angles with varying impact velocities keeping other parameters constant at erodent sizes of 300, 450 and 600 μm are shown in Fig. 5a–c. It can be concluded that erosion rate firstly increased with increment in impingement angle up to 45° but then decreased to minimum at 90° impingement angle. This trend is same across all velocities. Further, when the velocity is increased from 30 to 70 m/s, the erosive wear rate also increases for all type of erodent sizes.

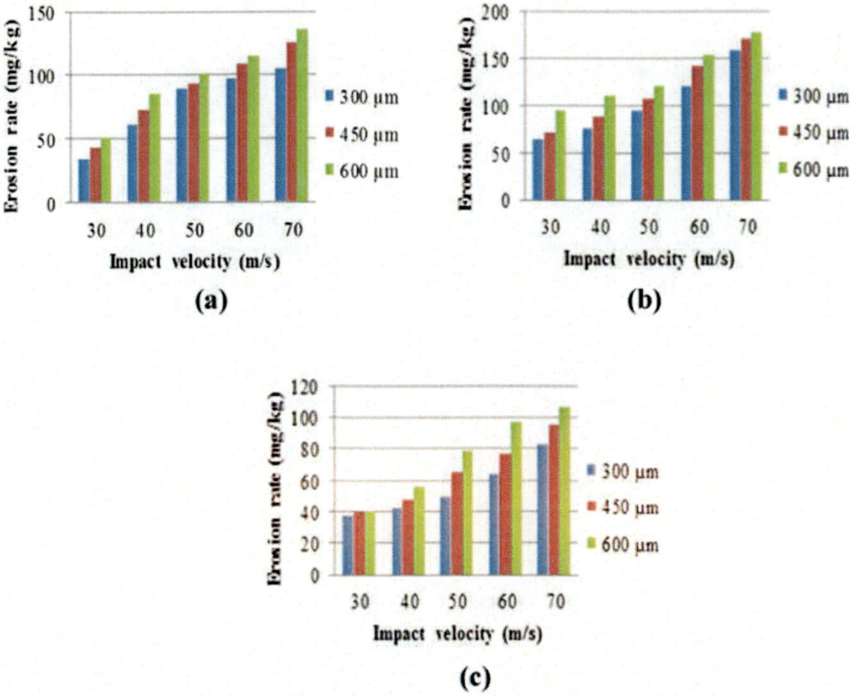

Fig. 4 Variation of erosion rate with variation in impact velocities and erodent size at standoff distance of 20 mm (impingement angle: **a** 30°, **b** 45° and **c** 75°)

4 Conclusion

1. Erosion wear is a serious problem in components made with Al alloys such as truss, cranes, high stress application, bridges, transport industry and beer barrels.
2. It was found that erosion wear of aluminum alloy is dependent upon velocity of impact, impingement angle, erodent size and standoff distance.
3. Taguchi design approach was used as design of experiment with the help of MINITAB 18 software, and ANOVA was used to investigate the erosion rate of alloy. It can be conclude that velocity of impact and impingement angle are significant factors for erosion rate of AA6082 T6 aluminum alloy.
4. The overall mean for S/N ratios of the erosion is found to be −36.6821 db for aluminum alloy. After analyzing the result, combination of factors A1 (Velocity of impact 30 m/s), B3 (Impingement angle 90°), C3 (Erodent size 600 µm) and D3 (Standoff distance 25 mm) was found for minimum wear rate.
5. Erosion rate of alloy is maximum at 45° impingement angle. Firstly, it increases up to 45°, and then, it decreases gradually up to 90° for all type of erodent size and impact velocity.

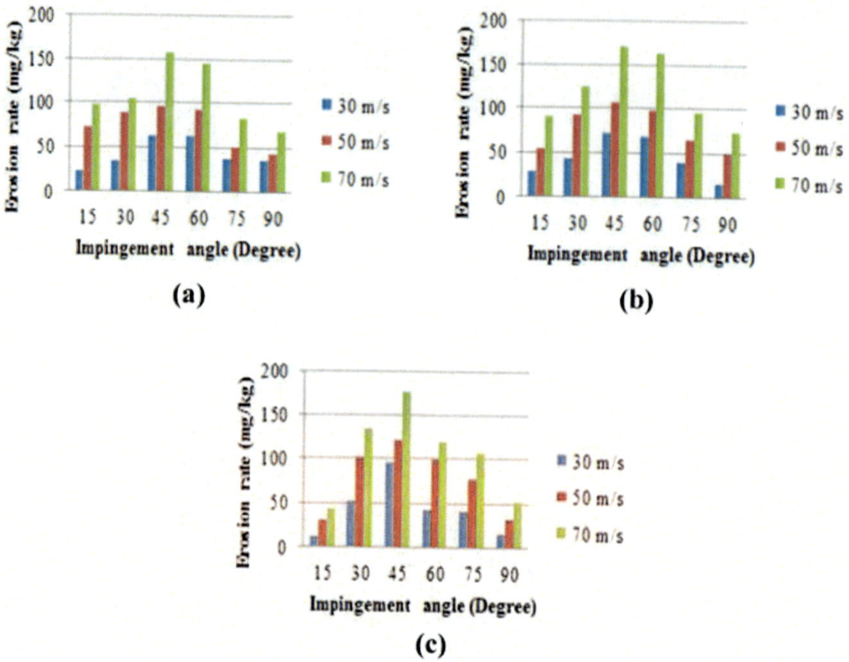

Fig. 5 Variation of erosion rate with variation in impingement angle and impact velocity at standoff distance of 20 mm (erodent size: **a** 300 μm, **b** 450 μm and **c** 600 μm

6. It can be concluded that the erosion wear continuously increases with increase in velocity of impact and erodent size.

References

1. Tisza, M., Budai, D., Kovacs, P.Z., Lukacs, Z.S.: Investigation of the formability of aluminium alloys at elevated temperature. IOP Conf. Ser. Mater. Sci. Eng. **159**(1), 22–29 (2016)
2. Liu, Y., Meng, G.Z., Cheng, Y.F.: Electronic structure and pitting behaviour of 3003 aluminum alloy passivated under various conditions. Electrochim. Acta **54**(1), 4155–4163 (2009)
3. Zhang, G.A., Xu, L.Y., Cheng, Y.F.: Investigation of erosion-corrosion of 3003 aluminum alloy in ethylene glycol-water solution by impingement jet system. Corr. Sci. **51**(1), 283–290 (2008)
4. Huang, H.W., Ou, B.L.: Evolution of precipitation during different homogenization treatments in a 3003 aluminum alloy. Mater. Des. **30**(1), 2685–2692 (2009)
5. Reddy, A.V., Sundararajan, G.: Erosion behaviour of ductile materials with a spherical non-friable erodent. Wear **111**(1), 313–323 (1986)
6. Finnie, I.: The mechanism of erosion of ductile metals. In: Proceedings of the Third National Congress on Applied Mechanics, pp. 527–532, New York (1958)
7. Tu, J.P., Pan, J., Matsumuran, M., Fukunaga, H.: The solid particle erosion behavior of Al18B4O33 whisker-reinforced AC4C Al alloy matrix composites. Wear **223**(1), 22–30 (1998)
8. Li, D.Y., Chen, Q., Cook, B.: A further simulation study on the dual role of porosity in solid-particle erosion of materials. Wear **271**(1), 325–1330 (2011)

9. Shimizu, K., Xinba, Y., Araya, S.: Solid particle erosion and mechanical properties of steel at elevated temperature. Wear **271**(1), 1357–1364 (2011)
10. Nsoesie, S., Liu, R., Chen, K.Y., Yao, M.X.: Analytical modeling of solid particle erosion of satellite alloys in combination with experimental investigation. Wear **309**(1), 226–232 (2014)
11. Harsha, A.P., Bhasker, D.K.: Solid particle erosion behavior of ferrous and non-ferrous materials and correlation of erosion data with erosion models. Mater. Des. **29**(1), 1745–1754 (2008)
12. Preece, C.M., Macmillan, N.H.: Erosion, annual review of materials science In: Huggins, R.A., (eds.) Annual Review of Material Science Conference 1977, Palo Alto (California), vol. 7, pp. 95–121, Stringer (1977)
13. Gilchrist, A., Neilson, J.H.: Erosion by a stream of solid particles. Wear **11**(1), 111–122 (1968)
14. Desale, G.R., Gandhi, B.K., Jain, S.C.: Effect of erodent properties on erosion wear of ductile type materials. Wear **261**, 914–921 (2005)
15. Riemsdijk, A.J.V., Bitter, J.G.A.: Erosion in gas-solid systems. In: 5th World Petroleum Congress, pp. 43–57, World Petroleum Congress, New York (1959)
16. Clark, H.M.: Specimen diameter, impact velocity, erosion rate and particle density in a slurry pot erosion tester. Wear **162–164** (B), 669–678 (1993)
17. Rao, K.V.S., Shivashankar, G.S., Sanman, S., Kumar, T.V.V.: Effect of standoff distance on solid particle erosion wear behavior of chill cast aluminum—boron carbide composites. Mater. Today Proc. **4**(1), 10015–10019 (2017)
18. Chowdhury, M.A., Debnath, U.K.: Experimental analysis of aluminum alloy under solid particle erosion process. Eng. Tribol. **230**(1), 01–26 (2016)

Muscular Discomfort in Occupational Motorcycle Riding

Mohd Parvez, Abid Ali Khan, and Siddharth Bhardwaj

1 Introduction

Motorcycles have recently gained prominence for the delivery of goods and services in the rapidly growing e-commerce sectors. The use of motorcycles in such occupations is increasing as motorcycles are fuel efficient, move quickly in congested traffic jams and require less parking space. This wider use of motorcycles has favored the genesis of occupational riders (delivery/pickup rider) in contrast to professional riders like racers and policemen [1]. Despite the increased occupational use of motorcycles, the health and safety issues of riders are however overlooked, thus imposing heath risk and even early retirement from the job [2].

Motorcycle riding has been linked with developing muscular discomfort, lower back and neck pain, forearm disorders, hand-arm vibration syndrome, sleepiness, etc. [3, 4]. The discomfort and health issues associated with riding are direct manifestation of the motorcycle design, the riding environment, and the physical and emotional state of the rider (Fig. 1) [5].

Vibration exposure during motorcycle riding is one major factor responsible for muscular discomfort and development of WMSDs like hand-arm vibration syndrome and lower back pain. Vibration transmission is dependent on the gripping force, rider-seat interface pressure, foot peg reaction force and ride duration. The exposure to vibration is further intensified by poor road profile, riding speed and poor riding skill [6–9]. Excessive gripping and large clutch forces are also documented for increasing the vibration transmissibility [10] as well as leading to chronic exertional compartmental syndrome in rider's forearm [11].

The estimation of muscular activity while riding is the direct manifestation of the physical and physiological demand in moterccyle riding. Studies have been

M. Parvez (✉) · A. A. Khan · S. Bhardwaj
Ergonomics Research Division, Department of Mechanical Engineering, Aligarh Muslim University, Aligarh, India
e-mail: parvez0527@gmail.com

M. Muzammil et al. (eds.), *Recent Advances in Mechanical Engineering*, Lecture Notes in Mechanical Engineering, https://doi.org/10.1007/978-981-15-8704-7_6

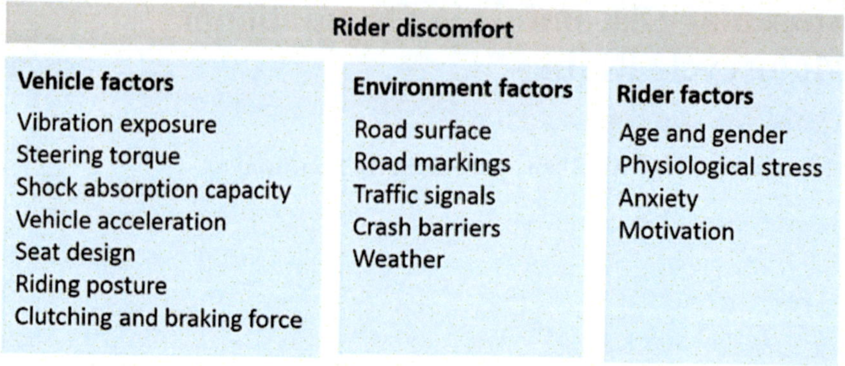

Fig. 1 Discomfort risk factors among motorcycle riders [5]

conducted to observe the effect of posture, riding conditions, clutching and gripping pattern on the muscular activity and fatigue of the riders using surface electromyography (sEMG) [12, 13]. Exposure to vibration has also been documented to cause muscle fatigue [14]; however, its effect on muscular discomfort is less studied in occupational riding.

The objective of the present study was to investigate the vibration exposure, sEMG activity of forearm muscles and subjective discomfort of the actively involved body parts in motorcycle riding. The study is a foreword step in learning the interactive aspect of vibration and muscular activity during motorcycle riding. The results can be used in improving the motorcycles' ergonomics.

2 Methodology

In the present study, conducted on two participants, vibration was recorded at rider-seat interface, while the sEMG activities of flexor carpi ulnaris (FCU) and biceps brachii (BB) were recorded during simulated ride experiment. Finally, the perceived discomfort score of the 13 body parts from the participants was obtained after the ride compilation. The recorded data were post-processed to evaluate the muscular fatigue and whole-body vibration (WBV) exposure. It was assumed that the muscular discomfort and fatigue of hand arm are caused solely by motorcycle riding, while the WBV exposure of the rider is caused by the combined effect of road and motorcycle vibration.

2.1 Design of Experiment

The experiment consisted of a participant riding the motorcycle in a closed loop circuit for 20 min on a good category bitumen road with no speed breakers and traffic at a speed of 30 ± 5 km/h. Vibration exposure from the seat, and bilateral measurement of sEMG activity of flexor carpi ulnaris (FCU) and biceps brachii (BB) were recorded during the ride. The subject rode their own motorcycles in the experimental trials. Both localized and overall discomfort rating were obtained by showing the Borg CR100 discomfort scale and 6 point ISO 2631-1:1997 discomfort scale before the start and after completion of the riding task in each experimental trial [9].

2.2 Participants

Two university post-graduation students (male, average age 27 years) were recruited for the study. Both the participants had at least 7 years of experience in riding motorcycles. The participants reported no prior history of injury and health condition. Prior to the experiment, a signed consent was obtained from the participants as per the ethics committee of the university.

2.3 Instrumentation

sEMG was recorded using bipolar EMG electrodes Biometrics REF SX230. A ground electrode was placed on the mastoid bone on the back of ear for attenuing high-frequency DC noise. Whole-body vibration (WBV) was recorded at the driver seat interface using a tri-axial seat pad accelerometer (Biometrics ACL 300) (standard protocol of ISO 2631-1:1997[15] was followed). Both sEMG and WBV data were recorded during experimental trial at a sampling rate of 1000 Hz using a portable Biometrics DataLOG (MWX8) logger. Figure 2 shows the sensor interfacing.

2.4 Procedure

First, the participant was briefed about the experiment and riding condition. Next, sEMG electrodes were placed at the respective muscles after cleaning the skin as per SENIAM standard (Fig. 1a). The tri-axial accelerometer was placed on the seat of the motorcycle (Fig. 1b) on which the participant was asked to sit. Once the participant sits comfortably, the sensors were connected to the respective channels of the data logger attached to the wrist belt of the participant as shown in Fig. 1c

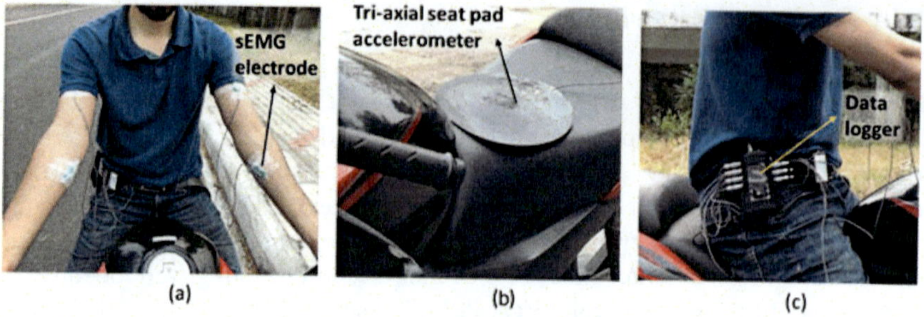

Fig. 2 Sensorial setup. **a** sEMG electrodes, **b** seat pad accelerometer and **c** data logger

and data acquisition was initiated. Following the sensorial setup, the participant was asked to provide discomfort rating after which the experimental trial was made. After completion of the ride, the participants were again interviewed for discomfort rating and the data acquisition was stopped.

2.5 Approach for Analysis

The logged data file for sEMG and accelerometer were first encoded in ASCII format for use in LabVIEW (ver 12, National Instrument, USA) environment for analysis. A custom-made LabVIEW program was used to evaluate median frequency slope (MDF slope) of the muscle groups for detecting muscle fatigue.

Vibration analysis was done as per ISO 2631-1 standard. Vibration dose value (VDV) of each individual axis (VDV_x, VDV_y and VDV_z) as well as total VDV was calculated as a measure of discomfort due to WBV exposure [16].

The overall discomfort and local discomfort rating were taken minimum of the rider's response before riding conditions and maximum of the rider's response after the riding to assess subjective discomfort.

3 Results and Discussions

The evaluated MDF slope of the sEMG data for the BB and FCU muscle groups of both arms is shown in Fig. 3a separately for each participant. For both the participants, the MDF slope was found to be negative which indicates that BB and FCU were fatigued during the riding task. However, the muscles of both the participant were fatigued differently. For participant 1, Right_FCU was most fatigued muscle while Right_BB was least fatigued. This observation was reverse in cases of participant 2, where the most fatigued muscle group was Right_BB and least fatigued was Right_FCU. This different pattern might be because of the different elbow flexion

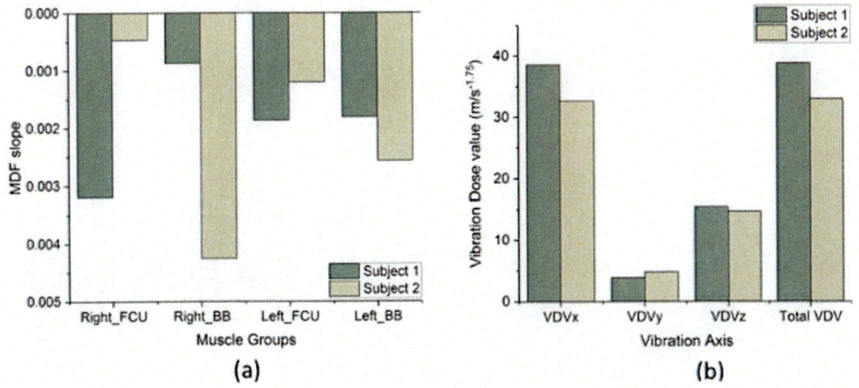

Fig. 3 **a** MDF slope of muscle groups and **b** vibration dose value of the rider

angles and wrist deviation while riding for the two participant, as it was shown previously that flexion of elbow and wrist deviation is related to the different work done by muscle group [17, 18]. Other reasons for this observation might be frequency of the use of front brake lever, wrist movement during throttling and hand-arm vibration transmissibility from the two motorcycles.

The calculated axial and total VDV of both the participants are shown in Fig. 3b. The VDV exposure value for both the riders was found to be more than 2 times higher the HGCZ (VDV= 17 m/s$^{1.75}$) as prescribed by ISO 2631-1:1997 [15], which demands to limit the vibration exposure by ergonomic interventions. Exposure to WBV is a serious concern in occupational riders, as they are exposed to such high levels of vibration for up to 8 h on daily basis, which may aggravate the risk of potential vibration hazard.

The overall discomfort (Fig. 4a) as well as local discomfort (Fig. 4b) was found to be increased after 20 min of riding. Most affected body parts were found to be wrist and palm followed by lower arm, ankle/foot, lower back, buttocks and neck. This is also supported by the fact that slope of MDF was found to be negative for both BB and FCU muscle group. This discomfort may lead to muscular fatigue and hence risk of musculoskeletal disorder and fatalities if occurs for prolonged duration of time like in case of occupational riding. Therefore, it demands the ergonomic interventions in motorcycle design to reduce the discomfort.

4 Conclusions

The present study showed that motorcycle riders suffer vibration exposure (VDV = 38 m/s$^{1.75}$) more than twice the limit set by Health Guideline Caution Zone (HGCZ: VDV = 17 m/s$^{1.75}$) of ISO-2631-1:1997. For occupational riding, such a high-level vibration is extremely hazardous that must be reduced to a permissible level. Further,

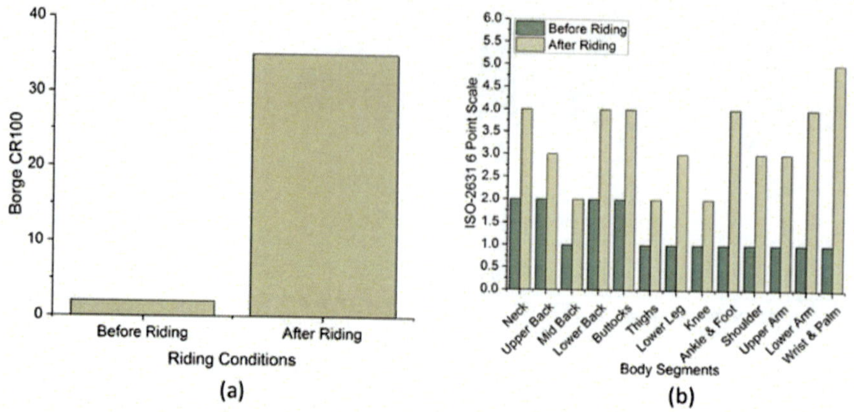

Fig. 4 **a** Overall discomfort of the rider and **b** local discomfort of different body parts before and after riding

the wrist and palm were the most reported discomfort site by the riders, and this was in positive agreement with the sEMG activity of the forearm muscles. Discomfort in the palm and wrist may be reduced by redesigning of the human coupling with clutch and brakes in occupational setting. The present study had observations only from two subjects, and further study is necessitated considering more samples from occupational and non-occupational groups to understand the effect of hand-arm vibration and manual controlling actions of the rider on discomfort in wrist and palm.

References

1. McCombe, A.W.: Hearing loss in motorcyclists: occupational and medicolegal aspects. J. R. Soc. Med. **96**, 7–9 (2003). https://doi.org/10.1258/jrsm.96.1.7
2. Richa Maheshwari, C.M.: Delivery boys suffer back problems: the backbone of Indian e-commerce boom faces serious breakdown. https://economictimes.indiatimes.com/industry/services/retail/delivery-boys-suffer-back-problems-the-backbone-of-indian-e-commerce-boom-faces-serious-breakdown/articleshow/50363413.cms?from=mdr (2015)
3. Arunachalam, M., Mondal, C., Singh, G., Karmakar, S.: Motorcycle riding posture: a review. Measurement **134**, 390–399 (2018). https://doi.org/10.1016/j.measurement.2018.10.019
4. Ospina-Mateus, H., Quintana Jiménez, L.A.: Understanding the impact of physical fatigue and postural comfort experienced during motorcycling: A systematic review. J. Transp. Heal. **12**, 290–318 (2019). https://doi.org/10.1016/j.jth.2019.02.003
5. Alias, A.N., Karuppiah, K., Tamrin, Shamsul Bahri Mohd Abidin, E.Z., Shafie, U.K.M., Sambasivam, S.: Risk factors of muscular discomfort among motorcyclist-review article (2016)
6. Diyana, N., Karuppiah, K., Rasdi, I., Sambasivam, S., Tamrin, S.M., Mani, K.C., Syahira, P., Azmi, I.: Vibration exposure and work-musculoskeletal disorders among traffic police riders in Malaysia: a review. Ann. Trop. Med. Public Heal. **10**, 334 (2017). https://doi.org/10.4103/ATMPH.ATMPH_91_17
7. Parvez, M., Khan, A.A.: Prediction of ride comfort of two-wheeler riders exposed to whole-body vibration. Presented at the (2010).https://doi.org/10.1007/978-981-13-6469-3_52

8. Parvez, M., Khan, A.A.: Simulation based discomfort assessment of two-wheeler riders. Int. J. Hum. Factors Model. Simul. 6, 314–332 (2018). https://doi.org/10.1504/IJHFMS.2018.096147
9. Mansfield, N., Sammonds, G., Nguyen, L.: Driver discomfort in vehicle seats—effect of changing road conditions and seat foam composition. Appl. Ergon. **50**, 153–159 (2015). https://doi.org/10.1016/j.apergo.2015.03.010
10. Wu, J.Z., Welcome, D.E., McDowell, T.W., Xu, X.S., Dong, R.G.: Modeling of the interaction between grip force and vibration transmissibility of a finger. Med. Eng. Phys. **45**, 61–70 (2017). https://doi.org/10.1016/j.medengphy.2017.04.008
11. Winkes, M.B., Teijink, J.A., Scheltinga, M.R.: Motorcycle racer with unilateral forearm flexor and extensor chronic exertional compartment syndrome. BMJ Case Rep. 2016, bcr2016214739 (2016). https://doi.org/10.1136/bcr-2016-214739
12. Marina, M., Rios, M., Torrado, P., Busquets, A., Angulo-Barroso, R.: Force-time course parameters and force fatigue model during an intermittent fatigue protocol in motorcycle race riders. Scand. J. Med. Sci. Sports. **25**, 406–416 (2015). https://doi.org/10.1111/sms.12220
13. Ma'arof, M.I.N., Omar, A.R., Haron, R., Husain, H., Wahab, M.A.: Muscular activities measurements of forward lean and upright sitting motorcycling postures via surface electromyography (sEMG). MATEC Web Conf. 135, 00046 (2017). https://doi.org/10.1051/matecconf/201713500046
14. Arpinar-Avsar, P., Birlik, G., Sezgin, O.C., Soylu, A.R.: The effects of surface-induced loads on forearm muscle activity during steering a bicycle. J. Sports Sci. Med. **12**, 512–520 (2013)
15. ISO 2631-1: Mechanical vibration and shock—Evaluation of human exposure to whole-body vibration—Part 1: General requirements (1997)
16. Khamis, N.K., Nuawi, M.Z., Deros, B.M., Ismail, F.R., Mohamad, D., Md Tahir, N.H.: Assessment of whole body vibration exposure among motorcyclist in Malaysia under different speeds and different road profiles: a preliminary study. Adv. Environ. Biol. **8**, 160–163 (2014)
17. Bhardwaj, S., Khan, A.A.: Ergonomics investigation for orientation of the handles of wood routers. Int. J. Occup. Saf. Ergon. **24**, 592–604 (2018). https://doi.org/10.1080/10803548.2017.1373479
18. Farooq, M., Khan, A.A.: Effects of shoulder rotation combined with elbow flexion on discomfort and EMG activity of ECRB muscle. Int. J. Ind. Ergon. **44**, 882–891 (2014). https://doi.org/10.1016/J.ERGON.2013.10.010

Group Preventive Maintenance Model for Multi-unit Series System: A TLBO Algorithm-Based Approach

Aseem K. Mishra, Divya Shrivastava, and Harsh Gupta

1 Introduction

Group maintenance planning decisions provide the economic benefits to the industrial systems in terms of reduced downtime and production losses. For instance, in a series production system, stoppage of one component due to scheduled preventive maintenance (PM) or failure may stop the whole system. This brings opportunity to perform PM on other components. This type of maintenance is generally referred as opportunistic maintenance (OM). To address this issue, several maintenance models have been proposed by researchers with an objective to minimize the total maintenance cost and to improve reliability and availability of the system [1]. However, researchers are facing several challenges in achieving toward optimum solution in this aspect. One such challenge is the stochastic behavior of machine failure. Mostly simulation and optimization techniques are utilized to deal with such situations.

A large amount of research in group maintenance optimization is available in literature which is beyond the scope of this article to include all contributions. Rather we refer the readers to the related review by Nicolai and Dekkar [2] and Samat and Kamaruddin [3, 4] for the comprehensive review of research works in this area. Since the group maintenance models involve large number of potential optimal solutions, researchers have frequently applied meta-heuristics in order to retrieve the near optimal solutions. For instance, Chalabi et al. [5] applied particle swarm optimization algorithm to solve a group PM model for multi-unit series system. An opportunistic group maintenance model with imperfect PM was proposed by Tambe et al. [6] to

A. K. Mishra (✉) · H. Gupta
Department of Mechanical Engineering, ITS Engineering College, 46, Knowledge Park-III, Greater Noida 201308, India
e-mail: aseem.me.engg@its.edu.in

D. Shrivastava
Department of Mechanical Engineering, Shiv Nadar University, NH 91 Tehsil Dadri, Greater Noida 201314, India

M. Muzammil et al. (eds.), *Recent Advances in Mechanical Engineering*, Lecture Notes in Mechanical Engineering, https://doi.org/10.1007/978-981-15-8704-7_7

minimize the expected total maintenance cost of the system. A genetic algorithm and simulated annealing algorithms were applied to obtain near optimum solution. Some of the latest research on group maintenance optimization are reported in literature [7, 8]. Although the group maintenance models proposed by researchers provide effective solutions, yet it appears from the literature review that most of approaches are intractable as the problem size increases. This is due to complex nature of optimization methods and grouping methodology which when combined leads to difficulties in achieving the near optimum solution. Furthermore, it is very selective to identify as to which group of components to be chosen for group maintenance as they have stochastic, economic, and structural interdependencies [9].

The present paper proposes a novel and effective methodology for grouping of components under PM intervals in a multi-unit series system. The objective is to retrieve the optimum PM interval and group of components, which minimizes the total maintenance cost per unit time of the system. A numerical example is adopted to identify the effectiveness of the proposed approach. A recently developed algorithm named teaching–learning-based optimization (TLBO) algorithm [10] is applied to optimize the objective function. The specialty of TLBO as against other evolutionary-based heuristics is that it is completely free from algorithm-specific parameters. This is perhaps the first attempt of TLBO for maintenance optimization problems. Computational results show the efficacy of the proposed approach as compared from the traditional maintenance practices.

2 Problem Statement

2.1 Description of the System

In this paper, we consider a system with n units arranged in series. Each unit $i \in \{1, 2, \ldots, n\}$ undergoes two types of maintenance, corrective maintenance (CM) which is performed upon failure and preventive maintenance (PM) which is done to avoid failure. An age reduction-based periodic imperfect preventive maintenance model is considered [11] which represent the practical situation that the age of the component reduces to some level but not all the way to new. On the other hand, CM is assumed to be minimally repaired. Each unit follows a two-parameter Weibull distribution with an increasing failure rate, which can be expressed as:

$$\lambda(t) = \frac{\beta}{\eta} \times \left(\frac{t}{\eta}\right)^{\beta-1} \tag{1}$$

where $\lambda(t)$ is the hazard rate function at time t, β and η are the respective shape and scale parameters for Weibull distribution. The expected number of failures are estimated based on the well-known theory that the number of minimal repairs between two consecutive PM follows a non-homogenous Poisson process (NHPP). It can be represented by a cumulative intensity function $\int_{r_i \times \mathrm{opm}_i}^{(r_i+1)\times \mathrm{opm}_i} \lambda(t)\mathrm{d}t$, where r_i is the rth PM interval of component i and opm_i is the optimum PM interval of the ith component. The expected total maintenance cost per unit time of the system is minimized pertaining to group PM activities which brings net savings in downtime cost of the system as compared to the individual maintenance models.

3 Mathematical Models

This section explains the development of various maintenance models, which provide the methodology of group maintenance planning in multi-unit series system.

3.1 *Single-Unit Model*

This model serves as initial solution to the proposed group maintenance model. It refers to the identification of optimum PM interval of each unit individually. It is based on the traditional approach of determining the optimum PM interval by strike of a balance between corrective and preventive maintenance cost [12]. The mathematical model of single-unit maintenance approach can be written as:

$$\text{Minimize } Z_i = \frac{\mathrm{Npm}_i * \left\{\mathrm{tr}_i^{\mathrm{pm}} * (\mathrm{pr} * \mathrm{lp_c} + \mathrm{lc}) + C_i^{\mathrm{pm}}\right\} + \mathrm{Nf}_i * \left\{tr_i^{\mathrm{cm}} * (\mathrm{pr} * \mathrm{lp_c} + \mathrm{lc}) + C_i^{\mathrm{cm}}\right\}}{T_\mathrm{m}} \quad (2)$$

where

Z_i	Total maintenance cost per unit time of the unit i (Rs./h).
Npm_i	Number of PM intervals of unit i.
$\mathrm{tr}_i^{\mathrm{pm}}$	Time to repair for PM action of unit i (h).
pr	Production rate (jobs/h).
$\mathrm{lp_c}$	Loss production cost (Rs./job).
lc	Labor cost (Rs./h).
C_i^{pm}	Fixed PM cost of unit i (Rs./unit).
Nf	Expected number of failures of unit i.
$\mathrm{tr}_i^{\mathrm{cm}}$	Time to repair for CM action of unit i (h).
C_i^{cm}	Fixed CM cost of unit i (in Rs./unit).
T_m	Mission time (in hours).

Thus, the optimum PM interval of unit i, $\mathrm{opm}_i = T_\mathrm{m}/\mathrm{Npm}_i$ can be obtained by minimizing Z_i as given in Eq. 2.

3.2 Group Maintenance Model

In a group maintenance approach, it is assumed that PM interval of each component will be nearest possible integer multiple of the minimum PM interval among the components of the system. For instance, let $\mathrm{opm}_{\min} = \min\{\mathrm{opm}_i\}_{i\in\{1,2,\ldots,n\}}$ be the minimum PM interval of all units in a system. The new PM intervals of units can be estimated based on the integer multiple of $\mathrm{opm}_{\min}$ as $\mathrm{opm}_i' = \alpha_i * \mathrm{opm}_{\min}$ where opm_i' are the new estimated PM interval of unit *i*. In this way, PM can be performed simultaneously on the grouped units having same PM intervals. However, this may not be the optimal solution as small shift in the value of minimum PM interval will create different integer multiples leading to large number of potentially optimal solutions. Furthermore, even at the fixed value of minimum PM interval, the rounding off to nearest possible integer (low or high value) may create multiple solutions. Therefore, the optimization problem becomes complex leading to decision variables as minimum PM interval ($\mathrm{opm}_{\min}$) and the integer multiples (α_i).

Hence, a mathematical model to obtain the optimum PM interval and grouping of components in order to minimize total maintenance cost can be formulated as follows:

$$\text{Minimize } Z_o = \left[\begin{array}{l} (T_\mathrm{m}/\mathrm{opm}_{\min}) \times \left\{\max\left(\mathrm{tr}_i^{\mathrm{pm}}\right)_{i\in G_i} \times (\mathrm{pr} * \mathrm{lp_c} + \mathrm{lc})\right\} + \\ \sum_{i=1}^{n} \frac{T_\mathrm{m}}{(\alpha_i * \mathrm{opm}_{\min})} \times C_i^{\mathrm{pm}} + \sum_{i=1}^{n} \mathrm{Nf}_i \times \left\{\mathrm{tr}_i^{\mathrm{cm}} \times (\mathrm{pr} * \mathrm{lp_c} + \mathrm{lc}) + C_i^{\mathrm{cm}}\right\} \end{array}\right] / T_\mathrm{m} \quad (3)$$

Subject to:

$$0 < \mathrm{opm}_{\min} \le \max\{\mathrm{opm}_i\} \forall i \in \{1, 2, \ldots, n\} \quad (4)$$

$$1 < \alpha_i \le \max\{\alpha_i\} \forall i \in \{1, 2, \ldots, n\} \quad (5)$$

where Z_o is the total group maintenance cost per unit time of the system. The first term in Eq. 3 represents the downtime cost of the system due to group PM which is reduced in group PM approach. $\max\left(\mathrm{tr}_i^{\mathrm{pm}}\right)_{i\in G_i}$ represents the maximum time to repair for PM among grouped components whereas G_i denotes the components grouped together. The second term of the equation signifies the fixed PM cost, which is independent of grouping of units. The third term indicates the downtime and fixed cost of corrective maintenance due to failures. Constraint 4 and 5 define the range of minimum PM interval and integer variables, respectively.

4 Solution Technique: TLBO Algorithm

This section presents the procedure of TLBO application to optimize the present objective function. Teaching–learning-based optimization (TLBO) is a novel and efficient meta-heuristic recently proposed by Venkata Rao [10]. It mimics the teaching–learning process of human beings in a classroom. TLBO has been successfully applied in various fields of engineering and sciences [13].

4.1 *Application of TLBO to the Group Maintenance Model*

In the present optimization problem, the decision variables are PM interval and integer multiples. The steps involved in the optimization of present objective function by TLBO algorithm are as follows:

Step 1 Initialize decision variables, i.e., PM interval and integer multiples (t_{pm} and α_i), population size (ps) and number of iterations (it).

Step 2 Update the values of decision variable and solution based according to Eq. 6.

$$x'_{j,l,k} = x_{j,l,k} + r * \left(x_{j\text{best},k} - T_f * \mu_{j,k}\right) \tag{6}$$

where

$x_{j,l,k}$	Previous value of jth variable of lth population in k^{th} iteration.
$x'_{j,l,k}$	Updated value of jth variable of lth population in k^{th} iteration.
$x_{j\text{best},k}$	Best value of jth variable in the kth iteration.
T_{f}	Teaching factor ($T_{\text{f}} = 1$).
$\mu_{j,k}$	Mean of all values of jth variable in each population in the kth iteration.
r	Random number in the range $U[0,1]$.

Step 3 Compare the solutions obtained from the updated variables in step 3 from previous variables. If the value of objective function is less as compared to the previous replace it otherwise retains the older one (This is the end of teacher phase).

Step 4 In the learner phase, each candidate in the population will interact with other candidate (selected randomly) to enhance the solution. The updated variables after interaction are obtained as per the following equations

$$x''_{j,l,k} = x'_{j,l,k} + r * \left(x'_{j,l,k} - x'_{j,l',k}\right) \quad \text{if } Z_{olk} < Z_{ol'k} \tag{7}$$

$$x''_{j,l,k} = x'_{j,l,k} + r * \left(x'_{j,l',k} - x'_{j,l,k}\right) \quad \text{if } Z_{ol'k} < Z_{olk} \tag{8}$$

where

$x''_{j,l,k}$ Updated value of jth variable of lth population in k^{th} iteration in the learner phase.

$x'_{j,l,k}$ Previous value of the variable j^{th} variable of lth population in k^{th} iteration (updated value in teacher phase).

$x'_{j,l',k}$ Previous value of the interacting variable of jth variable of lth population in k^{th} iteration (updated value in teacher phase).

Z_{olk} Previous Value of objective function of lth population in k^{th} iteration.

$Z_{ol'k}$ Previous Value of objective function of lth interacting population in k^{th} iteration.

Step 5 Compare the solutions obtained in the learner phase from the teacher phase and update the better values accordingly. This is the end of learner phase. This completes one iteration.

Step 6 Repeat the steps 2–6 until the number of iterations are completed.

Step 7 Report the near optimum solution.

Note that the above steps and equations are applied for both decision variables, i.e., optimum PM interval (opm_i) and integer multiples (α_i) in the evaluation of total expected maintenance cost per unit time of the system.

5 Results and Discussions

A numerical example of five components arranged in series is adopted to identify the effectiveness of the proposed methodology. The input parameters of the each component are given in Table 1. The common parameters include labor cost of 500 Rs./h, cost of lost production 40 Rs./job, and production rate of 20 jobs/h. Table 2 shows the results of optimum PM intervals and minimum expected maintenance cost per unit time obtained from the single-unit model. From Table 2, it can be seen that third component has minimum optimum PM interval of 46 h. The PM intervals of other components can be represented as integer multiple of 46. The new estimated PM intervals and integer multiples are shown in Table 2. The results indicate that

Table 1 Input parameters of multi-component system

S. No.	Weibull parameters		Cost parameters		Time parameters	
	η	β	C_i^{pm}	C_i^{cm}	$\mathrm{tr}_i^{\mathrm{pm}}$	$\mathrm{tr}_i^{\mathrm{cm}}$
1	480	1.8	1800	7460	4	11
2	500	1.9	2700	19,600	4	9
3	290	2.4	3700	22,500	7	17
4	900	2.5	4300	29,100	5	12
5	920	2.1	6800	36,600	8	20

Table 2 Optimum solutions of single-unit model

S. No.	Optimum PM interval	Number of failures	Total maintenance cost	PM ratio	Integer multiple	New PM intervals
	opm_i	Nf_i	Z_i	opm_i/opm_{min}	α_i	opm'_i
1	172	28.3	175	3.74	4	184
2	98	32.2	232	2.13	2	92
3	46	145.6	1534	1	1	46
4	183	13.6	214	3.98	4	184
5	232	8.8	256	5.04	5	230

components 1 and 4 form one group with optimum PM interval of 184 h. Units 2, 3, and 5 have optimum PM intervals of 92, 46, and 230 h, respectively. This grouping is known as initial group with total maintenance cost of Rs. 2156/h which is less than sum of costs obtained from single-unit model. However, this might not be the optimal solution as both PM interval and integer multiples are estimated values. To retrieve the near optimal solution, the problem is optimized by TLBO algorithm. The optimum solution is 4-1-1-4-4 with minimum PM interval of 54 h, i.e., components 1, 4, and 5 form a group with PM interval of 216 h and units 2 and 3 form another group with 54 h of PM time.

Figure 1a shows the comparison of maintenance cost with three different approaches. It is seen that the optimum group has a minimum cost of 1982 Rs./h which shows approximate 9% savings from the initial group and 21% single-unit model. Figure 1b depicts the grouped components with PM intervals. Thus, the proposed methodology and optimization methods can lead to direct cost reductions pertaining to maintenance optimization when applied to real industrial systems.

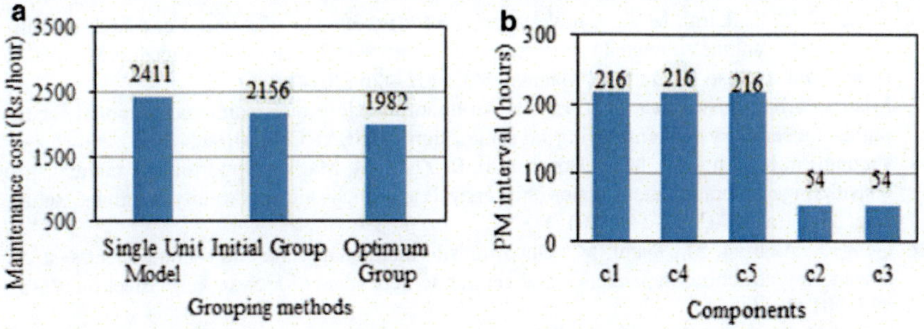

Fig. 1 **a** Cost comparison of different maintenance approaches and **b** optimum PM intervals of grouped components

6 Conclusions and Future Extensions

This paper proposed a group maintenance model for a multi-unit system to minimize the total maintenance cost of the system. The grouping of units is performed under PM intervals. A recently proposed TLBO algorithm was applied to optimize the objective function. Computational results show that the proposed approach yields 21% cost savings as compared to the traditional maintenance practices. Thus, the proposed methodology can bring economic benefits to the real manufacturing systems. Furthermore, group maintenance approach can be beneficial to continuous production systems such as chemical plants and electrical power plants where there is less scope of production stoppage due to maintenance. As a scope of future research, the present approach can be applied for multi-objective optimization such as increasing reliability and improve quality. Furthermore, TLBO algorithm can find its application to other maintenance optimization problems as well.

References

1. Cui, L.: Maintenance models and optimization. In: Handbook of Performability Engineering, pp. 789–805. Springer, London (2008)
2. Nicolai, R.P., Dekker, R.: Optimal maintenance of multi-component systems: a review. In: Complex System Maintenance Handbook, pp. 263–286. Springer, London (2008)
3. Ab-Samat, H., Kamaruddin, S.: Opportunistic maintenance (OM) as a new advancement in maintenance approaches. J. Qual. Maint. Eng. **20**(2), 98–121 (2014)
4. E. I. Basri, I. H. A. Razak, H. Ab-Samat, and S. Kamaruddin, "Preventive maintenance (PM) planning: A review," *Journal of Quality in Maintenance Engineering*, vol. 23, no. 2. Emerald Group Publishing Ltd., pp. 114–143, 2017.
5. Chalabi, N., Dahane, M., Beldjilali, B., Neki, A.: Optimisation of preventive maintenance grouping strategy for multi-component series systems: Particle swarm based approach. Comput. Ind. Eng. **102**, 440–451 (2016)
6. Tambe, P.P., Kulkarni, M.S.: A novel approach for production scheduling of a high pressure die casting machine subjected to selective maintenance and a sampling procedure for quality control. Int. J. Syst. Assur. Eng. Manage. **5**(3), 407–426 (2014)
7. Mishra, A.K., Shrivastava, D., Vrat, P.: An opportunistic group maintenance model for the multi-unit series system employing Jaya algorithm. OPSEARCH (2019)
8. Truong Ba, H., Cholette, M.E., Borghesani, P., Zhou, Y., Ma, L.: Opportunistic maintenance considering non-homogenous opportunity arrivals and stochastic opportunity durations. Reliab. Eng. Syst. Saf. **160**, 151–161 (2017)
9. Geng, J., Azarian, M., Pecht, M.: Opportunistic maintenance for multi-component systems considering structural dependence and economic dependence. J. Syst. Eng. Electron. **26**(3), 493–501 (2015)
10. Rao, R.V., Savsani, V.J., Vakharia, D.P.: Teaching–learning-based optimization: a novel method for constrained mechanical design optimization problems. Comput. Des. **43**(3), 303–315 (2011)
11. Zhou, X., Xi, L., Lee, J.: Opportunistic preventive maintenance scheduling for a multi-unit series system based on dynamic programming. Int. J. Prod. Econ. **118**(2), 361–366 (2009)
12. Pandey, D., Kulkarni, M.S., Vrat, P.: A methodology for joint optimization for maintenance planning, process quality and production scheduling. Comput. Ind. Eng. **61**(4), 1098–1106 (2011)

13. Venkata Rao, R.: Review of applications of TLBO algorithm and a tutorial for beginners to solve the unconstrained and constrained optimization problems. Decis. Sci. Lett. **5**, 1–30 (2016)

Parametric Investigation of a Rotational Parabolic-Tapered Cantilever with Elliptical Sectional Area Exposed to Variable Thermal Gradient and Axial Dynamic Load

Rakesh Ranjan Cahnd and Amit Tyagi

1 Introduction

Before going to design the blades for steam turbine and gas turbine, rotor blades, robotic manipulators and substrate beam for rotational vibration energy scavengers, etc., the stability study of the rotating beam is of much interest.

Rao and Carnegie [1] determined the natural frequencies of a cantilever using Galerkin's method under coupled torsional and bending vibration. Natural frequencies of an irregular cantilever structure mounted on a gyrating disk were investigated considering the polynomial frequency equation technique by Rao and Banerjee [2]. A finite element formulation is presented by Thomas and Sabuncu [3] for the stability analysis of a blade with irregular cross section. Shear center effects are also included to inspect the dynamic instability of revolving and non-revolving bladed disks. Banerjee [4] presented a dynamic stiffness matrix, and the three mode shapes and natural frequencies of a cantilever of Timoshenko type were inspected. The effects of the taper and stiffness of the foundation on the instability of an irregular beam fixed on a flexible foundation, subjected to a constant temperature grade, are analyzed by Kar and Sujata [5]. Nayak et al. [6] as well inspected the consequence of fixed temperature grade on the static stability of an irregular sandwich structure having viscoelastic boundary conditions at the ends.

But till now, no investigation has been conducted on a parabolic tape ring revolving system subjected to a variable temperature gradient. The elliptical cross section will definitely reduce the drag force on the system, and the taper profile will increase the strength to weight ratio of the system.

R. R. Cahnd (✉) · A. Tyagi
Mechanical Engineering Department, IIT(BHU), Varanasi 221005, India
e-mail: rakeshrchand.rs.mec17@itbhu.ac.in

A. Tyagi
e-mail: atyagi.mec@iitbhu.ac.in

M. Muzammil et al. (eds.), *Recent Advances in Mechanical Engineering*, Lecture Notes in Mechanical Engineering, https://doi.org/10.1007/978-981-15-8704-7_8

2 Modeling of the Problem

A revolving parabolic-tapered beam with an elliptical cross-sectional area of length l mounted on a hub of radius B_0, which rotates with a constant angular speed N. An amalgamation of static and transient external force $w(t) = w_0 + w_1\cos(\omega t)$ is applied at the free end along the centroid of the cross section, where w_0 and w_1 are the static and dynamic loads, respectively (Fig. 1).

With the assumption that the thermal gradient along the Y-direction is zero, and considering ρ is the mass density, $A(x)$ is the cross-sectional area at the standard section, $\delta(x, t)$ is the transverse deflection of the axis of the cantilever, $E(x)$ is the elasticity modulus and $I(x)$ is the area moment of inertia of the generic section of the system, using the total potential and kinetic energy, and the work exerted due to the external load in Hamilton's equation as in Chand et al. [8] and considering various non-dimensional parameters such as length $(\psi = x/l)$, transverse deflection $(\delta^* = \delta/l)$, time $(\tau = c \times t)$,non-dimensional excitation frequency (ω^*), the non-dimensional equation of motion can be written as;

$$\left[S(\psi)T(\psi)\delta^{*''}\right]'' + m(\psi)\ddot{\delta}^* + \left[r_g N_0^2 + w(\tau)\right]\delta^{*''} - N_0^2\left[q(\psi)\delta^{*'}\right]' = 0 \quad (1)$$

$\delta^*(\psi, \tau) = \sum_{r=1}^{N} \delta_r^*(\psi) f_r(\tau)$ In the above expression $I(\psi) = I * S(\psi)$, where $S(\psi)$ is the distribution of moment of inertia along the axis, $T(\psi)$ is Young's modulus distribution function, $m(\psi)$ is the mass distribution function,N_0 is the non-dimensional speed parameter. The series solution to the equation of motion can be considered as.

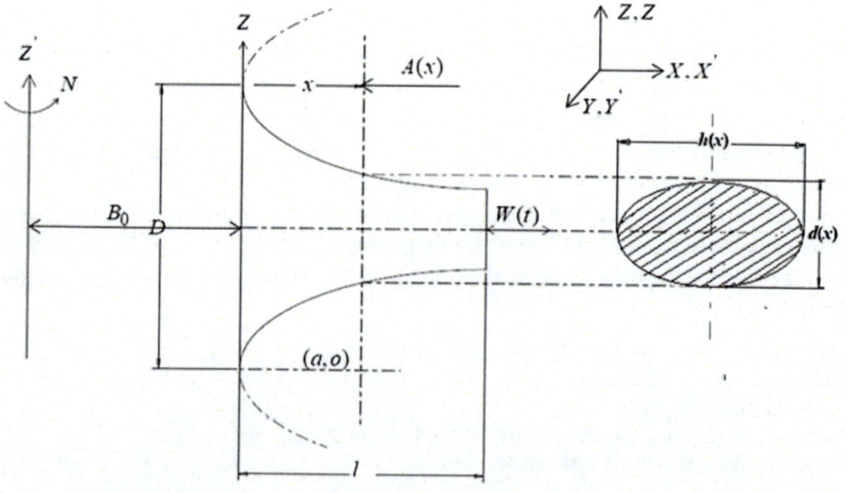

Fig. 1 System configuration

$$\delta^*(\psi,\tau) = \sum_{r=1}^{N} \delta_r^*(\psi) f_r(\tau) \tag{2}$$

where $f_r(\tau)$ and $\delta_r^*(\psi)$ are the time and coordinate function for the cantilever structure chosen from Parida and Dash [7] and given below

$$\zeta_r(\eta) = (j+2)(j+3)\zeta^{(j+1)} - 2j(j+3)\zeta^{(j+2)} + j(j+1)\zeta^{(j+3)} \tag{3}$$

With the help of Galerkin's principle and replacing the assumed series solution, the matrix form equation of motion can be acquired as

$$[M]\{\ddot{s}\} + [K]\{s\} - \{w_0[H] - w_1\cos(\omega^* \times \tau)[H]\}\{s\} = \{0\} \tag{4}$$

2.1 *Formulation for Regions of Dynamic Instability*

Introducing a linear coordinate transformation $\{s\} = [L]\{u\}$, where $\{u\}$ characterizes a group of novel generalized coordinates and bearing in mind, $[L]$ be the modal matrix of $[M]^{-1}[K]$, Eq. (4) can be rewritten as;

$$\{\ddot{u}\} + [\omega_n^2]\{u\} + w_1\cos(\omega^*\tau)[B]\{u\} = \{0\} \tag{5}$$

where the frequency matrix $[\omega_n^2]$ is equivalent to $[M]^{-1}[K]$ and $[B] = -[L]^{-1}[M]^{-1}[H][L]$. Equation (5) can be stated as;

$$\ddot{u}_n + \omega_n^2 + w_1\cos(\omega^*\tau)\sum_{m=1}^{N} b_{mn}u_n = 0 \tag{6}$$

The above equation illustrates a system of N-coupled complex Hill's equations. With the help of the Saito and Otomi formulations, the borders of the instability zones for higher modes, i.e., simple resonance ($\omega = 2\omega_n, n > 1$) and combination resonance ($\omega = \omega_n + \omega_m, n \neq m$), are obtained for the undamped case.

2.2 *Formulation for Static Buckling Load*

Substituting $\{\ddot{s}\} = \{0\}$ and $w_1 = 0$ in Eq. (4) will characterize an eigenvalue problem of $[K]^{-1}[H]\{s\} = \frac{1}{w_0}\{s\}$. The static buckling loads are found as the real parts of the reciprocal of the Eigenvalues of $[K]^{-1}[H]$. The buckling loads, w_0, for the first three modes are found as and plotted against various rotational speeds and taper parameters.

3 Computational Results and Analysis

Computational results are as simulated using MATLAB coding for a number of the performance affecting parameters like revolution speed, temperature grade and taper parameter and presented employing a series of diagrams.

3.1 Dynamic Instability Plots

The permanence of the system has been inspected as; if a definite variation in a parameter narrows the instability zones, moves them to the higher excitation frequency and decreases the total count of instability zones, then it is decided that the stability of the cantilever has enhanced or else reduced.

Increase in revolving speed stabilizes the system because the centrifugal resistance arises; as a result, the increased speed diminishes the consequence of the external excitation as in Fig. 2.

Figure 3 shows that with arise in taper parameter α_1, the stability zone moves leftward and the size of the unstable regions surges, which imply the deterioration instability with rising of α_1.The variation of α_2 has the same effect but less significant than the variation of α_1 as shown in Fig. 4. This is because of the stronger influence of thickness dimension than that of the width. The first three natural frequencies of the system for various parameters are given in Table 1.

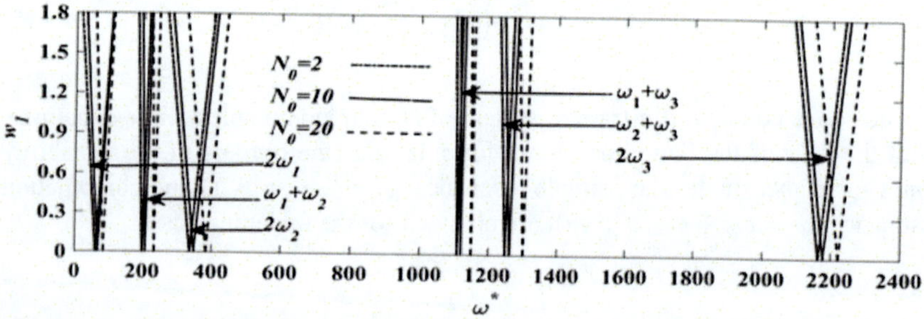

Fig. 2 Instability plot of the system with $D = 100$, $H = 50$, $l = 600$, $\alpha_1 = 0.001$, $\alpha_2 = 0.001$ and $\upsilon = 0.006$

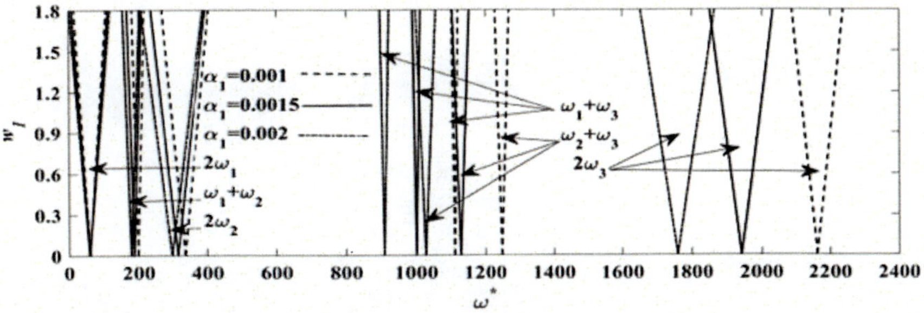

Fig. 3 Instability plot of the system with $D = 100$, $H = 50$, $l = 600$,$N_0 = 2$, $\alpha_2 = 0.001$ and $\upsilon =$ 0.006

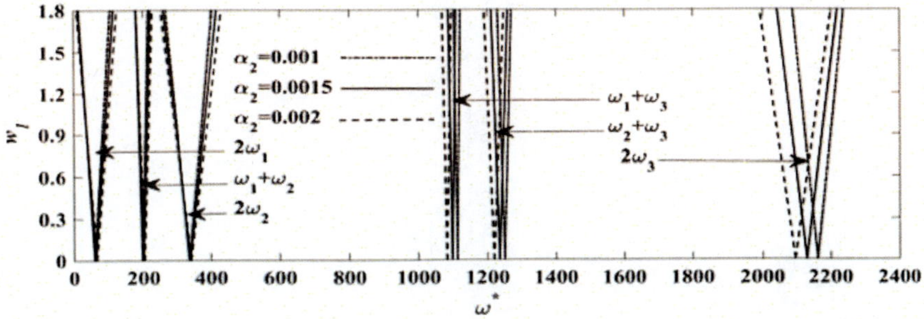

Fig. 4 Instability plot of the system with $D = 100$, $H = 50$, $l = 600$,$N_0 = 2$, $\alpha_1 = 0.001$ and $\upsilon =$ 0.006

3.2 *Static Stability Plots*

The results obtained for various parameters using a MATLAB code are presented through a series of graphs.

Figure 5a shows that there is a slight increase of static buckling load with an increase of revolving speed as with the increase of centrifugal force, the effect of static buckling load declines. Rise in both the variation parameters decreases the static buckling loads, as shown in Fig. 5b, c, and the rate of decrease of static buckling load is higher in case of increase in thickness taper parameter.

4 Conclusions

The computational analysis using MATLAB program leads to the following conclusions. The rise in revolving speed makes the system extra stable owing to the intensification in the centrifugal resistance force, which reduces the influence of the applied axial force. The surges in the taper parameters lead to a reduction in the stability of the

Table 1 First three natural frequencies of the system for various parameters

Modes	Rotational speed (N_0)			Taper parameter (α_1)			Taper parameter (α_2)		
Values	2	10	20	0.001	0.0015	0.002	0.001	0.0015	0.002
ω_1	30.796	34.240	42.212	30.796	30.475	30.230	30.796	32.365	34.070
ω_2	168.22	174.28	191.85	168.22	157.61	148.93	168.22	169.56	171.12
ω_3	1081.5	1088.6	1110.5	1081.5	971.84	879.90	1081.5	1065.6	1048.2

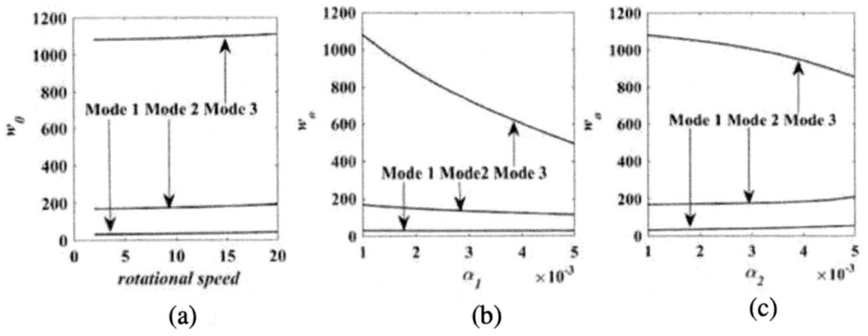

Fig. 5 Static buckling load plots for **a** different rotational speeds with $\alpha_1 = 0.001$ and $\alpha_2 = 0.001$, **b** various minor axis taper parameters with $N_0 = 2$ and $\alpha_2 = 0.001$, **c** various major axis taper parameters with $N_0 = 2$ and $\alpha_1 = 0.001$

rotating cantilever because of the decline in the flexural stiffness of the beam with the rise in taper parameters. For certain thermal gradient, stability is higher if it is taken as constant and declines if taken as a variable because of the reduction in stiffness of the cantilever structure under a variable temperature gradient. It is furthermore noticed that any change in the radius of the hub has a very negligible consequence on the transient instability of the structure as the centrifugal force generates predominantly because of the beam span. The buckling load factor decreases with arise in variation parameter and thermal gradient but surges with arise in rotational speed. This investigation can be valuable for vibration isolation of revolving irregular structures with high neighboring temperature and moderate revolving velocities and the design of the propeller and turbine blades, which requires a higher strength to weight proportion by selecting the appropriate value of the parameters acquired from the computational investigation.

References

1. Rao, J.S., Carnegie, W.: Solution of the equation of motion of coupled bending-torsion vibration of turbine blades by the method of Ritz-Galerkin. Int. J. Mech. Eng. **12**(10), 875–882 (1970)
2. Rao, J.S., Banerjee, S.: Coupled bending-torsional vibration of cantilever blades-method of polynomial frequency equation. Mech. Mach. Theor. **12**(4), 271–280 (1977)
3. Thomas, J., Subunca, M.: Finite element analysis of rotating pre-twisted asymmetric cross-section blades. ASME **1**(12), (1979)
4. Banerjee, J.R.: Development of an exact dynamic stiffness for free vibration analysis of a twisted Timoshenko beam. JSV **270**(1–2), 379–401 (2004)
5. Kar, R.C., Sujata, T.: Parametric instability of a non-uniform beam with thermal gradient resting on a Pasternak foundation. Comput. Struct. **29**(4), 591–599 (1988)
6. Nayak, S., Bisoi, A., Dash, P.R., Pradhan, P.K.: Static stability of a visco-elastically supported asymmetric sandwich beam with thermal gradient. IJSE **65**(6), 1–7 (2014)

7. Parida, R., Dash, P.R.: Dynamic stability analysis of a circularly tapered rotating beam subjected to axial pulsating load and thermal gradient under various boundary conditions. IJAV **21**(2), 137–144 (2016)
8. Chand, R.R., Behera, P.K., Pradhan, M., Dash, P.R.: Parametric stability analysis of a parabolic tapered rotating beam under variable temperature grade. JVET **7**(1), 23–31 (2019)

Analysis of E-Glass Fiber Wheel Rim by Using ANSYS

Pankaj Singh Bisht and Ankita Awasthi

1 Introduction

A wheel is circular made of the hard and durable material, and its center is a borehole circular in shape and connected directly with the axle of the vehicle about which wheel rotates. The wheel's moment is used to control the vehicle. When it is connected the crank, the wheel produces or transmits energy. Earlier wheel rims are made up of steel, but it has increased the weight and reduces the performance and speed of automobiles. After more improvising on materials, nowadays, aluminum is popularly used as wheel rim material. Aluminum is lightweight material, provides aesthetic look and supports the weight by giving appropriate strength. But in order to optimize the performance and strength to weight ratio, high-end automobiles are using carbon fiber.

Carbon fiber wheel rim: It is used by high-end automobile vehicles which require performance and more durability than other conventional such as sports cars, supercars, formula one racing cars use the carbon fiber wheel rim. Since the performance of carbon fiber-based wheel rim is best and more durable than convention wheel rims, but carbon fiber is costlier that is why for public comuters conventional wheel rims are used. The basic geometrical need of supercars is that they require high strength to weight ratio, toughness, resilience and great fatigue strength that is why the composite material is used in the manufacturing of wheel rim to full body manufacturing.

E-glass is a class of carbon fiber which has high strength and no corrosive nature, cheaper than the carbon fiber.

P. S. Bisht (✉) · A. Awasthi
IILM-CET-AHL, Greater Noida, India
e-mail: pankajbisht0195@gmail.com

A. Awasthi
e-mail: ankita.awasthi@iilmcet.ac.in

M. Muzammil et al. (eds.), *Recent Advances in Mechanical Engineering*, Lecture Notes in Mechanical Engineering, https://doi.org/10.1007/978-981-15-8704-7_9

1.1 Types of Wheel Rim

(a) **According to the dimensions, the wheel or rim is classified as follows:**

1. *Drop center wheel rim (DC)*
   It is the mounting which is shaped in slots or groove form in such a way that there is thin gap between the bead seat parts, which placed on both sides of the rim as shown in Fig. 1.
2. *Wide drop center wheel rim (WDC)*
   The design of wide drop center wheel rim is very similar with drop center wheel rim. In this, wheel's width is expanded little bit in comparison with DC wheel rim so that its lower flange is also elongated and made deeper. Such type of design for lower rim is applied to low aspect ratio tires.
3. *Wide drop center wheel rim with hump*
   This design has a hump and placed at the starting of the bead seat area of the tire (Fig. 2).

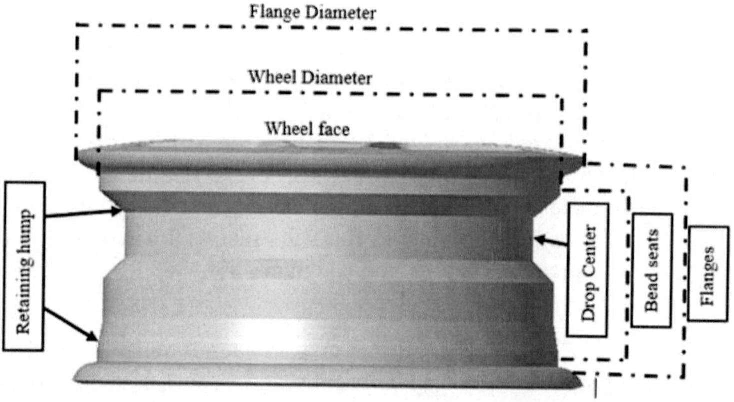

Fig. 1 Drop center (DC) rim

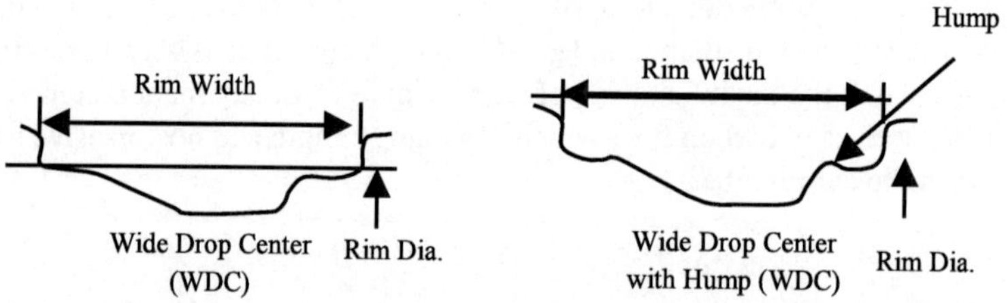

Fig. 2 Wide drop center (WDC) and wide drop center with hump wheel rim

2 Basic Nomenclature of Wheel Rim

1. **Wheel**: When disk and rim integrated with each other, then they form wheel.
2. **Rim**: It is the integral part of the wheel over which the tire is seated.
3. **Disk**: It is directly connected to wheel and mounted over the hub of vehicle axle.
4. **Bead Seat**: It is another integral part of the rim. The purpose of bead seat is to hold the tire in radial direction when radial forces applied toward the bead face.
5. **Hump**: The main purpose of hump is to prevent the tires for slipping from the rim when the vehicle is running and under the influence of high forces.
6. **Well**: The well is structured in order to facilitate the easy withdrawal and tire mounting over the rim (Fig. 3).

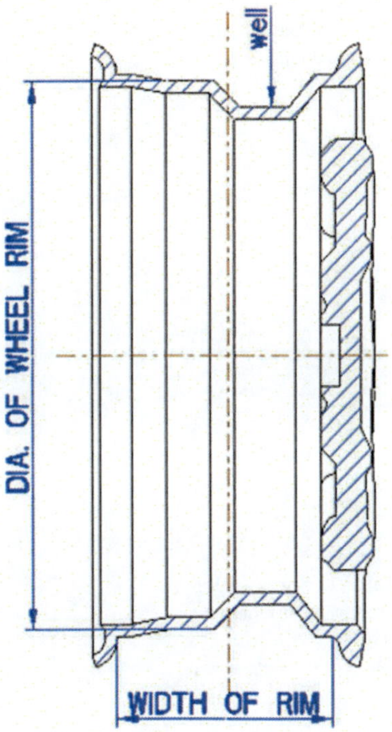

Fig. 3 Structure of wheel rim

3 Objective of Project

The main objectives of this study are as follows:

a. To show that multilayer wheel rim is suitable for high load and forces.
b. To show there is reduction in weight of wheel.
c. To show the stress in E-glass fiber in different orientation.
d. Wheel rim is designed as per JETMA.
e. The thickness of rim area of wheel is taken as 1 mm of seven layers.
f. The thickness of central part of wheel rim is taken as 1 mm of 14 Layers.

Problem Definition
To optimize the design of wheel rim, so that it will be more safe, lighter in weight with great strength than the conventional Al Alloy wheel rim used in present days. This paper has proposed the usage of E-glass fiber (Fig. 4; Table 1).

Fig. 4 Proposed 3D designed CAD

Table 1 Specification of proposed car wheel rim

S. No.	Specification	Value
1	Proposed width of rim (mm)	140 mm
2	Dia. of the rim (mm)	355 mm
3	Offset length (mm)	57 mm
4	Center hole dia. (mm)	40 mm
5	Thickness of the rim (mm)	7 mm
6	Dia. of stud hole (mm)	15 mm
7	Dia. of stud hole (mm)	68 mm
8	Number of stud holes	5 No's

4 Boundary Condition and Drawing

1. Fixed support is provided to set of five bolt-stud holes since wheel is seated over the vehicle axle with the help of bolts through stud holes.
2. The load is applied at the center of the hub in y-direction. As hub is fixed through the central hole of the rim so that maximum load is bore by this section. Total bearing load $(P) = 7769.5$ N.
3. On the tire allocation area, there is slight pressure of 0.241 Mpa which acts normally on the circumferential tread and flange portion of the wheel.
4. The diameter of tire as 550 mm, the proposed velocity of car rim taken 100 k mph, i.e., 965 rpm.
5. From annexure of AIS-073 (part-2), moment is taken as 5.0969e + 5 N mm.
6. Friction force is provided into the inner face of center hole which comes in action with axle (Fig. 5).

5 Static Analysis and Simulation of Composite Wheel Rim

The following load conditions will be shown in ANSYS Workbench 15.0 in analysis chapter (Fig. 6).

Variant-1
E-glass at [−25°, +25°] FW fiber orientation.

At ±25°, maximum deformation observed in flange area of the wheel rim shown in Fig. 7 is 0.2364 mm (Fig. 8).

At ±25° orientation of the lamina, there is maximum stress of 22.884 MPa at bolt area.

Layer stack up at ±25° of lamina is shown in Fig. 9

Variant-2
E-glass at [−35°, +35°] FW fiber orientation (Fig. 10).

At ±35° orientation of the lamina, there is maximum stress of 23.739 MPa at bolt area.

At ±35°, maximum deformation observed in flange area of the wheel rim shown in Fig. 11 is 0.2192 mm.

Layer stack up at ±35° of lamina is shown in Fig. 12.

Variant-3
E-glass at [−45°, +45°] FW fiber orientation.

At ±45°, maximum deformation observed in flange area of the wheel rim shown in Fig. 13 is 0.20061 mm (Fig. 14).

At ±45° orientation of the lamina, there is maximum stress of 21.736 MPa at bolt area.

Layer stack up at ±45° of lamina is shown in Fig. 15.

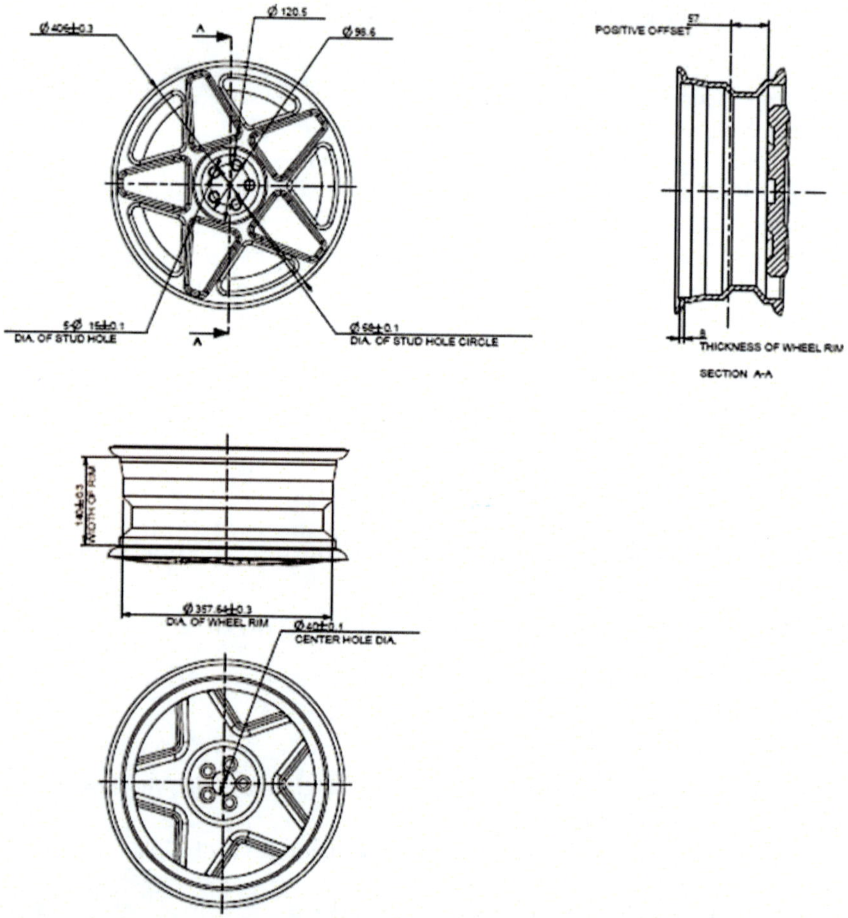

Fig. 5 Mechanical drawing

Variant-4

E-glass at [−55°, +55°] FW fiber orientation.

At ±55°, maximum deformation observed in flange area of the wheel rim shown in Fig. 16 is 0.19384 mm (Fig. 17).

At ±55° orientation of the lamina, there is maximum stress of 18.687 MPa at bolt area.

Layer stack up at ± 45° of lamina is shown in Fig. 18.

Fig. 6 Load condition for proposed wheel

Fig. 7 Total deformation of E-glass at [−25°, + 25°] FW orientation of fiber

6 Conclusion

After analyzing all the four variants of composite wheel rim, we have created a table to show the stresses and deformation at different orientation angles of composite material wheel rim at a specified mentioned boundary condition.

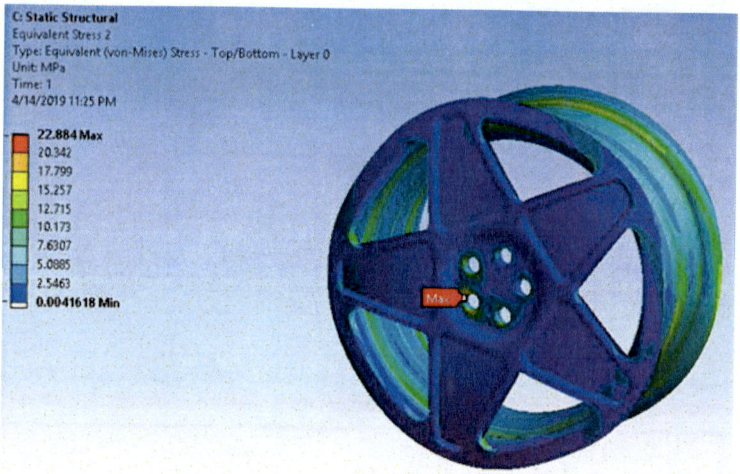

Fig. 8 Max stress is observed in bolt area

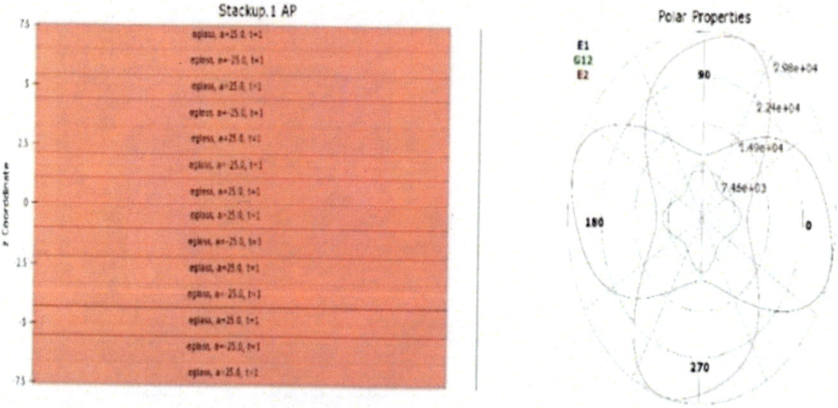

Fig. 9 Layer stack up at [±25°] FW

Variants	Orientation	Total deformation	Equivalent stress (MPa)
I	±25°	0.2364	22.884
II	±35°	0.2192	23.739
III	±45°	0.20061	21.736
IV	±55°	0.19384	18.687

- In current scenario, multilayered wheel rims are used in various automobile OEM's due to low weight and high strength along the wall thickness.
- Determined stress is less than allowable stress, so design is safe.

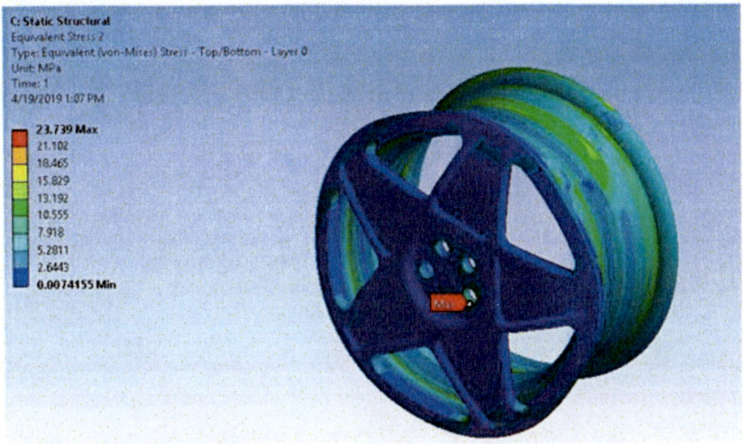

Fig. 10 Max stress is observed in bolt area

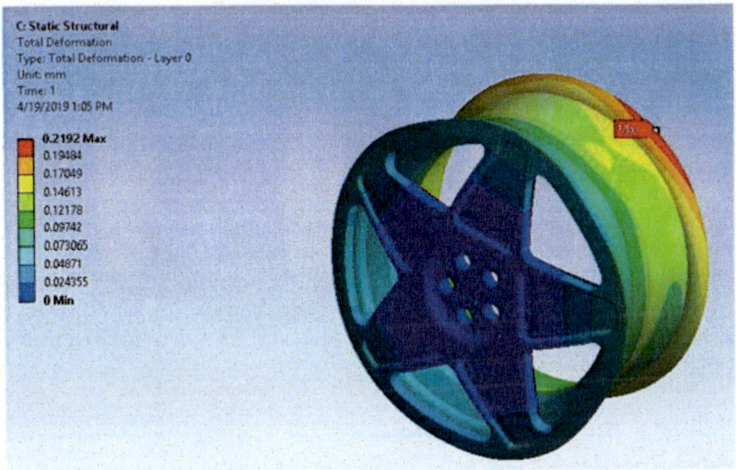

Fig. 11 Total deformation of E-glass at [−35°, +35°] FW orientation of fiber

- E-glass fiber having orientation [−55, +55] FW has less equivalent stress than currently using Al alloy wheel but greater deformation, which is allowed for safer design of particular geometry.
- It has good strength, and weight is around 7.8 kg.
- E-glass is 25% lesser in weight, and proposed design is safer.

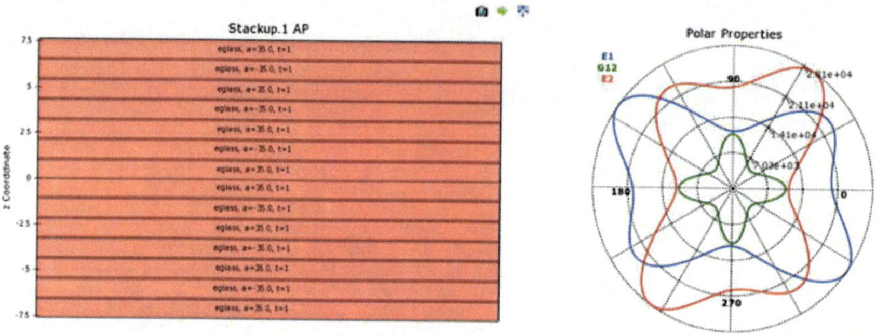

Fig. 12 Layer stack up at [±35°] FW

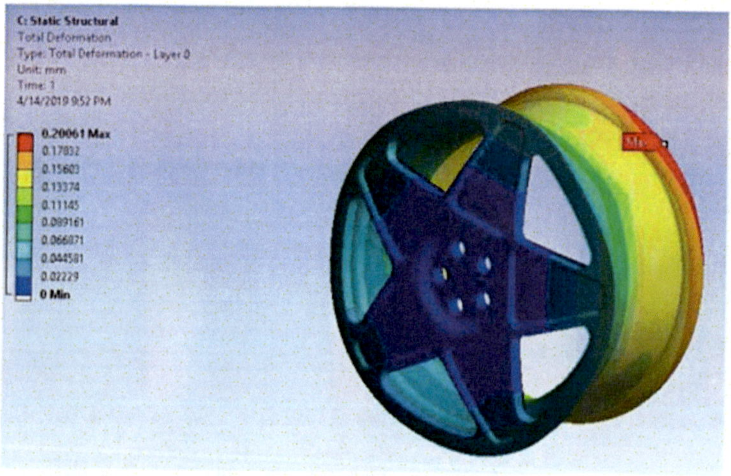

Fig. 13 Total deformation of E-glass at [−45°, +45°] FW orientation of fiber

7 Future Scope

The paper includes the static stress analysis, optimizes the design of composite wheel rim and reduces the weight of the wheel. In order to achieve greater strength to weight ratio, an E-glass is selected with different fiber orientation. But there is further scope in dynamic analysis of wheel rim, which includes the fatigue life of wheel rim.

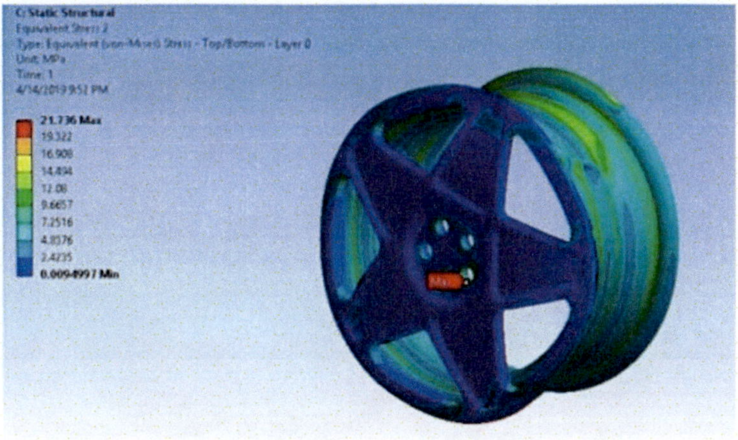

Fig. 14 Max stress is observed in bolt area

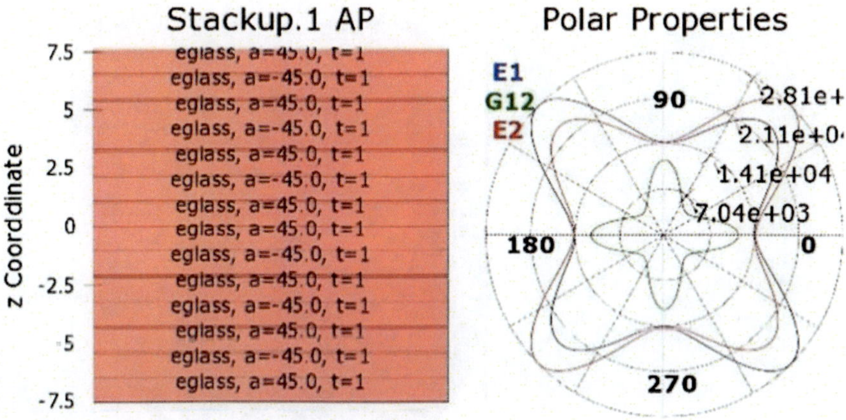

Fig. 15 Layer stack up at [±45°] FW

Fig. 16 Total deformation of E-glass at [−55°, +55°] FW orientation of fiber

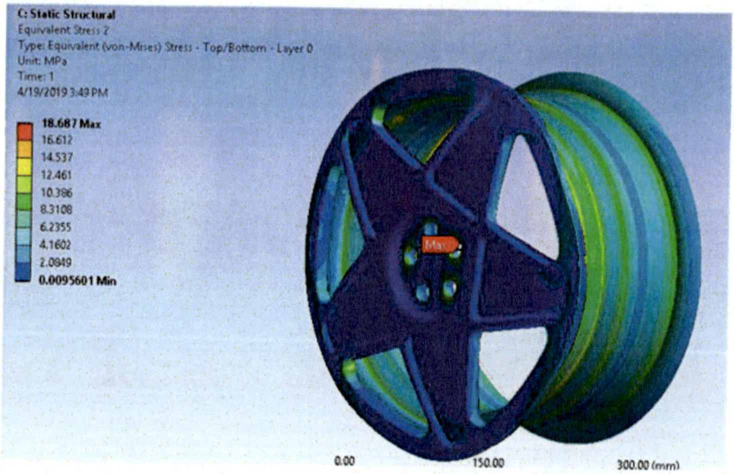

Fig. 17 Max stress is observed in bolt area

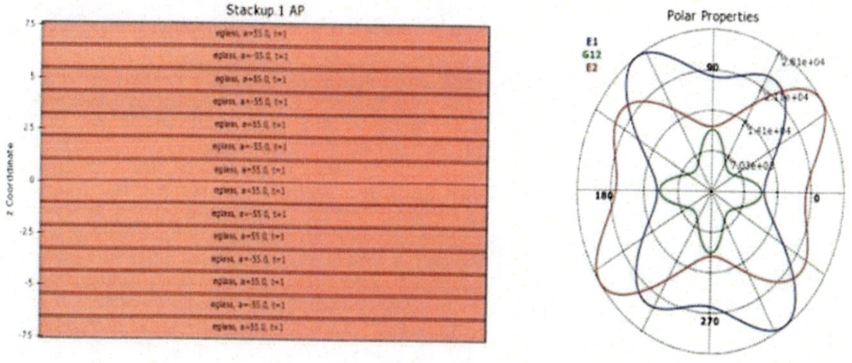

Fig. 18 Layer stack up at [±55°] FW

References

1. Bao, Y., Zhao, X.: Research of lightweight composite automobile wheel. World J. Eng. Technol. **5**, 675–683 (2017)
2. Nirala, S.K., Shankar, S., Sathishkumar, D., Kavivalluvan, V., Sivakumar, P.: Carbon fiber composites: a solution for light weight dynamic components of AFVs. Ref. Sci. J. **67**(4) (2017)
3. Schweizer, N., Giessl, A., Schwarzhaupt, O., Büter A.: Development of a composite wheel with integrated hub motor. In: ECCM15—15th European Conference on Composite Materials, Venice, Italy, 24–28 June 2012
4. Babu, M., Hariharan, V.S.: Modelling and analysis of automotive wheel rim. IJIRSET **5**(4) (2016)
5. Blasco, J., Valero, F., Besa, A., Rubio, F.: Design of a dynamometric wheel rim. Springer Science + Business Media Dordrecht (2014)
6. Tsai S.W., Wu E.M.: A general theory of strength for anisotropic materials. J. Compos. Mater. **5**, 58–80 (1971)
7. Gondhali, S.L., Dhale, A.D., Pagare, S.: Static structural analysis of car rim by finite element method. Springer Nature Singapore Pte Ltd. (2019)
8. Thakare, R.B.: Stress analysis in wheel rim by using dynamic cornering fatigue test under different conditions. IJARIIE **3**(2) (2017). ISSN(O)-2395-4396
9. Chopade, R.B., Pateriya, A., Shirbhate, A.D.: Review of design, analysis of four wheeler alloy material rim using FEA method under cornering fatigue test. IJIIRD **2**(2) (2018). ISSN: 2456-236X

Modeling and Simulation of a Distillation Column to Separate an Automotive Fuel

Asma Iqbal and Syed Akhlaq Ahmad

Nomenclature

L	Liquid stream
P	Pressure
F	Feed flow rate
K	Activity coefficient
V	Vapor stream flow rate
RR	Reflux ratio
Di	Internal diameter
Cp	Specific capacity
T_c	Critical temperature
T_j	Temperature at *j*th stage
T_r	Reduced temperature
V_j	Vapor flow rate from *j*th stage
L_j	Liquid flow rate from *j*th stage
PSD	Pressure swing distillation
LPC	Low-pressure column
HPC	High-pressure column
2-Methyl	2-Methylpyridine
$x_{i,j}$	Liquid mole fraction of *i*th component
$y_{i,j}$	Vapor mole fraction of *i*th component
h	Enthalpy
m	High- or low-pressure column
j	Stage number
z	Mole fraction at feed stage
λ	Latent heat

A. Iqbal (✉) · S. A. Ahmad
Department of Chemical Engineering, Aligarh Muslim University, Aligarh, India
e-mail: asmajawahira@gmail.com

M. Muzammil et al. (eds.), *Recent Advances in Mechanical Engineering*, Lecture Notes in Mechanical Engineering, https://doi.org/10.1007/978-981-15-8704-7_10

γ Activity coefficient of component
ϕ Partial fugacity coefficient

1 Introduction

The physical and chemical properties of working fluids strongly influence the plant efficiency, weight, and size of components, material of construction, etc. In some cases, it even affects the cost of power plants. Therefore, a research program has been initiated, which could help out to find the best working fluid for automotive engines. Although many guidelines have been outlined for the selection of an appropriate working fluid including thermal stability, availability, and efficiency, the most important being is that the fluid should be free from environmental hazards as well as health hazards. Earlier, only ideal solutions were permitted for the use of working fluids in external combustion engines, but due to the expansion of thermodynamic techniques, non-ideal solutions were also used. A report based on experimental and simulation studies suggests that 2-methylpyridine–water fulfills maximum criteria's to serve as a suitable automotive fuel for both combustion engines and reciprocating turbines. 2-methylpyridine has higher energy efficiency, high thermal conductivity, and is not susceptible to corrosion. Besides, it is compatible with most of the materials and is reasonably priced. However, the use of 2-methylpyridine–water as an automotive fuel requires strong corrosion-resistant steel or a more stable steel alloy. An economic evaluation of different candidate automotive fuels was carried out which suggest that 2-methylpyridine–water costs less in comparison to other fuels [1–4].

Since 2-methylpyridine serves as a good automotive fuel being non-hazardous to humans, its energy value is considerably lower due to the presence of water in it. Hence, a search based on the exploration of a separation method is a sought attribute. The process which can effectively separate 2-methylpyridine from water not only increases the fuel value of 2-methylpyridine–water but also eliminates the risk of health hazards caused by this azeotropic mixture. This reason has become the main motivation to carry out this work. This work is based on the exploration of an appropriate separation technique for the 2-methylpyridine–water azeotropic mixture.

Distillation, a separation technique, in process industries is used to separate impurities, and other by-products from various processes to recover highly pure products. It is a highly flexible process which involves low capital investment and low operational risk. However, the energy consumption of distillation columns is quite high, but that can be addressed either by the implementation of different heat integration techniques or by the optimization of the conceptual design of distillation columns [5, 6]. From past years, it has been observed that there is a tremendous rise in the applications of distillation-based separation techniques because of the advantages of distillation processes. One such distillation technique is based on the exploitation of pressure sensitivity of azeotropic mixtures. The process is known as pressure swing distillation

(PSD). It is quite similar to other distillation-based separation techniques; extractive and azeotropic distillation, in a way that it also utilizes two distillation columns. But the major difference of PSD technique concerning the other two techniques is that it doesn't require solvent (third component) to break the azeotropic point between two components of a binary azeotropic mixture. Rather, the PSD technique exploits the pressure-sensitive behavior of azeotropes. Therefore, the column arrangement in the PSD method is a low-pressure column (LPC) followed by a high-pressure column (HPC) or vice versa. The column arrangements can be decided based on the feed composition and azeotropic composition also [7, 8].

This work is primarily focused on the pressure swing distillation technique, which is used to separate 2-methylpyridine from water so that it can be used as an automotive fuel for external combustion engines. The distillation columns are modeled first and then simulated using simulation software of Aspen Plus V 8.8.

2 Modeling of Distillation Column

A steady-state mathematical model for the distillation columns has been developed. The equations include overall mass balances, energy balances, and summation equations.

2.1 *Total Mass Balance*

$$V_{j-1}^{m} - V_{i}^{m} + L_{j+1}^{m} - L_{j}^{m} + F_{j}^{m} = 0$$

2.2 *Component Balance*

$$V_{j-1}^{m} y_{i,j-1}^{m} + V_{j}^{m} y_{i,j}^{m} + L_{j+1}^{m} x_{i,j+1}^{m} + F_{j}^{m} z_{i,j}^{m} = 0$$

2.3 *Vapor–Liquid Equilibrium*

$$y_{i,j} = K_{i,j} x_{i,j}$$

where $K = \frac{\gamma_{i,j}\phi_{iL,j}}{\phi_{iV,j}}$

$$\sum_{i=1}^{NC} y_{j,i}^{m} = 1$$

$$\sum_{i=1}^{NC} x_{j,i}^{m} = 1.$$

2.4 Enthalpy Balance

$$V_{j-1}^{m} j_{j-1}^{v} + V_{j}^{m} j_{j}^{v} + L_{j+1}^{m} h_{j+1}^{l} - L_{j}^{m} h_{j}^{l} + F_{j}^{m} h_{j}^{f} - Q_{\text{loss}} = 0$$

The variation of enthalpy with temperature has been incorporated by use of the following correlation.

$$h = \int \text{Cpd}t$$

$$H = h + \lambda$$

$$\text{whereCp(J/kmol} - \text{K)} = C_1 + C_2T + C_3T^2 + C_4T^3 + C_5T^4$$

$$\lambda(\text{J/kmol}) = c_1(1 - T_r)^{(c_2 + c_3 T_r + c_4 T_r^2)}$$

$$T_r = \frac{T}{T_c}$$

The constants that appear in the specific heat (Cp) and latent heat (λ) equations are taken from [9].

3 Steady-State Design and Simulation

The distillation column for pressure swing distillation technique has been modeled using thermodynamic mass and energy balance equations. The next step is to design and simulate a suitable column configuration for pressure swing distillation scheme. In this design, a high-pressure column is followed by a low-pressure column to separate two components of an azeotropic mixture effectively. All the design parameters, including number of stages in two columns, feed stage locations, reflux ratio, etc., have been reflected in Fig. 1 (generated using a simulation software of Aspen Plus). Now, the synthesized flow sheet is run in Aspen Plus V8.8 to simulate the two columns. Once the simulation gets converged, the two products (2-methylpyridine and water) are obtained from the bottom of high-pressure column and low-pressure column, respectively.

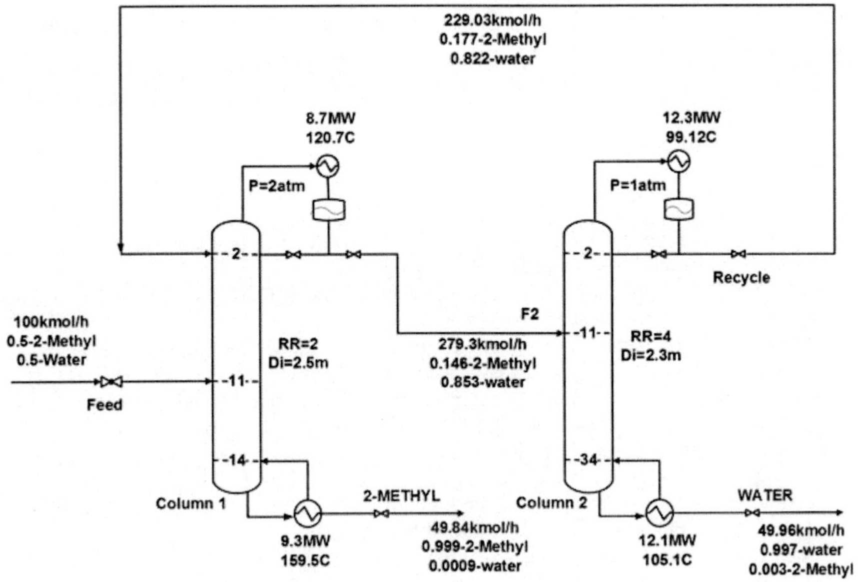

Fig. 1 Steady-state design flowsheet of PSD process

4 Results and Discussions

From Fig. 1, it can be observed that the proposed design of the pressure swing distillation technique has successfully separated 2-methylpyridine from water. This refers to the fact that the suggested scheme of separation has proved to enhance the fuel characteristics of 2-methylpyridine by separating water from it. High-purity 2-methylpyridine with 99.9% purity has been recovered from the bottom of the high-pressure column and requires only 15 stages for separation.

5 Conclusions

The work has explored an automotive fuel of 2-methylpyridine with its unique features of having low density and viscosity, readily available, higher freezing temperature than water, and higher energy efficiency. The potential of 2-methylpyridine–water as an automotive fuel has been exploited by proposing a competent separation scheme of pressure swing distillation. The distillation-based separation scheme of pressure swing distillation has been found to be capable of separating water from 2-methylpyridine effectively, such that 2-methylpyridine can be easily used as an automotive fuel in external combustion engines. The overall conclusion drawn from this work is that a high-purity 2-methylpyridine has been separated from water by the pressure swing distillation technique, to enhance energy potential of 2-methylpyridine, so that it can be readily used as a fuel for external combustion engines.

References

1. Cupper, R.A., Ridgefield, C., Somekh, G.S.: Thermally Stable Lubricants for External Combustion Engines. New Rochelle, N.Y. (1974)
2. Miller, D.R., Null, H.R., Thompson, Q.E.: Optimum Working Fluids for Automotive Rankine Engines Volume II-Technical Section (1973)
3. Petrillo, V.C., Springfield, P.: Motive Fluids for External Combustion Engines (1977)
4. George, S.S.: Working Fluids for External Combustion Engines. New Rochelle N.Y. (1974)
5. Jing, F., Yuqi, H., Chunli, L.: Energy-saving mechanism in heat transfer optimization of dividing wall column. Ind. Eng. Chem. Res. **52**, 18345–18355 (2013)
6. Kiss, A.A., Flores Landaeta, S.J., Infante Ferreira, C.A.: Towards energy efficient distillation technologies—Making the right choice. Energy **47**, 531–542 (2012)
7. Luyben, W.L.: Pressure-swing distillation for minimum- and maximum-boiling homogeneous azeotropes. Ind. Eng. Chem. Res. **51**, 10881–10886 (2012)
8. Wang, Y.H., Chien, I.L.: Unique design considerations for maximum-boiling azeotropic systems via extractive distillation: acetone/chloroform separation. Ind. Eng. Chem. Res. **57**, 12884–12894 (2018)
9. Perry, R.H., Green, D.: Perry's Chemical Engineers Handbook. McGraw Hill (2001)

Data Acquisition System for Measurement of Biomechanical Variables in Sit-to-Stand Task

Siddharth Bhardwaj, Abid Ali Khan, and Mohammad Muzammil

1 Introduction

The study on human motion has been widely conducted in research and clinical facilities to observe motor capabilities, improve athletic performance, and evaluate the outcome of rehabilitation programs [1, 2]. Biomechanical data concerning the kinematics and kinetics of human movement have been used successfully for motion reconstruction [3] and biofeedback applications [4]. Biomechanical data has also been used to study biomechanics in order to intervene in applications ranging from human movements [5] to the design of consumer products [6]. Moreover, in health-care systems, the fusion of physiological and biomechanical data has been argued to provide user complaint control of robotic devices [7, 8].

With the vast use of biomechanical data, it is therefore essential that the employed sensorial system should meet the data acquisition requirement. The sensors used for these purposes must be qualified in terms of handling large dynamic ranges, high sampling frequencies, precision and accuracy [9]. Various sensors are preferred for biomechanical measurements, namely force platform, accelerometer, inertial measurement unit, gyroscope, magnetometer, mechanomyography, etc. [10]. However, rather than complex instruments for studying human movement, simple, low-cost experimental design setups are found more appropriate for data collection in both local and multicenter repository context [11].

Understanding of biomechanics of different human movements lays the foundation of biomechatronic design criterion in rehabilitation robots [12]. Sit-to-stand (STS), i.e., standing from sitting posture, is an activity of daily life, involving transition of the posture through biomechanically unstable phases [8, 15]. STS task also serves as a therapeutic exercise for quadriceps strengthening and fall prevention in frail/elderly people [13, 14]. Hence, it is important to study the biomechanics of

S. Bhardwaj (✉) · A. A. Khan · M. Muzammil
Department of Mechanical Engineering, Aligarh Muslim University, Aligarh, UP, India
e-mail: siddharth.bhardwaj@live.com

M. Muzammil et al. (eds.), *Recent Advances in Mechanical Engineering*, Lecture Notes in Mechanical Engineering, https://doi.org/10.1007/978-981-15-8704-7_11

the STS task for assistive and biofeedback applications concerning such postural transfer.

In this article, a data acquisition system developed in the ergonomics laboratory of the Mechanical Engineering Department, AMU, is presented for the measurement of biomechanical variables associated with STS task. Sensorial setup and data acquisition software (coded in LabVIEW) for the measurement of three modalities, namely knee flexion/extension (K_{FE}), trunk flexion/extension (T_{FE}), and reaction forces at the right feet (R_{rf}), left feet (R_{lf}) and seat (R_s), are shown. For validation, a sample data has been collected to describe the STS movement from the developed data acquisition system, which is also presented in this article.

2 System Design

The proposed data acquisition system comprises (a) hardware, consisting of sensorial setup and instrumentation and (b) software, which helps to communicate with the hardware, to monitor the data in real time, and to save the acquired data in the PC.

2.1 Hardware

Trunk flexion/extension. An inertial measurement unit (IMU, MPU-6050) with onboard accelerometer (3 dof) and gyroscope (3 dof) was used for the measurement of T_{FE} data. Onboard digital motion processing (DMP) was exploited to fuse the computed quaternion data from the independent accelerometer and gyroscope. This fusion eliminated the inherent noise and drift errors associated with accelerometer and gyroscope. The fused quaternion (x, y, z, w) was used to estimate the pitch angle (θ) of IMU (Eq. (1)) which in turn gives the estimate of the T_{FE}.

$$T_{FE} = \theta = a\tan 2\big(2(x \times z - y \times w),\ 2 \times y^2 + 2 \times z^2 - 1\big) \times 180/\pi \quad (1)$$

The IMU was interfaced with Arduino Uno (ATmega328) and communicated to PC using serial communication protocol at 115,200 baud rate. The T_{FE} data was acquired at 50 Hz sampling frequency.

Knee flexion/extension. The K_{FE} data for the knee was measured with the help of 20 kΩ rotary potentiometer. Two potentiometers were used, each to acquire the data from both the knees (left and right). The potentiometric goniometer was designed to form a revolute joint at the lateral epicondyle of the knee. The potentiometer was secured to the knee using Velcro strap attached to the two metal strips placed at the thigh and shank. The potentiometers were pre-calibrated and tethered to MyDAQ data acquisition unit (National Instrument, USA) for interfacing with the PC. The T_{FE} data was acquired at 1000 Hz sampling frequency.

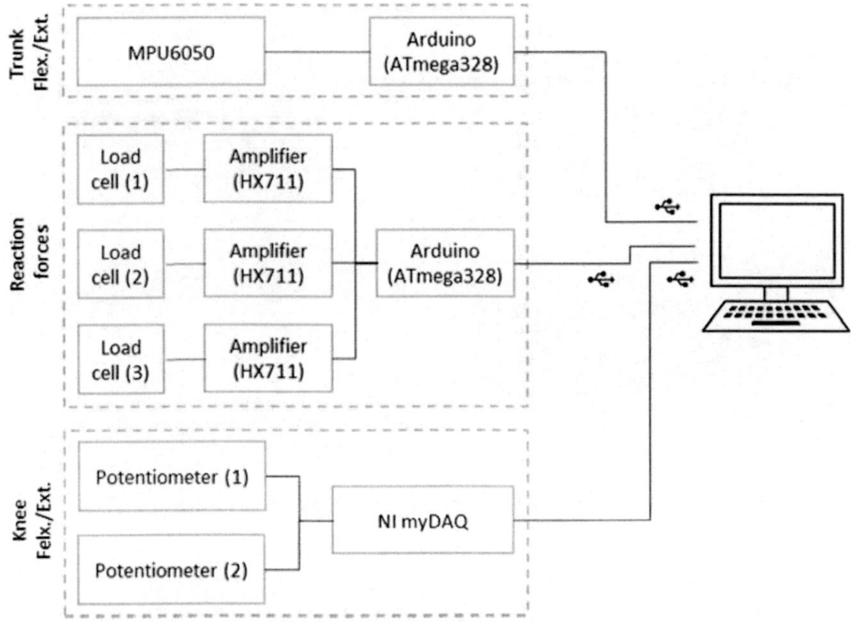

Fig. 1 Hardware and interfacing with PC

Reaction force. The normal reaction force at the feet (R_{rf},R_{lf}) and seat (R_s) was measured using single point load cells (rated 200 kg each). Foot plates, mounted on individual load cells, were used to measure ground reaction forces on the left and right foot, while a single load cell mounted beneath the stool seat was used to measure the seat reaction force. A 24-bit analog-to-digital converter, HX711, was employed for load cell signal amplification. The load cells were powered and tethered to PC using Arduino Uno (ATmega328) with serial communication protocol at 19,200 baud rate. The reaction force data was acquired at 90 Hz sampling frequency. Figure 1 shows the overall hardware interfacing.

2.2 *Software*

A LabVIEW program was developed for time-synchronized recording of T_{FE}, K_{FE}, R_{lf}, R_{rf}, and R_s data. Within the LabVIEW program, the data from different acquisition hardwares was recorded in separate loop structures with synchronized loop execution rate. Local variables were used to stop the loop execution while saving the acquired data. Digital filters were designed in the LabVIEW program to improve signal-to-noise ratio of the recorded variables. A low-pass Butterworth 3rd-order filter with 7 Hz cutoff frequency was used to reduce noise in potentiometer data. The data from the load cells were filtered using rank 2 median filter.

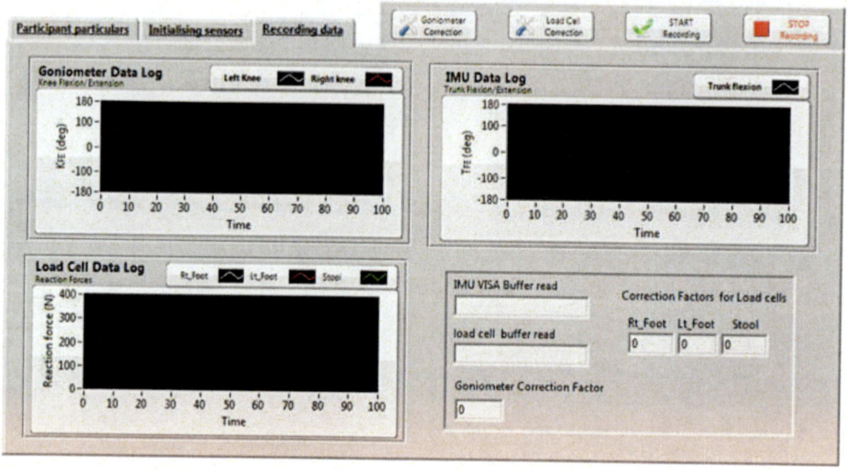

Fig. 2 Front panel of data acquisition program (build in NI LabVIEW version 12)

Figure 2 shows the front panel of the developed LabVIEW program. Graphical interface was provided for real-time data monitoring. Separate tabs were provided within the program's graphical interface to input the participant credentials, perform sensor initialization and start the data streaming, respectively. After the participant finished the task, the program was terminated using the STOP button provided in the front panel, which in turn saves the acquired data to the specified location on the PC hard drive.

3 Experimentation and Procedure

To demonstrate the utility of the proposed data acquisition system, a sample reading is presented for the recorded biomechanical variables during STS task. A participant (male, height = 1.58 m and weight = 58 kg) was recruited to volunteer the STS task. Before conducting the experimental trial, participant was briefed about the experiment and written consent was obtained as per university ethics board. Thereafter, all sensors were attached to the participant's body, and STS trials were conducted. IMU was placed at the chest of participant using a chest belt, while potentiometers were secured at the knees. Three consecutive trials of STS task were performed with a gap of 2 min in between the trials. The participant performed the task at his own comfortable speed after being verbally signaled to initiate the STS task. Figure 3 shows the experimental setup for the study.

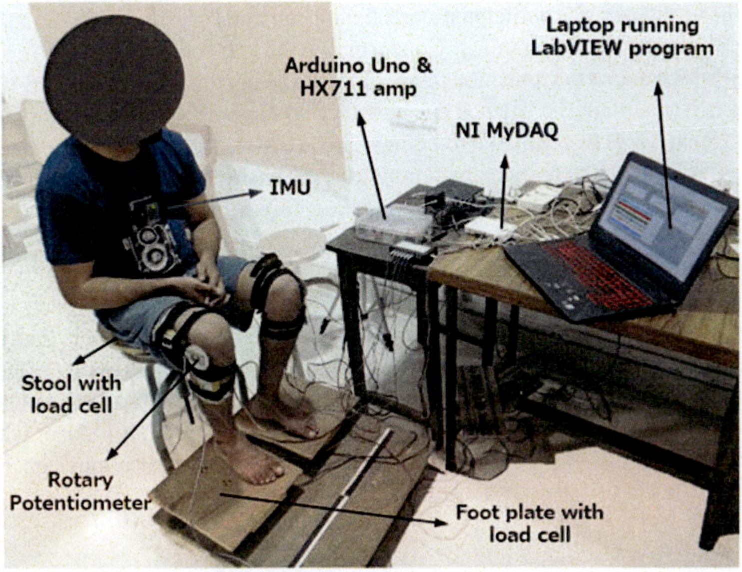

Fig. 3 Experimental setup

4 Test Results and Discussions

Figure 4 shows the mean of recorded data for the experimental trials. As evident from Fig. 4, the data is useful in describing the phases of STS task. The initiation of STS task is marked with the extension of trunk (increase in T_{FE} angle, Fig. 4b), and consecutive increase in the seat reaction force (increase in R_s) and decrease in feet reaction force (R_{lf} and R_{rf}) as shown in Fig. 4a. Extending the trunk helped the participant to gain momentum before lifting off from the seat, thereby reducing the torque requirement at the knees.

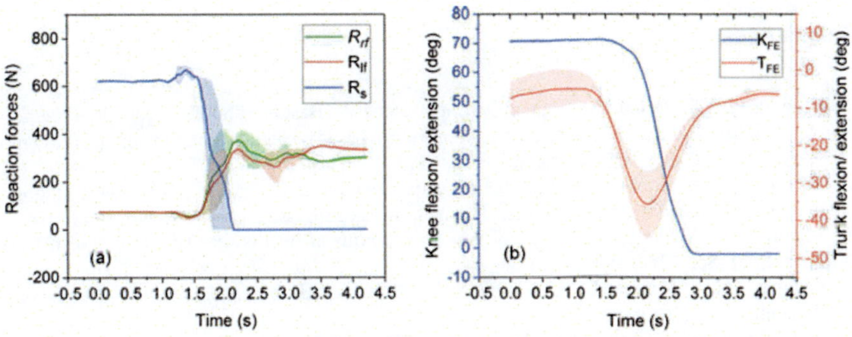

Fig. 4 Recorded data for sit-to-stand task

In the next phase, the participant starts flexing his trunk (decrease in T_{FE}) and initiated seat lift off (reduction in R_s). Concurrently, the knee extension starts (decrease in K_{FE}) which marks the transition phase of STS. During the transition phase, the feet struggles to maintain the balance as shown by continuous variation in the reaction forces (R_{lf} and R_{rf}) as compared to being seated (Fig. 4a). The STS phase marks the ending phase when the participant completes the transition phase and stands straight (knee in full extension, $K_{FE} = 0$ and reaction forces at the feet becomes almost constant).

The proposed data acquisition system provides a low-cost solution in measurement of biomechanical variables in STS task compared to camera-based motion tracking systems. The presented system provides capability of real-time data monitoring and can be augmented with algorithm-based phase markers to detect the STS phases. With the online phase detection, the data acquisition setup may be adopted for real-time diagnosis of pathologies and lower limb strength [16]. Also, with the online classification of phases, the system can be used in applications concerning rehabilitative and assistive devices for STS tasks [17].

5 Conclusions

The paper proposes a data acquisition system for online monitoring and recording of biomechanical variables, namely trunk flexion/extension, knee extension/flexion, and reaction forces at the feet and seat. The system is tested for sit-to-stand movement; however, it can be adapted to other movements such as gait and squatting. The system can be extended to rehabilitation and assistive applications, wherein the biomechanical data may be used for motion synthesis and control.

Acknowledgements This work was supported by the Council of Scientific and Industrial Research (CSIR), New Delhi, under Senior Research Fellow (SRF) scheme. File no. 09/112(0554)2K17.

References

1. Villa-Parra, A.C., Delisle-Rodríguez, D., Lóрez-Delis, A., Bastos-Filho, T., Sagaró, R., Frizera-Neto, A.: Towards a robotic knee exoskeleton control based on human motion intention through EEG and sEMGsignals. Procedia Manuf. **3**, 1379–1386 (2015)
2. Schindelholz, M., Stoller, O., Hunt, K.J.: A software module for cardiovascular rehabilitation in robotics-assisted treadmill exercise. Biomed. Signal Process. Control. **10**, 296–307 (2014)
3. Bhardwaj, S., Khan, A.A., Muzammil, M.: Knee torque estimation in sit to stand transfer. J. Phys. Conf. Ser. **1240**, 1–5 (2019)
4. Giggins, O.M., Persson, U.M., Caulfield, B.: Biofeedback in rehabilitation. J. Neuroeng. Rehabil. **10**, 60 (2013)
5. Law, M.H.C., Choi, E.M.F., Law, S.H.Y., Chan, S.S.C., Wong, S.M.S., Ching, E.C.K., Chan, Z.Y.S., Zhang, J.H., Lam, G.W.K., Lau, F.O.Y., Cheung, R.T.H.: Effects of footwear midsole thickness on running biomechanics. J. Sports Sci. **37**, 1004–1010 (2019)

6. Carse, B., Thomson, A., Stansfield, B.: Use of biomechanical data in the Inclusive Design process: Packaging design and the older adult. J. Eng. Des. **21**, 289–303 (2010)
7. Li, Z., Cheng, H., Guo, H., Sun, X.: Compliant training control of ankle joint by exoskeleton with human EMG-torque interface. Assem. Autom. **37**, 349–355 (2017)
8. Bhardwaj, S., Khan, A.A., Muzammil, M.: Lower limb rehabilitation using multimodal measurement of sit-to-stand and stand-to-sit task. Disabil. Rehabil. Assist. Technol. 1–8 (2019).
9. Kos, A., Umek, A.: Biomechanical biofeedback systems and applications. Springer Nature Switzerland AG, Swizerland (2018)
10. Howard, R.M., Conway, R., Harrison, A.J.: A survey of sensor devices: use in sports biomechanics. Sport. Biomech. **15**, 450–461 (2016)
11. Vannozzi, G., Della Croce, U., Starita, A., Benvenuti, F., Cappozzo, A.: Knowledge discovery in databases of biomechanical variables: application to the sit to stand motor task. J. Neuroeng. Rehabil. **1**, 1–10 (2004)
12. Simonetti, D., Tagliamonte, N.L., Zollo, L., Accoto, D., Guglielmelli, E.: Biomechatronic design criteria of systems for robot-mediated rehabilitation therapy. In: Colombo, R., Sanguineti, V. (eds.) Rehabilitation Robotics, pp. 29–46. Elsevier (2018)
13. Chumacero-polanco, E.A., Yang, J.: Fall prevention therapies for individuals with stroke: a survey. In: Proceedings of the ASME 2017 International Design Engineering Technical Conferences and Computers and Information in Engineering Conference, pp. 1–22. Ohio (2017)
14. McKay, C., Prapavessis, H., Doherty, T.: The effect of a prehabilitation exercise program on quadriceps strength for patients undergoing total knee arthroplasty: a randomized controlled pilot study. Phys. Med. Rehabil. **4**, 647–656 (2012)
15. Papa, E., Cappozzo, A.: Sit-to-stand motor strategies investigated in able-bodied young and elderly subjects. J. Biomech. **33**, 1113–1122 (2000)
16. Tsukahara, A., Kawanishi, R., Hasegawa, Y., Sankai, Y.: Sit-to-stand and stand-to-sit transfer support for complete paraplegic patients with robot suit HAL. Adv. Robot. **24**, 1615–1638 (2010)
17. Chen, B., Zhong, C.H., Ma, H., Guan, X., Qin, L.Y., Chan, K.M., Law, S.W., Qin, L., Liao, W.H.: Sit-to-stand and stand-to-sit assistance for paraplegic patients with CUHK-EXO exoskeleton. Robotica. **36**, 535–551 (2018)

Fabrication and Tribological Behavior of Copper Plain Bearing Lubricated with Biodegradable Oil

Uzma Iqbal, Asma Iqbal, and Siddharth Bhardwaj

1 Introduction

Most mechanical system failures occurring at the level of tribo-elements such as the bearing, gears are attributed to friction and poor tribological properties of the materials [1]. Mostly, one of the important tribo-elements in the mechanical system is plain bearing, where its functions are to support and guide the moving element against the direction of applied load. Since there is a relative motion between the bearing and the moving element, there will be a power or energy loss due to friction, and if their surfaces touch each other, wear may also take place. Hence, the bearing should perform its function with a minimum loss of energy while limiting the wear. There are certain applications where it is difficult to lubricate the bearing with synthetic lubricants, where the likelihood of leakage is higher and thus may pose an environment risk during operation [2]. Therefore, it becomes necessary to research eco-friendly biodegradable lubricants for such applications in particular [3]. Recently, ecological lubricants based on vegetable oil such as palm oil, soya bean oil, sunflower oil, and canola oil have tended to be used as lubricant in place of mineral oil.

In the present work, applicability of apricot oil as a lubricant is investigated while using copper bearings. The most common method of measurement of effectiveness of a lubricant is the measure of wear caused by relative movement between two mating surfaces. There are various methods of wear measurement, but the method which is used in this research for the calculation of wear measurement is mass loss

U. Iqbal (✉)
Division of Agricultural Engineering, SKUAST-K, Shalimar, JK, India
e-mail: uzmajawahira@gmail.com

A. Iqbal
Department of Chemical Engineering, Aligarh Muslim University, Aligarh, UP, India

S. Bhardwaj
Department of Mechanical Engineering, Aligarh Muslim University, Aligarh, UP, India

M. Muzammil et al. (eds.), *Recent Advances in Mechanical Engineering*, Lecture Notes in Mechanical Engineering, https://doi.org/10.1007/978-981-15-8704-7_12

measures of wear [4, 5]. The lubricating capacity of apricot oil on copper bearing in comparison with simple copper without lubricant is measured. The shaft material used for above tests is mild steel (MS). Tribological tests were carried out on pin on ring apparatus at room temperature. The use of apricot oil upholds the sustainable design methodology as the lubricant is not harmful to the environment as compared to synthetic lubricants and mineral oils.

2 Methodology

In the present work, pin on ring apparatus was used for measurement of coefficient of friction of copper specimen with apricot oil. Specimen was made pin, while rotating ring was made up of mild steel. The following two conditions were tested for the measurement of coefficient of friction:

- **Condition 1:** Sliding of copper specimen for 60 min at a load of 1.28 kg and sliding velocity (V_s) 0.1 m/s.
- **Condition 2:** Sliding of copper specimen and apricot oil (1 ml) for 60 min, at a load of 1.28 kg and sliding velocity (V_s) 0.1 m/s.

2.1 *Coefficient of Friction*

Readings for 60 min were recorded at a sliding velocity of 0.1 m/s, and deflections were taken after every 6 min. For 5 kg, the loading apparatus gives maximum deflection. For experimentation, the initial load applied was 1.28 kg, and maximum applied load (W) was 6.56 kg. Coefficient of friction (μ) was recorded for ten values of tangential force (F_t) making a deflection of 56, 57, 58, 59, 60, 61, 62, 63, 64, and 66 on the apparatus scale. A sample calculation shows the method or evaluating the value of μ for 56 deflection:

$$F_f = 5/100 \times 56 = 2.8\,\text{kg} = 27.44\,\text{N}$$

$$\text{Coefficient of friction } (\mu) = F_f/W$$

$$\mu = 27.44/(6.56 \times 9.81) = 0.426$$

$$\mu = 0.426$$

2.2 *Test Procedure for Wear*

Weight loss was calculated to measure the amount of wear produced in the two conditions: i.e., weight loss measurement for simple copper specimen and weight loss measurement for copper with 1 ml apricot oil (Cu + 1 ml apricot). Sample calculation provides the method for evaluating weight loss in different samples.

Sample A = (simple copper)
Sample B = (Cu + 1 ml apricot)

Original weight of samples:

Weight of sample A before experiment = 13.6045 g.
Weight of sample B before experiment = 13.6045 g.

Final weight of samples

Weight of sample A after experiment = 13.6000 g.
Weight of sample B after experiment = 13.6035 g.
Weight loss for simple copper = 4.5 mg.
Weight loss for (Cu + 1 ml apricot) = 1 mg.

3 Result and Discussion

It is evident from Figs. 1 and 2 that coefficient of friction increases with the increase in sliding distance; however, the increase was higher in case of copper without lubricant, as compared to apricot oil. Friction measurements against sliding distance are given in Table 1 for Cu and MS and in Table 2 when used with 1 ml of apricot oil.

It is evident from the results that use of apricot oil is effective in reducing the wear of material. Tribological behavior of apricot oil investigated on high load explains

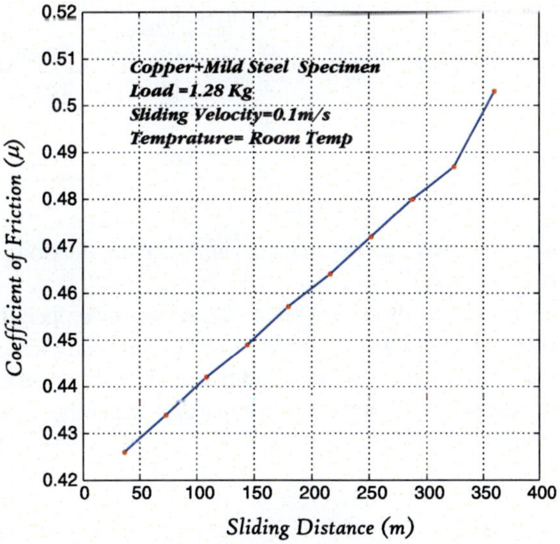

Fig. 1 Coefficient of friction and sliding distance without apricot oil (copper + MS). *Note* Load = 1.28 kg, V_s = 0.1 m/s

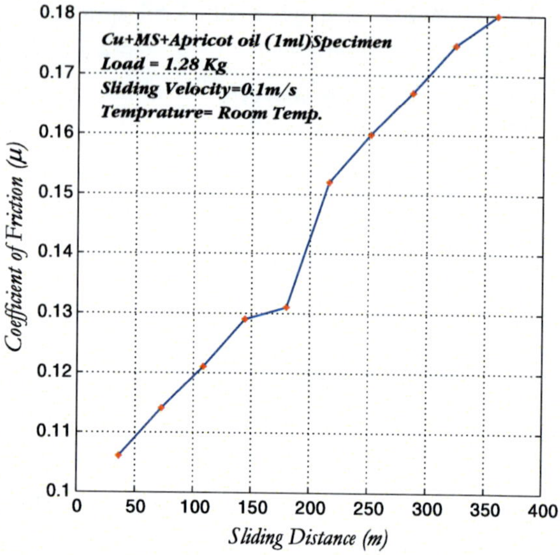

Fig. 2 Coefficient of friction and sliding distance with apricot oil (copper + MS + apricot oil 1 ml). *Note* Load = 1.28 kg, $V_s = 0.1$ m/s

Table 1 Coefficient of friction related to sliding distance for copper + MS at load of 1.28 kg for 60 min

S. No.	Deflection	Coefficient of friction (μ)	Sliding distance (m)
1	56	0.426	36
2	57	0.434	72
3	58	0.442	108
4	59	0.449	144
5	60	0.457	180
6	61	0.464	216
7	62	0.472	252
8	63	0.480	288
9	64	0.487	324
10	66	0.503	360

that apricot oil can be used for industrial applications as coefficient of friction and wear is reduced.

Lubrication by apricot oil in copper bearing reduces coefficient of friction by 75%, when compared to dry copper bearing. The study investigates the use of apricot oil as lubricant; however, comparative studies are necessary to be conducted to check the efficacy of apricot oil in comparison with other vegetable oils and mineral oils.

Table 2 Coefficient of friction related to sliding distance for Cu + Ms + apricot Oil 1 ml at load of 1.28 kg for 60 min

S. No.	Deflection	Coefficient of friction (μ)	Sliding distance (m)
1	14	0.106	36
2	15	0.114	72
3	16	0.121	108
4	17	0.129	144
5	18	0.131	180
6	19	0.152	216
7	20	0.160	252
8	21	0.167	288
9	22	0.175	324
10	23	0.503	360

4 Conclusion

The study investigates the use of apricot oil for lubrication purposes. From the bearing samples investigated, the one containing copper with apricot oil was shown to have improved performance characteristic with regard to friction and wear. The use of apricot oil was found to reduce the coefficient of friction by 75%. More studies are needed to draw the benefit of apricot oil in comparison with the available lubricants in the market.

References

1. Stachowiak, G.W.: How tribology has been helping us to advance and to survive. Friction. **5**, 233–247 (2017)
2. Panchal, T.M., Patel, A., Chauhan, D.D., Thomas, M., Patel, J.V.: A methodological review on bio-lubricants from vegetable oil based resources. Renew. Sustain. Energy Rev. **70**, 65–70 (2017)
3. Jain, A., Suhane, A.: Research Approach & Prospects of Non edible vegetable oil as a potential resource for biolubricant—a review. Adv. Eng. Appl. Sci. Int. J. **1**, 23–32 (2012)
4. Burris, D.L., Sawyer, W.G.: Measurement uncertainties in wear rates. Tribol. Lett. **36**, 81–87 (2009)
5. Iwai, Y., Honda, T., Miyajima, T., Yoshinaga, S., Higashi, M., Fuwa, Y.: Quantitative estimation of wear amounts by real time measurement of wear debris in lubricating oil. Tribol. Int. **43**, 388–394 (2010)

A Neural Network-Based Classification for Finger Motion While Grasping Different Sized Objects

Salman Mohd Khan, Abid Ali Khan, and Omar Farooq

1 Introduction

Generally, robotic assistive devices find their applications to redevelop or regenerate the motion or actions lost in limbs. Over the years, certain sophisticated devices have been developed [1]. However, these devices are rejected among users because of poor replicability and lesser control capabilities [2, 3].

The robotic assistive devices designed as the human–machine interface include the capture of electrical or mechanical signals in the human body [4, 5]. The electrical activities originating from the muscles of limbs or brain are more often used for human–machine interfaces. The muscle's electrical activity is acquired by the technique called electromyography (EMG), and these signals are referred as "EMG signals." The advantage of using these signals is the ease to use these signals into assistive devices. Presently, the EMG signals are more often processed by pattern recognition techniques, which involve processing and classification of signals for a different action–intention. Several studies have shown better results by pattern recognition techniques as compared to other techniques. The pattern recognition-based studies helped to evaluate the parameters named as muscle type; the number of muscles; processing parameters such as feature selection, feature extraction and classifications [1, 6, 7].

Recent studies have suggested that the use of EMGs is not enough for the design of devices. Certain studies have found the need to add a secondary set of sensors using EMGs for optimal information of the task. These sensors may involve accelerometers, potentiometers, flexi-force sensors, etc [8–10].

S. M. Khan (✉) · A. A. Khan
Department of Mechanical Engineering, Aligarh Muslim University, Aligarh, UP, India
e-mail: salmanmkhan225@gmail.com

O. Farooq
Department of Electronics Engineering, Aligarh Muslim University, Aligarh, UP, India

M. Muzammil et al. (eds.), *Recent Advances in Mechanical Engineering*, Lecture Notes in Mechanical Engineering, https://doi.org/10.1007/978-981-15-8704-7_13

One such issue in the design of an upper limb assistive devices is the finger motion capture using EMGs. For upper limbs, one of the important objectives of such devices is the replication of finger motion and manipulative grasping of objects [10, 11]. Often, using a certain number of muscles may cause loss of useful and intricate information, leading to inefficiency at times. This results in the need to record finger motion while grasping any object. The dual information allows researchers ease to identify the hand gesture while grasping objects of similar shape but different sizes [11]. Recently, an in-house research was published, in which knee angle, along with EMG, was used as input for sit-to-stand task-employed NN-based classification. The results were motivating to perform a similar analysis using neural network classification [12]. Based on these findings, in this study, the EMG data and finger joint data were fused together and then evaluated by comparing with the data of EMG signals only. This is done with the help of a neural network classification scheme.

This study is sequenced as follows: Section 2 details about experiment setup and its procedure. Section 3 discusses the techniques and methods applied for analysis. Further, the results and discussion are presented in Sect. 4.

2 Experimentation

The EMG was recorded using 4 bipolar EMG sensors (SX230) placed on 4 different muscles using a Biometric Data Acquisition (DAQ) System. The selected muscles were: flexor carpi radialis (FCR); brachioradialis; extensor carpi radialis (ECR); and extensor carpi ulnaris (ECU) based on previous studies [3, 11]. The standard protocol for surface electromyography for noninvasive assessment of muscles (SENIAM) was used for the positioning of electrodes. The reference electrode was placed behind the earlobe. The EMG signals were recorded using a Datalink software of Biometric DAQ system at a sampling rate of 1000 Hz.

The joint angle of the metacarpophalangeal (MCP) joint was measured by the flex sensor placed as shown in Fig. 1. The sampling rate of the flex sensor was 87 Hz. Besides, the subjects were asked to hold an object with a cylindrical surface. The object is designed such that the object span can be varied using an extra attachment on one side of object surface. The object spans were varied up to four levels, with initial being 6 cm and an increment of 1 cm at each level.

The experiment was repeatedly conducted on two abled-bodied male subjects (age: 26 & 27). Before experiments, both participants signed a written consent as approved by the university ethics committee. Each subject was asked to repeatedly perform the experiment 5 times. The subjects were seated on the chair at a fixed height. The subjects were asked to perform a simple pick and place task as shown in Fig. 2. The whole task is comprised of 12–15 s. It was divided into 5 phases: home position (Phase 1); transition phase (Phase 2); pick and place phase (Phase 3); transition phase (Phase 4); and home position (Phase 5). During Phase 3, the subjects were asked to make a fingertip contact with the surface of the object. Before recording, all subjects were allowed to practice on the experimental setup.

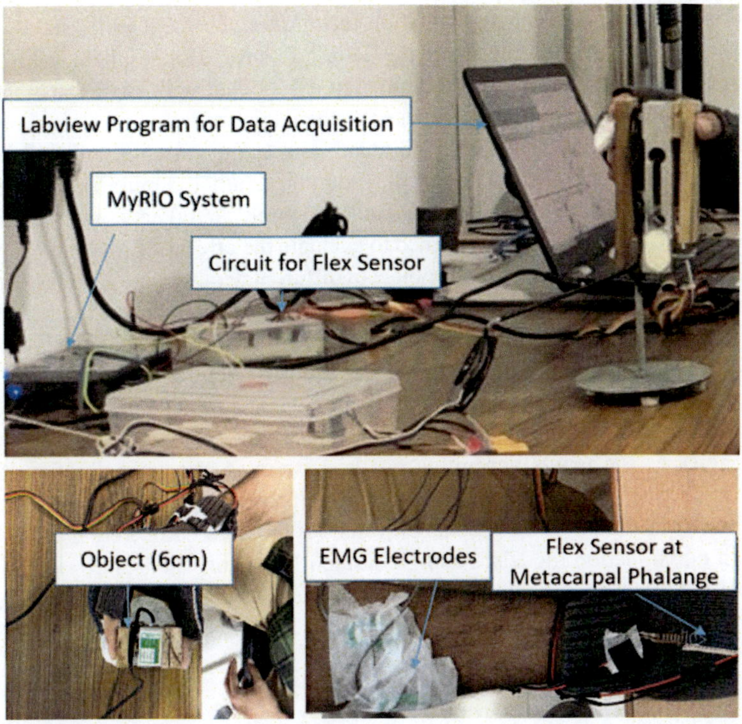

Fig. 1 Experiment setup design for data acquisition

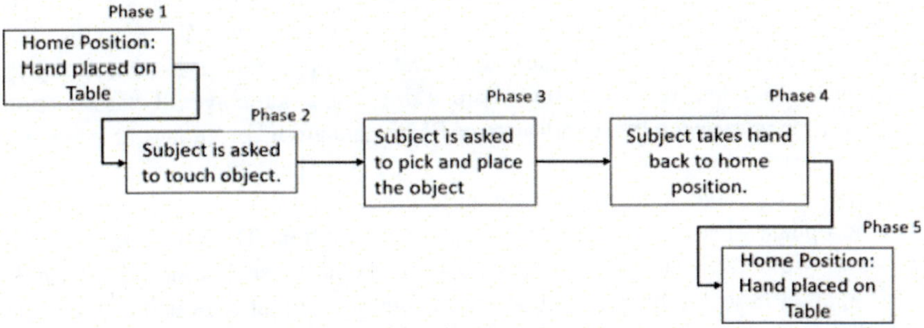

Fig. 2 Sequence of actions to be done during a task

3 Analysis

The raw EMG acquired from 4 muscles was subjected to data processing. It is comprised of data segmentation and feature extraction. The data-set was subjected to continuous data segmentation with a segment length of 250 ms. The continuous data segmentation helped to acquire better and detailed information as compared to overlapped and adjacent segmentation techniques. The segmented data-set was processed to acquire root-mean-square (RMS) feature. The resultant values were subjected to down-sampling to accommodate EMG data with angle data. The flex values incorporated in this study are used as the angle values.

The behavior of EMG signals and joint angles is evaluated using a neural network classification scheme. The scheme employed in this study is a multilayer perceptron (MLP). The MLP scheme is a conventional and simple neural network classification scheme. The classification is performed to evaluate the effect of EMGs and angles to perform a pick and place task for the object of different width spans. The classification models can be divided into sections. The first classification involves the testing of a system to identify continuous transition of finger motion. The second classification procedure was applied to test the system's capability to identify finger behavior, while the object span is varied. The MLP was applied using a module in SPSS software.

The NN models were designed using EMG data of the muscles and finger joint angles as the inputs depending on the system. At first, the RMS values of EMG activity were used as input for two classification models mentioned earlier. Later, the RMS feature values and angle values of the MCP joint were taken as input for both classification models. A single hidden layer consisted of 10 neurons and sigmoid functions as activation functions were used to design the models. The data percentage for training and testing was 70% and 30%, respectively.

4 Results and Discussion

In this section, the influence of MCP joint angles with EMG activities of muscles is compared. The analysis is performed for 5 phases of the task by MLP classification scheme. Further, another MLP classification scheme is applied to assess different grip spans with EMG data and EMG angle data. Table 1 shows the overall classification results for the task, when two conditions of EMG data are incorporated. It is evident that the classification largely improved when MCP joint behavior is used along with EMG data as the input in the study. The incapability of classification was found in the transition phases because of the continuous finger motion. The action recognition sequence with EMG data from highest to lowest was Phase 1; Phase 5; Phase 3; Phase 4; and Phase 2. Although this action recognition sequence remained the same when the angle data were added as input, the significant increase was found in the classification rate for all phases.

Table 1 Classification of task based on 2 conditions: (a) when only EMG signals are used as input and (b) when EMG and MCP joint angle values are used as input

Condition	Features	Classification percentage (%)
4 EMG muscles	RMS	61.3
4 EMG muscles and MCP angle	RMS, Angle	77.3

It was found that the extensor muscles were more active as compared to other muscles. The most prevalent muscle was ECR, followed by ECU and brachioradialis.

The second part of the study involved the identification of the different grip spans for 2 forms of sensing system using MLP classification schemes. The classification of each span is shown in Table 2. Table 2 clearly shows the effect of the MCP joint angle in detecting different grip spans, especially for larger grip spans. As the grip span increased, the activation of muscles became less differentiable. Besides this, the most prevalent muscle during Phase 3 of the task was found to be FCR, followed by ECU and ECR. Although the flexor muscles are regarded as important muscles, extensor muscles were also responsive to the grasping and gripping tasks. This analysis as shown in Fig. 3, can be conformed to the previous study by Hoozemans et al. in which extensor muscles were also capable to generate EMG and hand grip force relations [11].

Table 2 Classification result for different sizes of spans

Sample	Observed	For EMG and angle data: percent correct (%)	For EMG data: percent correct (%)
Testing	6 cm GS	95.8	90.4
	7 cm GS	99.7	87.8
	8 cm GS	81.4	53.1
	9 cm GS	91.8	55.1
	Overall %	93.9	79.4

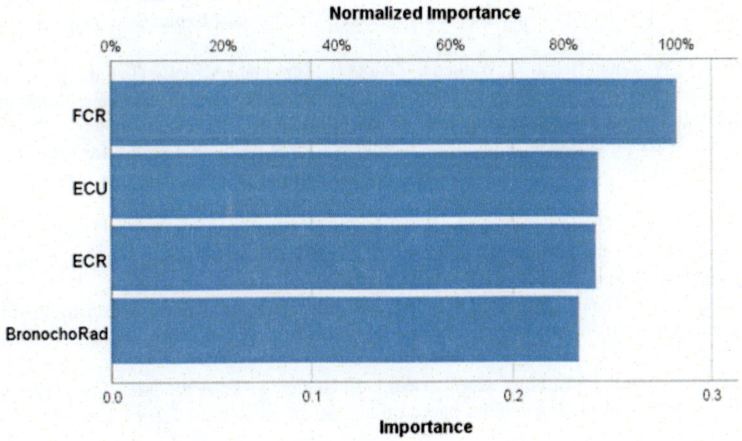

Fig. 3 Importance criteria of the muscles on change of object span. It shows the rate of influence of different muscles in the neural network

5 Conclusion

This study helped us to realize the need to fuse data of the joint angle motion along with the EMG signals for the design of biosignal-based assistive devices. In this study, the angle values of MCP joint of a finger along with EMG resulted in a marginally higher classification of action intentions. The modality of joint angles with EMGs also resulted in better recognition of the hand span for gripping different sized objects. This study motivates to add joint angles along with the data of EMG signals to recognize the activities for the similar objects of different dimensions.

References

1. Anam, K., Al-Jumaily, A.: Evaluation of extreme learning machine for classification of individual and combined finger movements using electromyography on amputees and non-amputees. Neural Netw. **85**, 51–68 (2017)
2. Biddiss, E.A., Chau, T.T.: Upper limb prosthesis use and abandonment: a survey of the last 25 years. Prosthet. Orthot. Int. **31**, 236–257 (2007)
3. Batzianoulis, I., Krausz, N.E., Simon, A.M., Hargrove, L., Billard, A.: Decoding the grasping intention from electromyography during reaching motions. J. Neuroeng. Rehabil. **15**, 1–13 (2018)
4. Ison, M., Artemiadis, P.: The role of muscle synergies in myoelectric control: Trends and challenges for simultaneous multifunction control (2014).
5. Guo, W., Sheng, X., Liu, H., Zhu, X.: Mechanomyography assisted myoelectric sensing for upper-extremity prostheses: a hybrid approach. IEEE Sens. J. **17**, 3100–3108 (2017)
6. Khushaba, R.N., Kodagoda, S., Takruri, M., Dissanayake, G.: Toward improved control of prosthetic fingers using surface electromyogram (EMG) signals. Expert Syst. Appl. **39**, 10731–10738 (2012)
7. Samuel, O.W., Zhou, H., Li, X., Wang, H., Zhang, H., Sangaiah, A.K., Li, G.: Pattern recognition of electromyography signals based on novel time domain features for amputees' limb motion classification. Comput. Electr. Eng. **67**, 646–655 (2018)
8. Roy, R., Sikdar, D., Mahadevappa, M.: A fingertip force prediction model for grasp patterns characterised from the chaotic behaviour of EEG. Med. Biol. Eng. Comput. **56**, 2095–2107 (2018)
9. Peng, L., Hou, Z., Chen, Y., Wang, W., Tong, L., Li, P.: Combined use of sEMG and accelerometer in hand motion classification considering forearm rotation. Proc. Annu. Int. Conf. IEEE Eng. Med. Biol. Soc. EMBS. 4227–4230 (2013).
10. Park, Y., Lee, J., Bae, J.: Development of a wearable sensing glove for measuring the motion of fingers using linear potentiometers and flexible wires. IEEE Trans. Ind. Informatics. **11**, 198–206 (2015)
11. Lin, B.-S., Chiang, P.-Y., Peng, C.-W., Lee, I.-J., Huang, S.-Y.: A modular data glove system for finger and hand motion capture based on inertial sensors. J. Med. Biol. Eng. **39**, 532–540 (2018)
12. Bhardwaj, S., Khan, A.A., Muzammil, M., Khan, S.M.: ANN based classification of sit to stand transfer. Mat. Today Proc. 1029–1034 (2020). https://doi.org/10.1016/j.matpr.2020.04.416
13. Powar, O.S., Chemmangat, K., Figarado, S.: A novel pre-processing procedure for enhanced feature extraction and characterization of electromyogram signals. Biomed. Signal Process. Control. **42**, 277–286 (2018)
14. Hoozemans, M.J.M., Van Dieën, J.H.: Prediction of handgrip forces using surface EMG of forearm muscles. J. Electromyogr. Kinesiol. **15**, 358–366 (2005)

Condition Monitoring in Additive Manufacturing Using Support Vector Machine

Durgesh Nainwal, Pavan Kumar Kankar, and Prashant Kumar Jain

1 Introduction

Additive Manufacturing is a name given to the various technologies that produce a three-dimensional object by layer-by-layer deposition of a material, viz. plastic, metal, concrete and so on in application ranging from the small button of the shirt to intricate part of aircraft, dental, medical implants, and fashion industry. FDM developed in 1989 is the second commonly used rapid prototyping technology today after stereolithography [1]. Parts manufactured using FDM-based printers are modeled in a computer-aided designing software and then converted into a surface geometry (e.g., .stl format) using a slicing software before finally sending to FDM machine for printing. In FDM, the gear mechanism powered by stepper motor feeds filament inside the liquefier, and then, solid filament next to it forces the molten matter inside the liquefier to finally extrude it out of the nozzle [2]. The deposition takes place in the form of a road which gets interlinked with each other on the successive deposition.

The linear motion of the hot end is controlled by a stepper motor by using a combination of lead screw and timing belt. AM has two vital characteristics; firstly, the printing time is usually long and secondly quality standards for dimensional accuracy ought to be maintained which motivates for developing an approach for health monitoring, fault detection and quality control for 3D printer. Many research efforts have been made to improve the FDM product quality. Yoon et al. [3] used acoustic emission (AE) sensor and piezoelectric strain sensor by creating slippage using metal strips in the timing belt to show the efficacy of such methodology. Rao et al. [4] used data from various sensors like thermocouple, accelerometers, etc. to identify FDM process failure using various models and theories. Wu et al. [5] used an

D. Nainwal (✉) · P. K. Jain
PDPM IIITDM, Jabalpur 482005, MP, India
e-mail: mail2durgesh14@gmail.com

P. K. Kankar
Discipline of Mechanical Engineering, IIT Indore, Indore 453331, MP, India

M. Muzammil et al. (eds.), *Recent Advances in Mechanical Engineering*, Lecture Notes in Mechanical Engineering, https://doi.org/10.1007/978-981-15-8704-7_14

acoustic AE sensor for monitoring purpose in which a manual fault of an improper air gap and layer thickness was created, the time domain and frequency domain signals from the AE sensor during printing condition were analyzed to identify failure modes. In a similar scope, Kim et al. [6] showed the efficacy of vibration sensor over AE sensor to monitor bolt fault using root mean square as a feature to monitor printing status using SVM technique to prevent poor printing.

1.1 Aim and Contribution of Work

The quality of an FDM component is usually inspected at the finishing point and results in a lot of wastage. Therefore, an effective technique should be developed that will not only help in monitoring the printing process in FDM, leading to fault diagnosis but also help in preventing the failure of the component at the earliest, thereby reducing the scrap rates [7]. This work aims at equipping an FDM machine (Creality Ender 3) with vibration sensors to capture time domain data, and using it for the construction of machine learning classification models, to show the ability for detection of part failure in FDM. This study demonstrates that the used methodology and monitoring setup for data collection is suitable for possible in-process monitoring applications.

1.2 Fish Bone Diagram

FDM printer failure causes were identified by Ishikawa diagram in Fig. 1 which is an essential visualization tools that help in identifying the root cause of a problem.

2 FDM Experiment

The common FDM printing problem includes material peeling off, filament buckling, elephant foot, warping syndrome, distortion, and part delimitation. While performing 3D printing, it was noticed that loosen bolt, a part of x-axis driving mechanism, is one reason for shifting of the layer. Therefore, loosen bolt is considered as a faulty condition of the printer in the present work. Figure 2b shows the essential mechanical component which aids in extruder head movement while printing. The pulley is connected to the stepper motor and tightened by a bolt, the timing belt over the pulley drives the extruder along the x-axis. When the bolt is loosened, the slipping occurs between the motor and the pulley, which gets misaligned causing intermittent motion of the driving belt. The printing head wobbles in its position giving rise to vibrations and shifting of subsequent layers resulting in the poor quality of the printed component. Thus, this failure condition was simulated by printing a

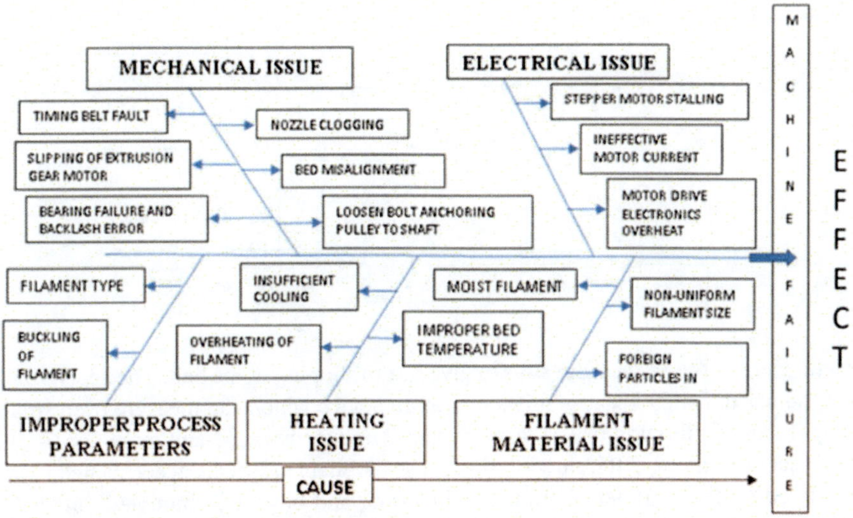

Fig. 1 Ishikawa cause-and-effect diagram

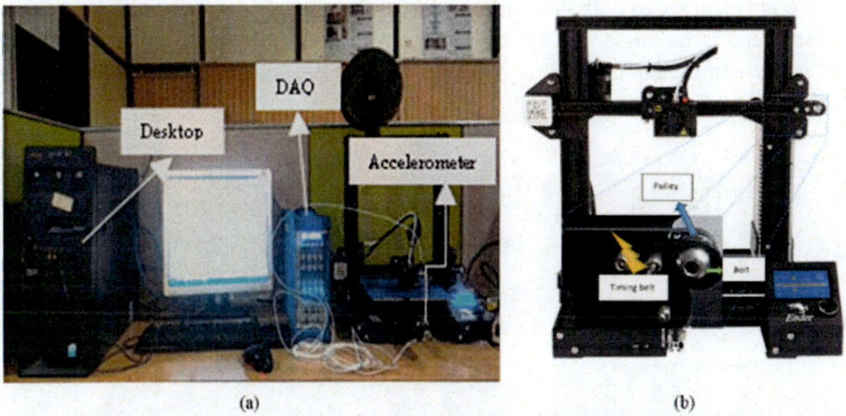

Fig. 2 **a** Experimental setup, **b** mechanical components affecting the motion of extruder head

standard specimen of ASTM D1708 in both the healthy and faulty state of printer. The experimental parameters are shown in Table 1. In a belt system, vibration problem generally occurs due to two reasons: (1) interference of the disruptive forces with belt and (2) improper belt like crack, pieces of chunks, etc.

In case of FDM printers, disruptive forces such as unbalance or the eccentricity of pulley (here) whenever the belt passes over the pulley a short pulse will be generated which will the disturbing frequency or the frequency of vibration of the belt. Previously, belt looseness was simulated by malfunctioning the toothed belt by inserting a

Table 1 Process parameters

Parameter	Description/value
	Creality Ender 3
Specimen model	ASTM D1708
Filament material	Polylactic acid
Filament diameter	1.75 mm
Extruder speed	60, 50, 40 mm/s
Infill	100%
Layer thickness	0.1 mm

metal piece between the teeth slots to give rise to slipping of the belt. The accelerometer comes in handy because it is a miniature device which can measure even small amplitudes of vibration over a large dynamic frequency range and more consistent and accurate among different vibration sensor available [8]. One accelerometer was mounted on the extruder along the x-direction, and other was mounted along the y-direction as shown in the experimental setup. Along the z-direction, negligible changes in vibration for the healthy and the faulty condition were observed. The vibration signals were taken corresponding to three different tool path of print head at different print speed both in healthy and faulty state of the printer. The infill pattern was line, and raster angle was 45 ° for the specimen. In Fig. 3, the healthy component and faulty component are shown; in faulty printed specimen, layer shifting can be easily observed.

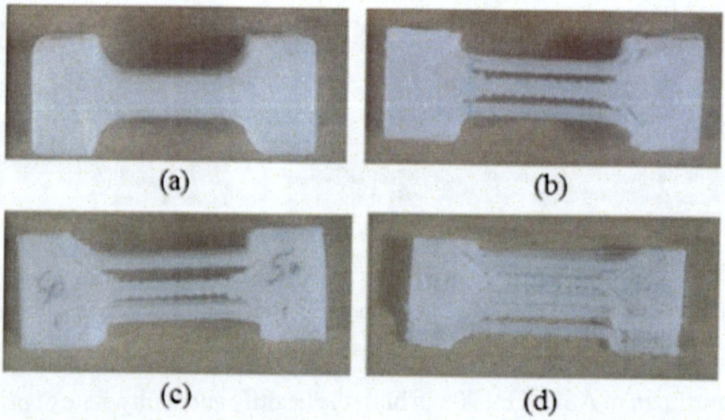

Fig. 3 Printed specimen at **a** healthy condition, **b** faulty condition at 40 mm/s, **c** faulty condition at 50 mm/s, **d** faulty condition at 60 mm/s

3 Feature Selection and Feature Vector

Feature selection is essential to choose key features; if properly selected, it helps in achieving fewer datasets when speed and storage are the concern. Out of the various features like kurtosis, standard deviation, crest factor, mean, etc. only root mean square, mean absolute deviation and interquartile range value of the signal are found to be significant for constructing different classification models.

Although there are many machine learning algorithms previously used by researchers for fault diagnosis, support vector machine (SVM) is found appropriate to construct the models for detecting the faulty and healthy state of the FDM-based printer. Both linear and quadratic SVM were applied in a different cases, and for each case, the discriminants were calculated and examined. In order to feed the model, 14 RMS data were captured from a FDM specimen, and the experiment was performed three times corresponding to healthy and faulty condition. So there were 42 data samples collected at different print speed and tool path. The features from the time data of the vibration signal which will be used in machine learning shown in Table 2.

3.1 Support Vector Machine and K-Cross Validation Approach

Support vector machine is mostly used supervised learning techniques which finds its application in both classification and regression analysis. For a linearly separable data, the two planes are drawn along the boundary of each type of data such that no

Table 2 Feature vector for classification model

RMS	MAD	IQR	Print speed (mm/s)	Class
0.9678	0.8864	1.5712	60	Healthy
1.1733	0.7744	1.3025	60	Healthy
0.6913	0.55	0.9243	50	Healthy
0.7266	0.5813	0.9606	50	Healthy
0.6506	0.5209	0.8914	40	Healthy
0.6456	0.5156	0.9707	40	Healthy
1.5377	1.2293	2.1654	60	Faulty
1.3229	1.0462	1.7764	60	Faulty
0.9216	0.709	1.1374	50	Faulty
0.9352	0.731	1.2327	50	Faulty
0.8465	0.6661	1.1097	40	Faulty
0.8471	0.6647	1.0854	40	Faulty

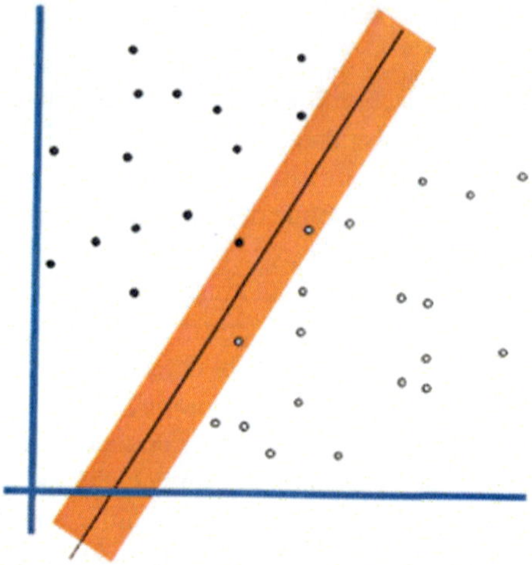

Fig. 4 SVM scatter plot showing maximum margin hyperplane

two points lie in between them. The distance between two margins is given by ||w||. For maximal margin, w is minimized.

The hyperplane has a general form

$$Y_i = \text{sign}\left(w^T x_i + b\right) \quad (1)$$

where w is the kernel plane to the hyperplane and b is the factor which determines the offset from the origin of the hyperplane and x is input vector.

The quadratic programming problem is expressed as:

$$\min_{w,b}\left(\frac{w^T w}{2}\right) \quad (2)$$

Subject to: $yi\left(w^T x_i + b\right) \geq 1$ (∀ data points x_i).

K-cross-validation is a resampling method mainly suitable when a limited number of the datasets are given for evaluating a machine learning model. The basic idea is to divide the dataset into k equal bins. In present case, there are 42 dataset and 5 bins, so each bin will consist of 8 or 9 values. One bin is used for testing, while remaining k minus 1 bins are used for training purpose. In this study, value of k is taken as 5 in order to prevent high bias and variance (Fig. 4).

4 Results

The experimental results show that the effect of the tool path is not considerable in the values of different features obtained from the time series data. Moreover, the

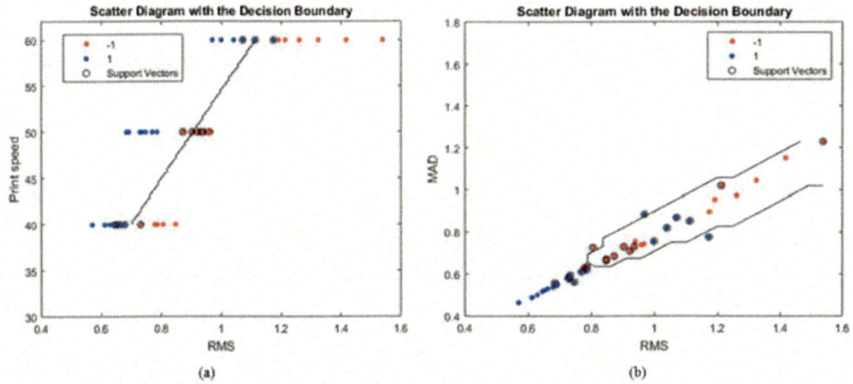

Fig. 5 Scatter plot of SVM. **a** Model 1, **b** Model 2

Table 3 Discriminant values

Model	Sensitivity	Specificity	Precision	Error rate	F1 Score
Model-1	0.9047	0.9047	0.9047	0.09523	0.9047
Model-2	0.80	0.7727	0.81	0.2142	0.7663

misalignment of the layer is more in case of specimen printed at 60 mm/s pertaining to the fact that slippage of the belt is more at a higher speed of stepper motor controlling the print head (Fig. 5).

The scatter plots were obtained for both the models, i.e., when RMS was used over a range of print speed using linear SVM and another when RMS and MAD were used as features using a quadratic kernel function. The classification accuracy of 90.5% is obtained for first model and 78.5% for second model. The values of various discriminants are shown in Table 3 for better understanding and predicting the performance of the model.

5 Conclusion

The methodology used here can be implemented in any FDM printer where such kind of faults occur more frequently, and the precision of the component is at risk. During the implementation phase, one can diagnose various printing cases and prevent any faulty specimen at the earliest helping in material saving.

References

1. Venuvinod, P. K., Ma, W.: Rapid Prototyping: Laser-Based and Other Technologies, vol. 1. Springer Science & Business Media, New York (2013)
2. Bellini, A., Guceri, S., Bertoldi, M.: Liquefier dynamics in fused deposition. J. Manuf. Sci. Eng. **126**(2), 237–246 (2004)
3. Yoon, J., He, D., Van Hecke, B.: A PHM approach to additive manufacturing equipment health monitoring, fault diagnosis, and quality control. In 29th Proceedings of the Prognostics and Health Management Society Conference, pp. 732–740, Fort Worth, TX, USA (2014)
4. Rao, P.K., Liu, J.P., Roberson, D., Kong, Z.J., Williams, C.: Online real-time quality monitoring in additive manufacturing processes using heterogeneous sensors. J. Manuf. Sci. Eng. **137**(6), 061007 (2015)
5. Wu, H., Yu, Z., Wang, Y.: Real-time FDM machine condition monitoring and diagnosis based on acoustic emission and hidden Semi-Markov model, **90**(5–8), 2027–2036 (2017)
6. Kim, J.S., Lee, C.S., Kim, S.M., Lee, S.W.: Development of data-driven in-situ monitoring and diagnosis system of fused deposition modeling (FDM) process based on support vector machine algorithm. Int. J. Precis. Eng. Manuf. Green Technol. **5**(4), 479–486 (2018)
7. Cheng, C., Sa-Ngasoongsong, A., Beyca, O., Le, T., Yang, H., Kong, Z., Bukkapatnam, S.T.: Time series forecasting for nonlinear and non-stationary processes: a review and comparative study. IIE Trans. **47**(10), 1053–1071 (2015)
8. Senldge, M., Llch, T.R.: Piezoelectric accelerometers and vibration preamplifiers, 1st edn. Bruel & Kjaer, Denmark (1987)

Analyses of Temperature and Thermal Stresses of a Ceramic-Coated Diesel Engine Valve

Subodh Kumar Sharma, Krishna V. Ojha, D. Pradhan, Pratibha Kumari, and Ajay kumar

1 Introduction

TBCs are applied on metal substrates to protect them by thermal distortion in order to work at elevated temperature. In diesel engines, the requirement is to enhance the thermal performance and reduce fuel consumption rate; hence, TBCs are proved to be promising in getting higher compression ratios [1]. It is used to protect ignition chamber components or close surfaces like piston head, valve head and cylinder crown. Due to low thermal conductivity and having high-quality strength, ceramic-based material is used for constructing TBCs, which is fit for supporting higher temperatures. A standout among the most utilized materials in TBC is zirconia, which is deposited by plasma-spray technique. Metallic surfaces are shielded from thermal fatigue, particularly from combustion stroke of engine cycle by using TBCs; by doing this, little bit of extra heat energy is saved and can be transmitted for useful work [2–7]. Additionally, it provides thermal insulation and decreases the necessity of cooling requirements. An intermetallic alloy, i.e., NiCrAl, is chosen as a bond coat (BC) material, which works as oxidation resistance at elevated temperature and helps to implant TBC over the substrate. It also sinks the internal stresses which may occur between the ceramic coating and substrate. Co-efficient of thermal expansion, conductivity and modulus of elasticity of TBCs are highly effective to work as thermal shock resistance [8]. It is understood that a thicker coating can give better assurance; however, residual thermal stress prompts spallation of thick coatings. So, choosing the most ideal thickness of the TBC is essential for exact assessment of the temperature drop [2–4, 9] and modeling of the valve temperature distribution for taking the thermal stresses within satisfactory levels at the interfaces between TC–BC and BC–SUBS. Modeling and simulations for temperature and stress investigations are extremely valuable and financially attainable for lessening the time as well as

S. K. Sharma (✉) · K. V. Ojha · D. Pradhan · P. Kumari · A. kumar
Mechanical Engineering Department, KIET Group of Institutions, Ghaziabad, India
e-mail: subodhmeerut@gmail.com

M. Muzammil et al. (eds.), *Recent Advances in Mechanical Engineering*, Lecture Notes in Mechanical Engineering, https://doi.org/10.1007/978-981-15-8704-7_15

cost at the production stage of engine components when prototype is developed [10]. This paper aim is to explore temperature and stress distribution in valve at different coating thicknesses to achieve higher engine performance.

2 Thermal Barrier Coating Materials

Magnesia-stabilized-zirconia ($MgZrO_3$) is widely used as top coat in TBC. It has a flexural strength and compressive strength nearly 520 and 1450 Mpa, respectively. In this study, thickness of the TC has been varied from 0.2 to 1.0 mm with a increment of 0.2 mm. An interlayer metallic bond coat NiCrAl of 0.2 mm thickness and substrate (valve material) of steel alloy is used to form TBC.

3 Temperature and Thermal Stress Analyses

To study the impact of the TBCs on diesel engine valves of different thicknesses of the $MgZrO_3$, steady-state thermal stress analyses are executed. Variations in temperature, strain and stresses on the valve are explored for both coated and uncoated conditions. Thermal stress examinations are executed by building up the PC code through Fortran FORCE 2.0. The engine picked up for this assessment is the AV1 single-cylinder, direct injection diesel engine with a 0.553-L cubic limit. The engine is rated at 5hp at 1500 rev/min and compression ratio of 19:1.

To diminish the number of triangular elements and computational time due to geometrical symmetry, boundary and loading condition, an axis-symmetric 2-D finite element model of quarter portion of the valve is utilized in the simulation. Efforts are used to make uniform shapes and kinds of components since it expects a basic part in giving the correct outcomes. While meshing of elements in the valve substrate, BC and TC are accomplished. In addition, line-to-line contact elements are characterized between the substrate and coating layer.

Certain point of confinement conditions is used to demonstrate the heat transfer in valve, valve seat and valve bush contact region with the accompanying presumptions: The effect of valve movement on the heat transfer is neglected, the valve and bush are completely immersed in oil, and there is no capitation, and furthermore, the conductive heat transfer in the oil film is disregarded [4]. From Fig. 1b, boundary conditions in the form of heat transfer coefficient and temperatures are shown.

4 Results and Discussion

Thermal stress investigations are performed using finite element technique, and evaluation of temperature and thermal stress field for uncoated and ceramic coated valve

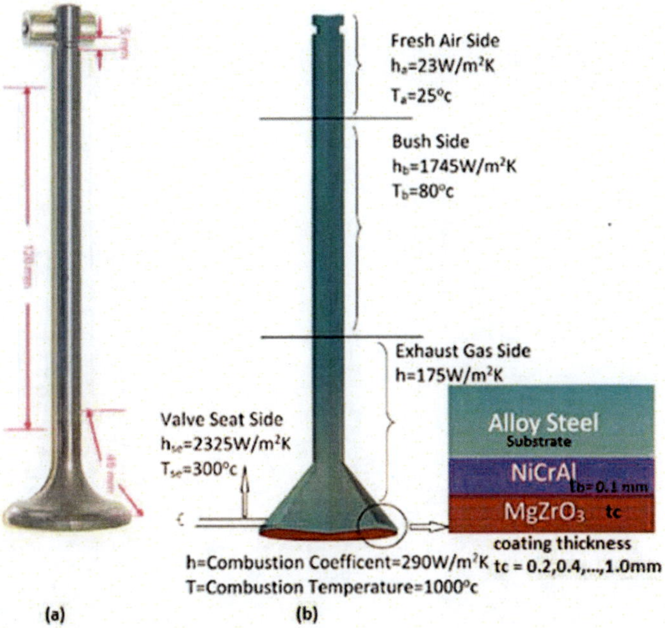

Fig. 1 **a** Valve used in diesel engine, **b** valve model

having different thicknesses of ceramic coating are investigation. Fig. 2a, b shows variation of temperature and radial stresses for the uncoated valve.

The extreme values of temperature and stress (normal stress) are 434.67 °C and 376.58 MPa in the valve, respectively. Analysis shows the decrement in temperature and radial stress from head center of valve head to the top valve stem (see Fig. 2). For the identical conditions, temperature distribution contours are presented in Fig. 3 (a-e) for coating thicknesses varying from 0.2 to 1.0 mm. Temperature at the center of valve head on the coating surface is in the range of 453.8–581.4 °C, respectively. Increased valve life is attributed to improved valve strength due to lowering of temperature. Therefore, radial stress distribution is shown in Fig. 4a–e. Also, maximum stress values at the center of valve head at the coating surface are ranging from 393.59 to 482.57 MPa, respectively, as shown in Fig. 4a–e. For 1.0 mm coating thickness, the maximum increase in temperature is observed at the valve's head surface, due to low thermal conductivity of ceramic coating. The temperature of the top coat (TC) surface versus radial distance (starting from valve head center to ending at the outer edge) for various thicknesses is plotted in Fig. 5.

Temperature of the bond coat (BC) surface and substrate surface (SUBS) versus radial distance for varying TBC thicknesses are plotted and presented in Figs. 6 and 7 respectively. The maximum temperature is shown at the surface of bond coat which ranges from 444.08 to 425.41 °C (see Fig. 6) and variations in maximum temperatures of substrate surface is 442.88–424.53 °C (see Fig. 7) respectively for

Fig. 2 Counterplots of the **a** temperature (°C) and **b** radial thermal stress (MPa) distributions of a quarter part of the uncoated valve

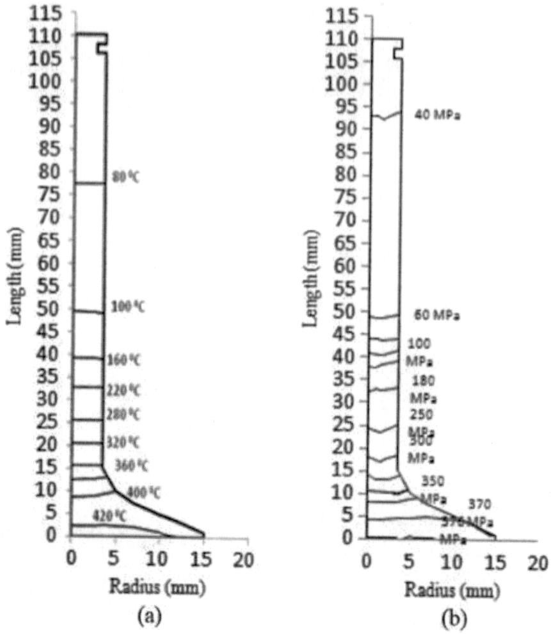

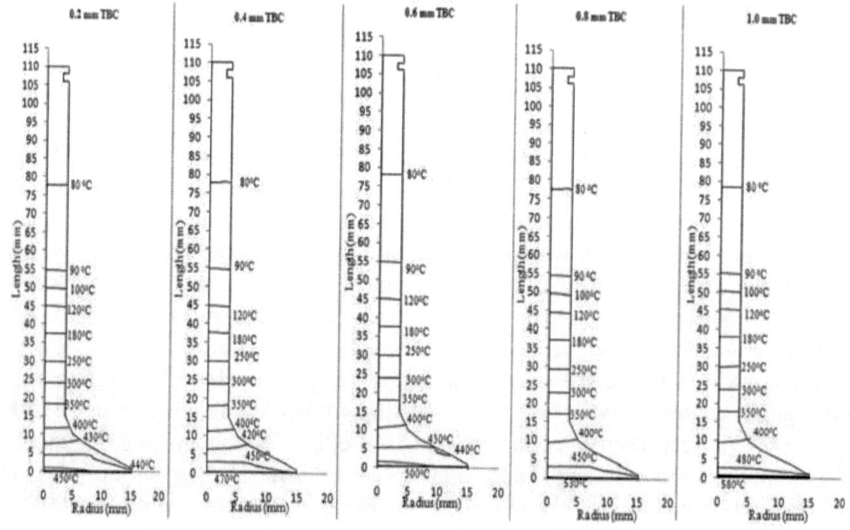

Fig. 3 Temperature (°C) distributions in the valve for various coating thicknesses

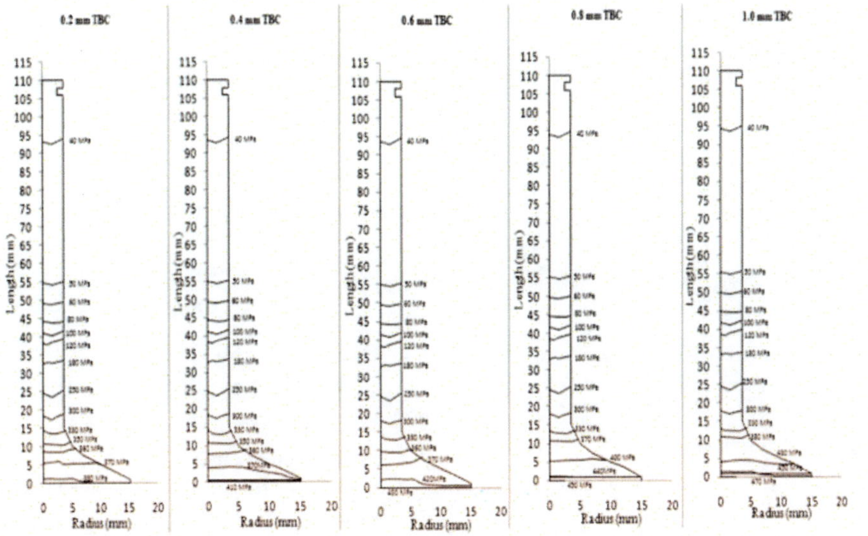

Fig. 4 Radial thermal stress (MPa) distributions in the valve for various coating thicknesses

Fig. 5 Top coat surface temperature

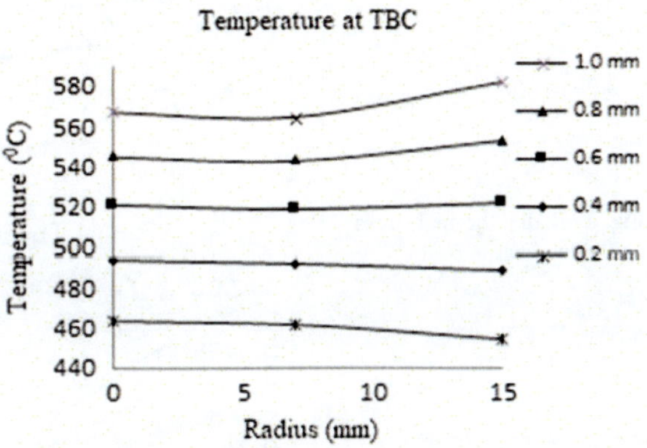

coatings with a thickness of 0.2–1.0 mm, respectively. Also, with the decrease in the coating thickness, the maximum temperature in the bond coat and substrate surface decreases, and no significant difference is shown in the temperature curves of substrate and bond coat material with marginal difference of 3 °C.

A point of highest thermal stress is the most susceptible point for the crack nucleation. Since TBC is brittle, so small imperfection may lead to rapid propagation of cracks, leading to the delamination of coating from the bond coat. Therefore, it is necessary to evaluate the distribution of interfacial thermal stresses for the stresses to be in operation limits. For this motive, a detailed analysis for the thermal stress distribution on the surfaces of top coat, bond coat and substrate is shown in Fig. 8.

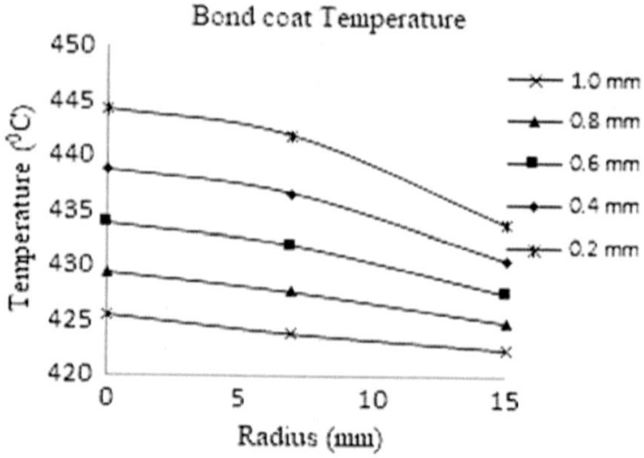

Fig. 6 Bond coat surface temperature

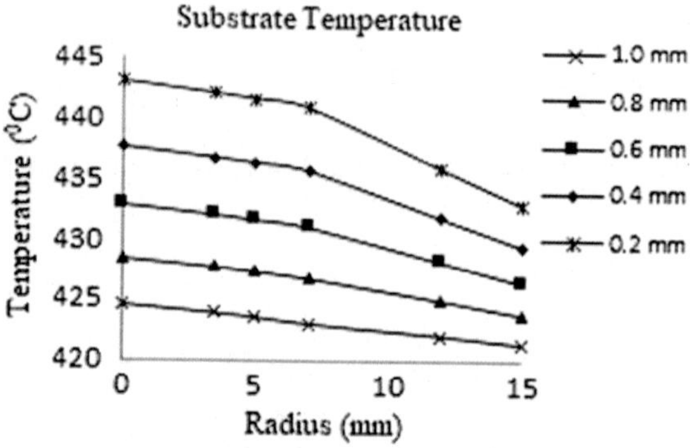

Fig. 7 Substrate surface temperature

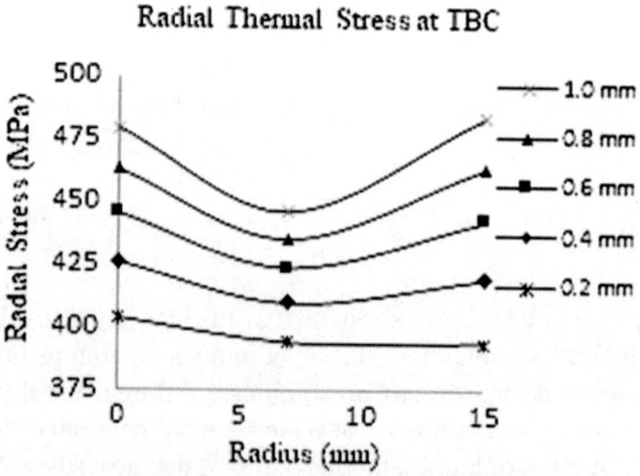

Fig. 8 Radial thermal stress on the top coat

The maximum normal stresses at near the edge of valve head edge are varied from 403.58 to 478.45 MPa for 0.2 to 1.0 mm coatings.

The stress is found to increases with the increase in the thickness of the coating, at the top coat surface. Figure 9 shows the normal stress distribution with the radial distance for different coating thickness, at the bond coat surface.

Figure 10 presents the distributions of normal stress vs radial distance at substrate surface. The maximum normal stress occurs at the valve head center as 385.17 to 367.83 MPa for 0.2 to 1 mm coating thicknesses, and finally decreases significantly with increase in the thickness of coating. The maximum thermal stress of 385.17 MPa for 0.2 mm coating thickness is observed on the surface of substrate, and with the gradual increase in coating thickness, the thermal stress decreases due to lower thermal conductivity of the ceramic material.

Since thermal stress is a function of coating thickness, hence when the normal stresses for different coating thicknesses are compared with the strength, lower values are observed.

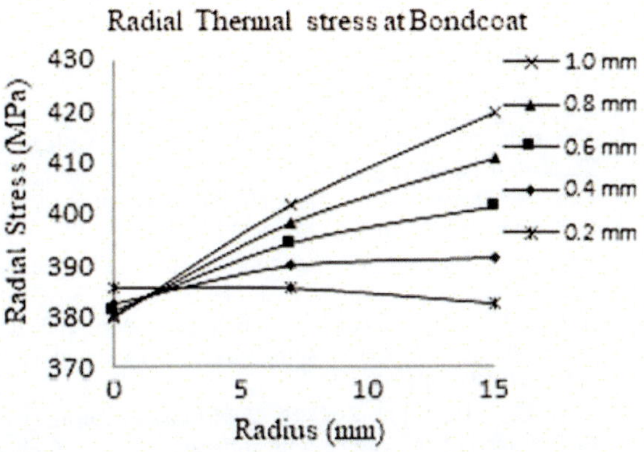

Fig. 9 Radial thermal stress on bond coat

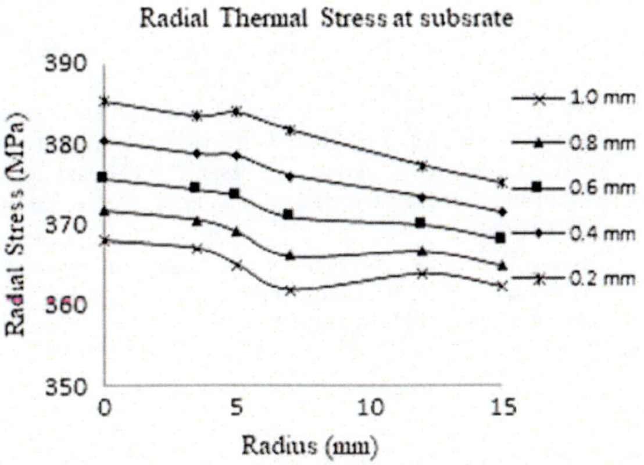

Fig. 10 Radial thermal stress on the substrate

5 Conclusions

Following are the conclusions drawn from the present study:

1. Temperature and thermal stress distribution are function of thickness of TBC coating established using numerical simulations.
2. Maximum temperature is at the center of valve base and at the edge of valve surface for all coating thickness.
3. Maximum temperatures for coated valve when compared with uncoated valve at the center of valve base are 6.67, 13.62, 19.97, 27.09 and 33.81% for 0.2, 0.4, 0.6, 0.8 and 1.0 mm thick thermal barrier coating, respectively.
4. TBC coating is proved to provide higher combustion chamber temperature; therefore, thermal efficiency of the engine increases. Also, reduction in the surface temperature of the valve (substrate) has improved effect on the performance of engine.

References

1. Yonushonis, T.M.: Overview of thermal barrier coatings for diesel engines. J. Therm. Spray Technol. **6**, 50–56 (1997)
2. Cerit, M.: Thermo mechanical analysis of a partially ceramic coated piston used in an SI engine. Surf. Coat Technol. **205**, 3499–3505 (2011)
3. Buyukkaya, E., Cerit, M.: Experimental study of NOx emission and injection timing of a low heat rejection diesel engine. Int. J. Therm. Sci. **47**, 1096–1106 (2008)
4. Sharma, S.K., Saini, P.K., Samria, N.K.: Modelling and analysis of radial thermal stresses and temperature field in diesel engine valves with and without air cavity. Int. J. Eng. Sci. Technol. **5**(3), 111–123 (2013)
5. Buyukkaya, E., Engin, T., Cerit, M.: Effects of thermal barrier coating on gas emissions and performance of the engine with different injection timings and valve adjustments. Energy Convers. Manage. **47**, 1298–1310 (2006)
6. Cerit, M., Coban, M.: Temperature and thermal stress analyses of a ceramic-coated aluminum alloy piston used in a diesel engine. Int. J. Therm. Sci. **77**, 11–18 (2014)
7. Parlak, A., Ayhan, V.: Effect of using a piston with a thermal barrier layer in a spark ignition engine. J. Energy Inst. **80**, 223–228 (2007)
8. Sharma, S. K., Saini P. K., Samria, N. K.: Experimental thermal analysis of diesel engine piston and cylinder wall. J. Eng. **2015**, 1–10 (2015). Article ID 178652
9. Cerit, M., Ayhan, V., Parlak, A., Yasar, H.: Thermal analysis of a partially ceramic coated piston: effect on cold start HC emission in a spark ignition engine. Appl. Therm. Eng. **31**(2–3), 336–341 (2011)
10. Sharma, S.K., Saini, P.K., Samria, N.K.: Computational modeling of temperature field and heat transfer analysis for the piston of diesel engine with and without air cavity. Jordan J. Mechan. Indus. Eng. **9**(2), 139–147 (2015)

Study on Optoelectronic Properties of Slurry Coated Binary Cadmium Chalcogenide Films

Vipin Kumar, Vandana Grace Masih, and V. K. Sachan

1 Introduction

Over the years, II–VI group compound semiconductors have gained substantial attraction of numerous investigators in view of their exhaustive applications in the solid-state physics. These semiconductors are widely used in optoelectronic and photovoltaic technologies [1]. Among all II–VI group compound semiconductors, the cadmium chalcogenides emerged as the potential candidates for optoelectronic technologies and other upcoming technologies such as light emitting diodes, photodetectors, solar cells and lasers [2]. The direct band gap nature and high absorption coefficients are the main reason of their popularity [3].

It is well recognized that any large-scale applications must depend on low-cost polycrystalline materials. The low production cost is the root cause of technical interest in polycrystalline-based devices [4]. Various techniques such as thermal evaporation [5], spray pyrolysis [6], screen-printing [7], sputtering [8], chemical bath deposition [9], successive ionic layer adsorption and reaction (SILAR) [10] and dip method [11] were reported to produce binary cadmium chalcogenide semiconducting thin films. Usually, in all of these methods, stable, uniform, adherent, hard and polycrystalline films are produced.

In the current work, binary cadmium chalcogenide (CdS, CdSe, CdTe) films were produced on glass substrate by using slurry coating method. This method is relatively low cost and appropriate for large area coating of the films. This is also suitable for

V. Kumar (✉) · V. G. Masih
Department of Applied Sciences, KIET Group of Institutions, Ghaziabad, India
e-mail: vipinkumar28@yahoo.co.in

V. K. Sachan
Department of Electronics and Communication Engineering, KIET Group of Institutions, Ghaziabad, India

M. Muzammil et al. (eds.), *Recent Advances in Mechanical Engineering*, Lecture Notes in Mechanical Engineering, https://doi.org/10.1007/978-981-15-8704-7_16

coating surface of any shape and morphology. The coated films were characterized for their optoelectronic properties to explore the application of these films in optoelectronic technologies.

2 Experimental Techniques

To prepare binary cadmium chalcogenide films, the commercially available AR grade powders of these cadmium chalcogenides (CdS, CdSe, CdTe) were taken as the base material. A slurry having this powder (CdS or CdSe or CdTe) with 10% of cadmium chloride (as adhesive) and suitable amount of ethylene glycol (as binder) were prepared and mixed properly in an agate and mortar. The slurry was screen printed on glass substrate through a silk screen. The coated films were then dried at 120 °C for three to four hours. Finally, the films were further fired for 10 min at 450–500 °C in a muffle furnace.

Structural characterization for the films was carried out by XRD studies, reflectance measurements were carried out to calculate the optical band gap (E_g). DC electrical conductivity measurements for the films were made on usual two-probe system.

3 Results and Discussion

XRD graphs of these binary cadmium chalcogenide films were recorded at room temperature and are presented in Fig. 1. The existence of intense peaks in these graphs authenticates the polycrystalline character of the coated films. The sharp and strong peaks specify the development of well-crystallized films. The additional details on the XRD graphs of these binary cadmium chalcogenide films have been described by the authors somewhere else [12, 13]

It is mostly stated that almost all binary cadmium chalcogenide semiconductors have direct transition of band gap [1]. The optical band gaps (Eg) of these slurry coated films were calculated by UV/Vis spectroscopy in reflection mode. As stated by Tauc [14], the absorption coefficient (α) of direct band gap semiconductors is expressed as

$$\alpha h\nu = C(h\nu - Eg)^{1/2} \tag{1}$$

where C is constant and other symbols has their usual meaning. For calculating the band gap from reflection data, a graph of $(\alpha h\nu)^2$ or $h\nu$ ln $[(R_{max} - R_{min})/(R - R_{min})]^2$ (on y-axis) and $(h\nu)$ (on x-axis) is plotted as here absorption coefficient (α) is proportional to ln $[(R_{max} - R_{min})/(R - R_{min})]$ [7]. The extrapolation of linear part of this graph to x-axis provides the value of band gap of material under investigation.

Fig. 1 XRD graphs of slurry coated binary cadmium chalcogenide films

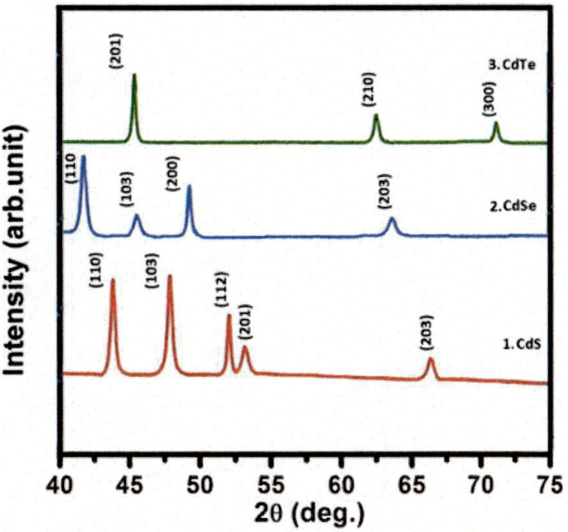

The plot of $h\nu$ ln $[(R_{\max} - R_{\min}) / (R - R_{\min})]^2$ versus $((h\nu)$ for the binary cadmium chalcogenide films is illustrated in Fig. 2.

The values of optical band gap (Eg) acquired form Fig. 2 is displayed in Table 1.

It is clear from the study of band gap that a wide range of band gap, i.e., from 1.48 to 2.47 eV can be obtained by these binary cadmium chalcogenides. This wide range of band gap is indicative of the fact that these films can be exploited in the optoelectronic technologies for specific requirements.

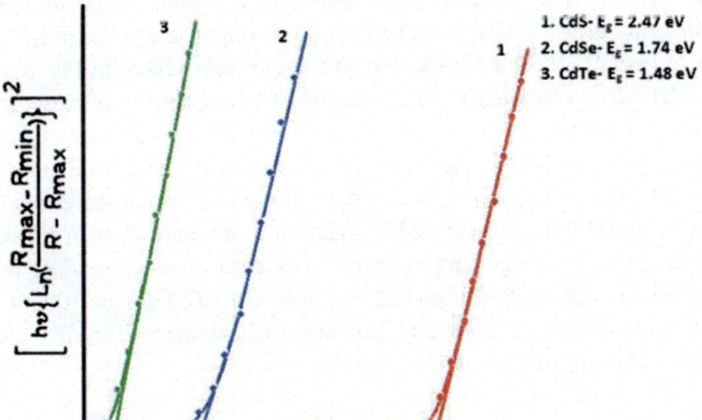

Fig. 2 Plot of $h\nu$ ln $[(R_{\max} - R_{\min})/(R - R_{\min})]^2$ versus $((h\nu)$ for the binary cadmium chalcogenide films

Table 1 Calculated values of band gap of binary cadmium chalcogenide slurry coated films

Composition	Optical band gap (eV)
CdS	2.47
CdSe	1.74
CdTe	1.48

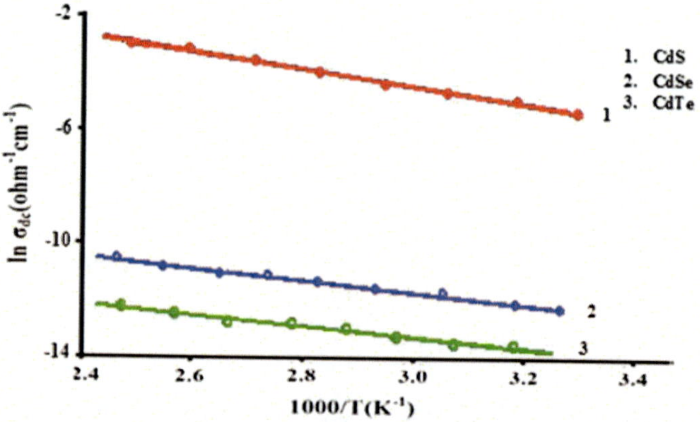

Fig. 3 Plot of ln σ_{DC} versus 1000/T for slurry coated binary cadmium chalcogenides

The DC electrical conductivities in dark for the coated films were measured in the range of temperature 300–400 K by two-probe system. A DC voltage was applied across the film and corresponding current was studied by a Keithley electrometer. A copper constantan thermocouple was employed to record the temperature of the films. The measurement of electrical conductivity was done by coating small film on glass substrate on which indium electrodes with small gap were pre-coated. The plot of log (σ) versus 1000/T for the coated films is represented in Fig. 3.

It was observed that conductivity increases with increase in temperature specifying the semiconducting nature for all the films. The increase in conductivity can be accounted for the growth of crystallite size and increase in carrier density [1]. The linear nature of plot of log (σ) versus 1000/T curve shows that the conductivity of these films is fitted with the Arrhenius equation and conduction in these films is due to thermal activation of charge carriers. The activation energies of the coated films were evaluated from the slope of the graph of log (σ) versus 1000/T (Fig. 3). The electrical conductivities and activation energies calculated for binary cadmium chalcogenide films are mentioned in Table 2

Table 2 Calculated values of electrical conductivity and activation energy of binary cadmium chalcogenide slurry coated films

Composition	Electrical conductivity (Ω^{-1} cm^{-1})	Activation energy (eV)
CdS	2.42×10^{-3}	0.16
CdSe	3.55×10^{-6}	0.25
CdTe	3.72×10^{-6}	0.21

4 Conclusions

It is hereby concluded that binary cadmium chalcogenide films (CdS, CdSe, CdTe) were synthesized by low cost and simple slurry coating method. The polycrystalline behavior of the coated films was ensured through XRD analysis. The optical band gap of the coated films alters from 1.48 to 2.47 eV. The DC electrical conductivity of these films is well fitted with the Arrhenius equation. The obtained results indicate the suitability of these slurry coated films in different optoelectronic technologies.

References

1. Kumar, V., Agarwal, S., Dwivedi, D.K.: Study on optical investigations and DC conduction mechanism in polycrystalline chalcogenide (Cd, Zn) semiconductor films grown by screen-printing method. J. Mater. Sci.: Mater. Electron. **28**, 1715–1719 (2017)
2. Goyal, S., Chauhan, R.P.: Substrate dependent variation in the properties of cadmium telluride thin films deposited on glass. J. Mater. Sci.: Mater. Electron. **30**, 1345–1353 (2019)
3. Kamble, V.K., Pujari, V.B.: Electrical and micro-structural properties of cadmium chalcogenides: A comparative study. Int. J. Emerg. Technol. Adv. Eng. **5**, 172–176 (2015)
4. Patidar, D., Saxena, N.S., Sharma, T.P.: Structural, optical and electrical properties of CdZnS thin films. J. Mod. Opt. **55**, 79–88 (2008)
5. Khairnar, O.P., Bhavsar, D.S., Vaidya, R.U., Bhavsar, G.P.: Optical properties of thermally evaporated Cadmium telluride thin films. Mater. Chem. Phys. **80**, 421–427 (2003)
6. Yadav, A.A., Barote, M.A., Masumdar, E.U.: Studies on Cadmium selenide (CdSe) thin films deposited by spray pyrolysis. Mater. Chem. Phys. **121**, 53–57 (2010)
7. Kumar, V., Sharma, D.K., Bansal, M.K., Dwivedi, D.K., Sharma, T.P.: Synthesis and characterization of screen-printed CdS films. Sci. Sinter. **43**, 335–341 (2011)
8. Ghoranneviss, Z., Akbarejad, E., Ghoranneviss, M.: Effects of various deposition times and RF powers on CdTe thin film growth using magnetron sputtering. J. Theor. Appl. Phys. **10**, 225–231 (2016)
9. Gopakumar, N., Anjana, P.S., Vidyadharan Pillai, P.K.: Chemical bath deposition and characterization of CdSe thin films for optoelectronic applications. J. Mater. Sci. **45**, 6653–6656 (2010)
10. Sankapal, B.R., Mane, R.S., Lokhande, C.D.: Deposition of CdS thin films by successive ionic layer adsorption and reaction (SILAR) method. Mater. Res. Bull. **35**, 177–184 (2000)
11. Hankare, P.P., Chate, P.A., Sathe, D.J., Jadhav, B.V.: Photoelectrochemical studies of CdSe thin films by dip method. J. Alloy. Compd. **474**, 347–350 (2009)
12. Kumar, V., Sharma, T.P.: Structural and optical properties of sintered CdS_xSe_{1-x} fiims. J. Phys. Chem. Solids **59**, 1321–1325 (1998)
13. Sirohi, S., Kumar, V., Sharma, T.P.: Optical, structural and electrical properties of CdTe sintered films. Opt. Mater. **12**, 121–125 (1999)
14. Tauc, J.: Amorphous and liquid semiconductors. Plenum Press, New York (1974)

An Integrated Maintenance Management: A Practical Approach

B. Hari Prasad and Mahesh Bhardwaj

1 Introduction

In recent years, there is considerable change in production and process industries. The market is driven by variety, quality and delivery performance. The product life cycles are shrinking day by day. The flexible manufacturing is the new advancements adopted by manufacturers. Just-in-time is the new material technology applied by industries to work without inventory. Industries have utilized state of the art technologies to minimize adjustment and set up time of machines as discussed by Prasad [1]. As compared to earlier culture of breakdown maintenance, manufacturers are focusing on early maintenance. The organizations or companies are more focused on preventive, predictive and productive maintenance. Even Internet of things (IoT) has been invoked by organizations for maintenance tasks. However, the production/process organizations/companies have invested heavily in new equipment in an effort to become less labour intensive.

2 Integrated Maintenance Management

An integrated approach to maintenance can be termed as integrated maintenance management. The three subsystems of maintenance are as under:

- Management of maintenance
- Maintenance of operations

B. Hari Prasad
Bharat Dynamics Limited, Hyderabad, India

M. Bhardwaj (✉)
KIET Group of Institutions, Ghaziabad, India
e-mail: mahesh.bhardwaj@kiet.edu

M. Muzammil et al. (eds.), *Recent Advances in Mechanical Engineering*, Lecture Notes in Mechanical Engineering, https://doi.org/10.1007/978-981-15-8704-7_17

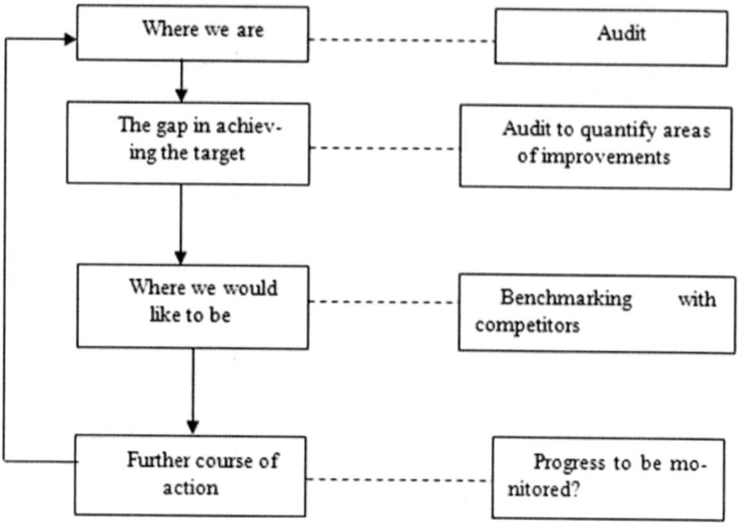

Fig. 1 Continuous improvement of maintenance effectiveness

- Management of equipment.

The aim of management through integrated maintenance approach is to provide a methodology for an effective and continuous maintenance. The philosophy behind this proposal is as enumerated below:

- There is a requirement of designing a comprehensive audit for maintenance department. This will tell us exact state of maintenance.
- Development of tools for correct analysis of audit results. This will help in establishing the current maintenance state and will help in identifying potential areas for improvement.
- The best maintenance practices help in setting up higher goals. This benchmarking will give improvement in a different dimension.

The proposed framework for an effective continuous improvement programme is shown in Fig. 1.

2.1 Management of Maintenance

2.1.1 Organization

The productive maintenance activity can only be implemented by a well-designed maintenance department organization. The organization chart should be complete and always current. It should also take the following issues into considerations:

- The ratio of supervisors to craftsmen should be appropriate.
- The number of planners should be proper.
- Necessary support functions to be in place.
- The maintenance response to be quick.

2.1.2 Employee Training and Motivation

Good quality maintenance work can only be ensured with proper training, good working environment and necessary support system. It requires proper planning, required material and accurate supervision. The result will be motivated the employee with good performance. The same was discussed by Misra [2].

2.1.3 Planner Training

Planning is a critical function of department handling maintenance function. Proper training of planner should be ensured. The planner will focus on the following activities:

- To determine the content in a job.
- The duration of all contents in a job.
- To plan the entire work process utilizing industrial engineering techniques.
- To determine the number of skilled workers required for the job utilizing industrial engineering techniques.
- Planner will estimate the material, spare parts and tools required for the job.
- Planner will plan and schedule the work orders.
- To estimate costs.

2.1.4 Maintenance Control

The thorough maintenance control system will help in identifying delays. The corrective action can be carried out and that will help in better labour and cost control. This is possible since it is based on factual data and identification of areas having high potential for improvement. Such a system as discussed by Prasad [1] and it includes:

- The status information of work orders. There estimated time can be compared to actual time.
- The productivity reports can be generated for maintenance.
- The overtime, backlog and emergent work can be depicted through charts and graphs.

2.1.5 Supervision by Maintenance Supervisor

The planner plays an important role here. He should relieve the supervisor accountable for maintenance from the role of planning and scheduling. This will help him to focus mainly on maintenance function.

2.2 *Management of Maintenance Operations*

2.2.1 Work Measurement

Good maintenance practice requires proper planning. Time standards are required to plan proper planning.

2.2.2 Maintenance Scheduling

The daily maintenance scheduling should match with planner function and accordingly day-to-day work orders should be scheduled. Apart from issuing daily schedules, this function determines the priority of work orders, follow-up their progress and track the backlogs.

2.2.3 System of Work Order

A well-designed works order form and procedure is a necessity for maintenance management system. All stakeholders involved in a maintenance job request and get the clear communication by this system. They include support function staff, planner, supervisor, requisitioner and craftsman. This also ensures proper documentation of maintenance work for analysis and exact corrective action. Thus, works orders are essential for all maintenance work.

2.2.4 Control of Materials and Tools

The timely availability of tools and spare parts are necessary to avoid any form of delays in the maintenance activity. It necessitates an installation of computerized inventory control system. This will help in deciding optimal order quantities. This will also help in having appropriate interface and co-ordination procedure between maintenance department and warehouse.

2.3 *Management of Equipment*

2.3.1 Equipment History and Preventive Maintenance

A comprehensive preventive maintenance system reduces maintenance costs. However, it requires well-trained preventive maintenance inspectors who are well versed with well-maintained equipment history and are dedicated to preventive maintenance work.

In preventive maintenance system, a checklist provides lists of preventive maintenance work for each part of equipment required to be maintained. The preventive maintenance route in its schedule provides the sequence of equipment to be worked in a given period. The actual preventive maintenance work carried out on a daily basis is covered in its reports.

Equipment history is required to be well maintained. It is an important part of preventive maintenance system. The regular analysis of this history is mandatory. It will help in adjusting the preventive maintenance schedule. The overhaul and replacement policies can also be developed based on this analysis.

2.3.2 Management of Predictive Maintenance

The management of predictive maintenance was introduced to Japan from America in the 1950s. This was the time when the Japanese process industries were beginning to get back on their feet after the war. Predictive maintenance management has been defined by Hartman [3]. According to him, it is the extrapolation of graphic trends of measured physical readiness of equipment when compared with its known engineering limits. It involves the periodic measurements of vibration velocity, tension, pressure, alignment, wear, heat, resistance, capacitance, etc. This is done for the purpose of detecting, analysing and correcting equipment problems before failure. The predictive maintenance management should be applied to all critical equipment. Mean time to repair should be calculated through analysis of equipment histories.

2.3.3 Equipment Management Programme

Flexible manufacturing systems warrant maintenance function with the increase in automation and acquisition of high-tech equipment by many organizations. Efficient equipment management programmes ensure peak performance and it also ensures high availability of equipment. The philosophy of total productive maintenance (TPM) is promising in this context since it focuses on reduction in equipment losses. The various losses in equipment identified by Nakajima [4] are as under:

- Failure of equipment.
- Losses during setting and adjustment of equipment.
- Losses due to idling of equipment.

- Losses due to minor stoppages in equipment.
- Losses due to reduction in speed of an equipment.
- Losses due to defects in process.
- Losses due to reduction in yield.

The availability of equipment, efficiency of equipment and quality of product are affected by these losses. Equipment effectiveness must take into consideration in all these aspects when measuring.

Both the Japanese or western style approach to TPM focuses on zero defects, zero accidents and zero breakdown. TPM experts *Nakajima, Hartmanm, Willmott and wireman* all agree on this aspect. The operational element of any TPM programme is based on five pillars of TPM development. These are summarized by Nakajima [5] as:

(i) The designed improvement activities to be implemented to increase equipment efficiency.
(ii) Equipment operators to propagate autonomous maintenance. The operators are to be trained in equipment skills.
(iii) The efficiency of the maintenance department will be increased by a planned maintenance system.
(iv) The skill levels of equipment operators will be enhanced with planned training courses.
(v) A system to be established where maintenance prevention is part of design. An early equipment management is part of maintenance activity.

2.3.4 Measuring and Improvement of Maintenance Productivity

The productivity needs to be measured to ensure effectiveness of any maintenance system. Then only areas can be identified where improvements can be made. The formulation of appropriate improvement programme is only feasible after audit of current status. The steps to be followed in audit of maintenance system are as under:

- A survey using a well-designed questionnaire to be carried out which will aim at comparing the current practices with what they should be.
- The analysis to be carried out of data gathered in above step.
- Based on the analysis improvement programmes to be formulated.

3 Benchmarking

The best strategies in maintenance be implemented through benchmarking by searching the best maintenance practice. There are many approaches to benchmarking. One of the simple benchmarking techniques is to priorities the list of activities to be benchmarked. This should also include the list of activities which are not required to be benchmarked.

4 Conclusion

The maintenance management is a requirement of current age organizations/companies due to tough competition and shrinking profit margins. The industries no longer can afford losses occurring for any mundane reason it may be. This is a reason enough for industries to adopt an integrated maintenance management system. Through this paper, an effort is made to present a methodology for an integrated maintenance management system which can effectively be implemented by any product/process industry.

References

1. Prasad, B.H.: An Integrated Approach to Mission Reliability for Strategic Equipment Management, PhD Thesis, JNTU, Hyderabad, India (2009)
2. Kiran, D.: Total productive maintenance. Total Qual. Manag. 177–192 (2017). doi: 10.1016/b978-0-12-811035-5.00013-1
3. Bouranta, N., Psomas, E., Suárez-Barraza, M. F., Jaca, C.: The key factors of total quality management in the service sector: a cross-cultural study. Benchmarking: Int. J. **26**(3), 893–921 (2019). doi: https://doi.org/10.1108/bij-09-2017-0240.
4. Sangode, P.B., Hedaoo, H.R.: Implementation of total quality management in manufacturing firms: an empirical study (February 11, 2019). IUP J. Oper. Manag. **XVIII**(1), 21–35 (2019). https://papers.ssrn.com/sol3/papers.cfm?abstract_id=3402308
5. Protzman, C., Whiton, F., Kerpchar, J., Lewandowski, C., Stenberg, S., Grounds, P., Bond, J.: Total Productive Maintenance (2018). 10.4324/9781315373843-23

Synthesis and Flexural Behavior of Silica Coated ABOw Reinforced Aluminum Matrix Composites for Structural Application

Neeraj Pandey, S. C. Ram, I. Chakrabarty, and M. R. Majhi

1 Introduction

Aluminum alloys and its composites have been extensively used in military industries, aerospace, auto industries and transport, marine industries, construction structures and electrical industries due to their lightweight, low density, high strength and ease of fabrication [1–3]. Reinforcements used may be continuous, discontinuous or particulate in nature. Whiskers are gaining their importance as reinforcements as they provide large surface area for bonding with matrix material. The whiskers mainly used as reinforcements are SiC, $Mg_2B_2O_5$, Mg_2Si and ABOw [2, 4–9]. Among these, ABOw is gaining importance as they are low-cost one-dimensional material. The ABOw ($Al_{18}B_4O_{33}$) samples may have potential oxidation resistance, and the physical properties such as bulk density 1.46 g/cm^3, open porosity 54.4 and linear shrinkage percentage −1.8% and flexural strength 51 ± 2 MPa at 1300 °C sintering temperature [10]. Several fabrication techniques using whiskers as reinforcements have been reported for production of such composites such as squeeze casting, stir casting and centrifugal casting [11]. However, due to lack of research work reported on powder metallurgy route for the fabrication of coated ABOw whiskers reinforced composite. Wang et.al. found that the ABOw was generally prepared by sol–gel method [12]. Gao et al. [13] reported that by coating of ABOw and the addition of soybean sacrificial filler in the matrix, the volume fraction of whiskers decreased, but the ultimate tensile strength remains unchanged. Yue et al. [9] also found that by coating a suitable layer of Zn, Cu, Ti, Al or Ni oxide on

N. Pandey (✉) · M. R. Majhi
Department of Ceramic Engineering, Indian Institute of Technology (BHU) Varanasi, Varanasi, UP, India
e-mail: neerajp.rs.cer16@itbhu.ac.in

S. C. Ram · I. Chakrabarty
Department of Metallurgical Engineering, Indian Institute of Technology (BHU) Varanasi, Varanasi, UP, India

M. Muzammil et al. (eds.), *Recent Advances in Mechanical Engineering*, Lecture Notes in Mechanical Engineering, https://doi.org/10.1007/978-981-15-8704-7_18

ABO whiskers, the interfacial reaction product Al_4C_3 reduces. Many researchers have studied ZrO_2, Al_2O_3 reinforced aluminum metal matrix composites but their tensile and fracture behavior needs much improvement. Roseline et al. [14] varied ZrO_2 percentage in ceramics and found that the impact strength and tensile strength are slightly enhanced with improved ductility. The higher porosities developed in the composites due to stir casting process causes low strength of components. The adaptation of powder metallurgy route shows better results in respect to lower porosities and defect-free ceramic reinforced composites fabrications. The desired shape component was obtained which requires minimum machining [15–17]. In the current work, the coating of SiO_2 particles was introduced into the preparation of ABOw/Al composites. The composites were fabricated by powder metallurgy route.

2 Experimental Procedure

2.1 Raw Materials Used

Aluminum borate whisker (ABOw) powder size (>10–20 μm) was synthesized by hydrolization route using alumina (Al_2O_3) and boric acid (H_3BO_3) and calcined at 1300 °C for 4 h and used as reinforcement [18], the mass ratio of SiO_2 to ABOw selected was 1:10 [18]. Firstly, fine SiO_2 size (>40 μm) and purity (99%) procured from CDH private limited New Delhi were taken as a starting material to form a clear sol from mixing with distilled water for 20 min. The aluminum powder (purity 99.5%) particle size (>45 μm) CAS: 1344-28-1 was used as matrix material procured from Poly-Chem private India Limited, Bangalore.

2.2 Coating of ABOw with Silica

As shown in Fig. 1, ABOw powder was mixed with the SiO_2 sol under a certain condition, and then filtered and dried at 110 °C for 2 h and dried powder was thermally heat treated at 500 °C for 4 h in air. Then cooled down to room temperature naturally

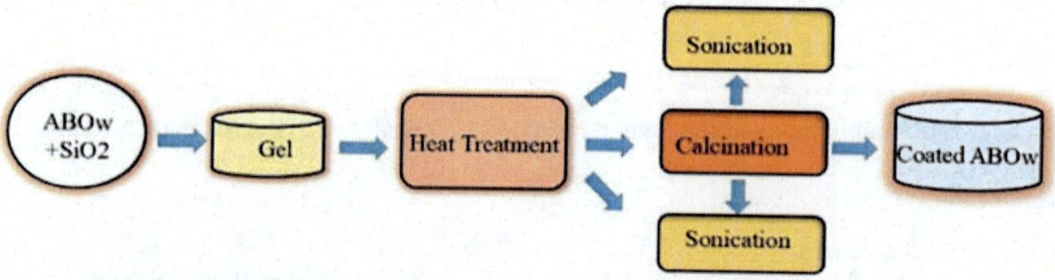

Fig. 1 Coating process of ABOw

in air, now whiskers could be easily dispersed in distilled water for sonication using an ultrasonic vibration apparatus for 45 min to prevent whisker agglomeration. Finally, the preform was calcined at 1100 ∘C for 1 h to obtain coated ABOw with sufficient strength.

2.3 Composite Sample Formation

The aluminum metal powder was vigorously mixed with ethanol in magnetic stirrer for 40 min, then coated ABOw was added in required percentage through a mechanical method, now the dried mixture was cold isostatic pressed at 250 MPa for 3 min to obtain green compacts, which would be kept at elevated temperature to remove residual gases. Rectangular pellets for flexural strength of dimension 40 × 10 × 10 were fired in argon gas environment in a sintering furnace at heating rate 3 °C/min followed by heating at 580 °C and 90 min soaking time then cooled to room temperature.

2.4 Characterization Methods

X-Ray Diffraction by Rigaku Smart Laboratory at 40 kV and 15 mA using CuKα ($\lambda =$ 0.154 nm) radiation was used to confirm the phases in sintered composite samples. Microstructures were studied using scanning electron microscopy (SEM) ZEISS EVO 18 and high resolution scanning electron microscopy (HRSEM) operating at 20 kV was used to identify the morphologies of sintered composites. Transmission electron microscopy (TEM TECHNAI G^2 T20) operated at 200 kV was carried out to study morphology and crystal structures of coated ABOw reinforcement. The TEM thin foils having 3 mm in diameter and 0.3 mm in thickness were prepared by a twin–jet electro polisher operated at – 30 °C with 20 V using of 30% HNO_3 and 70% methanol. Density of the composites was calculated using Archimedes' immersion method. The bending strength is measured by universal testing machine using Tinius Olsen (H10KL) with a crosshead speed of 0.15 mm/min. Flexural strength was calculated using the following expression:

$$\text{Flexural Strength} = \frac{3FL}{2bt^2}$$

where F is the load at failure, L is the distance between the supports, b is the sample width and t is the thickness of the sample.

3 Result and Discussion

3.1 Measurement of Density and Porosity

The density of three samples of each composite with different volume fractions was calculated using immersion method based on Archimedes' principle. Initially, the dry weight of sample has been measured then all samples have been suspended in water by using thread, and soaked weight was measured. Below mathematical equations are used for calculating the density and apparent porosity [18]. Finally, the average porosities and density were calculated for each composite.

$$\text{Apparent porosity (A.P \%)} = X100$$

$$\text{Bulk Density (B.D gm/cm}^3) = D/(W - S)$$

where W = suspended weight in gm. S = soaked weight in gm. D = dry weight in gm.

Figure 2 shows the density porosity variation with varying percentage of coated ABOw. It can be observed that the density of the specimens decreases with increasing the volume fraction of coated whiskers in the aluminum matrix. However, in 5 wt.% coated ABOw, the percentage porosity increased due increase in whiskers amount up to 5%. Also, the increased percentage of whiskers increases the average porosities in the sintered composites.

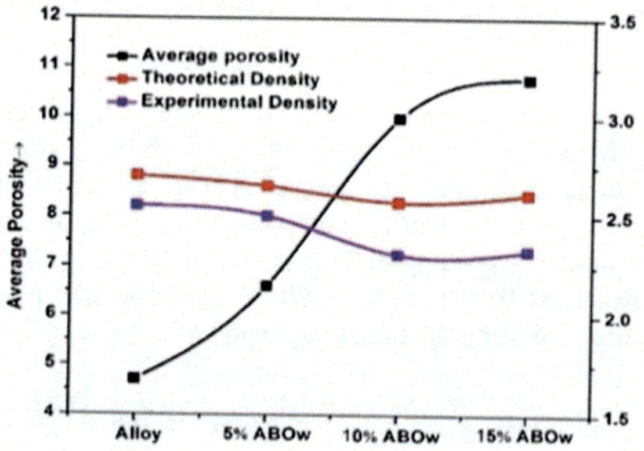

Fig. 2 Physical properties of base alloy and composite samples

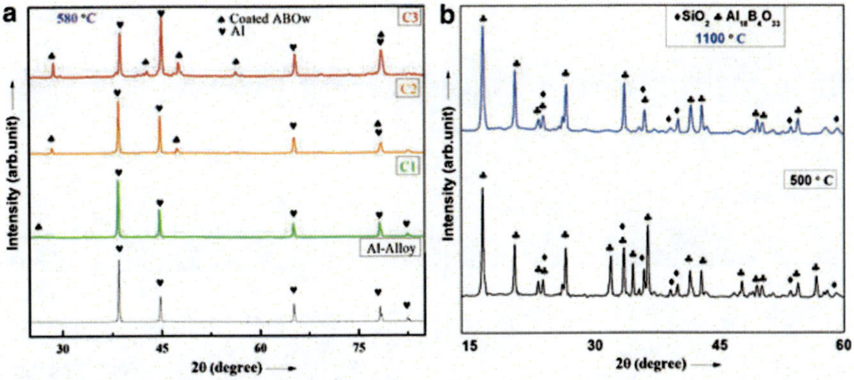

Fig. 3 XRD patterns of **a** SiO_2 coated ABOw sintered at 580 °C for 90 min, **b** ABOw calcined at 500 °C for 4 h and calcined at 1100 °C for 1 h

3.2 *Microstructural Examination and Phase Determination*

3.2.1 XRD Analysis

XRD patterns of aluminum matrix composites samples sintered in the range of 580 °C for 90 min are shown in Fig. 3a. It shows that except aluminum and coated ABOw, other phases were not detected. However, it can be observed that in sample C2 and C3, the coated ABOw volume percentages increases, the peaks height of aluminum matrix decreases.

Moreover, from XRD results as shown in Fig. 3b shows two phases of SiO_2 at low temperature (500 °C), low tridymite with ABOw (JCPDS, 11-0252) and at high temperature (1100 °C) coesite with ABOw (JCPDS, 83-1833) are present. It can be seen no reaction between SiO_2 and ABOw and it takes place upon calcination at 1100 °C for 1 h.

3.2.2 Electron Microscopy

The SEM images are shown in Fig. 4 for all sintered composites and reinforcement, respectively. The SEM image shows and it was found that, after sintering, the uniform distribution of reinforcement in matrix materials offered excellent flexural properties. However, some regions as shown in Fig. 4d with high whisker percentage are more porous due to agglomeration of whiskers.

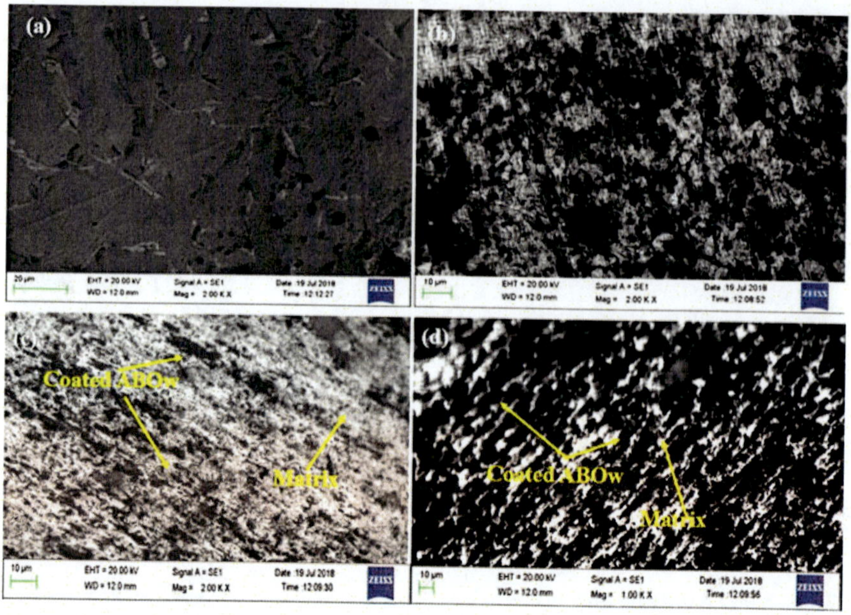

Fig. 4 SEM micrographs of **a** aluminum alloy, composite with SiO_2 coated, **b** 5 wt.%, **c** 10 wt.%, **d** 15 wt.% ABOw whiskers sintered at 580 °C

3.2.3 Characterization of Reinforcement and ABOw/Al composite

The TEM bright field images with corresponding selected area diffraction patterns (SADP) of ABOw powder are shown in Fig. 5. The crystal parameters of micro-whisker were determined as $a = 7.6874$ Å, $b = 15.0127$ Å and $c = 5.6643$ Å [12]. The parameters confirm the orthorhombic structure of ABOw reinforcements as

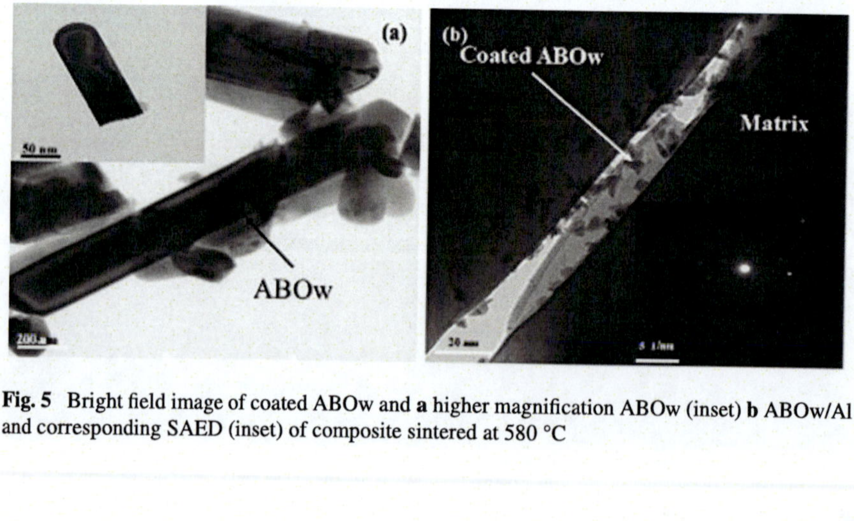

Fig. 5 Bright field image of coated ABOw and **a** higher magnification ABOw (inset) **b** ABOw/Al and corresponding SAED (inset) of composite sintered at 580 °C

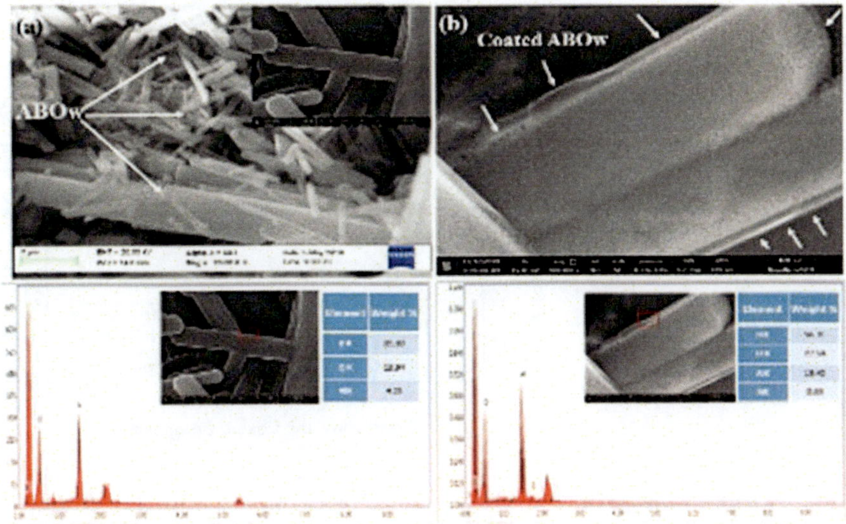

Fig. 6 HRSEM image of reinforcement, **a** calcined at 1300 °C without coated ABOw corresponding single whisker (inset) and **b** coated ABOw calcined at 1100 °C

shown in Fig. 5a. Figure 5b shows typical ABOw/Al composites reinforced with coated SiO_2 particles. It was observed that the surface of the whisker layer shows very smooth and also observed that zig-zag appearances due to clean and fine SiO_2 surfaces. In contrast, it was also observed that there is a clean interface between ABO whisker and pure aluminum and no interfacial reaction occurs with matrix.

The energy dispersive spectrometry specific region shows that the whisker surface is shown in Fig. 6a without coated alumina borate and Fig. 6b It clearly shows the presence of SiO_2 thin layer, confirming the coating on ABOw. It was confirmed that the tight interface bonding also occurred between the whiskers and SiO_2 particles.

3.3 *Mechanical Characterization*

3.3.1 Vickers Hardness Measurement

The measurements of microhardness of the each composite sample have been done using Vickers hardness tester of 1Kg load. The hardness and flexural strength are shown in Fig. 7. The hardness of the samples increases with increasing the percentage of coated ABOw in the matrix. The hardness of aluminum sample and composites C1, C2 and C3 were calculated as 27, 38.9, 45.2 and 40 Hv, respectively. This may be due to inhibition of the progress of plastic deformation by whiskers present in the matrix.

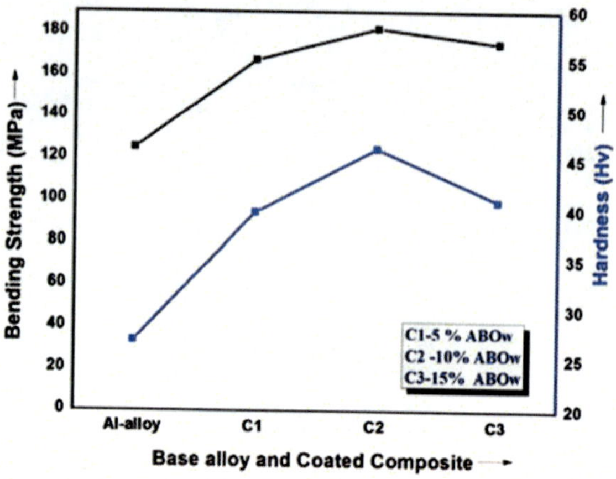

Fig. 7 Measurement of flexural strength and hardness properties of alloys and coated composites

3.3.2 Flexural Strength

The flexural strength of aluminum sample and composites can be determined on the basis of matrix type, whisker aspect ratio and volume fraction. The flexural strength of coated composite C1, C2 and C3, respectively as 125, 167, 178 and 173 MPa were observed. Resultant, it was found that in (C2) 10 wt.% coated ABOw-Al composite has more flexural strength, i.e., 178 MPa.

3.4 Fracture Surface Analysis of Flexural test-

In Fig. 8a, the fractured surface of aluminum sample after three-point flexural test at room temperature shows a ductile nature along the micro-holes and dimples on the surface and whisker-free weak areas in the composite. Thus, the resulting in a lower strength. The fractographs of coated ABOw/Al composites are shown in Fig. 8b–d reflect brittle nature of fracture specifically at whisker and matrix interface. However, Fig. 8d shows more agglomerates of aluminum on the fracture. The incorporation of SiO_2 coatings in the whisker-reinforced composite has a tendency to hinder the bond strength effect due to weakening of Al matrix alloys.

4 Conclusion

In this study, Al-SiO_2 coated (ABOw) metal matrix composite has been developed with the variations of SiO_2 coated ABOw reinforcements. On the basis of result and discussions, the following conclusions have been drawn:

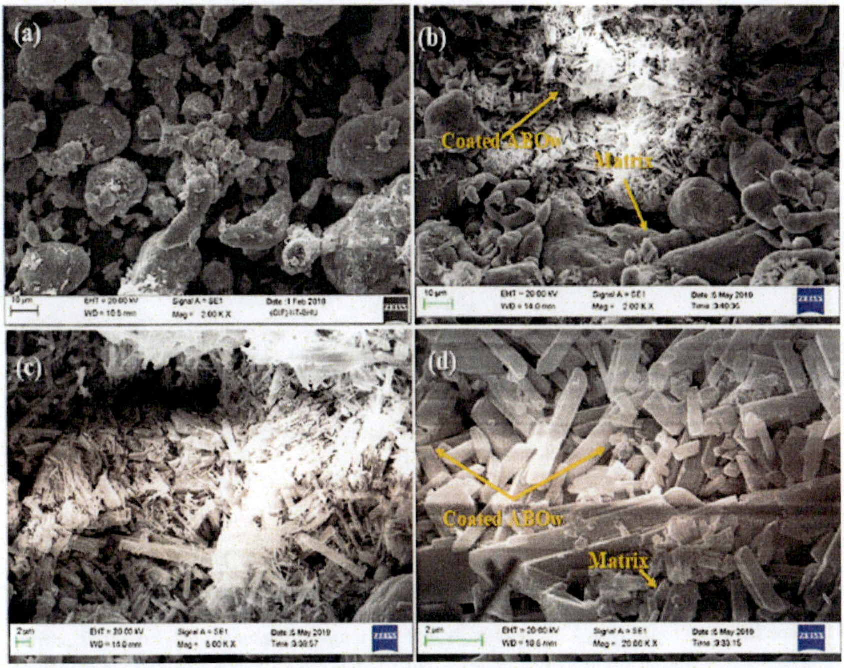

Fig. 8 Fractography images of **a** aluminum, composite SiO_2 coated ABOw with **b** 5 wt.% **c** 10 wt.%, **d** 15 wt.% sintered at 580°

The SEM, HRSEM were used for microstructures observations, i.e., phases as well as grain structure, and TEM analysis was done to characterized the morphological and discontinuous structure of the micro-whiskers with matrix. SiO_2 is the strongest because of high density and low porosity at 1100 °C coesite phase is formed which is very dense. The density of the composite decreases with increasing percentage of whisker in the matrix due to increase in porosity with increasing content of coated whisker. The microhardness of composite improved in 10 wt.% coated ABOw/Al with reduced porosity and more better bonding between matrix and reinforcement. The flexural strength around 178 MPa 10 wt.% coated ABOw reinforcement was observed due to lower porosity than 15 wt.% coated ABOw/Al sintered samples. The following factors pertaining to reinforcement can be attributed:

(a) The coated whiskers are shows the uniform distribution, (b) the debonding particle due to detachment of coated ABOw from the matrix shows whiskers are unbreakable under the application of load.

References

1. Torralba, J. M., Da Costa, C.E., Velasco, F.: P/M aluminum matrix composites: an overview. J. Mater. Pross. Tech. **133**(1–2), 203–206 (2003)
2. Mehta, N.S., Sahu, P.K., Tripathi, P., Pyare, R., Majhi, M.R.: Influence of alumina and silica addition on the physico-mechanical and dielectric behavior of ceramic porcelain insulator at high sintering temperature. Boletín de la Sociedad Española de Cerámica y Vidrio **57**(4), 151–159 (2018)
3. Hernândez, M.F., Suárez, G., Cipollone, M., Conconi, M.S., Aglietti, E.F., Rendtorff, N.M.: Formation, microstructure and properties of aluminum borate ceramics obtained from alumina and boric acid. Ceramics Inter **43**(2), 2188–2195 (2017)
4. Qu, X., Fucheng, W., Chunsheng, S., Naiqin, Z., Enzuo, L., Chunnian, H., Fang, H. In situ synthesis of a gamma-Al_2O_3 whisker reinforced aluminium matrix composite by cold pressing and sintering. Mater. Sci. Eng. A **709**, 223–231 (2018)
5. Yu, Z., Zhao, N., Enzuo, L., Chunsheng, S., Xiwen, D., Jian, W.: Low-temperature synthesis of aluminum borate nanowhiskers on the surface of aluminum powder promoted by ball-milling pretreatment. Powd. Tech. **212**(2), 310–315 (2011)
6. Jin, P.P., Geng, C.H.E.N., Li, H.A.N., Wang, J.-h.: Dry sliding friction and wear behaviors of $Mg_2B_2O_5$ whisker reinforced 6061Al matrix composites. Trans. Nonferr. Met. Soc. Chin. **24**(1), 49–57 (2014)
7. Gudlur, P., Adam, F., Jonathan, L., Miladin, R., Anastasia, M. Thermal and mechanical properties of Al/Al_2O_3 composites at elevated temperatures. Mat. Sci. Eng. A **531** 18–27 (2012)
8. Ram, S.C., Chattopadhyay, K., Chakrabarty, I.: High temperature tensile properties of centrifugally cast in-situ Al-Mg2Si functionally graded composites for automotive cylinder block liners. J. Alloys. Compd. **724**, 84–97 (2017)
9. Gao, X., Yue, H.Y., Lin, X.Y., Yao, L.H., Fei, W.D., Guo, E.J.: Effect of $ZnAl_2O_4$ coating on the compressive behaviors of aluminum borate whiskers-reinforced aluminum composites at elevated temperatures. Comp. Interf. **23**(5), 383–394 (2016)
10. Pandey, N., Chakrabarty, I., Singh, P., Majhi, M.: Effect of sintering temperature on microstructure and flexural strength of alumina borate whiskers. Mater. Res. Exp. doi.org/10.1088/2053-1591/ab4225 (2019)
11. Kala, H., Mer, K.K.S., Sandeep Kumar: A review on mechanical and tribological behaviors of stir cast aluminum matrix composites. Procedia Mater. Sci. **6**, 1951–1960 (2014)
12. Wang, J., Sha, J., Yang, Q., Wang, Y., Yang, D.: Synthesis of aluminium borate nanowires by sol–gel method. Mater. Res. Bullet. **40**(9), 1551–1557 (2005)
13. Gao, X., Yue, H.Y., Guo, E.J., Lin, X.Y., Yao, L.H., Wang, B.: Design and tensile properties of aluminum borate whiskers reinforced aluminum composite with low whisker volume fraction. Comp. Interf. **24**(4), 371–379 (2017)
14. Roseline, S., Paramasivam, V., Jayabal, S.: Evaluation of structure and mechanical properties of zirconia reinforced aluminium matrix composites. J. Comput. Theor. Nanosci. **15**(3), 830–838 (2018)
15. Hardy, D., Green, D.J.: Mechanical properties of a partially sintered alumina. Alumina J. Eur. Ceram. Soc. IS **15**, 769–715 (1995)
16. Pickens J.R.: Aluminium powder metallurgy technology for high-strength applications. J. Mater. Sci. **16**(6), 1437–1457 (1981)
17. Readey, M.J.: Formation and sintering characteristics of aluminum borate whiskers. J. Am. Ceram. Soc. **75**(12), 3452–3456 (1992)
18. Pandey, N., Ram, S.C., Chakrabarty, I., Majhi, M.R.: Physical and thermo-mechanical properties of coated and without coated alumina-boron based refrectory castable for petrochemical industry application. Mater. Today Proc. **28**(2), 698–703 (2020)

Advancement and Challenges in Latent Heat Thermal Energy Storage System

Jayesh Kumar, Pushpendra Singh, and Rajesh Kumar

1 Introduction

Heat is stored in the material by rising of the temperature of material recognized as sensible heat storage and transformation of phase of material at a constant temperature is considered as latent heat storage. The material which changes the phase during absorbing or releasing of heat is known as phase change material (PCM). Solid to liquid is more promising as high latent heat of phase transformation comparatively from solid to solid phase transformation and less volume change comparatively from liquid to vapor phase change. Mass and latent heat of enthalpy of PCM decide amount of heat stored in thermal energy system. Latent heat thermal energy density is nearly ten times higher of sensible thermal energy density [1]. PCM works as thermal energy storage system (TES) should manifest specific thermophysical, kinetic and chemical properties [2].

Thermal Properties

1. A liquefy temperature in the operating span of application and an elevated latent heat of enthalpy per unit volume (it is diminished material required for given amount of energy).
2. There should be an elevated specific heat to yield significant supplementary sensible heat capacity with elevated thermal conductivity, thermal reliability and stability.

Physical Properties

J. Kumar (✉)
Department of Applied Sciences, Maharaja Surajmal Institute of Technology, Delhi, India
e-mail: jayesh2117@gmail.com

J. Kumar · P. Singh · R. Kumar
Department of Mechanical Engineering, Delhi Technological University, Delhi, India

M. Muzammil et al. (eds.), *Recent Advances in Mechanical Engineering*, Lecture Notes in Mechanical Engineering, https://doi.org/10.1007/978-981-15-8704-7_19

1. Minimum volume change on phase transformation—It minimizes the complexity of design of heat exchanger.
2. A high density, congruent melting of the PCM and phase separation—A high density helps to minimize the size of container to store given amount of energy. The composition of the liquid that is formed after melting same as that of the solid known as congruent melting. Bottom sediment salt promotes the incongruent melting of salt. Stirring, encapsulating the PCM, addition of the thickening agents, rolling cylinder and addition of extra water in supersaturated salt solution are same methods suggested by the researchers [2]. Salt hydrate PCMs must be used in air-tight containers to prevent exposure to the atmosphere as TES [3]. Otherwise, there would be loss of water via evaporation which would critically affect the performance of the salt hydrate PCM.

Kinetic Properties

Supercooling—Solidification of PCM below the melting temperature (range generally 10–20 °C) due to poor nucleating properties and inadequate rate of crystallization is known as supercooling. Heat transfer rate from thermal energy storage system to application is markedly changed due to supercooling as temperature gradient shrink. Addition of nucleating agent, rough metallic heat exchanging surface in contact with the PCM and cold finger (Confined cold area) in the PCM can reduce the effect of supercooling [2].

Chemical Properties

It should be chemical stable. Flammability, toxicity, explosiveness and corrosiveness should not be exhibited by material.

2 Classification of PCM

PCM is classified generally in to three categories by the researchers mainly organic, inorganic and eutectic [2, 4].

2.1 Organic

Organic PCM compound has carbon atoms in its structure. It exhibits poor thermal conductivity in span of 0.1–0.7 W/mK [5]. Organic compounds manifest the volumetric phase transformation storage capacity in the span 125–200 kJ/dm^3 [3]. It takes lead from the non-organic by favorable physical properties of melting and solidification [3].

Paraffins—Paraffin consists of straight long chain alkanes CH_3–$(CH_2)_n$–CH_3. Chain length decides the phase transformation temperature and latent heat of phase transformation of the paraffin [2]. Alkane consisting melting temperature in the domain of 6–80 °C possess 14–40 carbon atoms [3]. Main challenges are poor thermal conductivity and moderate flammability of material. Generally, its melting temperature is low. Hence, it is used only for low temperature application. Low cost, chemically stable, little volume changes on melting, freeze without supercooling and low vapor pressure in the melting form, congruent melting, good nucleating properties and moderate latent heat of enthalpy are the favorable properties of paraffin [3].

Non-paraffins—Non-paraffins show highly varied thermophysical properties as consisting of numerous materials. It consists of number of esters, fatty acids, alcohol's and glycol's considerable for TES. Flammability is the major issue with poor thermal conductivity of non-paraffins.

Fatty acid—It shows corrosion with thermal energy storage system container [4]. It has higher latent heat of transformation than that of paraffin and low supercooling [2]. It is costlier about 2–2.5 times than the price of technical grade paraffin [2].

Sugar alcohols—Erythritol and xylitol exhibit considerable volumetric storage densities up to 450 MJ/m^3 [6]. But erythritol and xylitol are not suitable for mobile thermal energy storage applications. Erythritol has a wide range of melting temperature. High supercooling and high price of erythritol and xylitol are main drawback for thermal energy system [6]. D-Mannitol is combined with O_2 from the environment in open container and change into metastable material with degradation. Hence, the thermal energy system should be air-tight for sugar alcohols.

2.2 *Inorganic*

It is the material which does not accommodate carbon and is also not extracted from living material.

Salt hydrate—Salt hydrate of general formula $AB.nH_2O$ [2] is formed by addition of water in the anhydrous salt at room temperature. Bond strength of water molecules and salt decide the latent heat of phase transformation [5]. Major advantages are moderately higher thermal conductivity (0.4 and 0.7 W/mK), inexpensiveness and safe operation than that of paraffins. Decomposition, poor nucleating properties and corrosion to the container of salt hydrate can limit life of thermal energy storage system [2]. Incongruent melting and segregation may happen after several repeating cycles. Due to the phase separation, salt amount in the liquid medium will be diminished with each successive cycle. Salt hydrates are not compatible with aluminum, stainless steel and copper [5] but compatible with plastic and it is slightly toxic [2].

Metal and metals alloys—High heat convey capacity, high latent heat/volume and minor volume expansion at phase transformation are major advantage on the paraffin and salt hydrate. Weight, higher melting temperature as well as cost are major issues for selection of metal as thermal energy storage system [2].

2.3 *Eutectic*

It is explained as a combination of two or more elements which usually do not form a new chemical compound by interaction like $NH_2CONH_2 + NH_4Br$, $NH_2CONH_2 + NH_4NO_3$, etc. [2]. It is usually done to reduce the melting temperature of material to achieve the operating range of temperature of application or enhance the desirable properties of PCM. Magnesium chloride hexahydrate ($MgCl_2{\cdot}6H_2O$) has melting temperature 117 °C with enthalpy 150 kg/kJ which is not suitable for low temperature water heating system. However, $NH_4Al(SO_4)_2{\cdot}12H_2O$ has high supercooling with good latent heat of enthalpy (260 kJ/kg) with melting temperature 93 °C [7]. The addition of 30% mass fraction of $MgCl_2{\cdot}6H_2O$ in the eutectic mixture $MgCl_2{\cdot}6H_2O$—$NH_4Al(SO_4)_2{\cdot}12H_2O$ reduces the melting temperature up to 64.15 °C with a very low supercooling degree (0.74 °C).

3 Thermal Conductivity Enhancement of PCM

Organic and salt hydrates PCM have low thermal conductivity. Heat transfer rate from/to PCM depends upon heat transfer area, thermal gradient and thermal conductivity of PCM. Enhancement of thermal gradient increases the irreversible thermal loss of system. Hence, the heat transfer area and thermal conductivity of PCM are on the limelight of the researchers to attain the required heat transfer rate demanded by the application as demonstrated in Fig. 1.

3.1 *Dispersion of Nanoparticles in the PCM*

Researchers have experimented and simulated the utility of nanoparticles for thermal conductivity magnification in the PCM like silica, alumina, titania, exfoliated graphite nanoplatelets, multilayer graphene nanoplatelets, multiwalled carbon nanotubes, Ag nanowires and carbon black nanopowder, manganese dioxide nanowires and nanotubes, etc. [8, 9]. Exfoliated graphite nanoplatelets in the bio-based fatty acid improves the thermal conductivity from 0.154 (pure PCM) to 0.557 W/mK [8]. Graphite nanofiber (herringbone type) is added in the paraffin (n-tricosane) 10% by weight by Omar Sanusi et al. [9]. It is found that graphite nanofibers are effective additives for diffusion enhancement in all aspect ratios of thermal containment unit of PCM.

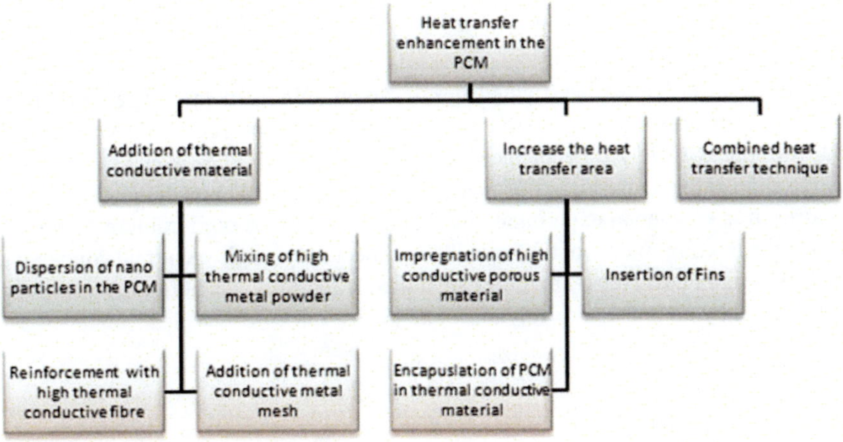

Fig. 1 Heat transfer enhancement method [4]

3.2 Heat Transfer Enhancement by Encapsulated PCM

Encapsulation sizes can be categorized according to the size of shell and named as macro, micro and nano. Appasaheb Raul et al. [10] examined the thermal properties of Hytherm 600 PCM encapsulated by stainless steel. It is found that smaller diameter has faster melting and solidification than that of larger diameter. Melting is not uniform throughout of PCM capsule during phase transformation. It is stated that the melting of the PCM in top portion is higher than the bottom portion in PCM capsules, which is achieved by predominantly convection [10]. Although, the conduction is the main approach of heat convey throughout the solidification. Sodium nitrate is encapsulated in the capsule and heat transfer characteristics examined by Selvan Bellan et al. [11]. It is concluded that when the thickness of capsules (polymer) is surged about 66%, the total melting time of PCM in the capsules is extended around 15% as thermal conductivity of polymer is poor.

3.3 Dispersion of Thermal Conductive Fibers

High thermal conductive fibers like ceramics, metals and carbon materials are dispersed in PCM to improve the thermal conductivity [12, 13]. Jun Fukai et al. [12] examined the random and brush types carbon fibers dispersion in the paraffin. They observed that the thermal conductivity improves with the volume fraction of carbon fibers. They further stated that brush type carbon fibers manifest the highest utility on the thermal conductivity enhancement of paraffin. This is happing because the fibers from the brush type are oriented in the heat flow direction in the paraffin.

3.4 Mixing of Metal and Nonmetal Powder in the PCM

Mettawee et al. [14] have investigated the effect of aluminum metal powder embedded in the wax (paraffin). It is stated that the aluminum metal powder in the wax (PCM) reduces approximately 60% of thermal charging time of TES. Wei Li et al. [15] also examined the consequence of higher mass fraction of Al powder (1–2%) in the paraffin. It is concluded experimentally that mass fractions of aluminum 1 and 2% reduce the melting time about 22 and 39%, respectively. Highly thermal conductive filler like graphite and nickel particles are mixed to magnify the thermal conductivity of erythritol by Okinaka and Akiyama et al. [16].

3.5 Metallic Fins

Al-Abidi et al. [17] have probed the consequences of number of fins, fin length and fin thickness in a triplex tube heat exchanger to explore thermal properties of PCM during melting process. The complete melting time is diminished to 34.7% by fins. It is stated that the melting time is diminished when the fin length and fins thickness are increased. Recently, Aly et al. [18] have investigated the corrugated fins consequences on the solidification rate of latent heat thermal energy storage by numerical modeling. Corrugated fin reduces the solidification times by 30–35% as compared to flat fins. Aluminum (Al) cross fins and stainless fins are also used in the salt hydrate (sodium acetate trihydrate) to enhance the thermal conductivity [19]. Thicker Al fins (2–4 mm) has not significant effect on the thermal charging time of PCM. Al cross fins is more effective in term of heat transfer during melting than that of stainless steel.

3.6 Impregnation of Porous Material in PCM

Inclusion of copper metal foam in the PCM increases the heat transfer rate 3 to 10 times than that of pure PCM [20]. Heat transfer rate is considerable effect by the metal foam cell size as the heat transfer area increases. PCM composite is fabricated by impregnating porous graphite matrices with the paraffin by Mills et al. [21]. It is stated that the thermal conductivity depends upon the compaction direction of composite. Thermal conductivity boosts 30–130 times higher in the perpendicular samples (heat flow perpendicular to compaction direction of composite) than that of pure paraffins [21]. Researchers have also successfully used metal foam for salt hydrate for thermal conductivity enhancement.

4 Conclusion

Maximum research of thermal conductivity enhancement is found on the organic material. There are very limited research papers available on the salt hydrate for thermal conductivity magnification. Nanoparticle is in the limelight of researchers for thermal conductivity magnification of PCM. Salt hydrates are more promising than that of paraffin due to more latent heat enthalpy density and low cost, but challenges for commercialization of salt hydrates as thermal energy storage system are more critical and severe than that of paraffin. So, the future research work is required to enhance the heat transfer rate in the salt hydrate by new techniques like heat pipe.

References

1. Nazir, H., Batool, M., et al.: Recent developments in phase change materials for energy storage applications. Int. J. Heat Mass Transf. **129**, 491–523 (2019)
2. Sharma, A., Tyagi, V.: Review on thermal energy storage with phase change materials and applications. Renew. Sustain. Energy Rev. **13**, 318–345 (2009)
3. Abhat, A.: Low temperature latent heat thermal energy storage: heat storage materials. Sol. Energy **30**, 313–332 (1983)
4. Ibrahim, N.I., Al-Sulaiman, F.A. et al.: Heat transfer enhancement of phase change materials for thermal energy storage applications. Renew. Sustain. Energy Rev. **74**, 26–50 (2017)
5. Pereira da Cunha, J., Eames, P.: Thermal energy storage for low and medium temperature applications using phase change materials—a review. Appl. Energy **177**, 227–238(2016).
6. Höhlein, S., König-Haagen, A. et al.: Thermophysical characterization of $MgCl_2 \cdot 6H_2O$, xylitol and erythritol as phase change materials (PCM) for (LHTES). Materials **10** (2017)
7. Sun, W., Zhou, Y. et al.: Compounding $MgCl_2 \cdot 6H_2O$ with $NH_4Al(SO_4)2 \cdot 12H_2O$ or $KAl(SO_4)2$ $\cdot 12H_2O$ to obtain binary hydrated salts as high-performance. Molecules MDPI (2019)
8. Jeong, S.G., Chung, O., et al.: Improvement of the thermal properties of Bio-based PCM using exfoliated graphite nanoplatelets. Sol. Energy Mater. Sol. Cells **117**, 87–92 (2013)
9. Sanusi, O., Warzoha, R., et al.: Energy storage and solidification of paraffin PCM embedded with graphite nanofibers. Int. J. Heat Mass Transf. **54**, 4429–4436 (2011)
10. Raul, A., Jain, M., et al.: Modelling and experimental study of latent heat thermal energy storage with encapsulated PCMs for solar thermal applications. Appl. Therm. Eng. **143**, 415–428 (2018)
11. Bellan, S., Alam, T.E., et al.: Numerical and experimental studies on heat transfer characteristics of thermal energy storage system packed with molten salt PCM capsules. Appl. Therm. Eng. **90**, 970–979 (2015)
12. Fukai, J., Kanou, M., et al.: Thermal conductivity enhancement of energy storage media using carbon fibers. Energy Convers. Manage. **41**, 1543–1556 (2000)
13. Nomura, T., Tabuchi, K., et al.: High thermal conductivity phase change composite with percolating carbon fiber network. Appl. Energy **154**, 678–685 (2015)
14. Mettawee, E.B.S., Assassa, G.M.R.: Thermal conductivity enhancement in a latent heat storage system. Solar Energy, **81**, 839–845 (2007)
15. Li, W., Wang, Y.H., et al.: Experimental study on melting/solidification and thermal conductivity enhancement of phase change material inside a sphere. Int. Commun. Heat Mass Transfer **68**, 276–282 (2015)
16. Okinaka, N., Akiyama, T., et al.: Thermal conductivity enhancement of erythritol as PCM by using graphite and nickel particles. Appl. Therm. Eng. **61**, 825–828 (2012)

17. Al-Abidi, A.A., Mat, S., et al.: Internal and external fin heat transfer enhancement technique for LHTES in triplex tube heat exchangers. Appl. Therm. Eng. **53**, 147–156 (2013)
18. Aly, K.A., El-Lathy, A.R., et al.: Enhancement of solidification rate of latent heat thermal energy storage using corrugated fins. J. Energy Storage **24**, 100785 (2019)
19. Wang, Y., Yu, K., et al.: Experimental and modeling study on thermal performance of hydrated salt LHTES. Energy Convers. Manage. **198**, 111796 (2019)
20. Zhao, C.Y., Lu, W., et al.: Heat transfer enhancement for thermal energy storage using metal foams embedded within PCMs. Sol. Energy **84**, 1402–1412 (2010)
21. Mills, A., Farid, M., et al.: Thermal conductivity enhancement of phase change materials using a graphite matrix. Appl. Therm. Eng. **26**, 1652–1661 (2006)

Modal Analysis of Twin S Shape Coriolis Mass Flow Sensor

Resham Taluja, Rushali Pant, and Pravin P. Patil

1 Introduction

Coriolis mass flow sensor is a device for flow measurement which depends on the fluid and solid interaction in order to sense the changes in the solid structure generated due to acceleration produced because of coriolis force acting on the geometry. Coriolis flow meters consist of two parts, one the flow sensor also known as flow transducer and second is the flow converter also known as flow transmitter. The flow sensors comprises of vibrating tube, sensors, and drive system to provide external vibration and structure to support the system. The flow converter gives input to the flow sensor, process the flow sensor signal, and then gives the output.

Pradeep Gupta [1] performed experiments on various shapes (U, S, and L) of mass flow sensor. The results indicate that the evaluated phase shift is directly proportional to the mass flow rate. The slope ratio of U tube is witnessed to be constant for different parameters and maximum sensitivity is deduced for S and L tubes by varying the sensor position. Sharma [2] carried out the modeling of omega shape CMFS made up of copper to check it using ANN. Tao Wang has performed the literature survey of development of CMFM in past years. Sharma [3] performed the analysis on U type copper tube and checked performance of CMFS. Patil [4] has used the response surface method (ESM) for optimization of copper omega type CMFS. Advanced coriolis mass flow sensors are independent of viscosity and density of flowing fluids.

R. Taluja (✉) · R. Pant · P. P. Patil
Graphic Era (Deemed to be University), Dehradun, India
e-mail: reshamtaluja11493@gmail.com

M. Muzammil et al. (eds.), *Recent Advances in Mechanical Engineering*, Lecture Notes in Mechanical Engineering, https://doi.org/10.1007/978-981-15-8704-7_20

2 Cad Model and Material Properties of Twin Tubes

FEA simulation of S shape tube is performed in ANSYS. The inner and outer diameter of the tube is kept constant of all the profiles of S tube. The length of both the profiles is 400 mm and the semi distance is 150 mm. Figure 1 shows the profile of the tube and Fig. 2 shows the mesh model generated in order to carry out the modal analysis of the tubes. The analysis is performed for two materials of the tube: copper and stainless steel. The properties of the materials of the tube are shown in Tables 1 and

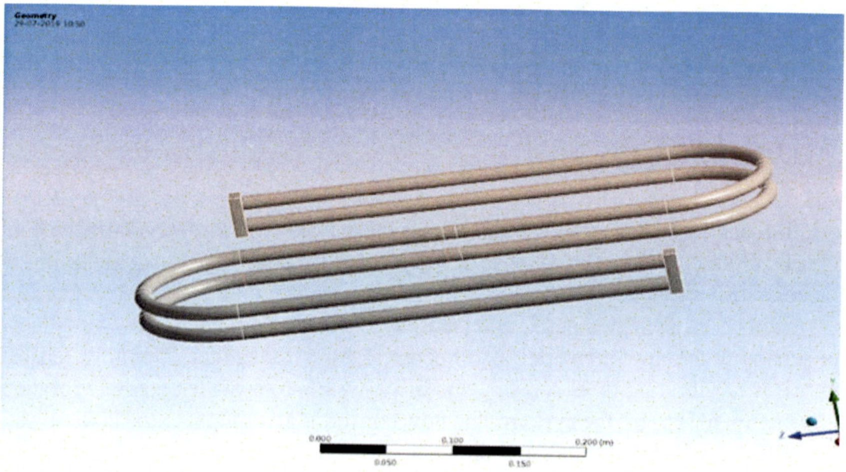

Fig. 1 Profile of twin tube

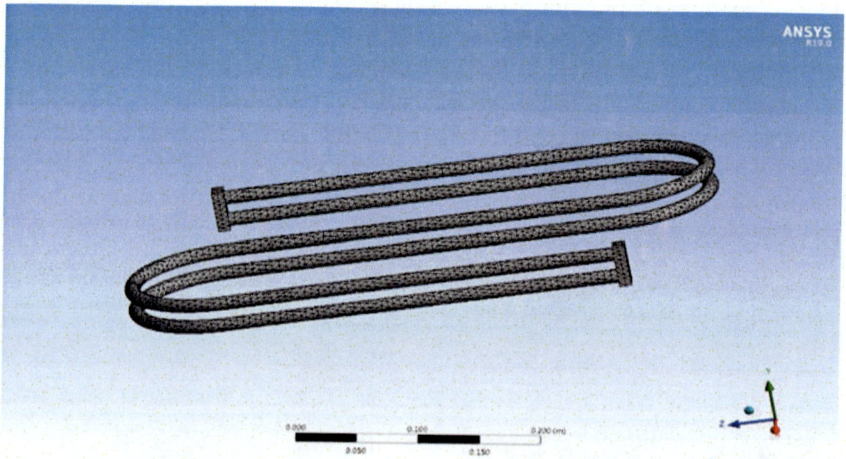

Fig. 2 Mesh model of the twin tube

Table 1 Material properties of tube materials

Tube material	Density (kg/m^3)	Young's modulus (GPa)	Poisson's ratio	Coefficient of thermal expansion
Copper	8940	128	0.360	17.10e−6
Stainless steel	8000	193	0.265	15e−6

Table 2 Material properties of water

Fluid	Density (kg/m^3)	Young's modulus (GPa)	Poisson's ratio	Coefficient of thermal expansion
Water	1000	2.2	0.2	2.14e−6

2 give the property of water which is used as a working fluid.

3 Results and Observations

In this study, two twin S shape coriolis mass flow sensors were designed and simulation was done using FEA method to calculate the frequency for all three profiles. First, the natural frequency achieved in modal analysis is known as fundamental frequency. It is the resonance frequency for twin S shape tube filled with the fluid (water).

Frequency of copper twin tube for mode 1 is 39.468 Hz and for mode 6 is 51.724 Hz. The frequency observed for copper twin s shape tubes is given in Table 3. Mode shape generated for six modes are shown in Fig. 3.

Frequency observed for stainless steel twin tube is 50.2 Hz for mode 1 and 65.348 Hz for mode 6. Table 4 shows the frequency variation of stainless twin S shape tube. Figure 4 shows the mode shape generated for all six modes for stainless steel. Figure 5 shows the frequency variation graph for six modes for two twin tubes.

Table 3 Frequency variation for copper twin S shape tube

Modes	Frequency
1	39.464
2	39.846
3	42.174
4	42.875
5	51.391
6	51.724

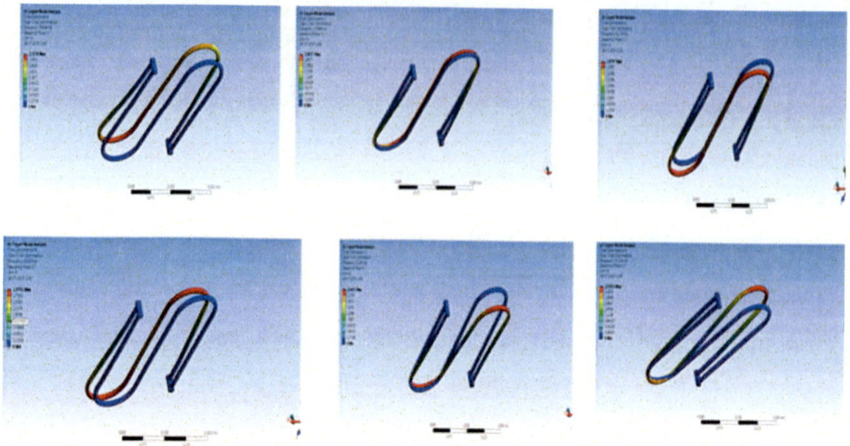

Fig. 3 Six mode shapes for copper twin s shape tube

Table 4 Frequency variation for stainless steel twin S shape tube

Modes	Frequency
1	50.2
2	50.658
3	53.299
4	54.174
5	64.925
6	65.348

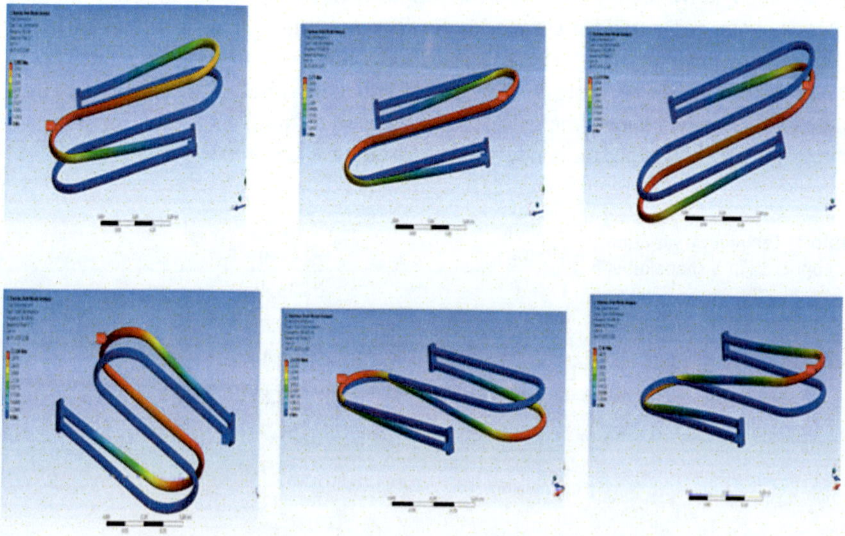

Fig. 4. Six mode shapes for stainless steel twin S shape tube

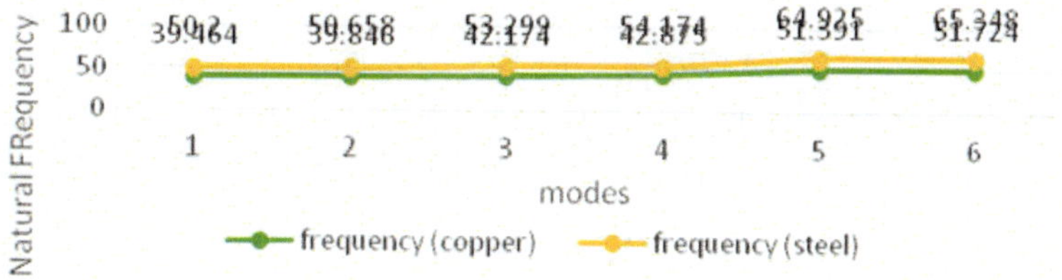

Fig. 5 Frequency variation graph for six modes for two twin tubes

4 Conclusion

Modeling of twin S configuration tubes is done in Catia v5 and finite element analysis for three profiles is performed in ANSYS 16.2 to assay the fundamental frequency and mode shape results for each tube. The frequency is measured in Hz. For each tube, resonant frequency is 39.464 and 50.2 Hz. By calculating the resonant frequency of the tubes, the initial step for this research has been completed. In the future, the results achieved in this analysis can be used for simulation as well as for experimental process to be carried out for coriolis mass flow sensor under varying sensor positions and mass flow rate.

References

1. Gupta, P., Srinivasan, K., Prabhu, S.V.: Tests on various configurations of Coriolis mass flowmeters. Measurement **39**(4), 296–307 (2006)
2. Patil, P., Sharma, S.C., Paliwal, V., Kumar, A.: ANN modelling of Cu type omega vibration based mass flow sensor. Procedia Technol **14**, 260–265 (2014)
3. Patil, P.P., Sharma, S.C., Mishra, R.: Performance optimization of aluminium (U) type vibration based electromechanical Coriolis mass flow sensor using response surface methodology. Mechanics **17**(3), 264–270 (2011)
4. Patil, P.P., Sharma, S.C., Jain, S.C.: Response surface modeling of vibrating omega tube (Copper) electromechanical Coriolis mass flow sensor. Expert Syst. Appl. **39**(4), 4418–4426 (2012)

Study of the Effect of Leak Location in Water Pipeline Using CFD

Nagendra Singh Ranawat and B. P. Nandwana

1 Introduction

Pipeline transportation system is most economic and efficient method to transport water, oil, gas, etc., from one place to another. This method of transportation and distribution system is growing continuously but is prone to various risks. Leakage in pipeline is one of the major problems worldwide [1]. Leak in pipe is caused mainly due to corrosion, pressure surges, engineering failure, earthquakes, movement of grounds or may be caused due to the human activities such as excavation, constructions or theft [2]. Leakage in the system is costly in terms of water loss, adverse water quality effect and energy consumed for supplying water through leak [3]. Therefore, leak must be detected and located as early as possible to reduce the consequences of leak.

Various leak detection systems (LDSs) are available which are used to detect and locate leak. These LDSs are broadly classified into two groups: external LDS and internal LDS [4]. External method uses sensing device to sense the escaping fluid outside the pipe through leak, while internal method uses the field sensor output that monitors the flow parameter continuously and on basis of changes in these flow parameters decides the occurrence of leak. Internal LDS is most commonly used as it is cheaper, can be easily installed in the existing system and can even monitor the pipeline continuously.

Leak can be spotted by observing the pressure at the outlet pipe and comparing it with the healthy state of the outlet pipe [5]. When the leak occurs, the sound is generated at leak location which can be captured using vibration/acoustic sensors

N. S. Ranawat (✉) · B. P. Nandwana
Mechanical Engineering Department, College of Technology and Engineering, Udaipur 313001, Rajasthan, India
e-mail: nagendra.ranawat007@gmail.com

B. P. Nandwana
e-mail: bpnand@gmail.com

M. Muzammil et al. (eds.), *Recent Advances in Mechanical Engineering*, Lecture Notes in Mechanical Engineering, https://doi.org/10.1007/978-981-15-8704-7_21

and leak can be detected on basis of the noise generated due to leak [6–8]. A drop-in pressure takes place at leak location which causes pressure oscillation in fluid which propagates as pressure signal at speed of sound known as negative pressure wave through the fluid and away from leak location in opposite direction. Leak can also be detected by analyzing the negative pressure wave which is generated due to occurrence of leak [9–11]. Leak can be located by using gradient intersection method which uses the phenomenon that when leak occurs pressure drops linearly for each point but at the same time pressure before leak drops significantly and pressure after leak drops slowly. The location of leak can be determined using intersection of these two curves [12].

Numerical modeling and simulation using CFD is a new area to analyze the effect of the flow parameters due to occurrence of leak. A very few literatures are available in this area which is discussed here. Ben-Mansour et al. carried a CFD simulation on a pipe with length 1 m and diameter 0.1 m with a leak of size 1 mm × 1 mm and found that leak causes fluctuation in both pressure and pressure gradient across pipeline at leak location [13]. Zeng and Luo performed simulation to analyze the effect of inlet mass flow rate and pipe length on total pressure drop due to occurrence of leak and found that mass flow rate has significant effect on pressure drop as compared to pipe length [14]. Araujo et al. conducted hydrodynamic study of leakage in pipeline using Ansys CFX to investigate the dynamic behavior of pressure and velocity in pipe containing one or two leaks, and they observed that pressure drop due to leak may vary based on the preexisting leak [15]. Jujuly et al. in their study found that flow rate through leak increases with increase in operating pressure [16]. Sousa and Romero found in their CFD simulation that leak can be identified by monitoring the pressure and velocity field as they are influenced by the perturbations in both upstream and downstream at leak location [17]. Wei and Masouri examined the effect of second leak in pipeline, and they observed that occurrence of second leak does not cause much effect in flow parameters but flow rate at first leak is always more as compared to the second leak [18].

In this study, effect on flow parameters such as pressure drop, leak flow rate and outlet velocity due to variation in location of leak is analyzed using Ansys Fluent 2019 R1.

2 Methodology

CFD simulation is an effective approach used to study the fluid flow characteristics of interest. In this study, pipe of diameter 150 mm and thickness 9.15 mm with a length of 3 m was taken. To model pipe with different leak locations, a circular leak of size 10 mm was selected at different locations such as 0.5, 1, 1.5, 2 and 2.5 m from the inlet. The geometry of pipe was modeled in SOLIDWORKS 2014 and was imported to Ansys Fluent 2019 R1 workbench for simulation under steady-state condition as shown in Fig. 1.

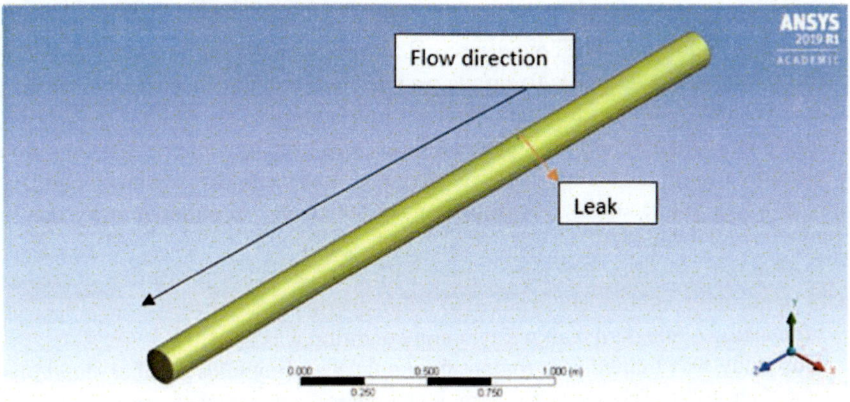

Fig. 1 Pipe model with leak

The computational domain was discretized into a small domain known as elements, and refined mesh was used around the leak to capture the flow variables as shown in Fig. 2. The element for mesh was selected on the basis of the mesh independency test conducted which is shown in Table1. The tetrahedral element with mesh Set 2 in Table 1 was used for this discretization of the domain.

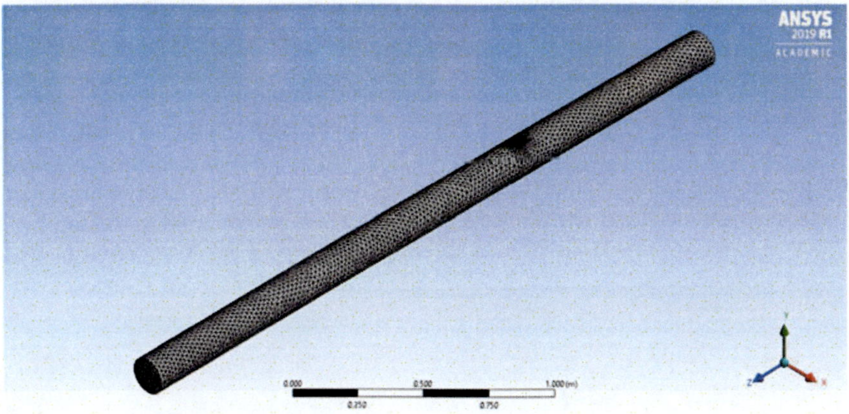

Fig. 2 Pipe model with leak after meshing

Table 1 Mesh independency test

S. No.	Total nodes	Total elements	Mass flow rate at inlet (kg/s)
1	11,223	9,984	17.3509
2	21,098	19,125	17.4388
3	1,66,754	1,60,200	17.5893

The boundary conditions used for the study were selected such that it simulates the real flow conditions. The boundary conditions used in the analysis are as follows:

Inlet boundary conditions: To simulate a real condition in which supply is from a source, velocity—inlet with water velocity 1 m/s is used.

Outlet boundary condition: Pressure—outlet was used to simulate real condition where flow in pipe at section has a 1 bar gauge pressure. The velocity boundary condition was not used as we want to analyze the change in outlet velocity due to change in leak location.

Leak outlet condition: Pressure—outlet with zero bar gauge pressure was used as leaks in normal condition are exposed to atmospheric condition.

Pipe walls: A standard wall with no-slip condition was used.

This study was limited to turbulent flow only; i.e., Reynolds number should be greater than 4000. The value for turbulent kinetic energy and dissipation rate at inlet was defined using intensity and viscosity ratio. The simulation was done using realizable k-ε model which is sufficient model to capture the turbulence features of flow, and pressure-based solver was selected to solve the problem as water is an incompressible fluid.

For pressure velocity coupling, PISO algorithm and second-order discretization scheme were selected for flow variables, turbulent kinetic energy and dissipation rate to discard any numerical errors and confirm accurate numerical solution.

The simulation was considered as converged when the maximum residual was less than at least three orders of magnitude as shown in Fig. 3.

The CFD modeling used in this study is validated with the models available in the literature. A CFD study on simulation of small leaks in water pipelines was performed by Ben-Mansour et al. (2012).

They performed the simulation on 2-m pipe length having outside diameter of 0.1 m and modeled a square-shaped leak with dimension 1 mm × 1 mm. They placed the leak at top middle section of the pipeline. The fluid velocity was taken as 1 m/s, and the centerline pressure was varied from 1 to 6 bar. The pressure outlet at leak hole was taken as atmospheric pressure, where gauge pressure is zero bar.

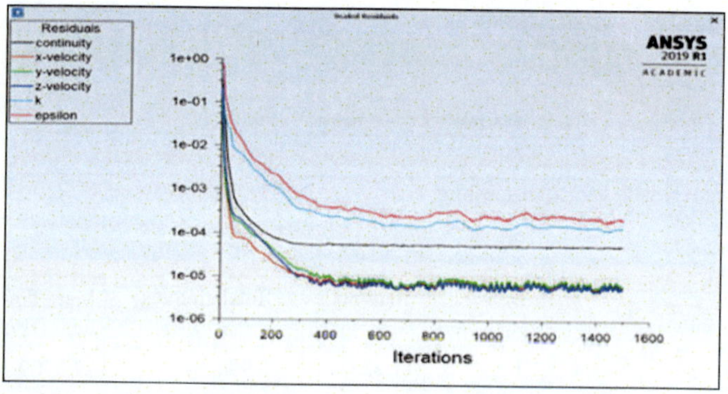

Fig. 3 Iteration for the converged solution

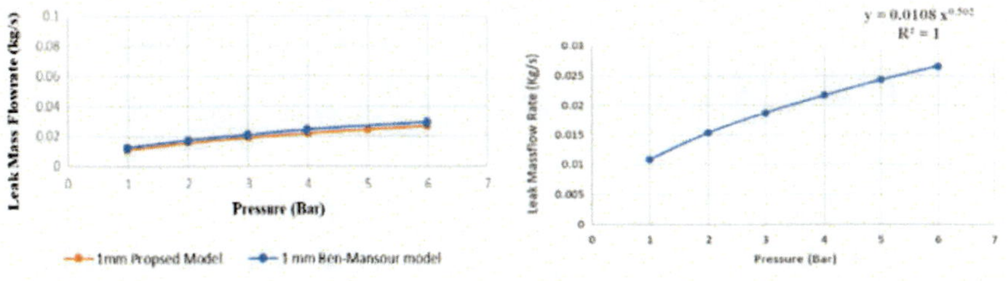

(a) (b)

Fig. 4 Validation of CFD model: **a** validation of proposed model with Ben-Mansour model. **b** CFD estimated effect of line pressure on leak mass flow rate

Figure 4a illustrates the close match with the results obtained by Ben-Mansour et al. (2012), thus validating the modeling adopted. The CFD simulation carried above and summarized in Fig. 4b reveals that $Q_{\text{leak}} = C(\Delta P)^{0.5}$ where Q_{leak} is leak mass flow rate (in kg/s), C is a constant and ΔP is difference between centerline pressure and atmospheric pressure which is in close match with orifice equation [13].

3 Results and Discussion

The results obtained from the CFD simulation of leakage in pipe and effect in the flow parameters due to the location of leak are presented below.

3.1 Variation of Pressure Drop

The pressure contours around the leak are shown in Fig. 5 which shows that there is a high fluctuation of pressure around the leak hole, and this pressure fluctuation first drops at upstream and then increases downstream till it matches the line pressure.

The pressure drops across the pipe with leak depend on the leak location in the pipeline as in Fig. 6 which shows that pressure drop is minimum when the leak is located close to the inlet but it increases as the position of leak goes away from inlet. This happens because when the location of leak moves away from inlet, mass flow rate at main outlet increases and mass flow rate at leak location decreases. This results in high velocity and low pressure near main outlet. This low pressure at main outlet results in increase in the pressure drop when the location of leak moves away from the inlet.

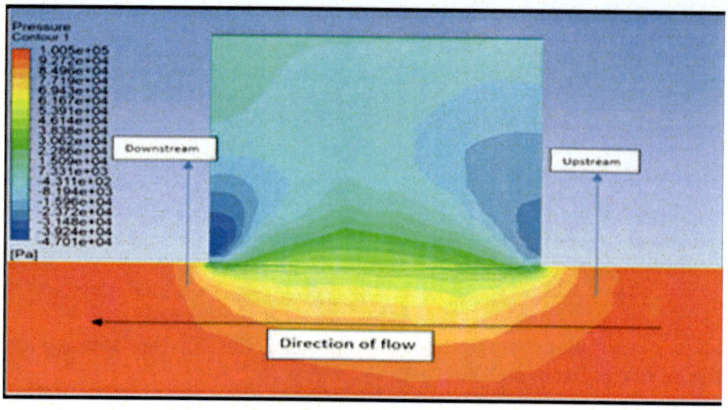

Fig. 5 Pressure contours at leak located at center (1.5 m from inlet)

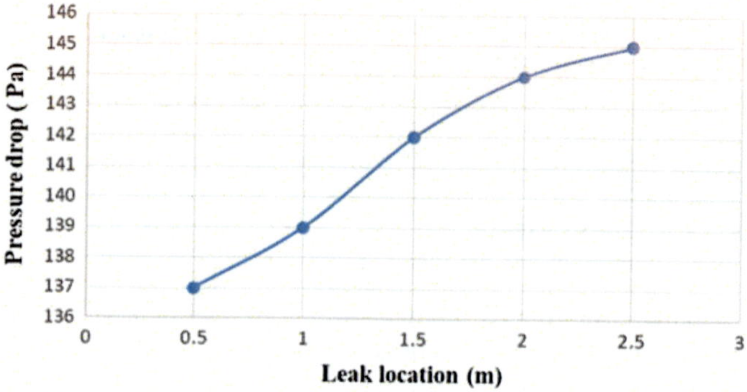

Fig. 6 Pressure drop due to change in location of leak

3.2 *Variation of Leak Mass Flow Rate*

The leak mass flow rate is more if the leak is located close to the inlet but it decreases as the location of leak increases from the inlet because the pressure at a location decreases with increase in its distance from inlet so as the leak location moves away from the leak the pressure at its location decreases. Due to this reduction, the pressure difference between the leak location and outlet pressure decreases slightly resulting in small reduction of mass flow rate at leak with increase in its distance from inlet as shown in Fig. 7.

3.3 *Variation of Outlet Velocity*

The occurrence of leak decreases the velocity at outlet of the pipe as one part of the mass flows through the leak. According to the conservation law,

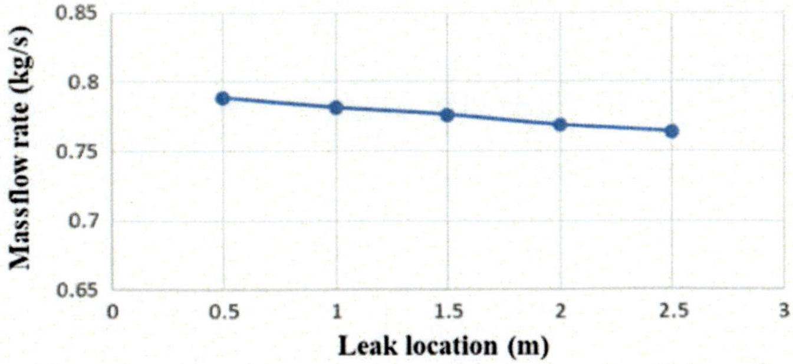

Fig. 7 Leak mass flow rate due to change in leak location

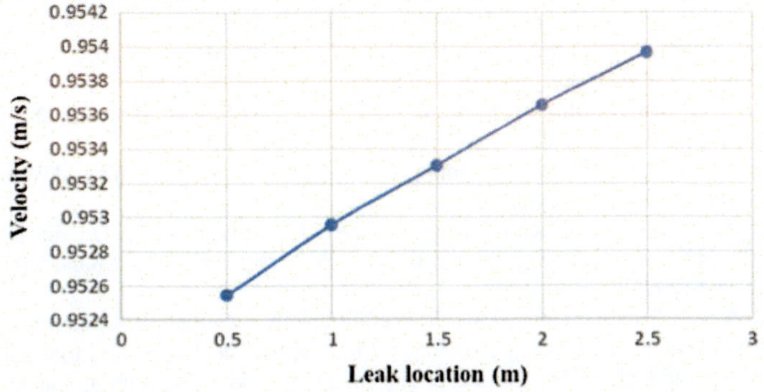

Fig. 8 Effect on outlet velocity due to change in leak location

$$Q_{\text{inlet}} = Q_{\text{leak}} + Q_{\text{outlet}}$$

Therefore, when leak mass flow rate changes then the flow rate at outlet also changes to balance the conservation of mass at inlet which changes the velocity at outlet.

Figure 8 shows the behavior of outlet velocity due to change in location of the leak. When the location of leak is near to inlet, then more mass flows through leak and as the location goes away from the inlet leak mass flow rate decreases resulting in more mass through outlet which increases the velocity at outlet.

4 Conclusions

In the present study, change in behavior of various flow parameters such as pressure drop, leak mass flow rate and outlet velocity due to the presence of leak and variation in them due to change in the location of leak were studied. The conclusions from the analysis are as follows:

- There is a huge pressure fluctuation around the leak taking place which changes the overall pressure drop of the pipe. The pressure drop increases with increase in distance of the leak from the inlet.
- The leak mass flow rate decreases with increase in the location of leak from inlet.
- The velocity at outlet increases with increase in distance of leak from inlet.

References

1. Burn, S., Desilva, D., Eiswirth, M., Hunaidi, O., Speers, A., Thornton, J.: Pipe leakage—future challenges and solutions. Pipes Wagga Wagga, Australia (1999)
2. Saghi, H., Ansari, A.: Effective factors in causing leakage in water supply systems and urban water distribution networks. Am. J. Civil Eng. **3**, 60–63 (2015)
3. Colombo Andrew, F., Karney Bryan, W.: Energy and costs of leaky pipes. J. Water Resour. Plan. Manage. **128**, 441–450 (2002)
4. Zhang, J., Hoffman, A., Murphy, K., Lewis, J.,Twomey, M.: Review of pipeline leak detection technologies, pp. 1–13. Pipeline Simulation Interest Group (2013)
5. Brunone, B.: Transient test-based techniques for leak detection in outfall pipes. J. Water Resour. Plan. Manage. **125**, 302–306 (1999)
6. Gao, Y., Brennan, M.J., Joseph, P.F., Muggleton, J.M., Hunaidi, O.: On the selection of acoustic/vibration sensors for leak detection in plastic water pipes. J. Sound Vibr. Elsevier **283**, 927–941 (2004)
7. Khalifa, A.E., Chatzigeorgiou, D.M., Youcef-Toumi, K., Khulief, Y.A., Ben-Mansour, R.: Quantifying acoustic and pressure sensing for in-pipe leak detection. In: Proceedings of the ASME 2010 International Mechanical Engineering Congress & Exposition, Sound, Vibration and Design, vol. 13, pp. 489–495 (2010)
8. Kiat, S.N., Pei-Yin, C., Yuan-Chi T.: A design of automatic water leak detection device. In: 2nd International Journal on Opto- Electronic Information Processing, IEEE, pp. 70–73 (2017)
9. Zhang, Y., Chen, S., Li, J., Jin, S.: Leak detection monitoring system of long-distance oil pipeline based on the dynamic pressure transmitter. Measurement **49**, 382–389 (2014)
10. Tuck, J., Lee, P.: Inverse transient analysis for classification of wall thickness variations in pipelines. Sensors **13**, 17057–17066 (2013)
11. Siebenaler, S., Tervo, E., Vinh P., Lewis, C.: Field testing of negative wave leak detection systems. In: Proceedings of the 2014 10th International Pipeline Conference, ASME, pp. 1–8 (2014)
12. Prihtiadi, H., Azwar, A., Jamal, M.: A simple method to determine leakage location in water distribution based on pressure profiles. AIP Conf. Proc. **1719**, 1–5 (2016)
13. Ben-Mansour, R., Habib, M.A., Khalifa, A., Youcef-Toumi, K., Chatzigeorgiou, D.: Computational fluid dynamics simulation of small leaks in water pipelines for direct leak pressure transduction. Comput. Fluids Elsevier **57**, 110–123 (2012)
14. Zeng, Y., Luo, R.: Numerical analysis of incompressible flow leakages in short pipes. In: Proceedings of 12th Pipeline Technology Conference, pp. 1–6 (2017)
15. Araújo, M.V., Neto, S.R.F., Antonio, G.B.L., Lavia, D.T.L.: Hydrodynamic study of oil leakage in pipeline via CFD. Adv. Mech. Eng. 1–9 (2014)
16. Jujuly, M., Thodi, P., Rahman, A., Khan, F.: Computation fluid dynamics modeling of subsea pipelines leak in arctic conditions. In: Offshore Technology Conference, pp. 1–19 (2016)

17. Sousa, C.A., Romero, O.J.: Influence of oil leakage in the pressure and flow rate behaviors in pipeline. Latin Am. J. Energy Res. LAJER **4**, 17–29 (2017)
18. Wei, O.Y., Masuri, S.U.: Computational fluid dynamics analysis on single leak and double leak subsea pipeline leakage. CFD Lett. **11**, 95–107 (2019)

Experimental Investigation of Machining Parameters on MRR of Biomaterial Stainless Steel 316L by EDM

Preeti Chauhan, M. A. Saloda, B. P. Nandwana, and S. Jindal

1 Introduction

EDM was introduced by physicist Joseph Priestley in 1770, but it was not much effective as more electrode material was eroded and there were manually operated feed systems which resulted in more arcing. N. I. Lazarenko and B. R. led the most important improvements, i.e. use of RC relaxation circuit in EDM because of which wear was reduced on electric power contacts [1].

EDM is one of the non-traditional machining methods which can machine workpiece with any hardness [2]. EDM can easily machine ceramics, super-alloys and composites which are not easy to machine [3]; even very delicate and fragile parts can be machined which cannot take the stress of conventional machines. For the last few years, to increase MRR, the tool is given ultrasonic-assisted vibration and tool vibrates, it increases flushing away of debris from workpiece which increases MRR [4]. The rotation of tool also increases MRR [5]. Some appropriate powder additives in the dielectric fluid are being used to improve surface roughness (SR) and decrease tool wear. Water as dielectric fluid is being used over other hydrocarbon dielectric fluids for the safe environment as decomposition of hydrocarbon oil releases carbon monoxide and methane which is harmful to the environment [6]. Figure 1 as shown classifies different types of EDM used for different purposes.

Normally, tool is connected with negative polarity and the workpiece is connected with positive polarity. Tool acts as cathode and workpiece as anode. The dielectric fluid has some dielectric constant. It is an electrically neutral liquid. It has some dielectric breakdown voltage. Potential applied between tool and workpiece should be greater than the breakdown voltage of dielectric fluid; otherwise, flow of current between tool and workpiece will not take place. When the capacitor discharges after

P. Chauhan (✉) · M. A. Saloda · B. P. Nandwana · S. Jindal
Mechanical Engineering Department, College of Technology and Engineering, Udaipur 313001, Rajasthan, India
e-mail: preeti.9950919824@gmail.com

M. Muzammil et al. (eds.), *Recent Advances in Mechanical Engineering*, Lecture Notes in Mechanical Engineering, https://doi.org/10.1007/978-981-15-8704-7_22

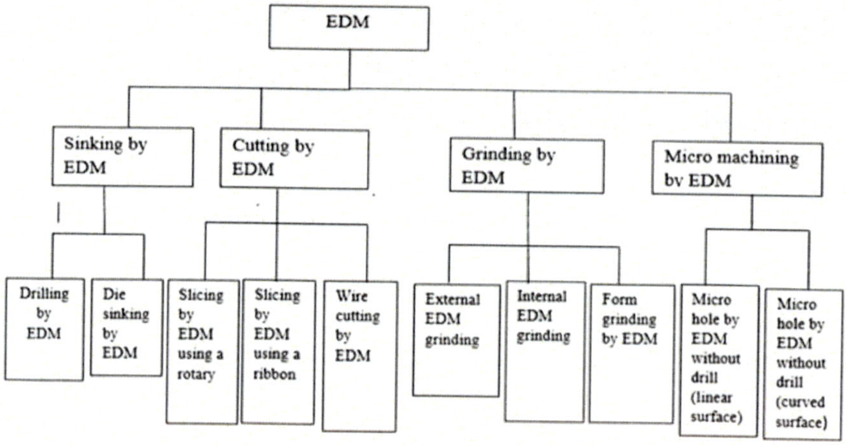

Fig. 1 Classification of spark erosion machining [7]

being charged, a huge amount of potential is supplied between tool and workpiece. The electrical field is established between them. Due to electrostatic force, electrons from tool are attracted towards workpiece. If the bond energy of electrons is less, then electrons move towards workpiece through the dielectric fluid. Electrons travel with very high velocity. The velocity of electrons depends on the potential supplied through capacitor. As electrons gain velocity, they move towards workpiece and collide with molecules of dielectric fluid. Collisions between them result in ionization of molecules. Collisions of electrons and molecules result in the generation of more positive ions and electrons. The high concentration of these ions between tool and workpiece within minimum inter-electrode gap takes place. Electrons move towards workpiece, and positive ions move towards tool resulting in sparking known as plasma channel. Electrons collide with workpiece with high velocity. The kinetic energy of electrons gets converted into thermal energy, and material gets removed from the workpiece. Material removal occurs due to vapourization and melting. Some material is also removed from tool due to collisions of positively charged ions with it but as the concentration of positive ions is less so material removal from tool is very less. As the supply of potential between tool and electrode is stopped, plasma channel collapses. The pressure is generated and removes molten material from workpiece.

There are different techniques for surface modification of biomaterials such as PVD, CVD, nitriding and anodizing, but due to weak adhesion, the bond cannot sustain under repeated loading conditions in corrosive environment and these techniques have become inefficient. Before these techniques, coating is required, i.e. shot peening and grit blasting which increases their production cost. Recently, EDM has been acknowledged as a promising technique for surface treatment of biomaterials for implants [8, 9]. When machining is done by EDM, dielectric fluid used for machining degenerates into oxides and carbides and gets deposited on workpiece which increases biocompatibility [10]. Nanoporous surfaces are produced when machined by EDM [11]; as a result, available pore spaces encourage the mineralized tissue growth into pores which allows the transportation of body fluid by interconnected pores [12]. As a result, cell proliferation and adhesion are improved [13]. Attainment of high temperature during sparking when machined by EDM improves the hardness of the surface. EDM machined surfaces improve corrosion [10, 14] and wear resistance [15].

Stainless steel 316L has been used as workpiece material. It is the most commonly used material for biomedical application as it is easily available and low cost and has excellent fabrication properties, good strength, high fracture toughness, good corrosive resistance and good biocompatibility [16, 17]. SS 316L is used in biomedical implants such as artificial valves, artificial joints, bone fixation (plate, screw, pin), orthodontic wire and artificial eardrum. In stainless steel 316L, L stands for low carbon percentage, i.e. 0.03%. Graphite produces more carbide phases on EDM machined surfaces which increases the corrosion resistance of workpiece.

2 Experimental Procedure

The workpiece of size 32 × 32 × 150 mm is used and graphite as an electrode material with dimensions 20 × 20 × 100 mm as shown in Fig. 2. Graphite is a non-metal electrode that has high melting point which increases its stability, and it has low density.

MRR has been calculated and examined using the Taguchi method by Minitab 18. Orthogonal array minimizes the number of trials, so it is most effective as it handles the number of variables and vast data. Three different parameters, i.e. T_{on}, I_m and V_s, are varied, and their effect on MRR has been studied as shown in Table 1. L_9 orthogonal array has been selected for evaluating maximum MRR. "Larger the better" option was selected.

Fig. 2 Machined workpiece and graphite electrode

Table 1 Independent parameters

S. No.	Independent parameters	Level 1	Level 2	Level 3	Units
1	I_m	10	14	18	Ampere
2	T_{on}	50	350	650	μs
3	V_s	3	5	7	Volt

$$\text{MRR} = \frac{\text{weight before maching} - \text{weight after machining}}{\text{time taken} \times \text{density}} (\text{mm}^3/\text{min})$$

The experiment was performed on ZNC 250 EDM, and only four parameters can be varied on this machine, i.e. I_m, T_{on}, V_s and T_{off}. From the literature survey and experiments performed, it was observed that I_m and T_{on} have major impact on MRR. T_{off} has less effect on MRR, and few researches have been carried out by varying parameter V_s. Hence, three parameters were varied I_m, T_{on}, V_s, and T_{off} was taken as constant. Values of parameters were selected such that to achieve maximum MRR and to observe variation in MRR by varying these parameters so their one lower value, one intermediate value and one higher value were selected.

3 Results and Discussion

The experimental investigation of SS 316L by EDM was performed, and the results obtained are discussed here.

Table 2 shows the experimental results of EDM machined surface of SS 316L. "Larger the better" option was used in Taguchi as the aim was to increase MRR of material.

It is concluded from the graph shown in Fig. 3 that maximum MRR was observed at the peak value of I_m 18 A, T_{on} 50 μs and V_s 5 V.

From Table 3, it can be seen that I_m is the most influencing factor (Rank 1), and then T_{on} (Rank 2) and V_s have a minimum effect (Rank 3). Here, I_m is increased from 10 to 18 A; 10 A current has a low value of MRR, as its value is increased from 10 to 14 A and then 18 A MRR keeps on increasing. T_{on} was ranged between 50 and 650 μs. At a lower value of T_{on}, MRR was high and as its value was increased, MRR decreased. V_s has very less effect on MRR. As V_s is increased from 3 to 7 V, MRR decreases.

Table 2 L_9 orthogonal array: independent parameters and response data

Independent parameters				Response	
S. No.	I_m	T_{on}	V_s	MRR	S/N ratio
1	10	50	3	2.417	6.8289
2	10	350	5	1.688	4.4407
3	10	650	7	0.917	0.1909
4	14	50	5	4.312	13.6370
5	14	350	7	3.278	9.4755
6	14	650	3	3.081	9.6670
7	18	50	7	5.702	15.0137
8	18	350	3	5.218	15.2935
9	18	650	5	4.116	12.8171

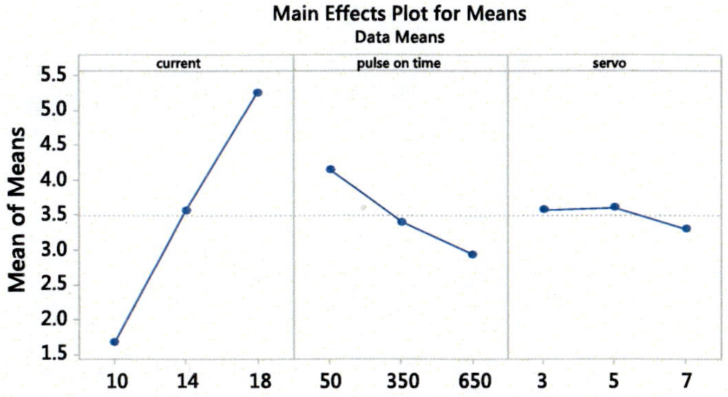

Fig. 3 Main effect plot for means of MRR

Table 3 Rank table for S/N ratio and with control machining parameters

Level	I_m	T_{on}	V_s
1	3.820	11.827	10.596
2	10.927	9.737	10.298
3	14.375	7.558	8.227
Delta	10.555	4.268	2.370
Rank	1	2	3

Results of ANOVA explained in Table 4 reveal the most significant factor which influences the material removal rate. *P*-value of these parameters is less than 0.05, so these parameters are significant and affect the response MRR.

Figure 4 shows interaction between different parameters I_m, T_{on} and V_s. It shows the effect of these parameters on MRR. Figure 5a shows the interaction between I_m and T_{on} and their effect on MRR. From the graph, it can be observed that as I_m increases MRR increases whereas with the increase in T_{on}, MRR decreases. Similarly, the interaction plot between I_m, and V_s as shown in Fig. 4b reveals that the slope of MRR increases with an increase in I_m and decreases with an increase in V_s. It is concluded from graphs that MRR increases as the value of I_m increases and decreases with an increase in the value of T_{on} and V_s.

Table 4 ANOVA for MRR

Source	DF	Adj SS	Adj MS	*F*-value	*P*-value
I_m	2	19.1506	9.57529	27,480.11	0.000
T_{on}	2	2.2232	1.11159	3190.15	0.000
V_s	2	0.1695	0.08474	243.20	0.004
Error	2	0.0007	0.00035		
Total	8	21.5439			

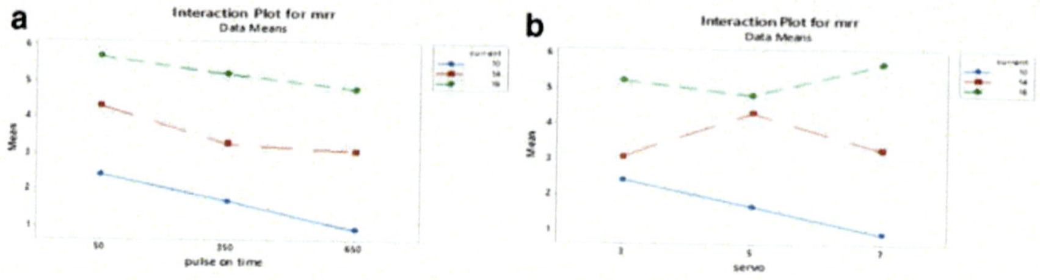

Fig. 4 Interaction plot for MRR: **a** T_{on} and I_m, and **b** V_S and I_m

As I_m increases, stronger spark is produced. More energy is transferred between tool electrode and workpiece. Causing high temperature between them and more material is eroded from workpiece, and as a result, MRR increases with an increase in I_m. Similar results were seen by [18] which showed that I_m was the most influencing factor.

Increase in T_{on} results in increases in time duration for supplied energy between tool and workpiece, and spark energy between tool and workpiece increases and results in more MRR. Up to some value of T_{on}, MRR keeps on increasing and then suddenly it starts decreasing; this is because at a higher value of T_{on}, the plasma channel between the inter-electrode gaps obstructs energy transfer and electrons jump out from plasma channel which reduces MRR. T_{on} here is 50, 350 and 650 μs which misses the peak value; hence, graph obtained here is decreasing. The peak value of T_{on} for maximum MRR lies in the range of 50–150 μs. Similar results were presented by [19] which shows that with an increase in T_{on} MRR drops.

4 Conclusion

On the basis of experiments, observations and calculations carried out, the following conclusions were drawn:

- MRR increases as the value of I_m rises from 10 to 18 A. An increase in I_m causes more energy supplied between tool and material which results in more material to be melted and evaporated which causes increase in MRR.
- MRR increases up to some value as T_{on} increases, and then it starts to decrease. Values selected were 50, 350 and 650 μs which misses the peak value; hence, the graph obtained here was decreasing.
- MRR decreases as V_s increases.
- I_m, T_{on} and V_s have *P*-value less than 0.05 so these are significant parameters and affect the response MRR.
- Maximum MRR is observed at a set of parameters I_m 18 A, T_{on} 50 μs and V_s 5 V.

References

1. Schumacher, B.M., Krampitz, R., Kruth, J.P.: Historical phases of EDM development driven by the dual influence of "market pull" and "science push". In: The Seventeeth CIRP Conference on Electro Physical and Chemical Machine (ISEM), vol. 6, pp. 5–12 (2013)
2. Mahmud, N., Yahya, A., Rafiq, M., Samion, S., Safura, N.L.: Electrical discharge machining pulse power generator to machine micropits of hip implant. (ICoBE), pp. 493–497 (2012)
3. Pandey, A., Singh, S.: Current research trends in variants of electrical discharge machining, a review. Int. J. Eng. Sci. Technol. **2**, 2172–2191 (2010)
4. Guo, Z.N., Lee, T.C., Yue, T.M., Lau, W.S.: A study of ultrasonicaided wire electrical discharge machining. J. Mater. Process. Technol. **63**, 823–828 (1997)
5. Yan, B.H., Chen, S.L.: Characteristics of SKD11 by complex process of electrical discharge machining using liquid suspended with alumina powder. Nippon Kinzoku Gakkaishi/J. Japan Inst. Metals **58**, 1067–1072 (1994)
6. Zhang, Q.H., Zhag, J.H., Ren, S.F., Niu, Z.W., Ai, X.: A theoretical model of surface roughness in ultrasonic assisted electrical discharge machining in gas. J. Mater. Process. Technol. **7**, 381–390 (2005)
7. Rajesh, R., Anand, M.D.: The optimization of electro discharge machining process using response surface morphology and genetic algorithms. Proc. Eng. **38**, 3941–3950 (2012)
8. Kumar, V., Beri, N., Kumar, A.: Electric discharge machining of titanium and alloys for biomedical implant applications: a review. Int. J. Res. Anal. Rev. **3**, 120–128 (2018)
9. Harcuba, P., Cákovάb, L.B., Strάskýa, J., Cákovάb, M.B., Novotnάb, K., Ceka, M.J.: Surface treatment by electric discharge machining of Ti–6Al–4V alloy for potential application in orthopaedics. J. Mech. Behav. Biomed. Mater. **7**, 96–105 (2011)
10. Prakash, C., Kansa, H.K., Pabla, B.S., Sanjeev, P., Aggarwal, A.: Electric discharge machining—a potential choice for surface modification of metallic implants for orthopedic applications. Rev. J. Eng. Manuf. **230**, 1–30 (2015)
11. Sidhom, H., Ghanem, F., Amadou, T.: Effect of electro discharge machining (EDM) on the AISI316L SS white layer microstructure and corrosion resistance, vol. 65, pp. 141–153. Springer, London (2012)
12. Dewidar, M.M., Khalil, K.A., Lim, J.K.: Processing and mechanical properties of porous 316L stainless steel for biomedical applications. Trans. Nonferrous Metals Soc. China **17**, 468–473 (2007)
13. Aliyu, A.A.A., Abdul-Rani, A.M., Ginta, T.L., Prakash, C., Axinte, E., Razak, M.A., Ali, S.: A review of additive mixed-electric discharge machining current status and future perspectives for surface modification of biomedical implants. Hindawi Adv. Mater. Sci. Eng. 1–23 (2017)
14. Jahan, M.P., Kakavand, P. and Alavi, F.: A comparative study on micro-electro-discharge-machined surface characteristics of Ni-Ti and Ti-6Al-4V with respect to biocompatibility. In: 45th SME North American Manufacturing Research Conference, NAMRC 45, vol. 10, pp. 232–242. LA, USA (2017)
15. Güngör, E., Ekmekçi, B.: Wear resistance of electrical discharge machined surfaces. In: Conference: 3rd International Symposium on Innovative Technologies and Science, vol. 1, pp. 1080–1089. At Universidad Politecnica de Valencia, Valencia, Spain (2015)
16. Simionescu, N., Benea, L.: Corrosion behavior of 316l stainless steel as biomaterial in physiological environment. In: 13th Annual Conference on Materials Science Metal and Manufacturing. Paris, France (2017)
17. Mariotto, S.F.F., Guido, V., Cho, L.Y., Soares, C.P., Cardoso, K.K.: Porous stainless steel for biomedical applications. Mater. Res. **14**, 146–154 (2011)
18. El-Taweel, T.A.: Multi-response optimization of EDM with Al–Cu–Si–TiC P/M composite electrode. Int. J. Adv. Manuf. Technol. **44**, 100–113 (2008)
19. Kakkar, K., Rawat, N., Janwal, A., Aggarwal, N.: Optimization of surface roughness, material removal rate and tool wear rate in EDM using Taguchi 72 method. Int. J. Adv. Res. Ideas Innov. Technol. **4**, 16–23 (2018)

Global Advancement in Hybrid Solar Drying with Using Thermal Energy Storage System

Narender and Pushpendra Singh

1 Introduction

In rapid changing world, the storage and preservation of food materials with requisite quality are the biggest challenging tasks. To meet the continuous increasing demand of food materials, the various new innovative techniques have been developed and research is still continuing in this area. In 1976, Everitt and Stanley prepared the first kind of solar dryer to make the alternate of open sun drying. He made transparent glass covered box-shaped drying unit. Midilli [1] describes the drawback of open sun drying which includes poor product quality due to rain, insects, animal, debris and wind blow. Drying of food materials is the best type of methods that not only meet the continuous increasing demand of food materials globally but also preserve and store food materials with requisite quality for the future. Thermal drying is one of the general methods used which involves heat and mass transfer simultaneously to dry the food products. We know that the harvest cycle of food materials specially fruits and vegetables has for a short time. Due to insufficient storage, improper preservation and unbalance supply chain, a large amount of food materials spoil every year. According to UN Food and Agriculture Organization (2019), approximately one trillion $ of food is lost or wasted every year and that is sufficient to preserve enough food to serve two billion people. In World Food Programme (WFP), the training on post-harvest crop management is provided for zero post-harvest loss missions. The five key areas of the training are harvesting, drying, threshing, solarization and storage. According to World Health Organization (2019), an estimated 600 million peoples

Narender (✉) · P. Singh
Department of Mechanical Engineering, Delhi Technological University, Delhi, India
e-mail: narendra_phd2k18@dtu.ac.in

P. Singh
e-mail: dr.psinghs@gmail.com

Narender
Department of Food Engineering, NIFTEM, Kundli, Sonepat, India

M. Muzammil et al. (eds.), *Recent Advances in Mechanical Engineering*, Lecture Notes in Mechanical Engineering, https://doi.org/10.1007/978-981-15-8704-7_23

(at most 1 in 10) in the world become ill and 4, 20,000 die every year due to the effect of contaminated food. Drying is the oldest, simple, safe and easy-to-learn method to preserve food materials. It removes water/moisture in food materials so bacteria, mold & yeast cannot grow and spoil food. Drying slows down the growth of enzymes and inactivates them for long period. The optimum temperature required for food drying is 60 °C. Drying requires large amount of energy for removal of moisture level in the food products by means of supplying hot air to achieve the desired amount of moisture. Ekechukwu and Nortan [2] developed an advance drying unit in which they used auxiliary heating source (i.e., electricity and fossil fuels) with natural and forced circulation to attain the desirable drying properties.

2 Description of Previous Work on Solar Dryer

In 1976, Everitt and Stanley prepared the first type of solar dryer to make the alternate of open sun drying. He fixes the transparent glass cover at the upper face top of drying unit. The transparent glass cover allows sunlight to pass through it and heats up the food materials to dry it. The descriptions of previous work are shown in the broad classification of solar dryer developed by various researchers.

2.1 Classification of Solar Dryer Developed by Various Researchers

Since from the invention of solar-based dryer, the various categories of solar-operated dryer have been developed. The detailed classification of solar dryer developed by researchers is given in Table 1.

3 Hybrid Solar Dryer

Hybrid system solar dryer is basically a joining of solar energy collectors, auxiliary heating devices, biomass and thermal energy storage systems. Hybrid solar drying not only facilitates quick and continuous drying but also protects the food materials form direct environment, insects, dust, smoke, etc. In hybrid solar dryer, the solar energy is converted into usable form by the help of solar collectors. The available converted solar energy is used in drying, in thermal energy storage and to run the auxiliary devices. Hybrid solar dryer provides continuous heating supply to the drying unit which reduces the drying time, improves quality and uniformity, increases overall efficiency of the system and reduces expenses. Thanaraj et al. [3] compare the result of solar hybrid dryer and copra kiln. In the study, they found that white copra with

Table 1 Classification of solar-operated dryer

Solar dryer	Direct solar dryer	(a) Direct solar-based dryer (b) Solar tunnel-based dryer (c) Solar-operated dryer with cylindrical collector (d) Forced air solar dryer (e) Solar dryer with photovoltaic cell
	Indirect solar dryer	(a) Indirect solar-operated dryer (b) Solar-assisted indirect solar dryer (c) Solar dryer with absorption systems (d) Indirect solar dryer with rectangular shape plate fins (e) Indirect solar dryer with movable mechanism (f) Roof-integrated indirect solar dryer (g) Prototype tray dryer (h) Indirect solar dryer by using thermic oil (i) Indirect forced convection-operated solar dryer (j) Solar dryer combined with air heater and heat accumulator (k) Solar-operated dryer coupled with V grooved collector (l) Solar dryer with microcontroller-based system (m) Forced air convection vegetable dryer (n) Solar dryer fitted with double-pass operated solar collector with fins (o) Forced convection solar dryer with evacuated tube collector
	Hybrid solar dryer	(a) Solar-assisted integrated drying system (b) Solar dryer with biomass burner (c) Indirect and mixed mode of solar dryer (d) Portable solar dryer with multi-shelf design (e) Mixed-mode forced convection solar dryer (f) Mixed-mode natural convection solar dryer (g) Hybrid system solar dryer integrated with auxiliary heating source

high-quality standard may be processed in hybrid solar drying, while white copra could not be formed in kiln drying. Ferreira et al. [4] made a hybrid solar dryer and conducted test on the banana chips drying. In his studies, he found that drying rate of the hybrid solar dryer was observed larger than the artificial and open dryer. Drying rate of the hybrid dryer was claimed 38% in comparison with artificial dryer. The drying time was observed lower for hybrid solar dryer in comparison with the artificial and open dryer at the same conditions of outlet air temperatures. Prasad et al. [5] experimentally analyzed and compared the results of ginger drying with open sun drying and hybrid solar biomass drying. Drying rate of hybrid dryer was found better in comparison with open sun drying. The diagram of hybrid solar dryer with biomass system is given in Fig. 1.

Ouassila et al. [6] experimentally studied tomato waste solar-operated drying by the use of new greenhouse-type solar drier. Aymen et al. [7] developed mixed-mode-type solar greenhouse dryer with forced convection system and study of solar drying of red pepper grown with them. Forson et al. [8] designed and developed a mixed-mode solar crop dryer with natural convection used for drying cassava and other

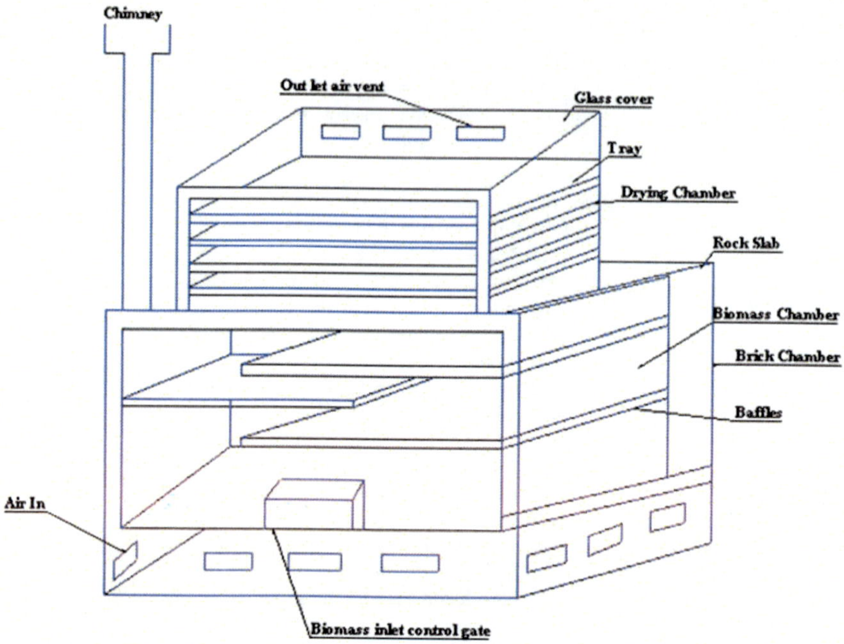

Fig. 1 Hybrid solar dryer with biomass system by Prasad et al. [5]

crops. Ramana et al. [9] showed in his review that various categories of solar dryer are used for grain drying, fruit drying, vegetable drying, grapes drying, medicine drying, etc. Saravanapriya et al. [10] developed a solar tunnel dryer of semicylindrical shape covered with UV stabilized polyethylene sheet. They perform experiments on it and compare the results with general open sun drying. In the solar tunnel, maximum temperature achieved was 65 °C. The estimated time for drying the coconut was 49.5 h recorded, and on the other hand open sun drying took 80 h for drying the same sample. In solar tunnel dryer, the moisture removal was uniform with drying rate throughout the drying period. The maximum drying rate for solar tunnel dryer was 0.042 g of moisture per hour per gm of moisture, whereas in case of open sun drying, the maximum drying rate was 0.34 g. per hour per gm of moisture achieved. Mehla and Yadav [11] developed an evacuated tube solar dryer and connected it with thermal energy storage for continuously producing hot air for clouds and night hours. Acetamide was used in the form of phase change material (PCM) for thermal energy accumulation.

The thermal energy storage efficiency of phase change material was varying from 14.43 to 17.65%. The pictorial view of evacuated tube solar dryer connected with storage of thermal energy is given in Fig. 2. Hossain et al. [12] made hybrid solar dryer fitted with auxiliary devices for tomato drying. The dryer efficiency was reported 17–29%, varying due to the change in operating conditions. The collector efficiency reported 10% increase due to attachment of plane reflector with the dryer. Baher et al.

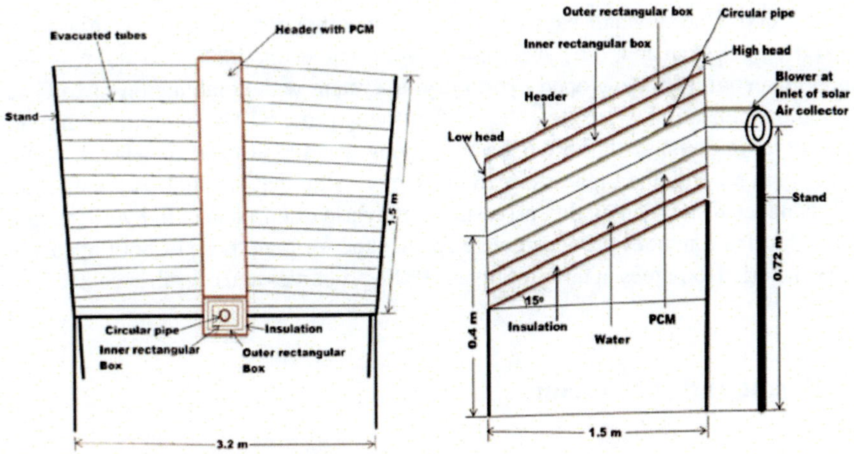

Fig. 2 Solar evacuated tube air collector connected with drying system and PCM unit [11]

[13] prepared hybrid solar dryer for chamomile drying. The drying unit consists of solar-operated collector, main drying unit, reflector, heat exchanger, additional drying unit, auxiliary heating source and water tank to store thermal energy (Fig. 3).

The integrated hybrid solar dryer decreases the moisture level of chamomile from 72–75 to 6% on wet basis in 30–33 h, while in open sun drying it takes 60 h for the moisture reduction from 72–75 to 9–10% on wet basis. Cipline et al. [14] developed hybrid solar drying for drying the medicinal plants. It comprises two solar-operated collectors, and air-type solar-operated collector was used for drying the medical plant whereas flat plate-based solar-operated collector was used to accumulate the thermal energy. A continuous drying was achieved by using hybrid drying system connected with auxiliary heating source. Chaudhari et al. [15] prepared a hybrid solar dryer for ginger drying and conducted economic analysis. The manufacturing cost of the hybrid solar dryer was Rs. 9000, and it takes 6 h' average drying time for 3 kg ginger. The annual repair and maintenance cost take 5%, and total benefit claimed

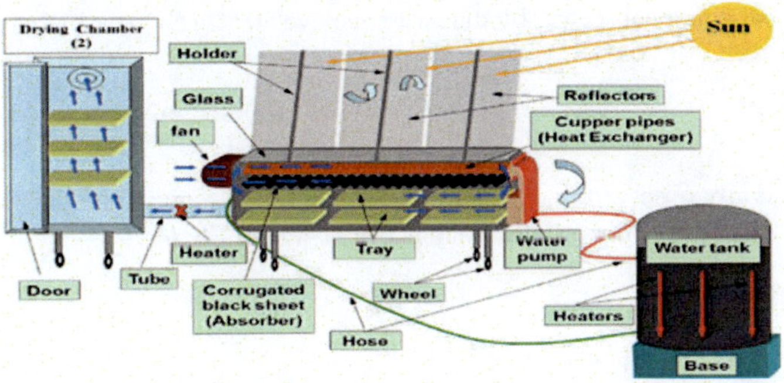

Fig. 3 Components of the integrated hybrid solar dryer system by Baher et al. [13]

Rs. 207,000. Due to small payback period (6 months) compared to the life of the solar drying system and, the hybrid solar dryer found economic viable and feasible. Suherman et al. [16] developed hybrid drying system with combination of LPG as an additional heating source to enhance the drying process. The outcome of hybrid solar dryer in comparison with open sun drying was found more beneficial. At 60 °C, system records the highest effectivity factor value that was 6.4. Hussein et al. [17] conducted study on quality assessment of hybrid dried tomato. In hybrid drying method, dried tomatoes produced were found superior than those of both open sun and solar dried tomatoes in terms of organoleptic properties and overall acceptability.

4 Result and Discussion

The general equations of solar dryer to calculate various parameters are given below.

4.1 Calculation of Energy Parameters [4]

The solar energy strikes on device which is calculated by E_s

$$E_s = \int (G \cdot A)\mathrm{d}t \tag{1}$$

where G is the incident global radiation and A denotes the dryer collector area.

The solar energy used to increase the airflow temperature (E_c) was calculated by mass airflow ($\mathrm{d}m/\mathrm{d}t$), airflow specific heat at constant temperature (C_p), ambient temperature (T_{amb}) and outlet airflow temperature (T_{out})

$$E_c = \int \frac{\mathrm{d}m}{\mathrm{d}t} C_p (T_{\mathrm{out}} - T_{\mathrm{amb}})\mathrm{d}t \tag{2}$$

The global energy absorbed by the air to achieve 50 °C (E_T) was calculated by

$$E_T = \int \frac{\mathrm{d}m}{\mathrm{d}t} C_p (50^\circ - T_{\mathrm{amb}})\mathrm{d}t \tag{3}$$

It is generally observed that 50 °C is the temperature required to airflow drying in experiments. The collector efficiency is calculated by Eq. (4).

$$\eta = \frac{E_c}{E_s} \tag{4}$$

4.2 Calculation of Dryer Efficiency [5]

The solar dryer overall thermal efficiency and solar biomass dryer are calculated by using the standard formula. It is the ratio of the amount of energy necessary to extract moisture content (through evaporation) from the product to the energy given to the dryer.

Mathematically, the overall thermal efficiency of solar dryer is calculated by:

$$E_o = \frac{m_e L}{IA} \tag{1}$$

Also, the solar biomass dryer overall thermal efficiency is calculated by:

$$E_{\text{bio}} = \frac{m_e L}{IA + CM} \tag{2}$$

where E_o = overall efficiency of solar dryer in percentage, E_{bio} = overall efficiency of the solar biomass dryer in percentage, m_e = moisture evaporated in kg, L = latent heat of vaporization in MJ/kg, C = biomass calorific value in MJ/kg, A = front sloping surface area in $m^2 M$ = mass of biomass substance used in kg, I = insulation on the dryer in MJ/m^2.

5 Conclusion

Hybrid solar dryer connected with auxiliary heating source provides continuous drying which increases the overall efficiency of the system and protects food materials from the dust, smoke, insects, debris, animals, etc. The quality of dried food materials observed quite better in hybrid solar dryer in comparison with the open sun drying techniques. It reduces the energy consumption, saves money, saves crucial time, reduces pollution and protects environment from smoke and other harmful gases. It is suitable for long time use and small-scale to large-scale industrial as well as residential drying applications. It requires very less maintenance cost, operational cost and handling cost. No extra skill or training is required for running the hybrid solar dryer.

References

1. Midilli, A.: Determination of pistachio drying behavior and conditions in a solar drying systems. Int. J. Energy Res. **25**, 715–725 (2001)
2. Ekechukwu, O., Norton, B.: Review of solar energy drying systems II: an over view solar drying technology. Energy Convers. Manag. **40**, 615–655 (1999)

3. Thanaraj, T., Dharmasena, N., Samarajeewa, U.: Comparison of drying behaviour quality and yield of copra processed in either a solar hybrid dryer on in an improved copra kiln. Int. J. Food Sci. Technol. **42**, 125–132 (2007)
4. Ferreira, A., Charbel, A., Pires, R., Silva, J., Maia, C.: Experimental analysis of a hybrid dryer. Therm. Eng. **6**(2), 03–07 (2007)
5. Prasad, J., Prasad, A., Vijay, V: Studies on the drying characteristics of zingiberofficinale (Ginger) under open sun and solar biomass Hybrid drying. Int. J. Green Energy **3**, 79–89 (2006)
6. Ouassila, B., Salah, H., Ahmed, D., Brahim, H., Amina, B.: Experimental and modelling study of tomato pomace waste drying in a new solar greenhouse: evaluation of new drying models. Renew. Energy **133**, 144–155 (2019)
7. EL Aymen, K., Ilhem, H., Sami, K., Amenallah, G.: Drying of red pepper slices in a solar greenhouse dryer and under open sun: experimental and mathematical investigations. Innov. Food Sci. Emer. Technol. **52**, 262–270 (2019)
8. Forson, K., Nazha, A., Akuffo, O., Rajakaruna, H.: Design of mixed-mode natural convection solar crop dryers: application of principles and rules of thumb. Renew. Energy **32**(14), 2306–2319 (2007)
9. Vijaya Venkata Ramana, S., Iniyan, S., Goic, R: A review of solar drying technologies. Renew. Sustain. Energy Rev. **16**(5), 2652–2670 (2012)
10. Saravanapriya, G., Mahendiran, R.: Design and development of solar tunnel dryer with control system for large scale drying of agro products. Int. J. Agric. Sci. **9**(40), 4626–4631 (2017)
11. Mehla, N., Yadav, A.: Experimental investigation of a desiccant dehumidifier based on evacuated tube solar collector with a PCM storage unit. Dry. Technol. **35**(4), 417–432 (2017)
12. Hossain, M., Amer, B., Gottschalk, K.: Hybrid solar dryer for quality dried tomato. Int J Dry Energy **26**(12), 1561–1601 (2008)
13. Baher, A., Klaus, G., Hossain, M.: Integrated hybrid solar drying system and its drying kinetics of chamomile. Renew. Energy **121**, 539–547 (2018)
14. Cipline, A., Novosinskas, H., Raila, A., Zvicevicius, E.: Usage of hybrid solar collector system in drying technologies of medical plants. Energy Convers. Manage. **93**, 399–405 (2015)
15. Chaudhari, R.H., Gora, A., Modi, V.M., Chaudhari, H.: Economic analysis of hybrid solar dryer for ginger drying. Int. J. Current Microbiol. Appl. Sci. **7**(11), 2725–2731 (2018)
16. Suherman, S., Susanto, E.E., Zardani, A.W., Dewi, N.H.R.: Performance study of hybrid solar dryer for cassava starch. AIP Conf. Proc. **2197**(080003), 1–10 (2020)
17. Hussein, J.B., Usman, M.A., Filli, K.B.: Effect of hybrid solar drying method on the functional and sensory properties of tomato. Am. J. Food Sci. Technol. **4**(5), 141–148 (2016)

Development of a Building Energy Model Based on State-Space Analysis and Determining the Performance Characteristics

V. S. K. V. Harish and Arun Kumar

1 Introduction

Buildings account to close to 33% of world's energy consumption and energy efficiency and conservation measures are being considered to bring this figure down. Major energy consuming services in a typical building are space heating/cooling, lighting and refrigeration. Energy control of systems responsible for delivery of these services requires development of a fairly accurate building energy models. Several literatures are available for building energy modeling approaches, and all can be categorized into forward and data-driven approach.[1, 2]. In forward approach, fundamental laws of building energy physics are used to develop the building energy system model. Output variables of the building energy model are predicted using detailed analysis subjected to a specific set of input variables. Data-driven building energy models are experimentally developed models where there is no detailed analysis of the model structure and input and output data obtained through measurements is used to construct the model. Experimentally measured data or historical data is used to develop such models. Such an approach is regarded as grey box approach. Due to limitations of computational complexity in white box models and accuracy of black box models, grey box approach is used for building energy modeling [3, 4, 6]. Remainder of the chapter is organized as follows. Section 2 involves describing the building energy transfer equations for the system under study. Section 3 present simulation results of the building energy system model. Model is validated using the

V. S. K. V. Harish (✉)
Department of Electrical Engineering, School of Technology, Pandit Deendayal Petroleum University Gandhinagar, Gujarat 382007, India
e-mail: harishvskv.iitr@gmail.com

V. S. K. V. Harish · A. Kumar
Department of Hydro and Renewable Energy, Indian Institute of Technology Roorkee, Uttarakhand 247667, India

M. Muzammil et al. (eds.), *Recent Advances in Mechanical Engineering*, Lecture Notes in Mechanical Engineering, https://doi.org/10.1007/978-981-15-8704-7_24

parameters being used currently by researchers round the world. Numerical values of the test case setup under study are adopted from [5].

2 Building Energy Systems

Systems which account for major consumption of energy delivering services in terms of comfort or any other functionality for a building are defined as building energy systems (BESs) [3]. Such systems include any equipment or machinery like HVAC system, ducts, lighting, etc., and/or processes like heat and mass transfer among occupants' and building construction elements or a combination of them. Services delivered by the equipment are majorly thermal and visual comfort and maintaining the building space indoor air quality depending upon the functionality of a building (Fig. 1).

Mathematical equations describing buildings' thermal energy transfer are a set of both time and space dependent. For analyzing the energy transfer processes, a test case is used with different layers of the multi-layered construction element as shown in Fig. 2. For the building space of BES under study, energy balance equation is given as Eq. (1).

$$\rho_a C_{p\_a} V_{bs} \frac{dT_{bs}}{dt} = \dot{Q}_{\mathrm{HVAC}} + \dot{Q}_{\mathrm{casual}} - \dot{V}_{bs} \rho_a C_{p\_a} (T_{bs} - T_{\mathrm{vent}}) - \ldots \frac{1}{\frac{1}{h_i} + \frac{x_1}{2k_{\mathrm{ins}}}} A_{\mathrm{ins}} (T_{bs} - T_{\mathrm{ins}}) - \frac{1}{\frac{1}{h_i} + \frac{x_{1,rr}}{2k_{\mathrm{ins},r}}} A_{\mathrm{ins},r} \left(T_{bs} - T_{\mathrm{ins},r}\right) \tag{1}$$

Following similar principles, energy balance equations for other layers of the building construction elements making up the building space are deduced in matrix

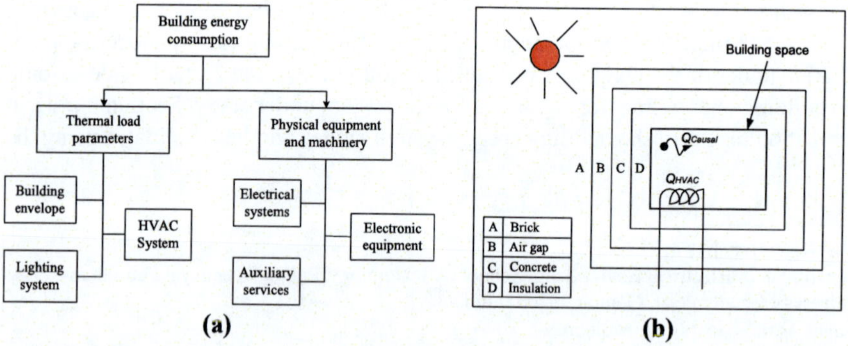

Fig. 1 **a** Services responsible for energy consumption in buildings, **b** building space with multi-layered construction elements (A, B, C, D)

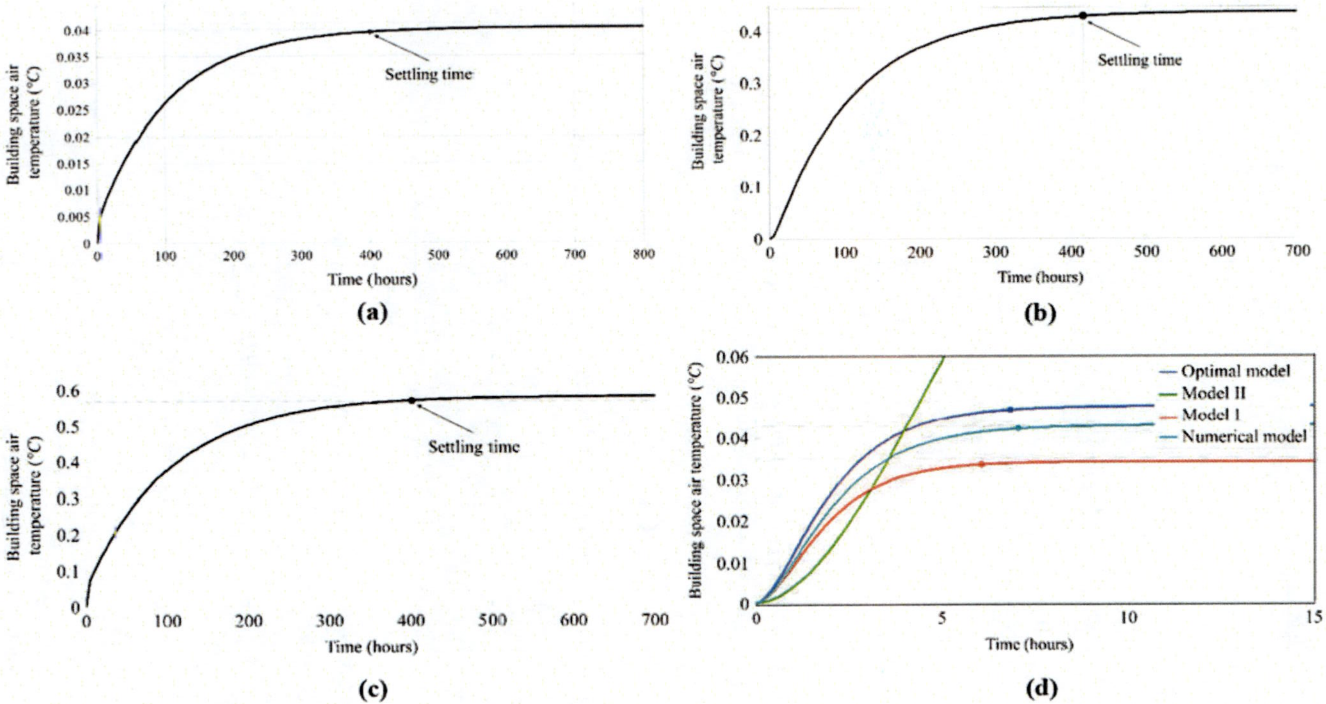

Fig. 2 Response of the BES model for step excitation of: **a** heat rate of HVAC system, **b** outdoor air temperature, **c** ventilated air temperature; **d** model validation against published research works

form as Eqs. (2) and (3).

$$\begin{pmatrix} \dot{T}_{bs} \\ \dot{T}_{\text{ins}} \\ \dot{T}_{\text{conc}} \\ \dot{T}_{b} \end{pmatrix} = \begin{bmatrix} \frac{-\dot{V}_{bs}\rho_a C_{p\_a} - \omega_1 A_1}{\rho_a C_{p\_a} V_{bs}} & \frac{\omega_1 A_1}{\rho_a C_{p\_a} V_{bs}} & 0 & 0 \\ \frac{\omega_1 A_1}{\rho_{\text{ins}} C_{p\_ins} V_{\text{ins}}} & \frac{-\omega_1 A_1 - \omega_2 A_2}{\rho_{\text{ins}} C_{p\_ins} V_{\text{ins}}} & \frac{\omega_2 A_2}{\rho_{\text{ins}} C_{p\_ins} V_{\text{ins}}} & 0 \\ 0 & \frac{\omega_2 A_2}{\rho_{\text{conc}} C_{p\_conc} V_{\text{conc}}} & \frac{-\omega_1 A_1 - \omega_3 A_3}{\rho_{\text{conc}} C_{p\_conc} V_{\text{conc}}} & \frac{\omega_3 A_3}{\rho_{\text{conc}} C_{p\_conc} V_{\text{conc}}} \\ 0 & 0 & \frac{\omega_3 A_3}{\rho_b C_{p\_b} V_b} & \frac{-\omega_3 A_3 - \omega_4 A_4}{\rho_b C_{p\_b} V_b} \end{bmatrix} \cdots \begin{pmatrix} T_{bs} \\ T_{\text{ins}} \\ T_{\text{conc}} \\ T_{b} \end{pmatrix}$$
$$+ \begin{bmatrix} \frac{1}{\rho_a C_{p\_a} V_{bs}} & 0 & \frac{\dot{V}_{bs}}{V_{bs}} \\ 0 & 0 & 0 \\ 0 & 0 & 0 \\ 0 & \frac{\omega_4 A_4}{\rho_b C_{p\_b} V_b} & 0 \end{bmatrix} \begin{pmatrix} \dot{Q}_{BS} \\ T_{\text{out}} \\ T_{bs} \end{pmatrix} \tag{2}$$

$$(T_{o/p}) = [1\,0\,0\,0] \begin{pmatrix} T_{bs} \\ T_{\text{ins}} \\ T_{\text{conc}} \\ T_{b} \end{pmatrix} + [0\,0\,0] \begin{pmatrix} \dot{Q}_{BS} \\ T_{\text{out}} \\ T_{bs} \end{pmatrix} \tag{3}$$

Also, net heat rate ($\dot{Q}_{BS}$) in the building space due to HVAC system ($\dot{Q}_{\text{HVAC}}$) and causal factors ($\dot{Q}_{\text{Casual}}$) like occupancy, furniture, electronic equipment, etc., is given as Eq. (4)

$$\dot{Q}_{BS} = \dot{Q}_{\text{HVAC}} + \dot{Q}_{\text{causal}} \tag{4}$$

Parameters (ω_1,ω_2, ω_3, and ω_4) used in Eq. (2) are defined as Eqs. (5) and (6)

$$\omega_1 = \frac{1}{\frac{1}{h_i} + \frac{x_1}{2k_{\text{ins}}}} \omega_4 = \frac{1}{\frac{x_3}{2k_b} + \frac{1}{h_o}} \tag{5}$$

$$\omega_2 = \frac{2k_{ins}k_{\text{conc}}}{x_1 k_{\text{conc}} + x_2 k_{\text{ins}}} \omega_3 = \frac{2k_b k_{\text{conc}}}{x_3 k_{\text{conc}} + x_2 k_b} \tag{6}$$

Above-mentioned equations can be represented in a state-space form of Eqs. (7) and (8)

$$\dot{T} = AT + BU \tag{7}$$

$$T_{\text{o/p}} = CT + DU \tag{8}$$

where

T	State vector.
U	Vector of inputs to the state-space model.
$T_{\text{o/p}}$	Output vector of the state-space model.

A, B, C, and D Coefficient matrices.

Transfer function for the developed state-space model is given as Eq. (9).

$$G(s) = C(sI - A)^{-1}B + D \tag{9}$$

where

s Laplace transform variable.
$G(s)$ System transfer function.

$|sI - A|$ term in Eq. (9) is the characteristic polynomial of $G(s)$ and the eigenvalues of matrix A are identical poles of $G(s)$. Transformation of a state-space model into transfer function enables to describe a dynamic relationship between inputs or excitations and outputs or responses of the system under study. Laplace transformation is applied on input–output differential equations with a condition that all initial values are zero or changing infinitesimally.

State-space model developed using Eqs. (7) and (8) does not take into consideration the effect of windows or glazing area within a building space. For the present study, 33% of the indoor space area is assumed to be exposed to opaque elements. The effect of window panes surrounding the glass has been neglected as window panes, and frames do not account for appreciable heat transfer within a building space [5].

3 Simulation Results and Model Validation

The developed state-space model is simulated in MATLAB/Simulink with step input excitations (Fig. 2a–c). The developed model is validated against: (1) The model parameters are equally distributed as a fraction of the building elements' total thermal resistance and capacitance (regarded as Model II in Fig. 2d), (2) the model parameters are tuned as a fraction of the total thermal resistance and capacitance depending upon the surface's exposure to outdoor climatic conditions (regarded as Model I in Fig. 2d).

4 Conclusion

A MATLAB/Simulink-based single-zone building energy model is developed using state-space approach. Developed building energy model is subjected to step excitations of HVAC heat rate, outdoor air temperature and building space indoor air temperature. Performance characteristics based on the system transfer functions and HVAC power have been analyzed for different scenarios by varying the excitation values. Developed model performance analysis is carried out by specifying the response characteristics of settling time, rise time, etc. BES model under study is excited with step excitations of outdoor air temperature, (T_{out}), ventilated air temperature, (T_{vent}) and HVAC plant heat rate ($\dot{Q}_{\text{HVAC}}$) is analyzed for different scenarios by varying the excitation values. In order to maintain the building space air temperature (T_{BS}) at 20 °C, 505 W of HVAC power is required under zero outdoor and ventilated air temperature values. When there is no HVAC power, then the building space temperature is 9.7 °C with ventilated air temperature of 17 °C and zero outdoor temperature. Under zero outdoor temperature condition, 255 W of HVAC power is required to maintain the building space air temperature and ventilated air temperature at 20 °C and 17 °C, respectively.

References

1. O'Neill, Z., O'Neill, C.: Development of a probabilistic graphical model for predicting building energy performance. Appl. Energy **164**, 650–658 (2016)
2. Schiefelbein, J., Rudnick, J., Scholl, A., Remmen, P., Fuchs, M., Müller, D.: Automated urban energy system modeling and thermal building simulation based on OpenStreetMap data sets. Build. Environ. **149**, 630–639 (2019)
3. Harish, V.S.K.V., Kumar, A.: A review on modeling and simulation of building energy systems. Renew. Sustain. Energy Rev. **56**, 1272–1292 (2016a)
4. Harish, V.S.K.V., Kumar, A.: Modeling and simulation of a simple building energy system. In: 2016 International Conference on Microelectronics, Computing and Communications (MicroCom), pp. 1–6. IEEE (2016)
5. American Society of Heating, Refrigerating, and Air-Conditioning Engineers (ASHRAE): ASHRAE handbook of fundamentals. Atlanta, U.S.A. (2013)
6. Harish, V.S.K.V., Kumar, A.: Reduced order modeling and parameter identification of a building energy system model through an optimization routine. Appl. Energy **162**, 1010–1023 (2016b)

Optimal Controller Design for Altitude Control of Modern Airship

Shahida Khatoon, Huma Khan, Prerna Gaur, and Mohd Faisal Jalil

1 Introduction

The evolution of flying machines has been impacted by the naturally born flying objects in the universe and encouraged the researchers for development of innovation of autonomous system of multiple uses and applications. The researchers have always faced the challenges in design of flying machines and simultaneously the control aspects for lack of existing solution. The motivation for the development for civilian and military needs led the researchers to the advancement in the autonomous system both on the ground and flying systems. In the last few years, the technology in the flying machines has been tremendously changed to meet the needs of the hour. The design of an autonomous system/flying machine/airship requires a numbers of various parameters to be selected based on a certain enabling decision to follow a validated methodology.

The first autonomous man-free controlled flight [1] was achieved as tendem like helicopter. In 1990s, projects on helicopters are presented in [2]. Nowadays, autonomous system is an advance technological-based flying machine to tackle the various challenges in terms of performances execution [3]. The modeling for AS system [4, 5] and analysis of dynamics performed based on state-space model [6] is presented. The Euler–Lagrange approach [7] is used for mathematical modeling of the system. The authors [8] proposed an optimal controller based on H2 and H∞,

S. Khatoon · H. Khan (✉)
Department of Electrical Engineering, Jamia Millia Islamia, New Delhi, Delhi 110025, India
e-mail: humazakhan@yahoo.com

P. Gaur
Division of Instrumentation & Control Engineering, NSUT, Dwarka, New Delhi, India

M. F. Jalil
Department of Electrical & Electronics Engineering, KIET Group of Institutions, Delhi-NCR, Ghaziabad, India

M. Muzammil et al. (eds.), *Recent Advances in Mechanical Engineering*, Lecture Notes in Mechanical Engineering, https://doi.org/10.1007/978-981-15-8704-7_25

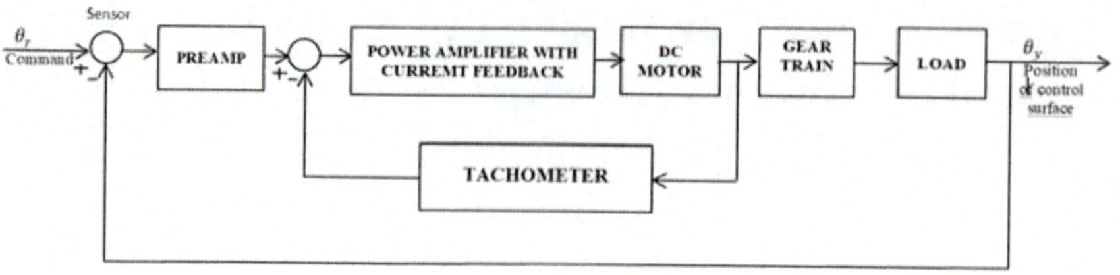

Fig. 1 Block diagram of an attitude control system of an aircraft

norms of the system were calculated, and the controller was designed for stabilization of the system. The altitude control, takeoff and landing [9] for performing autonomous hovering have been designed. The auto-lift and easier control are highly sought for the critical applications [10]. The researchers [11, 12] describe the design for trajectory tracking.

The impetus of the proposed work is on modeling and control approach for airship to invent a new way for controller's design based on certain variables [13, 14]. The present paper also proposed a new devised control strategy based on full-state feedback and pole-placement technique for the design of linear quadratic regulator to provide the stability to airship (Fig. 1).

2 Mathematical Modeling

The inertia and friction coefficient for the motor for considering the gear train ratio may be given as follows:

$$J_t = J_m + N^2 J_L \tag{1}$$

$$B_t = B_m + N^2 B_L \tag{2}$$

The forward-path transfer function of Fig. 2 by applying the SFG gain formula

$$\begin{aligned} G(s) &= \frac{\theta_y(s)}{\theta_e(s)} \\ &= \frac{K_s K_1 K_i K N}{s\left[L_a J_t s^2 + (R_a J_t + L_a B_t + K_1 K_2 J_t)s + R_a B_t + K_1 K_2 B_t + K_i K_b + K K_1 K_i K_t\right]} \end{aligned} \tag{3}$$

The system is of the third order, since the highest order term in $G(s)$ is s^3. The electrical time constant of the amplifier motor system is

$$\tau_\theta = \frac{L_a}{R_a + K_1 K_2} \tag{4}$$

The mechanical time constant of the motor-load system is:

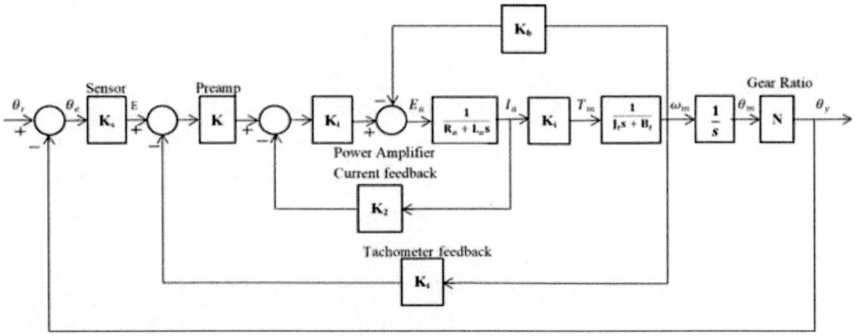

Fig. 2 Transfer-function block diagram of the system

$$\tau_T = \frac{J_1}{B_1} \tag{5}$$

The equations of the system are as follows:

$$\begin{bmatrix} \dot{x}_1 \\ \dot{x}_2 \\ \dot{x}_3 \end{bmatrix} = \begin{bmatrix} 0 & 1 & 0 \\ 0 & 0 & 1 \\ 0 & -\left(\frac{R_a B_t + K_1 K_2 B_t + K_i K_b + K K_1 K_i K_t}{L_a J_t}\right) & -\left(\frac{R_a J_t + L_a B_t + K_1 K_2 J_t}{L_a J_t}\right) \end{bmatrix} \begin{bmatrix} x_1 \\ x_2 \\ x_3 \end{bmatrix} + \begin{bmatrix} 0 \\ 0 \\ K K_1 K_i K_s N \end{bmatrix} r(t) \tag{6}$$

The A, B and C matrixes are given below:

$$A = \begin{bmatrix} 0 & 1 & 0 \\ 0 & 0 & 1 \\ 0 & -\left(\frac{R_a B_t + K_1 K_2 B_t + K_i K_b + K K_1 K_i K_t}{L_a J_t}\right) & -\left(\frac{R_a J_t + L_a B_t + K_1 K_2 J_t}{L_a J_t}\right) \end{bmatrix} \tag{7}$$

$$B = \begin{bmatrix} 0 \\ 0 \\ K K_1 K_i K_s N \end{bmatrix} \tag{8}$$

$$C = \begin{bmatrix} 1 \; 0 \; 0 \end{bmatrix} \tag{9}$$

The specifications of the parameters in the dc motor of airship are given in Table 1.

Table 1 Specification of the parameters of modern airship

Parameter	Description	Value	Units
K_s	Gain of encoder	1	V/rad
K	Gain of preamplifier	1	V/V
K_1	Gain of power amplifier	10	V/V
K_2	Gain of current feedback	0.5	V/A
K_t	Gain of tachometer feedback	0	V/rad/s
R_a	Motor armature resistance	5.0	Ω
L_a	Motor armature inductance	0.003	Henry
K_i	Motor torque constant	9.0	oz-in/A
K_b	Motor back-emf constant	0.0636	V/rad/s
J_m	Motor inertia	0.0001	oz-in-s^2
J_L	Inertia of load	0.01	oz-in-s^2
B_m	Motor friction coefficient	0.005	oz-in-s^2
B_L	Load friction coefficient	1.0	oz-in-s
T_m	Motor torque	1.0	N m
T_L	Load torque	1.0	N m
N	Gear train ratio	0.1	

3 State Feedback Controller Design Based on LQR Technique

The controller based on full-state feedback is suitable to handle systems of high order, and it is shown in Fig. 3 [13]. The proper placements of the poles are important for enhancement of margin of stability. The general expression is shown as;

$$x(k+1) = Ax(k) + Bu(k) \tag{10}$$

$x(k)$ state vector,
$u(k)$ control vector,
A System matrix,
B Control matrix.

Feedback control for the pole-placement technique

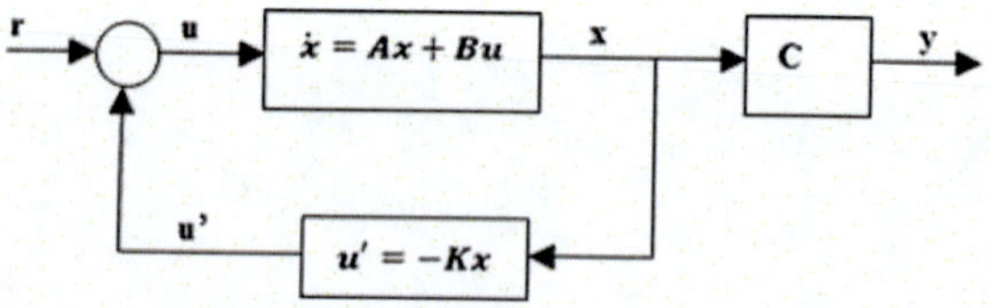

Fig. 3 State feedback controller

$$u(k) = -Kx(k) + r(k) \tag{11}$$

The performance of the airship control system is designed as per (12);

$$J = \frac{1}{2}x^T(N)S_N x(N) + \frac{1}{2}\sum_{k=0}^{N-1}\left[x^T Qx(k) + u^T(k)Ru(k)\right] = \text{Min} \tag{12}$$

Q, S and R are positive semi-definite symmetric matrices. K should be designed optimally subject to the minimization of performance index [5]. Riccati equation is given below for the solution of S.

$$S_k = Q + A^T \cdot S_{k+1} \cdot \left(I + B \cdot R^{-1} \cdot B^T \cdot S_{k+1}\right)^{-1} \cdot A, \tag{13}$$

and optimal control vector can be obtained as;

$$u_k = -\left[\left(B^T \cdot S_{k+1} \cdot B + R\right)^{-1} \cdot B^T \cdot S_{k+1} \cdot A\right] \cdot x_k \tag{14}$$

Using Eqs. (11–14) gain matrix can be obtained as;

$$K_k = \left(B^T \cdot S_{k+1} \cdot B + R\right)^{-1} \cdot B^T \cdot S_{k+1} \cdot A \tag{15}$$

LQR control is applied for the considered diagonal elements based on set procedure of the matrices Q, R and S.

4 Simulation Results

The full-state feedback controller is designed for the optimal feedback gain matrices as shown in Table 2. The system eigenvalues are computed to examine the system stability and are given in Table 3. The system response for the altitude control is shown in Fig. 4. And comparative analysis in done for both LQR and pole-placement-technique-based controller design. The matrices are computed for the controller design of airship and are given as follows;

Table 2 Feedback gain matrices

K_1	0.0029	0.0090	3.1623
K_2	0.0115	0.9839	0.0000

Table 3 System eigenvalues

E_1	−3008.1	−4003	0000
E_2	−450.3635	−350.7562	−150.0019

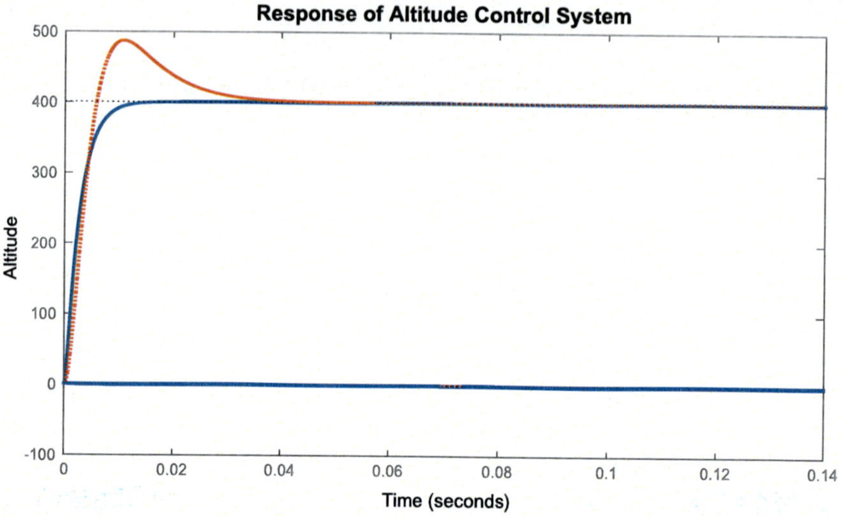

Fig. 4 Response of the altitude control

System matrix;

$$A = \begin{bmatrix} 0 & 1 & 0 \\ 0 & 0 & 1 \\ 0 & 1204000 & 3408.33 \end{bmatrix}$$

Control matrix;

$$B = \begin{bmatrix} 0 \\ 0 \\ 15000000 \end{bmatrix}$$

Measurement matrix;

$$C = [1\,0\,0]$$

The electrical time constant of the amplifier motor system is

$$\tau_\theta = \frac{L_a}{R_a + K_1 K_2} = 0.0003\text{s}$$

The mechanical time constant of the motor-load system is

$$\tau_T = \frac{J_1}{B_1} = 0.01333\text{s}$$

5 Conclusion

The mathematical model is developed, and a full-state feedback controller is designed based on linear quadratic optimization and pole-placement technique. LQR yields better result as compared to pole-placement controller and evident from the response plot for altitude stabilization. The main emphasis was given for proper selection and design of matrix Q and considered in the design of controller for airship.

References

1. Boulet, J.: Histoire de l'Hélicoptère. France-empire (1991)
2. Eck, C.: Navigation algorithms with applications to unmanned helicopters. Ph.D. Thesis, ETHZ (2001)
3. Aerovironment. https://www.aerovironment.com
4. Triputra, F.R. Trilaksono, B.R. Sasongko, R.A., Dahsyat, M.: Longitudinal dynamic system modeling of a fixed-wing UAV towards autonomous flight control system development a case study of BPPT Wulung UAV platform. In: Conference Proceedings, International Conference on System Engineering and Technology, pp: 1–6. Bandung, Indonesia (2012)
5. Done, G., Balmford, D.: Bramwell's Helicopter Dynamics. Oxford Butterworth-Heinemann (2001)
6. Rauf, A., Zafar, M.A., Ashraf, Z., Akhtar, H.: Aerodynamic modeling and State-Space model extraction of a UAV using DATCOM and Simulink. In: Conference Proceedings, Conference on Computer Research and Development (ICCRD), pp. 88–92. Shanghai (2011)
7. Mattio, A.: Modeling and control of the UAV sky-sailor. In: Master's Project Report, Swiss Federal Institute of Technology, Lausanne
8. Hussain, S.A., Kadri, M.B.: Optimal control synthesis for UAV. In: Conference Proceedings, IEEE International Conference on Computer, Control and Communication (IC4), pp. 1–6. Karachi, Pakistan (2013)
9. Krstic, M., et al.: Nonlinear and Adaptive Control Design. Wiley Interscience, New York, USA (1995)
10. Elfes, A.A., et al.: Robotic airship for exploration of planetary bodies with an atmosphere autonomy challenges. Auton. Robots (2003)
11. Low, C.B.: A trajectory tracking control design for fixed-wing unmanned aerial vehicles. In: Conference Proceedings, IEEE International Conference on Control Applications, Part of 2010 IEEE Multi-Conference on Systems and Control Yokohama, pp. 2118–2123. Japan (2010)
12. Ren, W., Beard, R.W.: A trajectory tracking control design for fixed-wing unmanned aerial vehicles. IEEE Trans. Control Syst. Technol. **12**(5), 706–716 (2004)
13. Nagarath, I.J., Gopal, M.: Control System Engineering. Wiley, Singapore (1982)
14. Golnaraghi, F., Kuo, B.C.: Automatic Control System, 9th edn. Wiley

Two-Wheeled Self-Balancing Mobile Robot Using Kalman Filter and LQG Regulator

Shahida Khatoon, Md. Istiyaque, Naimul Hasan, and D. K. Chaturvedi

1 Introduction

Two-wheel self-balancing mobile robot (TWSBMR) is based on two degrees of freedom (2 DOF) robot workstation. It is highly unstable and nonlinear system which is the keen interest of researchers to make it stable and linearize. Two-wheel inverted pendulum which is basically mounted on 2-DOF means two rotary servo base units with a 4-bar linkage system. Two-wheel self-balancing robot consists of a two-DOF instrumented joint, on which an almost 12 in. rod is bolted, which is free to rotate about two rectangular axes. The main aim of the 2-DOF inverted pendulum is to command and control position of the 2-DOF robot end effectors [1] to balance the 2-DOF inverted pendulum system module [2]. By measuring vertical position deviation, a controller is used to rotate the servo such that the end of the servo effectors balances the pendulum [3]. Designing a controller is necessary so that it can maintain the pendulum upright using the two servomotors. It provides so many concepts to the student for aerospace engineering application, such as rocket stabilization, self-balancing robot, earthquake resistant, and building construction [4–6].

Previously, several researches have been done which are actually based on inverted pendulum, and it is old secure and easy topic for the research to start and understand for the researchers, like a person upper body needs adjustment constantly to maintain and balance when we are standing or walking. From past few years, it has been found that the researchers design some controllers for the two-wheel self-balancing mobile

S. Khatoon · Md. Istiyaque (✉)
Jamia Millia Islamia, New Delhi, Delhi 110025, India
e-mail: istijmi@yahoo.com

N. Hasan
Qassim University, Buraidah, Kingdom of Saudi Arabia

D. K. Chaturvedi
Dayalbagh Educational Institute, Agra, India

M. Muzammil et al. (eds.), *Recent Advances in Mechanical Engineering*, Lecture Notes in Mechanical Engineering, https://doi.org/10.1007/978-981-15-8704-7_26

robot, and using PID controller and fuzzy logic, it is easy to control any system [7]; in this paper, LQR and PID controller have been proposed. Researchers have made two-wheel self-balancing mobile robot in laboratory at the industrial electronics. There are various studies based on stabilization and optimization of two-wheel self-balancing mobile robot. For this chapter I studied proportional and proportional derivative (PD) and go through it [8], proportional integral derivative (PID), linear quadratic regulator (LQR) [9], and model predictive control (MPC). And some more research work which is relevant and based on two-wheel mobile robot [10, 11] has been presented. Further, the chapter is categorized in following sections like Sect. 2 illustrates dynamic modeling of two-wheel self-balancing mobile robot, in Sect. 3 LQG and MPC controller configuration is modeled and designed, Sect. 4 discusses the simulated results and performance comparison of critical characteristics, followed by conclusion in last section.

2 System Modeling

The physical prototype of two-wheeled mobile robot consists of a rigid pole on a cart which is attached to two driving wheels. The mathematical formulation [12] of two-wheeled mobile robot is obtained assuming zero resistance in the air flow, negligible secondary friction, etc. The horizontal movement of cart is represented by distance x. M and m are used to define the mass of cart and pole, respectively. L_c represents the total pole length. The angle by which the pole is titled is described by φ. The wheel mobile robot is driven with the help of a wheel inverted pendulum [13, 14].

3 Model of the Wheel of Inverted Pendulum

The torque produced by left wheel is represented by D_L and that by right wheel is D_R as shown in Fig. 3. The reactive force exerted on left wheel is defined by R_{FL}, R_{FR} and that in right wheel by R_{FL}, R_{FR}. The equation of motion of left wheel can be represented by Eq. (7) and that of right wheel by Eq. (8).

For the left wheel

$$M_w\ddot{y} = -\frac{\tau_{mot}\tau_e}{Rr}\dot{\sigma}_w + \frac{\tau_{mot}}{Rr}V_a - \frac{I_w\ddot{\sigma}_w}{r} - R_{FL} \tag{1}$$

For the right wheel

$$M_w\ddot{y} = -\frac{\tau_{mot}\tau_e}{Rr}\dot{\sigma}_w + \frac{\tau_{mot}}{Rr}V_a - \frac{I_w\ddot{\sigma}_w}{r} - R_{FR} \tag{2}$$

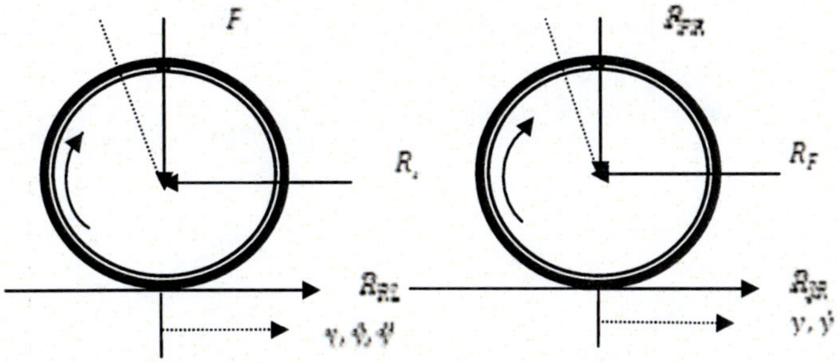

Fig. 1 Wheel of robot free body

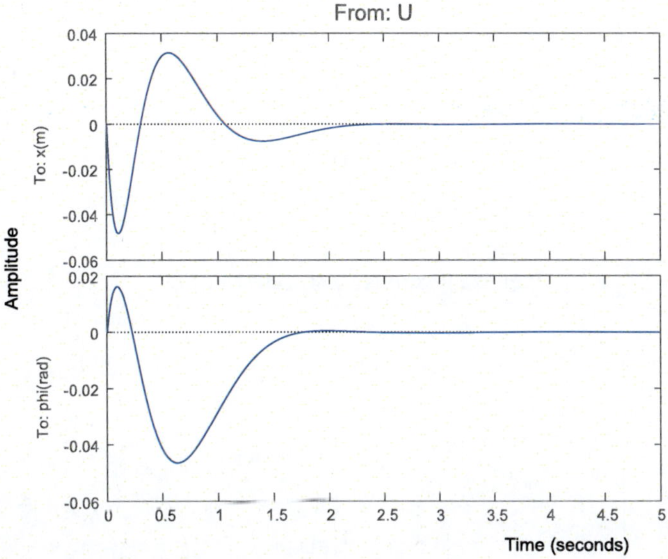

Fig. 2 LQG controller with plant

After converting angular velocity described by $\dot{\theta}$ into linear velocity $\dot{x}$ Eqs. (3) and (10) can easily be derived for left and right wheels, respectively,

For the left wheel

$$M_w\ddot{y} = -\frac{\tau_{mot}\tau_e}{Rr^2}\dot{y} + \frac{\tau_{mot}}{Rr}V_a - \frac{I_w\ddot{y}}{r^2} - R_{FL} \tag{3}$$

For the right wheel

$$M_w\ddot{y} = -\frac{\tau_{mot}\tau_e}{Rr^2}\dot{y} + \frac{\tau_{mot}}{Rr}V_a - \frac{I_w\ddot{y}}{r^2} - R_{FR} \tag{4}$$

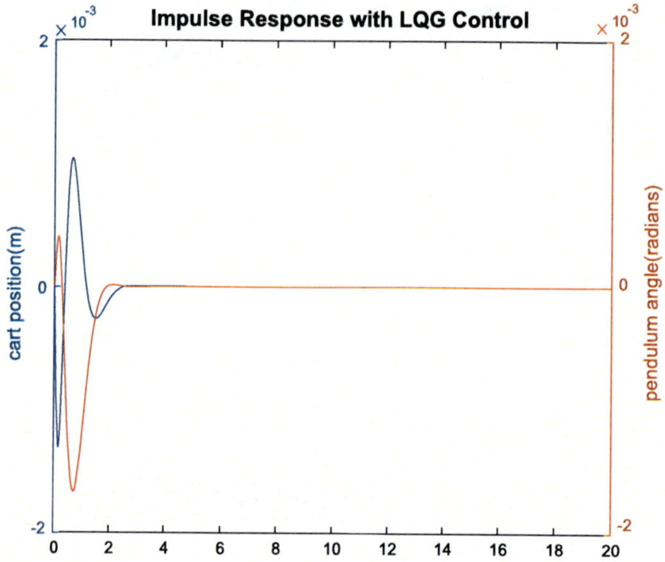

Fig. 3 Disturbance rejection capability of LQG controller with plant

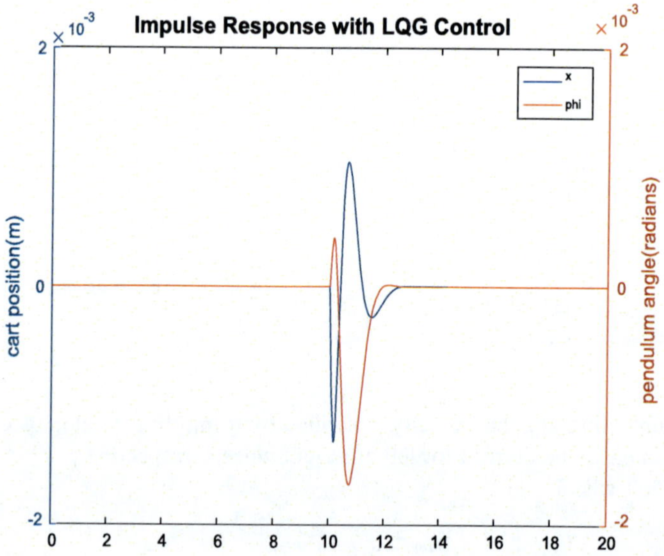

Fig. 4 Disturbance at $t = 10$ s and LQG controller with plant once disturbance produced

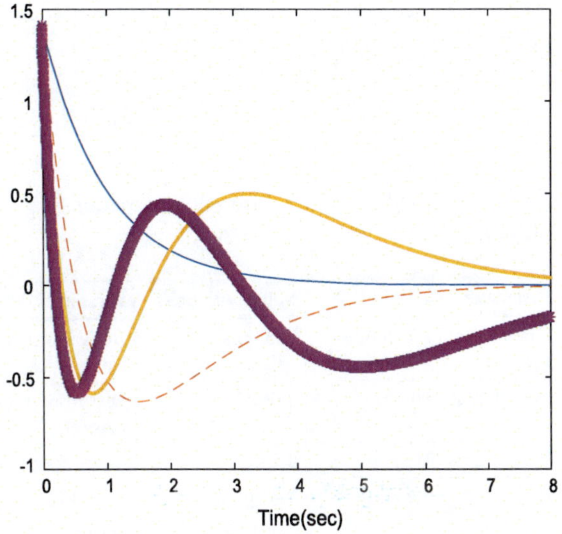

Fig. 5 Laguerre function for different values of *p* and *N*

By adding Eqs. (9) and (10), resultant equation will be

$$2\left(M_w + \frac{I_w}{r^2}\right)\ddot{y} = -2 - \frac{\tau_{mot}\tau_e}{Rr^2}\dot{y} + 2\frac{\tau_{mot}}{Rr}V_a - (R_{FL} + R_{FR}) \tag{5}$$

3.1 Cart Model

The mathematical formulation of the cart on pole is derived by many scientists [4–6] and is written as follows;

$$(I_p + l^2 M_b)\ddot{\sigma}_p - 2\frac{\tau_{mot\tau e}}{Rr}\dot{y} + 2\frac{\tau_{mot}}{Rr}V_a + M_p gl\sin\sigma_p = -M_p\ddot{y}l\cos\sigma_p \tag{6}$$

$$2\frac{\tau_{mot}}{Rr}V_a = \left(2M_w + 2\frac{I_w}{r^2} + M_p\right)\ddot{y} + 2\frac{\tau_{mot\tau e}}{Rr^2}\dot{y} + M_p l\ddot{y}_p\cos\ddot{y}_p\cos\sigma_p - M_p l\dot{\sigma}_p^2\sin\sigma_p \tag{7}$$

Since state-space representation of any given system requires first-order differential equations, above equations are linearized as follows;

$$\sigma_p = \pi + \emptyset \tag{8}$$

where $\emptyset$. The right angle which is upside

$$\cos\sigma = -1, \sin\sigma = -\emptyset, \frac{d\sigma_{p^2}}{dt} = 0$$

$$\ddot{y} = 2\frac{\tau_{mot}\tau_e\left(M_p l_r - I_p M_p l^2\right)}{Rr^2 a}\dot{y} + \frac{M_p^2 g l^2 \theta}{\alpha} + 2\frac{\tau_{mot}\left(I_p + M_p l^2 - M_p l_r\right)}{Rr\alpha}V_a \quad (9)$$

$$\ddot{y} = 2\frac{\tau_{mot}\tau_e\left(r\beta - M_p l\right)}{Rr^2 a}\dot{y} + \frac{M_p g l \rho \theta}{\alpha} + 2\frac{\tau_{mot}\left(M_p l - r\rho\right)}{Rr\alpha}V_a \quad (10)$$

The actual values of the different parameters of given pole and cart model are as follows;

Parameter description	Value with units	Parameter description	Value with units
Gravitational acceleration (g)	9.80 m/s^2	Body inertia (I_p)	0.21 Kgm^2
Wheel radius (r)	0.062 M	Dist from body's center of mass (l)	0.22 m
Wheel mass (M_w)	0.1 kg	Motor torque (K_m)	0.0335 N m/A
Body mass (M_p)	5.4 kg	Back EMF (K_e)	0.0435 V/(rad/s)
Wheel inertia (I_w)	0.0013 Kgm^2	Terminal resistance (R)	Ω

4 Designing of Controller

Any disturbance causes the pole on the cart to deviate from its upright position, and the role of the controller is to bring back the pole to its stable vertical position. The system specifications described above impose some constraints on the settling time of four seconds to regain its stable cart position and to the deviation in the angle of pole is ± 0.6 to ± 0.8 rad.

4.1 Linear Quadratic Regulator

The difference between the linear quadratic regulator and Gaussian controller is operated along with Kalman's filter, and this gives it an edge over LQR which is used for both time-variant as well as time-invariant systems.

$$\dot{y} = Ay + Bu \quad (11)$$

$$z = Cy \quad (12)$$

LQR and Kalman filter together formed linear quadratic controller, which can be used for linear time invariant and time variant. In this chapter, we design LQG controller and found satisfactory results, as the optimal controllers are distinguished from normal controllers in terms of performance measure, *J*, given by following quadratic relationship.

$$J = \frac{1}{2}\int_0^\infty \left(y^T Q y + U^T R U\right)\mathrm{d}t \tag{13}$$

where *Q* and *R* are state variable and control vector weighting matrix

$$U = -Ky \tag{14}$$

where '*K*' is the gain matrix given by

$$K = R^{-1} B^T P \tag{15}$$

P can be calculated using matrix Riccati equation

$$A^T P - PA + Q - PBR^{-1}B^T P = 0 \tag{16}$$

$$\dot{y}(t) = Ay(t) + BU + L(z(t) - Cy(t) - DU) \tag{17}$$

$$U = -Ky(t) \tag{18}$$

$$\dot{y}(t) = (A - BK - LC + LDK)y(t) + Lz(t) \tag{19}$$

5 Result and Discussion

All the simulation work has been done on MATLAB software, and we found that the impulse response using LQG controller is better what we have gone through. Settling time is 2.30 s position and the pendulum angle is 1.78 s and their peak amplitude is 0.166 rad. When the system based on LQG controller starts to stabilize, it will be easily stabilized in three seconds, when we produce disturbance system, it will take ten seconds to stabilize itself again. LQG controller found good when disturbance is produced in it and can stabilize itself quickly.

For describing the CMPC controller, Laguerre function is used. Different values of *P* and *N* are shown here in the given below.

6 Conclusion

Mathematical modeling of LQG and MPC controller is designed well. Both the controllers are designed using state space, and comparison has been shown. For the simulation work, we use MATLAB software, and all the simulated results have been produced well. System impulse response using LQG controller and their disturbance rejection capability with the plant have been produced. Design controller is found well, and Laguerre function with second- and fourth-order coefficient has been produced and found satisfactory. Controller designed for the two-wheeled self-balancing mobile robot is found easily used for the system.

References

1. Peng, X., Jia, M., He, L., Yu, X., Lv, Y.: Fuzzy sliding mode control based on longitudinal force estimation for electro-mechanical breaking systems using BLDC motor. CES Trans. Electric. Mach. Syst. **2**(1), 142–151 (2018)
2. Ghany, M.A., Bensenouci, A., Bensenouci, M.A., Syed-Ahmad, M.N.: Fuzzy fractional-order PID Tuned via relative rate observer for the Egyption load frequency regulation. In: Twentieth International Middle East Power Systems Conference (MEPCON), pp. 103–109. Cairo University, Egypt (2018)
3. Mahaboob Basha, S., Swarup, A.: Disturbance observer based fuzzy PID and robust PID autopilot design for bank-to-turn missile. In: 8th IEEE International Conference on Computing, Communication and Networking Technologies (ICCCNT), pp. 103–109
4. Huang, J., Ri, M.H., Wu, D., Songhyok: Interval type-2 fuzzy logic modeling and control of a mobile two-wheeled inverted pendulum. IEEE Trans. Fuzzy Syst. 1–9 (2017)
5. Kim, S., Kwon, S.: Nonlinear optimal control design for underactuated two-wheeled inverted pendulum mobile robot. IEEE/ASME Trans. Mechatron. 1–6 (2017)
6. Qiu, J., Hou, Z., Wang, W., Zhang, G., Li, Y., Feng, W., Han, C.: Two-wheeled self balancing robot modeling and nonlinear control. In: 14th IEEE International Conference on Ubiquitous Robots and Ambient Intelligence (URAI), pp. 734–739. IEEE (2017)
7. Khatoon, S., Chaturvedi, D.K., Hasan, N., Istiyaque, M.: Optimal control of double inverted pendulum by linearization technique. In: 2017 International Conference on Multimedia, Signal Processing and Communication Technologies (IMPACT), pp. 123–127. IEEE (2017)
8. Zheng, Z., Teng, M.: Modeling and decoupling for two-wheeled self-balancing robot. In: Chinese Control and Decision Conference (CCDC), pp. 5263–5269. IEEE (2016)
9. Zad, H.S., Ulasyar, A.: Adaptive control of self-balancing two-wheeled robot system based on online model estimation. In: International Conference on Electrical and Electronics Engineering (ELECO), pp. 876–880 (2017)
10. Jamil, O., Jamil, M., Ayaz, Y., Ahmad, K.: Modelling control of a two wheeled self balancing robot. In: International Conference on Robotics and Emerging Allied Technologies in Engineering (ICRESATE), vol. 10, pp. 191–199 (2014)
11. Nawawi, S.W., Ahmad, M.N., Osman, J.H.S.: Real-time control of a two-wheeled inverted pendulum mobile robot. World Acad. Sci. Eng. Technol. **39**, 214–220 (2008)
12. Nasir, A.N.K., Ahmad, M.A., Raja Ismail, R.M.T.: The control of a highly non-linear two-wheels balancing robot: a comparative assessment between LQR and PID-PID control schemes. World Acad. Sci. Eng. Technol. Int. J. Mech. Mechatron. Eng. **4**(10), 942–947 (2010)
13. Kun, L., Ming, B., Yuhua, N.: Two-wheel self-balanced car based on Kalman filtering and PID algorithm. In: IEEE 18th International Conference on Industrial Engineering and Engineering Management (IE&EM), vol. Part 1, pp. 281–285 (2011)

14. Pathak, K., Franch, J., Agrawal, S.K.: Velocity and position control of a wheeled inverted pendulum by partial feedback linearization. IEEE Trans. Robot. **21**(3), 505–513 (2005)

Microstructure and Mechanical Properties of Al–Cu–Mg Alloy Produced by Non-contact Ultrasonic Casting Route

Sneh Suman, Mukkollu Sambasiva Rao, Amitesh Kumar, and K. K. Singh

1 Introduction

Aluminium alloys are recently used in advanced applications because of their combination of high strength, low density, durability, machinability, availability and comparable cost compared to current applicable materials [1]. High specific strength, corrosion resistance and formability, Al–Cu–Mg alloys are potential materials for a number of industrial applications [2]. In conventional casting, aluminium–copper alloys solidify with the dendritic structure [3]. Generally, Al–Cu–Mg alloys solidify with dendrite–cellular structures and the dendrites of aluminium solid solution as the primary phase and the inter-dendrite space filled with eutectic mixture [4]. Castings with non-dendritic structure are better in mechanical properties than dendritic microstructure. The non-dendritic structure is obtained by inoculation, rapid solidification, agitation during solidification and other techniques [5–7]. Grain size can be reduced and compacted by mechanical mould vibration of hypoeutectic aluminium based alloy [8]. The mechanical mould vibrations increasingly improve the mechanical properties of the alloy [9, 10].

Based on literature review, it was observed that maximum works have been done to refine grain structure by rapid solidification, inoculation, mould vibration and semi-solid casting. However, very few researches have been carried out in the field of non-contact ultrasonic casting. In this work, non-contact ultrasonic casting of Al–4wt%Cu–2wt%Mg was carried out. Microstructure and mechanical properties of non-contact ultrasonic castings were compared with conventional castings. In ultrasonic casting, refined and better mechanical properties were obtained than the conventional casting.

S. Suman (✉) · M. S. Rao · A. Kumar · K. K. Singh
Foundry Technology Department, National Institute of Foundry and Forge Technology, Hatia, Ranchi, Jharkhand 834003, India
e-mail: Snehsuman891@gmail.com

M. Muzammil et al. (eds.), *Recent Advances in Mechanical Engineering*, Lecture Notes in Mechanical Engineering, https://doi.org/10.1007/978-981-15-8704-7_27

Fig. 1 Ultrasonic vibration set-up

2 Experimental Procedure

2.1 *Preparation of Master Alloy*

Master alloy of Al–4wt%Cu–2wt%Mg was prepared by melting in an induction furnace using commercially pure Al, Cu and Mg. Chemical composition of above composition is confirmed by optical emission spectrometer.

2.2 *Ultrasonic Casting*

The experimental set-up of ultrasonic casting is shown in Fig. 1. It consists of a steel mould which was surrounded by sufficient water so that effective transmission of ultrasonic waves to liquid metal could take place. The ultrasonic vibration was allowed till solidification. 20 kHz frequency was set for ultrasonic vibration. Conventional casting was made in the same mould. Water of ultrasonic chamber was heated to 70 °C. Melting of master alloy was done in the induction furnace. The molten metal was transferred in ultrasonic vibration-assisted mould and conventional casting mould. Liquid metal was poured into mould at 700 °C. Cast samples were characterized for microstructure study and mechanical properties.

2.3 *Hardness*

Hardness measurements were performed on the polished samples. Hardness was measured by using the Brinell hardness tester. The indenter ball of 10 mm diameter

Fig. 2 Tensile bar specimen prepared by using milling machine

and 500 Kgf load was applied. Hardness at five locations was taken and finally averaged it out.

2.4 Tensile Testing

Tensile test samples were made by the machining as per the ASTM B557M-02 a standard. ASTM A536 method (EN1563) was employed for tensile testing; the machine used was Zwick/Roell Universal Testing Machine Model BTC-EXOPTIC.005 (50 KN), at IIT Patna. A loading rate of 0.05 mm/min was set for the tensile test of specimens; meanwhile, for adjusting and stabilizing the experimental set-up, a preload of 50 N was applied to the specimen. Prepared tensile specimens are shown in Fig. 2.

3 Results and Discussion

3.1 Microstructure Analysis

Microstructures were taken at by using a NEOPHOT 32 metallographic microscope on surface etched with Keller's reagent (20 ml HF (48%), 30 ml HCl, 50 ml HNO_3 and 190 ml water). Microstructures are shown in Fig. 3. Figure 3a shows microstructure of castings with ultrasonic casting route. Very fine structures and free from porosity

Fig. 3 Microstructure of **a** ultrasonic casting and **b** conventional casting

(a) Microstructure of Ultrasonic Casting

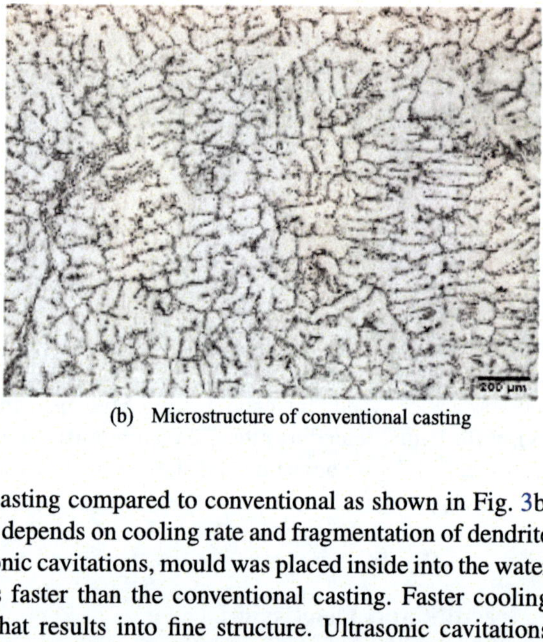

(b) Microstructure of conventional casting

were observed in ultrasonic casting compared to conventional as shown in Fig. 3b. Refinement of microstructure depends on cooling rate and fragmentation of dendrite or cellular structure. In ultrasonic cavitations, mould was placed inside into the water bath, where cooling rate was faster than the conventional casting. Faster cooling causes more under cooling that results into fine structure. Ultrasonic cavitations generate instantly huge shock waves dues to collapsing of bubbles. This shock wave fragments dendrites or cellular structure. Therefore, in ultrasonic casting, there is combined effect of cooling and vibration that causes grain refinement as observed in Fig. 3a.

Table 1 Hardness value of ultrasonic casting and conventional casting

Sample	Hardness (BHN)
Ultrasonic casting	64
Conventional casting	43

Table 2 Ultimate tensile strength of ultrasonic casting and conventional casting

Sample	UTS (MPa)
Ultrasonic casting	189
Conventional casting	148

3.2 Brinell Hardness

The Brinell hardness has been taken at different five locations on the polished surface. Average value was taken as the final results. Average hardness value is given in Table 1. The hardness of ultrasonic casting was obtained 64 BHN, while conventional casting was 43 BHN. Higher value of hardness in ultrasonic casting is due to grain fineness and least porosity comparing to conventional casting.

3.3 Ultimate Tensile Strength

The average ultimate tensile strength of ultrasonic casting and conventional casting is given in Table 2. Ultimate tensile strength of ultrasonic casting is higher than that of conventional casting due to finer microstructure of ultrasonic casting. Ultrasonic casting shows least porosity compared to conventional casting because ultrasonic vibration also helps in degassing. This may be also one of the causes of higher UTS.

4 Conclusions

Based on the above results and discussions, the following conclusions can be drawn:

- Ultrasonic mould vibration causes refinement of microstructure as well degassing due acoustic streaming in the molten metal and faster cooling as heat transfers from mould wall to water.
- Mechanical properties of casting can also be enhanced by ultrasonic casting due to refinement of microstructure and degassing of molten metal.
- Since casting was done non-contact type, i.e. in bath stead of probe, so bigger casting can be made with better mechanical properties compared to conventional casting. Therefore, ultrasonic casting will be an alternative casting method to make casting with minimal porosity and very fine microstructure.

References

1. Dursun, T., Soutis, C.: Recent developments in advanced aircraft aluminum alloys. Mater. Des. **56**, 862–871 (2014)
2. Salihu, S.A., Isah, A.: Polycarp evarastics: influence of magnesium addition on mechanical properties and microstructure of Al–Cu–Mg alloy. IOSR J. Pharm. Biol. Sci. (IOSR-JPBS) **4**(5), 15–20 (2012)
3. Olufemi, A.F., Ademola, I.S.: Effects of melt vibration during solidification on the mechanical property of Mg–Al–Zn alloy. Int. J. Metal. Eng. **1**(3), 40–43 (2012)
4. Komura, S., Horita, Z., Furukawa, M., Nemoto, M., Langdon, T.G.: An evaluation of the flow behavior during high strain rate super plasticity in an Al–Mg–Sc alloy. Metal. Mater. Trans. A **32A**, 707–716 (2001)
5. Muszka, K., Majta, J., Bienias, L.: Effect of grain refinement on mechanical properties of micro alloyed steels. Metall. Foundry Eng. **32**(2), 87–97 (2006)
6. Lee, Y., Dahle, A., Stjohn, D.: The role of solute in grain refinement of magnesium. Metal. Mater. Trans. A **31a**, 2895–2906 (2000)
7. Apelian, D.: Aluminum cast alloys: enabling tools for improved performance. North American Die Casting Association (2009)
8. Mondolfo, L.F.: Aluminium alloys: structure and properties. Butterworths, London, Boston (1976)
9. Girisha, H.N, Sharma, K.V.: Effect of magnesium on strength and microstructure of aluminium copper magnesium alloy. Int. J. Sci. Eng. Res. **3**(2), 1–4 (2012)
10. Kumar, R., Ansari, M.S., Mishra, S.S., Kumar, A.: Effect of mould vibration on microstructure and mechanical properties of casting during solidification. Int. J. Eng. Res. Technol. **3**(4), 90–92 (2014)

Performance Augmentation of Low-Temperature Sub-critical Organic Rankine Cycle Using First and Second Law-Based Analysis

Komal Madan and Omendra Kumar Singh

Nomenclature

T	Temperature
P	Pressure
M	Molecular Mass
$\tilde{n}$	Molar flow rate
$\hat{c}p$	Specific heat at constant pressure
η	Efficiency
$\dot{m}$	Mass flow rate
h	Specific enthalpy
Q	Heat
W	Work
ODP	Ozone depletion potential
GWP	Global warming potential

Subscript

cr	Critical temperature
NBP	Normal boiling point
i	Index
p	Pump
E	Expander
wf	Working fluid

K. Madan (✉) · O. K. Singh
Mechanical and Automation Dept, Indira Gandhi Delhi Technical University for Women, Kashmere Gate, Delhi 110006, India
e-mail: komal55_90@yahoo.co.in

M. Muzammil et al. (eds.), *Recent Advances in Mechanical Engineering*, Lecture Notes in Mechanical Engineering, https://doi.org/10.1007/978-981-15-8704-7_28

s	Saturation
Exe	Exergy
G_{in}	Gas in
G_{ex}	Gas exit ambient
o	Ambient

1 Introduction

The combined cycle power plant (CCPP) is the only solution to reduce fossil fuel consumption and to utilize the industrial emissions to meet the energy demand of general public. A number of thermodynamic cycles such as Kalina cycle (KC) [1], trilateral flash cycle (TFC) [2], organic Rankine cycle (ORC) [3, 4] and supercritical Rankine cycle (SRC) were studied and compared [5]. Among the all other thermodynamic cycles, the organic Rankine cycle is the highly efficient cycle to operate on low-temperature heat of biomass, solar, geothermal, waste heat of engine and other sources for generating an additional power to meet the increased demand and reduce the high-temperature emissions into the environment. The use of organic working fluid instead of water differs the conventional Rankine cycle from ORC. Looking at its advantages, this cycle is more reliable and chosen for the heat source temperature greater than or equal to 100 °C [3]. The ORCs are classified on the basis of the turbine inlet pressure as subcritical and supercritical cycles [4]. The selection of working fluid based on its thermo-physical properties, heat source condition and the expander design is the roots for performance enhancement of ORCs and is a challenge for researchers. Based on the slope of saturated vapour curve on temperature–entropy (T-s) diagram, the organic fluids can be classified as dry, wet and isentropic type [4]. Wang et al. [6] investigated the performance of nine working fluids for engine waste heat recovery application using a thermodynamic model built in MATLAB together with REFPROP and compared them based on its thermal efficiencies, exergy destruction and safety and environmental impacts. Uusitalo et al. [7] recommended fluorocarbons and low critical temperature hydrocarbon for low-temperature ORC system and siloxanes and high critical temperature hydrocarbon for high-temperature ORC applications. Although an immeasurable research is done on ORCs over wide range of application, this paper focuses on screening of seven organic working fluids based on its physical, thermal and environmental properties and evaluates the thermal efficiency, net power output and exergy efficiency at different conditions. The conclusion is done based on the results obtained and validated with the finding.

Table 1 Thermophysical properties of pure working fluid

Substance	Molecular structure	M. Mass	T_{cr} (K)	P_{cr} (Bar)	T_{NBP} (K)	ODP	GWP	Type
R245fa	CF_3-CH_2-CHF_2	134.05	427.2	36.40	288.05	0	950	Dry
R245ca	CF_3-CHF-CH_2F	134.05	447.57	39.25	298.28	0	610	Dry
R236ea	CF_3-CHF-CHF_2	152.04	412.44	35.02	279.34	0	1200	Dry
R236fa	CF_3-CH_2-CF_3	152.04	398.72	32.00	271.75	n.a	n.a	Dry
R227ea	CF_3-CHF-CF_3	170.02	374.89	29.29	256.81	n.a	n.a	Dry
R134a	CF_3-CH_2-F	102.03	374.18	40.56	247.08	0	1430	Wet
Toluene	C_7H_8	92.14	591.8	41.3	383.75	0	0	Isentropic

2 Working Fluid Postulant

The thermo-physical and environmental properties of working fluid postulant have been listed in Table 1 taken from REFPROP [8].

3 System Description and Assumption

The subcritical ORCs consist of following components: organic vapour generator (OVG), high-pressure organic vapour expander (HP OVE), low-pressure organic fluid condenser (LP OFC) and organic fluid feed pump (OFFP) as shown in Fig. 1. The working theory of ORC is similar to conventional Rankine cycle as shown in T-s diagram (Fig. 2).

For calculating the cycle performance, several assumptions are made to conceive the ORCs as follows.

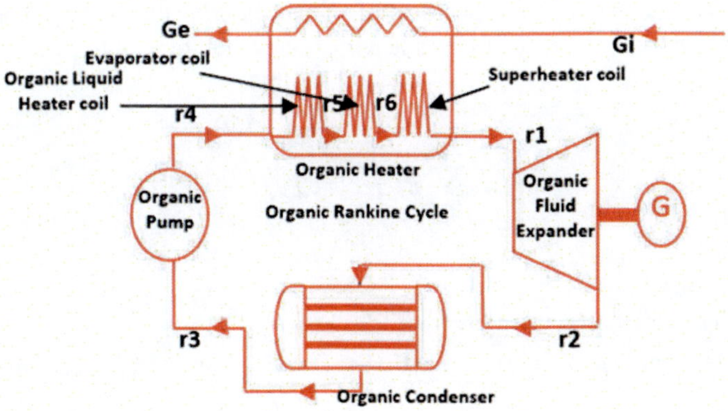

Fig. 1 Simplified ORC system

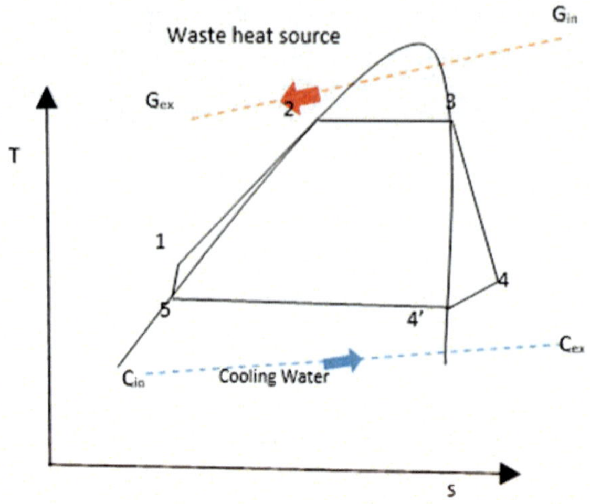

Fig. 2 T-s diagram of the ORC

- The analysis assumes system under steady-state condition.
- The heat losses and the drop in pressure in different heat exchangers and pipes are undervalued.
- The ambient temperature and pressure are 307.15 K and 100,900 Pa.
- The isentropic efficiency of turbine and pump is taken as 75% and 60%, respectively [7].

4 Modelling of ORC System

The heat source is the heat of waste gases coming from the stack of two heat recovery steam generators (HRSGs) with an average temperature of 371.15 K, and total heat drop of gases in the ORC vapour generator is 5012 kJ/s. The molar flow rate of components of waste gases and specification of heat entering the evaporator are given in Table 2. The total enthalpy of gases leaving the evaporator can have expressed as [9].

$$\sum_{i=1}^{i=5} \tilde{n}_i \int_{T_o}^{T} \hat{c}p_i \mathrm{d}T \quad (1)$$

Table 2 Molar flow rate of components of waste gases

Molar flow rate (kmol/s)	CO_2	H_2O	N_2	O_2	Ar
$\tilde{n}_1$	0.3285	0.8193	7.9743	1.5027	0.0954
$\tilde{n}_2$	0.3073	0.7663	7.4581	1.4054	0.0892

where $i =$ number of components of gases, $\tilde{n}_i$ is molar flow rate of components of gases, $\int_{T_o}^{T} \hat{c}p_i \mathrm{d}T$ is the molar physical enthalpy of each component of gases.

Process 5-1: The isentropic efficiency and rate of work done on pump is expressed as [10]

$$\eta_p = \frac{h_5 - h_{1s}}{h_5 - h_1} \tag{2}$$

$$W_p = \dot{m}_{wf} \cdot (h_1 - h_{5s}) \tag{3}$$

Process 1-3: The waste heat of exhaust absorbed by working fluid flowing in an organic vapour generator can be expressed as

Heat gain by ORC fluid = Heat loss by gases

$$Q_{\text{gases}} = m_{wf} \cdot (h_3 - h_1) \tag{4}$$

Process 3-4: The isentropic efficiency and rate of work done by organic expander can be expressed as [11]

$$\eta_E = \frac{h_3 - h_4}{h_3 - h_{4s}} \tag{5}$$

$$W_E = \dot{m}_{wf} \cdot (h_3 - h_4) \tag{6}$$

Process 4-5: The condensation of organic fluid in condenser can be expressed as [11]

$$Q_c = m_{wf} \cdot (h_5 - h_4) \tag{7}$$

4.1 Performance Parameters

The net power output can be expressed as [11]

$$P_{\text{net}} = W_E - W_p \tag{8}$$

The cycle efficiency can be expressed as [11]

$$\eta = \frac{W_E - W_p}{Q_{\text{gases}}} \tag{9}$$

The exergy efficiency can be expressed as [12]

$$\eta_{\text{Exe}} = \frac{W_E - W_p}{m_{wf} \cdot c_{pg}[T_{G,\text{in}} - T_{G,\text{ex}} - T_o \cdot Ln\frac{T_{G,\text{in}}}{T_{G,\text{ex}}}]} \quad (10)$$

5 Results and Discussion

The evaporator temperature varies from 75–85 °C and the condenser temperature from 20–45 °C with the difference of 5 °C. Figure 3 depicts the net power output with respect to critical temperature for different working fluid at an evaporator and condenser temperature of 85 °C and 20 °C, respectively. Figure 3 revealed for the same type of fluid there is increase in net power output with critical temperature arranged in the order of R134a(wet) > Toluene(isentropic) > R245ca(dry) > R245fa(dry) > R236ea(dry) > R236fa(dry) > R227ea(dry). Figure 4 shows the effect of net power output with condensation temperature at a given evaporator temperature. The maximum power output is obtained from R134a and minimum from R227ea with a power increase of 534.19 kW at 20 °C to 649.87 KW at 45 °C. If the condensation temperature increases, the specific enthalpy drop in the expander decreases and hence the net power output decreases. Figure 5 depicts the relationship between the exergy efficiency with evaporation temperature at a condenser temperature of 20 °C for different fluids. The maximum exergy efficiency is obtained for wet fluid at an evaporator temperature of 85 °C. Figure 6 depicts the condensation temperature effect on cycle efficiency for different working fluids at a given evaporation temperature of 85 °C. The cycle efficiency reduces with an increase in condensation temperature except for R134a. The maximum cycle efficiency among the dry and isentropic fluids is for Toluene which decreases from 12.153 to 7.735% when condensation temperature increases from 20–45 °C. Figure 7 depicts exergy efficiency for

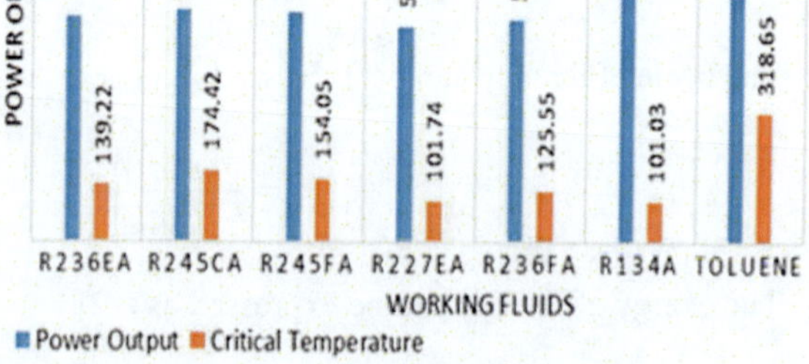

Fig. 3 Variation of Power output with critical temperature of organic fluids

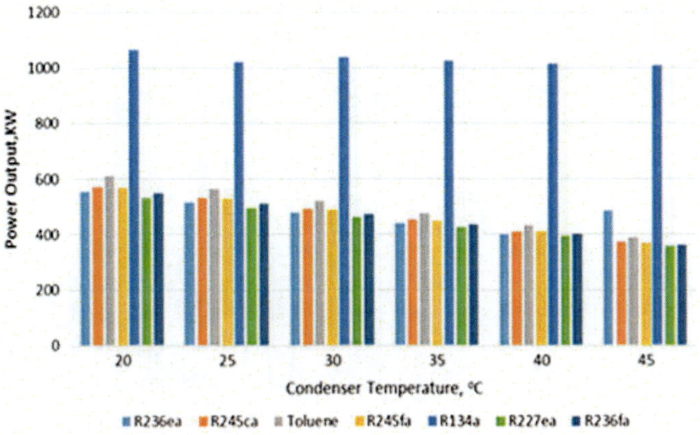

Fig. 4 Variation of power output at different condenser temperature

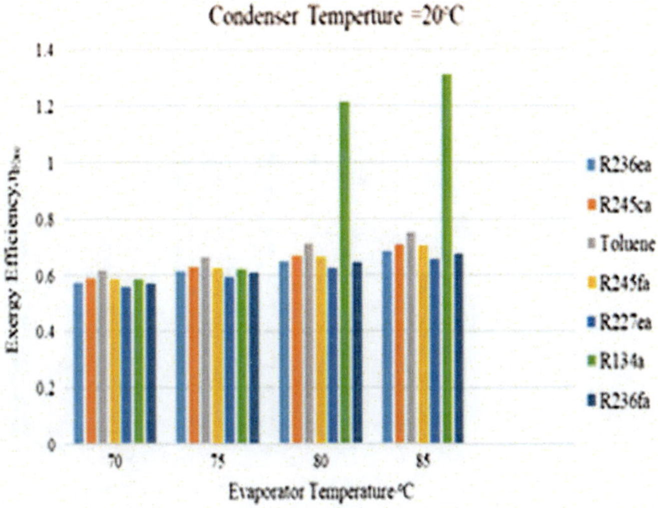

Fig. 5 Variation of exergy efficiency at different evaporator temperature

different condensation temperature. The maximum exergy efficiency is attained at a condensation temperature of 20 °C for toluene (75.282%) with a trend of decrease with increase in condensation temperature followed by R245ca > R245fa > R236ea > R236fa > R227ea. For working fluid R236ea, the exergy efficiency decreases first upto 40 °C and then increases with increase in condensation temperature. The heat of waste gases absorbed by organic fluid decreases when the condenser temperature increases due to increase in the specific enthalpy at evaporator inlet which results in cut in network output and hence cycle efficiency and exergy efficiency decrease [12].

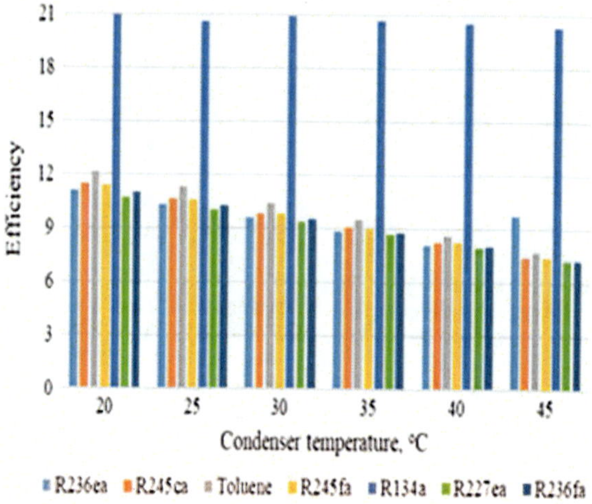

Fig. 6 Variation of efficiency at different condenser temperature

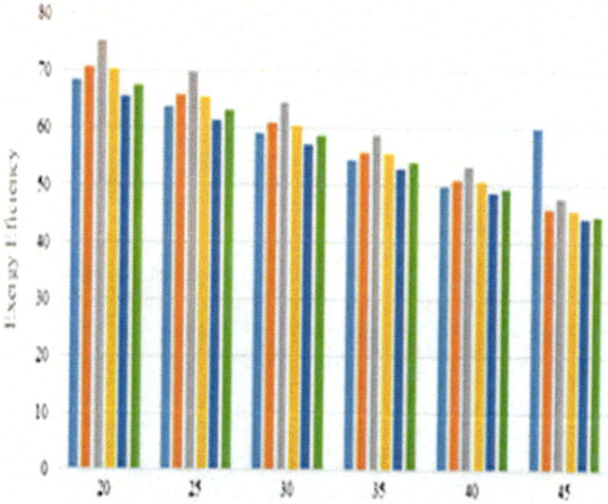

Fig. 7 Exergy efficiency for different condensation temperature

Figures 3 and 4 show power output with condensation temperature for different working fluids, respectively.

Figures 5 and 6 show the effect of evaporator temperature on exergy efficiency and thermal efficiency for different condensation temperature, respectively.

6 Conclusion

- Among the dry fluids, the maximum efficiency and highest power are achieved with R245ca ($T_e = 85$ °C and $T_c = 20$ °C) due to high critical temperature and pressure among them.
- The mass flow rate of working fluids among the dry fluids is maximum for R227ea with higher temperature of 70 °C and lower temperature of 45 °C due to high evaporator pressure for the same heat source condition.
- The cycle efficiency decreases with rise in condenser temperature at an evaporator temperature of 85 °C and is maximum at 20 °C.
- R245ca is considered as the highly efficient fluid among the studied fluids with a maximum power output of 762.7622 KW for winters.

Acknowledgement This work was sponsored by Indira Gandhi Delhi Technical University for Women under IGDTUW Research Fellowship Scheme (JRF/SRF).

References

1. Singh, O.K., Kaushik, S.C.: Reducing CO2 emission and improving exergy based performance of natural gas fired combined cycle power plants by coupling Kalina cycle. Energy **55**, 1002–1013 (2013a)
2. Lai, N.A., Fischer, J.: Efficiencies of power flash cycles. Energy **44**(1), 1017–1027 (2012)
3. Wang, Z.Q., Zhou, N.J., Guo, J., Wang, X.Y.: Fluid selection and parametric optimization of organic Rankine cycle using low temperature waste heat. Energy **40**(1), 107–115 (2012)
4. Saleh, B., Koglbauer, G., Wendland, M., Fischer, J.: Working fluids for low-temperature organic Rankine cycles. Energy **32**(7), 1210–1221 (2007)
5. Bao, J., Zhao, L.: A review of working fluid and expander selections for organic Rankine cycle. Renew. Sustain. Energy Rev. **24**, 325–342 (2013)
6. Wang, E.H., Zhang, H.G., Fan, B.Y., Ouyang, M.G., Zhao, Y., Mu, Q.H.: Study of working fluid selection of organic Rankine cycle (ORC) for engine waste heat recovery. Energy **36**(5), 3406–3418 (2011)
7. Uusitalo, A., Honkatukia, J., Turunen-Saaresti, T., Grönman, A.: Thermodynamic evaluation on the effect of working fluid type and fluids critical properties on design and performance of organic Rankine cycles. J. Clean. Prod. **188**, 253–263 (2018)
8. Lemmon, E.W., Huber, M.L., McLinden, M.O.: NIST standard reference database 23, reference fluid thermodynamic and transport properties (REFPROP), version 9.0, National Institute of Standards and Technology. R1234yf. fld file dated December, 22, 2010
9. Singh, O.K., Kaushik, S.C.: Variables influencing the exergy based performance of a steam power plant. Int. J. Green Energy **10**(3), 257–284 (2013b)
10. He, C., Liu, C., Gao, H., Xie, H., Li, Y., Wu, S., Xu, J.: The optimal evaporation temperature and working fluids for subcritical organic Rankine cycle. Energy **38**(1), 136–143 (2012)
11. Xu, W., Deng, S., Zhao, L., Su, W., Zhang, Y., Li, S., Ma, M.: How to quantitatively describe the role of the pure working fluids in subcritical organic Rankine cycle: a limitation on efficiency. Energy Convers. Manage. **172**, 316–327 (2018)
12. Sun, W., Yue, X., Wang, Y.: Exergy efficiency analysis of ORC (organic Rankine cycle) and ORC-based combined cycles driven by low-temperature waste heat. Energy Convers. Manage. **135**, 63–73 (2017)

Dynamic Analysis and Stability of Turbocharger Rotor Supported with Auxiliary Gas Foil Bearing Under Exhaust Emission Conditions

Rajasekhara Reddy Mutra, J. Srinivas, D. Mallikarjuna Reddy, and Gunji Balamurali

1 Introduction

Turbochargers have high-speed rotors with compressor and turbine disks on either end along with radial and axial bearing supports. The main reasons for the failure of turbochargers are due to casing—impeller rub and bearing wear events, etc. To avoid these failures, a reliable design of the rotors is necessary. In addition to high speeds, the rotor is subjected to high thermal heat flux exclusively on the turbine side and cavitated flow conditions on the compressor side. More commonly, turbocharger rotors are supported on floating ring bearings. Several recent studies [1–3] focused on the oil-free bearings for turbocharger rotors, and gas foil bearings are found to be successful alternatives in this regard. In the construction of the gas foils, various layers of sheet foils are used. The performance of the gas foil depends on the support structure. The foil structure is commonly consists of bump strips and the smooth foil rest on top of these strips [4]. The contamination is one of the major disadvantage in oil bearings to overcome this, and gas foil bearing technology is developed. However, due to their relatively low damping capabilities, these cannot be used alone in critical applications.

Often the temperatures at the exhaust manifolds are very high. Exhaust gas emission conditions are very essential in the analysis. Liang et al. [5] investigated a 3D thermo hydrodynamic model of the TC supported on the semi-floating ring bearings. The expansion of solid parts and heat conduction greatly influenced the temperature fields and oil film clearances [6]. Hoepke et al. [7] explained the conjugate heat transfer in TC components such as compressor, turbine and bearing housing. Li et al. [8] explained the thermo hydrodynamic analysis of the TC rotor floating ring bearing

R. R. Mutra (✉) · D. Mallikarjuna Reddy · G. Balamurali
Vellore Institute of Technology, Vellore, Tamil Nadu 632014, India
e-mail: rajmech03@gmail.com

J. Srinivas
National Institute of Technology, Rourkela, Odisha 769008, India

M. Muzammil et al. (eds.), *Recent Advances in Mechanical Engineering*, Lecture Notes in Mechanical Engineering, https://doi.org/10.1007/978-981-15-8704-7_29

system with numerical and experimental simulations. The heat transfers in the lubrication system, and the solid parts play a vital role. The thermal expansion influences the oil film clearances and temperature of the ring changes for large amplitude as rotor speed varies. More recent works (e.g., [9]) concerned with the use of auxiliary bearing for control of dynamic response of the rotors.

Present work focuses on the dynamic modeling and simulation studies of an on-board turbocharger rotor supported over two floating ring radial bearings with an auxiliary gas foil bearing. Initially, finite element method is used to model the rotor bearing system. The external gas excitation loads are modeled as axial harmonic forces expressed in terms of force magnitude, phase and number blades. By varying the auxiliary gas foil bearing parameters, the dynamic response is significantly affected, and this is used as an effective control technique in certain operating conditions of rotor.

2 Mathematical Model of TC

Both compressor and turbine disks are considered as rigid members, whereas the rotor is considered as flexible member. Each element has eight degrees of freedom (DOF). Figure 1 shows the FE model used in the present work.

The final equations of motion of the rotor are expressed as:

$$\mathbf{M}\ddot{q} + [\mathbf{C} + \Omega\mathbf{G}]\dot{q} + \mathbf{K}q = F \tag{1}$$

Here, the assembled mass, damping, gyroscopic and stiffness matrices are represented as **M**, **C**, **G** and **K**. Also, q is the displacement vector, and the unbalance and gravity force is denoted with F at the disks.

Fig. 1 FE model used

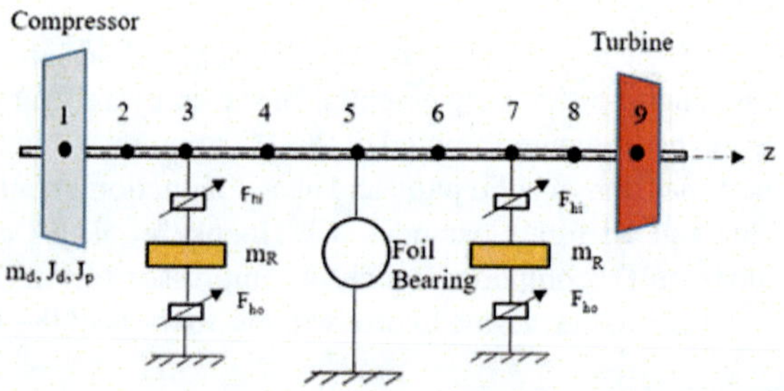

2.1 Bearing Forces

The Capone model in Refs. [10–12] is referred in this article to model the FRB forces. The inner and outer component forces from the pressure distribution are expressed as:

$$\begin{Bmatrix} F_{ix} \\ F_{iy} \end{Bmatrix} = \mu_i(\Omega_j + \Omega_r)R_j L_i \left(\frac{R_j}{C_1}\right)^2 \left(\frac{L_i}{2R_j}\right)^2 \begin{Bmatrix} f_{ix} \\ f_{iy} \end{Bmatrix} \tag{2}$$

$$\begin{Bmatrix} F_{ox} \\ F_{oy} \end{Bmatrix} = \mu_o \Omega_r R_{ro} L_o \left(\frac{R_o}{C_2}\right)^2 \left(\frac{L_o}{2R_{ro}}\right)^2 \begin{Bmatrix} f_{ox} \\ f_{oy} \end{Bmatrix} \tag{3}$$

where f_{ix},f_{iy} and f_{ox},f_{oy} indicates the journal and bearing component forces, respectively. The forces of GFB forces are evaluated from instantaneous bearing pressures predicted from finite difference method, and the force components are obtained according to the following equations [2].

$$F_x = p_a R^2 \int_0^{2\pi} \int_0^{L/R} (p - 1) \sin\theta \mathrm{d}\theta \mathrm{d}z \tag{4}$$

$$F_y = -p_a R^2 \int_0^{2\pi} \int_0^{L/R} (p - 1) \cos\theta \mathrm{d}\theta \mathrm{d}z \tag{5}$$

2.2 Exhaust Emission Forces

The exhaust gas excitation is considered as the external load in the axial direction at the turbine side. Under first blade frequency load excitation, the longitudinal primary resonance occurs. Here the harmonic excitation force is

$$F_g = F_0 \sin(n_b \Omega t + \varphi) \tag{6}$$

where F_0 is the magnitude, $n_b\Omega$ is first blade frequency of rotor,ϕ is the phase angle and n_b is the number of blades on turbine wheel. Thus, it is affected by number of impeller blades, speed and magnitude. This axial load along with radial Muszynska forces together constitutes one ideal representation of gas excitation.

3 Results and Discussion

Dimensional data of the system is selected from Ref. [13]. The validation of the present model was explained in the previous publication [14]. Initially, without considering an auxiliary gas foil bearing, the system response is identified. Figure 2 shows the time history and FFT of the rotor system without an auxiliary gas foil bearing at a rotor speed of 15,000 rpm.

It is identified that there are three dominant peaks obtained: First two are the sub-synchronous frequencies due to oil whirl/whip, whereas the third component peak is speed dependent. From time history, it is observed that the system is in periodic in nature. Further, the influence of the auxiliary GFB in the system is studied. The time histories and FFT of the system with auxiliary GFB in the system at the same rotor speed of 15,000 rpm are shown in Fig. 3.

The three peak amplitudes without auxiliary bearing in the system are found as 2.669×10^{-6} m, 4.306×10^{-6} m and 5.944×10^{-6} m, respectively. Whereas with auxiliary bearing in the system, the two peak amplitudes are found to be 3.95 $\times 10^{-6}$ and 8.01×10^{-6} m. It is observed that the first sub-synchronous peak is disappearing and the amplitude of the second sub-synchronous peak is also reducing with introducing the GFB in the system. The auxiliary gas foil bearing is therefore influencing the response of the system. Further, at two different rotor speeds ($N =$ 10,000 rpm and 25,000 rpm), the frequency response of the rotor system is predicted which is shown in Fig. 4. It is observed that the frequency of dominant peak is dependent on the rotor speed.

Further, keeping the floating ring bearings parameters constant, the response of the system is obtained by varying the GFB parameters. Figure 5 shows the FFT of the system at right bearing with different gas foil bearing clearances.

It is identified that, as the bearing clearance increases, the amplitude is increasing and the fundamental frequency of the system is also increasing.

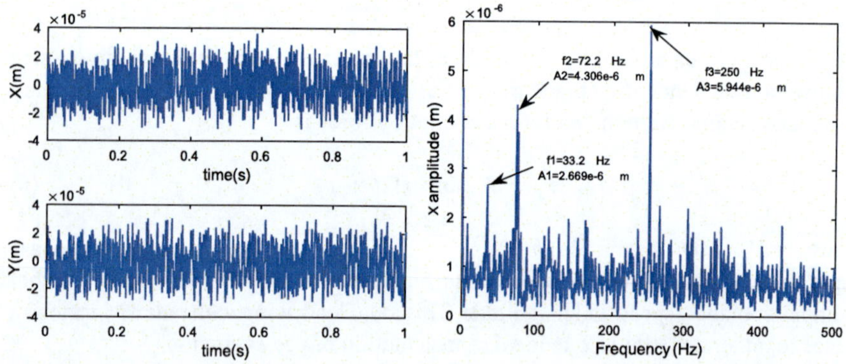

Fig. 2 Time and frequency response at right floating ring bearing without auxiliary gas foil bearing in the system

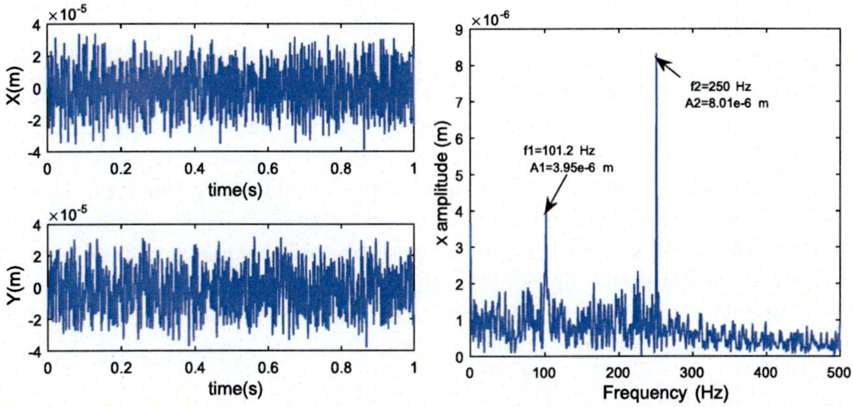

Fig. 3 Time and frequency response at right floating ring bearing with auxiliary GFB in the system

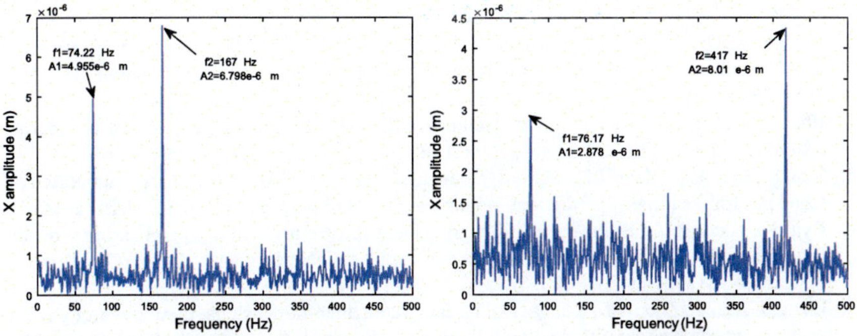

(a)N=10000 rpm (b) N=25000 rpm

Fig. 4 Frequency response at right floating ring bearing with auxiliary GFB in the system at two different rotor speeds

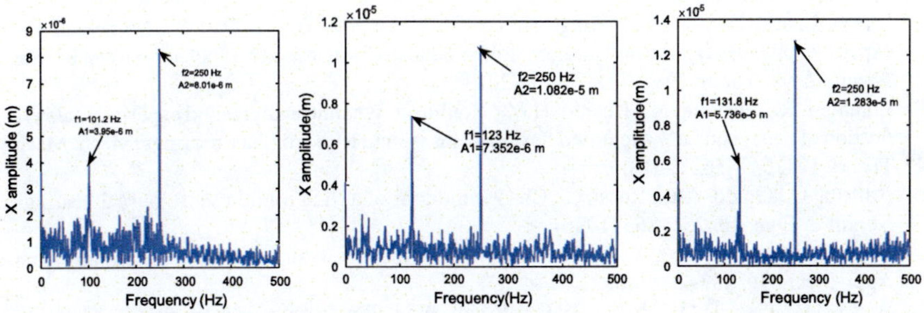

(a)0.0318 mm (b) 0.0418mm (c) 0.0518mm

Fig. 5 Frequency response at right floating ring bearing with various auxiliary GFB clearances

4 Conclusion

Dynamic analysis and stability studies of the turbocharger rotor system with an auxiliary GFB under exhaust gas excitations have been presented. The rotor model was developed with FEM, and the dynamic response of the system due to unbalance, gravity and exhaust gas excitation was obtained. The influence of the GFB parameters on the system response was studied. It was concluded that the auxiliary GFB influences the response of the system considerably and acts like amplitude control mechanism. Further studies are required to know the effects of GFB replacing one of the failed floating ring bearings. The optimum location and parameters of the GFB can also be predicted as future scope.

References

1. Hassan, M.F.B., Bonello, P.: A new modal-based approach for modelling the bump foil structure in the simultaneous solution of foil-air bearing rotor dynamic problems. J. Sound Vib. **396**, 255–273 (2017)
2. Bhore, S.P., Darpe, A.K.: Nonlinear dynamics of flexible rotor supported on the gas foil journal bearings. J. Sound Vib. **332**, 5135–5150 (2013)
3. Sim, K., Lee, Y.B., Kim, T.H.: Rotordynamic analysis of an oil-free turbocharger supported on lobed gas foil bearings—predictions versus test data. Tribol. Trans. **57**, 1086–1095 (2014)
4. Ryu, K., Ashton, Z.: Bump-type foil bearings and flexure pivot tilting pad bearings for oil-free automotive turbochargers: highlights in rotordynamic performance. J. Eng. Gas Turbines Power. **138**, 042501–042510 (2015)
5. Liang, F., Li, Y., Zhou, M., Xu, Q., Du, F.: Integrated three-dimensional thermohydrodynamic analysis of turbocharger rotor and semifloating ring bearings. J. Eng. Gas Turbines Power. **139**, 082501–082510 (2017)
6. Pinkus, O.: Thermal Aspects of Fluid Film Tribology, ASME Press (1990)
7. Hoepke, B., Vieweg, M., Pischinger, S.: Numerical analysis of energy flow paths in exhaust gas turbochargers by means of conjugate heat transfer. J. Eng. Gas Turbines Power. **139**, 061901–061909 (2017)
8. Li, Y., Liang, F., Zhou, Y., Ding, S., Du, F., Zhou, M., Bi, J., Cai, Y.: Numerical and experimental investigation on thermohydrodynamic performance of turbocharger rotor-bearing system. Appl. Therm. Eng. **121**, 27–38 (2017)
9. Ebrahimi, R., Ghayour, M., Khanlo, H.M.: Nonlinear dynamic analysis and experimental verification of a magnetically supported flexible rotor system with auxiliary bearings. Mech. Mach. Theory **121**, 547–562 (2018)
10. Adiletta, G., Guido, A.R., Rossi, C.: Chaotic motions of a rigid rotor in short journal bearings. Nonlinear Dyn. **10**, 251–269 (1996)
11. Ying, G., Meng, G., Jing, J.: Turbocharger rotor dynamics with foundation excitation. Arch. Appl. Mech. **79**, 287–299 (2009)
12. Wang, J.L., Cao, D.Q., Huang, W.H.: A new fluid film force model of elliptical bearing: Modelling and case studies. Proc. Inst. Mech. Eng. Part J J. Eng. Tribol. **224**,595–608 (2010)
13. Bonello, P.: Transient modal analysis of the non-linear dynamics of a turbocharger on floating ring bearings. Proc. Inst. Mech. Eng. Part J J. Eng. Tribol. **223**, 79–93 (2009)
14. Mutra, R.R., Srinivas, J.: Comparative studies on the dynamic performance of high speed turbocharger rotor supported on oil-free bearings versus conventional floating ring systems. In: Proceedings of ASME 2017 Gas Turbine India, vol. 2, pp. 1–8 (2017)

Optimization of Hardness Properties of Magnesium-Based Composites by Using Taguchi Method

Sakshi Singh and Nathi Ram Chauhan

1 Introduction

In the automobile sector, minimum cost and lightweight are some key issues. Exhaustive exercises have been performed by manufacturers to produce lightweight automotive parts. Meantime, purchasers' interest increases for comfort, interior design, safety, and navigations [1]. To justify these demands, automotive manufactures are switch to light materials and composites as purchasers' demands. After aluminum as light metal, magnesium is the correct preference to replace Al in automobile parts because of its low density, i.e., 1.72 g/cm^3. Valve covers, intake manifolds, transmission, and alternator housings are some capable components that can be taken over with Al [2]. Despite, Mg and its alloys are efficient in minimizing the mass of automobile parts, mechanical features of Mg, and its alloys that include immense strength, and hardness should be refined [3]. Many types of research have been executed to reach these conditions such as design, material, and processing modifications [4, 5].

Synthesizing of magnesium metal matrix composite (MgMC) by stir casting with the inclusion of reinforcements is one of the methods to improvise the mechanical characteristics. Several authors have studied carbide particulates that increase the mechanical characteristics of magnesium like graphite, TiC, SiC, B_4C, and alumina [6–10]. Silicon carbide is a hard carbide material which is used for synthesizing of MgMC [11]. Some researchers also have used TiC into the synthesizing of MgMC as reinforcements [12, 13]. The destructive properties of MgMC like TiC particles in MgMC increased [14, 15].

In order to minimize the high cost of reinforced materials, composites would be combined with graphite to common reinforcements such as alumina and silicon carbide to make it hybrid composites. This results high dislocation density to Mg-based composites and improves mechanical properties also [16]. It is also reported

S. Singh (✉) · N. R. Chauhan
MAE Department, Indira Gandhi Delhi Technical University for Women, New Delhi, Delhi, India
e-mail: singh.sakshi0408@gmail.com

M. Muzammil et al. (eds.), *Recent Advances in Mechanical Engineering*, Lecture Notes in Mechanical Engineering, https://doi.org/10.1007/978-981-15-8704-7_30

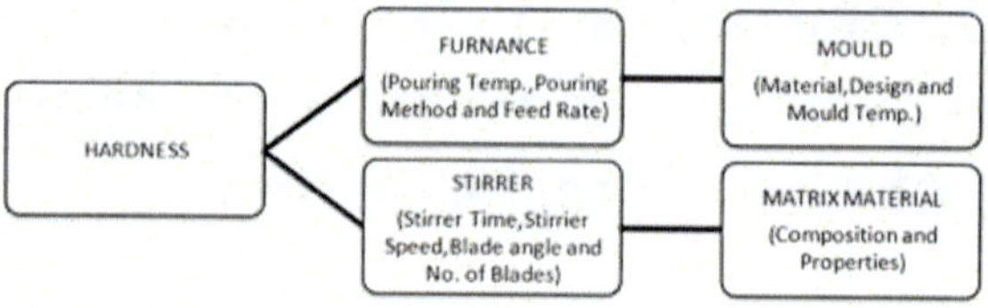

Fig. 1 Different parameters used to study the hardness effect of MgMC

that the weight percentage of graphite is low, then Mg-based composite shows low hardness as well as tensile and compressive strength [17–20]. The addition of graphite with silicon carbide in Mg-based composite plays a key role in improving the macro and microhardness properties of composites [21, 22]. The same trend also described improved tensile and hardness properties of Mg-based composites where B_4C and graphite are reinforced materials [23].

In this study, Mg with (B_4C + Al) hybrid composite has been processed. Also, evaluations of mechanical properties have been done by optimizing through Taguchi method.

2 Methodologies

Pure Mg ingot is a matrix material, with B_4C and Al as reinforcements. Mg-matrix has been blended in a crucible (made of steel) at 650 °C. Further, the B_4C and Al (reinforcements) have been added in different proportions. During this liquefying process, stirring has been performed.

MgMC synthesizing by VSP-SC has been associated with different specifications like furnace, mold, and composition. The specifications (stirrer, mold, furnace, and matrix alloy) that affect the Brinell hardness of MgMC during the VSP-SC process are shown in Fig. 1. The weight percentage of MgMC composition, liquefying temperature, stirring speed, and stirring time, i.e., processing parameters, have been preferred by Taguchi design where parameters have different levels as presented in Table 1.

3 Results and Discussion

The hardness values of MgMCs are measured by hardness number, i.e., BHN, and the indenter in Brinell hardness tester machine. Meanwhile, each of the experimental conditions is given in Table 2. Then, analysis of variance has been performed to investigate the factors which significantly support the Brinell hardness of MgMC. The average Brinell hardness values and the S/N ratio values are given in Table 3. In order to achieve the best Brinell data of MgMCs, "Larger is better" has been applied. Based on the investigation D3, C3, B2, and A3 values, the optimum condition of

Table 1 Processing parameters with their levels and elements for the stir casting process

Elements	Level		
	1	2	3
Composition of MgMC(A)	Mg + 5% wt. B_4C + 0% wt. Al	Mg + 0% wt. B_4C + 5% wt. Al	Mg + 5% wt. B_4C + 5% wt. Al
Stirring speed, rpm (B)	400	450	500
Stirring time, min (C)	9	12	15
Molten temp. of MgMC,°C (D)	750	800	850

hardness value has been concluded. It is observed that the increase in hardness value increases proportionally with the increased in the Element A level.

Table 4 shows the ANOVA results. Element A and element B improve the Brinell hardness. However, element C and element D indicate the minimal response to the hardness of MgMC. Therefore, great origination is observed in element D during the VSP-SC process of MgMCs that cannot be eliminated.

The 9.98% contribution error has been obtained as given in Table 4. This means that such a percentage factor affects all significant factors in the evaluation. Because of these factors in the adopted model, the percentage factor value must be ≤50%.

Equally, analysis of variance shows that element A and element B have a significant input to the hardness as given in Table 4. For the optimum mean surface, the confidence interval for hardness value is: 58.572 ± 1.30 kgf/mm^2, which equal to = 56.795 to 59.755 kgf/mm^2.

From the Taguchi method (by experimental analysis), the mixing of B_4C plays a key point in the hardness of MgMCs. Composite having the highest B_4C weight percentage shows the maximum hardness. Another element that devotes to the maximum hardness value is the 450 rpm stirring speed. It is observed that when stirring speed is more than 450 rpm, hardness value degenerated. It is because of the slight formation of porosity in MgMCs because of the blade (stirrer) when speed rose (Fig. 2).

4 Conclusions

Fabrication of MgMCs using the VSP-SC process has been successfully conducted. The Brinell hardness of MgMCs has been evaluated by the L9 orthogonal method. ANOVA evaluates that two processing factors (i.e., composition of MgMC and speed during stirring) influence the Brinell hardness of MgMCs prominently. The best level is observed at A3 and B2. As per ANOVA, element A (MgMC weight percentage

Table 2 Layout of experimental factors, their distribution, and hardness values of all MgMCs specimens by L9 orthogonal array

No	Factors				Experimental values				Replications		
	A	B	C	D	Composition of MgMCVA 3% v.1	Stirring speed, rpm CB)	Stirring time, min (*P*)	Molten temp of MelvfC	BHX	BE	BHX
1	1	1	1	1	BiC + 0% wt A1 3% wt	400	*9*	750	50.22	51.45	50 S4
2	1	2	2	2	BiC + 0% wt A1 3% wt	450	12	300	52.56	52.33	53.03
	1	3	3	3	BiC + 0% wt A1 Mg + 0% wt	300	15	350	55.14	55.35	55.41
4	2	1	3	3	BiC + 3% wt A1 Mg + 0% wt	400	12	350	43.10	47.79	43 66
	2	2	2	1	BX + 3% wt A1 Mg + 0% wt	450	15	750	37.55	33.94	39.03
6	2	3	1	2	BX + 3% wt A1 3% wt	500	*9*	300	45 4S	44 44	4433
7	3	1	3	2	BX + 3% wt A1 3% wt	400	15	300	60.57	61 59	60 92
3	3	2	1	3	BiC + 3% wt A1 3% wt	450	*9*	350	75. IP	75.26	75.21
9	3	3	2	1	BiC + 3% wt A1	500	12	750	65 04	64 62	64 S6

Table 3 S/N response values with their Brinell hardness

No	Factors					
	A	B	C	nz >	BH.N	S.-IST ratio
1	1	1	1	1	50.83	25.55
2	1	2	2	2	52.64	34.12
3	1	3	3	3	55.30	34.4–2
4	2	1	3	3	48.1 S	34.85
5	2	2	2	1	38.52	33.65
6	2	3	1	2	44.76	31.71
7	3	1	3	2	61 0.02	33.OL
8	3	2	1	3	75.22	35.70
9	3	3	2	1	64.84	37.52

Table 4 Mg-MCs Brinell hardness values from ANOVA

Source	ss	DF	MS	F-ratio	SS'	Ratio (%}	F-table
A	484.49	2	240.42	111.30	475.14	73.19	3.46
B	69.29	2	34.14	15.16	64.49	11.23	3.46
C	51.06	2	26.30	10.25	46.17	9.49	3.46
D	2.40						
Error	38.11						
Pooled	40.61	20	2.8050	1	52.81	6.09	
SSt	630.45	26	302.80		646.4	100	
Mean	72,100	1					
S Stntsl	72,544	*11*					

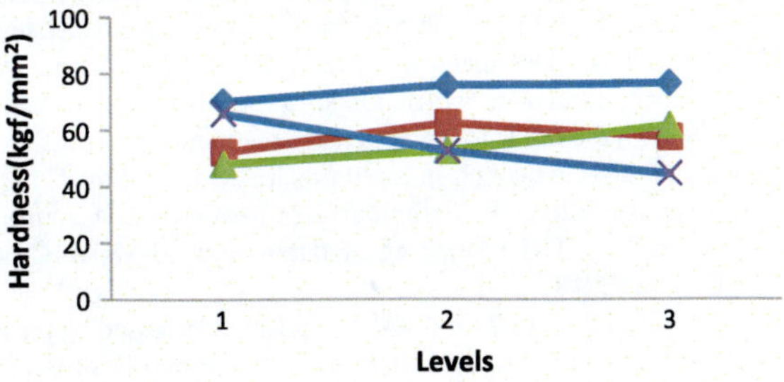

Fig. 2 Graphical response of mean hardness at different levels and their factors

composition) and element B (speed during stirring) influence the Brinell number prominently by 72.14% and 08.32%, respectively. Also, the temperature of MgMC, i.e., element D, gives no prominent influence on the Brinell hardness of MgMC.

References

1. Hegde, A.K.: Dry sliding wear of saffil short fibre reinforced AZ91D Mg alloy composite. Doctoral Dissertation (2018)
2. Arora, G., Sharma, S.: Production of hybrid reinforcement by ball milling for the development of aluminum matrix composites. World J. Eng. (2019)
3. Imran, M., Khan, A.A.: Characterization of Al-7075 metal matrix composites: a review. J. Mater. Res. Technol. (2019)
4. Crivello, J.C., Dam, B., Denys, R.V., Dornheim, M., Grant, D.M., Huot, J., Milčius, D.: Review of magnesium hydride-based materials: development and optimization. Appl. Phys. A **122**(2), 97 (2016)
5. Zhang, W., Liaw, P.K., Zhang, Y.: Science and technology in high-entropy alloys. Sci. China Mater. **61**(1), 2–22 (2018)
6. Rezayat, M., Parsa, M.H., Mirzadeh, H., Cabrera, J.M.: Microstructural investigation of Al-Mg/B4C composite deformed at elevated temperature. J. Alloy. Compd. **763**, 643–651 (2018)
7. Syahrial, A.Z., Salahuddin, J., Ahmad, H.E.: Characterization of Al-Si-Mg/Al2O3 nanocomposite produced by stir casting method. In: Materials Science Forum, vol. 827, pp. 294–299. Trans Tech Publications (2015)
8. Zhang, X., Fang, L., Xiong, B., Hu, H.: Microstructure and tensile properties of Mg (AM60)/Al 2 O 3 metal matrix composites with varying volume fractions of fiber reinforcement. J. Mater. Eng. Perform. **24**(12), 4601–4611 (2015)
9. Zhang, L., Luo, X., Liu, J., Leng, Y., An, L.: Dry sliding wear behavior of Mg-SiC nanocomposites with high volume fractions of reinforcement. Mater. Lett. **228**, 112–115 (2018)
10. Narayanasamy, P., Selvakumar, N., Balasundar, P.: Effect of weight percentage of TiC on their tribological properties of magnesium composites. Mater. Today Proc. Elsevier Publ. **5**(2), 6570–6578 (2018)
11. Ram, B., Deepak, D., Bala, N.: Role of friction stir processing in improving wear behavior of Mg/SiC composites produced by stir casting route. Mater. Res. Exp. **6**(2), 026577 (2018)
12. Selvakumar, N., Narayanasamy, P.: Optimization and effect of weight fraction of MoS2 on the tribological behavior of Mg-TiC-MoS2 hybrid composites. Tribol. Trans. **59**(4), 733–747 (2016)
13. Reyes, A., Bedolla, E., Perez, R., Contreras, A.: Effect of heat treatment on the mechanical and microstructural characterization of Mg-AZ91E/TiC composites. Compos. Interf. **24**(6), 593–609 (2017)
14. Sahoo, B.N., Panigrahi, S.K.: Synthesis, characterization and mechanical properties of in-situ (TiC-TiB2) reinforced magnesium matrix composite. Mater. Des. **109**, 300–313 (2016)
15. Navazani, M., Dehghani, K.: Investigation of microstructure and hardness of Mg/TiC surface composite fabricated by friction stir processing (FSP). Proc. Mater. Sci. **11**, 509–514 (2015)
16. Guo, R., Wu, J.: Dislocation density based model for Al-Cu-Mg alloy during quenching with considering the quench-induced precipitates. J. Alloy. Compd. **741**, 432–441 (2018)
17. Zabihi, A., Soltani, R.: Tribological properties of B4C reinforced aluminum composite coating produced by TIG re-melting of flame sprayed-Mg-B4C powder. Surf. Coat. Technol. **349**, 707–718 (2018)
18. Yao, Y.T., Chen, L.Q.: B4C/Al composites processed by metal-assisted pressureless infiltration technique and its characterization. Mater. Manuf. Process. **31**(10), 1286–1291 (2016)
19. Azizieh, M., Mazaheri, M., Balak, Z., Kafashan, H., Kim, H.S.: Fabrication of Mg/Al12Mg17 in-situ surface nanocomposite via friction stir processing. Mater. Sci. Eng. A **712**, 655–662 (2018)
20. Dash, D., Samanta, S., Rai, R.N.: Study on fabrication of magnesium based metal matrix composites and its improvement in mechanical and tribological properties—a review. In: IOP Conference Series: Materials Science and Engineering, vol. 377, No. 1, p. 012133. IOP Publishing, (2018)

21. Isfahani, M.J.N., Payami, F., Asadabad, M.A., Shokri, A.A.: Investigation of the effect of boron carbide nanoparticles on the structural, electrical and mechanical properties of Al-B4C nanocomposites. J. Alloy. Compd. **797**, 1348–1358 (2019)
22. Singh, A., Bala, N.: Synthesis and comparative sliding wear behavior of stir cast Mg and Mg/Al2O3 metal matrix composites. Mater. Res. Exp. **6**(7), 076512 (2019)
23. Aydin, F., Sun, Y., Turan, M.E.: Investigation of microstructure, mechanical and wear behavior of B4C particulate reinforced AZ91 matrix composites by powder metallurgy. Indian J. Chem. Technol. **26**, 351–354 (2019)

Grey Wolf Optimizer-Based PID Controller Design for Laser Beam Pointing Applications

Himanshu Chaudhary, Shahida Khatoon, Ravindra Singh, and Ashish Pandey

1 Introduction

Fast steering mirrors are becoming an integral part for laser beam steering and pointing in numerous applications; such as surveillance, tracking and pointing system, guided weaponry system, navigation system, ground and space communication, astronomy research and so on. They are generally employed to obtain much precise pointing, jitter attenuation and fast response by eliminating the effect of various internal disturbances such as frictional coupling, kinematic coupling, geometric coupling, mass unbalance, mechanical and electrical issues, thermal limits of actuators and external disturbances due to operating environment effect on LOS and platform disturbances [1]. There are several types of tracking and pointing configuration developed so far other than fast steering mirror approach [2]. In recent past, several configurations of fast steering mirrors with basic difference in their mechanical analysis and control configuration designs are proposed [3], such as four-quadrant-based piezoelectric fast steering mirror(PFSM) [4], magnetic suspended fast steering mirror (MSFSM), high bandwidth fast steering mirror (HBFSM) [5] and micro-electrical mechanical system (MEMS)-FSM [6]. A scientist or designer developing all such acquisition, tracking and pointing (ATP) system always faces many issues and performance limitation, which are broadly described in [1].

Several control law have been proposed so far to steer the line of sight and minimize the effect of disturbances. These includes PI and PID controller configuration [7], H infinity control [8], feed forward system design, real-time FPGA-based feedback control, Lyapunov stability approach [9], sliding mode control (SMC) [10] and

H. Chaudhary (✉) · S. Khatoon
Department of Electrical Engineering, Jamia Millia Islamia University, New Delhi, Delhi 110025, India
e-mail: himanshuchaudhary.h@gmail.com

R. Singh · A. Pandey
Defence Research and Development Organization, New Delhi, Delhi 110054, India

M. Muzammil et al. (eds.), *Recent Advances in Mechanical Engineering*, Lecture Notes in Mechanical Engineering, https://doi.org/10.1007/978-981-15-8704-7_31

many more. But the use of optimization algorithm and intelligent controllers is still very constrained with the FSM dynamics. Optimization algorithms [11, 12] proven to very efficient enough in recent past for complex dynamics problems to provide more robust and accurate solution.

2 FSM Modelling and Design

Consider FSM mirror as a standalone device with four piezo-actuators and with no coupling with gimbal assembly as shown in Fig. 1 [13]. Then, this FSM with piezo-actuator can be modelled as a spring mass system. Simplified form of spring mass system is represented as;

$$F = M\ddot{x} + D\dot{x} + Kx \tag{1}$$

$$G(s) = \frac{X(s)}{F(s)} = \frac{1}{MS^2 + DS + K} \tag{2}$$

Consider motion of FSM along azimuth axis as shown in Fig. 1b, the forces acting on the mirror are piezoforce, and spring forces are due to piezo-stiffness and inertial force.

For small angle of rotation $\delta\theta_x$ of the mirror, the equation of rotational motion about the X-axis is:

$$2Fr_1 = I_{xx}\delta\ddot{\theta}_x + B\delta\dot{\theta}_x + 2K\delta x r_1 \tag{3}$$

As the $\delta\theta_x$ is very small, it can be replaced by $(r_1.\delta\theta_x)$ and then equation becomes;

$$2Fr_1 = I_{xx}\delta\ddot{\theta}_x + B\delta\dot{\theta}_x + 2Kr_1^2\delta\theta_x \tag{4}$$

where F is the applied piezoforce, K represents piezo-actuator linear stiffness, B is viscous damping constant and Ixx is the moment of inertia. Considering Eq. 4 and taking its Laplace transformation, the angular rotation with respect to applied

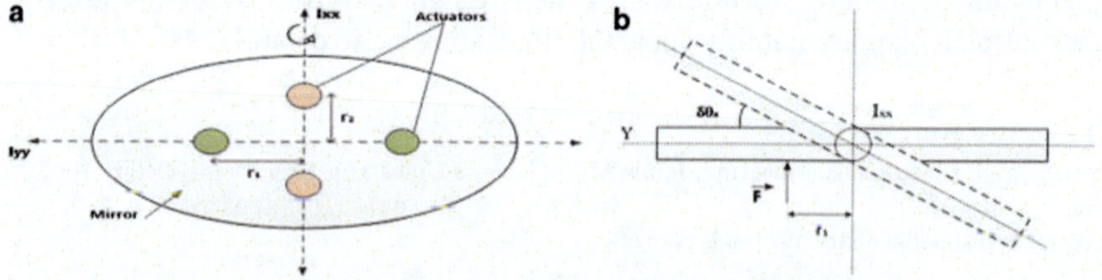

Fig. 1 **a** Piezo-actuated fast steering mirror, **b** FSM rotation about X-axis (azimuth)

piezoforce is represented in form of second-order transfer function as;

$$\frac{\delta\theta_x(s)}{F(s)} = \left(\frac{1}{I_{xx}S^2 + B.S + 2Kr_1^2}\right) \quad (5)$$

3 Performance Specifications

The control configuration is designed for the FSM pointing system, such that it must adhere to the following minimum specifications.

(1) Close loop bandwidth > 100 Hz.
(2) Accuracy of 30 μrad for commands up to 5 Hz.
(3) Step response shows minimal overshoot and rise time must be less than 5 ms.
(4) Disturbance frequencies 0.1–6 Hz.

Exact parametric values of FSM system is considered as follows:

Moment of inertia about the X- and Y-axis are 6.43e−3 kg m^2 and 3.663e−3 kg m^2; K (piezo-linear stiffness) = 80e6 N/m; A (maximum angular displacement of FSM) = 3 mrad; damping ratio $\zeta = 0.5$ and peak acceleration = 1650 rad/s^2.

Based on the above parameters, required peak azimuth torque can be calculated as:

$$Ixx * (\text{peak acceleration}) = 6.43\text{e} - 3 * 1650 = 10.6\,\text{N m}$$

Based on the above calculated specification, differential angle of rotation to applied piezoforce second-order transfer function can be rewritten as:

$$\frac{\delta\theta_x(s)}{F(s)} = \left(\frac{1}{6.4296\text{e} - 3S^2 + 2.03S + 2 * 80\text{e}6 * 0.05\hat{\ }2}\right)$$

Simplified uncompensated FSM open loop simulation is developed first for azimuth axis only. Figure 2a shows the frequency response of the uncompensated plant model with resonant frequency of 1.255 kHz. Considering the disadvantages and practical feasibility of open loop system, it is always requires to have close loop control system design. A PI compensator along with a lag compensator is designed, targeting a bandwidth of 100 Hz and open loop gain of 60 dB at 1 Hz:

$$\text{PI Compensator} = 2125 * \left(1 + \frac{1200}{s}\right)$$

$$\text{Lag Comensator} = \left(\frac{S + 2 * \text{pi} * 20}{S + 2 * \text{pi} * 1}\right),$$

Figure 2 shows the open loop, compensated open loop and closed loop responses.

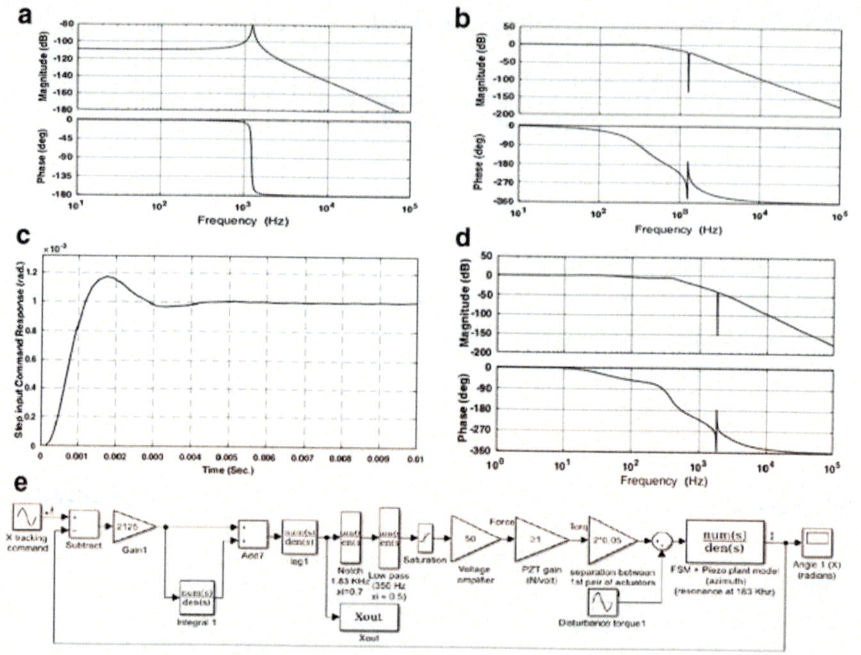

Fig. 2 **a** open loop uncompensated system bode plot, **b** open loop compensated system bode plot, **c** FSM open loop compensated system step response, **d** closed loop frequency response, **e** compensated closed loop FSM block diagram

4 Controller Design and Optimization

4.1 PID Controller Design

PID controllers are the most standard, readily available and flexible tool for every industrial automation application. This is the biggest advantage of using PID controller in every control-related real-time applications. PID controller can be used in different configurations such as serial, parallel, cascaded and in different combinations of P, I and D.

Standard form of the PID control algorithm in *S*-domain is:

$$U(s) = K_p\left(1 + \frac{1}{T_i s} + \frac{sT_d}{1 + s\frac{T_d}{N}}\right)E(s)$$

where K_p represents proportional gain, T_i and T_d are the integral and derivative time constants and N is the filter coefficient. The ultimate aim is the adjustment or tuning of K_c, T_i and T_d for designing of PID controller [14, 15]. Here, the good gain method is applied for determining the parameters of PID controller at the initial level.

The parameters obtained with the fine-tuning of PID controller are: $K_{PGG} = 300$; $T_{ou} = 0.0025$; $T_i = 1.5 * 0.0025 = 0.00375$; $K_p = 0.8\,K_{PGG} = 240$; $T_d = \frac{T_i}{4} = 0.0009375$. Further, for the optimization of the PID controller parameters, grey wolf optimization (GWO) algorithm is used with the minimization of the objective function.

4.2 Grey Wolf Optimizer Algorithm

Grey wolf approach is a metaheuristic approach that can provide solution to the various control optimization problem. It has been first introduced by Mirjali et al. [16]. GWO approaches the optimal solution based on the hunting behaviour and social hierarchy of the wolves.

Initially, the search starts with the random population generation of wolves followed by the arrangement of wolf in social hierarchy. Next, in the hunting process, these wolves try to estimate the optimum location of prey through a continuous iteration method. The objective of the optimization algorithm is to obtain finest values of PID controller parameter by minimizing integral time absolute error (ITAE) function.

In the present work, social hierarchy and hunting behaviour of grey wolves are mathematically modelled in order to develop GWO algorithm, and the optimization of ITAE is performed to obtain the parameter of PID controller. GWO algorithm for minimization of ITAE function is developed in MATLAB, and the optimized values of PID controller are obtained with number of simulation runs. The performance evaluation graph is shown in Fig. 3.

ITAE function is represented by;

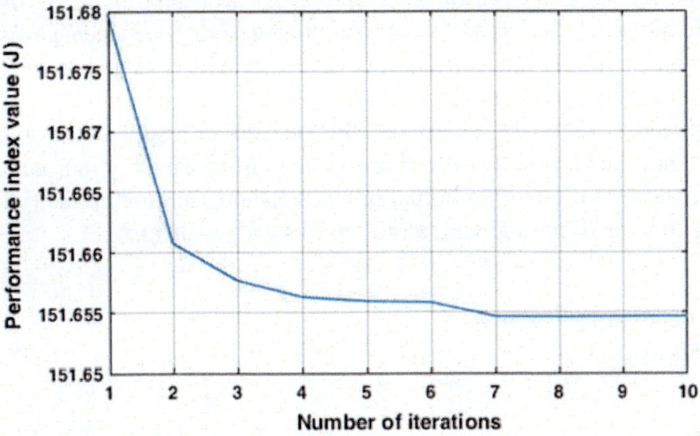

Fig. 3 Performance index evaluation

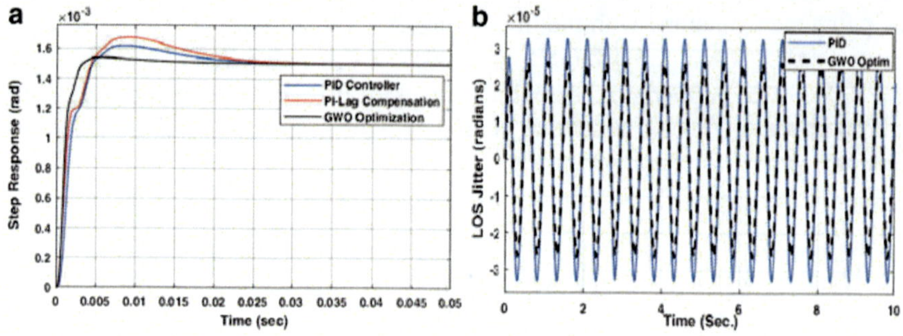

Fig. 4 **a** Step input command response, **b** line of sight jitter for disturbance signal

$$\mathrm{ITAE}(t) = \int_{t=0}^{t=\infty} t * |e(t)|\mathrm{d}t$$

The obtained values are given as;

total number of search agents (population): 100; number of iteration: 10; fitness function average: 1.51654e+02; elapsed time 1179.79228 s; parameters boundary limit; [0 10; 0 10; 0 10]; values: $K_{\mathrm{p}} = 204.1722$; $T_i = 0.045$; $T_{\mathrm{d}} = 0.075$.

5 Result and Discussion

Different design parameters like closed loop bandwidth, transient response characteristics, steady-state response and jitter attenuation performance have been taken into consideration for design of control loop strategy for fast steering mirror dynamics. The overall dynamics and control configuration design have been performed in MATLAB and Simulink. The performance characteristics obtained from the above discussion and design are shown in Fig. 4.

It is observed that the GWO optimized PID controller results in the ameliorated system dynamic performance with minimal jitter of about 25 μrad, and optimal input command response with minimal or no overshoot, minimum rise time, i.e., 2 ms, settling time of 20 ms and with almost zero steady-state error.

6 Conclusion

The kinematic relationship is established between angular rate responses to the input torque followed by simulation of the complete FSM dynamic model for azimuth axis. Standard compensator design approach is used to achieve the desired specifications, particularly, the bandwidth limitation. The performance characteristic of PI-Lag compensation-based FSM model validates the modelling and dynamics of the designed system. Further, for improving the performance of designed model, PID controller is designed based on good gain method and the gains of the PID controller are fine-tuned using grey wolf optimizer algorithm. Number of simulation runs validated the output response characteristics of both PID controller and GWO optimized PID controller, although the observed performance characteristics of all three controller configuration are satisfactory. But certainly, the response of GWO-tuned PID controller dominates the other two configurations in terms of both input command response and jitter attenuation performance.

References

1. Kennedy, P.J., Kennedy, R.L.: Direct versus indirect line of sight (LOS) stabilization. IEEE Trans. Control Syst. Technol. **11**, 3–15 (2003)
2. Sheikh, M., Zareinejad, M., Parsa, M., Sheibani, H.: Integral based sliding mode stabilizing a camera platform using Kalman filter attitude estimation R. Mechatronics **44**, 42–51 (2017)
3. Chaudhary, H.: Fast steering mirror for optical fine pointing applications : a review paper. In: IEEE Conference (CIPECH-2018), pp. 102–106 (2018)
4. Zhang, B., Zhang, L., Huang, G., Shu, R.: Research on pointing of piezoelectric fast steering mirror under vibration condition. Proc. SPIE Soc. Photo Instrum. Eng. **8191**, 1–7 (2011)
5. Li, Q., Liu, L., Ma, X., Chen, S., Yun, H., Tang, S.: Development of multitarget acquisition, pointing, and tracking system for airborne. IEEE Trans. Ind. Informatics. **15**, 1720–1729 (2019)
6. Tian, J., Yang, W., Peng, Z., Tang, T.: Inertial sensor-based multiloop control of fast steering mirror for line of sight stabilization. SPIE Opt. Eng. **55**, 1–6 (2019)
7. Zhou, Q., Ben-Tzvi, P., Fan, D., Goldenberg, A.A.: Design of fast steering mirror systems for precision laser beams steering. ROSE IEEE Int. Work. Robot. Sensors Environ. Proc. 144–149 (2008)
8. Yu, Z., Cui, N., Chen, X., Xu, C., Cao, K.: H ∞ control for fast steering mirror based on the incremental PI controller. Proc. SPIE. **9521**, 95210G–1–95210G–5 (2015)
9. Ramírez, L.F., Saldivar, B., Ávila Vilchis, J.C., Montes de Oca, S.: Lyapunov–Krasovskii approach to the stability analysis of the milling process. IET Control Theory Appl. **12**, 1332–1339 (2018)
10. Wang, H., Ye, X., Tian, Y., Zheng, G., Christov, N.: Model-free-based terminal SMC of quadrotor attitude and position. IEEE Trans. Aerosp. Electron. Syst. **52**, 2519–2528 (2016)
11. Jin, Z., Hou, Z., Yu, W., Wang, X.: Target tracking approach via quantum genetic algorithm. IET Comput. Vis. **12**, 241–251 (2018)
12. Fatihu, M., Jen, H., Ahmed, I.: Cuckoo search algorithm based design of interval type-2 fuzzy PID controller for Furuta pendulum system. Eng. Appl. Artif. Intell. **62**, 134–151 (2017)
13. Jones, C., Griffin, S.: State space representation of optical systems. Proc. SPIE. **7338**, 73380J–73380J–12 (2009)
14. Basilio, J.C., Matos, S.R.: Design of PI and PID controllers with transient performance specification—Education, IEEE Trans. **45**, 364–370 (2002). ieee-edu2002.pdf
15. Tang, K.Z., Huang, S.N., Tan, K.K., Lee, T.H.: Combined PID and adaptive nonlinear control for servo mechanical systems. Mechatronics **14**, 701–714 (2004)
16. Mirjalili, S., Mohammad, S., Lewis, A.: Grey wolf optimizer. Adv. Eng. Softw. **69**, 46–61 (2014)

Effect of Iron Content and Machining Parameters on Surface Roughness of Al–1V–1Si Alloys

Bibeka Nand Pathak and Arunesh Chandra

1 Introduction

Machinability is one of the major criteria for the selection of material for different applications. The simplicity with which a metal can be machined is an important variables influencing the utility, quality and cost of the item. On the basis of application, machinability is considered in terms of a tool wear rate, tool power consumption and surface quality. There are many ways that can be judged by the machinability of materials depending on the tool material, work material and machining operation for comparison purposes. The main criteria adopted for machinability assessment of the material are tool life, tool wear rate, cutting power and surface roughness created at the workpiece [1, 2]. Generally, it is related to the machining cost of production engineer, and they are serving to the actual behaviour of the material during machining [3].

Properties of the aluminium alloys can improve by addition of some alloying elements which also affect the machinability. Among all the alloying elements, silicon is mostly used and improves the fluidity and castability of Al alloys. Small amount of silicon (<0.8%) addition improves machinability but when silicon percentage is higher, it forms hard silicon particles which is quite abrasive to the tool [4]. Minimum 80 BHN (Brinell hardness number) hardness is required for good machinability; however, hardness is not the only judgement criteria for machinability. If there are higher content of iron in an Al–12Si–Cu–Mg alloy, then it results greater amount of coarser structure than lower content of iron alloy [5]. Modification of the internal structure with addition of alloying elements, selection of process, or subsequent

B. N. Pathak (✉)
Department of Mechanical Engineering, IMS Engineering College, Ghaziabad, India
e-mail: bibekanandp96@gmail.com; bnpathak2007@rediffmail.com

A. Chandra
Department of Mechanical Engineering, KIET Group of Institutions, Ghaziabad, India
e-mail: arunesh.chandra@kiet.edu

M. Muzammil et al. (eds.), *Recent Advances in Mechanical Engineering*, Lecture Notes in Mechanical Engineering, https://doi.org/10.1007/978-981-15-8704-7_32

heat treatment can improve machinability of Al alloys. It also observed that the internal structure of the alloys affected dramatically the machinability [6]. When tool life increases or hardness of workpiece material decreases, the machinability of the workpiece increases, and lower production cost and high productivity are possible [7]. Surface roughness is one of the imperative parameters which influence the machinability as well as service life of the component.

From the literature study, it shows that there are some relation between the machined surface quality and their microstructure of the alloy. From the literature survey, it was found that cutting force and the surface quality increases if the sizes of the soft matrix grains produced increases [8]. It has been seen from the cutting force value obtained that generally the cutting force lowers with cutting speed increasing for all the workpiece materials. There are two main factors to drop in the forces in the specimen. In the flow zone, shear strength decreases as cutting speed increasing, and the other factors are to decrease in tool–chip area for the responsible to drop in the forces as cutting speed increases [9]. From the experimental results, it has been shown that the surface quality is a function of cutting parameters, composition and alloy conditions [10].

The present work concerns the test investigation of machinability of high iron-containing aluminium alloys in terms of surface quality of machined samples. Therefore, it is important to analyse the machining behaviour of these alloys and compare them between different compositions of alloys and also to study the effect of the following various parameters like feed rate, depth of cut and cutting speed on machinability behaviour and their surface quality during milling of aluminium alloys.

2 Experimental Method

2.1 Alloy Preparation

The experimental specimen were made in an electric heating furnace in an earth-reinforced graphite pot under the cover of without Na transition flux. For the preparation of alloys, Al-21% Fe, Fe-50%V master alloys were utilised. At first, the graphite pot was preheated to about 600 °C. Then, weighted quantity of master alloys, 99.95% pure Al and 99.99% pure Si metallic were charged (Compositions are given in wt% unless otherwise mentioned.).

Just after melting, the molten alloy was secured with a sodium-free transition (2% of dissolved). Na free dissolved (flux) is required because Na free motion is required in light of the fact that Na addition expands the pin holing tendency and decreases fluidity [11]. Subsequent to melting, adequate time was given for complete mixing of the melt. The melt was every now and again unsettled with a graphite bar for complete homogenisation mixing. The cover flux through scaling and dross and so forth were skimmed off before the degassing treatment. The melt was degassed with Ar gas introduced through steel lancing tubes. After degassing, the melt was cast

Table 1 Alloys designation compositions of the alloys prepared

Alloy designation	Chemical composition (wt%)			
	Fe	V	Si	Al
BH1	1	1	1	Balance
BH2	2	1	1	Balance
BH3	3	1	1	Balance
BH4	4	1	1	Balance

in different moulds. The compositions of different alloys and different modification treatment are given in Table 1.

2.2 Machining of Specimen

For machinability study, a high-precision vertical CNC milling machine was used for machining the alloys developed. The machining procedure was completed without utilizing a cutting liquid. The samples size was kept 75 × 30 × 20 mm^3 for each experiment as shown in Fig. 1. A 16 mm diameter measurement HSS M35 type device (Addison organisation) was utilised for the processing tests. The machining parameters settled for examination as 0.5–1.5 mm depth of cut, 0.14–1.0 mm/rev feed rate and 25–175 m/min cutting speed. These parameters were taken on the basis of literature review and performance of the milling machine used for the experiment.

Fig. 1 Test pieces of samples used for machinability test

2.3 Estimation of Surface Roughness

To survey the surface roughness of the specimen, the average mean, Ra and maximum top-to-valley height, Rz were utilised for examination. The work piece surface quality were estimated with the assistance of Mitutoyo compact surface roughness analyser (model no. SJ-400). Surface roughness of the machined surface was estimated after each test. Surface roughness of each samples was studied under various conditions, and charts (graph) were plotted for all conditions. The average of four values was taken for each samples.

3 Result and Discussion

In this segment, the experimental analysis of machinability of aluminium alloys in terms of surface quality of machined alloys surfaces are reported and discussed. In order to detect the average surface roughness (R_a) and maximum top-to-valley height (R_z) value, experiments were carried out by changing the machining parameter one after another. The goal of the present work is, in this manner, to examine the impact of various parameters on machinability properties, viz. surface quality, the process of amid milling processing of Al–(1–4)Fe–1Si–1V alloys.

3.1 Surface Roughness

There are different surface roughness parameters might be utilised for the assessment of surface roughness testing, viz. average surface roughness (R_a), root mean square roughness (R_q) and greatest crest-to-bottom roughness (R_z). In the present examination, R_a and R_z parameters were utilised for the surface roughness analysis. The impact of machining parameters on average centre line height (R_a) and most extreme crest-to-bottom height (R_z) for Al–(1–4)Fe–1Si–1V alloys is appeared from Figs. 4, 5, and 6.

Figure 4a, b demonstrate the impact of depth of cut on R_a and R_z, individually, amid machining without utilisation of cooling fluid. From the graph of surface roughness and depth of cut, it was observed that the surface roughness height increases on addition of iron from 1 to 4% for Al–(1–4)Fe–1Si–1V alloy. It was also seen that in higher percentage of of iron in Al–(1–4)Fe–1Si–1V alloys, the structure was form blocky Al_3Fe precipitates which is brittle in nature of intermetallic phases and lowers properties of the alloys. Among all these intermetallic phases appeared in aluminium alloy, it is commonly observed that plate-like structure is not favourable to mechanical properties specially for elongation. Altering the state of intermetallic stages from plate-like structure to other reduced form enhances the properties and structure of Al alloys. Optical microstructures of the Al–(1–4)Fe–1 V-1Si alloys

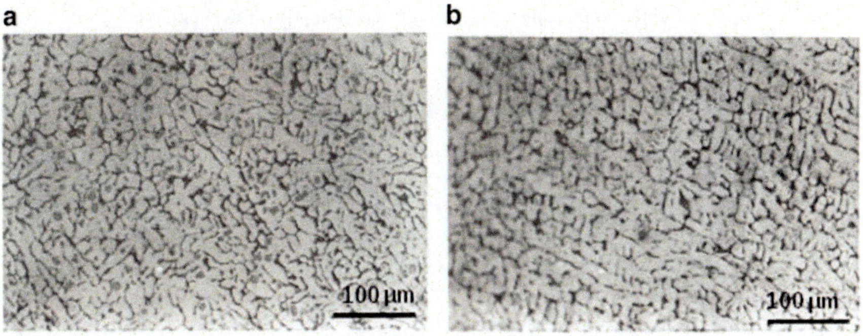

Fig. 2 Microstructures of **a** Al–1Fe–1V–1Si alloy and **b** Al–2Fe–1V–1Si alloy

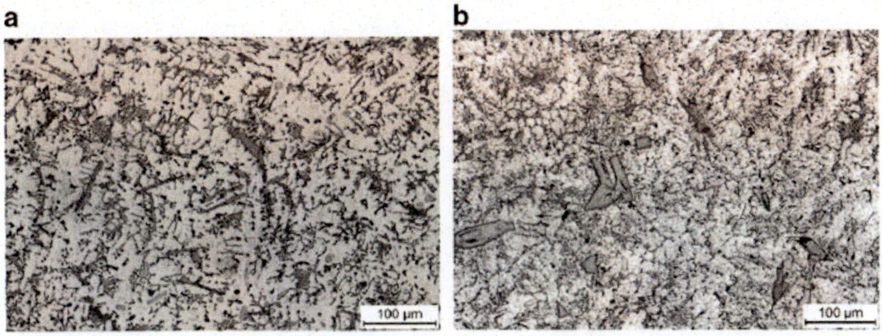

Fig. 3 Microstructures of **a** Al–3Fe–3V–1Si alloy and **b** Al–4Fe–1V–1Si alloy

are appeared in Figs. 2, 3 separately. The metallographic observation were done to see the morphological changes in the phases and its impact on the surface quality. Therefore, machinability reduces with increasing amount of iron.

Figure 5a, b demonstrate the impact of feed rate on surface roughness properties without utilisation of cooling fluid. The processing activities were performed at consistent cutting speed (75 m/min) and 0.5 mm depth of cut. It tends to be observed that when Fe content increments from 1 to 4%, surface roughness height increments for unmodified alloys.

Figure 6a, b demonstrate the impact of cutting speed on surface roughness qualities, i.e., R_a and R_z, amid processing without utilisation of cooling fluid. The milling processing were performed at 0.5 mm/rev constant feed rate and 0.5 mm depth of cut. The outcomes demonstrate that the estimation of both surface roughness height R_a and R_z increase with increment in Fe content from 1 to 4%.

The above discussion highlights the fact that machinability of alloys is directly related to the compositional changes and its modification treatment.

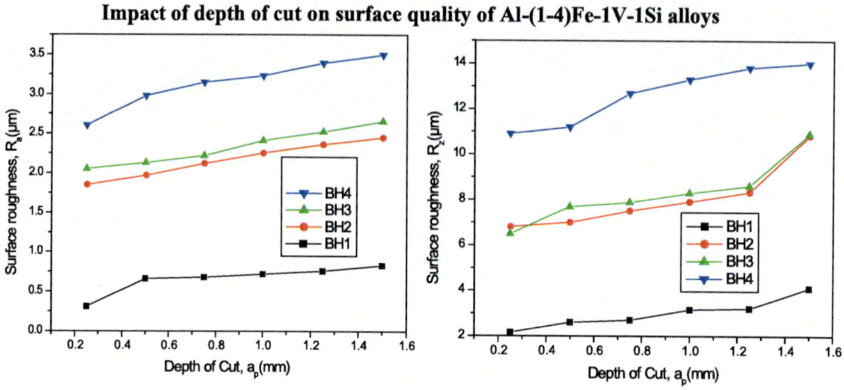

Fig. 4 Impact of depth of cut on surface roughness (R_a) and surface roughness (R_z) of Al–(1–4)Fe–1V–1Si alloy

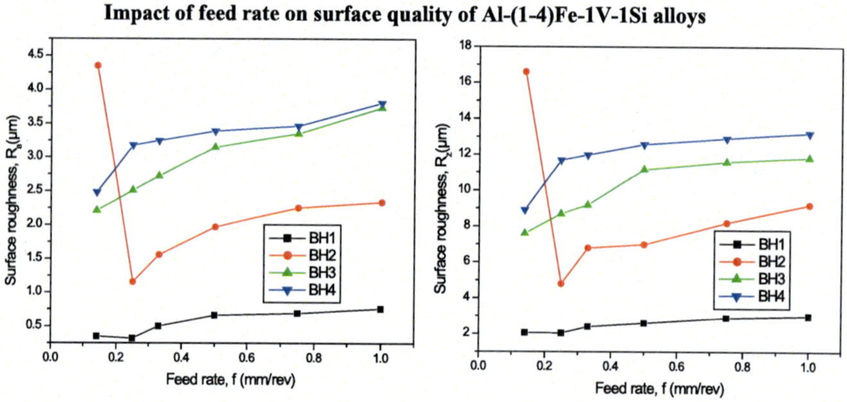

Fig. 5 Impact of feed rate on surface roughness (R_a) and surface roughness (R_z) of Al–(1–4)Fe–1V–1Si alloy

4 Conclusions

In view of above outcomes and discussion, it is concluded that,

1. The surface quality enhanced altogether with increment of cutting speed and decrement of feed rate of Al–(1–4) Fe–1V–1Si alloys in tried tested range.
2. It was suggested that Rapid speed, lower feed rate and lower depth of cut should be recommended for better surface quality.
3. Surface quality are influenced greater by depth of cut when compared with the other parameters like cutting speed and followed by the feed rate.
4. With increasing Fe percentages in Al–Fe–V–Si alloys (from 1 to 4%), surface roughness increases in all the conditions.

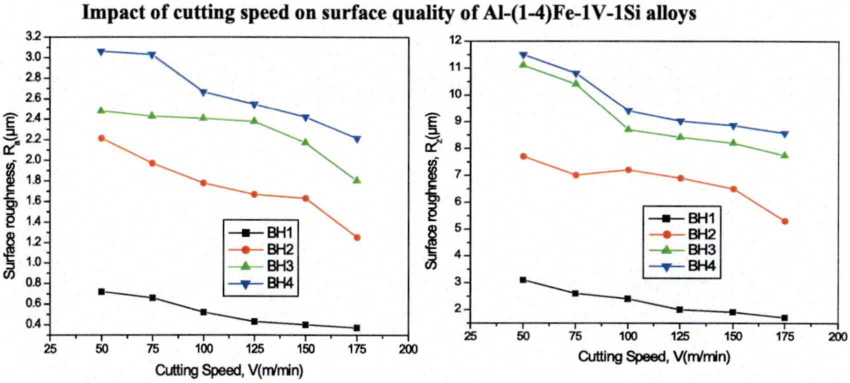

Fig. 6 Impact of cutting speed on surface roughness (R_a) and surface roughness (R_z) of Al–(1–4)Fe–1V–1Si alloy

5. The chip thickness also affect the surface quality. As chip thickness increases, surface quality adversely affected. Consequently, machinability unfavourably influenced by chip thickness.

References

1. Hossainy, T.M., EL-Zoghby, A.A., Bady, M.A., Maalawi, K.Y., Nasy, M.F.: Cutting parameters optimization when machining different materials. Mater. Manuf. Processes **25**, 1101–1114 (2010)
2. Coello, C.A.C., Becerra, R.L.: Evolutionary multi-objective optimization in materials science and engineering. Mater. Manuf. Processes **24**(2), 119–129 (2009)
3. Prasanna, N.D., et al.: Machinability characteristics of Austempered ductile iron. Indian Foundry J. TP51–TP56
4. Pathak B.N., Sahoo, K.L., Madhawanand, M.: Effect of machining parameters on cutting forces and surface roughness in Al–(1–2) Fe–1V–1Si alloys. Mater. Manuf. Processes **28**(4), 463–469 (2013)
5. Rodríguez, S.H., Rafael, E., Reyesa, G., Dwivedi, D.K., González, O.A., Hernández, B.: The effect of Al–5Ti–1B on microstructure and mechanical properties of Al–12Si–xFe alloy. Mater. Manuf. Process. **27**(6): 599–604 (2012)
6. Skingle, T.J., Thompson, R.W.: Machining of aluminium alloys. J. Appl. Metal Work. Am. Soc. Metals **1**(2), 76–78 (1980)
7. Chiranjeev, K., Prodhan, A.: Effect of DC plasma and strontium treatment of a hyoereutectic Al-Si alloy on its Microstructure and Machinability. Indian Foundry J. 45–52 (2007)
8. Basavakumar, K.G. et al.: Influence of grain refinement, modification and turning inserts on machinability and surface characteristics of hypoeutectic Al-Si cast Alloys. Indian Foundry J. **53**(6), 35–44 (2007)
9. Demir, H., Gunduz, S.: The effects of aging on machinability of 6061 aluminium alloy. Mater. Des. **30**, 1480–1483 (2009)

10. Dwivedi, D.K.: Influence of grain refinement, modification and heat treatment on surface roughness in machining of cast Al-Si Alloys. Indian Foundry J. **48**(1), 32–38 (2002)
11. Gourishankkar, N., Prabhakar, O.: Effect of magnesium on the cast microstructure Aluminum-Silicon Alloys. Trans. Indian Inst. Met. **47**(4), 229–237 (1994)

Optimization and Characterization of Low-Cost MR Fluids

Vivek Sharma, Kalidasan Rathinam, Gavendra Norkey, Swastik Pradhan, Mukthiar Singh, and Mandeep Singh

1 Introduction

The MR liquids have a one of a kind property of controllable rheological attributes with the utilization of attractive field [1].

The MR liquids will in general create stress of 27–48 kPa which can end up being valuable for applications in dampers and brakes [2]. Iron particles jumbled to get dispersion medium together with suitable additive like tetra methyl ammonium hydroxide lead to the formation of magneto-rheological fluids [3]. When an external magnetic field is carried out to MR fluid, the iron debris organizes along the magnetic discipline lines to form strong chains and increases fluid resistance to float as proven in Fig. 1 [4, 5]. This results in the development of a excessive yield stress that is proportional to the strength of the magnetic field [6]. An experimental installation including electromagnet that is able to producing an excessive magnetic field up to two Tesla has been developed.

2 Design of Experimental Setup

A trial setup having an electromagnet with 1900 turns has been created. The Perspex tube having an internal distance across of 15 mm is fixed between the shafts of electromagnet by drawing these posts nearer with locking system. This cylinder is loaded up with MR liquid and current is gradually expanded from 0.1 to 5 A to increment attractive field. The various estimations of attractive field comparing to include current are shown on the Gauss meter advanced yield screen, when its lobby test is drenched inside the cylinder containing the MR fluid. The post width

V. Sharma (✉) · K. Rathinam · G. Norkey · S. Pradhan · M. Singh · M. Singh
School of Mechanical Engineering, Lovely Professional University, Phagwara, India
e-mail: vivek.23319@lpu.co.in

M. Muzammil et al. (eds.), *Recent Advances in Mechanical Engineering*, Lecture Notes in Mechanical Engineering, https://doi.org/10.1007/978-981-15-8704-7_33

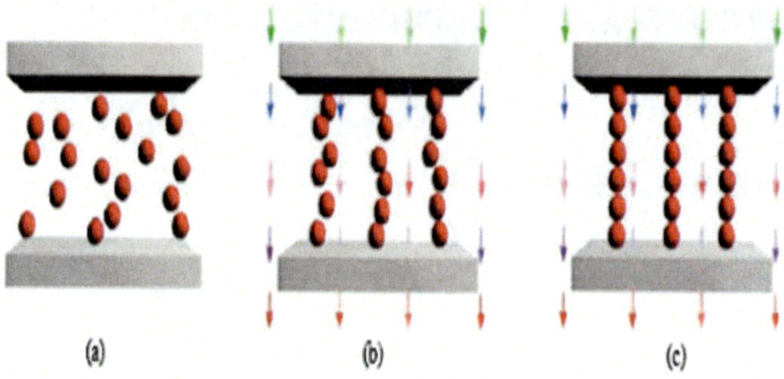

Fig. 1 Application of magnetic field to MR fluid particles

of the electromagnet is 75 mm. The length of Perspex tube utilized is 50 mm, and its inward width and external measurement will be 15 and 18 mm individually. The Perspex cylinder will be filled by MR liquid and will be put between the posts of electromagnet. Upon charge, the liquid will change its rheological properties from fluid to strong state and will be exceptionally solid. This on state energized liquid will in general hinder the pole speed, and we will fix the pole speed to 10 rpm so as to make thick impacts unimportant and to consider just yield pressure segment of torque. This is additionally done to acquire the torque esteems on the grounds that at high rpm the torque esteems will be immaterial. Presently, we acquire the three parameters relating to supply current from 0.1 to 5 A: (a) magnetic flux thickness, (b) torque esteems, and (c) RPM esteems. The opposing torque of the MR fluid on the part which is in contact with the pivoting shaft can be determined as pursues:

$$T_1 = \int_{R_{\text{int}}}^{R_{\text{ext}}} \int_{0}^{2\pi} \tau_y r^2 \mathrm{d}r \mathrm{d}\theta = 2\pi \tau_y \frac{R_{\text{ext}}^3 - R_{\text{int}}^3}{3}$$

$$T_2 = \int_{0}^{R_{\text{ext}}} \int_{0}^{2\pi} \tau_y r L \mathrm{d}r \mathrm{d}\theta = 2\pi \tau_y R_{\text{ext}}^2 L$$

$$T_{\text{total}} = T_1 + T_2 = 2\pi \tau_y \left(\frac{R_{\text{ext}}^3 - R_{\text{int}}^3}{3} + R_{\text{ext}}^2 L \right)$$

$$\tau_y = \frac{T_{\text{total}}}{2\pi \left(\frac{R_{\text{ext}}^3 - R_{\text{int}}^3}{3} + R_{\text{ext}}^2 L \right)}$$

The shear stress can be computed considering the effect of applied torque, over the two above-calculated surfaces.

Table 1 Levels of various components for MR fluid formation

Volume %	Levels		
Iron	22	27	32
Silicon oil	67	72	77
Tetra methyl ammonium hydroxide	0.5	0.6	0.7

3 Method of Fluid Formation

Carbonyl iron particles (80% by weight) are mixed with tetra methyl ammonium hydroxide (0.5% by volume) for 30 min at 400 rpm using a stirrer in a steel container. Further, silicon oil was added (67% by volume) in this mixture, and the whole liquid was stirred for around 3 h at 450 rpm using the stirrer inside the steel container. Table 1 shows the various levels fixed for various components of MR fluid formation.

4 Taguchi L-9 Orthogonal Array

Based on the various levels selected for L-9 orthogonal array as given in Table 2, a following orthogonal array was designed using response variable as yield stress. The yield stress was used as a rheological characteristic in the present research work.

Table 2 Orthogonal array for MR fluid formation

S. No.	Factor 1	Factor 2	Factor 3	Factor 4	Response
	Iron vol%	Silicon oil vol%	Oleic acid vol%	Tetra methyl vol%	Yield stress (kPa)
1	22	67	0.5	0.6	27.0388
2	22	72	0.6	0.7	24.786
3	22	77	0.7	0.8	27.021
4	27	67	0.6	0.8	34.321
5	27	72	0.7	0.6	33.32
6	27	77	0.5	0.7	35.465
7	32	67	0.7	0.7	45.684
8	32	72	0.5	0.8	47.7888
9	32	77	0.6	0.6	48.896

5 Anova Results for Yield Stress

Based on the Taguchi L-9 orthogonal array, following results were calculated for yield stress as a rheological characteristic as given in Table 3.

The model F-value of 97.93 implies that the model is significant. The "Pred. R-Squared" of 0.9313 is in a reasonable agreement with the "Adj. R-Squared" of 0.9864, i.e., the difference is less than 0.2. Further, the optimized and actual values are given in Tables 4, 5.

Table 3 ANOVA results for yield stress

Analysis of variance table [classical sum of squares—Type II]						
Source	Sum of squares	DOF	Mean square	F-value	p-value	Prob. > F
Model	693.66	6	115.61	97.93	0.0101	Significant
Iron vol%	685.03	2	342.52	290.13	0.0034	
Silicon oil vol%	5.58	2	2.79	2.36	0.2972	
Oleic acid vol%	3.04	2	1.52	1.29	0.4371	
Residue	2.36	2	1.18			
Total	696.02	8				

Table 4 R-squared results for ANOVA Table

Std. Dev.	1.09	R-squared	0.9966
Mean	36.04	Adj. R-squared	0.9864
C.V. %	3.02	Pred. R-squared	0.9313
Press	47.81	Adeq. precision	24.006

Table 5 Optimized and actual values for MR fluid

Run order	Actual value	Predicted value	Residual	Leverage
1	27.04	26.66	0.38	0.778
2	24.79	25.51	−0.72	0.778
3	27.02	26.68	0.34	0.778
4	34.32	33.98	0.34	0.778
5	33.32	32.94	0.38	0.778
6	35.47	36.19	−0.72	0.778
7	45.68	46.41	−0.72	0.778
8	47.79	47.45	0.34	0.778
9	48.90	48.51	0.38	0.778

6 Regression Analysis

Further the regression analysis has also been done in order to ascertain the relationship between the yield shear stress and magnetic field using an equation

$$\tau_y = 0.003B4 - 0.076B3 + 0.411B2 + 3.724B - 4.322$$

7 Conclusions

From this research on review of MR fluids characteristics and formulation, it can be seen that the yield stress of MR fluids depends mainly on the volume fraction of the iron particles of the fluid. The stiffness and consistency of MR fluid can be regained on removal of magnetic field from solid-to-liquid state and vice versa occurs on removal of applied magnetic field. With increase in the magnitude of applied magnetic field, different iron particles get aligned to form chains along the lines of stronger magnetic flux. This high value of yield stress can be used more effectively in formulation of higher capacity MR dampers and brakes.

References

1. Jolly, M.R., Bender, J.W., Carlson, J.D.: Properties and applications of commercial magnetorheological fluids. J. Intell. Mater. Syst. Struct. **10**(1), 5–13 (1999)
2. Kciu, M., Turczyn, R.: Properties and application of magneto-rheological fluid. J. Achiev. Mater. Manuf. Eng. **18**, 1–2 (2006)
3. Tang, X., Chen, Y., Conrad, H.: Structure and interaction force in a model magneto rheological system. J. Intell. Mater. Syst. Struct. **7**, 517–521 (1996)
4. Roupec, J., Mazůrek, I., Strecker, Z., Klapka, M.: J. Phys. Conf. Ser. **412**(1), 66–67
5. Varela-Jiménez, M.I., Vargas Luna, J.L., Cortés-Ramírez, J.A., Song, G.: J. Smart Mater. Struct. **24**, 39–45 (2015)
6. Bogdan, S., Wojciech, H.: Rheological properties of MR fluids recommended for use in shock absorbers. J. actamechanica et automatica **7**(2), 107–110 (2013)

A Simulation-Based Study for Multi-response Optimization of FMS Performance Measures Using Combined Grey Relational and Principal Component Analyses

Shafi Ahmad, Mohammed Ali, and Zahid A. Khan

1 Introduction

In the current scenario, manufacturing has become highly competitive, and therefore, manufacturing firms are moving toward flexible manufacturing systems (FMS). The performance of FMS system is mainly reliant on its machine layout. Using a proper layout for machines, it has been reported that manufacturing costs can be reduced by 10–30% [1]. Further, FMS systems operate in an asynchronous manner which results in substandard utilization of the machines and AGVs. Thus, proper scheduling of AGVs and CNC machines affects the performance of FMS [2]. For analyzing and designing of systems, simulation is a widely used tool which requires few simplifying assumptions. However, it does not provide any method for optimization. Deng in 1982 proposed grey relational analysis (GRA) which is a convenient method for resolving multi-response optimization problem [3]. In this method, all responses of a system are given equal weights, but in real-life problems, the weights of the responses may not be the same. Principal component analysis (PCA) is a method which can be used to determine weights of the responses. A combined GRA-PCA method can provide relatively better results as compared to only GRA method, and therefore, researchers have used combined GRA-PCA method for multi-response optimization [4, 5]. However, literature reveals a very few application of the combined GRA-PCA in simulation-based studies for multi-response optimization of FMS performance measures. Hence, a combined GRA-PCA method has been used in this work for multi-response optimization of the FMS performance measures.

S. Ahmad (✉) · Z. A. Khan
Department of Mechanical Engineering, Jamia Millia Islamia, New Delhi 110025, India
e-mail: Shafiahmad.amu@gmail.com

M. Ali
Department of Mechanical Engineering, Aligarh Muslim University, Aligarh 202002, India

M. Muzammil et al. (eds.), *Recent Advances in Mechanical Engineering*, Lecture Notes in Mechanical Engineering, https://doi.org/10.1007/978-981-15-8704-7_34

2 Methodology

The step-by-step procedure used in this research work is as follows:

1. Design of experiments.
2. Development of simulation model using ARENA.
3. GRA-PCA Method.
4. Analysis of results.
5. Conclusion.

2.1 Design of Experiments

Taguchi method has been used extensively for experimental designs. Orthogonal arrays given by Taguchi significantly decreases the number of experimental configurations to be performed, and it provides a best set of well-balanced experiments [6]. Four input parameters of FMS viz. type of layout (A), velocity of the AGV (B), number of AGVs (C), and sequencing rule (D) are used in this study. Types of layout considered are Inline, Ladder, Loop, and OpenField. Velocity of AGV is taken as 10, 20, 30, and 40 m/min. Number of AGVs is 1, 3, 5, and 7, and sequencing rule FCFS and HPT are considered. It has been found that L_{16} orthogonal array is the best suitable orthogonal array for this study.

Flexible manufacturing systems are known for the faster delivery of products with maximum utilization of machines and material handling systems. Hence, performance measures considered in this work are make span time (MST), AGV utilization (AU), and average machine utilization (AMU).

2.2 Development of Simulation Model Using ARENA Simulation Software

A hypothetical case problem based on a real-time data of a factory is considered [7]. ARENA simulation software which is a powerful tool for carrying simulation experiments is used in this study. Five different parts are generated according the inter arrival rate (i.e., expo (10) minutes). Figure 1 shows the logic module used in this work.

Verification is the process of checking that the model developed fulfills its intended purpose or not. With the help of animation of the developed simulation model, verification is done. If the real data is not available, explicit validation is impossible [8]. In this study also, it was difficult to collect real data because of several constraints, and therefore, explicit validation could not be done. The results obtained from simulation model are used to calculate signal-to-noise ratio.

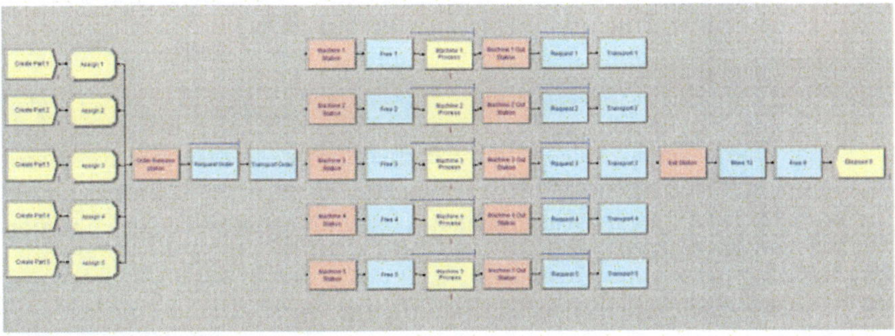

Fig. 1 Logic module used in this work

2.3 GRA-PCA Method

We have used combined GRA-PCA method in this study. The steps used for combined GRA-PCA method employed in this work are as follows:

Step1 Normalization of the data using Eq. 1.

$$u_{i,k}^{*} = \left\{ \begin{array}{ll} \frac{\max(u_k) - u_{i,k}}{\max(u_k) - \min(u_k)}, & \text{if characteristic is "lower-the-better"} \\ \frac{u_{i,k} - \min(u_k)}{\max(u_k) - \min(u_k)}, & \text{if characteristic is "lower-the-better"} \end{array} \right\} \tag{1}$$

where $u_{i,k}$ represents the S/N ratio for the kth performance measure in ith experiment.

Step 2 Determining grey relational coefficient using Eq. 2.

$$b_{i,k} = \frac{\Delta_{\min} + \tau.\Delta_{\max}}{\Delta_{i,k} + \tau.\Delta_{\max}} \tag{2}$$

where $\Delta_{i,k}$ is the deviation sequence, i.e., $\Delta_{i,k} = |\max(u_i{}^{*}{}_k) - u_i{}^{*}{}_k|$. τ is the distinguishing coefficient which is set to be 0.5 in this study.

Step 3 Determining weights for response variables using PCA as follows:

(a) Develop variance covariance matrix V using Eq. 3.

$$V = \left[b_{i,j}\right]_{m \times n} \tag{3}$$

(b) Develop correlation coefficient array using Eq. 4.

$$A_{j,l} = \left[\frac{\text{Cov}\left(b_j, b_l\right)}{\sigma_{b_j} * \sigma_{b_l}}\right] \tag{4}$$

(c) Calculate Eigen vectors and Eigen values using Eq. 5.

$$(A - \lambda_p I_p)X_{ip} = 0 \tag{5}$$

(d) The weighting value of the related performance measures is the squares of its corresponding Eigen vector. Hence, weight for make span time is 0.51123, for AGV utilization it is 0.02372, and for machine utilization 0.46512.

Step 4 Calculate grey relational grade (GRG) using Eq. 6.

$$\delta_i = \sum_{p=1}^{n} w_p y_i(p) \tag{6}$$

where n denotes number of response variables and w_p denotes weight of the performance measures. Grey relational grade values thus obtained along with the results obtained from simulation are given in Table 1.

Table 1 Experimental results and GRG values

Ex. No.	*A*	*B*	*C*	*D*	MST (in min)	AU	AMU	GRG
1	Inline	10	1	FIFO	19,369.162	0.9802	0.60528	0.6434
2	Inline	20	3	FIFO	18,492.73	0.4946	0.63638	0.8996
3	Inline	30	5	HPT	18,693.199	0.5761	0.62668	0.6168
4	Inline	40	7	HPT	18,669.124	0.6893	0.62854	0.6473
5	Ladder	10	3	HPT	18,761.252	0.5903	0.6256	0.5784
6	Ladder	20	1	HPT	18,925.212	0.7578	0.61952	0.4777
7	Ladder	30	7	FIFO	18,514.879	0.6892	0.63384	0.8215
8	Ladder	40	5	FIFO	18,421.758	0.5861	0.6372	0.9878
9	Loop	10	5	FIFO	18,453.742	0.6325	0.63464	0.8910
10	Loop	20	7	FIFO	18,463.421	0.7111	0.63482	0.8848
11	Loop	30	1	HPT	18,687.635	0.3232	0.62786	0.6445
12	Loop	40	3	HPT	18,647.891	0.3713	0.62944	0.6778
13	OpenField	10	7	HPT	18,678.285	0.7169	0.6275	0.6305
14	OpenField	20	5	HPT	18,616.661	0.5892	0.63	0.6940
15	OpenField	30	3	FIFO	18,442.937	0.4449	0.63702	0.9640
16	OpenField	40	1	FIFO	18,436.871	0.4086	0.6357	0.9377

3 Analysis of Results

The S/N ratio corresponding to each GRG value is used to search the best input parameter setting under various conditions. For every level of the input parameter, the mean value of the S/N ratio is given in Table 2. From Table 2, it can be observed that sequencing rule has the highest effect on grey relational grade (rank 1) followed by number of AGVs (rank 2), velocity of AGVs (rank 3), and type of layout (rank 4).

Figure 2 shows the main effects plot for the S/N ratios of the GRG. It is found that GRG value will be maximum at level 4 of type of layout (i.e., OpenField layout), level 4 of AGV velocity (i.e., 40 m/min), level 3 of number of AGVs (i.e., 5), and level 1 of sequencing rule (i.e., FCFS). Hence, the optimal parameter combination for multi-response optimization is $A_4B_4C_3D_1$.

ANOVA is performed for GRG with a significance level of 5%, and the results thus obtained are given in Table 3. Table 3 shows that only sequencing rule significantly affects the multi-performance characteristics (p-value ≤ 0.05).

Table 2 Response table for GRG

Input parameter	Level 1	Level 2	Level 3	Level 4	Δ	Rank
Type of layout	−3.181	−3.247	−2.315	−2.014	1.233	4
AGV velocity	−3.399	−2.893	−2.510	−1.955	1.443	3
Number of AGVs	−3.655	−2.343	−2.120	−2.639	1.535	2
Sequencing rule	−1.188	−4.190				1

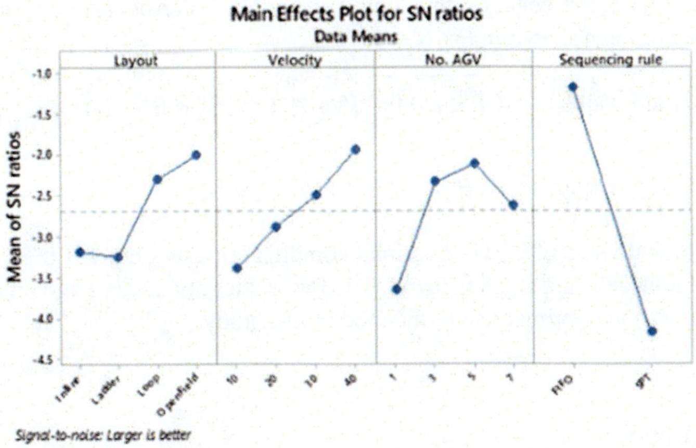

Fig. 2 Main effect plot for S/N ratio for grey relational grade

Table 3 Analysis of variance (ANOVA) for grade value

Source	DOF	SS	MS	F-value	p-value
Type of layout	3	0.02904	0.00968	2.8	0.148
AGV velocity	3	0.03321	0.01107	3.2	0.121
Number of AGV	3	0.03463	0.01155	3.34	0.114
Sequencing rule	1	0.26594	0.26594	76.91	0.0001
Error	5	0.01729	0.00346		
Total	15	0.38011			

3.1 Confirmatory Test

The correctness of the obtained results is ensured by comparing the experimental value of the GRG with that of its corresponding predicted value at optimal combination. The predicted value of the GRG at optimal input parameters setting is obtained by using Eq. 7.

$$\eta_0 = \eta_{\mathrm{m}} + \sum_{i=1}^{j} (\eta_i - \eta_{\mathrm{m}}) \tag{7}$$

where j is the number of input parameters, η_{m} represents the mean value of the GRG, and η_i represents the mean of the GRG corresponding to optimal input parameter levels.

To calculate the experimental value of GRG, regression analysis has been performed. Regression equation thus obtained for GRG considering a significance level of 5% has been shown in Eq. 8.

$$\text{Grade Value} = -69.5 + 0.00158 * A + 0.288 * B + 64.5 * C \tag{8}$$

where A = make span time, B = AGV utilization, and C = average machine utilization.

The predicted value of GRG at optimal combination using Eq. 7 is 0.79, and the experimental value using Eq. 8 is 0.768. The two values are in close agreement, and thus, it validates the optimal result obtained in this study.

4 Conclusion

This paper demonstrated the application of grey relational analyses combined with principal component analyses for multi-response optimization of performance measures of an FMS. ARENA simulation software is used to model the FMS and to extract values of three performance measures under the different levels of four input/control parameters. The conclusions of the present study are summarized below:

- Principal component analysis combined with grey relational analysis is simple, effective, and efficient method for multiple response optimization problems.
- Within the range investigated, it has been found that input/control FMS parameters with OpenField layout, 40 m/min AGV velocity, 5 AGVs, and first-come first-served sequencing rule yield optimal performance measures of the FMS.
- Sequencing rule has the maximum percentage contribution (69.96%) followed by number of AGVs (9.11%), AGV velocity (8.74%), and type of layout (7.64%) toward multi-performance measures of the FMS.

References

1. Tompkins, J.A., White, J.A., Bozer, Y.A., Tanchoco, J.M.A.: Facilities Planning, 4th Edn. John Wiley & Sons. New York (2010)
2. Ali, M.: Effect of scheduling and manufacturing flexibility on the performance of FMS. Glob. J. Flexible Syst. Manage. **11**, 21–38 (2010)
3. Kuo, Y., Yang, T., Huang, G.W.: The use of grey relational analysis in solving multiple attribute decision-making problems. Comput. Ind. Eng. **55**, 80–93 (2008)
4. Dewangan, S., Biswas, CK., Gangopadhyay, S.: Optimization of the surface integrity characteristics of EDM process using PCA based grey relation investigation. In: 3rd International Conference on Materials Processing and Characterization, pp. 1091–1096, Elsevier (2014)
5. Umamaheswarrao, P., Raju D.R., Suman, K.N.S., Sanka, B.R.: Multi objective optimization of process parameters for hard turning of AISI 52100 steel using hybrid GRA-PCA. In: International Conference on Robotics and Smart Manufacturing, pp. 703–710, Elsevier (2018)
6. Phadke, M.S.: Quality Engineering Using Robust Design. Prentice Hall, USA (1989)
7. Chang, G.A., Peterson, W.R.: Modeling and analysis of flexible manufacturing systems : a simulation study. In: 122nd ASEE annual conference and exposition, Seattle, WA (2015)
8. Sargent, R.G.: Verification and validation of simulation models. In: Proceedings of the 2010 Winter Simulation Conference, pp. 166–183, IEEE (2010)

Investigational Assessment of Heat Transfer and Pressure Drop in a Circular Tube

Mohan Gupta, Kamal Sharma, and Kunwar Mausam

1 Introduction

1.1 Heat Exchangers

Heat exchanger is an important bit of the refrigeration and potential series that empowers the trading of centrality beginning with one means then after onto the join by rectitude of temperature discrete. In the brilliance exchanger, the temperatures of all fluid variety as this go over the exchanger and from here time propelling the temperature of the pulling back curtain among the fluids in particular manner moves with the entire extent of the exchanger. This is consequently suggested to pass the perfect degree of warmth conformity as lively as could sensibly be regular. The inventors race to develop the capable and progressively minor warmth exchangers in any event theory and functioning cost. Heat pass refinement orders are pulled back into two sorts, the beginning is a operating system that it needs an outer potential center point, and the second is a blocked off technique which not requires any sort of external potential authority.

The pace of action of heat of the warmth exchangers is generally reliant on the surge trimming of deportation plane and the liquid attributes identified with its thermo-physicality. The procedure of heat move is hampered in light of the fact that the languid layer of the outskirt appended to the divider gave convective obstruction. To expand the action of heat, any procedure for advancing upheaval in the region of the warmed divider would be gainful presumably for helping with controlling the concurrent increment in frictional misfortunes for holding them under bearable point of confinement. Consequently, it is in every case exemplary to utilize embeds on the grounds that it is equipped for expanding the pace of heat move for recede increment in frictional misfortunes. Diverse cylinder embeds have been analyzed

M. Gupta (✉) · K. Sharma · K. Mausam
GLA University, Mathura, U.P, India
e-mail: mohanguptaucer@gmail.com

M. Muzammil et al. (eds.), *Recent Advances in Mechanical Engineering*, Lecture Notes in Mechanical Engineering, https://doi.org/10.1007/978-981-15-8704-7_35

with center around accomplishing improved thermo-water potentiated execution of warmth exchangers.

In the investigation, Eiamsa-ard et al. [1] completed audition for assessing the action of heat and nature of polish in a circuitous bower, i.e., added to complete-length "TTs" escorted by mutate degrees of "6 and 8" and the margin degrees of "1, 2, and 3." The assessment saying that utilizing the "TT" embeds in within housing of a bifold channel, HE can broadly lift the pace of action of heat. The entirety examines the rubbing factor of the housing at the same time increments with the use of "TT" inserts. Heat pass and contact are uplifted by ideals of the turning advancement occurring in perspective on the optional developments of the liquid.

A further assessment, Eiamsa-ard et al. [2] composed assessment for taking a gander at the action of vitality and surge liquid information of the gleam trading contraption that utilized a helical screw-strip built of hardened steel of 17 mm width and the respire margin of housing layer "$(D - W)/2 = 4$" mm. The strips are utilized in the absence of or with center rod introduces in the bifold funnel. In the sparkle exchanger, there are external and inward housing with partitions transversely more than 50 and 25 mm by way of warm and cool waters surge as they were utilized as appraisal liquids, and these liquids stay in tube and housing side, independently. The "Re" connected from "2000 to 12,000." The result of assessment portrayed that the action of heat was superior for the empty fit with no center shaft rather than a screw-shaped strip with center post. The sparkle trade rates were commonly 340 percentages more in spite of utilizing the screw-shaped strip with no center bar embed rather than empty housing. Notwithstanding the manner in which that contact diminished extensively for utilizing a helical screw-strip with an inner side insert, the sparkle trade rates were 25–60 percentage high for the tape with no center strip. In like manner, the effectiveness wound up being bifold for the helical screw-strip with no center post.

In his further trial, Eiamsa-ard et al. [3] explored the effect of utilizing delta-wingtip strips as augmentations in the round loads. For organizing the assessment, a crabwise delta-wingtip "TT" and a horizontal delta-wingtip "TT" are utilized. The delta-wingtip twisted strips utilized for assessment had three sagacity of wing slash degrees 0.11, 0.21, and 0.31 and curve degrees "3, 4, and 5." The "Re" span from "3000 to 27,000." On the basis of the founded outcomes, this is clear, by reducing in twine degree and enlargement in the essentialness of wing cut degree; there is expansion in the sparkle pass coefficient and scrub part. In addition, the slanted delta-wingtip "TT" indicated increment in the sparkle pass coefficient get up apart from horizontal delta-wingtip "TT."

It is found by Eiamsa-ard et al. [4], the propensities beyond the twin counters and co-turn twisted strips sway the pace of heat move. Principally, the twin counter new shaped funnels were utilized as opposite-turn surge generators; anyway, the twin co-wound strips contemplate the action of co-turn surge passing on inserts. The "CTs" and "CoTs" introduce alongside four different bend degrees "$y/w = 2.5, 3, 3.5, 4$" were utilized for organizing the assessment. The "Re" continued gushing from "3700 to 21,000." What is more, the examination delineated that there is modification of the shine pass coefficient and pounding changing with reducing in turn degree.

Again Eiamsa-ard et al. [5] investigate the presentation of opposite clockwise and exchange clockwise "TT" introduces "C–CC twisted tapes" contoured with a roundabout load and separated them and the introduction of an ordinary "TT." The nine "TT" inserts with three twist degrees 3, 4, and 5, and three breeze centers 30, 60, and 90 degrees were broke down with in assessment. The C–CC twine strips essentially broadened the pace of action of heat and crushing then again with the normal twirled strips under tantamount circumstances. Similarly, the starter information uncovered that the gleam move coefficient of the "C–CC" tapes increments with decreasing in the twirl degree and refinement of the curve tip.

1.2 *Mechanism of Heat Transfer in Heat Exchanger*

See Fig. 1.

1.3 *Merits of Inserts in Round Tube*

- As appeared differently in relation to the plain housing, the housing with adjunct gets for up heat transformation standard to the detriment of extended weight fall.
- In lower regularity fluids, additional roughness is made by the utilization of options.
- While the enhancements utilized in indirect housings, the distraction is induced in the surge within the housing in view of which the point of confinement layer surge is stimulated.
- The usage of different amounts of enhancements articulates colossal redesign in the glow transformation recognized of the housing.

The genuine objectives of this analysis task are according to the accompanying:

- To accumulate the probing data identifying with heat move in an indirect housing with one, twin, and four distorted strip installs.
- To discuss regarding glow pass and disintegration attribute of the examine pipe along with different augmentations (Figs. 2, 3; Table 1).

2 Experimental Methodology

The appraisals are made by round weight comprises along with one and certain "TT" implants beyond an enormous level of "Re" "4000 to 14,000" accompanied by the water used as mixing mechanism of fluid to mean the searching detail, identifying with the miracle pass rate and pressure fall [6]. Appraisal is also made on clear housing to check the introduction of various "TT" presented housing. The liquid from the

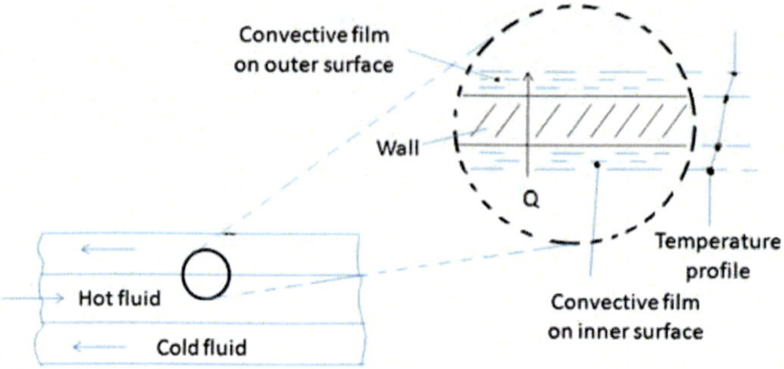

Fig. 1 Mechanism of heat transfer in heat exchanger [6]

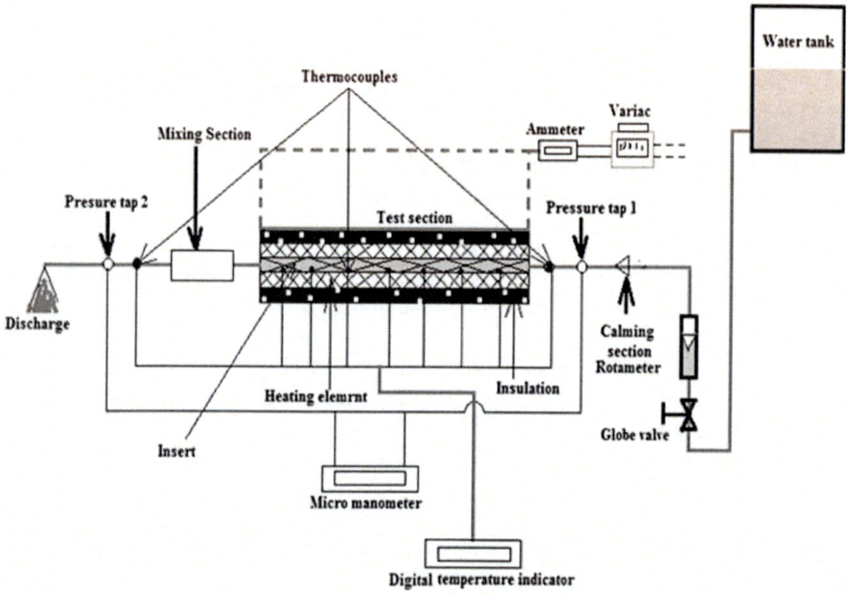

Fig. 2 Examination setup

enduring head tank passed at a perfect surge rate obliged by the glove valve, whereas the surge rate is surveyed by the rotameter. Occurrences to position the quality refinement and fluid surge rate, the temperatures of the appraisal housing surface and the fluid at passage and outward are notice. From the earliest starting point, arrange the housing surface and leave fluid temperature, rough approximations are unbalanced and approach to manage administer control undeviating higher traits with the time. It is noticed that all thing considered following "30 min," the temperature at each of the section winds up liberated from the time that stands up concentrated on that the pattern is developed the suffering state condition. Not long after the accomplishing

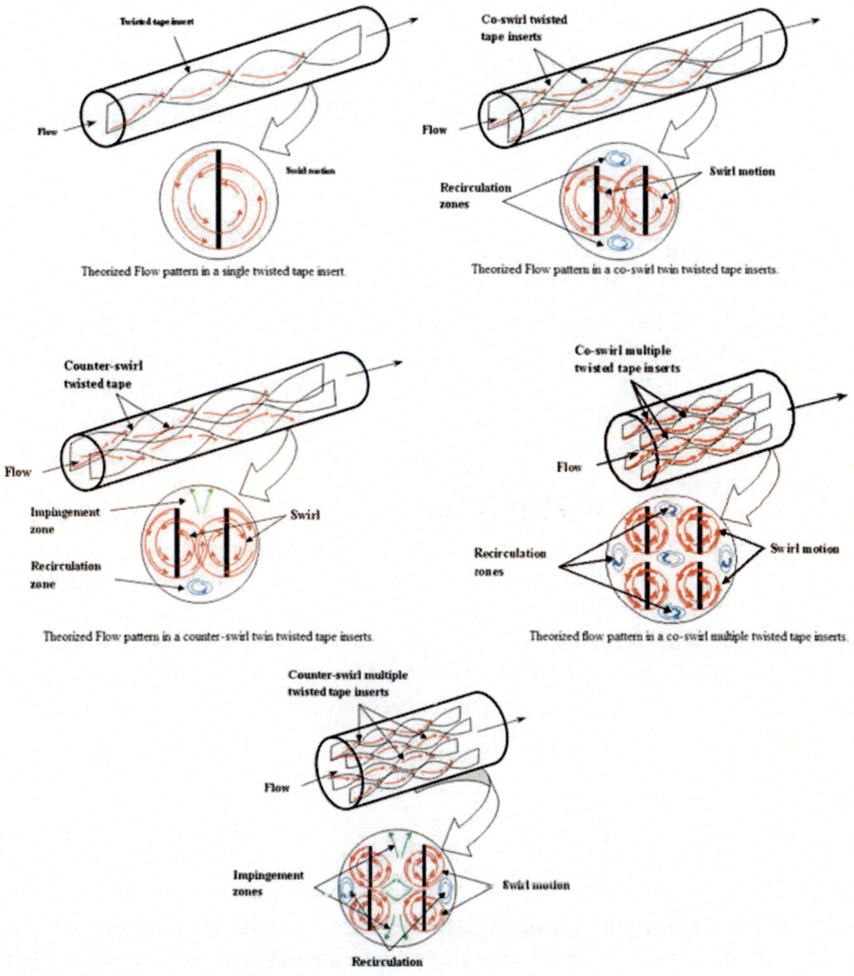

Fig. 3 Tube inserts configuration [6]

Table. 1 Scale of "TT" guidelines

Twisted tapes	ST	CoTs	CTs	4CoTs	4CTs
"Tape width" w	24	12	12	6	6
"Tape pitch" P	84,72, 60	42, 36, 30	Alike "CoTs"	21, 18, 15	Alike "4CTs"
"Twist ratio" "$y = P/w$"	2.50, 3, 3.50	Alike "ST"	Alike "ST"	Alike "ST"	Alike "ST"
"Tape thickness" t	0.6	Alike "ST"	Alike "ST"	Alike "ST"	Alike "ST"
Material	Aluminum	Al	Al	Al	Al

(all in mm)

the technique on identical state, the housing surface temperature at eight regions, closure centers, and departed fluid temperature is noted.

3 Results and Discussion

3.1 "Outcome" of "Re"

The refinement in the guesstimate of "Nu" and disintegration part exhibiting the assortment of the "Nu" overhaul extent and contact factor refinement extent escorted by alteration in "Re" for (*y*) of 2.5. Redesigns in "Nu" and scouring component move in the solicitation of one distorted strip insert, twin co-spin twirl strip implant, twin opposite-whirl wound strip install, four co-spin bended tape additive, and four opposite-spin tape implant.

The introduction of four opposite, spin wound strip "4CTs," outflanks the remainder of the arrangements paying little heed to the guesstimate of the "Re." The thermo-water controlled execution part had an irrelevant assortment regarding surge "Re" for the entire instance. The assortments of "Nu" and disintegration part concerning the "Re" for twist extent of 3 and 3.5 are as per the examples obtain if there ought to be an event of the breeze extent equal to 2.5 beside that the effect of "Re" on the "Nu" and scouring component is covered, and even the twist extent (*y*) is extended.

3.2 "Outcome of Twist Ratio" "y"

Twain "Nu" and the grinding changing are extended accompanied by abatement in the guesstimate of reshape extent offering slight appreciation to the guesstimate of "Re" in every single case. It may be accredited to the manner in which that if water skims outwardly of the twisted strip incidentally and the disconnected speed part starts an emanating potential that reestablishes the point of confinement coating stream. By the decline in the curve extent, the locale influenced by the spiral forces grows up and accordingly propels the wild potential of the fluid adjacent to the divider.

3.3 "Outcome of Counter-Swirl and Co-Swirl Twisted Tapes" "TTs"

The level of progress of the "Nu" and contact part if there ought to be an event of four opposite-spin twirl strips outperforms the update made by the remainder of the sorts of bended tapes at the distort extent of 2.5. On the off chance that there ought to emerge an event of co-spin perverted tapes, the unclear spins are made in the surge field; however, opposite-whirl wound strips start spins toward the way reverse to each other, along these lines forming all the all the more adjusting fluid streamlet which rises the turn paces of the fluid within the housing. The opposite-spin surge in like manner impels the discretionary surge due to which extremity of unsettling affects extends that propel the fluid merging and thusly engages as far as possible surface interior the exterior of the appraisal tube.

4 Conclusions

In perspective on the assessment given closures can be acquired:

- The "Nu" augmentations and crushing vacillating decays as "Re" increases for each kind of bowed strips. The "Re" has increasingly essential effect on "Nu" and disintegration factor regards even as amount of wound strip adjuncts is extended.
- The refinement degrees of "Nu" and rubbing part rot as "Re" enlarges in each case. The "Nu" raised degrees exist within "1.13–1.2, 1.3–1.4, 1.6–1.7, 1.9–2.1, and 2.3–2.4" for "ST," "CoT," "CT," "4CoT," and "4CT," self-bouncy, and the breeze degree is remain stays as 2.5. The thudding part refinement degrees are acknowledged to be in the degree of "2.6–3.4, 3.1–4.1, 4.3–5.34, 5.2–6.18, and 6.5–6.9" for the beginning late cited arrangements of redesigns.
- The "Nu" and rubbing factor regard is extended by decreasing the bend extent 'y' of twine strips free of the alteration in "Re."
- The counter-spin sort of twisted strip adjunct comes out higher heat move and granulating interestingly with the co-whirl reshaped strip installs.
- The most outrageous guesstimate of the thermo-water-driven enactment part is seen to be '1.26' for a great deal of four opposite-spin bended channels with twirl extent of '2.5.'

References

1. Eiamsa-ard, S., Thianpong, C., Promvonge, P.: Experimental investigation of heat transfer and flow friction in a circular tube fitted with regularly spaced TT elements. Int. Commun. Heat Mass Transf. **33**, 1225–1233 (2006)
2. Eiamsa-ard, S., Promvonge, P.: Heat transfer characteristics in a tube fitted with helical screw-tape with/without core-rod inserts. Int. Commun. Heat Mass Transf. **34**, 176–185 (2007)
3. Eiamsa-ard, S., Wongcharee, K., Eiamsa-ard, P., Thianpong, C.: Heat transfer enhancement in a tube using delta-winglet twisted tape inserts. Appl. Therm. Eng. **30**, 310–318 (2010)
4. Eiamsa-ard, S., Thianpong, C., Eiamsa-ard, P.: Turbulent heat transfer enhancement by counter/co-swirling flow in a tube fitted with twin twisted tapes. Exp. Thermal Fluid Sci. **34**, 53–62 (2010)

5. Eiamsa-ard, S., Promvonge, P.: Performance assessment in a heat exchanger tube with alternate clockwise and counter-clockwise twisted-tape inserts. Int. J. Heat Mass Transf. **53**, 1364–1372 (2010)
6. Vashistha, C., Patil, A.K., Kumar, M.: Experimental investigation of heat transfer and pressure drop in a circular tube with multiple inserts. Appl. Therm. Eng. **96**, 117–129 (2016)

A Comparative Analysis of Transmission Efficiency of Polyamide 66 Spur Gears Meshing with Similar and Dissimilar Gear Material

Akant Kumar Singh, Prashant Kumar Singh, Siddhartha, and Sanjay Yadav

1 Introduction

In general, the thermoplastic materials which are most frequently used for the manufacturing of plastic gears are acetal and nylon. Polymer gear possess some significant edges over the metal gears in terms of quieter operations, lower inertia, lower coefficient of friction, and most importantly the lower cost. Due to these characteristics, the plastic gears are now replacing the metal gears in plethora of industrial applications. However, their applications are limited to low and medium load conditions only. Utilization of the plastic gear enhances the efficiency of the mechanical systems due an overall reduction in the weight of the system [1]. Nowadays, there are various applications where plastic gears are used. Though, the study regarding TE of plastic gears are not explored much so far. It has been observed by Walton et al. [2, 3] that TE of polymer gears are very much influenced by load and speed. It is found that the material of gears and their geometries also affect the TE of these gears.

Lots of work has been done on plastic gears in the last decade [4]. It has been observed by Senthilvelan and Gnanamoorthy [5] that TE of carbon fiber-filled polymer composite gear improved as compared to neat gear. It happens because thermal peculiarity and gear tooth stiffness improved by reinforcing the fibers. It has also been observed that addition of nano-clay fillers also improves the TE of plastic gears [6]. Some researchers have used the compressed air at the mating surface to improve the TE of plastic gears, and almost 2% of enhancement is achieved in TE [7]. The effect of tooth deformation and gear tooth wear rate on transmission efficiency is investigated by Nagarajan and Rayudu [8]. Error in transmission due to the geometry of polymer gear is studied by Luscher et al. [9]. It has been observed in the study

A. K. Singh (✉) · S. Yadav
Department of Mechanical Engineering, I.T.S Engineering College, Greater Noida 201308, India
e-mail: akant.me@its.edu.in

P. K. Singh · Siddhartha
Department of Mechanical Engineering, NIT Hamirpur, Hamirpur, H.P. 177005, India

M. Muzammil et al. (eds.), *Recent Advances in Mechanical Engineering*, Lecture Notes in Mechanical Engineering, https://doi.org/10.1007/978-981-15-8704-7_36

of Mertens and Senthilvelan [10] that TE of polymer gears is enhanced by adding carbon nanotubes in gear material. They have done the experiment on nanotube-filled polypropylene spur. Kodeeswaran et al. [11] have done the experimental as well as numerical estimation of TE of polymer gears.

Reinforcement of fibers in polymer gear alters the performance of polymer gear very much. It improves the performance of the polymer gears by reducing the wear and temperature at the mating surfaces of polymer gear teeth [12–14]. Also, fiber-filled polymer composite gears have improved TE [2, 5]. Addition of fibers in polymer gears increases the damping characteristics of these gears which affect the transmission efficiency [15]. TE is also explored by the researchers for different materials [16–18]. A novel FGM based polymer spur gear has been fabricated by Singh and Siddhartha [19]. It has been observed that FGM based gears possess better TE in comparison with conventional polymer gears. Transmission efficiency has observed for various plastic gears fabricated with different materials. Maximum investigations are done for the polymer-to-metal gear pair. However, it has been found from literature that transmission efficiency has not been explored much for polymer-to-polymer gear pair. Therefore, the focus of this study is to investigate the TE of polymer–polymer gear pair and compared with the polymer–steel gear pair.

2 Materials and Methodology

2.1 Gear Manufacturing

In this study, the plastic gears are manufactured through injection molding technique. These gears are fabricated using PA 66 material. A dryer is used for heating PA66 materials at 90 °C for 4 h to eliminate the moisture content. Injection pressure of 90 MP is used to fabricate gears. A pictorial view of a fabricated gear is presented in Fig. 1.

2.2 Polymer Gear Test Rig

The testing of the gears is carried out by using a customized gear test rig. The testing of the polymer gears is done with similar as well as with their steel counterparts. Specifications of polymer and steel gears are same. Polymer test rig is manufactured by Bangalore-based company DUCOM Instruments. A pictorial view of the test rig is provided in Fig. 2. The torque is measured by torque sensors which are acting on driver and driven gears. Accuracy of the torque sensor is ±0.2%. Temperature was continuously measured with the help of an infrared sensor during experimentation. Temperature sensor is just fitted above the contact surface of driver and driven gear

Fig. 1 Fabricated PA66 spur gear

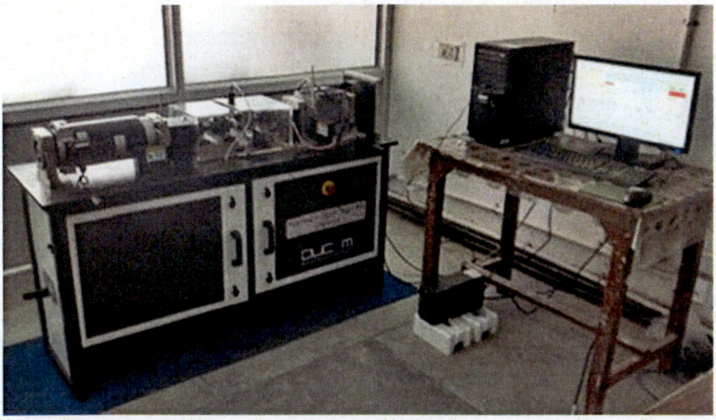

Fig. 2 Pictorial view of test rig

pairs. List count of the temperature sensor is 1 °C, and it can measure the gear tooth surface temperature up to 130 °C.

2.3 *Transmission Efficiency*

Transmission efficiencies of similar and dissimilar gear pairs are examined. The efficiency is computed with the help of Eq. 1 which is provided below. The loss of power which might be occurring at couplings and bearings is neglected in this computation.

$$TE = DGT_1/DGT_2 \tag{1}$$

DGT_1 Driven gear torque.
DRT_2 Driver gear torque.

3 Results and Discussions

3.1 Influence of Speed and Torque on TE of Polymer Gears

TE very much depends on the gear material of driver and driven gears pairs. Torque and speed are very much affected by the polymer gears. Deviation in the TE and surface temperature of polymer gear is shown in Fig. 3a, b. It is caused by the torque at different speeds of 600 and 1200 rpm. It is observed in Fig. 3 that when torque increases, it results in an increase in the temperature of gear tooth surface and a reduction of TE. Polymer gear tooth flexibility affects the temperature of the gear tooth very significantly. An increasing torque rises the temperature of gear tooth surface which ultimately results in the deflection of gear teeth. This deflection causes a reduction in the torque of the test gear that finally results in a reduced TE of the test gears [3] which is observed after accomplishment of 1.2 million cycles.

TE of polymer–steel gears pair is higher as compared to polymer–polymer pair. It occurs due to lower temperature of gear tooth surface in case of polymer–steel gears pair. Polymer–polymer and polymer–steel gear pairs have high TE at 600 rpm while decrement in TE is seen for both gear pair at 1200 rpm as shown in Fig. 3a, b. This may happen because of the higher temperature of gear tooth surface at the speed of 1200 rpm bases less TE for both gear pairs. The temperature at the surface of polymer–steel gear pair is less because steel gear has higher thermal conductivity. Whatever heat generated because of the meshing of the gear teeth surface of PA66 gear dissipates quickly. However, it does not happen for polymer–polymer gear pair. Both are the polymer material having less thermal conductivity causes high surface temperature. TE of polymer-to-polymer gear pairs and polymer-to-steel gear pairs is reduced by 19% and 13%, respectively, at 600 rpm and 22% and 17%, respectively, at 1200 rpm with an increase in the torque value from 0.8 Nm to 2 Nm.

TE of polymer-to-polymer gear pairs and polymer-to-steel gear pairs does not get influenced much with the variation of speed, as witnessed in Fig. 4. Thus, it is clear that TE of polymer-to-polymer gear pairs and polymer-to-steel gear pairs decreases with upsurge in speed. Same results are also observed by Walton et al. [2, 3] for various polymer materials used to manufacture gears. TE of polymer-to-polymer and polymer-to-steel gear pair is high for 0.8 Nm torque as compared to 2 Nm torque for each constant speed which is shown in Fig. 4. This may happen because of a higher temperature at the surface of both gear pairs at the torque of 2 Nm. TE of polymer-to-polymer gear pair and polymer-to-steel gear pair decreases marginally from 4 to 8% when the speed rises from the speed of 600–1200 rpm.

At 0.8 Nm, polymer-to-steel gear pair have 9% higher TE as comparison to polymer–polymer pair at the speed of 1200 rpm. However, at a constant torque of 2 Nm, polymer–steel gear pair have 13% higher TE as compared to polymer–polymer

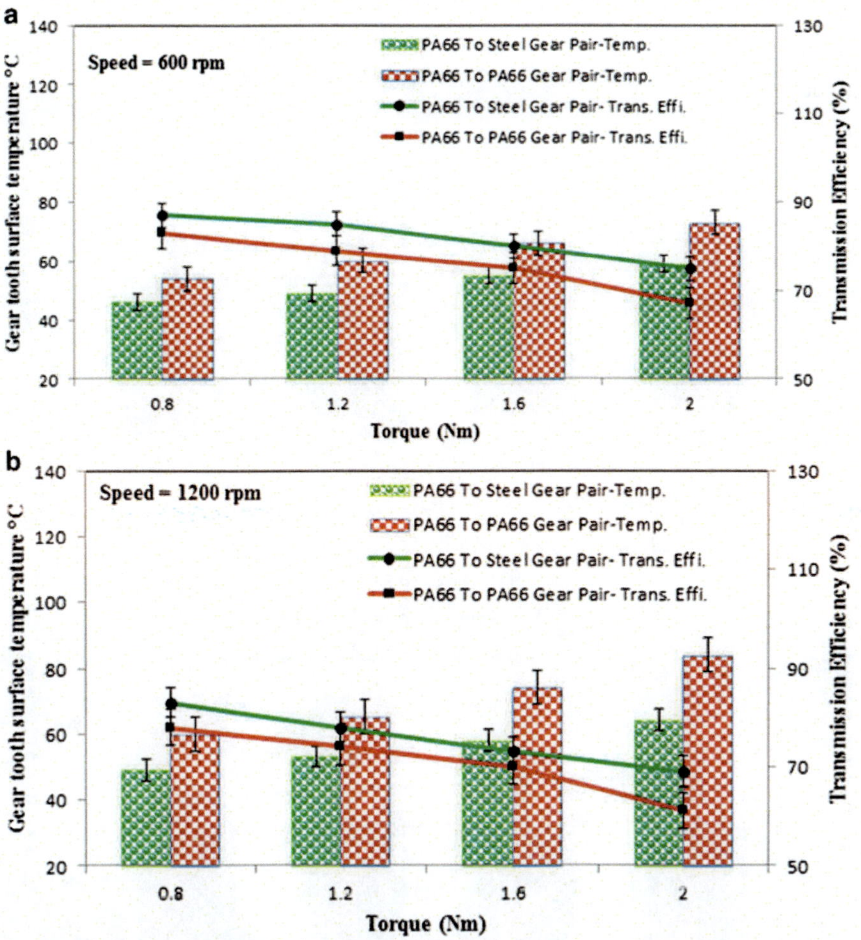

Fig. 3 **a** Transmission efficiency of gear pairs at speed of 600 rpm, **b** transmission efficiency of gear pairs at speed of 1200 rpm

pair at the speed of 1200 rpm. Hence, it is concluded that the polymer–steel gear pair performs better than the polymer–polymer gear pair.

4 Conclusions

Conclusions from this study are shown below:

1. TE is less influenced by speed rather than torque for both gear pairs. Transmission efficiency is reduced by 14% with increase in torque from 0.8 to 2.6 Nm. However, it is reduced only by 2% with increase in speed from 600 to 1200 rpm.

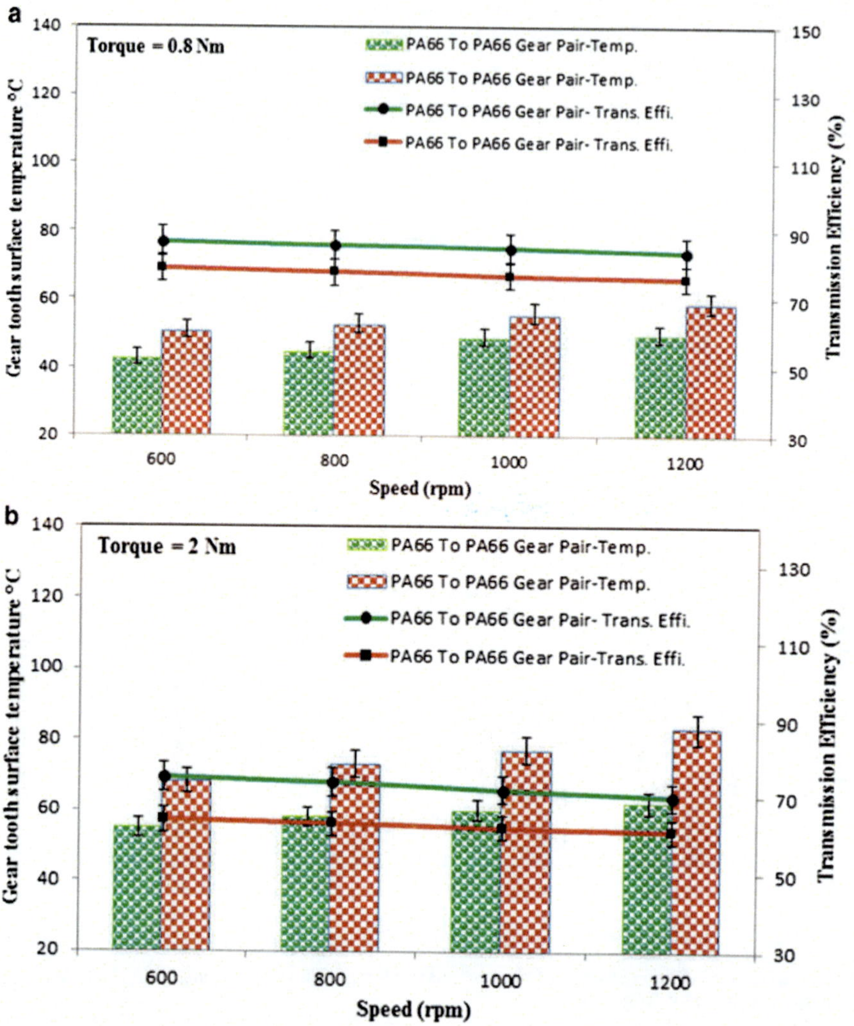

Fig. 4 **a** Transmission efficiency of gear pairs at 0.8 Nm torque, **b** transmission efficiency of gear pairs at 2 Nm torque

2. Polymer-to-steel gear pair has higher TE as compared to polymer-to-polymer gear pair. At 2 Nm torque, polymer-to-steel gear pair have 13% higher TE in comparison with polymer-to-polymer gear pair at 1200 rpm.
3. Polymer-to-polymer gear pair has minimum transmission efficiency at the torque of 2 Nm and the speed of 1200 rpm.

References

1. Adams, C.E.: Plastic gearing: Selection and application, 2st edn. Marcel Dekker, New York (1986)
2. Walton, D., Cropper, A.B., Weale, D.J., Meuleman, P.K.: J. Eng. Tribol. **216**, 75–92 (2002a)
3. Walton, D., Cropper, A.B., Weale, D.J., Meuleman, P.K.: J. Eng. Tribol. **216**, 93–103 (2002b)
4. Singh, A.K., Siddhartha, Singh, P.K.: J. Eng. Tribol. https://doi.org/10.1177/1350650117711595 (2017)
5. Senthilvelan, S., Gnanamoorthy, R.: J. Eng. Tribol. **223**, 925–928 (2009)
6. Kirupasankar, S., Gurunathan, C., Gnanamoorthy, R.: Maters. Des. **39**, 338–343 (2012)
7. Mertens, A.J., Senthilvelan, S.: J. Mat. Des. Appl. **230**, 515–525 (2015)
8. Nagarajan, G.V.N.R.: Proceedings of the 7th World Congress on the Theory of machines and mechanisms Sevilla, Spain 1335–1338 (1987)
9. Luscher, A., Houser, D., Snow, C.: J. Injection Molding Technol. **4**, 177–190 (2000)
10. Mertens, A.J., Senthilvelan, S.: J. Mat. Des. Appl. https://doi.org/10.1177/1464420716654308 (2016)
11. Kodeeswaran, M., Suresh, R., Senthilvelan, S.: Int. J. Power Trains **5**, 246–263 (2016)
12. Gurunathan, C., Kirupasankar, S., Gnanamoorthy, R.: J. Eng. Tribol. **225**, 299–306 (2011)
13. Kurokawa, M., Uchiyama, Y., Iwai, T., Nagai, S.: Wear **254**, 468–473 (2003)
14. Singh, P.K., Siddhartha, Singh, A.K.: Maters Today. **4**, 1606–1614 (2017)
15. Senthilvelan, S., Gnanamoorthy, R.: Polym. Test. **25**, 56–62 (2006)
16. Singh, A.K., Siddhartha, Singh, P.K.: Maters Today. **5**, 18038–18044 (2018)
17. Singh, A.K., Siddhartha, Yadav S, Singh, P.K.: Adv. Engg. Desig. **4**, 269–277 (2019)
18. Singh, A.K., Siddhartha, J.: Eng. Tribol. https://doi.org/10.1177/1350650119886233 (2019)
19. Singh, A.K.: Siddhartha Poly. Compos. https://doi.org/10.1002/pc.24682 (2017)

Design, Analysis and Fabrication of Semi-automatic Trash Masher

S. C. Sakthi Vigneshwaran and Chandran Arun Prakash

1 Introduction

In todays scenario, the amount of waste generated increases day by day. Starting from agricultural by-products to e-waste, waste management plays an important role to maintain an eco-friendly environment. Old waste management techniques such as collecting the waste, dumping it or burning the wastes create negative impact on the environment [1]. The waste-carrying trucks and the other waste management processes have a higher risk of causing accidents [2, 3]. Further, the waste can also cause infectious diseases in human beings and animals [4]. So, it is necessary to recycle the waste and reuse it for some other purpose [5].

Today, most of the industries use thermocol or Plaster of Paris to insulate the rooms. Better insulated buildings offer better energy economy. Also the industries have significant temperature variation which is likely to cause discomfort to the occupants. Therefore, insulation has become an integral part in industrial design. Thermocol, the other name of polystyrene, is a polymer made from the monomer styrene or phenylethene. The chemical properties of phenylethene are identical to polyethene. It responds very slowly to bacterial decomposition in the soil, thus making the soil infertile. In addition, it also releases sulphur dioxide on burning which can cause respiratory ailments or even death when inhaled. Plaster of Paris is generally regarded as a safe material for routine use, but it is considered dangerous if worked with repeatedly. It can cause various health hazards such as inhalation problems and skin irritation. Hence, thermocol and Plaster of Paris are regarded harmful both for the human body and the natural environment.

Therefore, it is necessary to go for an eco-friendly insulating materials. Rice husk, a by-product of rice industry, has various uses. In addition to protecting rice during

S. C. Sakthi Vigneshwaran · C. Arun Prakash (✉)
Department of Mechanical Engineering, Sri Sivasubramaniya Nadar College of Engineering, Chennai, Tamil nadu, India
e-mail: arunprakashc@ssn.edu.in

M. Muzammil et al. (eds.), *Recent Advances in Mechanical Engineering*, Lecture Notes in Mechanical Engineering, https://doi.org/10.1007/978-981-15-8704-7_37

Table 1 Chemical composition rice husk ash in weight percentage [10]

Elements	SiO_2	Al_2O_3	Fe_2O_3	CaO	Mg O	K_2O	Na_2O	P_2O_5
Percentage	93.4	0.05	0.06	0.31	0.35	0.14	0.1	0.8

the growing season, rice husks can be used as building material, fertilizer, insulation material or fuel. Rice husk ash (RHA), rich in silica content [6], can be produced from rice husk using appropriate combustion technique for use in buildings as a replacement of thermocol for the purpose of false ceiling. For this very purpose, it is necessary to design a machine which compresses the rice husk ash (RHA) material, so that it takes the form of cardboard which can be then used as false ceiling for insulating air-conditioned room. This paper discusses the design of such a machine. Trash masher is a machine used to compress eco-friendly waste products like rice husk ash into a finished cardboard-like form.

2 Rice Husk Ash

India is one among the largest rice-producing countries in the world. The major by-product of the rice-processing industries is rice husk. Rice husk has been a promising fuel over ages for processing the paddy or for producing energy through direct combustion. Incineration of rice husk produces RHA. The chemical composition of rice husk ash is given in Table 1. After removing cellulose and lignin from the rice husk under controlled combustion conditions, amorphous silica is obtained which is known as rice husk ash. Every year our country produces about 20 million tonnes of RHA. But the disposal of RHA has been a greatest problem since it is a great threat to the natural environment. Because of its excellent binding properties, it is used as a cement replacement material in recent days for constructions of buildings [7–9].

3 Design of Trash Masher

The schematic diagram of the trash masher is shown in Fig. 1.

It consists of a bin into which the waste material such as rice husk ash is dumped. The bin is closed using a top lid with suitable locking mechanism. The bottom plate of the bin is movable and the side walls are fixed. The movable bottom plate is fixed to a hydraulic cylinder. The hydraulic cylinder is controlled using a 4/3 lever-operated direction control valve. 4/3 valve is chosen because three types of operations have to be done by the cylinder. The operations are as follows:

Extension of the cylinder so that the plate moves up and compresses the waste to make it into the slab.

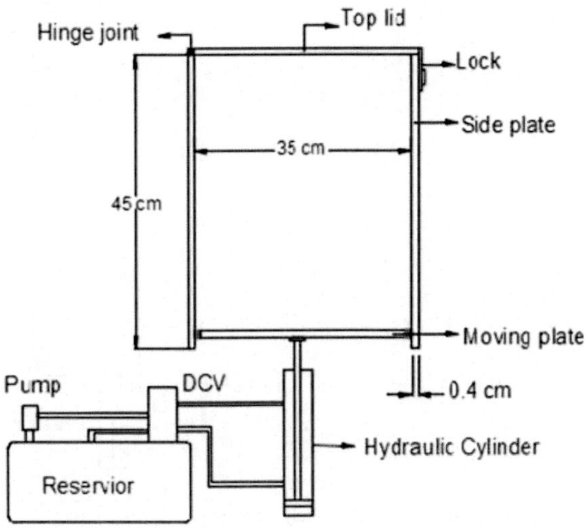

Fig. 1 Schematic diagram of trash masher

Retraction of the cylinder so that the plate comes down after the operation is done (i.e., after the slab is formed).
Idle position when the cylinder neither moves up nor down so that sufficient time is given for compressing the waste.

The circuit design for the hydraulic circuit is done using the software Automation studio and is shown in Fig. 2.

The material used for the fabrication of trash masher is mild steel. Mild steel is chosen to withstand the high stress during compression. The mild steel sheet is initially cut into four plates of size 45 cm height, 35 cm width and 0.4 cm thickness. The plates are then welded along the edges by keeping adjacent plates at right angle to each other. A square plate of dimensions 35 cm × 35 cm and thickness 0.4 cm is used as the top lid of the trash masher. The lid is attached to the masher by means of a clamp and a latch mechanism. This facilitates easy opening and closing of the lid. A square slab of dimensions 33 cm × 33 × cm and thickness 0.4 cm is used as the bottom moving plate. A cylinder of stroke length 450 mm is attached to the moving base plate by welding technique. The CAD model of the design is shown in Fig. 3.

4 Design Calculations

4.1 Maximum Force Exerted by the Cylinder

The mixture of rice husk ash and the bee wax has to be compressed with an adequate force so as to obtain a ceiling board of required dimension. Hence, the maximum force that can be exerted by the double-acting hydraulic setup is an important consideration.

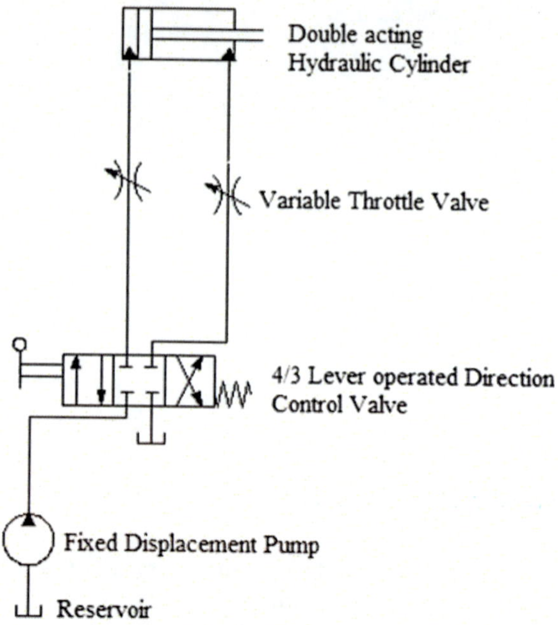

Fig. 2 Circuit design for hydraulic components

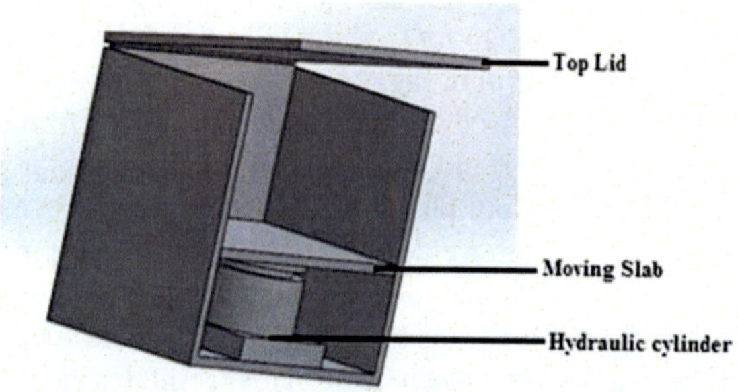

Fig. 3 Sectioned view of the trash masher

The chosen cylinder has a diameter of 4 cm and can exert a pressure of 40 kgf/(cm)^2

$$
\begin{aligned}
\text{Area}(A) &= \pi r^2 (\text{where radius} = 2\,\text{cm}) \\
&= \pi \times (2)^2 \\
&= 12.56\,\text{cm}^2 \\
\text{Pressure}(P) &= 40\,\text{kgf/cm}^2 \\
\text{Maximum force exerted} &= \text{Pressure} \times \text{Area} \\
&= 40 * 12.56 \\
&= 502.4\,\text{kgf}
\end{aligned}
$$

5 Maximum Force Withstand by Top Lid

During the forward stroke, the piston extends to compress the mixture in the bin. The movable bottom plate presses itself against the top lid due to the force exerted by the cylinder. The mixture which is placed between these two plates takes the form of a slab due to this movement. But the probability of top lid failure is high, since the entire force is transmitted to the top lid through the bottom plate. Hence, the material of the top lid should be selected carefully so as to minimise the failure costs.

Since the top lid is supported at only one end using a hinged support, it can be considered as a simply supported beam, where the component of force due to the hydraulic cylinder acts at the center. Hence, the force acting on the top lid can be calculated as follows

$$
\begin{aligned}
M/Y &= \sigma/I \\
M &= w/2 \times L/2 \\
&= 8.75 \times w\,\mathrm{N\,cm} \\
M &= 0.0875 \times w\,\mathrm{N\,m} \\
\sigma &= 125\,\mathrm{MPa} \\
\text{Deflection}\ Y &= 0.46/2 \\
&= 0.23\ \mathrm{cm} \\
\text{Moment of inertia}\ I &= (\mathrm{bt})^3/2 \\
&= (35 \times 0.46)^3/12 \\
&= 0.2838\ \mathrm{cm}^4 (\mathrm{or}) 28.38 \times 10^{-2}\mathrm{cm}^4 \\
&= 28.38 \times 10^{-10}\mathrm{m}^4
\end{aligned}
$$

Substituting the values, we get

$$
\begin{aligned}
(0.0875 \times w)/(28.38 \times 10)^{-10} &= (125 \times 10)^6/0.0023 \\
&= 1762.73\,\mathrm{N}\ (\mathrm{or})\ 1.762\,\mathrm{kN} (\mathrm{approx.}, 2\,\mathrm{kN})
\end{aligned}
$$

5.1 Maximum Induced Stress in the Slab

$$
\begin{aligned}
\text{Force} &= 2000\,\mathrm{N} \\
\text{Area} &= 0.35 \times 0.0036 \\
&= 1.26 \times 10^{-3}\,\mathrm{m} \\
\text{Stress} &= \text{Force}/\text{Area} \\
&= 2000/(1.26 \times 10^{-3})
\end{aligned}
$$

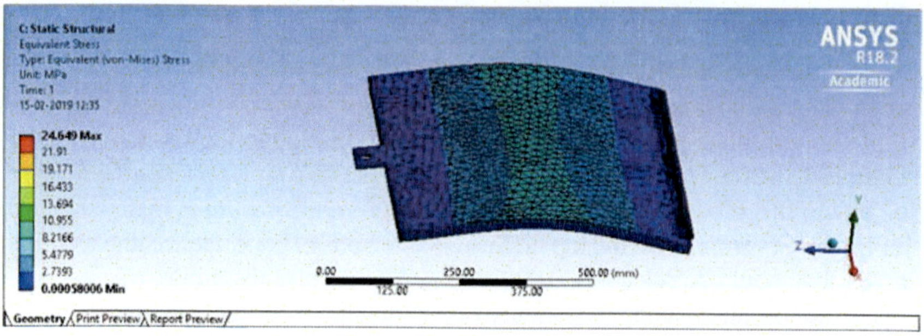

Fig. 4 Stress analysis of the top plate

$$= 1.58730\,\text{Mpa}.$$

But the ultimate shear strength of the mild steel is 400 MPa. So, the chance of failure of the top lid is very less. Hence, the design is safe.

5.2 ANSYS Modeling

Although the trash masher has five plates joined by welding joint, the probability of failure is high for the top plate. This is because the force exerted by the cylinder acts directly on the top lid through the moving plate, and hence, the chances of failure of top plate are very high when compared to side walls. But for the functioning of the trash masher, the top plate should be intact. Hence, ANSYS analysis of the top plate is done to analyse the stress concentration in the top lid. The top plate is modeled using the ANSYS software according to the dimensions shown in Fig. 4.

Mild steel is chosen as the material, and suitable mesh is chosen. The force conditions to which the plate is subjected are given. From the figure, it could be seen that the maximum stress induced in the top plate is around 5 MPa, and it occurs at the centre of the plate. But the ultimate tensile strength of steel is around 400 MPa. From the ANSYS analysis, it is found that the stress distribution is within the safe range. Hence, the chance of top lid failure is minimum for the above design of the trash masher.

Since the top lid is clamped at both the ends, there might be a deflection or bending of the plate at the centre portion. But from Fig. 4, it is clear that the stress at the centre is very less for causing bending of the plate. Therefore, the above design is suitable for the fabrication of the trash masher for efficient functioning.

6 Conclusion

In order to minimise the environmental impacts caused by the conventional ceiling boards and also to develop a safe disposal method for the RHA, harvesting its thermal insulation property, a hydraulically actuated trash masher made up of mild steel plate of thickness 4 mm was developed. For a mixture of 1.5 kg of RHA and 45 g of bee wax, a board of dimension 33 cm × 33 cm is developed using the trash masher. The board was dried in open sunlight for two days by swapping the sides to drive out the moisture. This leaves behind a well-binded hard RHA board with micropores which is then coated with an additional layer of varnish in order to prevent the boards from being attacked by insects such as termites. Also, it is important to check the leaks in the trash masher because the force exerted by the hydraulic cylinder causes the wet mixture of RHA and bee wax to squeeze out through the gap between the moving plate and side walls. Hence, an additional rubber lining was riveted around the edges of the plate to support the mixture. Additionally, a leak test was performed to ensure that no mixture gets out of the masher during the compression stroke. Also the RHA boards show a good strength due to the excellent binding properties of the RHA material and hence can be used as a substitute for plywood. Hence, in future, it has a high probability of bringing revolutionary changes in the field of plywood because the manufacture of plywood demands a massive deforestation which attributes to the current climatic change problems.

References

1. Management systems for domestic use. Energy **139**, 485–506 (2017)
2. Rosenfeld, J.: Statistics on truck accident fatalities n.d. https://www.rosenfeldinjurylawyers.Com/news/commercial-truck-fatality-statistics
3. Environment Agency: Regulating the waste industry, 1–8 (2015)
4. Pichtel, J.: Waste Management Practices: Municipal, Hazardous, And Industrial, 2nd edn. Taylor and Francis Group, Boca Raton (2014)
5. Bartl, A.: Moving from recycling to waste prevention: a review of barriers and enables. Waste Manage. Res. **32**, 3–18 (2014)
6. Siddique, R.: Waste materials and by-products in concrete. Chapter 7, pp. 235–265 (2008)
7. Adesina, P.A., Olutoge, F.A.: Structural properties of sustainable concrete developed using rice husk ash and hydrated lime. J. Build. Eng. **25** (2019)
8. Waswa-Sabuni, B., Syagga, P.M., Dulo, S.O., Kamau, G.N.: Rice husk ash cement—an alternative pozzolana cement for kenyan building industry. J. Civil Eng. **8**, 13–26 (2002)
9. Christopher, F., Bolatito, A., Ahmed, S.: Structure and properties of mortar and concrete with rice husk ash as partial replacement of ordinary Portland cement—a review Int. J. Sustain. Built Environ. **6**, 675–692 (2017)
10. Korotkova, T.G., Ksandopulo, S.J., Donenko, A.P., Bushumov, S.A. and Danilchenko, A.S.: Physical properties and chemical composition of the rice husk and dust. Orient J. Chem. **23**(6), 3213–3219 (2016)

Optimization of Cryogenic Turning Process Parameters Using Grey Relational Analysis (GRA) in Super-Duplex Stainless Steel (A479)

K. T. Akhil, Sanjivi Arul, and K. Shunmugesh

1 Introduction

Cryogenic turning process is a green manufacturing approach in which liquid nitrogen (LN2) is sprayed at a temperature of –196 °C to the tool–workpiece interface. Generally, almost all metal cutting industries use traditional cooling system to overcome the temperature rise during machining operation. There are many problems associated with usage of conventional fluid cooling systems; it is hazardous to health and its disposal causes environmental pollution [1, 2]. Cryogenic machining is a green manufacturing approach to overcome all above problems. Super-duplex stainless steel is extensively used in marine environment due to its mechanical properties and corrosion resistance. It is very difficult for machining because of the microstructure obtained by alloying various elements [3, 4]. Mohan et al. concentrate his study on cryogenic heat treatment and its effect on the mechanical properties [5]. This study results that mechanical properties can be improved by cryogenic heat treatment and it can further improved by carburizing material before heat treatment. This work analyses the improvement of mechanical properties in both shallow and deep cryogenic treatment. Work is concentrated in steel material. Shanmugam Murugappan et al. studied the relation between chip reduction coefficient and cryogenic precooling in turning process [6]. Since the cutting fore and chip reduction coefficient are interrelated, this study will help to reduce cutting force during turning

K. T. Akhil · S. Arul (✉)
Department of Mechanical Engineering, Amrita School of Engineering, Amrita Vishwa Vidyapeetham, Coimbatore 641112, India
e-mail: s_arul@cb.amrita.edu

K. T. Akhil
HMT Machine Tools Ltd., Kalamassery 683503, India

K. Shunmugesh
Department of Mechanical Engineering, Viswajyothi College of Engineering and Technology, Vazhakulam, Kochi, Kerala 686670, India

M. Muzammil et al. (eds.), *Recent Advances in Mechanical Engineering*, Lecture Notes in Mechanical Engineering, https://doi.org/10.1007/978-981-15-8704-7_38

process. Study reported that chip reduction coefficient can be reduced in turning process by precooling. Dry ice can be used for precooling the turning environment. They continued their study to determine the surface roughness after turning process [7]. This study results that surface roughness can be improved after machining by subzero temperature cooling. The quality of final product and cost of machining are determined by machining efficiency [8]. The machining efficiency can be improved by selecting optimum process parameters [9]. Optimum process parameters can be selected by using various multiresponse optimization methods [10]. Grey relational analysis is a multiobjective optimization technique used by various researchers for getting the optimum process parameters to various metal cutting processes at diverse environment. Tang et al. optimize the MRR and surface roughness value in electrical discharge machining (EDM) by means of grey relational analysis. Aggarwal et al. and Sivaiah et al. used multiobjective optimization technique for finding the optimum cutting parameters for various cryogenic machining [11, 12].

2 Experimental Methods

The workpiece selected for the turning process is super-duplex stainless steel (A479), and the cryogenic turning experiments were designed with Taguchi L27 orthogonal array with the aim of reducing experimental cost. The machining parameters selected for these experiments are speed, feed and depth of cut with three levels each, respectively. Table 1 shows the machining parameters and their levels used for experiment.

The experimental study was conducted to enhance the performance of material removal rate (MRR) and surface roughness (R_a and R_z). Since the experiment contains three process parameters with three levels, 27 numbers (n) of experiments with 26° of freedom ($n - 1$) were conducted. Table 2 shows the design of experiment and its results for studying the effect of response, MRR and surface roughness.

The machined surface roughness values Ra and Rz were measured using Mitutoyo Subtonic tester SI20IP. Material removal rate (MRR) is measured using difference in weight before and after machining (weight loss method) and the equation used is given below.

$$\text{MRR} = \frac{W_A - W_B}{T_M}$$

Table 1 Cryogenic machining parameters

Factors	Level 1	Level 2	Level 3
s (m/min)	80	160	240
f (mm/rev)	0.1	0.15	0.2
d (mm)	0.5	1	1.5

Table 2 Experimental design and its results

Exp. No.	v (m/min)	f (mm/rev)	d (mm)	R_a (μm)	R_z (μm)	MRR (g/min)
1	80	0.1	0.5	2.62	9.63	4.5
2	80	0.1	1	2.49	9.24	6.9
3	80	0.1	1.5	2.46	9.15	11.3
4	80	0.15	0.5	2.59	9.54	6.3
5	80	0.15	1	2.51	9.3	7.2
6	80	0.15	1.5	2.5	9.27	12.05
7	80	0.2	0.5	2.56	9.45	5.3
8	80	0.2	1	2.54	9.39	7.2
9	80	0.2	1.5	2.51	9.3	12.5
10	160	0.1	0.5	2.49	8.32	7.2
11	160	0.1	1	2.45	8.2	8.6
12	160	0.1	1.5	2.41	8.08	11.6
13	160	0.15	0.5	2.54	8.47	7.3
14	160	0.15	1	2.45	8.2	8.5
15	160	0.15	1.5	2.44	8.17	12.9
16	160	0.2	0.5	2.52	8.41	7.2
17	160	0.2	1	2.5	8.35	9.4
18	160	0.2	1.5	2.47	8.26	12.4
19	240	0.1	0.5	2.18	7.57	7.8
20	240	0.1	1	2.03	7.12	8.92
21	240	0.1	1.5	2.14	7.45	14.3
22	240	0.15	0.5	2.2	7.63	7.3
23	240	0.15	1	2.15	7.48	7.6
24	240	0.15	1.5	2.13	7.42	13.3
25	240	0.2	0.5	2.24	7.75	8.2
26	240	0.2	1	2.19	7.6	9.1
27	240	0.2	1.5	2.22	7.69	14.5

where

W_A Workpiece weight before turning (g).
W_B Workpiece weight after turning (g).
T_M Total turning time (min).

The statistical analyses were completed using Minitab 16 software. The objectives of the study were to maximize MRR and minimize the surface roughness value. Experimental results were converted into signal-to-noise (S/N) ratio using the Taguchi method.

3 Results and Discussion

Taguchi-based grey relational analysis was used to optimize the machining parameters. Table 3 shows coefficient and grade of grey relation for all experiment. The experiment 21 gives the optimized value which maximizes MRR and minimizes surface roughness (R_a and R_z).

Table 4 shows the rank of the input parameters, and it indicates that speed and depth of cut are the key parameters influencing the output response. The variation of grey relational grade (GRG) with speed, feed and depth of cut was shown in Fig. 1.

Table 3 Grey relational coefficient and grey relational grade

Exp. run	Grey relational coefficient			Grey relational grade	Rank
	R_a	R_z	MRR		
1	0.33333	0.33333	0.333333	0.333333333	27
2	0.39073	0.37185	0.396825	0.386468575	22
3	0.4069	0.38204	0.609756	0.466230741	17
4	0.34503	0.3415	0.378788	0.355104572	25
5	0.38065	0.36536	0.406504	0.384168616	23
6	0.38562	0.36858	0.671141	0.475112493	14
7	0.35758	0.35007	0.352113	0.353252723	26
8	0.36646	0.35603	0.406504	0.376330687	24
9	0.38065	0.36536	0.714286	0.4867625	13
10	0.39073	0.5112	0.406504	0.436144724	19
11	0.41259	0.53747	0.458716	0.469592081	15
12	0.43704	0.56659	0.632911	0.545513284	11
13	0.36646	0.48177	0.409836	0.419353843	21
14	0.41259	0.53747	0.454545	0.468202034	16
15	0.41844	0.54447	0.757576	0.573494674	9
16	0.3758	0.49312	0.406504	0.425141338	20
17	0.38562	0.50503	0.49505	0.4619002	18
18	0.40136	0.52401	0.704225	0.543198082	12
19	0.66292	0.73607	0.42735	0.608780719	7
20	1	1	0.47259	0.824196597	2
21	0.7284	0.7918	0.961538	0.827243877	1
22	0.63441	0.71105	0.409836	0.585097609	8
23	0.71084	0.77709	0.420168	0.636033741	5
24	0.74684	0.80707	0.806452	0.786787004	3
25	0.58416	0.66578	0.442478	0.564139595	10
26	0.64835	0.72334	0.480769	0.61748794	6
27	0.60825	0.68767	1	0.765306219	4

Table 4 Response table for grey relational grade

Level	v (m/min)	f (mm/rev)	d (mm)
1	0.4019	0.5442	0.4534
2	0.4825	0.5204	0.5138
3	0.6906	0.5104	0.6077
Delta	0.2887	0.0338	0.1544
Rank	1	3	2

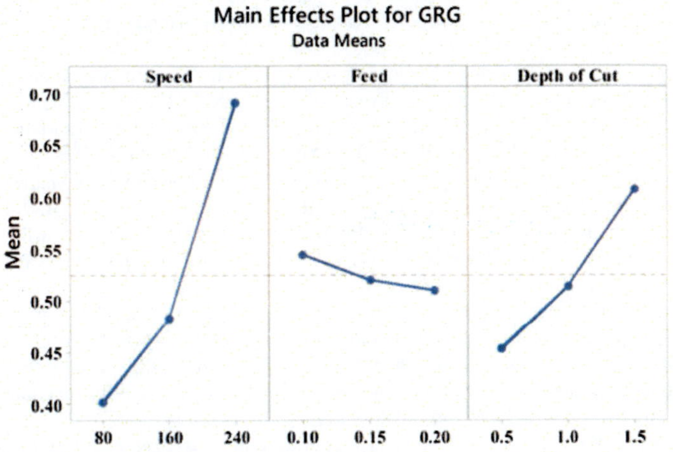

Fig. 1 Variation for GRG with machining parameters

GRG varies proportional to speed and depth of cut and inversely proportional to feed.

4 Conclusions

Taguchi-based grey relational analysis was successfully implemented to optimize cryogenic turning process parameters in super-duplex stainless steel. This machining experiment has led to the following conclusions. ANOVA analysis was also conducted to find the significant input parameters influencing each output response. The optimum process parameters of cryogenic turning process parameters which minimize the surface roughness and maximize the MRR are cutting speed 240 m/min, feed rate at 0.1 mm/rev and depth of cut 1.5 mm among the selected range of value for the study. Cutting speed is observed to be the most influencing parameters for surface roughness. The most momentous factor which influences the MRR is Depth of cut.

References

1. Hong, S.Y., Broomer, M.: Economical and ecological cryogenic machining of AISI 304 austenitic stainless steel. Clean Prod. Processes **2**(3), 157–166 (2000)
2. Hong, S.Y., Ding, Y., Jeong, J.: Experimental evaluation of friction coefficient and liquid nitrogen lubrication effect in cryogenic machining. Mach. Sci. Technol. **6**(2), 235–250 (2002)
3. Rajaguru, J., Arunachalam, N.: Coated tool performance in dry turning of super duplex stainless steel. Procedia Manuf. **10**, 601–611 (2017)
4. Samuel, A.P., Arul, S.: Effect of cryogenic treatment on the mechanical properties of low carbon steel IS 2062. Mater. Today Proc. **5**(11), 25065–25074 (2018)
5. Mohan, N., Arul, S.: Effect of cryogenic treatment on the mechanical properties of alloy steel 16MnCr5. Mater. Today Proc. **5**(11), 25265–25275 (2018)
6. Murugappan, S., Arul, S.: Effect of cryogenic pre cooling on chip reduction co-efficient during turning of EN8 steel rod. Mater. Today Proc. **4**(8), 8848–8855 (2017)
7. Murugappan, S.: Arul, A study on effect of sub zero temperature cooling on surface roughness of turned EN8 steel rod. Int. J. Appl. Eng. Res. **10**(8), 21549–21563 (2015)
8. Tomsun, N.: Determination of optimum parameters for multi-performance characteristics in drilling by using grey relational analysis. Int. J. Adv. Manuf. Technol. **28**(5–6), 450–455 (2006)
9. Akhil, K.T., Shunmugesh, K., Aravind, S., Pramodkumar, M.: Optimization of drilling characteristics using grey relational analysis (GRA) in glass fiber reinforced polymer (GFRP). Mater. Today Proc. **4**(2), 1812–1819 (2017)
10. Tang, L., Du, Y.T.: Multi-objective optimization of green electrical discharge machining Ti–6Al–4V in tap water via Grey-Taguchi method. Mater. Manuf. Processes **29**(5), 507–513 (2014)
11. Aggarwal, A., Singh, H., Kumar, P., Singh, M.: Optimization of multiple quality characteristics for CNC turning under cryogenic cutting environment using desirability function. J. Mater. Process. Technol. **205**(1–3), 42–50 (2008)
12. Sivaiah, P., Chakradhar, D.: Multi-objective optimisation of cryogenic turning process using Taguchi-based grey relational analysis. Int. J. Mach. Mach. Mater. **19**(4), 297–312 (2017)

Experimental Studies on Machining of SiC Under Different Assisted Electrode Conditions Using WEDM

Md Nadeem Alam, Arshad Noor Siddiquee, and Zahid A. Khan

1 Introduction

WEDM process is an advanced thermo-electric process in which the cutting tool is a conductive wire used to perform machining operations. The tool and workpiece are separated by a suitable dielectric fluid such as deionized water and kerosene oil which also acts as a coolant and flushes rubbish and detritus from the machining region. An electric spark is produced by maintaining a definite gap between tool and workpiece, and this gap is known as a spark gap. A series of spark causes a sudden rise in temperature of the workpiece up to approximately 10,000 °C due to which melting and evaporation of material take place and thus the material is removed. The schematic of the WEDM process is demonstrated in Fig. 1.

WEDM is a very effective process to machine hard materials up to 300 mm thickness with high precision including machining of die steel punches, ceramic tools and electronic components [1, 2]. WEDM is used for machining electrically conducting materials but nowadays it is also being used for machining electrically low conducting and non-conducting ceramics including silicon carbide (SiC), aluminium oxide (Al_2O_3), zirconium oxide (ZrO_2) and silicon nitride (Si_2N_4) by modifying the WEDM method with assisted electrode (AE) technique or forming composite of conductive and non-conductive materials [3–5]. For example, [6] successfully performed machining of a composite of Si_3N_4 and TiN ceramics using silver AE and in their study, and they explored the impact of clamping position, T_{ON} and T_{OFF} on MRR and R_a. Muttamara et al. [7] explored micromachining of Si_3N_4 by using the AE method on EDM. Their aim was to machine a hole of 0.05 mm by employing solid and hollow tools of copper and copper–tungsten alloy, kerosene as dielectric fluid and copper and TiN as an AE. From their results, they concluded that TiN-AE provided by physical vapour deposition enhances machining properties. However, the vapour

Md N. Alam (✉) · A. N. Siddiquee · Z. A. Khan
Department of Mechanical Engineering, Jamia Millia Islamia, New Delhi 110025, India
e-mail: mdnadeemalam7@gmail.com

M. Muzammil et al. (eds.), *Recent Advances in Mechanical Engineering*, Lecture Notes in Mechanical Engineering, https://doi.org/10.1007/978-981-15-8704-7_39

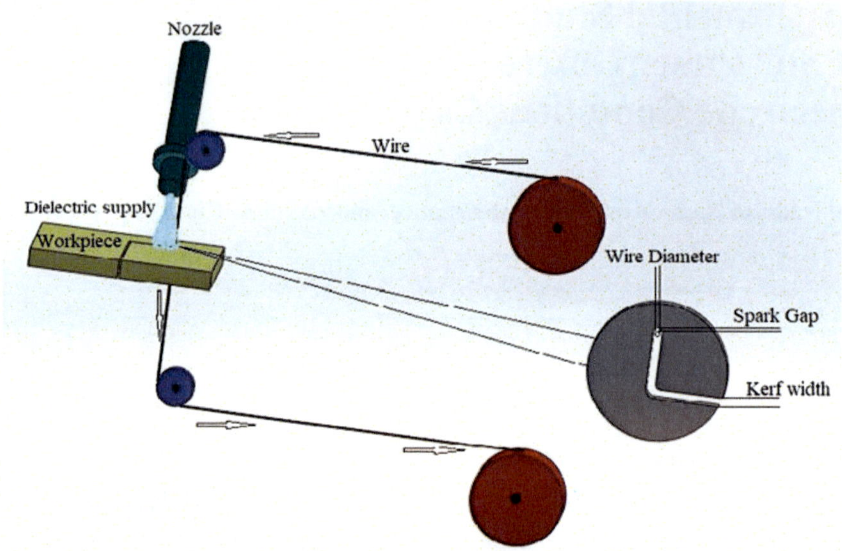

Fig. 1 Schematic of the WEDM process

deposition method is expensive and time-consuming, and hence, they developed a colloidal graphite solution and used it as an AE which drastically enhanced the machining properties and optimized the cost as well as time. According to Lal et al. [8], removal of material from the workpiece surface is done by spark erosion process. Higher value of current leads to produce high intensity spark that increases amount of discharge energy among the spark gap. The enhanced discharge energy leads to produce wider crater size, thus increasing KW and R_a. Pachaury et al. [9] examined the WEDM technique for machining of electro-conductive composite ceramics either by using AE method or by providing additional secondary impurities of the material to enhance the conductive property of the material which leads to continuous machining. Moudood et al. [10] adopted Cu foil as an AE to generate initial spark between tool and workpiece, and they used kerosene dielectric to create pyrolytic carbon layer over the workpiece to continue the machining operation. According to their study, the value of peak current ranged between 1.1 and 1.3 A is found as the compulsory condition to perform machining operation.

This study explores the machining of low conductive SiC by employing WEDM. The objective of the existing work is to scrutinize the influence of current on the response variables such as KW and Ra under three different types of AE conditions. During machining, T_{ON} and T_{OFF} were retained constant at 40 µs and 4 µs, respectively, and the current was varied to explore its effect on KW and R_a.

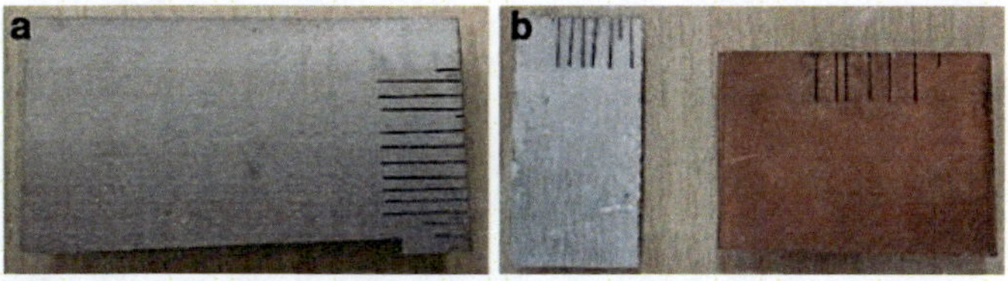

Fig. 2 **a** SiC ceramic, **b** Al and Cu-AE

2 Experimental Set-up and Material Selection

In the present study, SiC ceramic of dimension 50 × 50 × 1.3 mm is selected as a workpiece. Following AE conditions were adopted to perform the machining operations: (i) without AE, (ii) Cu–AE and (iii) Al–AE, where Cu–AE and Al–AE have a thickness of 1 mm. 0.18 mm diameter molybdenum wire is employed as a cutting tool electrode to cut the workpiece up to 10 mm depth. Deionized water is applied as a dielectric fluid to cool down the tool and to flush the debris from the machining region. Figure 2a, b shows the selected workpiece and AE, respectively.

During the experimental investigation, three different conditions of AE to machine low-conductive ceramic were taken. Three levels of the current value are chosen to investigate various output responses including KW and R_a. The three values of the current are chosen in such a way that the wire (tool) does not fail at these values.

3 Results and Discussion

3.1 Determination and Discussion of Kerf

Kerf demonstrates the machining of excessive part of the workpiece which is wasted during the WEDM process. It can be observed by estimating the difference between cutting width to wire diameter. Dimensional accuracy of the final product is significantly influenced by KW, and it is quantified in terms of μm. To measure KW, a stereozoom microscope was used, and three values of KW for each cut were taken with the help of the Image J software, and finally, the average value of KW was calculated. The three values were taken from three different regions, i.e. the starting side, intermediate portion and the end side. Figure 3 displays the images of the KW taken from stereozoom microscope.

Table 1 and Fig. 4 display the relationship between the average value of KW and the current for different AE conditions.

It can be understood from Table 1 and Fig. 4 that the KW increases with a rise in the current for all three AE conditions. Further, Cu-AE produces larger crater in WEDM by promoting the discharge between the electrode and work material. The formation of large-sized crater during machining resulted in higher KW.

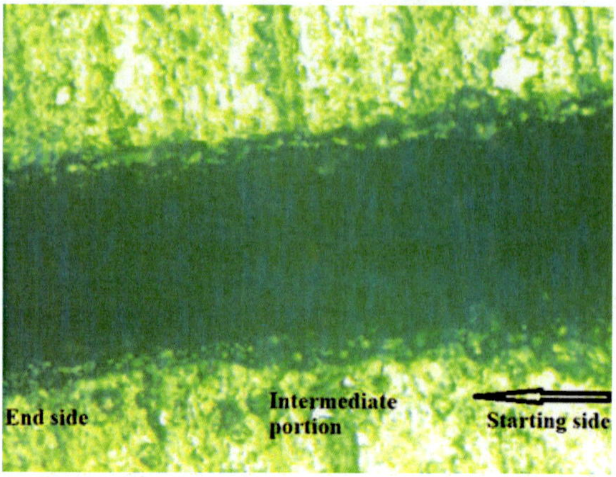

Fig. 3 KW image taken from stereozoom microscope

Table 1 Influence of current on KW for different AE conditions

Types of AE	T_{ON} (μs)	T_{OFF} (μs)	Current (A)	KW_{avg} (μm)
N-AE	40	4	2	251.00
			4	257.33
			6	263.66
Cu-AE	40	4	2	252.00
			4	259.00
			6	265.53
Al-AE	40	4	2	250.33
			4	258.00
			6	265.00

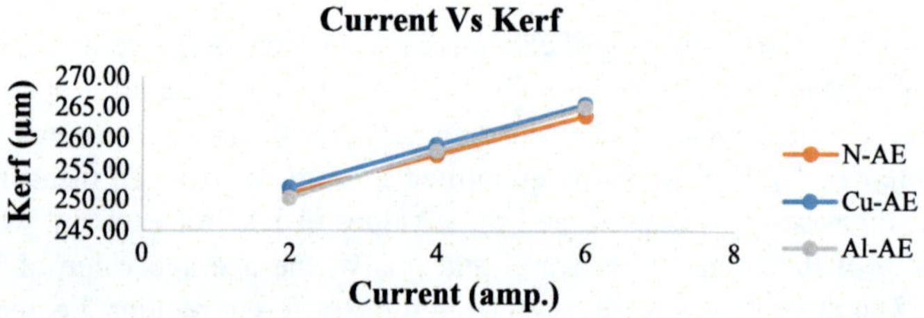

Fig. 4 Graph plotted between KW and current

3.2 Determination and Discussion of Ra

Ra is a constituent of surface texture which is enumerated by the aberrations along the normal vector of an actual surface as compared to its ideal or standard reference

Table 2 Impact of current on R_a for different AE conditions

Types of AE	T_{on} (μs)	T_{off} (μs)	Current (A)	R_a (μm)
N-AE	40	4	2	2.932
			4	2.952
			6	3.222
Cu-AE	40	4	2	2.815
			4	2.873
			6	3.618
Al-AE	40	4	2	2.642
			4	2.815
			6	3.301

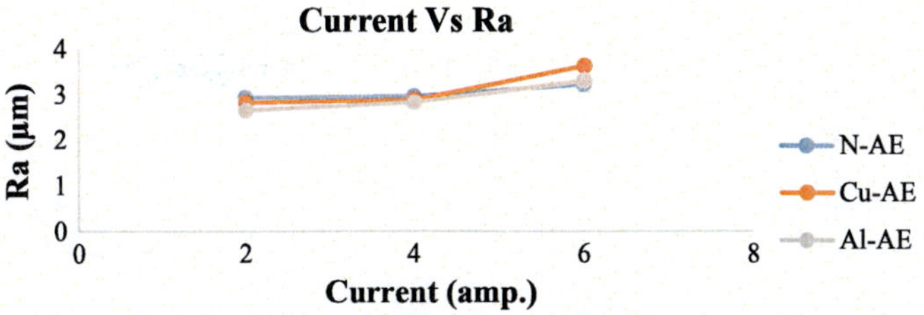

Fig. 5 Deviation in the R_a with current

surface. When these aberrations are comparatively large, the surface is considered as rough; but when these aberrations are comparatively small, the surface is relatively smooth. For any manufactured product, R_a value is one of the most significant response variables which demonstrates the quality of the product. Several functional properties including surface friction, surface wear, the reflectivity of light, heat transmissibility, the capability to distribute lubricant uniformly, etc., depend on Ra value. Table 2 and Fig. 5 display the impact of current on the R_a for different AE conditions.

It can be analysed in Table 2 and Fig. 5 that the value of R_a increases with a rise in the current for all three AE conditions. R_a value depends on the amount of power employed during discharge. The high value of the current leads to promote the amount of spark generated which causes rough cutting.

4 Conclusion

WEDM of low conductive SiC using three different AE conditions, i.e. without AE, Cu-AE and Al-AE using molybdenum wire of diameter 0.18 mm as tool electrode has been performed successfully. Among the three different AE conditions, maximum KW is found for Cu-AE at the highest current value. The Cu-AE produces larger crater in WEDM by promoting the discharge between the electrode and work material. The formation of large-sized crater during machining resulted in higher KW. Increasing the number of discharges, higher amount of energy is delivered to the working gap, which in turn, removes more material of the workpiece, thus leads to rough cutting. However, Al-AE produces a smooth surface compared to the other two AE conditions at low current.

References

1. Rao, M.S., Venkaiah, N.: Review on wire-cut EDM process. Int. J. Adv. Trends Comput. Sci. Eng. **2**(6), 12–17 (2013)
2. Kalpakjian, S., Steven R.S.: Manufacturing engineering and technology, 4th edn. Pearson Education, 7th floor, Knowledge Boulevard, A-8(A), Sector-62, Noida (U. P.), India (2013)
3. Menzies, I., Koshy, P.: Assessment of abrasion-assisted material removal in wire EDM. In: Menzies, I., Koshy, P. (eds.) CIRP Annals-Manufacturing Technology 2008, vol. 57, pp. 195–198. ELSEVIER, Hamilton, Canada (2008)
4. Mohri, N., Fukuzawa, Y., Tani, T., Sata, T.: Some considerations to machining characteristics of insulating ceramics. CIRP Ann. **51**(1), 161–164 (2002)
5. Gotoh, H., Tani, T., Mohri, N.: EDM of insulating ceramics by electrical conductive surface layer control. In: 18th CIRP Conference on Electro Physical and Chemical Machining, Procedia CIRP 42, pp. 201–205. Elsevier Procedia, Tokyo, Japan (2016)
6. Kozak, J., Rajurkar, K.P., Chandarana, N.: Machining of low thermal conductive materials by wire electric discharge machining (WEDM). J. Mater. Process. Technol. **149**, 266–271 (2004)
7. Muttamara, A., Fukuzawz, Y., Mohri, N., Tani, T.: Probability of precision micro-machining of insulating Si_2N_4 ceramics by EDM. J. Mater. Process. Technol. **140**, 243–247 (2003)
8. Lal, S., Kumar, S., Khan, Z.A., Siddiquee, A.N.: Multi-response optimization of wire electrical discharge machining process parameters for Al7075/Al_2O_3/SiC hybrid composite using Taguchi-based grey relational analysis. J. Eng. Manuf. 1–9 (2014)
9. Pachauri, Y., Tandon, P.: An overview of electric discharge machining of ceramics and ceramic-based composites. J. Manuf. Processes **25**, 369–390 (2017)
10. Maudood, M.A., Sabur, A., Ali, M.Y., Jaafar, I.H.: Effect of peak current on material removal rate for electrical discharge machining of non-conductive Al_2O_3 ceramic. Adv. Mater. Res. **845**, 730–734 (2014)

Optimization of Drilling Parameters of Glass Fiber-Reinforced Polymers Using Grey Relational Analysis

K. Shunmugesh, K. R. Arun, Arun Raphel, and K. T. Akhil

1 Introduction

Glass fiber composite is becoming a popular raw material for industrial applications such as the manufacturing of slurry and water tanks. Its high strength-to-weight ratio and low water absorption rate make it a preferred material over the natural fiber composite. The slight disadvantage being that glass-reinforced fibers are anisotropic. This dissimilarity of properties in different directions makes its machining difficult especially drilling. Paolo Priarone analyzed the effects of process parameters on machining carbon fiber-reinforced polymers (CFRP) [1]. The results showed that the tool feed has the strongest impact on the machining of CFRP, and therefore, a low value of feed rate is preferred. Lukas Herberger et al. discussed the effect of short amplitude drilling of CFRP [2]. This process is suitable for drilling holes having diameter more than 50 mm. DeFu Liu et al. studied about the mechanical drilling of composite laminates [3]. Hao et al. discussed the research of glass fibers (GFs) with carbon nanotubes (CNTs) and graphene by using an electrophoretic deposition (EPD) process. The incorporation of carbon-based nanomaterials made GFs sensitive to temperature and strain [4, 5]. Neeli et al. [6] has analyzed the effect of surface irregularity and delamination factor on the machining parameters. The authors concluded that the fiber orientation angle was the significant aspect for the decline of surface irregularity and delamination damage in drilling of GFRP.

K. Shunmugesh (✉) · K. R. Arun · A. Raphel
Department of Mechanical Engineering, Viswajyothi College of Engineering and Technology, Vazhakulam, Kerala 686670, India
e-mail: shunmugesh@vjcet.org

K. T. Akhil
HMT Machine Tools, Kalamassery 683503, India

M. Muzammil et al. (eds.), *Recent Advances in Mechanical Engineering*, Lecture Notes in Mechanical Engineering, https://doi.org/10.1007/978-981-15-8704-7_40

Table 1 Parameters and their level

Symbol	Parameters	Unit	Level 1	Level 2	Level 3
N	Cutting speed	m/min	30	50	70
F	Feed rate	mm/rev	0.04	0.08	0.12
Γ	Point angle	PA	90	118	140

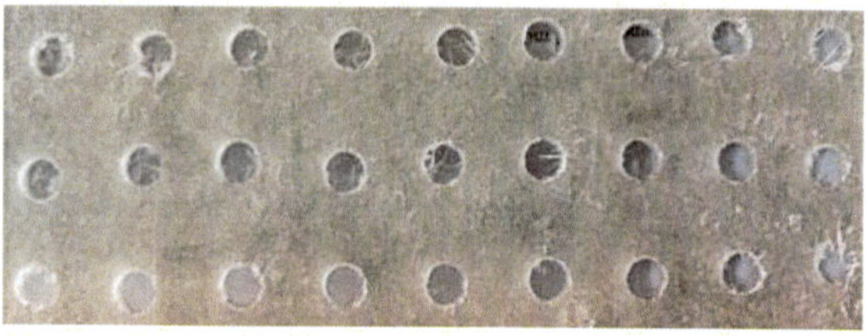

Fig. 1 Image of drilled GFRP composite

2 Experimental Setup and Optimization

The GFRP was manufactured by using simple hand layup method. The specimens have a weight ratio of 60:40 with the chopped stranded mat having an average thickness ranging from 0.8 to 0.9 mm. Seven layers of CSM were used to give a thickness of 3.4 mm to the test specimen. Twenty seven holes were drilled in the specimen to record the optimum machining parameters. The experiment for drilling operation was carried out in Vayu BMV 51 T20 having an output power of 7.5 KW. The point angles used were 90°, 118° and 140°. Table 1 shows the factors and their levels used for the experiment.

The principal objective is to find out the optimum drilling process factors to obtain minimum surface roughness, force and torque on the GFRP composite test specimen. The ANOVA for surface irregularity, torque and force was done by using the MINITAB 17 software. Figure 1 represents the image of the drilled GFRP composite.

3 Results and Discussion

The investigation was done by varying like cutting speed, feed rate and point angle. The recorded values of force, torque and surface roughness corresponding to varying input factors are shown in Table 2. The result of the ANOVA for torque is presented in Table 3. From the presented data, it is clear that the feed rate contributes 50.65% toward the torque followed by the cutting speed (42.99%).The lowest affecting factor is the point angle with a minimal contribution of 4.2%. The mean effect plot for torque is presented in Fig. 2.

The outcome of the ANOVA for force is displayed in Table 4. From the presented data, it is flawless that the feed rate contributes 51.2% toward the torque followed

Table 2 Experiment details

Exp. No.	Input process parameters			Responses		
	Cutting speed	Feed rate	Point angle	Force	Torque	Surface roughness
	m/min	mm/rev	degree	N	Nm	μm
1	30	0.04	90	54.5	21.45	3.65
2	30	0.04	118	43.4	19.83	3.6
3	30	0.04	140	44.4	21.35	3.83
4	30	0.08	90	60.5	22.93	3.77
5	30	0.08	118	52	22.04	3.8
6	30	0.08	140	57.5	22.34	3.9
7	30	0.12	90	68.5	28.28	4.44
8	30	0.12	118	65.5	25.51	4.33
9	30	0.12	140	69.5	27.19	4.57
10	50	0.04	90	46.4	17.09	3.43
11	50	0.04	118	34.4	16.65	3.25
12	50	0.04	140	37.4	16.96	3.4
13	50	0.08	90	50.5	21.55	3.6
14	50	0.08	118	41.4	18.18	3.47
15	50	0.08	140	45.4	20.66	3.8
16	50	0.12	90	67.5	24.92	4.02
17	50	0.12	118	56.4	23.29	3.99
18	50	0.12	140	59	23.59	4.04
19	70	0.04	90	31.4	16.1	3.2
20	70	0.04	118	24.8	13.03	3.1
21	70	0.04	140	27.6	13.82	3.3
22	70	0.08	90	45.4	16.99	3.3
23	70	0.08	118	31.4	15.5	3.24
24	70	0.08	140	35.6	16	3.48
25	70	0.12	90	57.5	22.54	3.77
26	70	0.12	118	46.4	20.05	3.9
27	70	0.12	140	49.4	20.75	4.05

by the cutting speed (36.4%). The lowest affecting factor is the point angle with a minimal contribution of 10.15%. The mean influence plot for force is shown in Fig. 3 . The result of the ANOVA for surface roughness is shown in Table 5. From the presented data, it is clear that the feed rate contributes 62.55% toward the surface coarseness followed by the cutting speed (30.43%). The lowest affecting factor is the point angle with a minimal contribution of 4.32%. The mean effect plot for surface roughness is shown in Fig. 4. It is observed that the surface roughness which has a variable effects with respect to the input factors.

Table 3 ANOVA for torque

Source	Degrees of freedom (DF)	Sequential sum of squares (Seq SS)	Adjusted sum of squares (Adj SS)	Adjusted mean square (Adj MS)	*F* value	Probability (*P*) (%)
N	2	175.195	175.195	87.597	209.69	42.99
F	2	206.437	206.437	103.218	247.08	50.65
Γ	2	17.519	17.519	8.76	20.97	4.20
Error	20	8.355	8.355	0.418		
Total	26	407.505				

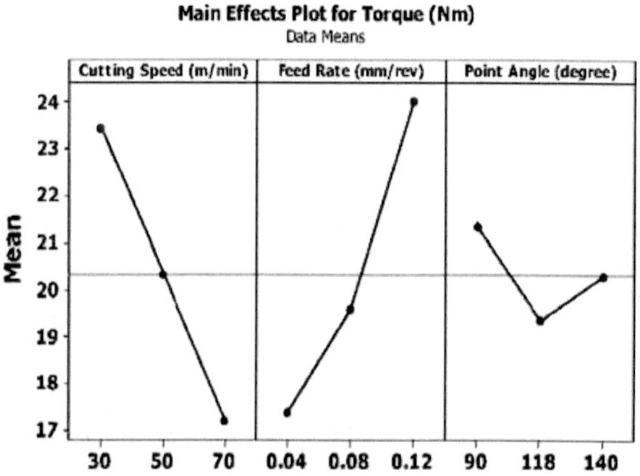

Fig. 2 Variation of torque

Table 4 ANOVA for force

Source	Degrees of freedom (DF)	Sequential sum of squares (Seq SS)	Adjusted sum of squares (Adj SS)	Adjusted mean square (Adj MS)	*F* value	Probability (*P*) (%)
N	2	1533.94	1533.94	766.97	172.58	36.40
F	2	2153.61	2153.61	1076.81	242.29	51.20
Γ	2	427.03	427.03	213.51	48.04	10.15
Error	20	88.88	88.88	4.44		
Total	26	4203.46				

3.1 Grey Relational Analysis

Grey relational analysis is a statistical analysis model, which makes use of various models of Grey Systems Theory [7].

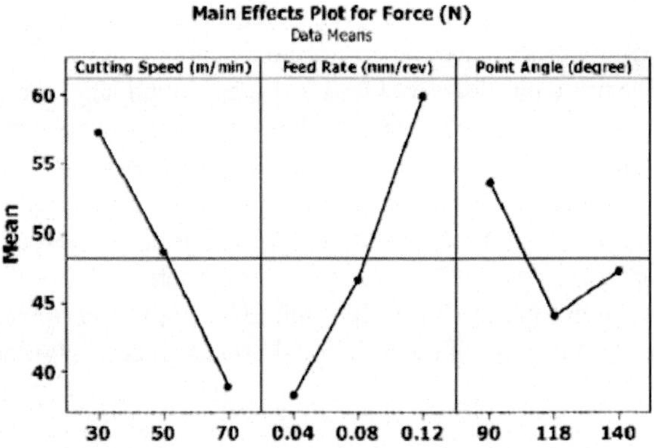

Fig. 3 Variation of force

Table 5 ANOVA for surface roughness

Source	Degrees of freedom (DF)	Sequential sum of squares (Seq SS)	Adjusted sum of squares (Adj SS)	Adjusted mean square (Adj MS)	*F* value	Probability (*P*) (%)
N	2	1.17816	1.17816	0.58908	107.17	30.43
F	2	2.42389	2.42389	1.21194	220.49	62.55
Γ	2	0.16749	0.16749	0.08374	15.24	4.32
Error	20	0.10993	0.10993	0.0055		
Total	26	3.87947				

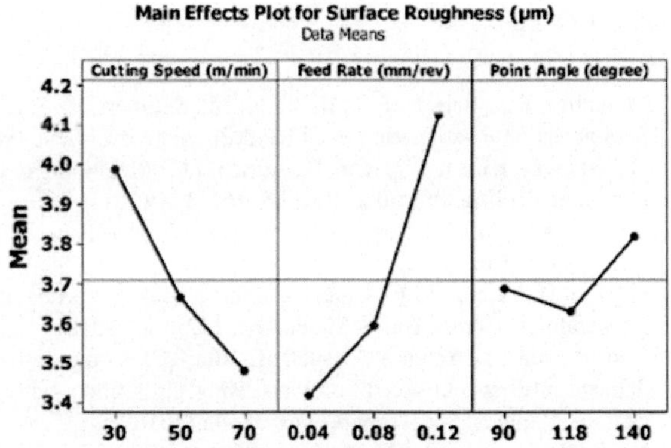

Fig. 4 Variation of surface roughness

Table 6 Response of grade

S. No.	*N*	*F*	*Γ*
1	0.441	0.692	0.523
2	0.544	0.57	0.608
3	0.69	0.414	0.545
Delta	0.25	0.28	0.09
Rank	2	1	3

GRA was applied to the experimental investigation to filter out the optimum drilling process parameter for the machining of the glass-reinforced fiber composite. The principle stages of GRA involve normalization of the recorded data followed by the preprocessing of the data. The weighted grey relational grade (WGRG) is derived from the calculated GRA, and the grades serve as a clear measure of the effect of the selected process factors on the surface irregularity, force and torque of the GFRP composite. A higher value of WGRG indicates excellent output characteristics at optimum levels of input process parameters. Table 6 shows the reaction for grey relational grade. From the analysis of the experimental data using GRG, it was learned though all the individual process factors significantly affect the output.

4 Conclusions

Drilling of GFRP composite having a thickness of 3.4 mm using three different point angles (90, 118, 140) by varying feed rate (0.04, 0.08, 0.12 mm/rev) and cutting speed (30, 50, 70 m/min) was conducted. It is noted that the surface roughness was comparatively low at intermediate point angle (118°). Maximum thrust force and torque were obtained at small cutting speed (30 m/min) and at higher tool rate (0.12 mm/rev). Thrust force and torque value establish lower value at intermediate point angle (118°).

References

1. Melentiev, R., Priarone, P.C., Robiglio, M., Settineri, L.: Effects of tool geometry and process parameters on delamination in CFRP drilling: an overview. Procedia CIRP **45**, 31–34 (2016)
2. Heberger, L., Kirsch, B., Henn, B., Aurich, J.C.: Machining of CFRP via short amplitude torsion pendulum drilling. Procedia CIRP **66**, 161–174 (2017)
3. Liu, D., YongJun, T., Cong, W.L.: A review of mechanical drilling for composite laminates. Composite Struct. **94**(4), 1265–1279 (2012)
4. Hajiloo, H., Green, M.F., Gales, J.: Mechanical properties of GFRP reinforcing bars at high temperatures. Constr. Build. Mater. **162**, 142–154 (2018)
5. Hao, B., Ma, Q., Yang, S., Mäder, E., Ma, P.C.: Comparative study on monitoring structural damage in fiber-reinforced polymers using glass fibers with carbon nanotubes and graphene coating. Compos. Sci. Technol. **129**, 38–45 (2016)
6. Neeli, N., Jenarthanan, M.P., Dileep Kumar, G.: Multi-response optimization for machining GFRP composites using GRA and DFA. Multidiscipline Model. Mater. Struct. **14**(3), 482–496 (2018)
7. Deng, J.L.: Introduction to grey system theory (1989)

Highly Productive Solar Stills for Efficient Water Utilization in Agriculture

Apurv Yadav, Vineet Kumar Vashishtha, and Harsha Yadav

1 Introduction

Energy demand worldwide is increasing day by day and the major reason for it is industrial expansion. Most of energy production technologies also require substantial water intake. These technologies are also the cause of two of the world's major concern, i.e., pollution and water scarcity. To reduce pollution, several policies are being implemented by various governments [1]. Adoption of more and more solar- or wind-based energy generation techniques is being considered [2]. Exploring various biodiesels is also a major step toward reducing vehicular emission pollutants [3]. Non-conventional energy-based water purification steps are being used to try to solve the problem of water scarcity. Scarcity of water also poses a problem for agriculture growth. In an economy like India, where agriculture produces 42% of employment and contributes to 14% of the country's GDP, this poses a serious threat [4]. Apart from water scarcity, agriculture also faces a threat from salt accumulation in soil due to improper drainage. Shallow water table and salt accumulation in a region's soil drive down the agricultural productivity of that region [5]. There is a need to utilize all possible recourse to retain all existing water resources and search for new sources.

Globally, nearly 66% of the water supplied for irrigation purposes is lost as run-off or agriculture drainage water (ADW) [6]. Availability of ADW in abundance has opened lots of doors for recovery of water [7]. Although the overall drainage volume could be reduced by a conventional approach, the methods would not be sustainable. Desalination of drainage water is beneficial than other re-utilization approaches, such

A. Yadav
Amity University Dubai, Dubai 345019, UAE

V. K. Vashishtha
Krishna Institute of Engineering & Technology, Ghaziabad, UP 201206, India

H. Yadav (✉)
Indian Institute of Technology, New Delhi, Delhi 110016, India
e-mail: harsha.civil32@gmail.com

M. Muzammil et al. (eds.), *Recent Advances in Mechanical Engineering*, Lecture Notes in Mechanical Engineering, https://doi.org/10.1007/978-981-15-8704-7_41

as irrigation of crops that tolerates salinity, as not only it recovers freshwater but also isolates potentially harmful salts [8]. Also, the water utilization efficiency for the irrigation increases [9]. The salinity-affected soils can be recovered, and the quality of agricultural products increases [10]. The nature of ADW limits the application of reverse osmosis due to high energy and chemical costs.

Also, recovery percentage or percentage reduction in drainage discharge volume is limited due to fouling and scaling which can be defined as

$$R = \frac{\text{FPR}}{\text{SFR}} \tag{1}$$

where R is the recovery percentage, FPR is freshwater production rate, and SFR is saltwater feed rate. This limitation requires additional measures which are more energy-intensive. The overall environmental impact of reverse osmosis does not make it a feasible solution for desalination of ADW [11].

2 Solar Stills

Solar stills are found to be the most promising and economical method of treating ADW among all solar desalination techniques. This desalination process uses the sun as their energy source, which offers a feasible solution for the treatment of ADW. Economical high-quality water is produced which can be used for irrigation or even for drinking purposes. Solar distillation has long been regarded as an easy to implement and economical way for the treatment of brackish water. Evaporation of water requires an energy of about 2.3 MJ/kg. Although it is more strenuous than reverse osmosis, it is advantageous that it does not need energy in terms of electrical power but as heat; dense medium distillation can be also conducted and it is almost no maintenance device.

In practical application, distillation by solar stills is much less expensive than reverse osmosis. The basic working principle is the same for all types of solar stills. A fundamental schematic of solar still is shown in Fig. 1. Solar radiation incidence on the glass cover is transmitted through it and absorbed by the black inner surface, specifically bottom, of the still. The water then gets heated and evaporates. Condensation of the evaporated vapors takes places on the inner side of the glass cover surface. From there, droplets trickle along the slope and get collected in a tray. Solar still efficiency, E, can be expressed as a ratio of the amount of solar radiation utilized for evaporation to the total amount of radiation incident on the still. The approximate daily output (l/day) from solar still can be found by

$$Q = \frac{EGA}{2.3} \tag{2}$$

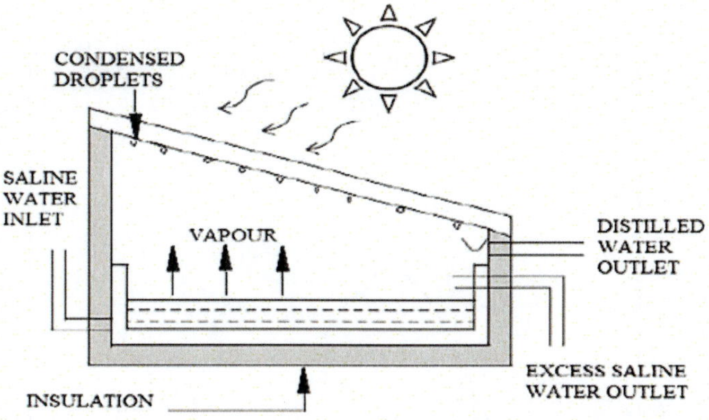

Fig. 1 Single basin solar still printed with permission from Elsevier [12]

where Q is the daily output, A is the aperture area of the solar still in $m^{2,}$ G denotes the global solar irradiation in MJ/m^2, and E denotes overall efficiency [13]. Although it is a reliable and simple device, it has low productivity. Therefore, many researchers have focused on limiting this drawback [14, 15].

3 Phase Change Materials Integrated Solar Stills

The most effective method of improving productivity was found out to be storing of sun's heat energy during the day and its release at night when there is no sun. El-Sebaii et al. [16] examined the usage of phase change material (PCM) as energy-absorbing material on the performance of fundamental solar still. It was observed that overnight productivity was significantly increased and a daily efficiency of 85.2% was obtained. Radhawan [17] performed a similar experiment on stepped solar still. The still temperature was found to be more uniform with PCM, and 57% of daily efficiency was achieved. The system utilizing PCM was found to be more effective in both cases, especially during the night. This happens as the temperature difference between the glass cover and water increases, facilitating higher heat transfer rates. Also, the PCM has accumulated a significant amount of heat during the day, as opposed to heat dissipation to surroundings. As PCM solidifies, the stored heat is released to the water for its evaporation and the still also works at night. However, there are very few works which have used PCM in solar stills. Shalaby et al. [18] reported that the utilization of paraffin PCM in solar still improves its yield by 12%. Mousa and Gujarathi [19] achieved an increase of 49% in the solar still productivity by the use of paraffin PCM. Chaichan and Kazem [20] added paraffin wax to a solar still and experience 10.38% improvement in the daily yield output. PCM integration in a solar still is shown in Fig. 2.

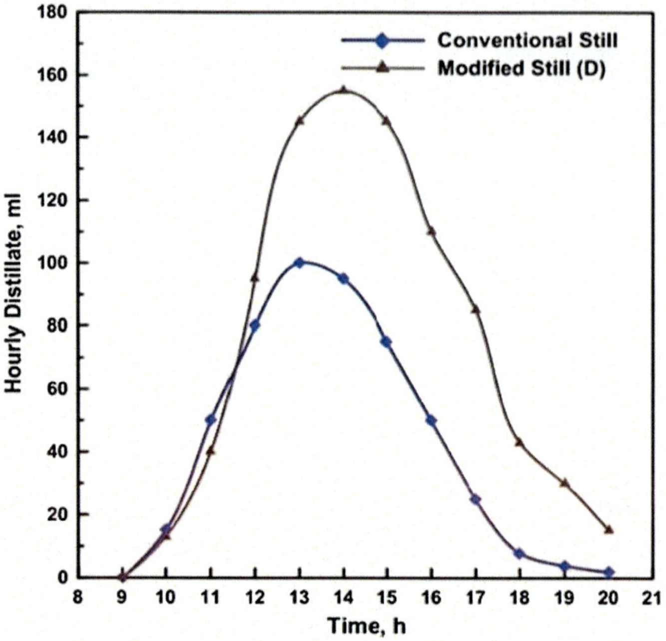

Fig. 2 Solar still with PCM integration printed with permission from Elsevier [20]

4 Nanoparticle-Enhanced Phase Change Materials

Although PCM has a high density of energy storage, their poor thermal conductivity limits their application due to poor heat release rate. Enhancement of thermophysical properties of PCM by incorporation of nanoparticles may overcome this limitation [21]. Nanoparticle addition improves the thermal conductivity and decreases the time taken for the melting and solidification process in comparison to pristine PCM [22, 23]. Elango et al. [24] reported an improvement of 29.95% in the overall productivity of a still by insertion of aluminum oxide nanoparticles in energy-absorbing PCM. Sahota and Tiwari [25] reported a 12.2% increment in the productivity of a dual-slope solar still by using aluminum oxide added PCM. Carbon-based nanoparticles have been found more suitable as they have high thermal conductivity with the additional advantage of low density [26]. Sharshir et al. [27] used film cooling and phase change nanoparticles with flake graphite nanoparticles in a solar still and observed 73.8% increment in the solar still productivity (see Fig. 3).

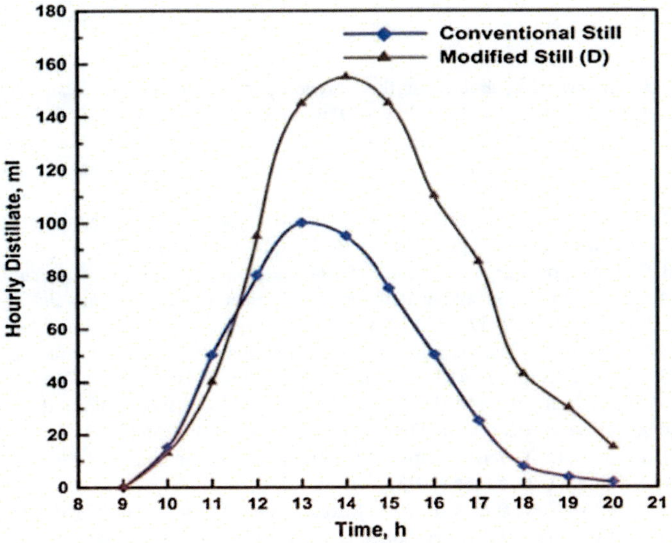

Fig. 3 Increased productivity of solar still by nanoparticle-enhanced PCM printed with permission from Elsevier [27]

5 Conclusion

This work discusses the modifications in solar stills which could provide freshwater for irrigation purposes and to maintain the fertility of the soil with negligible cost addition. In near future, potable supply of water in desert regions will be difficult without the use of solar energy. However, establishing solar stills utilizing appropriate technology should be carefully considered to generate interest in the technology. To decrease freshwater cost, efficiency needs to be increased. Integration of nanoparticle-enhanced PCM in solar stills has a huge potential to improve the efficiency and the yield of the solar stills. ADW can be used as feed water for solar stills, and produced excess freshwater can be reused again for irrigation purposes. This will not only increase solar still effectiveness but provide for low salinity water for agricultural purpose. Both water wastage and water scarcity problems could be solved by simple and economical design aspects of the solar still. Soil salinity also can be reduced by the utilization of the discussed approach. Further research on the integration of some more techniques like evacuated tubes, solar collector, etc., is required to generate a more effective process and explore more applications of solar stills for agricultural purposes. Also, research is required for the selection techniques of appropriate combination of PCM and nanoparticles yielding maximum productivity.

References

1. Stokes, L.C., Breetz, H.L.: Politics in the US energy transition: case studies of solar, wind, biofuels and electric vehicles policy. Energy Policy **113**, 76–86 (2018)
2. Khan, F.A., Pal, N., Saeed, S.H.: Review of solar photovoltaic and wind hybrid energy systems for sizing strategies optimization techniques and cost analysis methodologies. Renew. Sustain. Energy Rev. **92**, 937–947 (2018)
3. Kumar, S., Yadav, A.: Comparative experimental investigation of preheated thumba oil for its performance testing on a CI engine. Energy Environ. **29**(4), 533–542 (2018)
4. Indian GDP from agriculture https://tradingeconomics.com/india/gdp-from-agriculture
5. Singh, A.: Soil salinization and waterlogging: A threat to environment and agricultural sustainability. Ecol. Ind. **57**, 128–130 (2015)
6. Gregory, P.J., Hester, R., Harrison, R.: Soils and food security: challenges and opportunities. Soils and food security, pp. 1–30. RSC Publishing, London (2012)
7. Sorour, M.H., El Defrawy, N.M.H., Shaalan, H.F.: Treatment of agricultural drainage water via lagoon/reverse osmosis system. Desalination **152**(1–3), 359–366 (2003)
8. Zarzo, D., Campos, E., Terrero, P.: Spanish experience in desalination for agriculture. Desalin Water Treatment **51**(1–3), 53–66 (2013)
9. Rahardianto, A., McCool, B.C., Cohen, Y.: Reverse osmosis desalting of inland brackish water of high gypsum scaling propensity: kinetics and mitigation of membrane mineral scaling. Environ. Sci. Technol. **42**(12), 4292–4297 (2008)
10. Burn, S., Hoang, M., Zarzo, D., Olewniak, F., Campos, E., Bolto, B., Barron, O.: Desalination techniques—a review of the opportunities for desalination in agriculture. Desalination **364**, 2–16 (2015)
11. Stuber, M.D., Sullivan, C., Kirk, S.A., Farrand, J.A., Schillaci, P.V., Fojtasek, B.D., Mandell, A.H.: Pilot demonstration of concentrated solar-powered desalination of subsurface agricultural drainage water and other brackish groundwater sources. Desalination **355**, 186–196 (2015)
12. Selvaraj, K., Natarajan, A.: Factors influencing the performance and productivity of solar stills-A review. Desalination **435**, 181–187 (2018)
13. Scrivani, A., El Asmar, T., Bardi, U.: Solar trough concentration for fresh water production and waste water treatment. Desalination **206**(1–3), 485–493 (2007)
14. Samuel, D.H., Nagarajan, P.K., Arunkumar, T., Kannan, E., Sathyamurthy, R.: Enhancing the solar still yield by increasing the surface area of water—A review. Environ.l Progress Sustain. Energy **35**(3), 815–822 (2016)
15. Rajaseenivasan, T., Tinnokesh, A.P., Kumar, G.R., Srithar, K.: Glass basin solar still with integrated preheated water supply–theoretical and experimental investigation. Desalination **398**, 214–221 (2016)
16. El-Sebaii, A.A., Al-Ghamdi, A.A., Al-Hazmi, F.S., Faidah, A.S.: Thermal performance of a single basin solar still with PCM as a storage medium. Appl. Energy **86**(7–8), 1187–1195 (2009)
17. Radhwan, A.M.: Transient performance of a stepped solar still withbuilt-in latent heat thermal energy storage. Desalination **171**(1), 61–76 (2005)
18. Shalaby, S.M., El-Bialy, E., El-Sebaii, A.A.: An experimental investigation of a v-corrugated absorber single-basin solar still using PCM. Desalination **398**, 247–255 (2016)
19. Mousa, H., Gujarathi, A.M.: Modeling and analysis the productivity of solar desalination units with phase change materials. Renewable Energy **95**, 225–232 (2016)
20. Chaichan, M.T., Kazem, H.A.: Single slope solar distillator productivity improvement using phase change material and Al_2O_3 nanoparticle. Sol. Energy **164**, 370–381 (2018)
21. Yadav, A., Barman, B., Kumar, V., Kardam, A., Narayanan, S.S., Verma, A., et al.: A review on thermophysical properties of nanoparticle-enhanced phase change materials for thermal energy storage. In: Recent Trends in Materials and Devices pp. 37–47. Springer, Cham (2007)
22. Yadav, A., Barman, B., Kumar, V., Kardam, A., Narayanan, S.S., Verma, A., et al.:. Solar thermal charging properties of graphene oxide embedded myristic acid composites phase

change material. In: AIP Conference Proceedings, vol. 1731, no. 1, p. 030030. AIP Publishing (2013)

23. Yadav, A., Barman, B., Kardam, A., Narayanan, S.S., Verma, A., Jain, V.K.: Thermal properties of nano-graphite-embedded magnesium chloride hexahydrate phase change composites. Energy Environ. **28**(7), 651–660 (2017)
24. Elango, T., Kannan, A., Murugavel, K.K.: Performance study on single basin single slope solar still with different water nanofluids. Desalination **360**, 45–51 (2015)
25. Sahota, L., Tiwari, G.N.: Effect of Al2O3 nanoparticles on the performance of passive double slope solar still. Sol. Energy **130**, 260–272 (2016)
26. Yadav, A., Verma, A., Narayanan, S.S., Jain, V.K., Bhatnagar, P.K.:. Carbon based phase change nanocomposites for solar energy storage. In: AGU Fall Meeting Abstracts (2013)
27. Sharshir, S.W., Peng, G., Wu, L., Essa, F.A., Kabeel, A.E., Yang, N.: The effects of flake graphite nanoparticles, phase change material, and film cooling on the solar still performance. Appl. Energy **191**, 358–366 (2017)

Dimensional Stability Analysis of Teak Sawdust and Polypropylene Composite

Anil Kumar Yadav and Rajeev Srivastava

1 Introduction

Wood plastic composites (WPCs) are the mixture of wood and plastics having a ratio 50:50. WPCs are manufactured by using polymer and wood sawdust/fiber. Polymer (virgin/recycled) used as matrix and wood fibers/sawdust is used as a reinforcing material. The WPCs are presently used in outdoor/indoor applications like tiles railing door and door frame, etc. The properties of wood and plastics are entirely different; woods are hydrophilic and plastics are hydrophobic in nature. This affects the performance of composites. The properties of WPCs depends upon the bonding of wood and plastic [1, 2]. Water absorption in the WPCs occurs at the interface of wood and plastics. Water absorptions process in the WPCs occurs according to the Ficks law of diffusions [3, 4]. The body of the WPCs becomes swell due to water absorption, and other mechanical properties also affected by water absorption. The effect of virginity and recycling of polymer on the mechanical and physical properties of WPCs have been studied, and it has been observed that recycled polypropylene (RPP) matrix-based composite exhibits higher density, low porosity, low crystallinity, and high-dimensional stability [5]. The WPCs are produced by RPP and pine sawdust, RPP composites exhibit low water absorption and thickness of swell than VPP composites, and incorporation of MAPP reduces the water uptake and thickness of swell [6]. It has been reported in literature that reprocessing of polymer produces additional functional group in the recycled polymer and increases the chain length of the polymer. Reprocessing improves the mechanical properties and reduces water absorption and thickness of the swell of the WPCs. [7]. Literature shows that the composite produced by recycled polymer matrix possesses good strength and low water absorption than composite produced by virgin polymer of similar grade.

A. K. Yadav (✉) · R. Srivastava
Mechanical Engineering Department, MNNIT, Allahabad, India
e-mail: anilhbtik@gmail.com

M. Muzammil et al. (eds.), *Recent Advances in Mechanical Engineering*, Lecture Notes in Mechanical Engineering, https://doi.org/10.1007/978-981-15-8704-7_42

The addition of MAPP in the same formulation improves the mechanical and hygroscopic properties of the composites [8, 9]. Increase in the wood content reduces the weight of the composites and affects the mechanical hygroscopic and rheological of the composite. In the present work, dimensional stability of teak sawdust and polypropylene composites has been assessed for their end applications.

2 Material and Method

2.1 *Polymer*

Virgin polypropylene is collected from a local dealer of Repole (a unit of Reliance Petroleum India), and recycled polypropylene is collected from local recycler. The granules are dried in air circulatory oven for 12 h at 65 °C to remove the moisture content.

2.2 *Wood Sawdust*

Fresh teak wood sawdust is collected from a local sawmill. Sawdust is dried in an air circulatory oven for 24 h at 110 °C to remove the moisture of it. Sawdust particles of size 200–250 μm have been separated by the sieve separation method.

2.3 *Coupling Agent*

Coupling agent maleated polypropylene (MAPP) is used in the fabrication of composite.

2.4 *Composite Preparation*

Virgin PP/recycled PP and wood particles were compounded by the twin-screw extruder. The twin-screw extruder was PLC controlled and had four temperature zone of 160 °C, 170 °C, 180 °C, 190 °C, respectively. The screws speed was 90 r.p.m. at barrel pressure 40 bar (Fig. 1a). The materials were compounded in the batch of one kg for all formulations as given in Table 1. The compounded material comes in the form of wire through a bathtub and cuts in the small granules by a cutting machine as shown in Fig. 1b, c.

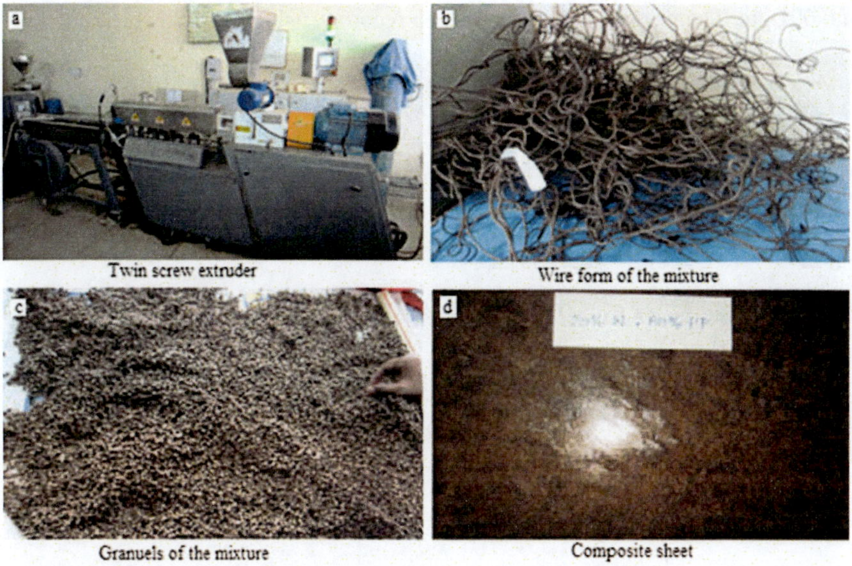

Fig. 1 Manufacturing process of composite

Table 1 Composites formulation and coding

S. No.	Composite code		Wood sawdust (%Weight)	Polypropylene (%Weight)	Coupling agent (%Weight)
	Virgin polypropylene	Recycled polypropylene			
1	V1	R1	00	100	00
2	V2	R2	20	80	00
3	V3	R3	30	70	00
4	V4	R4	40	60	00
5	V5	R5	50	50	00
6	V6	R6	50	47	03
7	V7	R7	50	45	05

3 Experiments

3.1 Water Absorption Test

The water absorption test is conducted according to ASTM D570-98 [10]. The specimens were heated in air circulatory oven for 24 h at 105 °C to remove the moisture content and cooled in desiccators. The test specimens were placed in the container of distilled water at a maintained temperature of 23 ± 1 °C for 24 h. The specimens were wiped properly and weighted at an interval of 2 h and 24 h with the accuracy of 0.001 gm. The water absorption has been recorded using the following equation-

$$WA(\%) = \frac{W_f - W_i}{W_i} \times 100 \quad (1)$$

where W_i and W_f are the initial and final weight of the specimen after immersion in the water.

3.2 Thickness of Swell Examination

Thickness swelling of the composite is measured with a micrometer having the least count of 0.01 mm. The thickness swell (TS) of the specimen was measured by using the following equation-

$$TS(\%) = \frac{\delta_f - \delta_i}{\delta_i} \times 100 \quad (2)$$

where δ_i and δ_f are the initial and final thickness of the specimen after immersion in the water.

3.3 Microstructure Examination

The microstructure of the composite is examined at 500X and 300X.

4 Result and Discussion

4.1 Water Absorption Test

The interfacial adhesion of wood and polymer is poor so water absorption takes place at the interface of wood and plastic. Adhesion between wood and polymers is poor due to hydrogen bonding. The poor adhesion wood and the polymer are responsible for water uptake, and water absorption occurs at the interface of wood and plastic. Water absorption tests of 2 h and 24 h for all formulations are conducted and plotted in Figs. 2 and 3. The water absorption percentage is calculated by using Eq. (1). The water absorbed by VPP and RPP polymer is 0.03% and 0.02% in 2 h, similarly 0.06% and 0.05% for 24 h during this experiment. The water absorption of VPP and RPP matrix composite varies from 0.09 to 1.65% and 0.08 to 1.46%, respectively. Similarly, water uptake by VPP and RPP matrix-based composite for

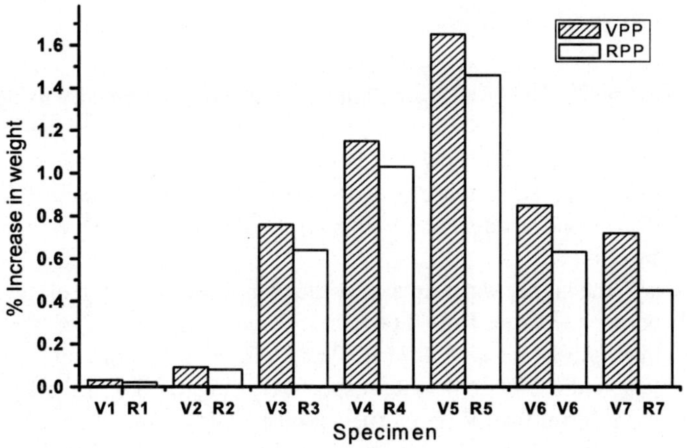

Fig. 2 Water absorption test for 2 h

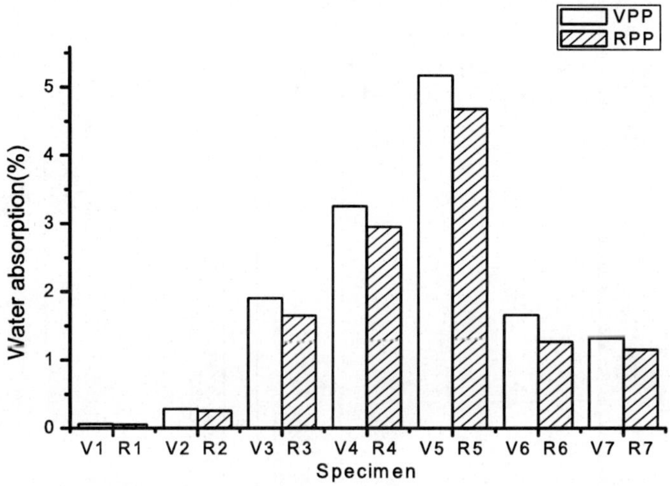

Fig. 3 Water absorption test for 24 h

24 h period varies from 0.28 to 5.17% and 0.25 to 4.68%, respectively. Water uptake increases with the increase of wood content in the composites, VPP with 50% of wood sawdust (V5) and RPP with 50% wood sawdust show maximum water absorption. The composites of RPP show low water absorption than VPP composite at the same level of wood sawdust loading. Incorporation of 3% and 5% of MAPP in the polymer matrix reduces the water uptake in both types of the composite as shown in V6 and R6 and V7 and R7.

4.2 Thickness of Swell Test

Thicknesses of swell (TS) of all formulations are given calculated by using Eq. (2) and plotted in Fig. 4(for 2 h) and Fig. 5(24 h), respectively. TS of the composite follows the trend of WA and shows TS is initially high. Thickness swelling of the composites rises with the increase of wood content in the composite. The results show that 50%wt wood sawdust and VPP matrix composite (V5) show a maximum thickness of swelling.

This is due to the higher water uptake by the composite. TS in 2 h of the RPP and VPP matrix composite varies from 0.06 to 1.35% and 0.05% to 1.48%, respectively (Fig. 4). Similarly, in 24 h test, TS for RPP and VPP matrix composite varies from 0.21% to 4.35% and 0.24% to 4.94%, respectively (Fig. 5). TS of the RPP matrix composite (R5) is lower than VPP matrix composite (V5) at the same wood level of wood content.

TS of RPP matrix (R6) and VPP matrix composite (V6) reduces with incorporation of 3% of MAPP as shown in Fig. 5. TS of RPP and VPP matrix composite reduces from 1.35% (R5) to 0.61% (R6) and 1.48% (V5) to 0.68% (V6) for 2 h test and decreases from 4.35% (R5) to 1.13% (R6) and 4.94% (V5)to 1.45% (V6) for 24 h test. TS reduces from 1.35%(R5) to 0.57% (R7) and 1.48 (V5) to 0.62% (V7) for 2 h and 4.35% (R5) to 1.05% (R7) and 4.94% (V5) to1.21% (V7) for 24 h test.

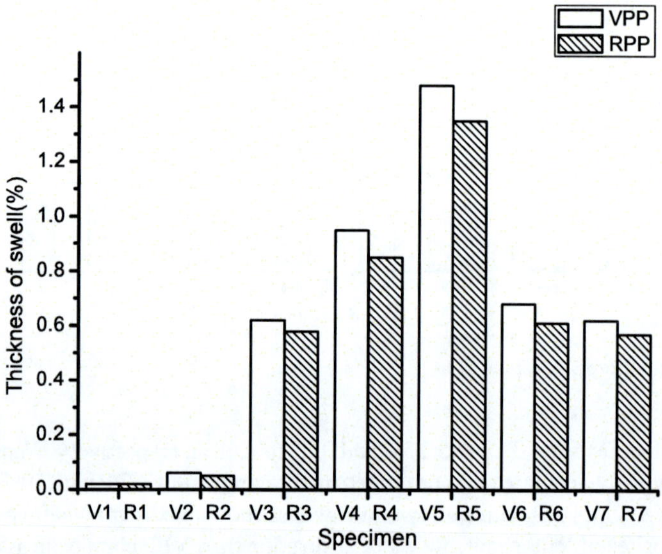

Fig. 4 Thickness of swell test for 2 h

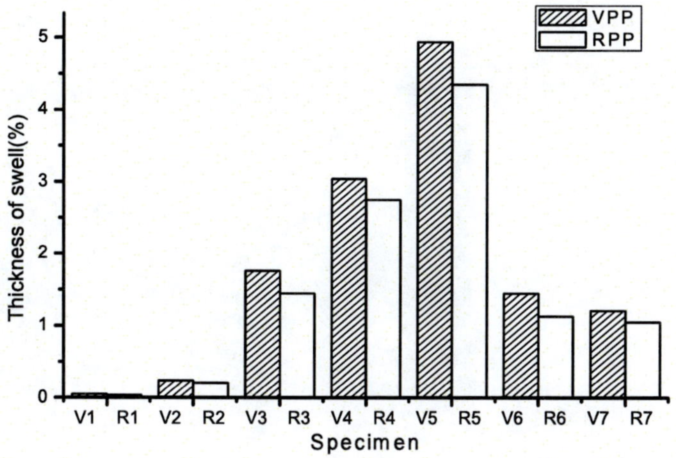

Fig. 5 Thickness of swell in 24 h

4.3 *Microstructure Characterization*

The microstructure of specimens is examined at 500X and 300 X magnifications. Scan 6(a) of R5 shows that wood particles are not properly encapsulated by the polymer, and poor encapsulation is responsible for higher water uptake in the composite.

Voids are present in the scans, which were formed due to the pulling of wood particles, and light flairs show the flaw of the polymer composite. The presence of a cluster and small gaps in the image confirms the inept adhesion of wood and polymer. This poor adhesion of wood and polymer exhibits poor mechanical, rheological, and hygroscopic properties.

The wood particles are comparatively fine mixed with the polymer in scan Fig. 6b R7 which is due to polymer ratio increase. The dispersion wood particle becomes fine as the polymer ratio increases in the composite. Coupling MAPP agent increases the interfacial bonding between wood and polymer, which reduces the water absorption of the composite Fig. 6b. The proper interfacial adhesion improves the mechanical strength of the composite and reduces water uptake (Fig. 7).

5 Conclusion

The composite having 20% wt wood sawdust shows the properties similar to polypropylene as the wood content increases the composite properties tends toward wood properties. The composites V5 and R5 absorb maximum water due to 50% wt wood content in their matrix. The water absorption by the composite is dependent upon the interfacial adhesion of wood particles and polymer matrix. SEM scans

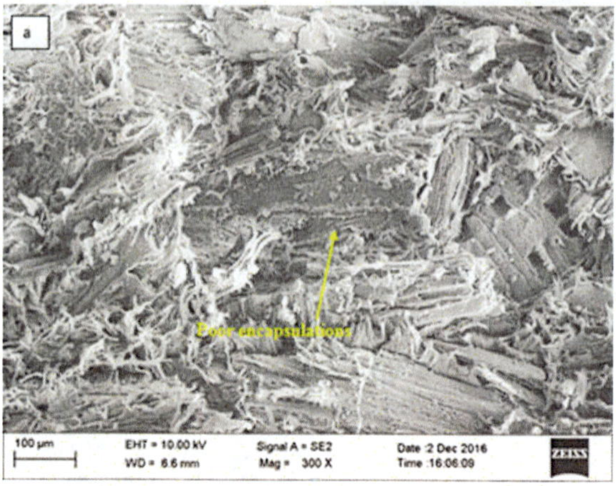

Fig. 6 SEM scan of R7

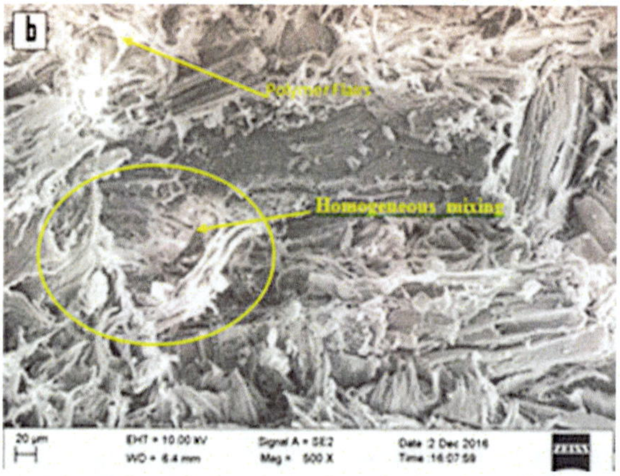

Fig. 7 SEM scan of R7

confirm that the encapsulation of wood particles becomes poor as wood content increases in the matrix. The coupling agent MAPP increases the bonding strength of wood and polymer which reduces the water uptake of the composites. The results show that the composite having maximum wood content (50%wt) in the presence of 3 and 5% wt MAPP has low water absorption. The composite R7 (50%wt wood content and 5%wt MAPP) shows minimum water absorption and maximum dimensional stability. Composite R7 is lighter in weight and possesses good dimensional stability which is useful in outdoor as well as indoor applications.

References

1. Ashori, A., Nourbakhsh, A.: Characteristics of wood-fiber plastic composites made of recycled materials. J. Waste Manage. **29**(4), 1291–1295 (2009)
2. Kaboorani, A.: Characterizing water sorption and diffusion properties of wood/plastic composites as a function of formulation design. J. Constr. Building Mater. **136**, 164–172 (2017)
3. Abedini, M.A., Najafabadi, N., Khorasani, S., Moftakharian Esfahani, J.: Water absorption behaviour and mechanical properties of high density polyethylene/pistachio shell flour nano-composites in presence of two different uv stabilizers J. Poly. Poly. Compos. 22(4), 409–416 (2014)
4. Adhikary, K.B., Pang, S., Staiger, M.P.: Dimensional stability and mechanical behavior of wood–plastic composite based on recycled and virgin high density polyethylene (HDPE). J. Compos. Part B **39**(5), 807–815 (2008)
5. Homkhiew, C., Ratanawilai, T., Thongruang, W.: Long-term water absorption and dimensional stability of composites from recycled polypropylene and rubber wood flour. J. Thermoplas. Compos. Mater. **29**(1), 74–91 (2014)
6. Shakeri, A., Ghasemian, A.: Water absorption and thickness swelling behavior of polypropylene reinforced with hybrid recycled newspaper and glass fiber. J. Appl. Compos. Mater. **17**(2), 183–193 (2010)
7. Mohebby, B., Younesi, H., Ghotbifar, A., Najafi, S.K.: Water and moisture absorption and thickness swelling behavior in polypropylene/wood flour/glass fiber hybrid composites. Reinforced Plast. Compos. **29**(6), 830–839 (2009)
8. Kord, B.: The impact of plastics virginity on water absorption and thickness swelling of wood plastic composites. J. World Appl. Sci. **17**(2), 168–171 (2012)
9. Bhaskar, J., Singh, V.K.: Water absorption and compressive properties of coconut shell particle reinforced-epoxy composite. J. Mater. Environ. Sci. **4**(1), 113–118 (2013)
10. Standard Test Method for Water Absorption of Plastics ASTM D570–98 (2018)

Parametric Optimization in the Laser Cutting of Titanium Alloy Sheet (Grade-II)

Tarun Singh and Arun Kumar Pandey

1 Introduction

Titanium and its alloys are highly demanded due to their advance properties like high strength-to-weight ratio, high resistance to corrosion, high resistance to fatigue, and high stiffness and capacity to bear high temperatures without any creep. These materials have low thermal conductivity, high reaction capability, and low elastic modulus. It is not easy to cut these alloys properly by using traditional machining methods due to such properties [1]. The dissipation of heat generated during the machining may be accumulated at a point because of very low thermal conductivity of these materials which results in melting of tool tip. There is also a chemical reaction that takes place between the tool and titanium alloy at elevated temperature, due to which tool is dissolved and hence reduces the tool life [1].

Machining by the laser beam is a unconventional machining process and uses the thermal energy for machining. The cutting of difficult-to-cut and latest developed materials for specific purposes materials such as superalloys, hardened steel, composites and ceramics is done by using laser beam. LBM is being employed for both macro- and micromachining [2]. The material removal mechanism in machining by laser beam has the stages as melting, vaporization, and degradation by chemical reaction. The assisted gas with high pressure is utilized to remove the melted material from machining area. The variation of operations and hybridization in laser beam machining are also possible [3, 4].

As discussed earlier, Nd: YAG laser is a solid laser which uses optically pumped excitation energy. It has 1.06 μm of wavelength and low average power in CW mode but it gives high peak power in pulsating mode that is responsible for small HAZ and small KW (kerf width). CO_2 laser is gas laser that uses electrical pumping source and has working wavelength 10.6 μm, and it has high average beam power [5]. The air is

T. Singh · A. K. Pandey (✉)
Bundelkhand Institute of Engineering and Technology, Jhansi, UP, India
e-mail: arunpandey76@gmail.com

M. Muzammil et al. (eds.), *Recent Advances in Mechanical Engineering*, Lecture Notes in Mechanical Engineering, https://doi.org/10.1007/978-981-15-8704-7_43

most commonly used as assisted gas due to low cost and easily availability. But HAZ by using air as assist gas gives more as compared to that of N_2 and Ar as assisted gasses [6]. In previous studies, Pandey and Dubey have found that the oxygen can give the satisfactory results by reducing the kerf deviation at top as well as bottom [7]. There is a need for several parameters to do the process successful. Optimization of the process parameters is done by using the proper optimization methodology [8].

The geometrical accuracy in laser cutting of material depends mainly upon the kerf qualities of cut sections. The deviation in length of cut shows its importance in the materials those are very difficult to cut by laser. The kerf taper always exists in cut specimens due to the inherent characteristics of beam profile [9]. Ghany and Newishy studied the laser cutting of austenitic stainless steel sheet and observed that upon increasing the frequency and cutting speed the decrease in the kerf width was observed while increase in kerf width was observed with enhancing the power and pressure of gas [10]. The cut quality study on 4130 steel by N. Rajaram et al. taking CO_2 laser system has performed and found the major effect of power and moderate effect of feed rate on kerf width [11].

A study conducted by Mondal et al. on Al7075 sheet of 2 mm thickness in laser micromachining found that the kerf width (KW) and other qualities such as HAZ and SR are increased with laser power and frequency while decreased with scanning speed [12]. G. Thawari et al. presented a specifically cutting parametric regime for pulsed laser cutting of Hastelloy X sheet having the thickness of 1 mm. It has been found that upon increasing the pulse energy, the spot overlap decreases and also found that width of kerf increases upon increasing the overlaps of spot [13].

Many authors have used the Taguchi methodology to improve the cut qualities of laser cut material in the different operating environment and they observed an improvement on different quality characteristics [14, 15]. Mathew et al. have analyzed the cutting by laser beam of 2 mm thick (CFRP) sheet and cut successfully performed [16].. Biswas, et al. have investigated the laser microdrilling for gamma titanium aluminide (Ti–44.5%, Al–2%, Cr–2) [17]. Saini et al. used zirconia-toughened alumina (ZTA) for studying the effects of process parameters on hole quality characteristics [18]. Kotadiya and Pandya have applied the RSM and ANOVA to effectively analyze the cutting variables and find out the optimum values for surface roughness [19]. Almeida et al. perform the factorial design (FD) approach during Nd: YAG laser cutting of titanium alloy grade-V and pure titanium to determine the effects on surface roughness and dross formation by pulse energy, overlapping rate, and gas pressure [20]. Shrivastava and Pandey have applied a genetic algorithm and multiple regression analysis to predict the optimum parameters of minimum width of kerf and minimum deviation of kerf in the cutting of Ti–6Al–4V sheet by using laser beam of solid state laser [21].

There is very less work that has been found on cutting by using the beam of solid-state laser of the Ti-alloy (grade-II) for optimizing the quality characteristics. So, it is decided to take this advanced material for the experimental study by using solid-state laser. For the conduction of experiments and optimization, it is decided to apply (TM) Taguchi methodology with L_{27} OA. The multiple order models are developed by using the data of the experiments. The settings of parameters for minimum width of kerf and deviation of kerf have been found by using TM. The confirmation tests have been conducted for the confirmation of the results found at parameters setting obtained by TM. The confirmation results have been found very near to the results found at optimum setting and also these results follow almost the same trend as predicted.

2 Methodology and Parameters Design

2.1 Experimental Design

Taguchi methodology is a technique for designing the robustness to improve the manufacturing productivity by reducing experiments required [22]. The designed matrixes for experimentation generally known as orthogonal array (OA) are based on process parameters and levels tabulated in Taguchi method [23]. In Taguchi methodology, the signal-to-noise (S/N) ratio ($\acute{\eta}$) represents the quality loss for the observed data. The word 'signal' is representing the value of desireness and word 'noise' is representing the undesirable values, and scatter around the desired values is expressed by ratio of these two. The signal-to-noise ratio for smaller the better type is computed by using following equation;

$$\acute{\eta} = -10\log_{10}(\text{MD}) \tag{1}$$

where MD is mean square deviation from desired values.

2.2 Experimental Work Planning

In this research work, parametric optimization in cutting by laser beam using solid state laser of 1.5 mm thick titanium alloy grade-II sheet which having chemical composition (weight percentage of nitrogen-0.015, carbon-0.022, oxygen-0.101, hydrogen-0.001, iron-0.118, and titanium-99.39) has been done. Since the aim is to optimize overall output of production, the air is used as the assist gas, which is easily available at very less cost after processing. The specifications of the machine are as 250 W average power, 240 A constant current, CNC (*X*–*Y*) travel operating mechanism with lead screw, assist gas-compressed air, 600 μm nozzle spot diameter, 1.5 mm nozzle tip distance, 40 mm focal length of focusing and collimation lances, sheet metal dimension of 200 × 1000 × 1.5 (mm), and optical measuring microscope 4X-zoom (Leica SAPO). The process parameters and their levels have been selected based on the results of pilot experiments, importance, and machine capability. The process parameters and their levels are given in Table 1. By using the values of different process parameters shown in Table 1, the through cuts of 20 mm

Table 1 Control factors and their levels used in the experiment

Symbol	Factors	Units	Level 1	Level 2	Level 3
A	Assisted gas pressure	Kg/cm^2	5.0	7.0	9.0
B	Pulse width	ms	2.0	3.0	4.0
C	Pulse frequency	Hz	12	14	16
D	Cutting speed	mm/min	10	20	30

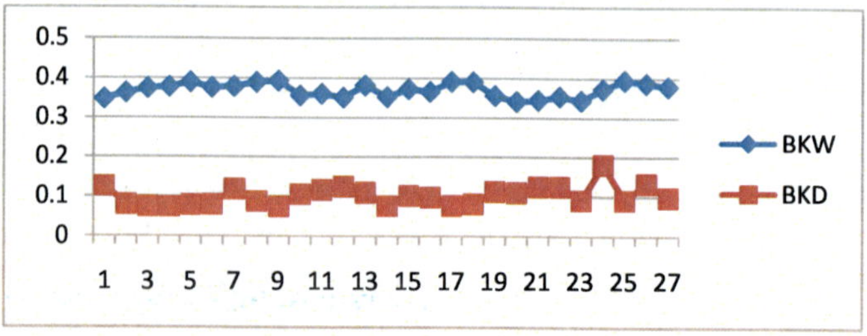

Fig. 1 Experimental values of quality characteristics

in a straight line are cut. The widths of kerf are measured at many places on the bottom side and then take mathematical average. Kerf deviation (KD) is measured by taking difference between the maximum and minimum values of kerf widths at bottom. The observed values of width of kerf (KW) and deviation of kerf (KD) at bottom side are shown by graph in Fig. 1.

3 Second-Order Regression Models

3.1 Modeling

The second-order regression models for responses BKW and BKD are developed by regression analysis using MINITAB software. The interactions between different control factors are taken into consideration to know the effect on responses. The uncoded units (actual experimental values) are taken for regression analysis. The final developed models are shown as below;

$$\begin{aligned}\mathbf{BKW} = & -0.118 + 0.0197\,Y_1 + 0.0516\,Y_2 + 0.0291\,Y_3 + 0.00942\,Y_4 + 0.000281\,Y_1^2 \\ & - 0.00269\,Y_2^2 - 0.000224Y_3^2 - 0.000080Y_4^2 + 0.00014\,Y_1 * Y_2 - 0.00032\,Y_2 * Y_3 \\ & - 0.000262\,Y_3 * Y_4 + 0.000025\,Y_1 * Y_4 - 0.00196\,Y_1 * Y_3 - 0.000592\,Y_2 * Y_4 \end{aligned} \quad (2)$$

$$S = 0.00911578 R - Sq = 92.5\% R - Sq(\mathrm{adj}) = 83.9\%$$

Table 2 ANOVA results

Source	DF	SS	MS	F-value	P-value
BKW					
Regression	14	0.01238185	0.00088442	10.64	0.000
Residual error	12	0.00099717	0.00008310		
Total	26	0.01337902			
BKD					
Regression	14	0.0170861	0.0012204	11.29	0.000
Residual error	12	0.0012975	0.0001081		
Total	26	0.0183836			

DF degree of freedom, *SS* sum of squares, and *MS* mean of squares

$$\begin{aligned}\textbf{BKD} = {} & 0.916 - 0.0269\,Y_1 - 0.0187\,Y_2 - 0.0725\,Y_3 - 0.0222\,Y_4 \\ & - 0.00018Y_1^2 + 0.00848Y_2^2 + 0.00204Y_3^2 + 0.000210Y_4^2 \\ & - 0.00385\,Y_1 * Y_2 - 0.00287\,Y_2 * Y_3 + 0.000347\,Y_3 * Y_4 + 0.000808\,Y_1 * Y_4 \\ & + 0.00225\,Y_1 * Y_3 + 0.00140\,Y_2 * Y_4 \end{aligned} \quad (3)$$

$$S = 0.0103982R - Sq = 92.9\%R - Sq(\text{adj}) = 84.7\%$$

where Y_1, Y_2, Y_3 and Y_4 are pressure of assist gas, width of the pulse, frequency of the pulse, and speed of the cutting, respectively.

3.2 *Adequacy of the Developed Models*

The S-values, R^2-values, and adjusted R^2-values have been calculated for models of BKW and BKD to check well fitting of predicted data using regression models. The coefficients of correlation (*R*-values) for BKW and BKD are found 0.962 and 0.964, respectively, which are found in the range that will be accepted. The data predicted by regression models are adequate. The adequacy may also be checked by conducting ANOVA analysis. The results found by this analysis are shown in Table 2. The *p*-values of all regression models are less than 0.01, i.e., the developed regression models are significant and adequate for all the responses [8].

4 Parametric Optimization

The prediction of optimum parameters for single quality has been analyzed by using the Taguchi methodology (TM). The mean values of S/N ratio for bottom kerf width

Table 3 Response table for bottom kerf width (BKW)

Level	A	B	C	D	
1		8.499	9.007*	8.695*	8.842*
2	8.670	8.674	8.685	8.636	
3	8.797*	8.285	8.586	8.488	
Delta	0.298	0.722	0.108	0.354	
Rank	3	1	4	2	

Table 4 Response table for bottom kerf deviation (BKD)

Level	A	B	C	D
1	21.24*	19.42	19.53	19.90
2	20.21	20.19	20.65*	20.93*
3	18.71	20.54*	19.97	19.32
Delta	2.53	1.11	1.12	1.60
Rank	1	4	3	2

Vs the process parameters are shown in response Table 3, and all mean values of S/N ratios for bottom kerf deviation versus process parameters are shown in response Table 4.

From the observations, it can be concluded by (TM) that the highest value of S/N ratio for respective process parameters represents the optimum parametric condition. The rank is estimated by delta which shows the sequence of contribution of control factors to achieve optimized quality levels. Thus, from Table 3, the optimum parametric condition for bottom kerf width (BKW) is $A_3B_1C_1D_1$ that is gas pressure of 9 (Kg/cm^2), width of pulse 2 (ms), frequency of pulse 12 (Hz), and speed of cutting 10 (mm/min). From Table 4, it is clear that the optimum parametric levels for bottom kerf deviation (BKD) are $A_1B_3C_2D_2$ that is assisted gas pressure of 5 (Kg/cm^2), pulse width of 4 (ms), pulse frequency of 14 (Hz), and cutting speed of 20 (mm/min).

5 Confirmation Experiments

To check the adequacy of predicted results and improvement at optimized parameters levels, an experimental test is done which is given in Table 5. The results of confirmation experiments show that the width of kerf and deviation of kerf both reduce at optimum parameters setting, i.e., an improvement on both characteristics has been observed at optimum setting as shown in Figs. 2 and Fig. 3.

Table 5 Confirmation experiments for quality characteristics

Initial cutting parameter	Optimum parameters for laser cutting		
	Setting for minimum value		
	Prediction	Experiment	
(BKW)			
Level	$A_3B_1C_2D_1$	$A_3B_1C_1D_1$	$A_3B_1C_1D_1$
Bottom kerf width (mm)	0.3408	0.33983	0.33693
S/N ratio (dB)	9.3500	9.37472	9.44920
Predicted error of S/N ratio (dB)	0.07448		
(BKD)			
Level	$A_1B_2C_1D_2$	$A_1B_3C_2D_2$	$A_1B_3C_2D_2$
Bottom kerf deviation (mm)	0.0740	0.06918	0.06871
S/N ratio (dB)	22.6154	23.20030	23.25960
Predicted error of S/N ratio (dB)		0.05930	

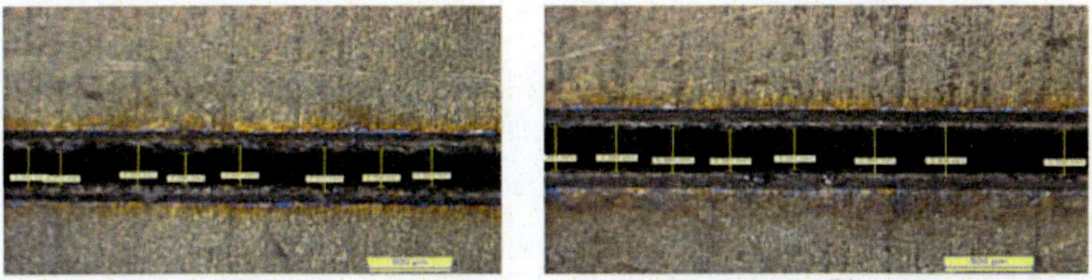

(a) (b)

Fig. 2 Optical images of laser cut bottom kerf width at **a** initial parameters setting having average width of 0.3408 mm and **b** optimal experimental parameter setting having average width of 0.33693 mm

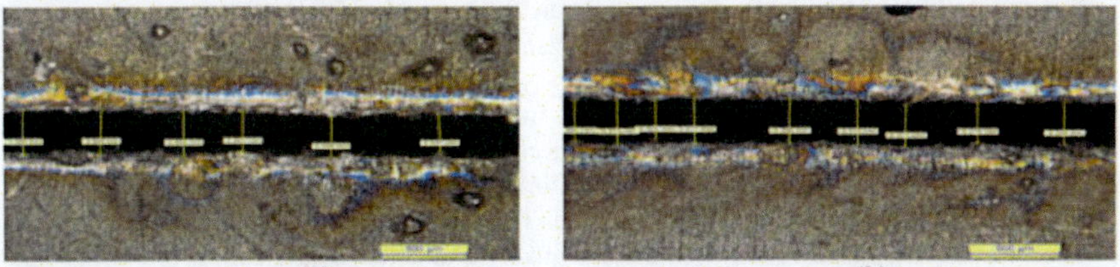

(a) (b)

Fig. 3 Optical images of laser cut bottom kerf deviation at **a** initial parameters setting having mean deviation of 0.0740 mm **b** optimal experimental parameter setting having mean deviation of 0.06871 mm

6 Conclusions

An experimental study on the cutting of titanium (grade-II) sheet by using the beam of solid-state laser has been carried out successfully. The main conclusions found from the research are as follows;

(i) The regression models of second order for width of kerf and deviation of kerf at the bottom side have been developed.
(ii) The regression analysis shows that the correlation coefficient for width of kerf and deviation of kerf at the bottom side is observed as 0.962 and 0.964, respectively, which shows that the developed models are reliable and adequate.
(iii) The statistical analysis shows that the regression models of second order for width of kerf and deviation of kerf at the bottom side are found to be adequate and reliable for predicting the above responses.
(iv) At optimal parametric levels, the improvement in S/N ratio as well as quality characteristics values can be seen. It has also been observed in the confirmation experiments as the BKW and BKD are reduced from 0.3408 mm and 0.0740 mm to 0.33693 mm and 0.06871 mm, respectively.

References

1. Dornfeld, D.A., Kim, J.S., Dochow, H., Hewsen, J., Chen, L.J.: Drilling burr formation in Titanium alloy (Ti-6Al-4V). Ann. CRIP **48**, 73–76 (1999)
2. Jain, V.K.: Advance Machining Process, 4th edn. Allied Publishers Private Limited, New Delhi (2005)
3. Majumdar, J.D., Manna, I.: Laser processing of materials. Sadhana **28**(3–4), 495–562 (2003)
4. Gautam, G.D., Pandey, A.K.: Pulsed Nd:YAG laser beam drilling: A review. Opt. Laser Technol. **100**, 183–215 (2018)
5. Steen, W.M.: laser Material Processing, 3rd edn. Springer, New York (1991)
6. Shanjin, L., Yang, W.: An investigation of pulsed laser cutting of Titanium alloy sheet. Opt. Lasers Eng. **44**, 1067–1077 (2006)
7. Pandey, A.K., Dubey, A.K.: Taguchi based fuzzy logic optimization of multiple quality characteristics in laser cutting of duralumin sheet. Opt. Laser Eng. **50**, 328–335 (2012)
8. Pandey, A.K., Dubey, A.K.: Simultaneous optimization of multiple quality characteristics in laser cutting of titanium alloy sheet. Opt. Laser Technol. **44**, 1858–1865 (2012)
9. Shrivastava, P.K., Pandey, A.K.: Geometrical quality evaluation in laser cutting of inconel-718 sheet by using Taguchi based regression analysis and particle swarm optimization. Inf. Phys. Technol. **89**, 369–380 (2018)
10. Ghany, K.A., Newishy, M.: Cutting of 1.2 mm thick austenitic stainless steel sheet using pulsed and CW Nd: YAG laser. J. Mater. Process. Technol. (168), 438–447 (2005)
11. Rajaram, N., Sheikh-Ahmad, J., Cheraghi, S.H.: CO_2 laser cut quality of 4130 steel. Int. J. Mach. Tools Manuf **43**, 351–358 (2003)
12. Mandal, K.K., Kuar, A.S., Mitra, S.: Experimental investigation on laser micro-machining of Al 7075. Opt. Laser Technol. **107**, 260–267 (2018)
13. Thawari, G., Sarin Sundar, J.K., Sundararajan, G., Joshi, S.V.: Influence of process parameters during pulsed Nd: YAG laser cutting of nickel-base superalloys. J. Mater. Process. Technol. **170**, 229–239 (2005)

14. Tam, S.C., Lim, L.E.N., Quek, K.Y.: Application of Taguchi method in the optimization of the laser-cutting process. J. Mater. Process. Technol. **29**, 63–74 (1992)
15. Li, C.-H., Tsai, M.-J., Yang, C.-D.: Study of optimal laser parameters for cutting QFN packages by Taguchi's matrix method. Opt. Laser Technol. **39**(4), 786–795 (2007)
16. Jose, M., Goswami, G.L., Ramakrishnan, N., Naik, N.K.: Parametric studies on pulsed Nd: YAG laser cutting of carbon fibre reinforced plastic composites. J. Mater. Process. Technol. (89–90), 198–203 (1999)
17. Biswas, R., Kaur, A.S., Mitra, S.: Multi-objective optimization of hole characteristics during pulsed Nd: YAG laser microdrilling of gamma-titanium aluminide alloy sheet. Optics. Laser Eng. **60**, 1–11 (2014)
18. Saini, S.K., Dubey, A.K., Upadhyay, B.N., Choubey, A.: Study of hole characteristics in laser Trepan Drilling of ZTA. Opt. Laser Technol. **103**, 330–339 (2018)
19. Kotadiya, D.J., Pandya, D.H.: Parametric analysis of laser machining with response surface method on SS-304. Procedia Technol. **23**, 376–382 (2016)
20. Almedia, A.I., Rossi, W., Lima, M.S.F., Beretta, J.R., Nogueria, G.E.C., Wetter, N.U., Vieira Jr., N.D.: Optimization of titanium cutting by factorial analysis of the pulsed Nd:YAG laser parameters. J. Mater. Process. Technol. **179**, 105–110 (2006)
21. Shrivastava, P.K., Pandey, A.K.: Multi-objective optimization of cutting parameters during laser cutting of Titanium alloy sheet using hybrid approach of genetic algorithm and multiple regression analysis. Mater. Today **5**, 24710–24719 (2018)
22. Phadke, M.S.: Quality Engineering Using Robust Design, 1st edn. Prentice-Hall, Englewood Cliffs, New Jersey (1989)
23. Ross, P.J.: Taguchi Techniques for Quality Engineering, 2nd edn. McGraw Hill, New York (1988)

Development of Hybrid Nanocomposites by Stir Casting Technique Using Two-Step Mixing

Rakesh Kumar Yadav, Zahir Hasan, and Akhter Husain Ansari

1 Introduction

The composite metal matrix (MMC) is designed to achieve a specific property by combining metal (matrix) with difficult particle/ceramic (reinforcement). For space shuttles, business aerospace, electronic equipment, bikes, cars and a range of wide application, MMCs are used or prototyped [1]. Aluminium matrix composites are not like conventional composite material but belong to a family of composites which can be apportioned to its rigidity, strength, density, heat and electrical qualities. To accomplish the necessary features, various parameters, like matrix alloy, strengthening material, reinforcement volume and shape, the place of reinforcement and manufacturing processes, etc., can all be varied. The goal is to enhance the combined properties of metals and ceramics in the design of composite metal matrix [2]. Stir casting is generally practiced and is known for an especially favourable process for discontinuous metal materials composites, which is presently practiced commercially. Its benefits are its simplicity, flexibility and manufacturability for mass production. This is also appealing as it enables a traditional metal handling route to be used in theory and therefore reduces the final price of the item to a minimum [3].

The most common matrix in MMCs is aluminium. The low density, strength, good forgeability, high hardness and greater resistance to wear make Al alloys quite fit for use [4]. The aluminium matrix composites (AMCs) are now being widely considered in the sports, electronics, automobile, aerospace, bicycle and other industries [5]. Furthermore, AMCs are lightweight material with high mechanical characteristics

R. K. Yadav (✉)
APJ Abdul Kalam Technical University, Lucknow, India
e-mail: rkymech@gmail.com

Z. Hasan
Department of Mechanical Engineering, NSAKCET, Hyderabad, India

A. H. Ansari
Department of Mechanical Engineering, Aligarh Muslim University, Aligarh, India

M. Muzammil et al. (eds.), *Recent Advances in Mechanical Engineering*, Lecture Notes in Mechanical Engineering, https://doi.org/10.1007/978-981-15-8704-7_44

with respect to weight, and also they can be cast with stir casting processes in all complex forms. The mechanical characteristics like strength during compression and during tension of nanocomposites Al and Al–AlN were examined [6] which by adding AlN nanoparticles results in an improved compressive strength. The final strength is projected to be 2610 bar at a percentage elongation of 39–41%. On the other hand, if Al_2O_3 has reinforcement of aluminium and magnesium in pulverized form, the corresponding strength characteristics are improved. Furthermore, if Al-A206 is used as matrix material with reinforcement of Al_2O_3 particles in solid or semisolid state, corresponding mechanical strength characteristics are improved [7].

It has been observed that Al-SiCp's pressure resistance, produced using ball-mounted and ball milling powders, has grown from 80 to 172 N/mm^2, with a growth in SiC_p weight. The bending strength of Al-composites, manufactured by the liquid aluminium infiltration method, with a ceramic reinforcement up to 13.0 vol. per cent, increased during third-point bend tests [8], extremely allowing dual-ceramic (aluminium oxide and silicon carbide)-reinforced ceramic compounds, which were further reduced [9]. The literature review shows that the main issue was the use of standard low-cost machinery for business purposes to achieve homogenous dispersion of ceramic particles. From the studies released by stir casting on Al/SiC composites, restricted work has been performed by two-step mixing in stir casting. As it is reviewed, by stir casting technique using two-step mixing, no work is done on aluminium alloy (A356) matrix hybrid composites. Therefore, A356 as matrix material and reinforced with 10 wt.% SiC microparticles along with different weight fractions of SiC nanoparticles (0.0–4.0 wt.%) were fabricated by stir casting process using two-stage mixing.

2 Experimentation

2.1 Matrix Material

A356 is widely used as the matrix material. The chemical composition of the aluminium matrix is shown in Table 1 and Fig. 1a show the A356 used as the matrix material. As-received A356 image of scanning electron microscopy is shown in Fig. 1b.

Table 1 Chemical composition of A356

Element	Si	Cu	Fe	Zn	Mg	Mn	Ti	Al
(%) weightage	6.5	0.2	0.25	0.15	0.4	0.1	0.2	Bal

Fig. 1 **a** Ingots A356, **b** SEM image of purchased aluminium powder

2.2 *Reinforcing Material*

The material used as the reinforcement phases were SiC microparticles and SiC nanoparticles. The size of micro- and nanoparticles is 38 μm and 48 nm, respectively. Figure 2a and b shows the SiC micro- and nanoparticle powder used as reinforcement material. The scanning electron microscopy of as-received silicon carbide micro- and nanoparticle is shown in Fig. 2c and d.

3 Mixing of Reinforcing Material

Ball mill is a mixing system for fine aluminium powder consisting of repetitive cold bonding, rupture and powder particle rewelding [10]. Micro-SiC powder, nano-SiC powder and aluminium powder have been obtained from Otto chemie Pvt Ltd. Mumbai, and A356 ingots have been purchased from Purshottam Brothers Kanpur.

Through the use of ball mill, pure aluminium fine powder and SiC nanoparticles are mixed together. According to the past study [7], the ball mill cylinder velocity was set at 100 rpm for 10 h ball milling length. The ratio between powder and ball was 1:10 [11]. Ball milling method was performed to safeguard the oxidation in the argon gas setting. The degasification prearrangement was also made from the cylinder of the ball mill. The primary benefits of using this method are to improve the wettability of aluminium fine powder nanoparticles. For making 10–10 g powder palette, we used 25 mm diameter die and on this we apply a load of nearly one tonne. The palette is used to preserve nanopowder from evaporation. Now, the pallets are wrapped with aluminium foil in direct molten metal with the help of tong to preserve it from oxidation [12]. Figure 3a shows pictorial view of palette and Fig. 3 b shows die and punch.

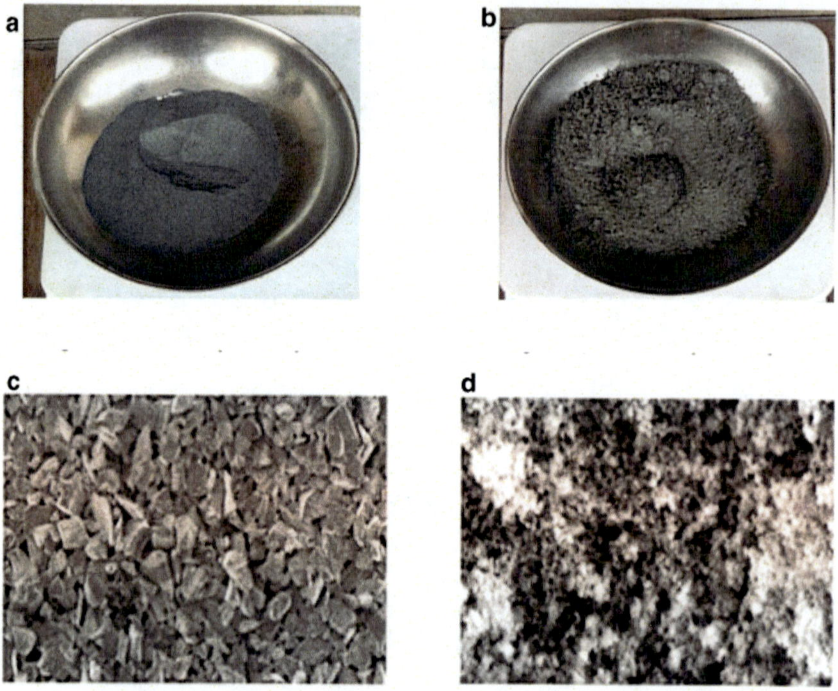

Fig. 2 **a** SiC microparticles powder, **b** SiC nanoparticles powder, **c** SEM image of purchased SiC microparticles, **d** SEM image of purchased SiC nanoparticles

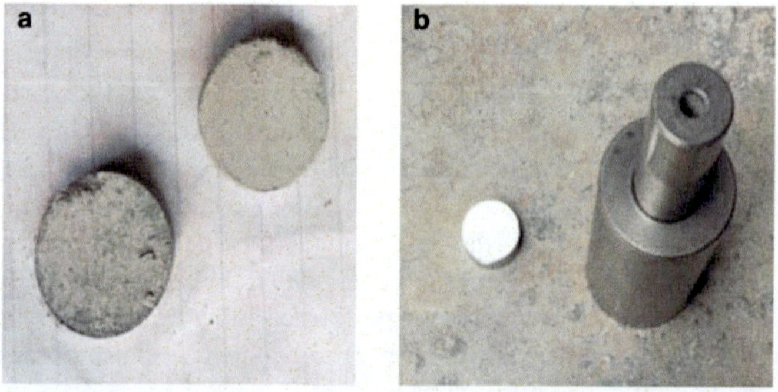

Fig. 3 **a** Pictorial view of pallets and **b** punch and die

4 Stir Casting of Hybrid Nanocomposites—The Stir Cast Product

Stir casting is commonly used in the fluid stage method to make the composites. Because it has different benefits compared to other manufacturing techniques such as easy, low cost, less manufacturing time and its applicability to continuous production. The melting was carried in a electric furnace in a range of 750–850 °C. A schematic view of the electric furnace has been shown in Fig. 4a

Initially, A356 ingots are heated in a electric furnace between temperature limit 750–850 °C and after melting the alloy a flux (coverall) weighted 20–25 g is added. The flux removes the impurities and also provides the blanket against heat radiation. Thereafter an inert argon gas is inserted for five minutes in furnace to remove the inherited oxygen. After the cleaning process, palette is mixed with molten metal and allowed to heat for 15 min and maintaining the temperature of furnace at 750 °C. Now, the temperature of furnace is reduced to 725–750 °C. After completing melting, the SiC microparticles are inserted in the molten metal and along with it stirring is done. The molten metal is stirred with the help of stirrerring at 400–500 rpm for five minutes. The process is carried out to avoid jurling. Now, the same operation is repeated after five minutes so that the molten metal reaches to liquid state and then again the stirring operation is performed to repeat the same operation as discussed above. Afterwards waiting for one minute, mixing solution is observed whether the

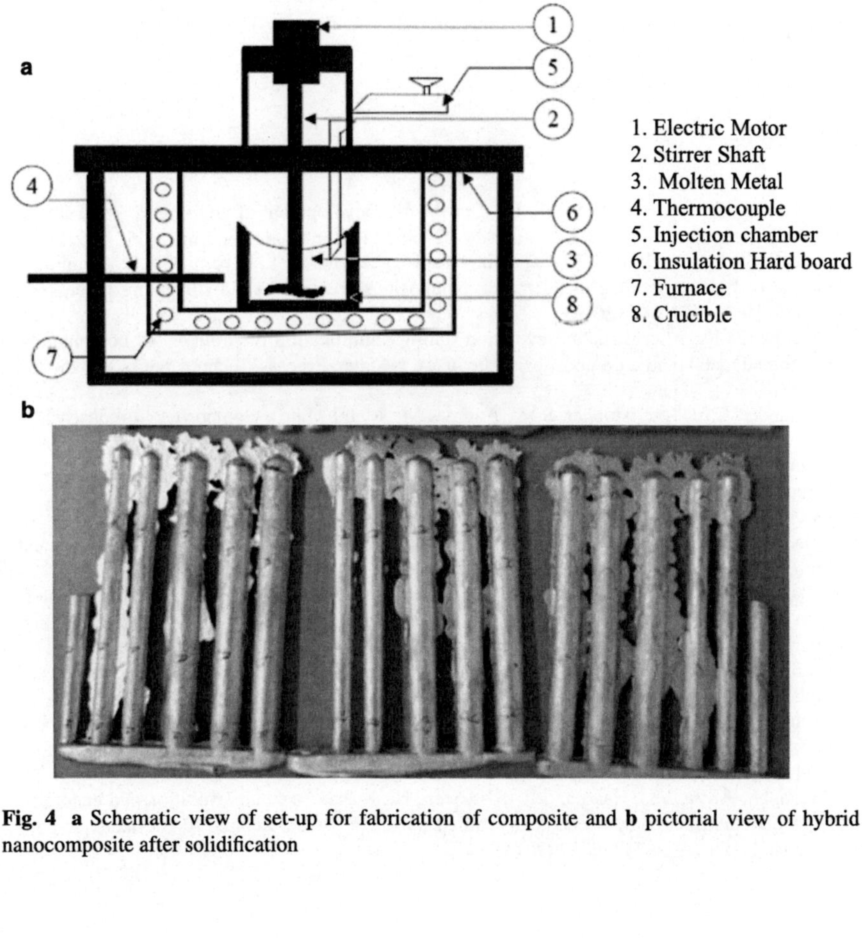

Fig. 4 **a** Schematic view of set-up for fabrication of composite and **b** pictorial view of hybrid nanocomposite after solidification

powder is properly mixed or not. If the prepared solution is successful, then we obtain shiny and homogeneous structure of molten metal. Thereafter, stirrer is removed and after heating for 15 min the prepared slurry is poured in the permanent mould. At the same time, permanent mould is also heated upto a temperature of 200–300 °C. So, the finished (solidified) product can be easily removed from mould. Finally, we obtain the final product as shown in Fig. 4b.

5 Conclusion

In the above operation, we obtain hybrid nanocomposite material which was fabricated successfully after following two-step stirring operation by using stir casting. The obtained product has wide application in engineering structures, aerospace and in automobile industry, etc. The obtained product is also very light in weight along with good strength. The stir casting process is very economical and simple in operation that is why it is practice in modern engineering applications. The hybrid nanocomposite material is found to be better than conventional material as the former is better in strength and high mechanical properties suitable for engineering purposes.

References

1. Singla, M., Dwivedi, D.D., Singh, L., Chawla, V.: Development of aluminium based silicon carbide particulate metal matrix composite. J. Miner. Mater. Character. Eng. **8**(06), 455 (2009)
2. Inegbenebor, A.O., Bolu, C.A., Babalola, P.O., Inegbenebor, A.I., Fayomi, O.S.I.: Aluminium silicon carbide particulate metal matrix composite development via stir cating processing. Silicon **10**(2), 343–347 (2018)
3. Surappa, M.K.: Microstructure evolution during solidification of drmmcs (discontinuously reinforced metal matrix composites): state of art. J. Mater. Process. Technol. **63**(1–3), 325–333 (1997)
4. Alexander, E.A., Charistopher, S.M., Andreas M.: Metal Matrix Composites in Industry: An Introduction and a Survey. Springer Science+Business Media, LLC, New York, pp 1–65 (2003)
5. Alam, M.Y., Ansari, A.H.: Mechanical behavior of A356 matrix nanocomposites fabricated using two step casting technique. Int. Conf. Adv. Mech. Ind. Automat. Manage. Syst. (AMIAMS), 196–201 (2017)
6. Fale, S., Likhite, A., Bhatt, J.: Compressive, tensile and wear behavior of ex situ Al/AlN metal matrix nanocomposites. J. Compos. Mater. **49**(16), 1917–1928 (2015)
7. Tahamtan, S., Emamy, M., Halvaee, A.: Effects of reinforcing particle size and interface bonding strength on tensile properties and fracture behavior of Al-A206/alumina micro/nanocomposites. J. Compos. Mater. **48**(27), 3331–3346 (2014)
8. Purohit, R., Rana, R.S., Verma, C.S.: Fabrication of Al-SiCp composites through powder metallurgy process and testing of properties. Int. J. Eng. Res. Appl. **2**(3), 420–437 (2012)
9. Demir, A., Altinkok, N.: Effect of gas pressure infiltration on microstructure and bending strength of porous Al_2O_3/SiC-reinforced aluminium matrix composites. Compos. Sci. Technol. **64**(13–14), 2067–2074 (2004)
10. Saheb, N., Aliyu, I.K., Hassan, S.F., Al-Aqeeli, N.: Matrix structure evolution and nanoreinforcement distribution in mechanically milled and spark plasma sintered Al-SiC nanocomposites. Materials **7**(9), 6748–6767 (2014)

11. Mazahery, A., Abdizadeh, H., Baharvandi, H.R.: Development of high-performance A356/nanoAl_2O_3 composites. Mater. Sci. Eng. **5**, 61–64 (2009)
12. Alam, M.T., Ansari, A.K., Arif, S., Alam, M.N.: Mechanical properties and morphology of aluminium metal matrix nanocomposites-stir cast products. Adv. Mater. Process. Technol. 1–16 (2017)

Review on Studying Effects of Enhancing Turbulence by Modifying the Intake Manifold of Internal Combustion Engine

Shahim Haider Abidi and M. M. Hasan

1 Introduction

As power plants in the current scenario is one of the important examples, in terms of durability and efficiency, CI engines have leverage on SI engines, but with these advantages, it has a problem in terms of emissions. Various techniques related to in-cylinder and also after combustion treatment are being developed to minimize the effects of harmful output. In the IC engine, mainly there are three types of efficiency: thermal, volumetric, and mechanical. As far as mechanical efficiency is concerned, a significant amount of optimization has been achieved during past research work, but still there is a scope of work in the other two (thermal and volumetric). In the field of designing intake manifold, various geometries were used by researchers to increase the mixing of fuel and air as it plays important role in combustion. This parameter of design is a major deciding factor for an efficient engine in terms of power and emissions. In the paper, an effort is done to discuss all possible arrangement of the intake manifold to serve the purpose of enhanced mixing and the possible ways to further increase the turbulence of inlet air.

2 Literature Review

Paul and Ganesan [1] in his work have taken three arrangement spiral, helical, and their combination intake manifold for analyzing two phenomena, air motion and turbulence inside direct injection diesel engine at 3000 rpm. In this study, RNG k–ε turbulence modeled was used to obtain results related to volumetric efficiency, swirl, and turbulent in the engine. It was found that the spiral arrangement of manifold

S. H. Abidi (✉) · M. M. Hasan
Department of Mechanical Engineering, Jamia Millia Islamia, 110025, New Delhi, India
e-mail: shahim24@gmail.com

M. Muzammil et al. (eds.), *Recent Advances in Mechanical Engineering*, Lecture Notes in Mechanical Engineering, https://doi.org/10.1007/978-981-15-8704-7_45

yields higher swirl ratio and turbulent kinetic energy than the other two arrangements. In terms of volumetric efficiency, spiral–helical combination leads with by 10% as compared to spiral one, but highest was found for a helical manifold. Also, it was derived that manifold configuration has minimal effect on the fluctuation of average RMS of turbulent swirl velocity during TDC of compression inside piston bowl. Rahiman et al. [2] also worked on the same arrangement used by Paul and Ganesan to compare the parameters like volumetric efficiency, swirl ratio, tumble ratio, and turbulence for diesel engine at 1000 rpm. Results for helical manifold are higher volumetric efficiency, tumble ratio, and turbulent kinetic energy than other two configurations. As swirl is concerned, helical–spiral combination was best one among all. Bari and Saad [3] in this work in-cylinder motion of CI engine with the use of guide vane swirl and tumble device, model was developed with SolidWorks and for CFD analysis ANSYS was used. The recorded increase of turbulence kinetic energy, velocity, and in-cylinder pressure, during injection were about 2%, 22%, and 1.3%, and, respectively, for six guide vane swirl and tumble devices. Mandloi and Verma [4] studied the variation of design parameters like guide curve angle, guide curve radius, etc., of an intake manifold, and the main objective was to optimize brake-specific fuel consumption by maximizing the flow area. Martins et al. [5] in his project used software programs like Gambit, SolidWorks, and Fluent, and these simulations were carried out to find out swirl in spark ignition engine. Jemni et al. [6] used six cylinders, heavy duty, and IVECO engine which are used for powering diesel engine of bus. k–ε turbulence model was used for solving Navier–Stokes and energy equation with help of 3D CFD code FLOWORK. In results, increase of brake torque, brake power, and brake thermal efficiency by 13.9%, 16%, and 12.5%, respectively, was found and also reduction of 28% brake-specific fuel consumption was recorded. Xu [7] conducted numerical simulation on three kinds of intake manifolds using software Fluent. The analysis laid stressed that two major evaluation indicators are pressure loss and flow uniformity. Sabale and Sanap [8] proposed a methodology to achieve 1.8 swirl numbers by opting for helical intake in a diesel engine. This study emphasizes the correlation between the throat height and the swirl ratio; for lower throat height value, strong swirl is achieved and for higher value, streamlines are focused toward the center in a cylinder having smaller swirl ratio. Vichi et al. [9] in this work have main focus on the design of the engine intake manifold in which two parameters, the maximum power and the drivability of the vehicle, were optimized, and the main objective is to locate the place of a plenum chamber between the restrictor and the engine. It allows one to minimize the pulsating nature flow through the restrictor. As a result, decrease in the pressure drop at the restrictor was observed and also rise in the mass flow rate elaborated by the engine was noticed. This enhances growth of the engine performance. It can be concluded that the increase in plenum volume leads to higher power that the engine can provide. Shenghua et al. [10] used several curvilinear diversion blades which are installed near the inlet valve. This system was able to generate different strength of swirl, and variable induced swirl has a considerable effect on CI engine fuel consumption. Triqui et al. [11] analyzed two-valve spark engine to calculate two parameters torque and power which are controlled by engine air intake capacity. In this work, various techniques were used to find a trade-off between the engine flow resistance and its ability to generate large-scale in-cylinder motion. Loong and Salim [12] had used the K–ε model in which standard wall function was incorporated by simulation and quantify the modification in design. With improved intake port, higher mass flow rates were achieved which are decisive performance factors of the engine. Hushim et al. [13] used computational fluid dynamics to verify that intake airflow plays a decisive role in efficient mixing in the intake manifold for port fuel injection. Six angles of manifold were investigated which were 30°, 60°, 90°, 120°, 150°, and 180°, among all these last arrangements, was found with the best performance. Martinas et al. [14] had done optimization of valve diameter which is trade-off between the need for bigger one for high mass of air filling in-cylinder and smaller one to reduce blind zone; more the shape of inlet port is smother and piston diameter is bigger, smaller will be the aerodynamic resistance of geometry due to smaller pressure difference of inlet and near to piston. Anilkumar and Elia [15] with the help of CFD tool three-dimensional flow within the manifold runners were simulated. The nature of flow is highly three-dimensional and have strong dependence on the valve lift except for upstream of the port bend. Flow separation is critical at higher valve lift. From the analysis, it can be concluded that the uniform pressure inside the runners is obtained and also a reduction in smoke level is reported for improved inlet manifold system. Mohiuddin [16] in this work investigation was carried out on double overhead camshaft engine having capacity 1597 cc with 16 valves installed on it, and this engine was developed by Malaysian car manufacturer PROTON. In this engine, swirl adapter is used inside the intake manifold. The objective of this investigation is to study the effect of swirl with respect to the normal turbulence mixing process in the engine. The GT-SUITE software was used which has a standard swirl flow embedded in it. It is found that the swirl factor is important tool in reduction of the fuel consumption and influence the volumetric efficiency improvement. The experimental analysis derives that the influence of swirl increases two parameters power as well as torque in both the idle and cruising speed conditions with respect to normal turbulence. But these two parameters decreases rapidly which is due to the inability of generating swirl to at higher wind flow velocity by the swirl adapter during the higher throttle opening condition. Singla et al. [17] in this study Maruti Wagnor engine was used to solve the problem of Unequal distribution of charge which leads to a reduction of efficiency of the engine. Modification in a manifold is done to achieve equal velocity at the end of each runner. Two models were analyzed by making some improvements in the actual manifold. This work focuses on various conclusions like the model has optimized design of runners and equal velocity at outlet is achieved through the curves at the end, the arrangement of runners with respect to the inlet is also important factor and the plenum faulty design is the reason for the variation in outlet velocity, due to restriction in the passage of flow leading low velocity and high-pressure loss at outlet-1 in initial model. Absence of unwanted projections and depth cuts in plenum chamber of intake manifold geometry shows satisfactory results. An increase of 16% in flow velocity at outlet-1, and approximately 5% to 7% in other outlets improved by modified design in which nearly equal velocity is achieved at all runner outlets with respect to actual intake manifold system. Raj Kumar et al. [18] the main objective was to study the ability of an air enhancement device for improving the performance of diesel engine. Four different types of vanes profiles of convergent nozzle were placed in four-stroke single-cylinder diesel engine at downstream of the manifold. Analysis was carried out at different loads in normal conditions and also by using air swirl devices. The emission analysis for both actual and modified design is carried out and it was derived that it decreases for all types of vane nozzle as compared to actual one. At 9 Kgs load, the 3 vanes nozzle emits 61.9% less hydrocarbon than normal engine, also at peak load there is decrease in NO_x and smoke density by 17.18% and 12.58%, respectively. At 9 kg, all vanes nozzle gives slight increase in brake power, i.e., 4.19% for 3 vanes nozzle and also specific fuel consumption is decreased by 9.09% with respect to normal condition. At peak load, highest brake thermal efficiency is achieved for 5 vanes nozzle, i.e., 8.85% increase as compared to normal engine. Shrirao and sambhe [19] in their paper aimed to design different types of inlet manifolds for direct injection (DI) single-cylinder diesel engine in order to create the turbulence by swirl. The swirl is one the important factors which leads to optimized combustion and also higher efficiency. At low speed, low mechanical losses and good combustion result in higher combustion efficiency. This study aims at the effect of swirl generation in intake system on engine performance and exhaust emissions. The enhanced turbulence is achieved in the intake manifold with different types of internal threads of constant pitch that are acme, buttress, and knuckle threads. It is inferred that the inlet manifold with buttress threads gives better performance and yields fewer exhaust emissions compared to inlet manifolds with acme and knuckle threads. This is because inlet manifold with buttress threads achieved a higher swirl coefficient and swirl ratio compared with inlet manifolds having acme and knuckle threads. Among all configurations of inlet manifold, buttress internal thread yields best air–fuel mixing due to enhancement in turbulence. At 2.5 KW load for engine, a reduction of 11.62% was noted in BSFC. The turbulence in inlet manifold configuration with buttress internal threads was enhanced and hence resulting this one best air–fuel mixing process among all the configurations of inlet manifolds. As exhaust gas temperature is concerned for different configuration of intake manifold, it is higher than normal engine. In terms of emission, reduction of 3.6%, 26.66%, and 12.32% of the reduction in NOx, CO, and HC was at 2.5 KW load. Gupta and Mishra [20] designed manifold for one of the truck engines, which has less plenum volume which is not suitable for air requirement of the engine, resulting reduction in the volumetric efficiency. Good airflow efficiency implies low restriction than effective utilization of the intake manifold runner area. A good guideline for judging an intake manifold is the amount of flow loss it provides from the cylinder head port flow. Generally, flow loss of 10% v occurs in good intake manifolds or less on the flow bench in the low-speed range; with this study, it was derived that the valve events play the most significant role in engine output. Also in the mid-range, the intake manifold tuning mechanisms are most effective in boosting volumetric efficiency. Optimum design was achieved by reducing the sharp corners in runner outlet area, also the vortex flow to be reduced by reducing dead plenum and collector volume which causes less uniform streamline resulting in an effect on increasing pressure drop.

Proposed Scope of Research

Still there is scope for applying new techniques to enhance the turbulence of intake air. One of them can be combination of airfoil and dimpled surface.

From above Fig. 1 (2 mm radius and $h/d = 0.5$ dimple) and Fig. 2 (NACA-0012 Airfoil).

This arrangement can be used for the intake manifold of diesel engine as shown below (Figs. 3 and 4).

Fig. 1 Dimple with h/d=0.5

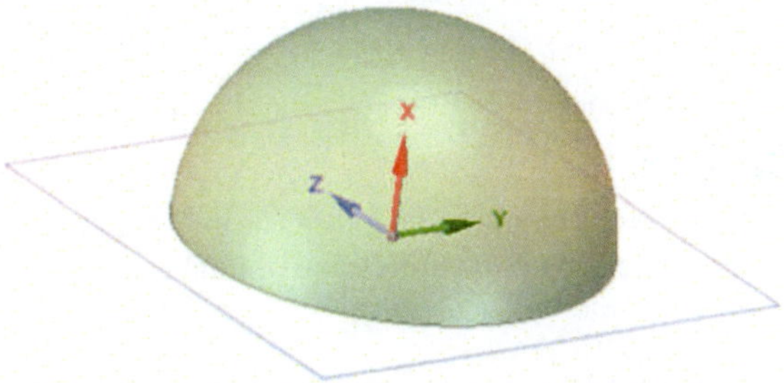

Fig. 2 3D Airfoil NACA0012

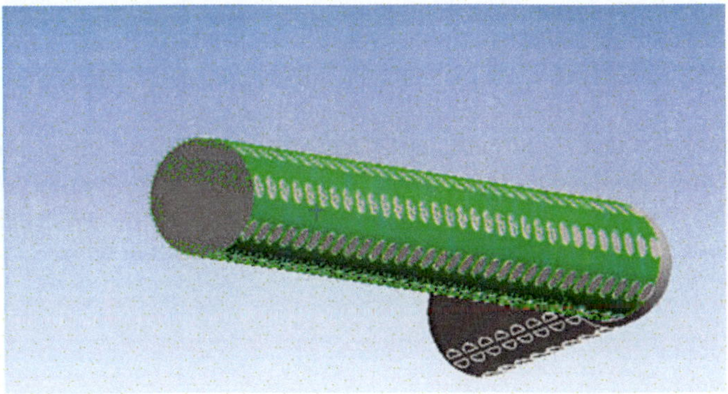

Fig. 3 (Dimpled intake manifold)

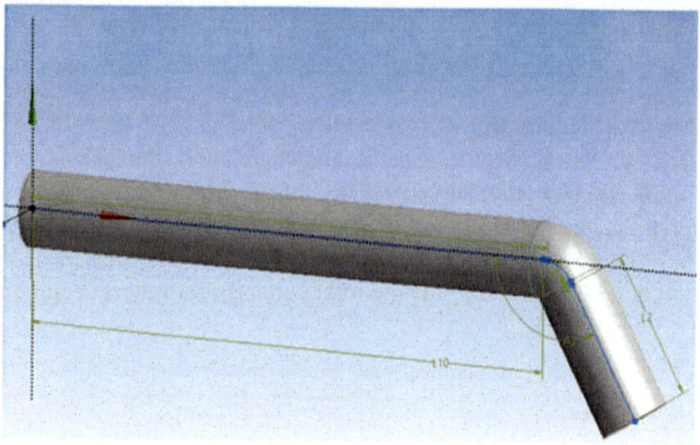

Fig. 4 (Intake manifold)

In above arrangement, airfoil can be incorporated in inlet of the dimpled manifold, which will enhance the turbulence of intake air.

3 Conclusion

From the literature review, it can be concluded that methods used like orientation of valve, placing guide vane swirl and tumble device (GVSTD) to guide the flow of air entering which lead enhancement in turbulence inside the combustion chamber, i.e., piston, using helical groove of different pitches, optimizing plenum volume, different geometry of internal thread of same pitch, using three arrangements helical, spiral, and helical–spiral manifold. Optimizing design parameters like guide curve angle and guide curve radius, installing several curvilinear diversion blades near inlet valve, using different types of vanes profiles of convergent nozzle, optimization of valve diameter increases the mixing in engine specifically in diesel engine which provides scope to analyze various parameters like brake-specific fuel consumption, tumble ratio and swirl ratio, volumetric efficiency and turbulence kinetic energy, brake thermal efficiency and emissions. Due to increase in pollution, norms are becoming tighter, so there is always scope for finding new techniques by enhancing turbulence which results improvement in mixing of air and fuel in diesel engines like the application of airfoils and dimple surface in intake manifold which is yet to be explored.

References

1. Paul, P., Ganesan V.: Flow field development in a direct injection diesel engine with different manifolds. Int. J. Eng. Sci. Technol. **2**(1), 80–91 (2010)
2. Paul, B., Ganesan, V.: Flow field development in a direct injection diesel engine with different manifolds. Int. J. Eng. Sci. Technol. **2**(1), 80-91 (2014). https://doi.org/10.5923/j.ajfd.20140403.03
3. Saad, I., Bari, S.: CFD investigation of in-cylinder air flow to optimize number of guide vanes to improve ci engine performance using higher viscous fuel. Int. J. Automot. Mech. Eng. **8**, 1096–1107 (2013). ISSN: 2229–8649 (Print); ISSN: 2180–1606 (Online). ©Universiti Malaysia Pahang
4. Mandloi, P., Gunjan, V.: Design optimization of an in-cylinder engine intake port. In: Nafems World Congress 2009 June 16th–19th, 2009 CRETE, GREECE
5. Martins, J., Senhorinha, T., Stijn, C.: Design of an inlet track of a small I. C. engine for swirl enhancement. In: Proceedings of COBEM 2009 20th International Congress of Mechanical Engineering Copyright © 2009 by ABCM November 15–20, 2009, Gramado, RS, Brazil
6. Jemni, M.A., Kantchev, G., Abid, M.S.: Influence of intake manifold design on in-cylinder flow and engine performances in a bus diesel engine converted to LPG gas fuelled, using CFD analyses and experimental investigations. Energy, **36**(5), 2701–2715 (2011)
7. Xu, J. (2017). Flow analysis of engine intake manifold based on computational fluid dynamics. In: Journal of Physics: Conference Series vol. 916, No. 1, p. 012043 (2017). https://doi.org/10.1088/1742-6596/916/1/012043
8. Sabale, S.K., Sanap, S.B.: Design and analysis of intake port of diesel engine for target value of swirl. Am. J. Mech. Eng. **1**(5), 138–142 (2013). https://doi.org/10.12691/ajme-1-5-6
9. Giovanni, V., Romania, L., Ferrarib, L., Ferrara, G.: Development of an engine variable geometry intake system for a formula SAE application. In: Science Direct 69th Conference of the Italian Thermal Machines Engineering Association, ATI2014
10. Shenghua, L., Ou, C., Won, H.: Development of New Swirl System and Its Effect on DI Diesel Engine Economy, SAE Technical Paper 1999-01-2889, https://doi.org/10.4271/1999-01-2889
11. Trigui, N., Griaznov, V., Affes, H., Smith, D.: CFD based shape optimization of IC engine oil & gas science and technology. Rev IFP **54**(2), 297–307 (1999)
12. Loong, Y.K., Salim, M.S.: Experimentation and Simulation on the Design of Intake Manifold Port on Engine Performance. EUREKA (2013)
13. Hushim, M.F., Ahmad, J.A., Mohd, A.R., Akmal, N.M. Azwan, S., Julio, C.M.C.: Air flow behaviour on different intake manifold angles for small 4-stroke Pfi retrofit kit system. ARPN J. Eng. Appl. Sci. **11**(12) (2016). ISSN 1819–6608
14. Martinas, G., Cupsa, O.S., Stan, L.C., Arsenie, A.: Cold flow simulation of an internal combustion engine with vertical valves using layering approach. IOP Conf. Ser. Mater. Sci. Eng. **95**, 012043 (2015). https://doi.org/10.1088/1757-899X/95/1/012043
15. Anilkumar, D.B., Elia, A.K.: Computational analysis of intake manifold design of a four cylinder diesel engine. (IJCESR) **5**(4) (2018). ISSN (Print): 2393–8374, (Online): 2394–0697
16. Mohiuddin, A.K.M.: Investigation of the Swirl effect on engine using designed Swirl adapter IIUM engineering. J. Special Iss. Mech. Eng (2011)
17. Singla, S., Sharma, S., Gangacharyulu, D.: Study of Design Improvement of Intake Manifold of Internal Combustion Engine, vol. 3 (2015). Special Issue, ISSN 2349–4476 234
18. Kumar, A.R., Raju, G.J., Reddy, K.H.: Performance evaluation of a diesel engine the presence of a convergent nozzle W Internal Blades in the air intake manifold. Int. J. Inn. Res. Sci. Eng. Technol. **4**(7) (2015)
19. Pankaj, N.S., Rajeshkumar U.S.: Effect of swirl induction by internally threaded inlet manifolds on exhaust emissions of single cylinder (DI) diesel engine. Int. J. Sci. Res. (IJSR) **3**(7) (2014). ISSN (Online): 2319–7064 www.ijsr.net
20. Gupta, A.K., Mishra, A.: Design and development of inlet manifold for six cylinder engine for truck application. Paripex Indian J. Res. **3**(7), (2014). ISSN 2250–1991

21. Prasad, S.L.V., Pandurangadu V.: Reduction of emissions by intensifying air swirl in a single cylinder di diesel engine with modified inlet manifold. Int. J. Appl. Eng. Tech. Int. J. **1**(1), 18–23. ISSN: 2277–212X (Online). Available at https://www.cibtech.org/jet.htm. Prasad and Pandurangadu
22. Ramasamy, D., Zamri, M., Mahendran, S., Vijayan, S.: Design optimization of air intake system (AIS) of 1.6L engine by adding guide vane. In: Proceedings of the International MultiConference of Engineers and computer scientists 2010 vol. II, IMECS 2010 March 17–19, 2010 Hong Kong
23. Sulaiman, S.A., Murad, S.H.M., Ibrahim, I., Abdul Karim, Z.A.: Study of flow in air-intake system for a single-cylinder go-kart engine. Int. J. Automot. Mech. Eng. (IJAME) **1**, 91–104 (2010). ISSN: 2229–8649 (Print); ISSN: 2180–1606 (Online). ©Universiti Malaysia Pahang https://doi.org/10.15282/ijame.1.2010.8.0008
24. Saad, I., Bari, S.: Optimizations of vane height of guide vane swirl and tumble device to improve in cylinder airflow characteristics of a diesel engine running with vegetable oil. Adv. Automobile Eng. **3**(106). https://doi.org/10.4172/2167-7670.1000106
25. John, A.T., Alfred, O.E.: Application of computational fluid dynamics code on flow process through inlet port of internal combustion engines. J. Multidis. Eng. Sci. Stud. (JMESS) **2**(2) (2016). ISSN: 2912–1309
26. Zhou, W., Yu, R., Hu, H.: An experimental investigation on the characteristics of turbulent boundary layer flows over a dimpled surface. J. Fluids Eng. **138**, 02104-1 (2016). Copyright VC 2016 by ASME
27. Qu, H., Shen, Z., Xie, Y.: Numerical Investigation of Flow and Heat Transfer in a Dimpled Channel among Transitional Reynolds Numbers. Hindawi Publishing Corporation Mathematical Problems in Engineering. Article ID 989237, 10 pages https://doi.org/10.1155/2013/989237.

Waste to Energy (WTE) by Incineration: Current and Future Practices in India

Prateek Verma, Oshi Jain, and Anurag Gupta

1 Introduction

Municipal solid waste comprises of all the private and business wastes; however, it does exclude any of the dangerous waste created in ventures with the prohibition of biotherapeutic waste that had been pretreated. MSW additionally incorporates waste created in family units, hotels, and restaurants. It likewise incorporates garbage from the civil works, sanitation buildup, and degradable waste (counting biomass). The non-recyclable piece of MSW can either be separated or be incorporated as fuel. Practically, all MSW scums post-material retrieval and reutilizing are treated as sustainable [1]. The energy created from the waste in WTE procedure can be recouped through direct ignition, (e.g., incineration, pyrolysis, and gasification) or ignition of inflammable fills as methane, hydrogen, or other artificial fuels (e.g., burning, pyrolysis and anaerobic absorption, mechanical, organic treatment and waste-determined fuel). The two principle WTE techniques effectively being used all through the world are incineration and gasification. According to the 2012 information of Central Pollution Control Board (CPCB), municipal experts have just set up 279 compost plants. There are 72 biomethanation plants, 29 refuse determined energizes (RDF) plants, and 8 WTE plants, out of these, few apparently do not work [2]. It is assessed that around 68% of the MSW produced all through India is gathered for processing and only 28% is treated by local municipal departments. But small amount, i.e., 19% of the total waste created is converted to energy appropriately. The majority of the unprocessed waste is disbursed in landfills or dump destinations. If country is able to manage the total amount (about 62 million tons per annum) of MSW produced, it can effectively altogether lessen the changes of maladies while upgrading standard of life. Up till 2013, out of the 29 Indian states/UTs, just 22 states/UTs have set up appropriate waste handling and transfer amenitics, while the rest have not even

P. Verma (✉) · O. Jain · A. Gupta
Department of Mechanical Engineering, KIET Group of Institutions, Ghaziabad, India
e-mail: Prateek.shanu007@gmail.com

M. Muzammil et al. (eds.), *Recent Advances in Mechanical Engineering*, Lecture Notes in Mechanical Engineering, https://doi.org/10.1007/978-981-15-8704-7_46

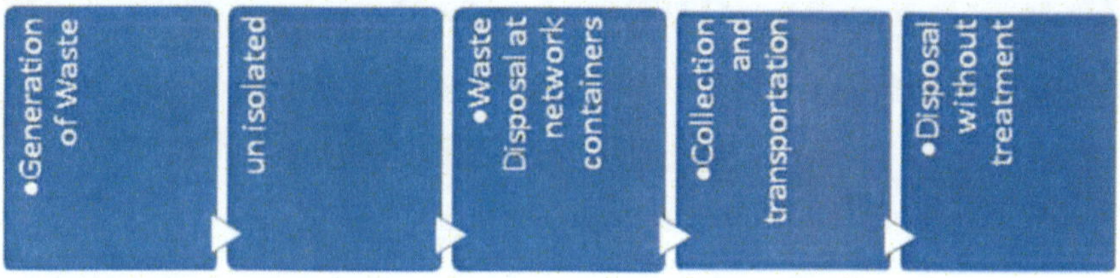

Fig. 1 Steps for untreated MSW disposal

attempted any effort in this regard [3]. The disbursal of MSW through WTE plants has an added advantage that this method diminishes the possibility of aqueous and air pollution, which are the drawbacks of landfill or composting methods. MSW is practically an energy wellspring of electricity generation in an inexorably carbon-compelled world. But unfortunately, in India, it has not picked up as much notoriety as it did in different nations like USA, China, Sweden, etc.

2 Urban Expansion in India and MSW Challenge

A number of jumbo-sized urban groups, with population more than 50 lakhs, have remained mostly unchanged over the last couple of years. But, a number of smaller urban areas group have increased very rapidly due to the changing pattern of industrialization [4]. As per the world urbanization report published in 2018, the urbanization in India increases more than 3% in the tenure of 2011–2017. This urbanization also causes an increasing industrial consumption and henceforth increase in MSW generation [1, 5]. Several states of India are now aiming at a new generation idea, such as IT/ITES, Biotech, SEZs, and Financial Hubs, but unfortunately, so far a very few are thinking efficiently about MSW disposal. So far not a single efficiently designed and competently operated MSW disposal facility is functioning in the country. The generalized steps of untreated MSW disposal in India are shown in Fig. 1.

Outmost favored three methods of MSW disbursal in Indian scenario, the landfill is not being used word wide as it may cause the soil and groundwater contamination and while composting need a large investment for segregation of MSW in dump yards due to diversified social and education level in India. A sample of quantity and quality of MSW is generated at different cities in India [6, 7] (Table 1).

3 Design Criteria of Incinerator for MSW to WTE

Increasing gathering and disbursal unit costs, aggravated by expanding refuse generation rates, has driven government to inspire industry, huge shopping malls, and hotels to introduce small incinerator unit for disbursal of their own waste and to generate

Table 1 A sample of quantity and quality of MSW generated at different cities in India [7]

S. No.	City	Population	Area (Km^2)	Waste quantity (TPD)	Waste generation rate (kg/c/day)	Compostable (%)	Recyclables (%)	C/N ratio	HCV[a] (Kcal/kg)
1	Kanpur (U.P)	2,551,337	267	1100	0.43	47.52	11.93	27.4	1571
2	Ahmedabad (Gujrat)	3,520,085	191	1302	0.37	40.81	11.65	29.6	1180
3	Hyderabad (Andhra Pradesh)	3,843,585	169	2187	0.57	54.20	21.60	25.9	1969
4	Bangalore (Karnataka)	4,301,326	226	1669	0.39	51.84	22.43	35.1	2386
5	Chennai (Tamil Nadu)	4,343,645	174	3036	0.62	41.34	16.34	29.2	2594
6	Kolkata (West Bengal)	4,572,876	187	2653	0.58	50.56	11.48	31.8	1201
7	Delhi (Union Territory)	10,306,452	1483	5922	0.57	54.42	15.52	34.8	1802
8	Mumbai (Maharastra)	1,437,354	286	574	0.40	52.44	22.33	21.5	1421

Note The data of the table have been taken from the Bhat et al. [7]

energy for their own. This thought much of the time may help, the investment funds from less reject stockpiling area and less incessant decline accumulations, together with full-filament of their own energy requirement and also the end of the unaesthetic and unhealthful collection of waste. These reasons are adequate to legitimize establishment and activity of a private transfer framework. Such an establishment must meet certain **presentation** prerequisites.

- A incineration system must most likely deal with a wide assortment of waste, including the capacity to devour reject of high dampness content, (e.g., trash or food wastage).
- While designing the incinerator unit for WTE, per day waste available for incineration and calorific value of waste must be well defined.
- The incinerator must almost certainly devour a wide assortment of waste materials including plastics, metal foils, wood, cardboard, glass, paper, etc.
- With the expanding stringency of air contamination codes in regard to both the particulate substance and foul aggravation qualities of the vent gas, new burning provisions must incorporate sufficient air contamination control frameworks.
- In request to legitimize an incinerator establishment versus the continuing with the scavenger for waste disbursal, the incineration system ought to have moderately minimal effort (capital and working expenses).
- High ignition effectiveness (great burnout of both flue gas and residue) must be acknowledged in an incinerator.
- The ash handling in the incinerator should be easy and must not pollute the nearby area.
- Based on earlier investigations by AGA [8] and others, the pilot burner structure for incinerator units must accommodate simple igniting system along with high level of pilot stability.
- Maintenance must be easy and inexpensive.

4 Waste-to-Energy Conversion

The basic process of waste-to-energy conversion by incinerator involves direct burning of waste in presence of excess oxygen at temperatures nearly 800 °C which produces heat energy, inert gases, and ash. 65–80% of heat content of combustion of the organic matter transfer to hot air, steam, and hot water which can further be used to run turbine, etc. Basic process of incineration is shown by following Fig. 2 [9].

5 India's First EURO Norm Compliant WTE Plant

India's first WTE plant was established in Ghazipur, Delhi. Delhi generates over 10,000 tons of municipal solid waste every day and it is often not dealt with properly, instead just pushed out of sight and eventually out of mind. The initiative was

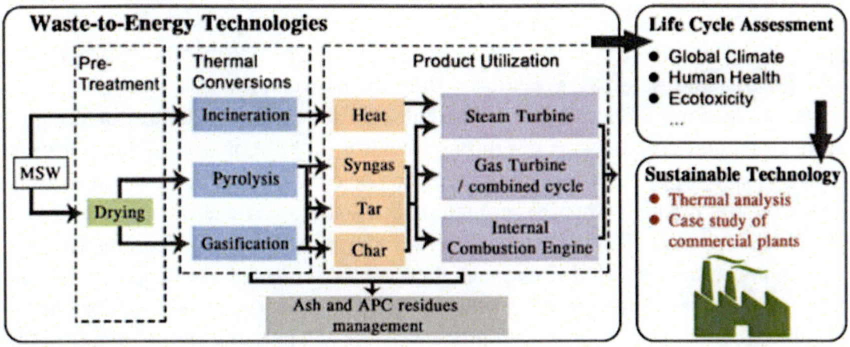

Fig. 2 Process of waste to energy [9]

taken by Delhi Municipal Corporation (DMC) with Private Sector Corporation—they set up a WTE plant at Ghazipur with the help of IL and FS environment. The plant can generate 12 MW of electric power at 11 kV by processing 2000 tons per day. The municipal solid waste is transported to the plant site by East Delhi Municipal Corporation. The plant has a detailed facility comprising of seven stages for preprocessing. This is done to prepare the waste to ensure high calorific value for the RDF. The RDF has a calorific value over 3000 kcal per kg, and such high value eliminates the need for any additional fuel. The state-of-the-art boiler is designed to burn around 550 tons per day of RDF at a temperature of 1100 °C. To eliminate carcinogenic dioxins, the flue gas produced from the combustion of RDF is passed through the real-time gas cleaning process comprising of semiwet reactor and bag filter and only after all this are the flue gases discharged in the atmosphere [10]. Some other incineration plants built (or in pipeline) are in Narela (Delhi), Jabalpur (Madhya Pradesh), Pallavapuram (Tamil Nadu), Surat (Gujarat), and Jawaharnagar (Hyderabad).

6 Air Pollution Control in WTE Incineration

Majority of the WTE plants those are currently working successfully in all over the word are incineration type which proved to be a simpler technique than others [11]. Prime benefits of WTE are conversion of heterogeneous waste into consistent residues such as flue gas, bottom ash, and fly ash, thereby reducing the bulk of waste up to 75% and volume up to 90%. The design chamber for incinerating the MSW is specially designed to sustain high temperatures of about 1100 °C. The air supply is kept uninterrupted to ensure turbulence which in turn guarantees complete combustion of the waste down to its most stable (or natural) molecular form. The solid residue can be as repurposed (after cleaning) construction material or simply be transported to landfills. A significant portion of the flue gases consists of pollutants (sulfur dioxide (SO_2), particulate matter (PM), heavy metals, hydrochloric acid (H_2SO_4), and dioxins). Previously, the dioxins were severe concern from the environmental point of view. Now, with significant advancements in design of incinerators and modified processes of emission control (in light of strict regulations put in place by developed nations), incineration can be operated with virtually no dioxin emissions. Apart from them, other contaminants (in air) can be efficiently controlled during combustion, and they can also be removed by the system for cleaning flue gas. The heat generated due to MSW incineration can be collected via steam generation which in turn is used for generating power. The sale of electricity or steam produced as the by-product of the combustion process partly balances the capital required for incineration [12].

7 Problems of WTE Incineration

- WTE needs high capital investments and functioning costs in comparison to other MSW management technologies and power generation by other renewable sources of energy. Incineration equipment is very expensive and has high operating cost and maintenance costs; it has been found that equipment are modeled on domestic technologies which are cheaper but largely restricted to relatively low capacities.
- Some air pollutants' emission.
- Management of fly ash produced during MSW incineration is a key factor which has to be taken care of [13].

8 Conclusion

Having gone over all the date, we can safely conclude that WTE efficiently deals with the problem of MSW disposal all the while recovering energy from the waste materials. Additionally, it lowers the level of the pollutants emitted. MSW is increasingly being accepted as one of the clean sources of energy, primarily due to the significant environmental quality benefits such as reduction of greenhouse gas (GHG) emissions. As the research and technology develop further, especially dealing with flue gas control, fly ash management, and corrosion phenomena, possible reuse of residues will accelerate the growth of the WTE industry even more. The task of MSW disposal and the call for alternative energy sources are common in several developing countries such as India. It is expected that the Indian experience with WTE will offer some helpful lessons to other developing nations as well.

References

1. Ghosh, R., Arun K.: Urban challenges in India and the mission for a sustainable habitat. Interdisciplinary **2**(2) (2014)
2. Central Pollution Control Board (CPCB): Management of Municipal Solid Waste. Ministry of Environment and Forests, New Delhi, India (2013)
3. A report by Abhishek Chandra (Municipal Commissioner Agartala Municipal Corporation) (https://iced.cag.gov.in/wp-content/uploads/C-05%20Shri%20Abhishek%20Chandra.pdf)
4. Asnani, P.U.: Solid Waste Management. India Infrastructure Report 570 (2006)
5. Shyamala, M., Satpal, S.: Sustainable Municipal Solid Waste Management in India: A Policy Agenda (2015)
6. Central Pollution Control Board (CPCB): Management of Municipal Solid Waste. Ministry of Environment and Forests, New Delhi, India (2004)
7. Bhat, R.A., Shabeer, A.D., Davood, A.D., Gowhar, H.D.: Municipal Solid Waste Generation and current Scenario of its Management in India (2018)
8. American Gas Association: USAS, Z21.6, Approval Requirements for Domestic Gas-Fired Incinerators, p. 16 (1966)
9. Dong, J., Tang, Y., Nzihou, A., Chi, Y., Elsa Weiss-Hortala, Ni, M.: Life cycle assessment of pyrolysis, gasification and incineration waste-to-energy technologies: theoretical analysis and case study of commercial plants
10. IL & FS waste to energy plant (https://www.ilfsindia.com/our-work/environment/waste-to-energy-plant-ghazipur/)
11. Gupta, N., Krishna, K.Y., Vinit, K.: A review on current status of municipal solid waste management in India. J. Environ. Sci. **37**, 206–217 (2015)
12. Gupta, S., Mohan, K., Prasad, R., Gupta, S., Kansal, A.: Solid waste management in India: options and opportunities. Resour. Conserv. Recycl. **24**(2), 137–154 (1998)
13. American Society of Mechanical Engineers: Waste-to-Energy: A Renewable Energy Source from Municipal Solid Waste. New York, NY, White Paper Submitted to Congress (2008)

Enablers for Lean Six Sigma and Agile Implementation: An Interpretive Structural Modeling Approach

Nidhi Mundra, Rajesh P. Mishra, and Abhishesh Mishra

1 Introduction

Nowadays, synergy of Lean and Six Sigma seems to be a renowned continuous improvement (CI) business strategy to reduce the cost, increase the productivity, enhance the product quality in the manufacturing sector.

Lean and Six Sigma are complementary approaches as Six Sigma will eliminate defects in processes, but it can not optimize process flow [1] and Lean cannot bring the process into statistically control [2–4]. To achieve contradictory goals, i.e., reducing the cost and enhancing the quality of products, LSS can be implemented; however, neither of LM nor Six Sigma addresses the total requirements demanded by the current market, which includes a simultaneous focus on efficiency and quality, as well as flexibility [5].

AM can cope up with such a situation [6]. In philosophies, AM is the capability to reconfigure and behavioral flexibility to operate in volatile environments [7].

Yet none of these approaches gives competitive advantages in today's market place when practicing individually. Although AM brings responsiveness and flexibility, it is far behind from LM in terms of efficiency. On the other hand, LM cannot handle the dynamic market demand due to lack of flexibility. While both production systems address basic improvement issues but they do not encompasses problem-solving approach, which is Six Sigma's main strength [5]. AM, when integrated with LSS methodology, can enable an organization to offer desired products and services quickly to its customers without compromising the quality at an optimum price. Several researchers have developed Lean, Six Sigma, LSS, AM, Lean agile enablers in their research individually but a comprehensive set of enablers which integrates LSS and AM is not found. There is a need to develop a framework which not only possesses combination of LSS and AM enablers but also gives step-by-step

N. Mundra · R. P. Mishra (✉) · A. Mishra
Department of Mechanical Engineering, BITS Pilani, Pilani Campus, Pilani, Rajasthan, India
e-mail: rpm@pilani.bits-pilani.ac.in

M. Muzammil et al. (eds.), *Recent Advances in Mechanical Engineering*, Lecture Notes in Mechanical Engineering, https://doi.org/10.1007/978-981-15-8704-7_47

Table 1 Literature review

S. No	Enablers	Authors
1	Knowledge management/intellectual capital management	[12, 14, 15, 21, 23, 29]
2	Virtual enterprise	[8, 9, 11, 12, 21]
3	Customer relationship management and market information	[11, 14–18, 21–25, 27]
4	Existing technology advancement and augment of new technology	[8–13, 20, 25]
5	Systematic learning and development of employees	[15–19, 24, 25, 27, 29, 31]
6	Employee empowerment	[9, 12, 15–17, 24, 25, 27]
7	Use of information technology for better communication	[12–19, 21, 22, 27, 28];
8	Top management commitment toward LSS & AM	[8, 10, 14–20, 23–29, 31]
9	Organization culture	[14, 15, 18, 19, 25, 26]
10	Supplier relationship	[8, 10–12, 21, 22, 25, 29]
11	Multi-skilled workforce	[9, 11, 20, 25, 29, 30]
12	Alignment of strategies	[15–17, 23, 24, 26, 33]
13	Project selection and prioritization	[11, 15–20, 24, 25, 27]
14	Financial Capabilities	[28, 30, 31]
15	Link to human resources-based actions (promotions, bonuses, etc.)	[16, 29–31]
16	Change management	[17, 25, 28–31];
17	Organization infrastructure	[16, 17, 25, 28–31]
18	Communication	[16, 17, 25]

guidelines to implement LSS and AM to practitioners. This paper is organized into four sections; Section 2 describes the ISM methodology while Section 3 discusses features of the proposed ISM frameworks. Finally, Section 4 presents the conclusion. All the enablers for LSS and AM implementation are identified through vast literature review and opinion of industry and academia experts are shown in Table 1.

2 Interpretive Structural Modeling

ISM methodology helps in identifying the mutual relationship among enablers and their driving power and inter-dependencies. Jharkharia and Shankar [32] have given the following steps for ISM technique (see Figs. 1 and 2) (Tables 2, 3 and 4).

2.1 Identify the Variables and Development of SSIM Matrix:

Enablers are identified from the collected data; the group of industries experts and researchers identified the kind of inter-relationships among enablers. Four symbols V; A; X and O represent the direction of the relationship between two enablers i and j in SSIM (shown in Table 2)

↓

2.2 Reachability Matrix

Now to get the initial reachability matrix, SSIM has been changed into a binary matrix. The instructions for V, A, X and O substitution by 1 and 0 are adopted from Ravi & Shankar [32]. In the final reachability matrix transitivity condition is considered. (See Table 3)

↓

2.3 Level Partitions:

The enabler having identical reachability and intersection set is placed at the topmost level in the echelon of ISM . Now the identified top-level enabler is segregated from the remaining enablers. Then, the reiteration of the same process is carried out to identify the level of each enabler (See Table 4).

↓

2.4 Formation of the ISM model:

A directed graph (digraph) is drawn from the final reachability matrix. It is transformed into the ISM-based model (Figure 2) by removing the transitivity's

↓

2.5 MIC-MAC Analysis:

In MIC-MAC analysis enablers are classified into four clusters: Autonomous enablers (both weak driving power & dependence); Dependent enabler (weak driving power & strong dependence); Linkage enablers(both strong driving power & dependence) and Independent enablers (strong driving power & weak dependence)

Fig. 1 Steps of ISM

Table 2 Structured self-interaction matrix (SSIM)

Enablers	18	17	16	15	14	13	12	11	10	9	8	7	6	5	4	3	2	1
1	A	A	A	X	A	X	A	V	V	A	A	X	X	X	V	V	V	
2	A	A	A	A	A	A	A	O	A	A	A	A	A	A	X	A		
3	A	A	A	A	A	A	A	X	X	A	A	A	A	A	V			
4	A	A	A	A	A	A	A	A	A	A	A	A	A	A				
5	A	A	A	X	A	X	A	X	V	A	A	X	X					
6	A	A	A	X	A	X	X	V	V	A	A	X						
7	A	A	A	X	A	X	A	V	V	A	A							
8	V	V	V	V	V	V	V	V	V	X								
9	V	V	V	V	O	V	V	V	V									
10	A	A	A	A	A	A	A	X										
11	X	X	X	X	V	X	V											
12	X	X	X	X	X	V												
13	A	A	A	X	A													
14	X	X	X	V														
15	O	A	A															
16	X	X																
17	X																	
18																		

Table 3 Final reachability matrix

	1	2	3	4	5	6	7	8	9	10	11	12	13	14	15	16	17	18	DRIVER
1	1	1	1	1	1	1	1	0	0	1	1	0	1	0	1	0	0	0	11
2	0	1	0	1	0	0	0	0	0	0	0	0	0	0	0	0	0	0	2
3	0	1	1	1	0	0	0	0	0	1	1	0	0	0	0	0	0	0	5
4	0	1	0	1	0	0	0	0	0	0	0	0	0	0	0	0	0	0	2
5	1	1	1	1	1	1	1	0	0	1	1	0	1	0	1	0	0	0	11
6	1	1	1	1	1	1	1	0	0	1	1	0	1	0	1	0	0	0	11
7	1	1	1	1	1	1	1	0	0	1	1	0	1	0	1	0	0	0	11
8	1	1	1	1	1	1	1	1	1	1	1	1	1	1	1	1	1	1	18
9	1	1	1	1	1	1	1	1	1	1	1	1	1	1*	1	1	1	1	18
10	0	1	1	1	0	0	0	0	0	1	1	0	0	0	0	0	0	0	5
11	0	1*	1	1	0	0	0	0	0	1	1	0	0	0	0	0	0	0	5
12	1	1	1	1	1	1	1	0	0	1	1	1	1	1	1	1	1	1	16
13	1	1	1	1	1	1	1	0	0	1	1	0	1	0	1	0	0	0	11
14	1	1	1	1	1	1	1	0	0	1	1	1	1	1	1	1	1	1	16
15	1	1	1	1	1	1	1	0	0	1	1	0	1	0	1	0	0	0	11
16	1	1	1	1	1	1	1	0	0	1	1	1	1	1	1	1	1	1	16
17	1	1	1	1	1	1	1	0	0	1	1*	1	1	1	1	1	1	1	16
18	1	1	1	1	1	1	1	0	0	1	1	1	1	1	1*	1	1	1	16
	13	18	16	18	13	13	13	2	2	16	16	7	13	7	13	7	7	7	201/201

*denotes the transitivity condition between i an j enablers

Table 4 Level partition of each enablers

S. No.	Enablers	Level No.
1	2, 4	First
2	3, 10, 11	Second
3	1, 5, 6, 7, 13, 15	Third
4	12, 14, 16, 17, 18	Fourth
5	8, 9	Fifth

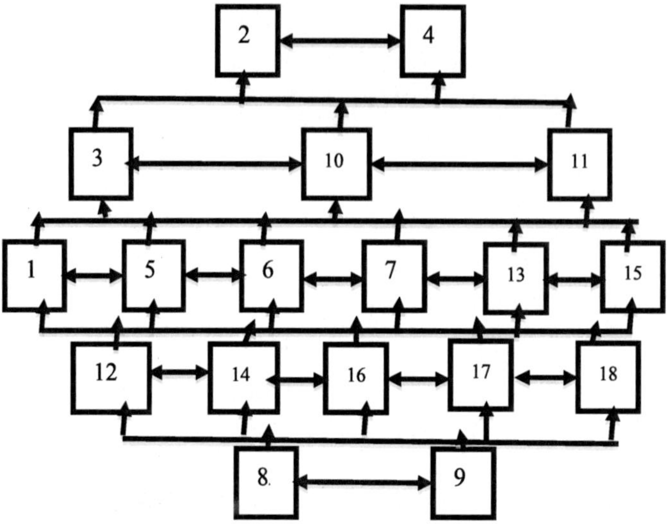

Fig. 2 Integrated framework for LSS and AM implementation

3 Result and Discussion

In ISM, the MIC-MAC analysis gives a classification of enablers. It has been obtained that no enablers present in the autonomous cluster, so no enabler is disconnected from the system, that means all the enablers obtained from the extensive literature survey are essential and organizations must pay attention to all of them. Enablers' top management commitment toward LSS & AM and organization culture has the highest driving power and the least dependence formed the base of the model. So the integration of these enablers is the key to the efficacious enactment of LSS and AM in several organizations.

4 Conclusion

In recent years, the manufacturing segment of the industry has acquired itself with modern and advanced technologies; however, in the very volatile market, customers are more demanding than ever. The fusion of LSS and Agile can give a competitive edge to the manufacturing industry in the upcoming era. This is one of the prior efforts made to present a comprehensive analysis of LSS and AM framework which also give step-by-step guidelines to implement LSS and AM to practitioners. Although this framework is not validated statistically. For statistically validation, structural equation modeling can be seen as groundwork for the future.

References

1. Antony J.: Six Sigma vs Lean: some perspectives from leading academics and practitioners. Int. J. Product. Perform. Manage. **2**, **60**(2), 185–90 (2011)
2. George, M., Rowlands, D., Kastle, B.: What is Lean Six Sigma?. McGraw-Hill Companies, New York, NY (2003)
3. Bhuiyan, N., Baghel, A.: An overview of continuous improvement: from the past to the present. Manag. Decis. **43**(5), 771–781 (2005)
4. Corbett, L.M.: Lean Six Sigma: the contribution to business excellence. Int. J. Lean Six Sigma **2**(2), 118–131 (2011)
5. Yaghoubi, N.M., Banihashemi, S.A.: Conceptual model of achieving competitive advantages by combination of production and quality systems. Eur. J. Sci. Res. **46**(4), 523–530 (2010)
6. Elmoselhy, S.A.: Hybrid Lean–agile manufacturing system technical facet, in automotive sector. J. Manuf. Syst. **32**(4), 598–619 (2013)
7. Leite, M., Braz, V.: Agile manufacturing practices for new product development: industrial case studies. J. Manuf. Technol. Manage. **27**(4), 560–576 (2016)
8. Gunasekaran, A.: Agile manufacturing: a framework for research and development. Int. J. Prod. Econ. **62**(1), 87–105 (1999)
9. Sharp, J.M., Irani, Z., Desai, S.: Working towards agile manufacturing in the UK industry. Int. J. Prod. Econ. **62**(1–2), 155–169 (1999)
10. Power, D.J., Sohal, A.S., Rahman, S.U.: Critical success factors in agile supply chain management-An empirical study. Int. J. Phys. Distrib. Logist. Manage. **31**(4), 247–265 (2001)
11. Kumar, P., Shankar, R., Yadav, S.S.: Modeling the enablers of agile manufacturing system. Proc. GLOGIFT **8**, 1–8 (2008)
12. Hasan, M.A., Shankar, R., Sarkis, J., Suhail, A., Asif, S.: A study of enablers of agile manufacturing. Int. J. Ind. Syst. Eng. **4**(4), 407–430 (2009)
13. Bottani, E.: Profile and enablers of agile companies: An empirical investigation. Int. J. Prod. Econ. **125**(2), 251–261 (2010)
14. Lean Six Sigma for higher education institutions (HEIs) Challenges, barriers, success factors, tools/techniques. Int. J. Product. Perform. Manag. **61**(8), 940–948 (2012)
15. Laureani, Antony.: Critical success factors for the effective implementation of Lean Sigma. Int. J. Lean Six Sigma **3**(4), 274–283 (2012)
16. Manville, G., Greatbanks, R., Krishnasamy, R., Parker, D: Critical success factors for Lean Six Sigma programmes: a view from middle management. Int. J. Q. Reliabi. Manage. **29**(1), 7–20 (2012)
17. Timans, W., Antony, J., Ahaus, K., van Solingen, R.: Implementation of Lean Six Sigma in small-and medium-sized manufacturing enterprises in the Netherlands. J. Oper. Res. Soc. **63**(3), 339–353 (2012)
18. Psychogios, A.G., Atanasovski, J., Tsironis, L.K.: Lean Six Sigma in a service context: a multi-factor application approach in the telecommunications industry. Int. J. Q. Reliabil. Manage. **29**(1), 122–139 (2012)
19. Jeyaraman, K., Teo, L.K.: A conceptual framework for critical success factors of lean Six Sigma. Int. J. Lean Six Sigma **1**(3), 191–215 (2010)
20. Aravind Raj, S., Sudheer, A., Vinodh, S., Anand, G.: A mathematical model to evaluate the role of agility enablers and criteria in a manufacturing environment. Int. J. Product. Res. **51**(19), 5971–5984 (2013)
21. Matawale, C.R., Datta. S., Mahapatra. S.S.: Interrelationship of capabilities/enablers for lean, agile and leagile manufacturing: an ISM approach. Int. J. Process Manage. Benchmark. **3**(3) (2013)
22. Habidin, Yusof.: Critical success factors of Lean Six Sigma for the Malaysian automotive industry. Int. J. Lean Six Sigma **4**(1), 60–82 (2013)
23. Antony, J.: Readiness factors for the Lean Six Sigma journey in the higher education sector. Int. J. Product. Perform. Manage. **63**(2), 257–264 (2014)

24. Abu Bakar, F.A., Subari, K., Mohd Daril, M.A.:. Critical success factors of Lean Six Sigma deployment: a current review. Int. J. Lean Six Sigma **6**(4), 339–348 (2015)
25. Dubey, R., Gunasekaran, A.: Agile manufacturing: framework and its empirical validation. Int. J. Adv. Manuf. Technol. **76**(9–12), 2147–2157 (2015)
26. Dev, C.A., Kumar V.S.S.: Analysis on critical success factors for agile manufacturing evaluation in original equipment manufacturing industry-an AHP approach (2016) https://doi.org/10.3901/cjme.2016.0608.071, available online at www.springerlink.com; www.cjmenet.com
27. Lande, M., Shrivastava, R.L., Seth, D.: Critical success factors for Lean Six Sigma in SMEs (small and medium enterprises). TQM J. **28**(4), 613–635 (2016)
28. Henderson, K., Evans, J.: Successful implementation of Six Sigma: benchmarking General Electric Company. Benchmark. Int. J. **7**(4), 260–81 (2000)
29. Coronado, R.B., Antony, J.: Critical success factors for the successful implementation of six Sigma projects in organisations. TQM Mag. **14**(2), 92–99 (2002)
30. Yi-Zhong, M., Gang, Y., Li-Lin, W., Ree, S.: The critical success factors of Six Sigma in China manufacturing industry. Asian J. Q. **9**(2), 39–56 (2008)
31. Vasileios, I., Odysseas, M.: Six Sigma's critical success factors and toolbox. Int. J. Lean Six Sigma **4**(2), 108–117 (2013). https://doi.org/10.1108/20401461311319310
32. Jharkharia, S., Shankar, R.: IT enablement of supply chains: modeling the enablers. Int. J. Product. Perform. Manage. **53**(8), 700–712 (2004)

Experimental Investigation and Analysis of Stainless Steel 316L by Salt Spray Test Method for Corrosion Behavior

Mohd Dawood and Prabhat Kumar Sinha

1 Introduction

In present scenario, the usage of the stainless steel is increasing in different fields such as chemical and allied industries, household utensils and marine structures. Stainless steel is also known as corrosion-resistant steel, because it is an iron-based steel alloy, which contains minimum 11% chromium. Chromium present in it prevents it from getting corroded. When ordinary carbon steel is exposed to rainwater, it corrodes easily due to formation of a brown iron oxide on the surface, which is commonly called as rust. But when more than about 10% chromium is added to ordinary steel, the oxide on the surface is transformed. Stainless steel generally has high ductility, weldability and cryogenic toughness properties.

Grade 316 is the standard molybdenum-bearing grade, second in importance to 304 amongst the austenitic stainless steels. Stainless steel differs from carbon steel by the amount of chromium present. When exposed to air and moisture, unprotected carbon steel rusts easily. This iron oxide film(the rust) is active and accelerates corrosion by forming more iron oxide [1]. The numerous and fruitful uses of stainless steel 316L make it a material to be researched upon. In this view, it was decided to take a study upon the following objectives:

1. To conduct an experiment of stainless steel SS316L by salt spray test method for corrosive atmospheric contamination.
2. To measure the corrosion of the SS 316L numerically by corrosion analysis.

M. Dawood · P. K. Sinha (✉)
Department of Mechanical Engineering, SHUATS, Prayagraj, UP, India
e-mail: prabhat.sinha@shiats.edu.in

M. Muzammil et al. (eds.), *Recent Advances in Mechanical Engineering*, Lecture Notes in Mechanical Engineering, https://doi.org/10.1007/978-981-15-8704-7_48

2 Review of Literature

Kumar et al. (2014) studied the austenitic stainless grade 316 corrosion resistances in one mole hydrochloric acid solution. The electrochemical potentiostatic polarization method was used to measure the corrosion behavior of 316 SS samples in corrosion environment.

Ribeiro et al. (2015) studied the electrochemical technique EIS used for the evaluation and investigate the corrosion behavior of metal in reinforced concrete.

Alar et al. (2016) investigated the corrosion behavior of SS 304, SS 316 and SS 316 Ti in red wine, white wine and beer. The cyclic potentiodynamic polarization method and gravometric method was used to determine the corrosion rate and its behaviors of the sample. [2]

Suresh(2017) performed the corrosion and wear characteristics of the propeller shaft material as 316LSS with increasing the hardness properties. Author found that increase in hardness and corrosion resistance can be achieved by cryogenic treatment.

3 Materials and Methods

The composition of the SS316 is given in the following Table 1.

Stainless steel 316L is taken in the form of the sheet of size 320*160*3 mm. In the sample of stainless steel 316L, salt spray test was performed. In salt spray test chamber, a standard solution of 5% NaCl (sodium chloride) is used to create a highly corrosive atmosphere. The spraying of salt is produced continuously over a surface of the stainless steel 316L sample, and corrosion over the surface of the sample depends on the resistance of the surface coating. Details of the sample taken are given in Table 2.

Table 1 Chemical composition of AISI 316L stainless steel

Element	% of elements
Chromium (Cr)	17.00–18.00
Sulfur (S)	0.0–0.04
Molybdenum (Mo)	2.00–2.50
Manganese (Mn)	0.00–2.00
Silicon (Si)	0.0–1.00
Iron (Fe)	Balance
Phosphorous(P)	0.0–0.05
Carbon (C)	0.0–0.04
Nickel	9.00–14.00
Nitrogen	0.0–0.10

Table 2 Detail are as follows of the Testing method

Reference to protocol	ASTM B117
Description	Stainless steel 316L (3 mm)
Test required	Salt spray test
Test Method/specification	ASTM B117
Salt solution	5% solution of NaCl
pH of solution	6.8–7.2
Temperature	35 ± 1 °C
Quantity of fog	1.0 ml/80cm^2/H
Exposing duration	120 years

3.1 *Corrosion Analysis Model*

The main purpose of this current research is the analysis of the corrosion over the stainless steel sample that taken the variation in the concentration of the stainless steel 316L. Chloride ions along with the oxygen ion called the pure diffusion. In this method, both the ions diffusion coefficient is taken as input value for the operation [4].

Corrosion of SS 316 L consists of two electrodes reactions that taken place (Fig. 1).

(1) Reaction at anode—conversion of iron to Fe2 + ions;
(2) Reaction at cathode—conversion of oxygen to *O* ions.

There are three causes of polarization, they are as follows:

Fig. 1 Description of the localized corrosion

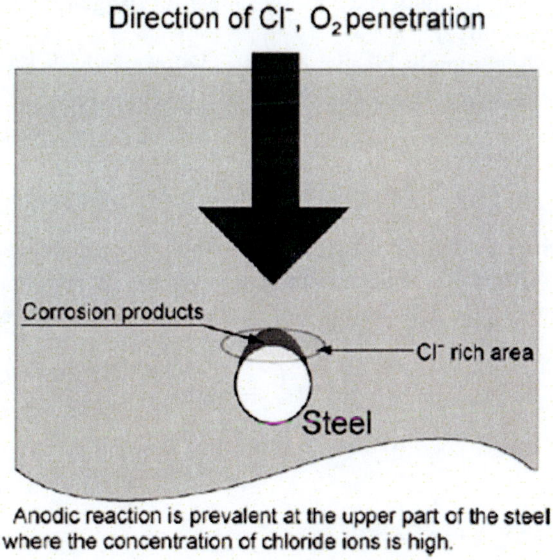

1. Concentration,
2. Ohm potential drop and
3. Activation.

3.1.1 Polarization at Cathodes

Concentration polarization for cathodes, gc, c (V), could be computed by the given below formula [5].

$$\eta_{c,c} = -\frac{RT}{zF}\ \text{In}\ \frac{i_L}{i_L - i_c} \tag{1}$$

Here,

R is universal gas constant (= 8.314 J/K mol),
F give Faraday's constant (= 9.65 × 104 C/mol),
T taken as absolute temperature (K)
Z defined the number of electrons exchanged in the reaction at cathode (= 4),
i_L taken the limiting density of current of the reaction at cathode (A/cm_2).
i_c defined the current density at cathode (A/cm^2)
The current density (limiting) can be obtained by the profile of oxygen on the steel surface.

$$\mathbf{i_L = zFJO_2,\quad max} \tag{2}$$

where JO_2, max taken as the supply for oxygen (maximum) for steel 316L sample ($mol/(cm^2\ s)$).

Accordingly, kinetics of Butler Volmer, cathode activation polarization$\eta_{c,a}$ could be found out by the use of the Tafel equation [6].

$$\eta_{c.a.} = \beta_c \log \frac{i_c}{i_\infty} \tag{3}$$

where β_c defined the slope of the Tafel for cathode reaction (V/decade) and

i_c 0 is exchange current density for the cathode reaction (A/cm^2). From Eq. (1) and (3), the polarization at the cathodes, η_c, could be expressed as Eq. (4).

$$\eta_{c.a.} = \beta_c \log \frac{i_c}{i_\infty} \frac{RT}{zF}\ 1n\ \frac{i_L}{i_L - i_c} \tag{4}$$

3.1.2 Polarization at the Anodes

polarization at anodes, η_a, could be determined by the following equation.

$$\eta_a = \beta_a \log \frac{i_a}{i_{a0}} \tag{5}$$

Where β_a is the Tafel slope of reaction at anode(V/decade),
i_a is the current density (A/cm^2) at anode and
i_a 0 taken as the exchange current density at the anode (A/cm^2). In this study, β is taken as 91.7 mV/decade for de passive steel and infinity for passive steel, and taken as to be 176.8 mV/decade for all other states of steel [7].

3.1.3 Current Density and Electrode Potential for Corrosive Sample

Using polarization, electrode potentials for the anodes and the cathodes reaction could be calculated by the following equations.

$$f\mathbf{a} = f\mathbf{a0} + \mathbf{ha} + \mathbf{ba}\,\mathbf{log}\frac{i_a}{i_{a0}}\,\textbf{(for the anodes)} \tag{6}$$

$$f\mathbf{c} = f\mathbf{c0} + \mathbf{hc}$$

$$= f\mathbf{c0} - \mathbf{bclog}\frac{i_c}{i_{c0}} - \frac{RT}{zF}\,\mathrm{In}\,\frac{i_L}{i_L - i_c}\,\textbf{(for the cathode)} \tag{7}$$

where $\phi a, \phi c$ taken as potentials (V) at the anode and cathode, whileϕa0, ϕc0 defined the potentials equilibrium at the anodes and cathodes (V) under an environment condition. For clarity taken for the analysis for corrosion, it has been taken in the research thatϕa0, ϕc0 remain constant throughout the analysis process. While in actual, these values depend on PH value for the amount of available oxygen [8].

Macro-cell current density at the point, I macro, could be calculated by the following

$$\mathbf{i_{macro}} = \mathbf{ia} - \mathbf{ic} \tag{8}$$

By Ohm's law, the normal gradient at the steel sample is proportional to the macro-cell current.

This means

$$\frac{1}{r}\frac{\partial\varphi}{\partial\varphi} = i_{\mathrm{macro}} \tag{9}$$

where r taken to be electrical resistance of concrete (Ωm);
n determine the direction normal for the steel surface.

3.2 Corrosion Analysis

Assuming electrical resistance for SS 316L isotropic, then electric potential could be found out by Laplace's equation.

$$\nabla 2\phi = 0 \tag{10}$$

where ϕ taken as electric potential (V) and ∇2 is the Laplace sign. For finding potential distribution for the sample, the calculation of Eq. (10) under boundary condition of Eq. (9) should be determined. As both sides of Eq. (9) are dependent on the distribution of potential, this problem could not be solved by any other ways. The solution to this problem is made in this study as given below [9]:

1. Obtain suitable point of pivot; also, give it an essential boundary condition with pivot value of potential (= ϕpivot).
2. Suppose the value of distribution of initial potential.
3. Find the macro-cell currents for the boundary by the use of Eqs. (6)-(8).
4. Find the natural boundary condition by the use of Eq. (9).
5. Find out potential distribution by using the above boundary conditions.
6. Do all the steps from (3) to (5) unless norm for convergence test. $\|\eta\|$, found to be lower than allowable limit value $\|\eta\|_{allow}$.

$$\|\eta\| = \sqrt{\frac{\sum_{i=0}^{nn}\left[\{\varphi i\}^{j} - \{\varphi i\}^{j-1}\right]^{2}}{nn}} \leq \|n\|_{allow} \tag{11}$$

Here, $\{\varphi i\}^{j}$ determine the potential value at the *j*th iteration, and *n* gives the number of nodes.

7. Change the value of the potential of pivot as well as do steps 1–6 unless the difference between the corrosion current given by the normal gradient of potential and the macro-cell. Current obtained from the potential at the point of pivot is lower than that of the allowable value.

α_{allow}.

$$\left|\frac{1}{r}\frac{\partial\varphi}{\partial n}\right| - i_{\text{macro}} \leq \alpha_{\text{allow}} \tag{12}$$

By current density at anode, depth of corrosion can be found out by the given below equation.

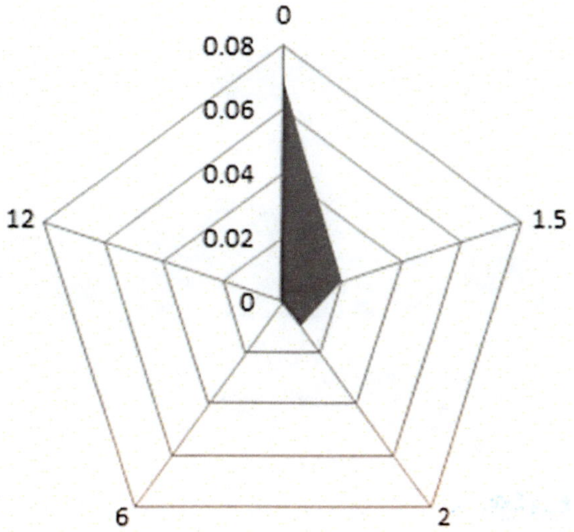

Fig. 2 Effect of immersion time on corrosion rate for SS 316L

$$d_{corr} = \int 1.15 \times 10^4 \, i_a \, dt \tag{13}$$

Where d_{corr} is the corrosion depth (mm), t is the time (year).

After the analysis, same boundary condition is applied to the analysis of oxygen to consider the consumption of oxygen during reaction at cathode (Fig. 2).

$$\frac{\partial Co_2}{\partial n} = \frac{i_c}{4FDo_2} \tag{14}$$

where CO_2 is the concentration for oxygen and Do_2 is taken as the coefficient of diffusion for oxygen.

Effect of immersion time on the corrosion rate of SS 316L subjected to 35 g/l NaCl solution is shown in the Fig. 3 [10].

1. 0–0.08: corrosion rate (in mmpy)
2. 0–12: immersion time (in hours)

4 Results

The result was demonstrated, and various mechanical properties were investigated on the basis of the exposure time and its corrosion behavior. Sample having less exposure time generally exhibits the higher corrosion resistance and having higher tensile strength, while the sample having more exposure time exhibits the less corrosion resistance and having the less tensile strength. Salt spray test was achieved under 5% NaCl solution conditions, and the results are shown in Table. The corrosion

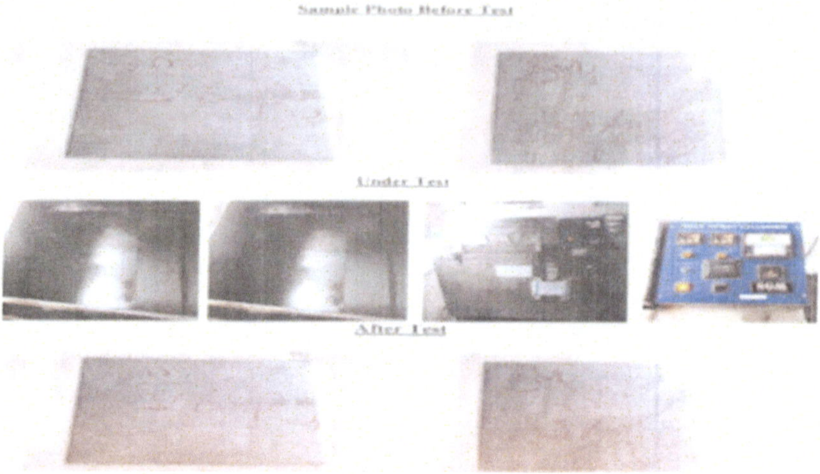

Fig. 3 Result after the salt spray test

Table 3 Examination (in Hours)

Salt spray test	24	48	72	96	120
Result	No effect	–	–	–	–

resistance seems to be the same, and the total duration of the test, 120 h, was certainly too short to point out differences. After completing the salt spray test under the salt spray test method chamber for the exposure of 120 h, it is observed that there is no corrosion effect is seen over the surface of the sample stainless steel 316L; also, numerical corrosion analysis shows that for the stainless 316L, corrosion resistance for the sample is inversely proportional to the exposure time (in years). The effect on the surface of SS 316L after every 24 h up to 120 h is shown in the Table no 0.3. From the given Table no 3, it is clear that we cannot determine and observe the corrosion on the surface of SS 316 L under such small time interval. The effect of the corrosion over the surface of SS 316 L depends mainly on the time (Table 3).

5 Conclusion

In this experiment, the condition for varying the immersion time of S.S 316L subjected to behaviors of stainless steel 316L has been studied, and it is found that under the corrosion test of above material sheet of 3 mm thickness has been done under 120 h and justified by numerical analysis which shows that order of corrosion resistance of SS 316L is suitable for the present atmospheric condition and optimized their utilization in different application in the industries. This study can be provided for acidic condition and analyzed for their best result.

References

1. Ahmadian, Danaee, Golozar, M.A.: Effect of surface treatment on corrosion resistance of 304 stainless steel implants in tyrode solution. Arch. Metall. Mater. **59**(1), 25–30 (2014)
2. Alar, Vesna: Corrosion behaviour of stainless steel in contact with wine and beer. Metalurgija **55**(3), 437–440 (2016)
3. Suresh, J., Suthagar S.: A study on wear and corrosion characterstics of martine propeller shaft material-AISI 316L stainless steel subjected to cryogenic treatment. In: Proceedings of ICSEM, 14–2nd International Conference on Science, Engineering and Management, vol. 4(6), pp 1049–1055 (2014)
4. Kumar, N.: Corrosion resistance of austenitic Cr–Ni stainless steel in 1M HCL. Int. J. Mech. Eng. Robotics Res. **3**(3), 21–26 (2014)
5. Lins, V., de Freitas Cunha, : Corrosion resistance of AISI 304 and 444 stainless steel pipes in sanitizing solutions of clean-in-place process. Mater. Res. **19**(2), 333–38 (2016)
6. Natishan, P.M.: Carbon surface modification for enhanced corrosion resistance. Mater. Sci. Technol. **4**(7), 89–95 (2008)
7. Sudjatmoko, Lely Susita, R.M., Wirjoadi, Bambang, S.: Corrosion resistance improvement of AISI 316L Stainless steel using nitrogen ion implantaion. Ganendra J. Nucl. Sci. Technol. **16**(2), 67–75 (2015)
8. Singh, G., Singh, H., Sidhu, B.S.: Corrosion behavior of plasma sprayed hydroxyapatite and hydroxyapatite-silicon oxide coatings on AISI 304 for biomedical application. Appl. Surf. Sci. **284**, 811–818 (2013)
9. Uflame U.G., da Silva Jr., J.F., Bezerra, G., Pinto, F.: Effect of the additives of nanosized Nb and Ta carbides on microstructure and properties of sintered stainless steel. **21**, 281–290
10. Umoru, L.E., Afonja, A.A., Ademodi, B.: Corrosion study of AISI 304, AISI 321 and AISI 430 stainless steels in a tar sand digester. J. Min. Mater. Charact. Eng. **7**(04), 291–99 (2008)

Experimental Investigation of Evacuated Tube Solar Air Collectors for Drying Application

Santanu Malakar and Vinkel Kumar Arora

1 Introduction

Solar energy is considered as most favorable renewable energy sources around world. Solar energy is harvested through the different types of collector which convert solar energy into heat energy and transfer it to a fluid. It is also extensively used in many countries for drying application [1]. Drying of agricultural products using direct sun rays under the open sky degrades the quality of the dried products due to dust, rodents and insects attack. Solar cabinet drying technology is the foremost substitute to all the problems discussed, and these can be resolved by the development of solar drying system, consisting of solar collector, air flow blower assembly, drying chamber [2]. The postharvest processing of garlic clove poses a great challenge because of high moisture content and bio-compound which causes browning and quick deterioration [3]. To overcome the spoilage, the garlic clove is dried and its moisture content is reduced up to the desired level for safe storage. For drying of garlic clove, different researcher applied different drying technique and some have worked on the garlic slice [4].

The biggest challenge lies in conversion of radiation energy to heat energy in efficient way for application of drying of agricultural produces. Mainly PV thermal collectors and flat plate collectors are used to collect the heat energy and proceed in the drying chamber by natural or forced convention methods in solar cabinet dryer and green house dryer [5, 6]. Numerous researchers have worked on free convection greenhouse dryer and flat plate collector to enhance the convective transfer coefficient, heat absorption, thermal efficiency and energy efficiency of the dryer [7, 8]. Multimode drying operation and phase change materials could be used for maintaining the better quality of food products and continue the drying process after sunshine, respectively [9, 10]. Due to low thermal efficiency and unsteady energy

S. Malakar · V. K. Arora (✉)
Department of Food Engineering, NIFTEM, Sonepat, Haryana 131028, India
e-mail: vinkelarora17@gmail.com

M. Muzammil et al. (eds.), *Recent Advances in Mechanical Engineering*, Lecture Notes in Mechanical Engineering, https://doi.org/10.1007/978-981-15-8704-7_49

absorption in flat plate collectors, ETSC is mostly used in air heating system for different applications. Also, the ETSC has much higher efficiencies than the flat plate collectors [11, 12]. Also, [13] has reported the performance and efficiency of evacuated vacuum tube solar drying system for different agricultural produces and confirmed that the air temperature increases up to 80 °C which is sufficient for drying of agricultural produces. The efficiency of the evacuated vacuum tube solar drying systems for particular vegetables and fruits was found around 70%–79% and 55%–78%, respectively. Evacuate tube-based solar dryer was developed by [14] for the drying of amla fruit.

The objective of the present research is to design and develop on evacuated tube solar dryer (ETSD) consideration on the design parameters. The thermal performance of ETSC and drying system is investigated on load and no load condition. Garlic clove are dried in ETSD and sun dried in order to investigate the average drying rate, percentage moisture content removed and also compared with sun drying.

2 Materials and Methods

2.1 Design Consideration

The ETSD is designed for approximately 10 kg of garlic clove to reduce the moisture content from 68% to 8% (*wb*) within 8 h. Technical specifications of evacuated tube is made of borosilicate of length 1.8 m with copper heat pipe of diameter 58 mm and placed at tilt angle 45 °C. Based on the assumptions and design criteria, the different parameters of evacuated tube solar dryer are determined by the following equations.

Amount of moisture content to be removed (M_w) and average drying rate (D_R) for the desired level of moisture from given quantity of garlic clove is given by: [15]

$$M_w = \frac{M_T \times (m_i - m_f)}{100 - m_f} \tag{1}$$

where, M_T is total mass of garlic clove, m_i and m_f *are* initial and final moisture content, respectively, and t_d is total drying time (h).

Amount of heat required (Q) for drying of garlic clove considering sensible heat and latent heat vaporization can be determined by: [16]

$$Q = M_T \times C_{pr} \times (T_p - T_a) + M_w \times h_{fg} \tag{2}$$

$$h_{fg} = 4.186 \times 10^3 (597 - 0.56T_p) \tag{3}$$

where C_{pr} is the sp. heat of produce (KJ/kg°C), T_p and T_a are product drying temperature and ambient air temperature, respectively, and h_{fg} is latent heat of vaporization (kJ/kg).

The required volume of heated air (V_{air}) for the decrease of moisture content from the garlic clove can be calculated by the energy balance equation [12].

$$M_w \times h_{fg} = \rho_a \times C_{pa} \times V_{air} \times \left(T_{co} - T_f\right) \quad (4)$$

where ρ_a (kg/m^3) is the density of air, T_{co} and T_f are collector outlet temperature and final temperature (°C), respectively.

The mass flow rate M_f(kg/s) and volume flow rate of air V_f (m^3/s) is requirement for effective drying of garlic clove, and it can be calculated as follows:

$$V_f = \frac{V_{air}}{t_d} \quad (5)$$

$$M_f = V_f \times \rho_a \quad (6)$$

The amount of heat absorbed by the evacuated tube collector should be equal to the amount of heat required for the designing the total solar collector area and the number of evacuated tube requirements. Therefore, total collector area and number tube can be estimated as follows: [17]

$$Q_{abs} = Q_{req} = A_c \times I_G \times \eta_{eff} \quad (7)$$

$$A_c = \frac{\pi}{2} \times D \times L \times n \quad (8)$$

where Q_{abs} and Q_{req} (kJ) are amount of heat absorption and heat requirement, respectively, A_c collector area (m^2), I_G is solar radiation (W/m^2), and η_{eff} is efficiency of collector. *D*, *L* and *n* are the diameter (mm), length (m) and number of evacuated tubes, respectively.

2.2 Drying Chamber and Blower Selection Procedure

The drying chamber is considered as convective batch—type and parallel hot air flows to the perforated stainless-steel trays of the product within the closed compartment. The dimensions of the drying chamber are length, breath and height estimated based on the product drying bed area, bulk density and porosity of the product and distance between trays. The drying bed area $\left(A_{dry}\right)$ can be calculated as:

$$A_{dry} = \frac{M_T}{x \times \rho_p \times \varepsilon} \quad (9)$$

The blower capacity for the effective drying was selected based on the requirements of air flow rate and pressure desired within the drying chamber. The power requirement of the blower for transferring of heated air to the drying chamber is calculated by this equation:

$$P = V_f \times \rho_a \times h \tag{10}$$

ρ_a (kg/m^3) is the density of air, x (mm) is the thickness of drying bed ε is the porosity of the product; V_f air volume flow rate (m^3/s), and h is the pressure head (m).

2.3 *Testing and Performance Investigation of ETSD*

The preliminary trails at no load condition were conducted at the NIFTEM, Kundli, Haryana, India, located at 28.8676°N, 77.1180°E in the month of July 2019. The study is focused to determine, the thermal performance and drying characteristics of garlic clove on ETSD.

2.3.1 Performance Analysis of Solar Collector

The ETSC performance is evaluated at no load condition based on the temperature raised in air at collector outlet. Different parameters like temperature, solar intensity and relative humidity were measured at an interval of one hour in a clear sunny day. Observations were measured by using different digital instrument like digital pyranometer (±0.05), digital anemometer (±0.2 m/s), thermometer (±0.01 °C), humidity meter(±0.02%), etc. The experiment of the ETSD is conducted from 10 AM to 5 PM, and data was recorded on hourly basis. The variation of solar radiation (I_T), ambient temperature (T_a), inlet temperature of drying chamber (T_{di}), outlet temperature of drying chamber (T_{do}) and relative humidity (H_{rel}) is recorded on hourly basis.

2.3.2 Performance Analysis of Drying Chamber

The ETSD dryer was investigated based on the moisture content removed (%), average drying rate and drying efficiency of the developed ETSD. The following constraints are taken into attention for the performance investigation of evacuate tube solar dryer and also comparison with sun drying of garlic clove.

a. ***Amount of moisture removed and drying rate***

Percentage of moisture content (MC) removed from the initial moisture content (*wb*) and drying rate of the garlic clove is expressed on and given by [18]

$$MC = \frac{w_i - w_f}{w_i} \times 100\% \quad (11)$$

$$D_r = \frac{w_i - w_f}{t_d} \quad (12)$$

w_i and w_f are initial and final weight of the product, respectively, and t_d is the drying time (h).

2.4 *Evacuated Tubes Solar Collector Efficiency*

Collector efficiency (η_c) is the ratio of the rate of heat gained by the fluid for effective drying to the total solar radiation incident on collectors. It is expressed by equation: [14]

$$\eta_c = \frac{M_f \times C_{pa}(T_{co} - T_a)}{A_c \times I_T} \times 100\% \quad (13)$$

where M_f (kg/s) is the air mass flow rate,T_{co} and T_a is the outlet temperature of the collector and ambient air temperature, respectively.

2.5 *Evacuated Tube Solar Dryer Efficiency*

Dryer efficiency (η_{dryer}) of the developed evacuated solar dryer is the vital factor for evaluating the performance of this dryer which can be estimated as: [12]

$$\eta_{dryer} = \frac{M_w \times L_w}{I_T \times A_c \times t_d} \quad (14)$$

I_T is solar radiation (W/m^2), L_w is latent heat of vaporization of H_2O (kJ/kg), and A_c is the collector area (m^2)

3 Results and Discussion

3.1 *Experimental Setup*

Evacuated tube solar dryer was developed based on the design parameters for the drying of garlic clove. As per the design estimation, the ETSD was fabricated. The 15 number of evacuated tubes solar collector was mounted parallel through header

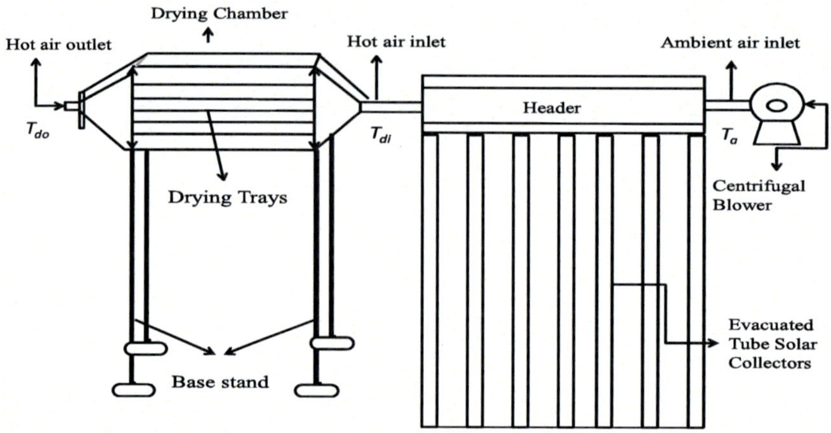

Fig. 1 Schematic representation of developed ETSD

which was made of 40-mm-diameter copper pipe through proper insulation to reduce the heat loss. Specifications, i.e. collector area, 1.34 m/s^2, air flow rate 3.7 m/s are estimated. The drying chamber area 4 m^2 with six number of trays is made up plywood with inner wall covered with aluminum sheet to diminish the heat loss.

The internal dimension of the chamber is (0.78 × 0.70 × 1.36) m^3. The developed drying chamber contains of six perforated trays for loading the garlic clove. A centrifugal fan of 0.2 HP was attached in inlet of header for the maintaining the air flow rate during drying processed. The fabricated dryer consists of evacuated tube solar collector, blower, drying chamber as displayed in Fig. 1.

3.2 *Performance Investigation of Solar Collector*

The drying system was developed for heating the air to supply in drying chamber for drying of garlic clove. The temperature profile of the evacuated tube solar dryer was recorded at no load condition by measuring the inlet zone and outlet zone temperature of the drying chamber. Also, the thermal profile of the drying system is presented in Fig. 2.

It is clearly observed that the maximum temperature 82.5 °C was achieved at the inlet zone of drying chamber when the solar intensity 1440 W/m^2, while ambient temperature was 39 °C, relative humidity 42% at mid-noon (1–2 pm). During the experiments, the ambient temperature changes from 32 to 39 °C although the highest temperature at collector outlet was recorded 82.5 °C at inlet and 62.8 °C at outlet temperature of drying chamber. The solar radiation was recorded from 650 to 1495 W/m^2. It was recorded that the intensity of solar radiation was higher during mid-noon, which resulted in higher air temperature inside the drying chamber. Similar trend was reported for the flat plate collector [12].

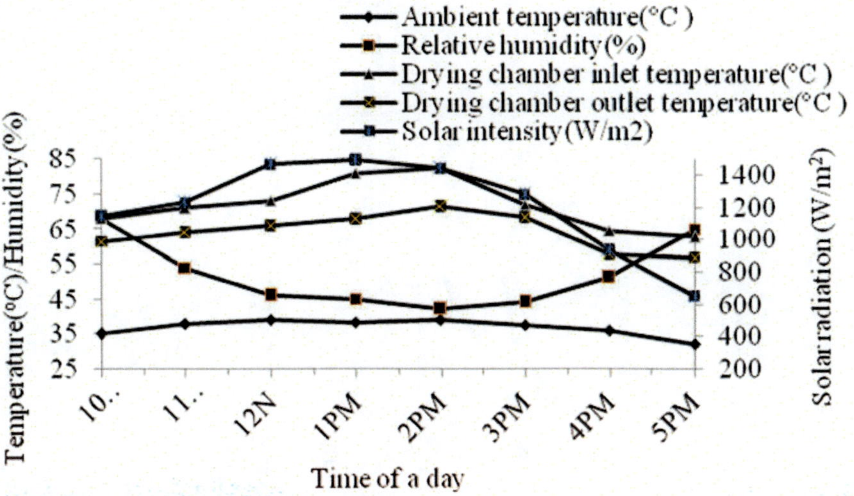

Fig. 2 Thermal profile of evacuated tube solar dryer

3.3 Performance of Drying System

3.3.1 Amount of Moisture Removed

The variation of moisture content using ETSD and sun drying is presented graphically in Fig. 3a. It was clearly observed that the initial amount of moisture of garlic clove was found to be 68% which decreases to 8% in ETSD and 24% in sun drying for same drying time. It concludes that the reduction of moisture content in ETSD was higher as compare to sun drying. Similarly, the reduction of moisture content was reported by [19].

3.3.2 Drying Rate of Garlic Clove

The drying rate curve for ETSD and sun drying is depicted in the Fig. 3b. It is clearly observed in ETSD, the initial drying rate was found to be 0.81 kg H_2O/kg dry matter/h which increases initially up to 0.96 kg H_2O/kg dry matter/hand; thereafter, it decreases to 0.58 kg H_2O/kg dry matter/h during drying time of 7 h. Similarly, during sun drying, the initial drying rate was 0.65 kg H_2O/kg dry matter/h which increased 0.68 kg H_2O/kg dry matter/h and thereafter continuously decreased to 0.23 kg H_2O/kg dry matter/h during drying period. The results show that the drying rate in evacuated tube solar drying system was higher than the sun drying at same environment conditions which concluded that the required drying time in ETSD is reduced as compared to sun drying.

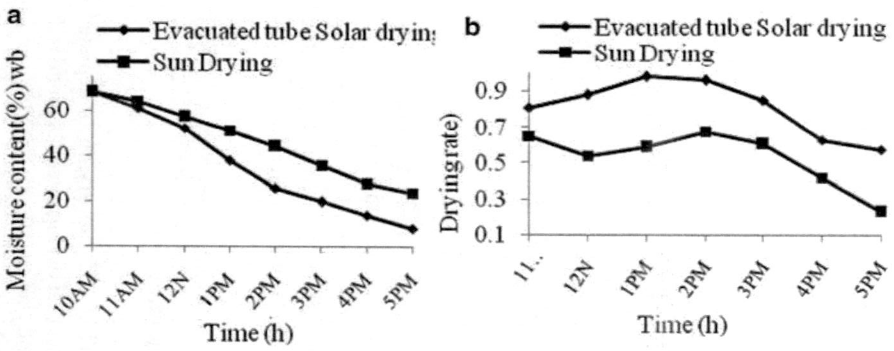

Fig. 3 **a** Moisture content (%) versus time and **b** drying rate (D_r) versus time, plots for ETSD and sun drying

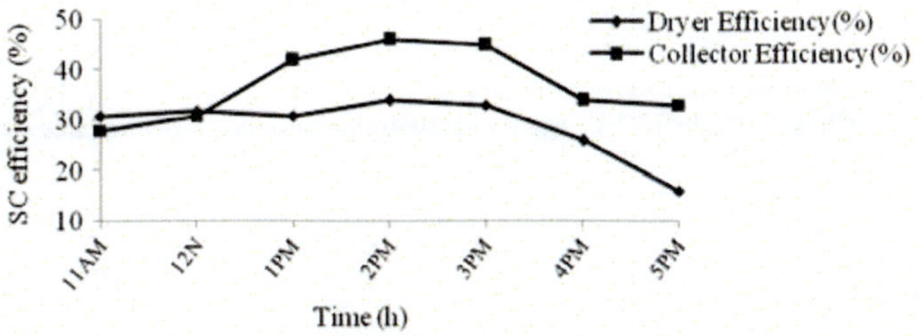

Fig. 4 Efficiency variations of the evacuated tubes solar collector and solar dryer

3.4 Evacuated Tube Solar Collector Efficiency

The performance of ETSC was evaluated through the efficiency of collectors during the experiment. The ETSC efficiency is graphically presented in Fig. 4. Initially, the efficiency of ETSC was recorded 28% which increase up to 46% with the increase of solar intensity 1440 W/m^2 during 1 pm to 2 pm and thereafter decreased to 33% with decreasing solar intensity 650 W/m^2 at 5 pm.

3.5 Evacuated Tube Solar Dryer Efficiency

The variation of overall dryer efficiency of the ETSD during the experiment is graphically as shown in Fig. 4. Initially, the dryer efficiency was observed 31% which increases 34% when solar intensity was maximum, and after that, the efficiency was decreased to 16% during the drying period. The average efficiency of ETSD was measured 29%.

4 Conclusion

Evacuated tube solar dryer was designed and fabricated for drying of garlic clove. The maximum collector outlet temperature was measured to be 82.5 °C during mid-noon at maximum solar radiation 1440 W/m^2. It is concluded that the amount of moisture content reduced by 60% in ETSD while 48% in sun drying for drying time. The average drying rate was found to be 0.81 kg H_2O/kg dry matter/h in ETSD which is higher than in sun drying. Also, the average collector efficiency and dryer efficiency were found 37% and 29%, respectively. So, it is evident that the developed ETSD is better and efficient and less time is required for drying of garlic clove as comparison to sun drying.

References

1. Sharma, A., Chen, C.R., Lan, N.V.: Solar-energy drying systems: a review. Renew. Sustain. Energy Rev. **13**, 1185–1210 (2009)
2. Pakhare, V.V., Salve, S.P.: Design and development of solar dryer cabinet with thermal energy storage for drying chilies. Int. J. Curr. Eng. Technol. **5**, 358–362 (2016)
3. Bisnoi, N., Kumari, P., Yadav, Y.K.: Study of dehydration characteristics of garlic. J Dairying, Foods. **27**, 238–240 (2008)
4. Wongsa, P., Spreer, W., Müller, J., Sruamsiri, P.: Effect of drying methods on anti-oxidative potential of garlic (*Allium sativum L.*). Acta Hortic. **1125**, 209–213 (2016)
5. Tiwari, S., Tiwari, G.N.: Grapes (*Vitis vinifera*) drying by semitransparent photovoltaic module (SPVM) integrated solar dryer: an experimental study. Heat Mass Trans. **54**, 1637–1651 (2018)
6. Tiwari, S., Agrawal, S., Tiwari, G.N.: PVT air collector integrated greenhouse dryers. Renew. Sustain. Energy Rev. **90**, 142–159 (2018)
7. Kumar, M., Sahdev, R.K., Tiwari, S., Panchal, H., Manchanda, H.: Experimental free convection thin layer groundnut greenhouse drying. Agric. Eng. Int. CIGR J. **21**, 203–211 (2019)
8. Montero, I., Blanco, J., Miranda, T., Rojas, S., Celma, A.R.: Design, construction and performance testing of a solar dryer for agro-industrial by-products. Energy Convers. Manag. **51**, 1510–1521 (2010)
9. Sehrawat, R., Chandra, A., Nema, P.K., Arora, V.K.: Drying of fruits and vegetables in a developed multimode drying unit and comparison with commercially available systems. J. Inst. Eng. Ser. A. **100**, 381–386 (2019)
10. Babar, O.A., Arora, V.K., Nema, P.K.: Selection of phase change material for solar thermal storage application: a comparative study. J. Brazilian Soc. Mech. Sci. Eng. **41** (2019)
11. Sabiha, M.A., Saidur, R., Mekhilef, S., Mahian, O.: Progress and latest developments of evacuated tube solar collectors. Renew. Sustain. Energy Rev. **51**, 1038–1054 (2015)
12. Wang, W., Li, M., Hassanien, R.H.E., Wang, Y., Yang, L.: Thermal performance of indirect forced convection solar dryer and kinetics analysis of mango. Appl. Therm. Eng. **134**, 310–321 (2018)
13. Mahesh, A., Sooriamoorthi, C.E., Kumaraguru, A.K.: Performance study of solar vacuum tubes type dryer. J. Renew. Sustain. Energy. **4**, 121–128 (2012)
14. UmayalSundari, A.R., Neelamegam, P., Subramanian, C.V.: Performance evaluation of a forced convection solar drier with evacuated tube collector for drying amla. Int. J. Eng. Technol. **5**, 2853–2858 (2013)
15. Hossain, M.Z., Hossain, M.A., Awal, M.A., Alam, M.M., Rabbani, A.H.M.M.: Design and development of solar dryer for chilli drying. Int. J. Res. **2**, 63–78 (2015)
16. Abubakar, S.: Design and performance evaluation of a mixed-mode solar crop dryer. FUOYE J. Eng. Technol. **3**, 22–26 (2018)
17. Ubale, A.B., Pangavhane, D., Auti, A.: Performance analysis of forced convection evacuated tube solar collector used for grape dryer. J. Eng. Sci. Technol. **12**, 42–53 (2017)
18. Demiray, E., Tulek, Y.: Drying characteristics of garlic (*Allium sativum L*) slices in a convective hot air dryer. Heat Mass Transf. **50**, 779–786 (2014)
19. Shringi, V., Kothari, S., Panwar, N.L.: Experimental investigation of drying of garlic clove in solar dryer using phase change material as energy storage. J. Therm. Anal. Calorim. **118**, 29–41 (2014)

Appliances Energy Prediction Using Random Forest Classifier

V. Vakharia, S. Vaishnani, and H. Thakker

1 Introduction

Energy consumption of world is increasing with time. Being the pure and high-grade form of energy, nowaday's share of electrical energy in total energy demand is increasing rapidly. Understanding the energy usage in the form of electricity has been a subject of various studies [1–3]. The supply of electrical energy is quite less than the demand. Moreover, electrical energy is to be supplied in real time as it cannot be stored in its pure form and even if stored in devices such as battery packs, the conversion efficiency of the equipment comes into play. One way to meet the demand is to supply only the real time required amount of electrical energy. This can be possible if somehow prediction of electrical energy load can be done. Understanding the energy use in the form of electricity has been the subject of various studies. One of the significant sectors in electrical energy consumption is appliances. It represents a considerable portion of about 20–30% of the electrical energy demand [4]. Considering the high demand and less supply of electrical energy, it is required to predict energy load. The consumption of electricity in domestic households depends mainly on two factors: the number and type of appliances and their use by the residents. Many factors such as building construction, weather conditions, dry-bulb temperature, wet-bulb temperature occupancy, behavior of occupancy, lighting, heating ventilation and airconditioning systems, their usage pattern and performance, etc., influences the energy behavior of a building [2]. Considering the complexity of the problem, accurate energy consumption prediction is not an easy task. Traditional analytical methods do not seem to be dominating here, as they become computationally very intensive and also very difficult to consider a huge number of variables [5]. Thus, it has become necessary to move toward data-driven machine learning techniques

V. Vakharia (✉) · S. Vaishnani · H. Thakker
School of Technology, Pandit Deendayal Petroleum University, Gandhinagar, Gujarat 382007, India
e-mail: vinayvakharia4343@gmail.com

M. Muzammil et al. (eds.), *Recent Advances in Mechanical Engineering*, Lecture Notes in Mechanical Engineering, https://doi.org/10.1007/978-981-15-8704-7_50

[6–9]. Regression methods of machine learning techniques can be used to build a regression model of energy use. It can also be used to understand the relationships between various variables and their importance [3, 10]. This energy load prediction models can be used for various applications like to detect unusual energy consumption, in demand side management, to take decision input for building performance simulation analysis, etc. In the methodology adopted, authors have compared three machine learning models, random forest, SVM and ANN, to predict the appliances energy prediction from the publicly available dataset provided by [1].

2 Data Acquisition

2.1 *Data Description*

The house of which energy consumption data are collected is located in Stambruges in Belgium. It is a low energy consumption house which has an annual heating and cooling load of less than 15 kWh/m^2. Total floor area of the house is 280 m^2. The house was occupied by four people consisting of two adults and two teenagers. The energy (Wh) data were logged every 10 min for the appliances. To capture the varying energy consumption more accurately, 10 min reporting interval was chosen. Lights combined with relative humidity seems to be good predictor of room occupancy. Thus, another sub-metered lights load is included. Data are recorded for 137 days, and high variability can be seen in appliance energy consumption as can be observed from Fig. 1. Weather data are taken from the nearest airport weather station (Chièvres Airport, Belgium) which is 12 km from the Stambruges house. Linear interpolation is done at 10-minute intervals for this data [1]. A ZigBee wireless sensor network used for monitoring the house temperature and humidity conditions with an accuracy of ±0.5 °C for temperature and ±3% for relative humidity. In total, 31 features are used for the prediction of household energy appliances out of which some statistical parameters are generated from the date/time variable like the number of seconds from midnight for each day (NSM), the week status (weekend or workday), and the day of the week.

2.2 *Exploratory Analysis*

Some important features are generated from other features. The week status (whether the day is weekday or weekend), the day of the week, and the number of seconds from midnight (NSM) are generated from date time feature. In this analysis, our target variable is appliance energy consumption with three conditions. It is very important to understand the trend of variation of appliance energy with respect to time for selecting the model to establish relation between target variable and other

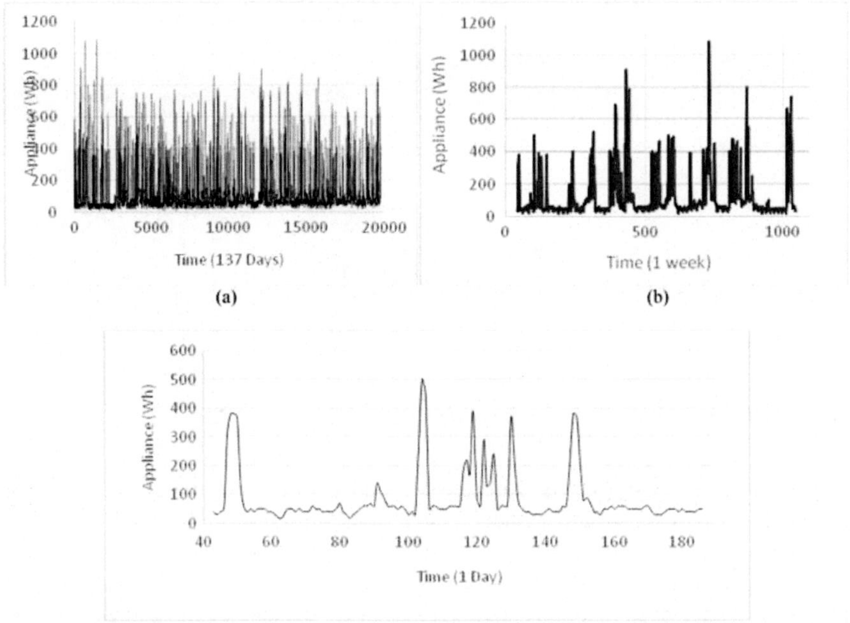

Fig. 1 Appliance energy consumption for **a** Whole period **b** 1 Week **c** 1 day [1]

predicting features. Figures 1 show how appliance energy consumption varies with different time intervals.

3 Regression Modeling Using Machine Learning Techniques

3.1 Performance Measurement

Various performance metrics are used for the purpose of comparison of each of the regression models [1, 12]. Metrics considered are coefficient of correlation (r), Mean Absolute error (MAE), root mean squared error (RMSE), relative absolute error (RAE), root relative squared error (R^2).

$$r = \frac{\sum(x-\bar{x})(y-\bar{y})}{\sqrt{\sum(x-\bar{x})^2\sum(y-\bar{y})^2}} \quad (1)$$

$$\text{MAE} = \frac{\sum|x-y|}{n}. \quad (2)$$

$$\mathrm{RMSE} = \sqrt{\frac{\sum (x-y)^2}{n}} \tag{3}$$

$$\mathrm{RAE} = \frac{\sum |x-y|}{\sum |x-\bar{x}|} \tag{4}$$

$$\mathrm{R}^2 = \sqrt{\frac{\sum (x-y)^2}{\sum (x-\bar{x})^2}} \tag{5}$$

3.2 Data Division and Model Generation

Different subsets were created from the original dataset and used for evaluating the performance of regression models with various subset of predictors. Three data subsets were created:

(1) All records (2) No light (3) Light. These data subsets are created based on the absent or presence of feature light. In first data subset, all the records are considered. In the second data subset, records with zero value of light feature are only included. While in third data subset, records with zero value of Light features are excluded from all data records. The data sets were divided into two parts: training and testing data in which 70% of data used for training and 30% data used for testing.

Three machine learning algorithms are used for model training and testing: Random forest, artificial neural network, and support vector machine. The random forest model is a tree-based classifier and the final predictions are based on the weighted output of several regression trees. Each tree is built with a random sample of selected features. This is because to decorrelate the trees and to improve prediction. Support vector machine (SVM) can use different kernels. Here, radial basis function kernel is used, as it has some numerical advantages. Artificial neural network (ANN) is inspired by the biological neuron present in human and animal brains. This type of systems iteratively improves the performance generally without task-specific programming. In our cas, multilayer perceptron (MLP) is used which belongs to feed forward artificial neural network (Table 1).

Table 1 Training and testing data set

	Training	Testing	Total
All Records	13,814	5920	19,734
No lights	5231	1373	6604
Lights	3138	1345	4483

4 Results and Discussion

Regression models are trained using three different techniques: Random forest, artificial neural network, and support vector machine. After training the regression models, testing is done. All three regression models are assessed for both training as well as testing by using five various statistical parameters [1, 4, 11, 12]. In the case of coefficient of correlation, higher is better considered for model evaluation as it represents the trend of regression model with actual data. For training of model, random forest gives highest value of correlation coefficient. Also, all the four types of errors are less for random forest as compare to ANN and SVM for all the three cases, as seen from Tables 2, 3 and 4. Thus, it can be observed that in our experiment, for training of random forest, better results are observed than ANN and SVM. For testing of regression models, again random forest gives highest correlation coefficient among all the three regression models. Also, all errors are less in case of random forest method. Thus, it can be observed that in the methodology used, random forest method is proven to be better for both training and testing as compare to artificial neural network and support vector machine. It should be considered that for training the regression model, support vector machine technique takes highest time among the above three methods being computationally more intensive and artificial neural network is the fastest one, while random forest remains in between the above two methods for time requirement.

Table 2 Model performance considering all records subset

Performance parameters	Random forest		ANN (MLP)		SVM (RBF)	
	Training	Testing	Training	Testing	Training	Testing
Correlation coefficient (r)	0.98	0.98	0.64	0.63	0.41	0.40
Mean absolute error (MAE)	9.47	10.28	42.15	45.76	36.60	39.92
Root mean squared error (RMSE)	21.40	22.65	74.79	80.46	88.62	94.35
Relative absolute error (RAE)	18.06	18.25	80.35	81.22	69.77	70.86
Root relative squared error (R^2)	23.37	23.24	81.66	82.57	96.75	96.82

Table 3 Model performance considering no light subset

Performance parameters	Random forest		ANN (MLP)		SVM (RBF)	
	Training	Testing	Training	Testing	Training	Testing
Correlation coefficient (r)	0.98	0.97	0.69	0.68	0.40	0.42
Mean absolute error (MAE)	9.23	9.43	39.77	34.43	35.22	39.19
Root mean squared error (RMSE)	21.89	21.46	75.75	67.71	87.08	96.61
Relative absolute error (RAE)	17.34	18.33	73.07	68.36	69.93	72.0
Root relative squared error (R^2)	22.08	23.90	75.40	76.40	96.97	97.44

Table 4 Model performance considering lights subset

Performance parameters	Random forest		ANN (MLP)		SVM (RBF)	
	Training	Testing	Training	Testing	Training	Testing
Correlation coefficient (r)	0.97	0.97	0.69	0.69	0.36	0.35
Mean absolute error (MAE)	15.89	15.95	46.54	45.94	51.67	52.04
Root mean squared error (RMSE)	29.98	30.63	81.69	82.68	109.90	111.66
Relative absolute error (RAE)	23.44	23.22	67.10	68.40	67.10	75.93
Root relative squared error (R^2)	27.02.	27.20	73.42	73.63	99.16	99.05

References

1. Candanedo, L.M., Feldheim, V., Deramaix, D.: Data driven prediction models of energy use of appliances in a low-energy house. Energy Build. **140**, 81–97 (2017)
2. Fan, C., Xiao, F., Wang, S.: Development of prediction models for next day building energy consumption and peak power demand using data mining techniques. Appl. Energy **127**, 1–10 (2014)
3. Zhao, H.-X., Magoulès, F.: A review on the prediction of building energy consumption. Renew. Sustain. Energy Rev. **16**(6), 3586–3592 (2012)
4. Cetin, K.S.: Characterizing large residential appliance peak load reduction potential utilizing a probabilistic approach. Sci. Technol. Built Environ. **22**(6), 720–732 (2016)
5. Kavousian, A., Rajagopal, R., Fischer, M.: Ranking appliance energy efficiency in households: utilizing smart meter data and energy efficiency frontiers to estimate and identify the determinants of appliance energy efficiency in residential buildings. Energy Build. **99**, 220–230 (2015)
6. Kusiak, A., Li, M., Zhang, Z.: A data-driven approach for steam load prediction in buildings. Appl. Energy **87**(3), 925–933 (2010)
7. Long, H., Zhang, Z., Su, Y.: Analysis of daily solar power prediction with data-driven approaches. Appl. Energy **126**, 29–37 (2014)
8. Oliver, E., Keshav S.: Data driven smart phone energy level prediction University of Waterloo Technical Report (2010)
9. Zhang, G., Patuwo, B.E., Hu, M.Y.: Forecasting with artificial neural networks: The state of the art. Int. J. Forecast. **14**(1), 35–62 (1998)
10. Jovanović, R.Ž., Sretenović, A.A., Živković, B.D.: Ensemble of various neural networks for prediction of heating energy consumption. Energy Build. **94**, 189–199 (2015)
11. Gujar, R., Vakharia, V.: Prediction and validation of alternative fillers used in micro surfacing mix-design using machine learning techniques. Construct. Build. Mater. **207**, 519–527 (2019)
12. Vakharia, V., Gujar, R.: Prediction of compressive strength and portland cement composition using cross-validation and feature ranking techniques. Construct. Build. Mater. **225**, 292–301 (2019)

Review of Effect of Nanofillers on FRP Composites

Tanmay Bansal, Suraj Malik, Tushar Batra, Munna Shah, Anurag Gupta, and K. L. A. Khan

1 Introduction

Manufacturing engineering materials with high performance from nanoparticles is one of the aspiring goals currently being followed by researchers across the globe. Nowadays, polymer nanocomposite materials are developed with the embodiment of nanofillers, such as nanoparticles, nanotubes, nanofibers. In addition, better diffusion of these fillers within the matrix gives high-performance nanocomposites; moreover, the properties of the nanoscale filler are remarkably higher apart from the base matrix. Composites are strong as well as light; this is highest advantage of composites. The two discrete components are mixed together in which one is the binder and the other is the matrix. Reinforcement surrounds and binds together fibers and fragments of a composite material. Matrix is normally less hard and more ductile. It detains dispersed phase and contributes a load with it. Matrix has dispersed phase implanted in an intermittent form. The dispersed phase is the secondary phase. Dispersed phase is usually longer then the matrix; consequently, it is sometimes known as reinforcing phase [1]. Fillers are added to the matrix in order to improve surface finish of composite materials as surfaces without fillers could become coarse and cause decrease in mechanical properties. Adding $CaCO_3$, glass fiber, and carbon black improves stiffness, elastic modulus. Adding silica, $CaCO_3$ as fillers in polyester resin matrix lessens tensile

T. Bansal (✉) · S. Malik · T. Batra · M. Shah
B.Tech. Student, Department of Mechanical Engineering, KIET Group of Institutions, Ghaziabad, India
e-mail: tbansal.5186004@gmail.com

A. Gupta
Associate Professor, Department of Mechanical Engineering, KIET Group of Institutions, Ghaziabad, India

K. L. A. Khan
Professor & Head, Department of Mechanical Engineering, KIET Group of Institutions, Ghaziabad, India

M. Muzammil et al. (eds.), *Recent Advances in Mechanical Engineering*, Lecture Notes in Mechanical Engineering, https://doi.org/10.1007/978-981-15-8704-7_51

strength in the span of 3–15% for 25–50% fillers in hand layup technique. Tensile strength as well as modulus of composites doped with carbon black and carbon fiber rises with increasing the filler contents; however, impact strength as well as elongation at break is seen to be reduced [2].

2 Nano-SiO_2-Based Epoxy Composites

Modified nano-SiO_2 is more effective than the standard SiO_2 as it has better tensile properties as well as impact properties because of the large specific area with active groups on surface of nanoparticles. Comparable behavior has been exhibited for the toughness of epoxy resin-based composites with nano SiO_2 as well as standard SiO_2 [3]. On mixing nanosilica powder with resin, there is a 23% increment of ultimate tensile strength at 5 wt.% and impact strength rises to 10 J/mm^2 at 10 wt.% Also, the fatigue life cycle increment to 14×10^5 cycles at 6 wt.% It demonstrated some significant impact on the electrical breakdown strength of the epoxy composites. Nanoparticles with saline coupling agent enhance the molecular interactions between the epoxy and nanoparticles. The softening point has been discovered to rise however coefficient of thermal expansion, and dielectric constant reduces with rise in SiO_2 content. Therefore, the breakdown strength of the modified nanoparticles is more than that of the unmodified one [4]. It was found that the softening point T_s rose by raising SiO_2 content; however, dielectric constant as well as coefficient of thermal expansion reduced [5].

3 TiO_2 Nanofiller-Based FRP Composites

Nano-TiO_2 is inorganic filler which when added to FRP composites improves the mechanical properties of the composite. Wisnom did several explorations on residual strength, thermal properties, and water sorption of FRP composites by addition of TiO_2 particles. The outcome showed that by adding 0.1 wt.% TiO_2 the water diffusion coefficient got decreased by 9%, residual flexural strength increased by 19%, and as well as residual interlaminar shear strength increased by 18% in the number of all TiO_2-based composites. This provides opportunity for the improvement of mechanical properties in many engineering applications [6]. According to Nayak and Ray for composites having 0.1 wt.% of the nano-TiO_2, after seawater aging, the flexural strength got improved by 15%. This occurs because of the improved interface/interphase bond which occurs between fiber and the matrix. Thereafter, increasing the nano-TiO_2 content results in agglomeration of nanoparticles within the matrix because of the van der Waal's forces acting between two nanoparticles [7].

4 Al_2O_3 Nanofiller-Based Composites

Useful procedures for the preparation of Al_2O_3 and other nano-oxide-based nanocomposites of conducting polymers from water-soluble/insoluble monomers and of binary polymers were developed. The hardness of the composite got better by the presence of Al_2O_3 fillers [8]. These materials were characterized by improved thermal stability compared to base polymer. All these composites exhibited DC conductivities in the order of 10^{-5} to 10^{-7} s/cm for the PNVC (poly-N-vinylcarbazole)-based composites and $10^{-2}$to 10^{-4} s/cm for the PPY (polypyrrole), PANI (polyaniline) and PTP (polythiophene)-based systems. TEM analyses confirmed the formation of particles in the nano-order (20–500 nm) range depending upon the nature of the oxide and polymer used. These composites produced aqueous dispersion of varying stabilities depending upon the nature of the polymeric component in the composites [9]. The impact strength and hardness Rockwell were increased up to 300 and 400%, respectively, as compared to the pure HDPE. The nano-Al_2O_3 is generally more efficient for the betterment of the impact strength [10].

5 Carbon Nanotubes (CNT)-Based Composites

According to Zhu et al. [11, 12] using single-walled carbon nanotubes (SWCNT) as nanofiller to reinforce the epoxy resin, it was noted that it exhibited some remarkable outcomes. An improvement in the mechanical properties and dispersion was exhibited by the addition of 1.0 wt.% of functionalized nanotubes to the composites. This resulted in 30% rise in modulus as well as 18% rise in tensile strength. According to Feng et al. [13], the tensile and dynamic properties got improved by the addition of SWCNT into the matrix. SWCNT when added on 39.1 wt.% loading, the tensile as well as Young's modulus was raised by 183 and 408%, respectively, when kept side by side to neat epoxy. Hsieh et al. explored the effect on the fatigue behavior as well as fracture toughness of a thermosetting epoxy polymer by the volume content of multi-walled carbon nanotubes (MWCNT). Several uniaxial tensile tests were done to found the mechanical properties of nanocomposites. It was found that mode I fracture toughness as well as stiffness behavior got raised by raising MWCNT content value [14].

6 $CaCO_3$ Nanofiller Composites

The tensile strength of composite increases with the decrease in the content of$CaCO_3$, because of the void formation owing to clustering of nano-$CaCO_3$ particles. Impact strength also decreases with increase in concentration because agglomeration of fillers, thereby generating the increase in void space which is responsible for stress propagation [15]. The highest erosion wear rate is at 90 impingement angles. With addition of $CaCO_3$ nanofiller, the resistance to impact loading increases and it also increases the hardness of the material. Hardness of composites increases by 116.16 HV. By increasing the weight percentage of the filler quantity, the ultimate tensile strength as well as flexural strength of the composites reduces. A new variety of $CaCO_3$-filled epoxy-based composites has been efficiently fabricated by the help of hand layup technique. When $CaCO_3$ is added in 10 wt.%, it increases the impact energy to 6 J because of the interfacial bonding in between matrix material and particulate fillers [16].

7 SiC Nanofiller Composite

Kaundal et al. [17] studied the consequences on physical properties, erosion behavior, and impact strength of nano- SiC- loaded short glass fiber on reinforced polymer composites. Polymer composite reinforced with nano-silica carbide showed enhanced erosion resistance when kept side by side with unfilled polymer composites. Silica carbide was added in 10 and 20 wt.% to polymer composites. Unfilled polymer composites show the maximum erosion taking place at 60^0 impingement angle, whereas silica carbide filled composites shows the maximum erosion taking place at 75° impingement angle. It can be concluded from the investigation made by Kaundal et al. that nano-silica carbide fillers was able to enhance the erosion rate of the material when compared to the unfilled FRP composites. This allows further improvement and advancement to occur in this field of nanofillers in future.

8 Graphene Nanofillers-Based FRP Composites

According to Qiu et al. [18] the crack propagation was prohibited because a large value of plastic deformation was generated by the presence of grapheme oxide (GO). The deformation and the structure of grapheme were explored thoroughly by using Raman spectroscopy [19]. The number of layers in graphene films was discovered by this method. Tang et al. [20] explained that Young's modulus was increased by 2.3% by incorporating 0.05 wt.% of reduced graphene oxide (RGO), for the good dispersion level. According to Bordz et al. [21], Young's modulus got improved by 6% on 0.5 wt.% loading. The stiffness was also improved by rising the loading content; also, the maximum value of UTS was seen at 0.5 wt.% loading. The highest increase was seen at 0.5 wt.% graphene content which showed the maximum value for stiffness as well as the strength.

9 Carbon Black Nanofiller Composite

The epoxy resin utilized in this work is diglycidyl ether of bisphenol A (DGEBA) which is a liquid epoxy resin which shows more than 75% of the resin usage in applications. It is heat treated. Tensile strength of composites reduces with rise in carbon black content due to increase of hardness with rise in carbon black percentage [22]. It increases the thermal properties of the composite and makes the composite fire resistant up to some extent. The carbon black can retard fire in less than 30 s when added in appropriate weight percentage with respect to the amount of composite. By consolidating an alternative high surface area or lower surface area carbon black in the resin, the electrical conductivity was achieved in the inner core [2]. Appropriate self–monitoring properties was showed by only one type of material, since at increased loading rise in electrical resistance was seen. Compared to epoxy/inorganic nanofiller, composite carbon fiber composites are employed more frequently and effectively in defense-related applications [23].

10 Conclusion

- Nanoparticle type affects the flexural, interfacial as well as vibration properties of composites which enhance the flexure strength as well as interfacial bonding. Al_2O_3-based composites produced aqueous dispersion with varying stabilities depending upon the nature of the polymeric components in the composites.
- In silica-based epoxy nanocomposites for electrical insulation dielectric properties has been improved. Ultimate tensile strength as well as flexural strength of composites reduces with rise in weight percentage of filler content also these values are less than unfilled epoxy composites.
- The enhancement of the properties depends upon the amount or concentration of nanofillers added and also on the distribution of the nanofillers on the composites. Carbon black fillers increase the hardness, and the fire-resistant properties of the composites are greatly enhanced.
- The tensile strength of the composites can possibly be enhanced by utilizing smaller particle magnitude of $CaCO_3$ and carbon black.
- Increase in flexural strength was remarkably seen in FRP composites after adding TiO_2 nanofillers. It also showed reduction in water diffusion coefficient as well as increase in the residual interlaminar shear strength.
- Several tests results show that by adding CNT nanofillers in different forms, improvement in mechanical properties like tensile strength was observed. Also, dispersion of the composite was improved.
- An increase in Young's modulus was observed by adding graphene nanofillers. The increase in stiffness and the maximum value of UTS was observed at 0.5 wt.% loading of graphene nanofillers.

- It can be concluded from the investigation made by Kaundal et al. that nanosilica carbide fillers were able to enhance the erosion rate of the material when compared to the unfilled FRP composites. This allows further improvement and advancement to occur in this field of nanofillers in future.

Acknowledgements Authors are grateful to have been supported by the Royal Academy of Engineering under the auspices of "UK-India Industry-Academia Partnership Programme" between Cranfield University, UK, and KIET, India (Grant No. IAPP18-19\295).

References

1. Jose, J. P., Thomas, S., Kuruvilla, J., Malhotra, S.K., Goda, K., Sreekala, M.S.: Advances in polymer composites: macro-and microcomposites—State of the art, new challenges, and opportunities. Polym. Compos. **1**, 3–16 (2012)
2. Bera, T., Acharya, S.K., Mishra, P.: Synthesis, mechanical and thermal properties of carbon black/epoxy composites. Int. J. Eng. Sci. Technol. 10(4), 12–20 (2018)
3. Nagalingam, R., Sundaram, S., Stanly, B., Retnam, J.: Effect of nanoparticles on tensile, impact and fatigue properties of fibre reinforced plastics. Bull. Mater. Sci. **33**(5), 525–528 (2010)
4. Mahrholz, T., Stängle, J., Sinapius, M.: Quantitation of the reinforcement effect of silica nanoparticles in epoxy resins used in liquid composite moulding processes. Compos. A Appl. Sci. Manuf. **40**(3), 235–243 (2009)
5. Borkar, S. P., Senthil Kumar, V., Mantha, S.S.: Effect of silica and calcium carbonate fillers on the properties of woven glass fibre composites (2007)
6. Nayak, R.K., Mahato, K.K., Ray, B.C.: Water absorption behavior, mechanical and thermal properties of nano TiO_2 enhanced glass fiber reinforced polymer composites. Composit. Part A: Appl. Sci. Manuf. **90**, 736–747 (2016)
7. Nayak, R.K., Ray, B.C.: Influence of seawater absorption on retention of mechanical properties of nano-TiO_2 embedded glass fiber reinforced epoxy polymer matrix composites. Archives Civil Mechan. Eng. **18**(4), 1597–1607 (2018)
8. Das, M, Singh, S.P., Pal, K., Jena, S., Karmakar, B.: Influence of combined Al_2O_3-SiO_2 filler on thermal and dielectric properties of barium zinc borate glass microcomposites for barrier ribs of plasma display panels. Ind. J. Eng. Mater. Sci. **17**(3), 199–207 (2010)
9. Ballav, N., Maity, A., Biswas, M.: Comparative studies on Al_2O_3 based nanocomposites of some speciality polymers and binary polymers with the corresponding SiO_2, MnO_2 and ZrO_2 based nanocomposites (2003)
10. Koyejo O. et al. Enhancement of high-density polyethylene properties by impregnation with inorganic alumina filler. Am. J. Chem. Eng. **5**(3–1), 49–54 (2017)
11. Zhu, J., Kim, J.D., Peng, H., Margrave, J.L., Khabashesku, V.N., Barrera, E.V.: Improving the dispersion and integration of single-walled carbon nanotubes in epoxy composites through functionalization. Nano Lett. **3**(8), 1107–1113 (2003)
12. Zhu, J., Peng, H., Rodriguez-Macias, F., Margrave, J.L., Khabashesku, V.N., Imam, A.M., Lozano, K., Barrera, E.V.: Reinforcing epoxy polymer composites through covalent integration of functionalized nanotubes. Adv. Funct. Mater. **14**(7), 643–648 (2004)
13. Feng, Q.P., Shen, X.J., Yang, J.P., Fu, S.Y., Mai, Y.W., Friedrich, K.: Synthesis of epoxy composites with high carbon nanotube loading and effects of tubular and wavy morphology on composite strength and modulus. Polymer **52**(26), 6037–6045 (2011)
14. Hsieh, T.H., Kinloch, A.J., Taylor, A.C., Kinloch, I.A.: The effect of carbon nanotubes on the fracture toughness and fatigue performance of a thermosetting epoxy polymer. J. Mater. Sci. **46**(23), 7525 (2011)

15. Yilmaz, M.G., Unal, H., Mimaroglu, A.: Study of the strength and erosive behavior of $CaCO_3$/glass fiber reinforced polyester composite. Express Polym. Lett. **2**, 890–895 (2008)
16. Sai Sravani, K., Ram Gopal Reddy B., Mohammed, R.: Effect of $CaCO_3$ and Al_2O_3 fillers on mechanical properties of glass/epoxy composites 3(6), 207–214 (2017). ISSN: 2455–3778
17. Kaundal, R., Patnaik, A., Satapathy, A.: Effect of SiC particulate on short glass fiber reinforced polyester composite in erosive wear environment. Walailak J. Sci. Technol. (WJST) **9**, (1) (2012): 49–64.
18. Qiu, J., Wang, S.: Enhancing polymer performance through graphene sheets. J. Appl. Polym. Sci. **119**(6), 3670–3674 (2011)
19. Gong, L., Kinloch, I.A., Young, R.J., Riaz, I., Jalil, R., Novoselov, K.S.: Interfacial stress transfer in a graphene monolayer nano composite. Adv. Mater. **22**(24), 2694–2697 (2010)
20. Tang, L.C., Wan, Y.J., Yan, D., Pei, Y.B., Li, Y.B., Wu, L.B., Jiang, J.X., Lai, G.Q.: The effect of graphene dispersion on the mechanical properties of graphene/epoxy composites. Carbon **60**, 16–27 (2013)
21. Bortz, D.R., Heras, E.G., Ignacio, M.G.: Impressive fatigue life and fracture toughness improvements in graphene oxide/epoxy composites. Macromolecules **45**(1), 238–245 (2012)
22. Gungor, S., Bakis, C.E.: Anisotropic networking of carbon black in glass/epoxy composites using electric field. J. Compos. Mater. **49**(5), 535–544 (2015)
23. Gupta, A., Singh, H., Walia, R.S.: Effect of fillers on tensile strength of pultruded glass fiber reinforced polymer composite (2015)

A Review of Nanofiller Coating on FRP Composites

Vijay Singh Parihar, Ankit Baranwal, Vikas Gautam, Shikhar Bajpai, Anurag Gupta, and K. L. A. Khan

1 Introduction

In the present times, the composite material incorporated with nanofillers like nanotubes, nanoparticles, nano-clays and incorporation of nano-reinforcement with elastomers, which improve the mechanical properties along with thermal properties and adhesion rheological behaviors [1, 2]. When we use these fillers with nanocomposites, higher results are produced than that of base matrix. McCrum gives a suggestion that it is a very important description for showing the effects of reinforcement, i.e., the ratio of surface area to volume reinforcement [3]. Recently in epoxy coating adhesive behavior of nanomaterials enhance the properties of composites. We know that nanoparticles cover more active surface composition. By the study, we can say that nanoparticles improves the strength, stiffness and toughness of resin composites [4]. Pavia and Cartin [5] suggested that "the development of nanofiber and nano-/micro-hybrid CMCs ought to be steered by an accurate understanding of the underlying toughening/strengthening mechanisms imparted by the nanofibers because these properties rely upon several system and material parameters." Composite materials are finding wide spread applications in commercial enterprises in light of their excellent properties like ease of creation, environment friendly, chemical and erosion resistance, high quality, solidness, light in weight and economically productive properties. Researchers always try to enhance the properties of materials. Epoxy composites are widely used in mechanical engineering, electronics/electrical engineering applications, aerospace and aviation fields and also in chemical industries [6, 7]. There are many reinforcements that can be added to polymers for the enhancement of properties like glass particles, ceramic particles, layered silicates, metal particles

V. S. Parihar (✉) · A. Baranwal · V. Gautam · S. Bajpai · A. Gupta
KIET Group of Institutions, Ghaziabad, India
e-mail: vijaysingh140298@gmail.com

K. L. A. Khan
Department of Mechanical Engineering, KIET Group of Institutions, Ghaziabad, India

M. Muzammil et al. (eds.), *Recent Advances in Mechanical Engineering*, Lecture Notes in Mechanical Engineering, https://doi.org/10.1007/978-981-15-8704-7_52

and thermoplastics. Sometimes ceramic nanoparticles like clay, graphene and carbon nanotube (CNT) are also added to toughen the epoxy resin by plastic void growth, crack pinning, etc., with a simultaneous increase in strength and modulus with no drastic effect on glass transition temperature. If fillers are added in larger weight percentage, then a noticeable improvement in toughness of epoxy matrix will also be seen [8]. When fiber-reinforced polymers (FRP) is used as a reinforcement in concrete, then it is considered as a good alternative over steel bars. The interface bond of steel bar reinforcement is different than the FRPs reinforcement bars. The interface bond of FRPs reinforcement with concrete is very complex [9].

2 Coating of Cellulose Nano-crystals on Short Glass Fibers

Automotive and marine organizations are using short glass fiber (SGF) polymer matrix composites (PMC) in various form because of their stiffness and high specific strength. Contribution of lightweight PMC in automobile sector is such that a little amount of weight reduction increases the fuel efficiency of that vehicle so these materials are used as a step toward reducing the weight of the vehicle [10]. As we increase the density of nanoparticles in the surface area, then as a result the mechanical property of that composite with other constituents is also increased [11]. We use cellulose nanomaterials (CNs) as an alternative nanoparticles that are used to increased the properties of glass fiber polymer matrix composites. CNs are cellulose-based nanoparticles which can be found from marine animals, plant and algae. We have discovered many cellulose nanomaterial (CNs) that can be divided on the basis of cellulose source and their extraction methods. There are various property of many cellulose nanomaterials (CNs) are similar to each other because alike arrangement of cellulose chains along the crystal length [12]. Various characteristics of CNs like elastic modulus, tensile strength, low density and high surface area are unique and make them attractive for other materials [13].

2.1 Enhanced Properties

2.1.1 Fracture Surface Morphology

Field emission scanning electron microscopy (FE-SEM) technique was used to analyze the fracture surface of the composites which are failed in tensile testing. Interfacial debonding mechanism is the most failure mechanism for uncoated glass fiber epoxy composites. Fiber breakage, matrix cracking, interfacial debonding and concurrent matrix cracking were the main failure mechanism for CNC-coated epoxy composites [14]. A rough fracture surface for coated glass fiber composites compared to same uncoated glass fiber composites with smooth fracture surface gives a better adhesion between fiber and matrix.

2.1.2 FElastic modulus of CNC-Coates

Elastic modulus of CNC-coated GF was enhanced by 10% for 1S-30GF/epoxy composites in comparison of uncoated GF/epoxy. According to Hashin "the presence of CNC increases the stiffness of the GF/epoxy interphase" [15]. Stress transfer with a faster rate across the matrix interface results the higher modulus for the imperfect interface composite [16]. Composite macroscopic modulus is increased by the increment in the apparent modulus of the GF/epoxy interphase. In our current study, we can say that higher value of macroscopic modulus of the composites is the result of enhancement in the apparent modulus of the GF/epoxy interface [17]. Increase in the tensile strength reflects the higher value of interfacial shear strength (IFSS) which results the better stress transfer across the fiber/epoxy interface [18]. It is observed that at the same CNC content enhancement in the flexural properties was larger than the enhancement in the tensile property. It is noted that in most of the epoxy systems, flexural strength is higher but flexural modulus is lower than tensile modulus [19].

2.1.3 Interfacial Properties

The load transfer across the glass fiber (GF)/cellulose nanocrystals (CNC) can be modified by changing the interfacial shear strength (IFSS) and fiber critical length. There is a reduction in interfacial shear strength (IFSS) due to which the concentration of cellulose nanocrystals (CNC) suspension increases [20]. Reduction in the value of IFSS indicates that stress transfer efficiency of CNC coating has been reduced.

2.1.4 Dynamic Thermo-mechanical Properties

Table 1 contains the mechanical and dynamic thermal properties of the composites above and below the Tg. storage modulus (E) of 1S-30GF/epoxy, and 1.5S-30 GF/epoxy composites can be enhanced by CNC coating on the GF roving at 25 °C. Due to the presence of CNC particles, the increment of storage modulus at 25 °C attributes to stabilize the glass fiber interphase [21]. At the temperature of 90 °C,

Table 1 Viscoelastic properties of CNC-GF/epoxy composites in three point bending mode [23]

Composite	E' at 25 °C (Mpa)	E_r at 90 °C (MPa)	T_g °C	Tanδ at T_g
30GF/epoxy	4932 ± 586	250 ± 21	50.3 ± 0.7	0.61 ± 0.06
1S-30GF/epoxy	5213 ± 543	224 ± 31	49.4 ± 0.5	0.61 ± 0.02
1.5-30GF/epoxy	5614 ± 695	228 ± 22	50.2 ± 0.8	0.65 ± 0.07
2–30GF/epoxy	4634 ± 257	243 ± 38	49.5 ± 1.1	0.59 ± 0.02

E Storage modulus, E_T Rubbery modulus, T_g Glass transition temperature measured in tanδ peak, Tanδ Value of tanδ peak. *Note* Error bars are 1 standard deviation

CNC-coated GF has lower value of rubbery moduli (Er) than that of composite having uncoated GF. Rubbery modulus is not affected by coating of GF with CNC [22].

3 Sand Coating on GFRP Bar

This coating contains GFRP bars which are made of orthophthalic polyester resin reinforced with glass fiber. In this coating, used rods were nominal diameters of 9 and 16 mm. According to ASTM 391, Youngs modulus and tensile strength of 9 mm bars are 44.5 GPa and 770 MPa and 41.4 GPa and 680 MPa for 16 mm bar, respectively [24]. Two different granulometries were used for the coating of smooth rods. ASTM C 778-02 was used for determining the sand granulometry. To check bond ability of concrete with FRP bars, we use two concrete named as conventional concrete (CC) and high resistance concrete (HRC).

3.1 Effect of the Concrete Strength

Conventional concrete (CC) presents an initial linear increase in the bond stress for smooth bar upto 1 MPa. Friction and a weak chemical bond provide load transfer. Pressure load transfer also depends on transverse pressure. Friction resists the slip by which bond stress increases slowly [25]. Sanded bars included in HRC show maximum bond strength than that in CC.

3.2 Effect of Length Embedded

Surface of bar shows an important character because bond strength depends on length of bar. Smooth bar shows same bond strength for different lengths [26].

3.3 Influence of bar Diameter

The effect of bar dia on bond strength is evaluated by the average bond stress relation obtained from the bars of 9 and 16 mm including both conventional concrete and high resistance concrete. Stress distribution depends on the pull-out test, and it is nonlinear along the bar [27]. Low bond strength is found in the larger bar diameters.

3.4 *Effect of Surface Modification*

Sand provides higher friction and interlock forces. Coating of sand on the surface of bar improves the mechanical property. Sand on the surface of bar also improves the bondability between concrete and bar surface [28]. Higher bond strength is found from coarse sand as it results more interlocking force and friction. In pull-out test, coating upon coarse sand is remained attached to the bar surface while in coating of fine sand upon the bar is removed completely from bar surface. This shows that bond between the concrete and bar is weaker in coarse sand than the fine sand [29].

4 Conclusion

The study of coating of cellulose nanocrystals on short glass fiber introduces coating of CNC on glass fiber (GF) that improves the IFSS, and it also improves mechanical property of or epoxy composites and short glass fiber. All the properties are enhanced without increasing the weight. Due to CNC coating, the stress transfer ability and interfacial adhesion of composites are enhanced. Tensile and flexural properties of GF/epoxy composites are increased by CNC coating. All the results shows that mechanical properties of GF/epoxy composites can be enhanced by coating of CNC on GF.

The study of sand coating on GFRP bar is done to find out the possibility of using GFRP bar with concrete structure. From this coating, we found that bond strength for short length bar doubled that of long length bar. Bond strength is not clear for the thicker diameters of bar. When bar is sand coated, strength of matrix results to a higher bond strength and it is independent of bar diameter. Higher bond strength between the concrete and the bar achieved due to interlocking force and friction when coating of coarse sand is done on bar.

Acknowledgements Authors are grateful to have been supported by the Royal Academy of Engineering under the auspices of "UK-India Industry-Academia Partnership Programme" between Crankfield University, UK, and KIET, India (Grant No. IAPP18-19295).

References

1. Pinnavaia, T.J., Beall, G.W. (eds.) Polymer-Clay Nanocomposites. Wiley (2000)
2. Anandhan, S., Bandyopadhyay, S.: Nanocomposites and Polymers with Analytical Methods, pp. 3–29. Rijeka, Croatia, Intech (2011)
3. McCrum, N.G., Buckley, C.P., Bucknall, C.B., Bucknall, C.B.: Principles of Polymer Engineering. Oxford University Press, USA (1997)
4. Giese, R.F., Van Oss, C.J.: Colloid and Surface Properties of Clays and Related Minerals, vol. 105. CRC Press (2002)

5. Kumar, S., Sundara Raju, Mohana, N., Sampath, P.S., Jayakumari, L.S.: Effects of nanomaterials on polymer composites—an expatiate view. Rev Adv Mater Sci 38(1) (2014)
6. Van Den Einde, L., Zhao, L., Seible, F.: Use of FRP composites in civil structural applications. Constr. Build Mater. **17**(6–7), 389–403 (2003)
7. Robert, M., Benmokrane, B.: Effect of aging on bond of GFRP bars embedded in concrete. Cement Concr. Compos. **32**(6), 461–467 (2010)
8. Davalos, J.F., Chen, Yi., Ray, I.: Effect of FRP bar degradation on interface bond with high strength concrete. Cement Concr. Compos. **30**(8), 722–730 (2008)
9. Sunter, D.A., Morrow III, W.R., Cresko, J.W., Liddell, H.P.H.: The manufacturing energy intensity of carbon fiber reinforced polymer composites and its effect on life cycle energy use for vehicle door lightweighting. In: Proceedings of the 20th International Conference on Composite Materials (ICCM), Copenhagen, Denmark (2015)
10. Dorigato, A., Morandi, S., Pegoretti, A.: Effect of nanoclay addition on the fiber/matrix adhesion in epoxy/glass composites. J. Compos. Mater. **46**(12), 1439–1451 (2012)
11. Luo, J.J., Daniel, I.M.: Characterization and modeling of mechanical behavior of polymer/clay nanocomposites. Compos. Sci. Technol. **63**(11), 1607–1616 (2003)
12. Moon, R.J., Martini, A., Nairn, J., Simonsen, J., Youngblood, J.: Cellulose nanomaterials review: structure, properties and nanocomposites. Chem. Soc. Rev. **40**(7), 3941–3994 (2011)
13. Roman, M.: Toxicity of cellulose nanocrystals: a review. Indus. Biotechnol. **11**(1), 25–33 (2015)
14. Hansen, F., Brun, V., Keller, E., Wegner, T., Meador, M., Friedersdorf, L.: Cellulose nanomaterials—a path towards commercialization workshop report (2014): 1–44
15. Hashin, Z.: Thermoelastic properties of fiber composites with imperfect interface. Mech. Mater. **8**(4), 333–348 (1990)
16. Nairn, J.A.: Generalized shear-lag analysis including imperfect interfaces. Adv. Composit. Lett. **13**(6), (2004): 096369350401300601.
17. Gao, S.L., Mäder, E.: Characterisation of interphase nanoscale property variations in glass fibre reinforced polypropylene and epoxy resin composites. Compos. A Appl. Sci. Manuf. **33**(4), 559 (2002)
18. Madhukar, M.S., Drzal, L.T.: Fiber-matrix adhesion and its effect on composite mechanical properties: II. longitudinal (0) and transverse (90) tensile and flexure behavior of graphite/epoxy composites. J. Compos. Mater. **25**(8), 958–991 (1991)
19. Kinsella, M., Murray, D., Crane, D., Mancinelli, J., Kranjc, M.: Mechanical properties of polymeric composites reinforced with high strength glass fibers. Int. SAMPE Tech. Conf. **33**, 1644–1657 (2001)
20. Tang, L., Weder, C.: Cellulose whisker/epoxy resin nanocomposites. ACS Appl. Mater. Interf. 2(4), 1073–1080 (2010)
21. Girouard, N., Schueneman, G.T., Shofner, M.L., Carson Meredith, J.: Exploiting colloidal interfaces to increase dispersion, performance, and pot-life in cellulose nanocrystal/waterborne epoxy composites. Polymer **68**, 111–121 (2015)
22. Ansari, F., Galland, S., Johansson, M., Plummer, C.J.G., Berglund, L.A.: Cellulose nanofiber network for moisture stable, strong and ductile biocomposites and increased epoxy curing rate. Compos. Part A: Appl. Sci. Manuf. **63**, 35–44 (2014)
23. Ahmad, F.S., Foret, G., Le Roy, R.: Bond between carbon fibre-reinforced polymer (CFRP) bars and ultra high performance fibre reinforced concrete (UHPFRC): Experimental study. Constr. Build. Mater. **25**(2), 479–485 (2011)
24. Achillides, Z. Bond Behaviour of FRP Bars in Concrete. PhD diss., University of Sheffield (1998)
25. Tighiouart, B., Benmokrane, B., Gao, D.: Investigation of bond in concrete member with fibre reinforced polymer (FRP) bars. Constr. Build. Mater. **12**(8), 453–462 (1998)
26. Cosenza, E., Manfredi, G., Realfonzo, R.: Behavior and modeling of bond of FRP rebars to concrete. J. Compos. Constr. **1**(2), 40–51 (1997)
27. Achillides, Z., Pilakoutas, K.: Bond behavior of fiber reinforced polymer bars under direct pullout conditions. J. Compos. Constr. **8**(2), 173–181 (2004)

28. Chang, J.J., Yeih, W., Tsai, C.L.: Enhancement of bond strength for epoxy-coated rebar using river sand. Constr. Build. Mater. **16**(8), 465–472 (2002)
29. Tasong, W.A., Cripps, J.C., Lynsdale, C.J.: Aggregate-cement chemical interactions. Cem. Concr. Res. **28**(7), 1037–1048 (1998)

Active Suspension System Modeling for a Passenger Car Subjected to Random Road Profile Inputs

Mohd Avesh and Rajeev Srivastava

1 Introduction

Suspension system is an important part of the road vehicle that had been extensively using since the beginning of 1800 [1]. The consistent technical developments resulted in advance suspension system with complex control algorithms [2, 3]. Its prime task is to isolate the vehicle body from road disturbances, jolts and the engine produced vibrations to help to provide the safe and a comfortable ride with better driving control [4, 5]. Road roughness, bumps and breakers create the disturbance to the vehicle that has to be overcome at maximum as possible [6]. Ride comfort and road holding are two main factors that have to be improved for better safety and durability of the vehicle [7]. Thus, suspension system design is an art of compromise between these two goals that fairly reduces the disturbance to the outputs [3, 8].

The proposed active suspension system is expected to exhibit better characteristics than the conventional one. A hydraulic actuator is placed vertically between sprung and unsprung mass, which power is governed by some disturbance feedback-based control laws to generate the required force in the system. The actuator is a force-generating element of active suspension that compensates the road transmitted disturbances by developing active vibration control phenomena; an obligatory force (u_f) is developed by the pressure difference across the hydraulic actuator. Further, a traditional PID controller has to model to implement in the suspension system to control the actuator's force. A two degree of freedom quarter-car suspension system has been constructed mathematically using nonlinear approach to apply the phenomena. A simplest quarter-car theoretical model is shown in Fig. 1, in which the hydraulic actuator with conventional spring and damper have been supported the car body. MATLAB SIMULINK platform would be utilized to simulate the model under the applied design parameters [9]. Ride comfort and stability criteria

M. Avesh (✉) · R. Srivastava
Motilal Nehru National Institute of Technology, Allahabad, Prayagraj, India
e-mail: mail2avesh@gmail.com

M. Muzammil et al. (eds.), *Recent Advances in Mechanical Engineering*, Lecture Notes in Mechanical Engineering, https://doi.org/10.1007/978-981-15-8704-7_53

Fig. 1 Simplest form of quarter-car model

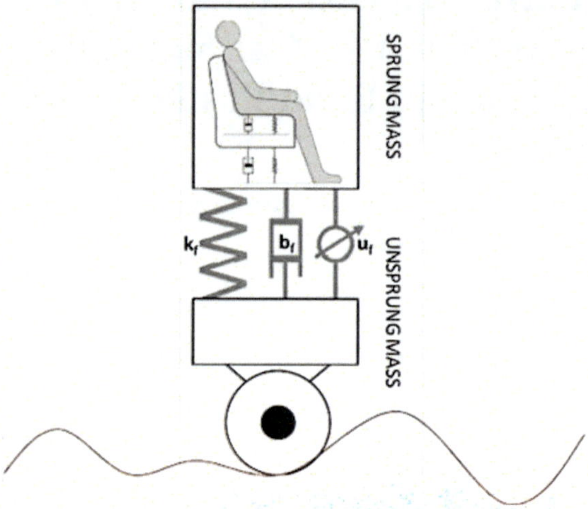

are applied to measure the performance improvements. The road disturbance model developed mathematically for the road surface of class D, for which the international standard roughness criteria has been followed [10]. The main inputs to the model are inertial force, road disturbance and spool displacement [11].

2 Model Construction

A commonly used two degree of freedom quarter-car model has been developed with system nonlinearities. The elemental, as well as entire system response, has been perceived through the experiments and numerical simulations. Passive suspension components are naturally modeled as numerical function of motion; for example, spring is the function of displacement and the damper is a function of velocity. The vertical displacement has been obtained from the difference between wheel travel and the sprung mass movement. The main disturbing inputs to the system are road excitations that typically develop the vertical dynamic forces and displacement in the vehicle body. The resulted model excited under a very poor road of class D. To get the time response under the applied inputs, the mathematical model transformed into a SIMULINK model and further simulated for the variable operating conditions. The blocks for various parameters are available in the SIMULINK library.

2.1 Quarter-Car Model

The different motions and degree of freedom of the discrete quarter-car suspension system signify through graphical representation in Fig. 2. Key parameters like sprung mass, unsprung mass, spring stiffness, damping coefficient are denoted by m_s, m_u, k_s and c_s, respectively. The tire behavior is also simplified as spring and damper. The nonlinear behavior of the model is governed by the equations of the motion. Equations 1–6 are derived by balancing the vertical forces in which Eqs. 1 and 2 describe the behavior of sprung and unsprung mass, respectively.

$$m_s\ddot{z}_s = F_{ss}(z_s, z_u, t) + F_{sd}(\dot{z}_s, \dot{z}_u, t) \tag{1}$$

$$m_u\ddot{z}_u = -F_{ss}(z_s, z_u, t) - F_{sd}(\dot{z}_s, \dot{z}_u, t) + F_{ts}(z_u, z_r, t) + F_{td}(\dot{z}_u, \dot{z}_r, t) \tag{2}$$

$$F_{ss}(z_s, z_u, t) = k_s(z_s - z_{us}) + \zeta k_s(z_s - z_u)^3 \tag{3}$$

$$F_{sd}(\dot{z}_s, \dot{z}_u, t) = c_s(\dot{z}_s - \dot{z}_u) + \zeta c_s(\dot{z}_s - \dot{z}_u)^2 sgn(\dot{z}_s - \dot{z}_u) \tag{4}$$

$$F_{ts}(z_u, z_t, t) = k_t(z_t - z_u) \tag{5}$$

$$F_{td}(\dot{z}_u, \dot{z}_t, t) = c_t(\dot{z}_t - \dot{z}_u) \tag{6}$$

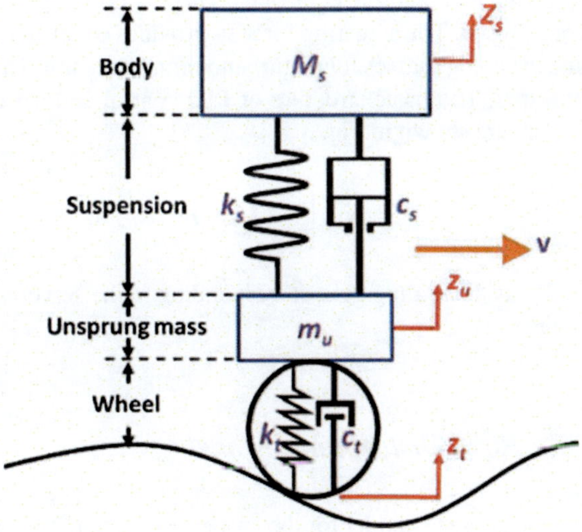

Fig. 2 Proposed quarter-car model of suspension system

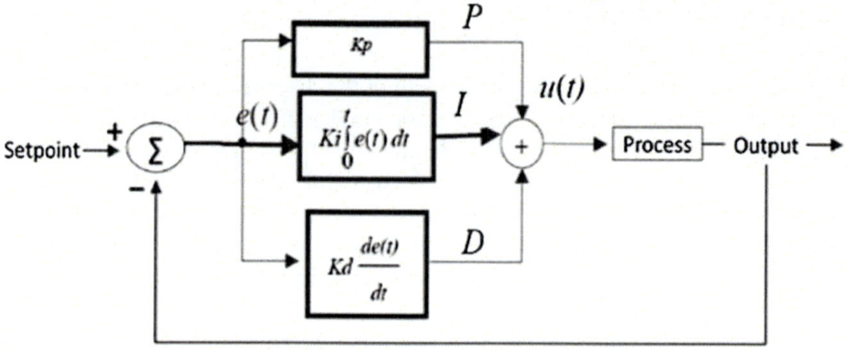

Fig. 3 PID controller structure

The vertical movement of the sprung mass, unsprung mass and tire is represented by z_s, z_u and z_t, respectively. The term ζ represents the nonlinearity of the spring and damper.

2.2 *PID Controller*

The proportional integral derivative (PID) control scheme is applied to the derived model of suspension system to regulate the actuation input–output in order to reduce the car body displacement. The road disturbance and the car cabin load are the feedback inputs to the controller. There is no separate reference input added in the control loop. The actuating force u_f provides equal and opposite reaction into sprung and unsprung masses for disturbance compensation. The control law is characterized by Eq. 7. The basic structure of PID control and the control loop of the proposed system are shown in Fig. 3.

$$r = K_{\mathrm{SV}}\left[-K_P x_c(t) - K_i \int x_c(t)\mathrm{d}t - K_D \frac{\mathrm{d}x_c(t)}{\mathrm{d}t}\right] \quad (7)$$

K_{sv} is the scaling gain corresponding to the servo valve that is tuned by trial and error.

2.3 *Road Disturbance Model*

The disturbance model for the suspension system has been created for Class D road profile. Car is considered to be run at a constant velocity (v) of 60 km/h. The vehicle velocity is a function of road profile excitations [12]. The resulted road surface inputs are governed by Eq. 8.

$$\dot{z}_t = -2\pi f_0 z_r + 2\pi\sqrt{\Phi_0 v w_0} \tag{8}$$

For class D surface, $\Phi(n_0) = 1024 * 10^{-6}$ m²/(m/cycle)

3 Simulation Outcomes

The time–domain analysis at MATLAB SIMULINK platform is carried out to make the performance comparison of active and passive suspension systems. Several parameters have been adopted to check the consistency of the results in different working conditions. The random road inputs at roughness quality under class D are applied that yields the dense responses of the measuring parameters. Thus, the root mean square (RMS) values of the parameters have to be computed for clear analysis of the responses under the applied parameters. The evaluation of system performance has been carried out using comfort and stability performance indices [11].

From Fig. 4, the suspension defection (SD) peak values of ACTIVE-PID and PASSIVE system are noted as 0.04512 m and 0.1267 m, respectively. The corresponding wheel deflection values for ACTIVE-PID and PASSIVE design are

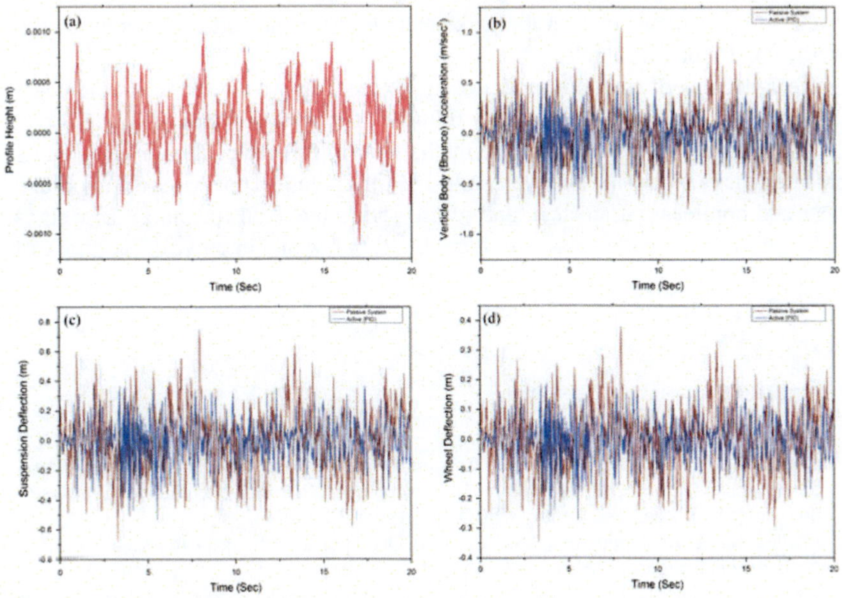

Fig. 4 Time response, **(a)** road input, **(b)** vertical body acceleration, **(c)** suspension deflection, **(d)** wheel deflection

0.1316 m and 0.1522 m. It shows that ACTIVE-PID design demonstrates the superior attitude toward vehicle road holding over the PASSIVE design. Similarly, for comfort analysis, vertical body acceleration (VBA) time responses are compared and the lower value of VBA is discovered for ACTIVE-PID design, i.e., 1.01 m/s^2. The pitch acceleration could not be discovered as the analysis carried out at quarter-car model. Apparently, the ACTIVE-PID design has better control in suppressing the acceleration of the vehicle body and in reducing the tire deflection.

4 Conclusions

The ride characteristics of a light passenger vehicle have been examined through a simulation study, using a quarter-car model of suspension system subjected to random road conditions. The wheel response, suspension displacement and body acceleration are the main factors of consideration that help to measure the performance in terms of stability and comfort. The vehicle response has been plotted in the time-domain and examined for characteristics comparison of proposed active suspension over the passive suspension characteristics. A notable improvement in ride comfort and vehicle stability has been observed as it reduces the maximum body acceleration and travel by 74.3% and 62.8% of that of the passive suspension system, respectively.

The percentage reduction in passenger cabin acceleration and suspension travel justifies the superiority of active suspension system over the passive suspension system; improvements in both the parameters indicate the better ride comfort and driving control of the vehicle.

The 2 degrees of freedom model used in the current study does not include a few important aspects that potentially influence the system response. Therefore, it is recommended to construct and use the half-car or full car model for vehicle ride characteristics analysis in future research. The full car model possesses the important linear and nonlinear parameters, and it also shows the axial (pitching), as well as rotational (rolling) characteristics of the vehicle, that would improve the accuracy of the analysis.

References

1. Avesh, M., Srivastava, R.: Parametric study on the performance of active suspension system for variable passenger size and repeated road bumps. In: 2016 10th International Conference on Intelligent Systems and Control (ISCO). pp. 1–6. IEEE (2016)
2. Fijalkowski, B.T.: Automotive Mechatronics: Operational and Practical Issues. Springer, Netherlands, Dordrecht (2011)
3. Mohd, A., Srivastava, R.: Parametric optimization and experimental validation for nonlinear characteristics of passenger car suspension system. Period. Polytech. Transp. Eng. (2019). https://doi.org/10.3311/PPtr.12999

4. Shehata, A., Metered, H., Oraby, W.A.H.: Vibration control of active vehicle suspension system using fuzzy logic controller. In: Vibration Engineering and Technology of Machinery, Mechanisms and Machine Science, pp. 389–399 (2015)
5. Aldair, A., Wang, W.: The energy regeneration of electromagnetic energy saving active suspension in full vehicle with Neurofuzzy Controller. Int. J. Artif. Intell. Appl. **2**, 32–43 (2011). https://doi.org/10.5121/ijaia.2011.2203
6. Sharma, S.K., Kumar, A.: Impact of electric locomotive traction of the passenger vehicle Ride quality in longitudinal train dynamics in the context of Indian railways. Mech. Ind. **18**, 222 (2017). https://doi.org/10.1051/meca/2016047
7. Nagarkar, M.P., Vikhe Patil, G.J., Zaware Patil, R.N.: Optimization of nonlinear quarter car suspension–seat–driver model. J. Adv. Res. **7**, 991–1007 (2016). https://doi.org/10.1016/j.jare.2016.04.003
8. Avesh, M., Srivastava, R.: Parametric optimization to design a passenger car suspension system for better dynamic performance. Eur. Transp. Trasp. Eur. 1–14 (2019)
9. Avesh, M., Srivastava, R.: Modeling simulation and control of active suspension system in Matlab Simulink environment. In: 2012 Students Conference on Engineering and Systems, pp. 1–6. IEEE, Allahabad (2012)
10. Mechanical Vibration-Road Surface Profiles-Reporting of Measured Data. Int. Organ. Stand. ISO 8608 (1995)
11. Avesh, M., Srivastava, R.: Passenger car active suspension system model for better dynamic characteristics. Natl. Acad. Sci. Lett. 1–5 (2019). https://doi.org/10.1007/s40009-019-00807-z
12. Pedro, J.O., Dangor, M., Dahunsi, O.A., Ali, M.M.: Differential evolution-based PID control of nonlinear full-car electrohydraulic suspensions. Math. Probl. Eng. 2013 (2013). doi: http://dx.doi.org/10.1155/2013/261582

Fabrication of the Composites (AA6082-T6/SiC) by Using Friction Stir Processing

Amit Kumar Gupta and Madan Mohan Puram

1 Introduction

There is a great demand for reduction of weight of material used in manufacturing in aerospace and automobile industries. It shows that industries want to increase strength-to-weight ratio. A number of research scholars and professor are working on this to find out the best material. Metals like aluminum, copper, magnesium, etc., are found out, but these having very low strength. These metals directly cannot be used, and processing of these metals should be done; after processing, these metals convert into metal matrix composite [1, 2]. In stir casting process, we reinforce the particles after melting of metal; in this process, distribution of particles in metal matrix is very difficult. This reaction takes place between reinforcement and metal matrix. So, uniform distribution of particles is not taking place [3]. An FSP work has been done on various numbers of materials like AA6061, AA7075, AA5083, AA6082, etc. [4]. In this paper, AA6082-T651 material is used. SiC particles are used to reinforce in AA6082T651 to improve its properties [5].

Tool can be of HSS(M2), carbide and H13 material [6]. The first tool is fitted in milling machine in jaw, and workpiece is fixed in fixture. FSP can be done on whole surface, and it can be done on a groove made in the workpiece. In the groove, reinforcement particles are filled. By FSP, particles reinforced in the metal matrix. By FSP, properties of materials are improved [7].

After the literature survey, it was decided that AA6082-T651 material will be used in the research. SiC particles will be used to reinforce in the matrix. Taguchi L-27 orthogonal array will be used in the research. Output parameter will be microhardness. Minitab software will be used for ANOVA analysis (Table 1; Fig. 1).

A. K. Gupta (✉) · M. M. Puram
ITS Engineering College, Greater Noida, India
e-mail: amitgupta.me@its.edu.in

M. M. Puram
e-mail: madanmpuram.me@its.edu.in

M. Muzammil et al. (eds.), *Recent Advances in Mechanical Engineering*, Lecture Notes in Mechanical Engineering, https://doi.org/10.1007/978-981-15-8704-7_54

Table 1 Working levels of chosen parameters

Parameters	Levels		
	1	2	3
Tool rotation speed (RPM)	520	970	1420
Groove width (mm)	2.5	3	3.5
Tool tilt angle (degree)	1	2	3
Pin profile	Triangular	Circular	Pentagonal

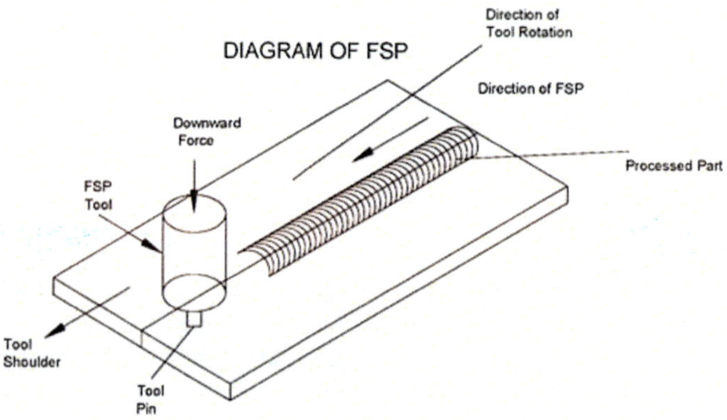

Fig. 1 Schematic diagram of friction stir processing setup

The workpiece diagram is shown below, which was used for making FSP samples (Figs. 2, 3, 4 and 5).

The following are graphs generated using ANOVA (Minitab software) (Figs. 6, 7 and 8).

Now, using the above analysis, optimum value of tool rotational speed, tool tilt angle, groove width and tool pin profile can be found out. By these values, a optimum composite can be made, and its tensile strength, impact strength and hardness and structure will be optimum.

Fig. 2 Workpiece without FSP

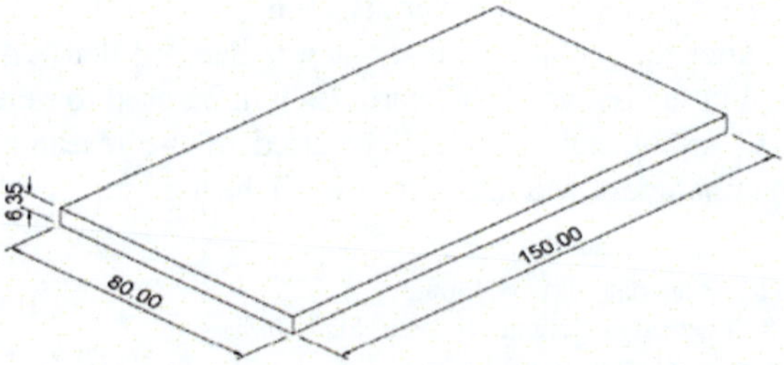

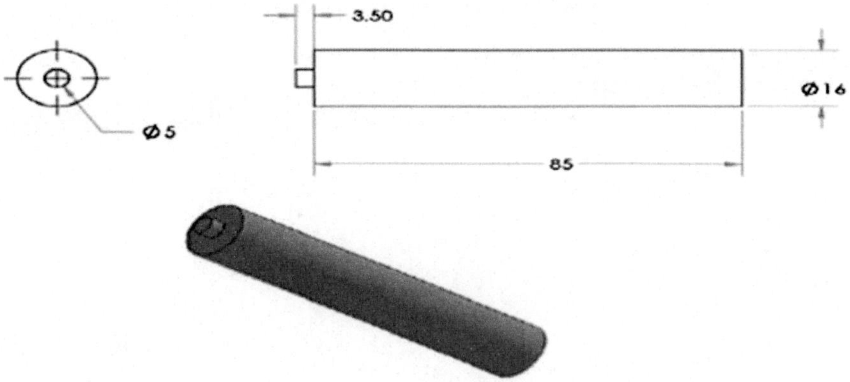

Fig. 3 Circular tool

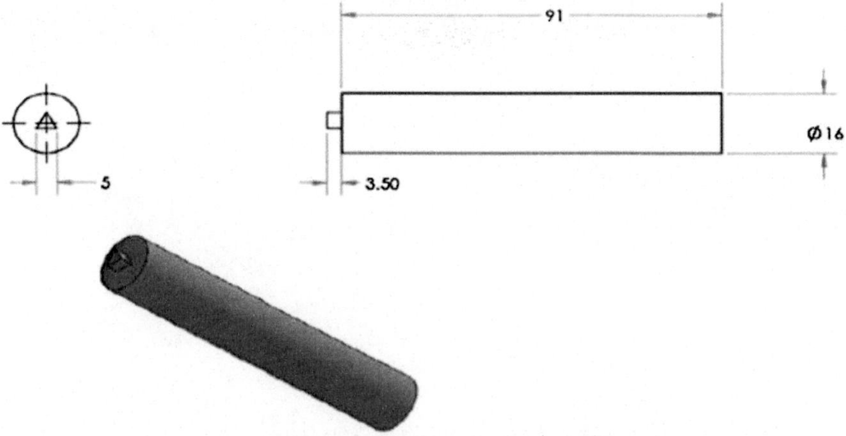

Fig. 4 Triangular tool

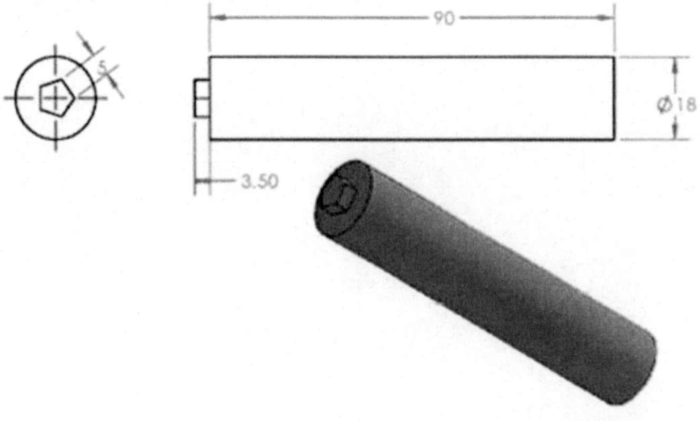

Fig. 5 Pentagonal tool

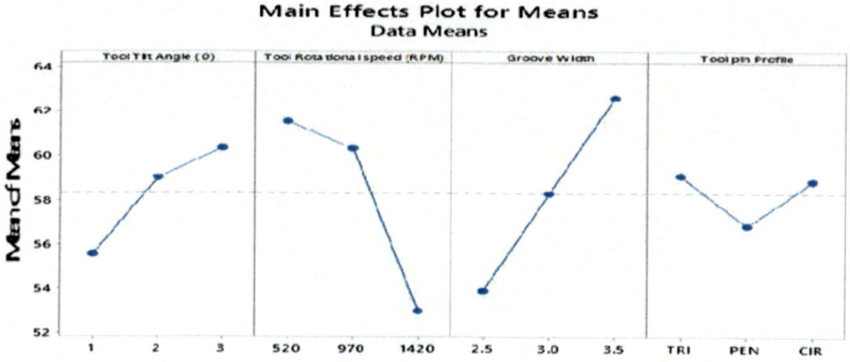

Fig. 6 .

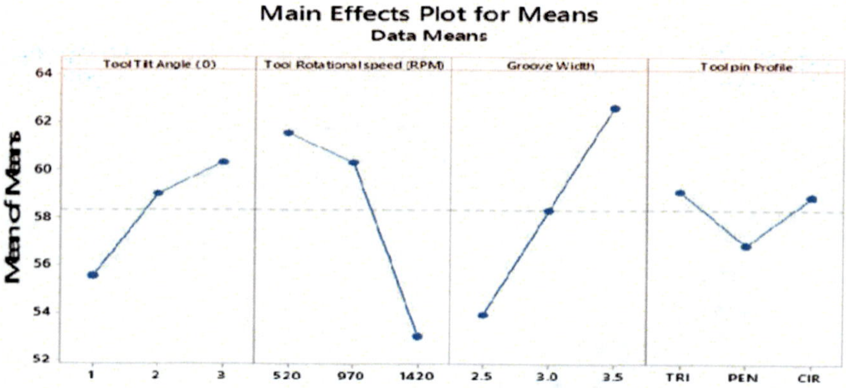

Fig. 7 .

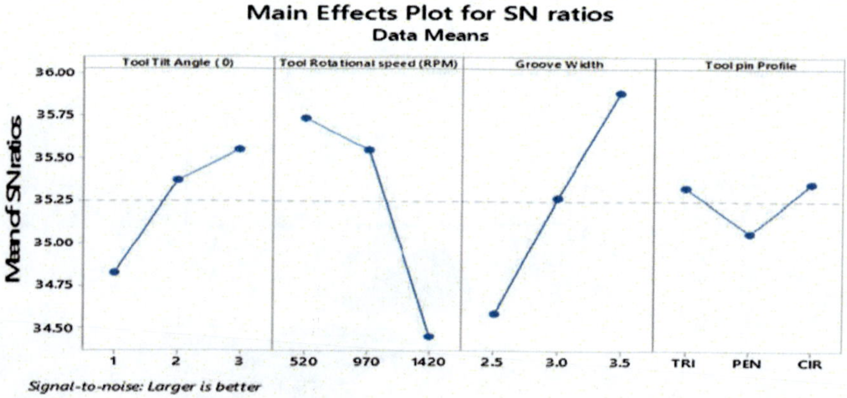

Fig. 8 .

2 Conclusion

Finally, it is concluded that AA6082-T651 material will be used in the research. SiC particles will be used to reinforce in the matrix. Taguchi L-27 orthogonal array will be very useful and recommended in the research. Four parameters, namely tool rotation speed, groove width, tool tilt angle and pin profile with three levels, are used. Outcome of the research is the latest parameters on the microhardness. Minitab software will be used for ANOVA analysis. For this, we have discovered the new thing which is the optimization of process parameters to enhance the microhardness of AA6082T651/SiC composites fabricated via friction stir processing [8, 18].

References

1. Hosseini Zeidabadi, S.R., Daneshmanesh. H.: Fabrication and characterization of in-situ Al/Nb metal/intermetallic surface composite by friction stir processing. **PII:S0921–5093**(17), 30302–30307. https://doi.org/10.1016/j.msea.2017.03.014.
2. Pilchak, A.L., Juhas, M.C., Williams, J.C.: Microstructural changes due to friction stir processing of investment-cast Ti-6Al-4V. Miner. Metal Mater. Soc. ASM Int (2007). doi: https://doi.org/10.1007/s11661-0069061-x
3. Khosravi, J., Beshrati Givi, M.K., Barmouz, M., Rahi, A.:. Microstructural, mechanical, and thermophysical characterization of Cu/WC composite layers fabricated via friction stir processing. Int. J. Adv. Manuf. Technol. doi: https://doi.org/10.1007/s00170-014-6050-x
4. Thangarasu, A., Murugan, N., Dinaharan, I., Vijay, S.J.: Microstructure and microhardness of AA1050/TiC surface composite fabricated using friction stir processing. Sadhana **37**(5), 579–586 (2012)
5. Khodabakhshi, F., Simchi, A., Kokabi, A.H., Nosko, M., Simancik, F., Svec, P.: Microstructure and texture development during friction stir processing of Al–Mg alloysheets with TiO_2 nanoparticles. Mater. Sci. Eng. A **605**, 108–118 (2014)
6. Li, J., Zhang, D.-T., Chai, F., Zhang, W.: Microstructures and mechanical properties of WE43 magnesium alloy prepared by friction stir processing. Rare Met. Rare Metals doi: https://doi.org/10.1007/s12598-014-0306-3
7. Gan, W.Y., Zhou, Z., Zhang, H., Peng, T.: Evolution of microstructure and hardness of aluminium after friction stir processing. Trans. Nonferrous Met. Soc. China **24**, 975–981 (2014)
8. Rathee, S., Maheshwari, S., Siddiquee, A.N., Srivastava, M., Sharma, S.K.: Process parameters optimization for enhanced microhardness of AA 6061/SiC surface composites fabricated via Friction Stir Processing (FSP). Mater. Today: Proc. **3**, 4151–4156 (2016)
9. Kumar, S., Kumar, S., Kumar, A.: Optimization of process parameters for friction stir welding of joining A6061 and A6082 alloys by Taguchi Method. In: Proceedings of the Institution of Mechanical Engineers, Part C: Journal of Mechanical Engineering Science, vol. 227, (2013), 1150 originally published online 13 September 2012. doi: https://doi.org/10.1177/0954406212459448
10. Dinaharan, I., Murugan, N., Thangarasu, A.: Development of empirical relationships for prediction of mechanical and wear properties of AA6082 aluminum matrix composites produced using friction stir processing. Eng. Sci. Technol. Int. J. **19**, 1132–1144 (2016)
11. Puviyarasan, M., Senthil Kumar, V.S.: Optimization of friction stir process parameters in fabricating AA6061/SiC_p Composites. Procedia Eng. **38**, 1094–1103 (2012)
12. Dakarapu, S.R., Nallu, R.: Process Parameters optimization for producing AA6061/TiB2 composites by friction stir processing. J. Mech. Eng. **67**(1), 101–118 (2017)

13. Singh, D.P., Singh, V., Kumar, S.:. Optimization of process parameters for friction Stirwelded AA7075–SIC composite joints by Taguchi method. YMCAUST Int. J. Res. **4**(I) (2016). ISSN: 2319-9377
14. Raweni, A., Majstorovic, V., Sedmak, A., Tadic, S., Kirin, S.: Optimization of AA5083 Friction Stir Welding Parameters Using Taguchi Method. ISSN 1330-3651 (Print), ISSN 1848-6339 (Online)
15. Anil Kumar, K.S., Karur, A.S., Chipli, S., Singh, A.: Optimization of FSW parameters to improve the mechanical properties of AA2024-T351 similar joints using Taguchi method. J. Mech. Eng. Autom. **5**(3B), 27–32 (2015). doi: https://doi.org/10.5923/c.jmea.201502.06
16. Ranjit Bauri, Devinder Yadav, C.N. Shyam Kumar, G.D. Janaki Ram. Optimized process parameters for fabricating metal particles reinforced 5083 Al composite by friction stir processing. Data Brief **5**, 309–313 (2015)
17. Joyson Abraham, S., Chandra Rao Madane, S., Dinaharan, I., John Baruch, L.: Development of quartz particulate reinforced AA6063 aluminummatrix composites via friction stir processing. J. Asian Ceram. Soc. (2016)
18. Baurio, R., Janaki Ram, G.D., Yadav, D., Shyam Kumar, C.N.: Effect of process parameters and tool geometry on fabrication of Ni particles reinforced 5083 Al composite by friction stir processing. Mater. Today: Proc. **2**, 3203–3211 (2015)

Assessment of Barriers of Green Supply Chain Management Using Structural Equation Modeling

Somesh Agarwal, Mohit Tyagi, and R. K. Garg

1 Introduction

In today's era, the major global problem arising is environmental degradation. Due to the upsurge in the raw material and energy consumption, natural resources are depleting at a very rapid rate. Moreover, pollution and waste production have also been increased significantly. Therefore, concurrently, there are twin challenges for the industries viz. competitive and environmental demands. The concept of GSCM is thus generated due to the alarming environmental issues and emerging economic opportunities. The assimilation of environmental concerns and the values of SC give rise to GSCM [1].

In the last two decades, most of the multinational industry and SME are turning toward GSCM for the sake of alarming environmental conditions. The demand for GSCM has risen nowadays as public awareness, environmental, economic or governmental reasons. But it is not an easy assignment. To make a model of various interrelated barriers to implement GSCM in Indian Rubber Industries and then validate that model using SEM is the main objective of this research.

1.1 Motivation

GSCM is a very good practice to be done for dealing with and minimizing the environmental issues that we are facing today. The boundary of GSCM is dependent on the researcher's goals and the problems which are at tips similar to the concept of SCM. The very initial green SC came into the framework by Kelle and Silver's

S. Agarwal · M. Tyagi (✉) · R. K. Garg
Department of Industrial and Production Engineering, Dr. B. R. Ambedkar National Institute of Technology, Jalandhar, Punjab, India
e-mail: mohitmied@gmail.com

M. Muzammil et al. (eds.), *Recent Advances in Mechanical Engineering*, Lecture Notes in Mechanical Engineering, https://doi.org/10.1007/978-981-15-8704-7_55

[2]. Apart from the operations perspective, considerable work has been done to facilitate the GSCM implementation in the industries at various stages by analyzing the barriers for the implementation of GSCM. GSCM activities are constrained on part of the suppliers by their unwillingness to support and lack of attentiveness. The outdated mindset of the supplier firm is the main key, which reflects suppliers' and the total network interests as being. Government guidelines can inspire or depress the acceptance of alterations.

Customers, investors and non-governmental organizations can be treated as external drivers. Rao [3] explored the idea of GSCM by empirical survey initiative in the Philippine context.

2 Review of the Literature

2.1 Supply Chain (SC)

SC is the progress of products as they move from their source to the end customer. Regularly, a SC is additionally depicted by the term worth chain, which mirrors the concept that worth is included along the chain.

2.2 Green Supply Chain Management (GSCM)

To minimize or reduce waste in the form of energy, emission, dangerous solid and waste chemicals is the main aim of GSCM. The addition of the 'green' component to SCM includes addressing the influence and connection between SCM and the natural environment [4, 5].

The main aim of the organizations is to adopt GSCM practices to enhance their environmental and financial performance [6, 7].

2.3 Structural Equation Modeling (SEM)

For testing and estimating the fundamental relationship, SEM approach is used by a combination of statistical data and qualitative triggered assumptions. Unlike other methods, number of variables and limitations are not a case for SEM; hence, it is considered as the best approach. Jiang et al. [8] presented a new Bayesian nonlinear SEM approach for the hierarchical valuation of dynamic systems, considering uncertainty in predicted and measured time series. The proposed model and assumptions were validated using SEM. Although the SEM approach has been used by researchers for numerous problems, the application of SEM to model a GSCM barrier system

Fig. 1 Generic SEM model

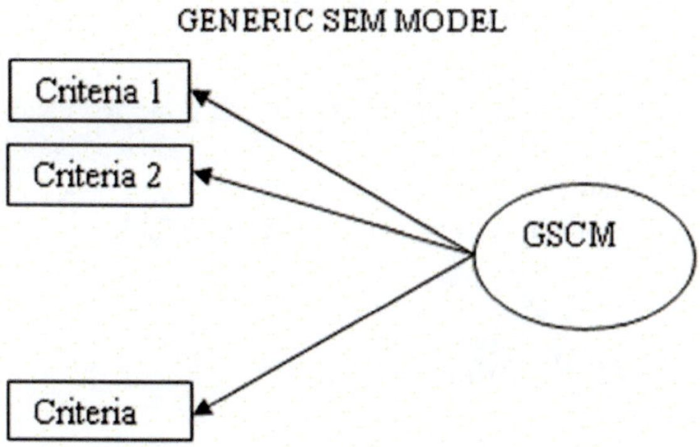

has been found to be very limited. The application of SEM to a GSCM barrier model has not been found in more practices. This research study was initiated in this context (Fig. 1).

3 Problem Description

Due to environmental issues, GSCM finds a great need in society. For that purpose, to study the hindrances or barriers that organizations are facing in the implementation of GSCM is the main objective of this research. For that, firstly, we should know about which barriers are influencing the SC. In view, a lot of pollution is caused by disposing of the waste rubber as they are burned after use which causes a lot of smoke and pollution. So, there is an utmost need to reduce the wastes of rubber. For that, recycling of rubber should be done. For that purpose, rubber industries are chosen for reference. In this research, it aims to find the barriers which cause hindrances in the rubber industry and then to validate these barriers using SEM.

3.1 Research Methodology

The objective of this research is to create a model by considering different barriers of GSCM and evaluate them using structural equation modeling (SEM). For this, barrier is identified which hinders rubber industries to adopt GSCM using survey and expert opinion. Then, hypothesis has been developed relating to these barriers. Data was collected using questionnaire-based survey and was analyzed using SEM, and then, validation of hypothesis has been done. The result thus obtained will show that the validated hypothesis will influence the system and vice versa.

3.2 Identification of GSCM Barriers

GSCM is a very good practice to be done for dealing with and minimizing the environmental issues that we are facing today. Many researchers contribute to this topic, and they had found barriers related to different kinds of industries, namely plastic, manufacturing, automobile, etc. (Table 1).

3.3 Methodology

Firstly, the research hypothesis was prepared based on the barriers related to the implementation of GSCM in the rubber industry. Then, a questionnaire was prepared to keep in mind the exhaustive and extensive factors in order to get the awareness of the industry regarding GSCM. The data attained from the responses to the questionnaire was analyzed using SEM with the help of LISERAL software.

3.4 Research Hypothesis

According to the available literature, the hypothesis has been synthesized considering the barriers to the implementation of GSCM in mind. Each major barrier found by the literature gives augmentation to each hypothesis as follows:

H1: Lack of top management support has a positive relationship with GSCM and they hinder in GSCM practices.

H2: Inability or unwillingness to share information has a positive relationship with GSCM and they hinder in GSCM practices.

H3: Inability or unwillingness to share risk and rewards has a positive relationship with GSCM and they hinder in GSCM practices.

H4: Inflexible organizational system and processes has a positive relationship with GSCM and they hinder in GSCM practices.

H5: Communication gap between management and shop floor workers has a positive relationship with GSCM and they hinder in GSCM practices.

H6: Unawareness of customers has a positive relationship with GSCM and they hinder in GSCM practices.

Table 1 Identification of barriers

S. No.	Barrier name	Description	Author
1	Lack of top management support	The top management is the only key for modifying the existing system from all points of consideration, but it should not have any negative impact on the system the main objective	Luthra et al. [9], Tyagi et al. [10]
2	Inability or unwillingness to share information	When the information flow is exaggerated by negligence or unwillingness, a big disorder is created in the process flow and chances of further development will decreases	Chen et al. [11], Roarty [12], Jose [13]
3	An unwillingness to share risks and rewards	If organizations are failing to take risks, it implies failing to progress and sometimes even failing to grow. The rewards should also be shared, not enjoyed selfishly	Sharma [14]
4	Inflexible organizational systems and processes	The organization should be so designed that it can rearrange itself for any forthcoming changes	Jose [13], Klassen and Whybark [15]
5	Communication gap between management and shop floor workers	There must not be proper interaction and proper understanding between management and workers. Also, it is due to under-educated workers	Ravi and Shankar [16]
6	Unawareness of customers	A most important type of external pressure is customer demand as they have to be strictly followed by the company	Zhu et al. [17]
7	Lack of training of workshop floor workers	Continuous improvement is a must objective which can be achieved by training for survival and excel in the field	Yu Lin et al. [18]
8	Incapable of achieving the exact design standard	They are incapable of achieving the same design as that obtained by the simple process	Hosseini [19]

(continued)

Table 1 (continued)

S. No.	Barrier name	Description	Author
9	Lack of financial supports	High investment is needed by green methods such as green design, green manufacturing and green labeling of packing which in turn upsurges the product cost	Klassen and Whybark [15]
10	Cost implications	The high investment requirement by green methodologies such as green design, green manufacturing and green labeling of packing which in turn increases the product cost	Hosseini [19], Mudgal et al. [20]
11	High initial capital cost	For the first implementation of the setup, the high initial cost is required	Hosseini [19], Mudgal et al. [20]
12	High cost in managing hazardous wastes	For the disposal of different types of wastes, different methods are used, which in turn is an expenditure of a bundle of money	Mudgal et al. [20]
13	Administrative and financial burden of tax	The direct and indirect taxes should be properly planned and managed for better financial consideration within the reverse chain	Chen et al. [11], Jose [13]
14	Market competition and uncertainty	Global competitiveness and customer's requirements give rise to market uncertainties and competency	Zhu et al. [17], Mudgal et al. [21]
15	Lack of Government support systems	No compensation to industries from the government for GSCM implementation	Yu Lin et al. [18], Tyagi et al. [22]
16	Lack of demand for GSCM	The market demand for the product is less	Mudgal et al. [21]

H7: Lack of training of workshop floor workers has a positive relationship with GSCM and they hinder in GSCM practices.

H8: Incapable of achieving the exact design standard has a positive relationship with GSCM and they hinder in GSCM practices.

H9: Lack of financial supports has a positive relationship with GSCM and they hinder in GSCM practices.

H10: Cost implications have a positive relationship with GSCM and they hinder in GSCM practices.

H11: High initial capital cost has a positive relationship with GSCM and they hinder in GSCM practices.

H12: High cost in managing hazardous wastes has a positive relationship with GSCM and they hinder in GSCM practices.

H13: Administrative and financial burden of tax has a positive relationship with GSCM and they hinder in GSCM practices.

H14: Market competition and uncertainty have a positive relationship with GSCM and they hinder in GSCM practices.

H15: Market lack of government support systems has a positive relationship with GSCM and they hinder in GSCM practices.

H16: Lack of demand for GSCM has a positive relationship with GSCM and they hinder in GSCM practices.

Then, for importance and the scale of each measure, the factor loading of each measure on GSCM is verified, irrespective of the sign will give the influence of that barrier of GSCM. These values are used for arriving at the relative weight of attributes. To bring out the above process, the SEM model with LISREL symbolization can be established by assuming the connection between the observed variables and their underlying factors. Using LISREL notations, the hypothesized conceptual models are presented.

We can summarize its configuration by writing the series of Equations/statements.

The equations for model factor structure are as:

$$Y_1 = \lambda_1 h_1 + e_1,\ Y_2 = \lambda_2 h_1 + e_2,\ Y_3 = \lambda_3 h_1 + e_3,\ Y_4 = \lambda_4 h_1 + e_4 \ldots Y_n = \lambda_n h_n + e_n$$

And vector form of the above equations can be written as:

$$Y \quad \Lambda \quad \varepsilon$$

$$\begin{bmatrix} Y_1 \\ Y_2 \\ Y_3 \\ Y_4 \\ . \\ . \\ . \\ Y_n \end{bmatrix} = \begin{bmatrix} \lambda_1 \\ \lambda_2 \\ \lambda_3 \\ \lambda_4 \\ . \\ . \\ . \\ \lambda_n \end{bmatrix} \begin{bmatrix} \eta \end{bmatrix} + \begin{bmatrix} \varepsilon_1 \\ \varepsilon_2 \\ \varepsilon_3 \\ \varepsilon_4 \\ . \\ . \\ . \\ \varepsilon_n \end{bmatrix} \tag{1}$$

Also, summarization of the above lower-order structure can be:

$$Y = \Lambda_Y \eta + \varepsilon \tag{2}$$

where model factor loadings are Λ the and error terms measurement is ε; thus, relative influence or weight of attributes on barriers can be established by SEM.

The achieved results from LISREL for the model are depicted in Fig. 2

4 Results and Discussion

According to the results obtained from the model demonstrates optimistic values for the entire path coefficient; at $p < 0.05$, all the t-values of the variables are statistically important. Thus, according to Table 2, the structural model supports all the 16 hypotheses of the projected model. The influence of all the listed barriers for GSCM has been proved by hypotheses H1, H2, H3, H4, H5, H6, H7, H8, H9, H10, H11, H12, H13, H14, H15 and H16. So, the projected model explicated a substantial percentage of variance in the barriers of GSCM. Thus, the SEM model confirms the consistency of the proposed model and achieves an acceptable level.

5 Conclusion

In the research, SEM approaches were proposed to implement green SC in the Indian rubber industry. This research is focused on the empirical validation of a real working environment using the SEM approach. SEM examines a set of associations between one or more independent variables (IV) and one or more dependent variables (DV). Here, a model is presented for which all the barriers are validated using SEM. This model validated the hypothesis made for the barriers, and it gives the right path for

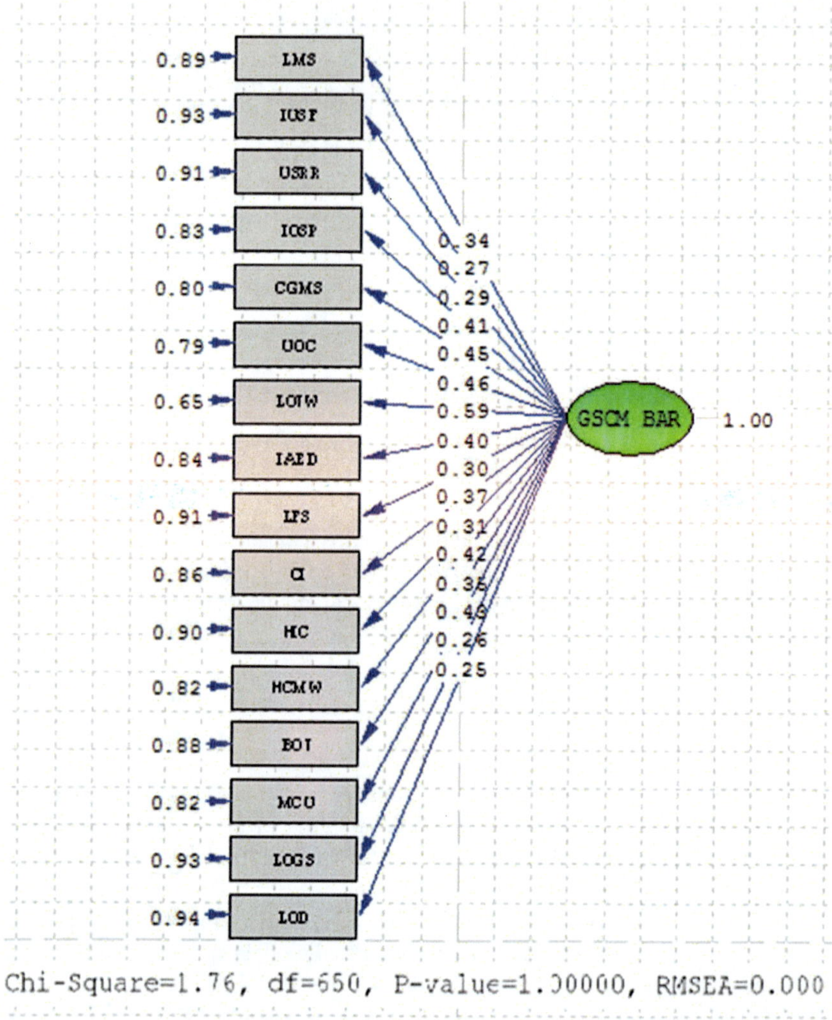

Fig. 2 Higher-order SEM model

the industries to work on. The Cronbach's alpha value demonstrates the reliability standard, which is greater than 0.96 for all criteria.

6 Managerial Implications

Researchers are confronted to provide a guide for selecting the technique that is both theoretically well originated and almost operational to crack the real problems.

Table 2 Results of hypothesis table

Hypothesis	Casual-path	Point estimates	t-values	Hypothesis support
H1	LMS	0.48	8.46	Yes
H2	IUCF	0.38	9.27	Yes
H3	USRR	0.42	8.88	Yes
H4	IOSP	0.58	12.42	Yes
H5	CGMS	0.63	10.9	Yes
H6	UOC	0.65	9.98	Yes
H7	LOTW	0.83	8.97	Yes
H8	IAED	0.57	9.8	Yes
H9	LFS	0.43	12.23	Yes
H10	CI	0.52	11.11	Yes
H11	HIC	0.44	10.23	Yes
H12	HCMW	0.53	9.89	Yes
H13	BOT	0.5	7.87	Yes
H14	MCU	0.6	8.86	Yes
H15	LOGS	0.36	8.85	Yes
H16	LOD	0.36	9.89	Yes

Result obtained shows that the barriers are hindering the supply chain performance for rubber industries. Suggested framework provides the barriers which are influencing the GSCM practices. This research provides an in-depth study of barriers and their implication on supply chain performance. Managers and decision makers are required to implement GSCM keeping the above-mentioned barriers in mind.

References

1. Hall, J.: Environmental supply chain dynamics. J. Cleaner Prod. **8**(1), 455–471 (2000)
2. Kelle, P., Silver, E.A.: Forecasting the returns of reusable containers. J. Oper. Manage. **8**(1), 17–35 (1989)
3. Rao, P.: Greening of the supply chain: an empirical study for SME's in the Philippine context. J. Asia Bus. Stud. **1**(2), 55–66 (2007)
4. Tyagi, M., Kumar, P., Kumar, D.: Modeling and analysis of barriers for supply chain performance measurement system. Int. J. Oper. Res. **28**(3), 392–414 (2017)
5. Srivastava, S.K.: Green supply-chain management: a state-of-the-art literature review. Int. J. Manage. Rev. **9**(1), 53–80 (2007)
6. Kumar, D., Jain, S., Tyagi, M., Kumar, P: Quantitative assessment of mutual relationship of issues experienced in greening supply chain using ISM-fuzzy MICMAC approach. Int. J. Logistics Syst. Manage. **30**(2), 162–178 (2018)
7. Zhu, Q., Sarkis, J.: Relationships between operational practices and performance among early adopters of green supply chain management practices in Chinese manufacturing enterprises. J. Oper. Manage. **22**, 150–159 (2004)

8. Jiang, X., Mahadevan, S., Urbina, A.: Bayesian nonlinear structural equation modeling for hierarchical validation of dynamical systems. Mechan. Syst. Signal Proces. **24**(4), 957–975 (2010)
9. Luthra, S., Kumar, V., Kumar, S., Haleem, A.: Barriers to implement green supply chain management in the automobile industry using interpretive structural modeling technique—an Indian perspective. J. Indus. Eng. Manage. **4**(2), 231–257 (2011)
10. Tyagi, M., Kumar, P., Kumar, D.: Analysis of interaction among the drivers of green supply chain management. Int. J. Bus. Perform. Supply Chain Model. **7**(1), 92–108 (2015)
11. Chen, Y.-S., Lai, S.-B., Wen, C.-T.: The influence of green innovation performance on corporate advantage in Taiwan. J. Bus. Ethics, **67**(4), 331–339 (2006)
12. Roarty, M.: Greening business in a market economy. Eur. Bus. Rev. **97**(5), 244–254 (1997)
13. Jose, P.D.: Getting serious about green. Real CIO World **3**(8), 26–28 (2008)
14. Sharma, S.: Managerial interpretations and organizational context as predictors of corporate choice of environmental strategy. Acad. Manag. J. **43**(4), 681–697 (2000)
15. Klassen, R.D., Whybark, D.C.: The impact of environmental technologies on manufacturing performance. Acad. Manag. J. **42**, 599–615 (1999)
16. Ravi, V., Shankar, R.: Analysis of interactions among the barriers of reverse logistics. Technol. Forecast. Soc. Chang. **72**, 1011–1029 (2005)
17. Zhu, Q., Sarkis, J., Geng, Y.: Green supply chain management in China: pressures, practices, and performance. Int. J. Oper. Prod. Manage. **25**(5–6), 449–468 (2008)
18. Lin, Y., Hills, P., Welford, R.: Extended producer responsibility and eco-design changes: perspectives from China. Corporate Soc. Respons. Environ. Manage. **15**, 111–124 (2008)
19. Hosseini, A.: Identification of green management of system's factors—a conceptualized model. Int. J. Manage. Sci. Eng. Manage. **2**(3), 221–228 (2007)
20. Mudgal, R.K., Shankar, R., Talib, P., Raj, T.: Greening the supply chain practices: an Indian perspective of enablers' relationship. Int. J. Adv. Oper. Manage. **2**(3), 151–176 (2009)
21. Mudgal, R.K., Shankar, R., Talib, P., Raj, T.: Modeling the barriers of green supply chain practices: An Indian perspective. Int. J. Logist. Syst. Manage. **7**(1), 81–107 (2010)
22. Tyagi, M., Kumar, D., Kumar, P.: Assessment of CSR based supply chain performance system using an integrated fuzzy AHP-TOPSIS approach. Int. J. Logist. Res. Appl. **21**(4), 378–406 (2018)
23. Lamming, R., Hamapson, J.: The environmental as a supply chain management issue. Br. J. Manage. **7**(Special Issue), 45–62 (1996)

Microstructural and Wear Characteristic of Fe-Based Nanostructured Hardfacing Alloy

Pratibha Kumari, Mohd. Parvez, Kumari Archana, Subodh Kumar Sharma, Dhananjay Pradhan, and Krishna Vijay Ojha

1 Introduction

To improve the wear resistant properties of any metal or alloy, the most prominent technique used is hardfacing. Reduction in inventory, repair and remanufacturing is some of the major benefits of hardfacing [1–3]. Among the commercial hardfacing alloys, Fe-based alloys are extensively used because of its excellent mechanical properties, good wear and corrosion resistance and optimum cost [4–6]. GMAW is used mostly in fabrication processes due to its all position welding capability, easy automation and can weld wide variety of steel and alloys [7], and one of the major applications of surfacing by gas metal arc welding is hardfacing.

Therefore, the present investigation aims to ensure the acceptability of the deposited Fe-based hardfacing alloy using gas metal arc welding GMAW by extensive microstructure, hardness and wear study.

2 Materials and Methods

2.1 Materials

GMAW process was employed to deposit hardfacing alloy on low carbon steel substrate of dimension 250 mm × 150 mm × 10 mm. The chemical composition of

P. Kumari (✉) · K. Archana · S. K. Sharma · D. Pradhan · K. V. Ojha
Mechanical Engineering Department, KIET Group of Institutions, Ghaziabad, UP, India
e-mail: pratibha.kumari@kiet.edu

Mohd. Parvez
Al-Falah School of Engineering and Technology, Dhauj, Faridabad, Haryana, India

M. Muzammil et al. (eds.), *Recent Advances in Mechanical Engineering*, Lecture Notes in Mechanical Engineering, https://doi.org/10.1007/978-981-15-8704-7_56

Table 1 Welding process parameters

Parameters	Wire diameter	Wire feed rate	Welding speed	Voltage	Nozzle to plate distance	Torch angle
Units	mm	m/min	cm/min	Volts	mm	Degree
Values	1.6	8	29	22	7	94

the hardfacing alloy (wt %) is 19.9 Cr, 2.75 C, 11 Mo, 0.62 B, 9.68 W, 0.53 Si, 0.79 Nb and balance Fe.

The hardfacing process parameters detail are listed in Table 1.

2.2 Microhardness Testing

Buehler micromet 5103 microhardness tester was used to record microhardness of samples at 30 N load, across the cross section of deposited hardfaced layer along vertical axis, near the surface of plate from centreline of the weld. Microhardness measurements in the weldzone were carried out in interval of 0.5 and 0.1 mm in heat affected zone and 0.5 mm in the non-affected zone.

2.3 Microstructural Study

After deposition, the air-cooled samples and substrate were sectioned for metallurgical study. The test pieces were polished in accordance with standard test technique used for steel. A.S.I. Sales inverted optical microscope and Zeiss Sigma HR SEM were used for microstructural characterization.

2.4 Wear Study

Wear resistance of the hardfacing alloy and the substrate was determined by pin-on-disc wear testing machine according to ASTM G99 standard. The samples of size 10×10 mm were welded to 7-mm-diameter cylindrical pin to ease the clamping of specimen on wear test machine. The disc of diameter 166 mm was hardened using electric furnace tempering.

The specimens were polished by 220 grit SiC emery paper followed by ultrasonic cleaning to maintain the uniformity of the surface [8]. Thereafter, specimens were subjected to wear test for the test parameters shown in Table 2. After wear test, samples were cleaned in Acetone and dried. The initial weight before and final

Table 2 Parameters for wear test

Load (L)	Sliding velocity (SV)	Sliding distance (SD)
50 N	3 m/s	1800

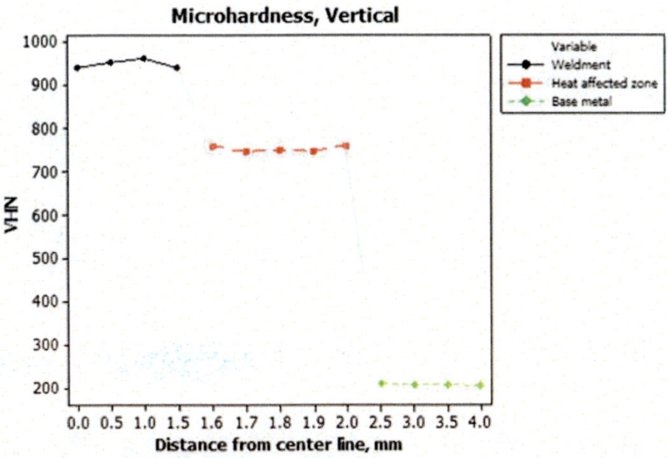

Fig. 1 Microhardness along vertical axis

weight after wear test was recorded using microbalance. The difference of initial and final weight divided by density was calculated for determination of wear volume [9].

3 Result and Discussion

3.1 Microhardness

Microhardness of weld zone and substrate are shown in Figs. 1 and 2, respectively. It is evident from Figs. 1 and 2, microhardness of weld zone was higher as compared to substrate due to nanoparticles present in the hardfaced coating and resulted in approximately five times increase in microhardness. This appreciable increase in microhardness of weld zone was due to the presence of martensite. Further, moving towards the base metal, a dip in microhardness was seen due to the presence of pearlite [10, 11]. The average microhardness in horizontal direction was 1025VHN.

3.2 Microstructure Characterization

The optical micrograph of weld metal and weld interface is shown in Fig. 3. Tempered

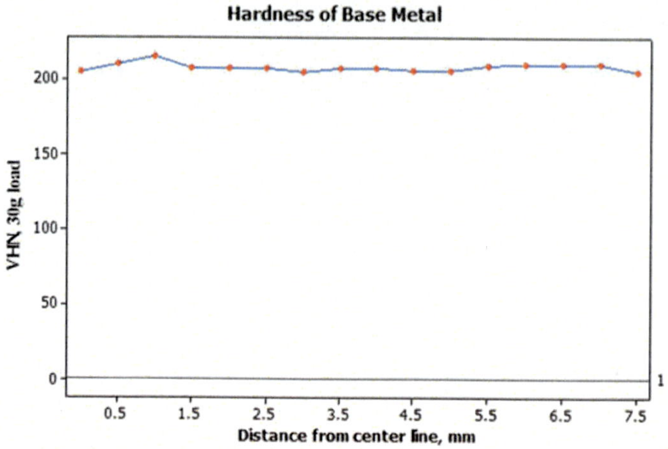

Fig. 2 Microhardness of substrate

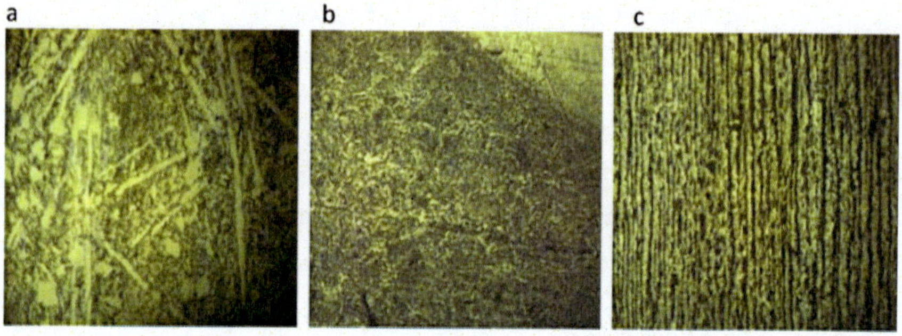

Fig. 3 Optical micrograph showing **a** weld zone, **b** HAZ and **c** parent metal

martensite with formation of metal carbide (Fig. 3a, b) was seen in the weldment and HAZ. Pearlite and ferrite in the substrate were observed in Fig. 3c.

Figure 4a shows SEM micrograph of the hardfacing alloys. Flatten disc-like structure shown in Fig. 4b represents perfect fusion with excellent continuity with no microcracks. Fine particles which do not get dissolved in the metal matrix were mainly due to local accumulation of nanoparticles (Fig. 4c). Nanostructured zone also facilitates the crack arrest, which might get propagated through weaker zones[12–14]. Figure 4d shows the perfect unity of hardfaced alloy with substrate at the weld interface.

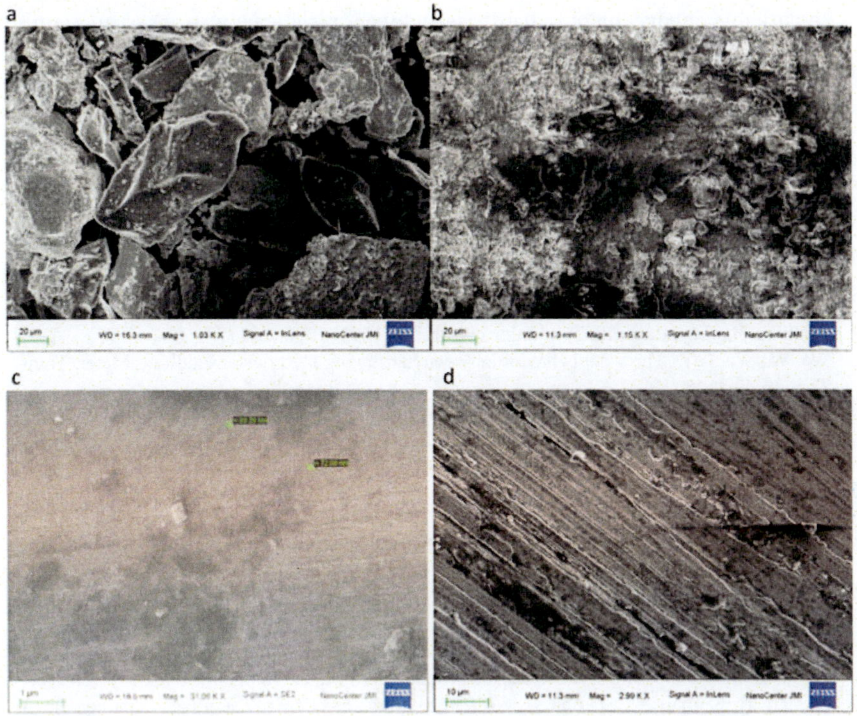

Fig. 4 SEM micrograph, **a** hardfacing alloy, **b** weld zone, **c** weld zone with nanoparticles, **d** weld metal interface

3.3 Wear Characteristics

The wear volume of hardfaced alloy and substrate was 135.69 and 785.5 mm^3, respectively. Abrasive wear of substrate was found to be higher as compared to the coating. The developed bead shows excellent wear-resistant properties due to the presence of tempered martensite matrix with formation of carbide.

4 Conclusion

Following are the conclusions drawn from this study.

1 Tempered martensite structure with formation of carbide was observed by optical micrograph.
2 Pearlite and ferrite were observed in optical microscope in base metal.
3 No discontinuity and cracks were seen in weld zone.
4 Perfect coherence of the deposited material with substrate was observed.

5 Remarkable enhancement in hardness and wear resistance was observed mainly by the formation of carbide.

References

1. Kumari, P., Singh, R.P.: Development of mathematical models for prediction of weld bead geometry of hardfaced steel. Int. J. Appl. Eng. Res. **10**(17), 38509–38525 (2015)
2. Yang, K., Gao, Y., Yang, K., Bao, Y., Jiang, Y.: Microstructure and wear resistance of Fe-Cr13-C-Nb hardfacing alloy with Ti addition. Wear **376**, 1091–1096 (2017)
3. Wo, W., Wu, L.T.: The wear behavior between hardfacing materials. Metallurg. Mater. Trans. **27A**, 3639–3648 (1996)
4. Kumari, P.: Optimization of welding parameters of nanostructured hardfacing. In: International Conference on Advances in Nanomaterials and Nanotechnology (2016)
5. Badisch, E., Kirchgaßner, M., Polak, R., Franek, F.: The comparison of wear properties of different Fe-based hardfacing alloys in four kinds of testing methods. Triobotest **14**, 225–233 (2008)
6. Coronado, J.J., Caicedo, H.F., Gómez, A.L.: The effects of welding processes on abrasive wear resistance for hardfacing deposits. Tribol. Int. **5**(42), 745–749 (2009)
7. Wani, P., Dutta, C. K.: Some studies of hardfacing and metallurgical structures of mild steel using arc welding processes. Delhi Coll. Eng. Weld. J. 22–23 (2004)
8. Pawar, S., Jha, A.K., Mukhopadhyay, G.: Effect of different carbides on the wear resistance of Fe-based hardfacing alloys. Int. J. Refract. Metal. Hard Mater. **78**, 288–295 (2019)
9. Kumari, P., Kumari A., Sharma.S.: Wear Behaviour of FeCrMoW hardfaced coating. Int. J. Appl. Eng. Res. **13**(6), 194–197 (2018)
10. Chatterjee, S., Pal, T.K.: Wear behaviour of hardfacing deposits on cast iron. Wear **255**, 417–425 (2003)
11. Raza, M. A., Kashyap, S.K., Rakesh.: The effect of welding on mechanical and microstructural properties of materials—A critical review. Elk Asia Pacific J. Manuf. Sci. Eng. **1**(2), (2016)
12. Miao, P., Odette, G.R., Gould, J., Bernath, J., Miller, R., Alinger, M., Zanis, C.: The microstructure and strength properties of MA957 nano-structured ferritic alloy joints produced by friction stir and electro-spark deposition welding. J. Nucl. Mater. **367**, 1197–1202 (2007)
13. Correa, E.O., Alcântara, N.G., Tecco, D.G., Kumar, R.V.: The relationship between the microstructure and abrasive resistance of a hardfacing alloy in the Fe-Cr-C-Nb-V system. Metallurg. Mater. Trans. **8**(38), 1671–1680 (2007)
14. Wang, Q.B., Li, X.Y.: Effects of Nb, V, and W on microstructure and abrasion resistance of Fe-Cr-C hardfacing alloys. Weld. J. **89**, 133–139 (2010)

Thermo-mechanical Peculiarity of TiC Filler Filled Glass Fiber-Based Hybrid Composites

Anant Krishan Pun and Siddhartha

1 Introduction

Polymer composite filled with particles has been widely studied because of their logical and scientific significance. Two-phase material composites are divided into two main categories that are particulate filled composites and fiber-reinforced composites. Particulate filled composites are the composites in which particles having different shapes and sizes are scattered inside a matrix in an arbitrary manner. The composite filled with particulate fillers has been utilized widely in different fields because of their low manufacturing expenses and ease to be formed into complex shapes. Also, they are isotropical in behavior and sensitive as compared to long fiber composites to the thermal expansion mismatch between the reinforcement and the matrix [1, 2]. In polymers, the fillers are generally used for various reasons like density control, price reduction, better processing, thermal conductivity, optical effects, control over thermal expansion, wear resistance, magnetic properties, better hardness, and electrical properties.

As opposed to a couple of exploratory examinations, an enormous number of theoretical and numerical examinations have been completed so as to make sense of how the interphase properties impact the by and large mechanical behavior of a fiber/filler-reinforced composite. For sake of simplicity, the interphase is considered to be a homogeneous material in most of the micromechanical models [3]. The metal particles reinforced polymer composites have been used in several industrial applications like heaters, electrodes [4], and thermally stable composites for high-temperature applications [5]. Hard particles as reinforcing materials played an

A. K. Pun (✉)
Department of Mechanical Engineering, Chandigarh University, Chandigarh 160036, Punjab, India
e-mail: anantkrishanpun.nith@gmail.com

Siddhartha
Department of Mechanical Engineering, NIT Hamirpur, Hamirpur 177005, HP, India

M. Muzammil et al. (eds.), *Recent Advances in Mechanical Engineering*, Lecture Notes in Mechanical Engineering, https://doi.org/10.1007/978-981-15-8704-7_57

important role in developing commercially viable polymer composites, but certain fillers may improve processing, physical and optical properties as well as mechanical behavior [6] and tribolgical applications [7]. Similarly, ceramic-based polymer composites are also widely studied in the last twenty years or so. The silica particles introduced in a polymer matrix significantly enhance the thermal, electrical, and mechanical properties of the composites [8, 9]. The polymer used for commercial application is filled with inorganic filler for the reduction in cost and improvement in the stiffness of polymer composites [10, 11]. In several real operational conditions, besides the fiber-reinforced composites, the particulate filled composites have also been observed to perform well. Bonner [12] found that the addition of micro-sized particulates in the polymer matrix requires a higher filler content (typically higher than 20 vol.%) to bring the above mentioned positive effects but it may cause some negative effects on some of the essential properties of the polymer matrix like appearance, processability, aging performance, and density. Mechanical performance of composites is of more interest to the component designer. The mechanical properties of polymer composites are dependent on the constituent materials properties (type, loading, distribution and orientation of fiber, void contents) other than these properties, the kind of interfacial bond formed between the matrix and the fiber strongly influences the performance of the composite [10]. Short fiber-reinforced polymer matrix (SFRP) composites are appealing as they can be easily fabricated, are economical, and have excellent mechanical properties. Generally, high fiber contents are necessary to attain high-performance short fiber-reinforced polymer composite [12]. Thus, the influence of the amount of fiber on mechanical properties of composites is of specific interest and importance. They also reported that by increasing the fiber content the strength and modulus as well as toughness increased if the matrix is of low toughness. Work has been done much for micro-size fillers. However, the performance of nano-size filler filled composites has not been explored much. Literature on TiC filler filled composites is negligible. Therefore, nano-TiC filler filled composites are fabricated in this work and their performance are compared with micro-TiC filler filled composites and unfilled composites.

2 Materials and Methods

2.1 Sample Fabrication

Nano- and micro-TiC filler filled glass fiber-reinforced hybrid polymeric composites are fabricated. Vinyl ester resin is used to fabricate composites. Unfilled composite is also fabricated for comparative study. The designation and composite compositions are given in Table 1. Layer-by-layer composite fabrication technique is used in this study. Cobalt naphthenate as accelerator is mix in vinyl ester resin within the amount of 2%. MEKP with 2% amount is also used to solidify the resin within 24 h. Twelve layers of glass fiber are used to fabricate the composites up to desired thickness.

Table 1 Void fraction of fabricated composites

Designation of composites	Composition of composites	Experimental density (ρ_e) g/cm^3	Theoretical density (ρ_t) g/cm^3	Void fraction (%)
VEGF	Resin + 50% glass fiber	2.0687	2.1138	2.1343
TiC-5 M	Resin + 50% glass fiber + 5 wt.% micro-TiC	2.0707	2.1968	5.7391
TiC-5 N	Resin + 50% glass fiber + 5 wt.% nano-TiC	2.1134	2.1973	3.8163
TiC-10 M	Resin + 50% glass fiber + 10 wt.% micro-TiC	2.0479	2.2967	10.8351
TiC-10 N	Resin + 50% glass fiber + 10 wt.% nano-TiC	2.1220	2.2973	7.6295

TiC filler with an amount of 5 and 10 wt.% is used to fabricate the composites. The fabricated composite was kept at room temperature under a load of about 20 kg for a time period of 24 h before removing from the mold. Vinyl ester resin and bidirectional E-glass fiber are supplied by Sakshi chemicals, Delhi. Nano- and micro-TiC fillers are provided by Nano Sell, Chandigarh, India.

2.2 Mechanical Characterization

Agarwal and Broutman's formula is adopted for measuring the theoretical density of nano- and microfiller filled composites [13]. However, experimental density of fabricated specimens is measured as per the ASTM D792 standard. The void content (%) of the fabricated samples is calculated with the help of theoretical and experimental densities as shown in Table 1. Dynamic mechanical analyzer machine is used for the analysis of thermo-mechanical property of all fabricated composites. The tensile as well as flexural tests are conducted on a UTM. Both the tests are executed as per the ASTM D3039 and D2344 standards, respectively.

3 Results and Discussion

3.1 Thermo-mechanical Analysis of Composites

The storage modulus is a key term in thermo-mechanical analysis of composites. It measures the maximum energy that is stored in a material for each cycle of oscillation. It provides important information about the elastic behavior of the material through the insight to fiber stiffness [14]. Generally, polymer materials have three physical states, viz. glassy, glass transition region, and rubbery. The glassy region represents a tight packing that results in the frozen movement of chain segment. Thus, the material tends to store the energy which leads to a high storage modulus [15].

The variation in storage modulus with varying temperature is shown in Fig. 1. It is observed that there is an increase in storage modulus value with the addition of TiC fillers. With the increase in the wt.% of micro- and nano-fillers, the storage modulus of composites increases. Nano-filled composites have more storage modulus than micro-filled composites. When temperature increases, the composites attain a dynamic nature and lose their close packing procedure. It results in a reduced storage modulus in the glass transition region of 25–50 °C. The temperature reliant reduction in storage modulus of the composite filled with both type of TiC fillers remains uniform in the glass temperature range which causes a meager change in storage modulus in the rubbery region.

The variation of loss modulus with temperature as shown in Fig. 2, with respect to energy lost due to friction as temperature and inner cues imitating viscous nature [16], shows that the magnitude of loss modulus decreased with the addition of both type of TiC fillers. The loss modulus peaks are higher for unfilled composite than those with filler filled composites material. It happens because of the hindrance of the molecular motion [15]. Loss modulus of micro-TiC filled composites is high in comparison with nano-TiC filled composites.

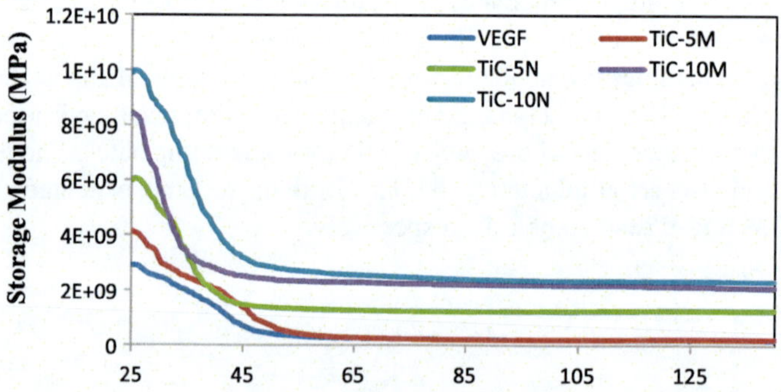

Fig. 1 Storage modulus of fabricated composites

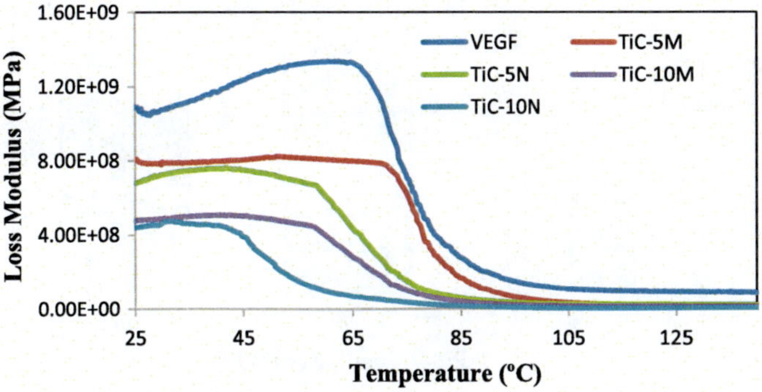

Fig. 2 Loss modulus of fabricated composites

Damping factor (tan δ) is loss modulus divided by storage modulus, also known as internal damping or dynamic composite behavior. Generally, the damping of the composites is dependent on the following factors: matrix and filler material type, nature of the interface, frictional damping due to slippage in the liberated region of filler–matrix interface and the delamination and energy dissipation in matrix crack area [15].

It can be deduced from Fig. 3 that 5 wt.% of both types of filler filled hybrid polymeric composites have higher damping factor than 10 wt.% of filler-reinforced composites. It may happen due to the fact that resin molecules have deformed in-between the cross-links [17]. However, damping properties remained same after 125 °C and were independent of the weight fraction of TiC particles in the composites.

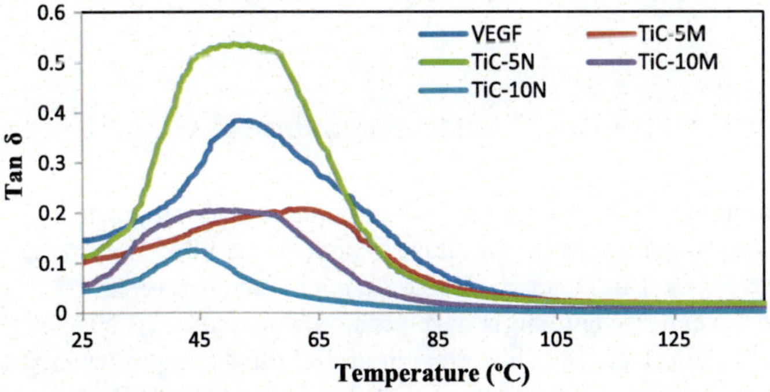

Fig. 3 Damping factor fabricated composites

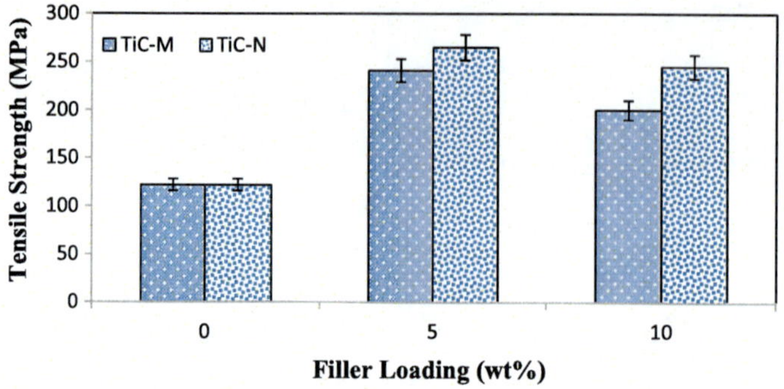

Fig. 4 Tensile strength of fabricated composites

4 Tensile Strength of Fabricated Composites

The tensile test results are shown in Fig. 4. It is evident from Fig. 4 that tensile strength increases for 5 wt.% of TiC filled composites and it further decreases for 10 wt.% of TiC composites. Tensile strength is higher for nano-TiC composites at both filler loading conditions. It is observed that the tensile property of the composite gets deteriorated at higher filler loading. It may happen due to the increased void sizes and the occurrence of agglomeration of filler particles [18]. In fact, an increase in the filler content makes the void sizes critically large due to the detachment of polymer matrix from the fillers which ultimately lead to the initiation of the main crack [13]. Also, a lower strength of the agglomeration causes a reduced tensile strength [19]. Composite filled with nano-filler has less void fraction which results in higher tensile strength in comparison with micro-size TiC composites.

5 Flexural Strength of Fabricated Composites

Flexural strength values of fabricated test samples are displayed in Fig. 5. These composites show an improved flexural strength when filler loading increases as evident from Fig. 5. This pattern is similar for micro- and nano-fillers. This improvement in the flexural strength might have occurred due to the existence of TiC fillers at fiber–matrix interface. Bending strength of TiC filled composites in micron size is less than that of nano-TiC composites at 5 and 10 wt.% of filler inclusion. This may happen due to the incomplete fusion of the matrix, weak interfacial adhesion, and due to the presence of voids in composites.

It is obvious from Table 1 that hybrid polymeric composites filled with micron size TiC have higher void fraction as compared to nano-size filler filled hybrid polymeric composites. These voids might have resulted due to the incomplete fusion of the matrix material. The flexural properties have a high significance in structural

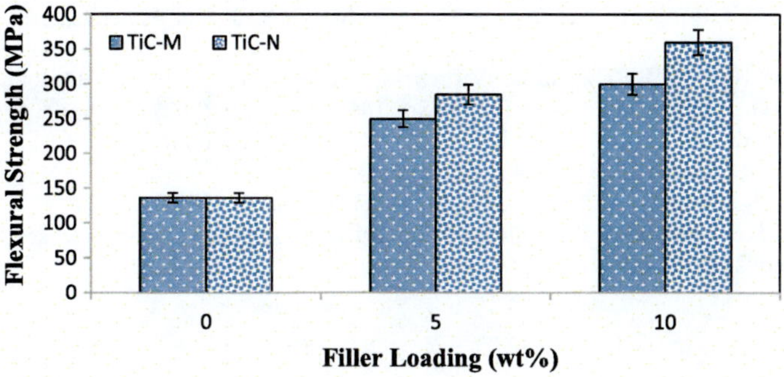

Fig. 5 Flexural strength fabricated composites

elements. Composite materials have a tendency to fail under bending conditions, and hence, it is important to create new composites having better flexural attributes.

6 Conclusions

1. TiC filled glass fiber-based hybrid composites are successfully fabricated via hand layup technique.
2. Nano-composite with 10 wt.% filler addition has high storage modulus as well as low damping factor and loss modulus among all the fabricated composites.
3. Tensile strength is maximum for 5 wt.% filler loading. Tensile strength decreases for 10 wt.% of both type of TiC filler loading.
4. Nano-TiC filled composites have high flexural strength than micro-TiC filler filled composites. Flexural strength is higher for 10 wt.% of filler loading.

References

1. Bae, J.W., Kim, W.H., Cho, S.H.: J. Mater. Sci. **35**, 5907–5913 (2000)
2. He, H., Fu, R.L., Shen, Y., Han, Y.C., Song, X.F., Compos. Sci. Technol. **67**, 2493–2499 (2007)
3. Lee, W.S., Yu, J.: Diamond Relat. Mater. **14**, 1647–165 (2005)
4. Chen, Y.M., Ting, J.M.: Carbon **40**, 359–362 (2002)
5. Fu, S.Y., Mai, Y.W.: J. Appl. Polym. Sci. **88**, 1497–1505 (2003)
6. Georje, J., Joseph, K., Bhagavan, S.S., Thomas, S.: Mater. Lett. **18**, 163 (1993)
7. Kubat, J., Rigdahl, M., Welander, M.J.: Appl. Polym. Sci. **39**, 1527 (1990)
8. Schledjewski, R., Karger-Kocsis, J.J.: Thermoplast Compos. Mater. **7**, 270–277 (1994)
9. Joseph, K., Thomas, S., Pavithran, C.J.: Reinf. Plast. Compos. **12**, 139–155 (1993)
10. Joseph, K., Thomas, S., Pavithran, C.: Mater. Lett. **15**, 224–228 (1992)
11. Ghosh, P., Bose, N.R., Mithra, B.C., Das, S.J.: Appl. Polym. Sci. **64**, 2467–2472 (1997)
12. Gassan, J., Bledzki, A.K., Composites processing and microstructure. In: Proceedings of ICCM-II, 4, 1997 Gold Goast, Australia, pp. 762–770 (1997)

13. Parvaiz, M.R., Mohanty, S., Sanjay, K.N., Mahanwar, P.A.: J. Min. Mater. Charact. Eng. **9**(1), 25–41 (2010)
14. Jacob, M., Francis, B., Thomas, S., Varughese, K.T.: Polym. Compos. **27**, 671–680 (2006)
15. Wu, G., Gu, J., Zhao, X.: J. App. Polym. Sci. **105**, 1118–1126 (2007)
16. Jiang, H., Kamdem, D.P.: J, App. Polym. Sci. **107**, 951–957 (2008)
17. Kaundal, R.: Silicon **10**, 2439–2452 (2018)
18. Sandi, A.R., Prasad, S.V., Rohatgi, P.K.: J. Mater. Sci. **21**, 81 (1986)
19. Nielsen, L.E. Mechanical Properties of Polymers and Composites. Marcel Dekker, New York, vol. 1, pp. 254–385 (1974)

Performance Analysis of Hybrid PCM by Doping Graphene

Ravi Kumar Samadhiya, Devendra Singh, Manoj Kumar Shukla, Kamal Sharma, and Ravindra Pratap Singh

1 Introduction

With the exhaustion of fossil fuel sources and amplification of environmental pollution, renewable energy consumption and waste heat recuperation have been an area of interest in research and development [1]. Meanwhile, according to the energy expenditure category in China, inhabited energy utilization accounted for 20% of the total in 2017 [2], in which household hot water production and spacing heating in a residential building explanation for a large quantity. Thermal energy storage (TES) is a successful solution to the intermittency and volatility of energy supply and forms a bridge between energy supply and end-users [3, 4] (Fig. 1).

Energy storage device helps in smooth supply and reliability of energy by synchronizing load leveling of power generating plant [5]. This makes the system more energy efficient though sun has clean safe and abundant source of energy, but the supply of energy is intermittent. Therefore, the requirement of energy source device in the case of solar device is needed. Shortage of fossil fuels and increasing consciousness toward the human health has made researchers to contemplate about the alternatives of fossils. Combustion of fossil fuel leads to harmful greenhouse gases like carbon monoxide and carbon dioxide, which causes incurable diseases like cancer and asthma which is a great cause of concern. For domestic hot water (DHW) and some industrial applications, FPSWs are mostly used. But due to the absence of solar radiation and disproportion between the intervals of hot water, utilization of FPSWs is limited. As a result, LHS system with PCM is incorporated into FPSWCs, and

R. K. Samadhiya · M. K. Shukla · K. Sharma (✉) · R. P. Singh
Department of Mechanical Engineering, GLA University, Mathura 281406, UP, India
e-mail: sharmakamal1978@gmail.com

D. Singh
Department of Mechanical Engineering, Malaya University, Kuala Lumpur, Malaysia

M. K. Shukla
KNPC (Rajiv Gandhi Technical University), Jabalpur, India

M. Muzammil et al. (eds.), *Recent Advances in Mechanical Engineering*, Lecture Notes in Mechanical Engineering, https://doi.org/10.1007/978-981-15-8704-7_58

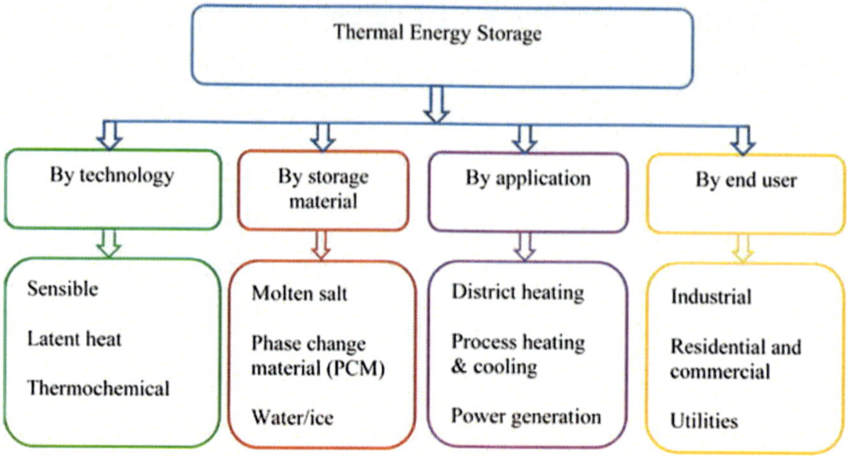

Fig. 1 Classification of thermal energy storage

this helps in increasing the operating time of FPSWCs which increases its thermal performance. Categories of PCMs used in LHS are shown in Fig. 2.

These PCMs have several advantages: (a) high latent heat, (b) the ability to store a huge amount of energy at a constant fusion temperature, (c) appropriate phase change temperature, (d) cost effective and availability.

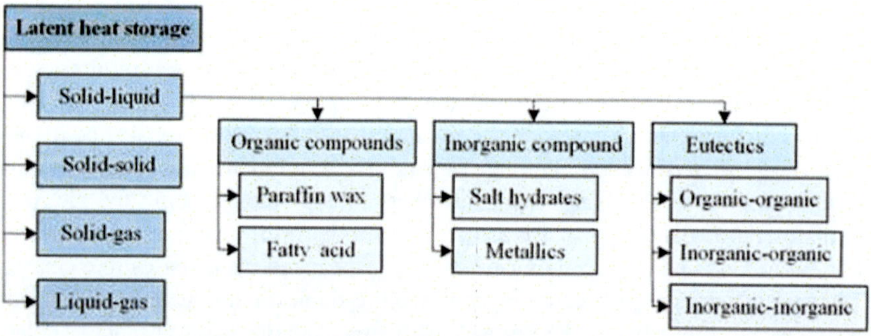

Fig. 2 Categories of LHS and classification of PCMs

2 Heat Transfer Fluid

2.1 Water

In latent heat storage system, the heat energy from the sun is transferred to the phase change material through a fluid. Generally, water is used as a heat transferring fluid as it has good thermal conductivity; it is easily available and economical. It used having high thermal conductivity.

2.2 Cerium Oxide

A colloidal suspension of cerium oxide having a concentration of 18% by weight as a nanofluid is used to enhance the properties of heat transfer fluid. Heat transfer fluid is prepared having concentration of one percent by weight of nanoparticles in water [6].

3 Experimental Set up

3.1 Working of the System

The working of the system starts with filling of water in the cold HTF storage tank. Once this tank is full, the water is uplifted to the HTF hot/cold source tank by switching the pump on. When water overflows, the pump is switched off and heat is supplied to the water by switching on the electric heaters. The hot water gives heat to the PCM by the process of convection and conduction on its way of flow from one end to other of the heat exchangers. This stage in which the PCMs accrued energy is known as charging stage. After giving heat to PCM, the water gets accumulated in the used HTF storage tank. The second step of the experiment is an extraction of heat from the PCM. This step is also known as discharging of PCM. After absorbing heat, the water gets accumulated in the used HTF storage tank [7, 8].

There is a control unit in the system. All required meters and control knots are fit in this unit. From this unit, the user can take the values of different parameters such as water flow rate and temperature at different conditions of the system (Fig. 3).

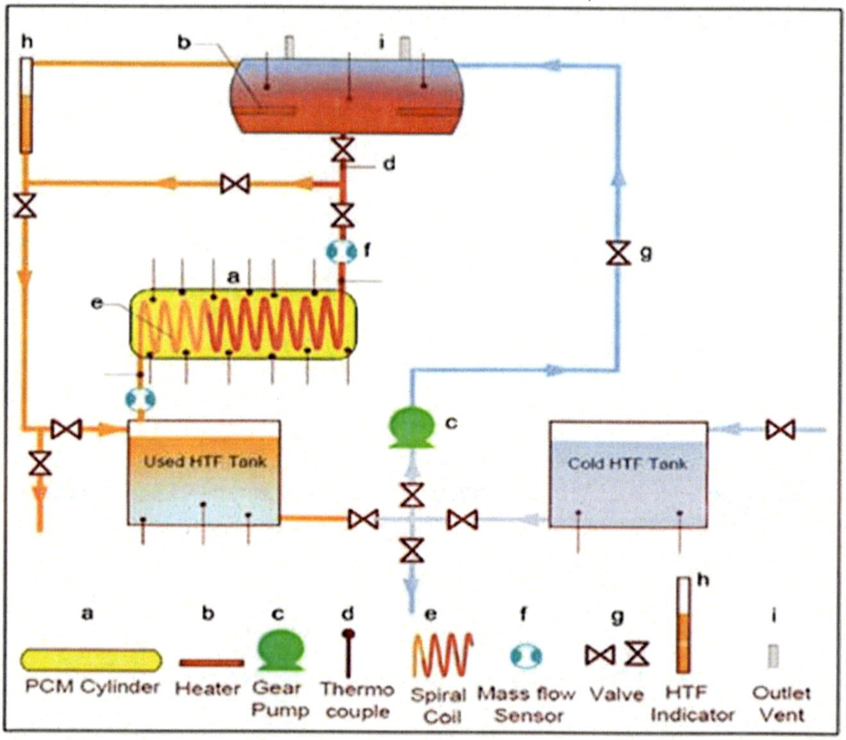

Fig. 3 Schematic diagram of system. (This figure shows all the internal connection of the system elements like valves, thermocouple, etc.)

4 Results and Analysis

Varying the HTF temperature, the charging and discharge behaviors are plotted on the graph for obtaining the performance of the LHESS by using water as an HTF flowing through PCM cylinder.

4.1 Water as Heat Transfer Fluid

The water has been used as an HTF in the first stage of the experiment. Water has many merits to be chosen as an HTF such as free availability and average specific heat capacity. Under this section, the inlet HTF temperature has been varied from 60 to 85 °C. At different temperatures, the charging time has been recorded and plotted to showcase the variation. The figure shows that the charging time decreases by 19.81% on increasing the temperature from 60 to 85 °C; the great reduction in time has been observed in the beginning of charging because the sensible heat is responsible for

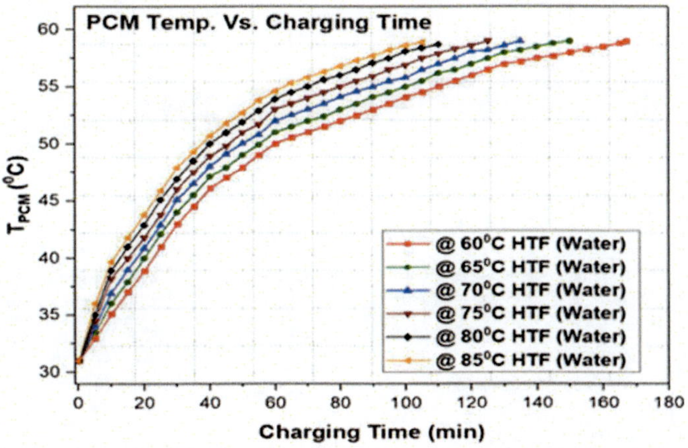

Fig. 4 Charging with water as an HTF

giving the rise in temperature till melting temperature. A large amount of heat transfer due to temperature difference occurs from HTF to PCM through conduction phase. After achieving the melting point of 58 °C, the latent heat transfer phase starts which is responsible for phase transformation process. Convention heat transfer mode starts when the PCM around the heat exchanger melts. Optimum results have been achieved at 80 °C where the reduction in time was 18.67%, as shown in Fig. 4.

4.2 *Water as HTF and PCM Induced with Graphene*

4.2.1 Effect of Variation at Inlet Temperature

In this section, the temperature has been varied from the range 60–85 °C. Figures 5a–f shows the charging time variation with respect to different HTF temperatures. The charging time has been recorded as 147, 140, 132, 127, 123, and 119 min at different HTF temperatures ranging from 60 to 85 °C at the interval of 5 °C as shown in figure a. The overall charging time is reduced by 26.98% as compared to HTF with respective to the time at 60 °C and at 0.25% concentration. This is because of the rate at which heat is advanced/lost and the time for phase changing process during charging process is also enhanced

4.2.2 PCM Average Temperature Versus Discharging Time

At the end of the charging process, the PCM is in the liquid phase at nearly 60 °C. The system was kept untouched for 4 h to store the energy [9]. The temperature was

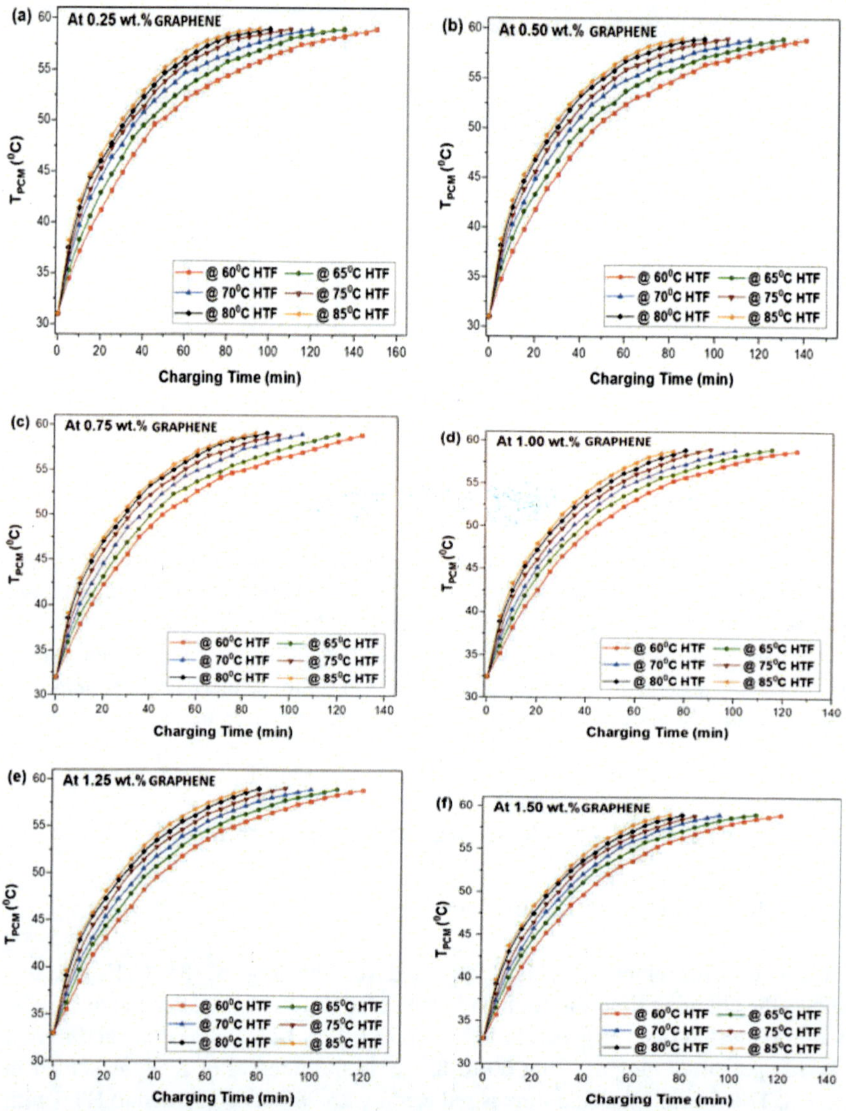

Fig. 5 **a**, **f** Charging using nanofluid at different concentrations

observed to get reduced to 54 °C at the starting of discharge process; this is due to the heat loss through insulation, as shown in Fig. 6.

During the discharge process, the cold HTF has passed through the heat exchanger to extract the energy from the system. The temperature profile during the discharging process is shown in figure. The mass flow rate is kept constant at 0.3 Lpm and temperature of cold HTF is 30 °C, which is known that temperature difference due

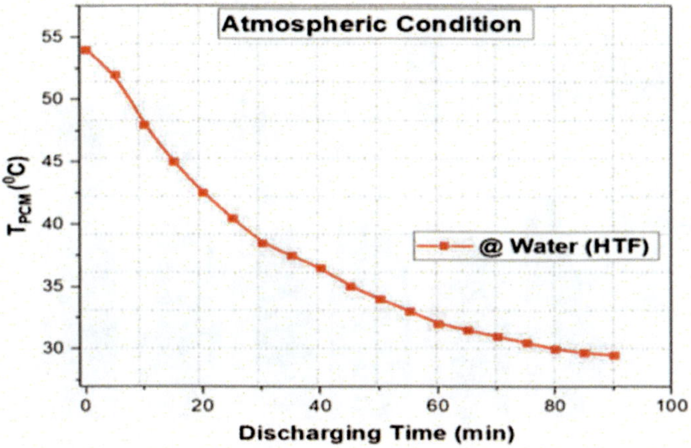

Fig. 6 Discharge using water

to driving force for heat transfer therefore initially temperature decreases drastically from large temperature difference between cold HTF and PCM.

4.3 *Efficiency*

This system is working efficiently because we are using the hybrid PCM with doping graphene. This is the most important portion of any study that will provide an information about working condition of the system after refining all thermal properties with the help of various alternatives such as extended fins and encapsulated PCM. It can be measured in terms of charging, exergy, and overall energy efficiency [10]. The efficiency results are based on the comparison of the outcome of water and nanofluid. On the basis of these analyses, the effective change is noticed and shown by Fig. 7.

4.3.1 Charging Efficiency

It can be observed that in the case of enhancing the HTF temperature, the charging efficiency is continuously improving. The reason behind the thermal conductivity of nanofluid is enhancing because its energetic nanoparticle makes the Brownian motion. The result of water is being compared with the consequences of nanofluid.

When water is used as an HTF, the charging efficiency is enhanced by 27.51% at variation in inlet HTF temperature. By applying the nanofluid as an HTF by varying HTF temperature and at constant concentration, the enhancement is noticed as follows. At 0.25% concentration, the charging efficiency is enhanced by 29.65% on the variation in HTF temperature from 60 to 85 °C with respect to water. Similarly, at 0.5, 0.75, 1.25, and 1.5% concentration, charging efficiency is enhanced by 32.56,

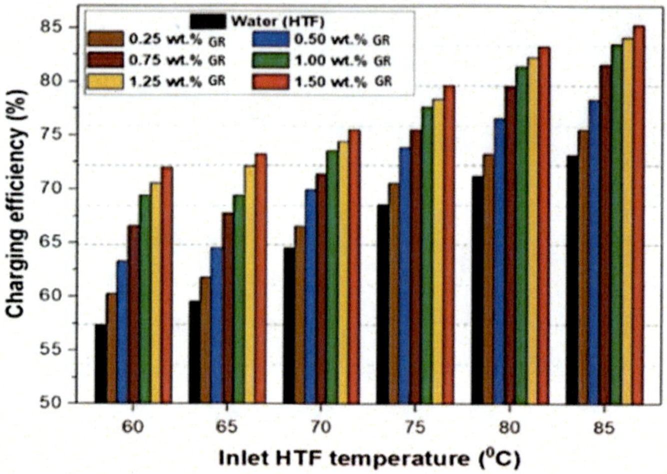

Fig. 7 Charging efficiency at different HTF temperatures

35.49, 39.12, 40.14, and 40.65%, respectively. The overall improvement of charging efficiency, approximately 17%, is recorded at varying HTF temperature, as illustrated in Fig. 7.

4.3.2 Discharging Efficiency

As the charging process gives the information about capability of the system, in the same manner, the discharging process also has a significant role for determining the storing capacity of the system. The storing of energy for a longtime eliminates mismatching of demand and supply. The energy is stored up to the end of the charging from the beginning of charging by disregarding all the losses during charging process. This stored energy is accumulated several times after discharging phenomenon takes place. The discharging efficiency depends upon recovered as well as available energy. It has been observed that there is no effective impact of nanofluid for improving the storing or discharging efficiency because it is depended on insulation, overall heat transfer coefficient of heat exchanger, and surface of cylinder [11, 12].

In this stage, the energy stored in the PCM is extracted to use in the load.

$$E_{\mathrm{st,begin}} = E_{\mathrm{in,chg}} - E_{\mathrm{loss,chg,i}} - E_{\mathrm{loss,chg,ii}} - E_{\mathrm{st,loss,i}} \cong E_{\mathrm{acc,chg}}$$

$$E_{\mathrm{st}}, =, -E_{\mathrm{st,loss,ii}}$$

$$E_{\mathrm{st,loss,ii}} = q_{\dot{3}} \times L_{\mathrm{ins}} \times t_i$$

$$E_{\text{rec, dischg}} = \dot{m}_{\text{HTF}}\, C_{\text{HTF}}\,(\text{T}_{o,\text{HTF,dischg}} - T_{i,\text{HTF,dischg}}) \times \text{timedischg}$$

$$E_{\text{av,ischg}} = E_{\text{st}}, -\left\{E_{\text{Losses,ischg(i + ii)}}\right\}$$

$$E_{\text{loss hg,i}} = q_{\dot{2}} \times L_{\text{ins}} \times \text{time}_{\text{dischg}}$$

$$E_{\text{loss,kg,ii}} = \dot{m}_{\text{HTF}} C_{\text{HTF}} (T_{\text{PCM}} - T_{\text{amb.}}) \times \Delta t_{\text{dischg - atm}}$$

$$\dot{q}_2 = \frac{(T_{\text{PCM}} - T_{\text{amb.}})}{R_1}$$

$$\dot{q}_3 = \frac{(T_{\text{PCM}} - T_{\text{amb.}})}{R_2}$$

$$R_2 = R_1 - \frac{1}{h_i A_i} + \frac{\ln\left(\frac{r_2}{r_1}\right)}{2\pi k_1 L_1} + \frac{\ln\left(\frac{r_3}{r_2}\right)}{2\pi k_2 L_2}$$

The efficiency for the discharging period can be expressed as follows

$$\eta_{\text{dis}} = \frac{\text{Energy recovered from TES during discharging}}{\text{Amount of energy available to discharge}}$$

$$\eta_{\text{st}} = \frac{E_{\text{st,end}}}{E_{\text{st,begin}}}$$

$$\eta_{\text{dischg}} = \frac{E_{\text{rec,dischg}}}{E_{\text{av,dischg}}}$$

4.3.3 Overall Efficiency

The enhancement in HTF temperature at the constant concentration and the change in overall efficiency of nanofluid are obtained as compared to that of base fluid. As in case of water, the enhanced efficiency 16.41% is to be observed, but as in nanofluid at 60 °C HTF temperature, the change in efficiency is obtained 20.72%. Similarly, at 65 to 85 °C, the change in efficiency is obtained 23.92, 25.76, 27.56, 29.73, and 30.26%, respectively. The overall change in energy efficiency is recorded approx. 12% that is more drastic change compared than that of water. But enhancing in concentration at constant inlet HTF temperature, the change in overall efficiency of nanofluid is lower than enhancing the HTF temperature [13, 14]. The optimum result of overall efficiency is obtained at 80 °C HTF temperature and 1% concentration

5 Conclusions

The experimental study is carried out for improving the thermal performance of thermal energy storage system using nanofluid as an HTF under varying concentration from 0.25 to 1.5%, as a recess of 0.25% and inlet HTF temperature from 600 to 850 °C, as a recess of 50 °C. The thermo-physical properties of nanofluid are determined from the laboratory. Based on performance factors, the outcomes of the water and nanofluid are compared to each other under varying concentration and inlet HTF temperature.

- The reduction in charging time is found up to **19.81%** while using water as an HTF, whereas the reduction in charging time is found up to **26.98%** when using Graphene as an HTF under varying inlet HTF temperature. Under varying concentration, the charging time has been reduced but that is significantly not as much of as variation in HTF temperature. However, the discharging time is not considerable affected as using nanofluid as an HTF.
- Enhancement in charging efficiency is found up to **27.51%** when using water as an HTF but the deviation in HTF temperature at a constant concentration, and the enhancement in charging efficiency is found up to **40.65%** while using nanofluid as an HTF.
- The enhancement in overall efficiency is found up to **30.26%** as variation in inlet HTF temperature of nanofluid at constant concentration. The enhancement in available energy (exergy efficiency) is recorded up to **17.02%** under varying inlet HTF temperature and constant concentration.

References

1. Pereira da Cunha, J., Eames, P.: Thermal energy storage for low and medium temperature applications using phase change materials—A review. Appl. Energy **177**, 227–238 (2016)
2. National Bureau of Statistics of China. China Statistical Yearbook 2019, China (2019)
3. Sahan, N., Paksoy, H.: Novel shapeable phase change material (PCM) composites for thermal energy storage (TES) applications. Sol. Energy Mater. Sol. Cells **174**, 380–387 (2018)
4. Amaral, C., Vicente, R., Marques, P.A.A.P., et al.: Phase change materials and carbon nanostructures for thermal energy storage: A literature review. Renew. Sust. Energy. Rev. **79**, 1212–1228 (2017)
5. Abhat, A.: Low temperature latent heat thermal energy storage: heat storage materials. Sol. Energy **30**, 313–332 (1983)
6. M.J. Hosseini, A.A. Ranjbar, K. Sedighi, M. Rahimi, A combined experimental and computational study on the melting behavior of a medium temperature phase change
7. Feldman, D., Shapiro, M.M., Banu, D.: Organic phase change materials for thermal energy storage. Solar Energy Mater. **13**, 1–10 (1986)
8. Hasan, A., Sayigh, A.A.M.: Some fatty acids as phase change thermal energy storage materials. Renew Energy **4**, 69–76 (1994)
9. Ahmet, A.K.: Thermal conductivity and latent heat thermal energy storage characteristics of paraffin/expanded graphite composite as phase change material. Appl. Therm. Eng. **27**, 1271–7 (2007)

10. Zalba, B., Man, J.M., Cabeza, L.F., Mehling, H.: Review on thermal energy storage with phase change: materials, heat transfer analysis and applications. Appl. Therm. Eng. **23**, 251–83 (2003)
11. Jiang, L., Li, Q., Li, D., Chen, Z., Hu, W., et al.: Aqueous preparation of polyethylene glycol/sulfonated graphene phase change composite with enhanced thermal performance. Energy Converse Manage. **75**, 482–487 (2013)
12. Jegadheeswaran, S., Pohekar, S.D.: Performance enhancement in latent heat thermal storage system: a review. Renew. Sustain. Energy Rev. **13**, 2225–44 (2009)
13. Sharma, A., Tyagi, V.V., Chen, C.R., Buddhi, D.: Review on thermal energy storage with phase change material sand applications. Renew. Sustain. Energy Rev. **13**, 318–345 (2009). Abhat. Low temperature latent heat thermal energy storage: heat storage materials. Solar Energy **30**(4), 313–332 (1983)
14. Liu, M., Saman, W., Bruno, F.: Review on storage materials and thermal performance enhancement techniques for high temperature phase change thermal storage systems. Renew. Sustain. Energy Rev. **16**, 2118–32 (2012)

Identify the Rational Performance and Effect of End State Temperature on a Regenerative–Reheat Coal-Based Thermal Power Plant in India

Shailendra Pratap Singh, Subrata Kumar Ghosh, and Vijay Kumar Dwivedi

1 Introduction

As per World Economic Forum Data, India's energy consumption has doubled since the year 2000 and is expected to be more than double by 2040 whereas its electricity consumption per capita is even less than that of Africa and is one-tenth that of America. Sixth largest consumer of energy in the world is India, consuming around 3.9% of the total energy production. Presently the installed capacity is 367280 MW and 56% of which, i.e., 205254 MW is produced by coal (CEA, December 2019). Energy is the most important parameter of a country's development status. In the wake of global climate concerns and shrinking coal reserves, it is substantial not just to perform minor modifications and retrofits at the energy generating facilities but to utilize coal in a more effective, efficient, and clean process. This can be done by extending the conventional energy analysis to the second-law-based exergy analysis useful for measuring the efficiency of an energy conversion process.

This study is carried out on a 210 MW LMW coal-based unit and provides an in-depth study of the performance of individual energy systems and their components. There are a number of papers, which have reflected upon the energy–exergy performance of thermal power plants. For instance, in a study done by Bhatt et al. [1], coal-fired power plants of 30-500 MW output, it was concluded that maximum potential for refinement is in 500 and 210 MW units followed by 120 and 110 MW units. Aljundi [2] estimated exergy and energy losses by conducting componentwise modeling and referred thermal power generation in Jordan. Datta et al. [3] spliced the power plant in three zones and conducted exergy analysis. Rosen [4] compared nuclear and thermal steam power plants and conducted energy as well as exergy

S. P. Singh (✉) · S. K. Ghosh
Indian Institute of Technology (ISM), Dhanbad 826004, India
e-mail: spsinghr@gmail.com

V. K. Dwivedi
Galgotia College of Engineering and Technology, Greater Noida 201306, India

M. Muzammil et al. (eds.), *Recent Advances in Mechanical Engineering*, Lecture Notes in Mechanical Engineering, https://doi.org/10.1007/978-981-15-8704-7_59

analysis. Hasan et al. [5] identified and discussed the thermal inefficiencies and compared different coal-based thermal power generation in Turkey. Naterer et al. [6] conducted the analysis on the coal-based thermal power generation on referred turbine and boiler losses. Zubair and Habib [7] analyzed the regenerative–reheat Rankine cycle power generation based on second law of thermodynamics. Ganapathy et al. [8] determined energy as well as exergy losses componentwise of the lignite-based thermal power plant. Reddy and Butcher [9] analyzed power generation based on waste heat recovery. Srinivas et al. [10] analyzed performance on the basis of first and second laws of power plant by using alternative fuel. Suresh et al. [11] performed exergy-based performance of the coal-fired thermal power plants using different steam conditions. Oktay [12] conducted exergy loss analysis and described the technique to reduce these losses by using Fluidized bed combustion technique in Turkey. Regulagadda et al. [6] performed both exergy and energy analysis of boiler and turbo alternator for a 32 MW subcritical, coal-based power unit and developed formulation for entire system. A study based on different operating parameter is being conducted for the plant. Chao Zhang et al. [13] conducted analysis on 300 MW thermal power generation in China. The study diverted the analysis from conventional techniques to thermo-economic study. Jassim et al. [14] conducted the exergy analysis of regenerative air heater of the thermal power plant. Habib et al. [15] performed analysis for coal-fired unit with reheating and regeneration on the basis of second law. Reddy et al. [16] conducted the energy and exergy analysis with main emphasis on the boiler and combustion chamber in cogeneration power plant. They also described the analysis procedure of the systems based on exegetical approach. Kaushik et al. [17] performed first and second-law analysis of power plants operated by gas and coal. They showed that first law of thermodynamics is not enough to make a right analysis, as the difference between quantity and quality of energy. Kopac and Hilalci [18] studied the exergy and energy analysis of a power generating unit in Turkey. The exergy analysis was done to determine the irreversibility rates and the heat loss at different temperatures. Ahmadi and Toghraie [19] analyzed the 200 MW Power Plant of Isfahan using energy, mass, and exergy balance equations. They used EES for analysis. Exergy clearly demonstrated the area of efficiency improvements and reductions in losses. Exergy provided comparative better economics and recognized environmental benefits of energy by Rosen et al. [20]. Several books written over the years have been based on the concept of exergy such as that written by Kotas [21], Moran and Shapiro [22]. This study aims to provide minor improvements in the exergy efficiency and helps to remove the error between the achieved values (which do not indicate the performance deterioration as it is understood only through the limited analysis of the conventional first law) and the desired values.

2 Descriptions of Reference Plant

The Unit-IV (210 MW), based on the Russian Technology, in the coal-fired power generating station of NTPC Limited, Badarpur, Delhi, India, has been selected for

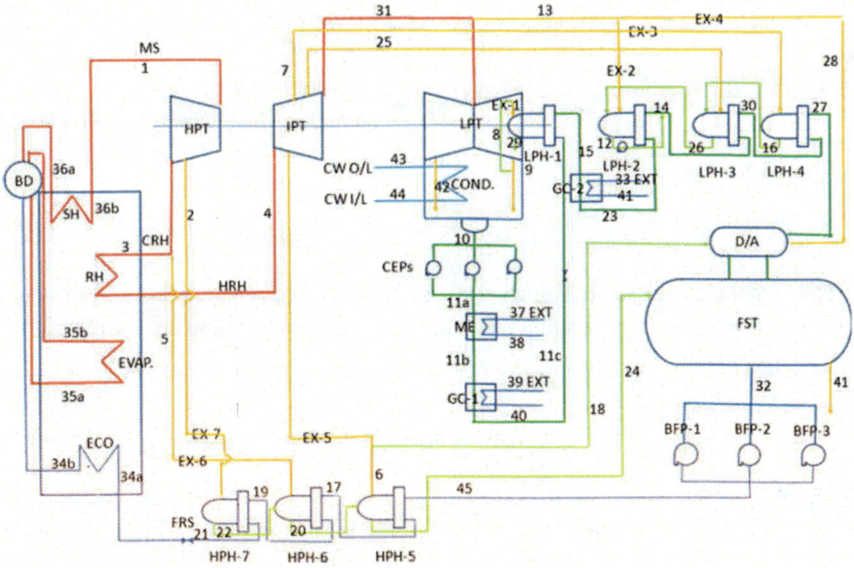

Fig. 1 Typical layout of 210 MW LMW thermal power plant

the purposes of study. This unit operates on a subcritical pressure, which is based on a Rankine cycle with reheating and regeneration. Figure 1 illustrates the entire cycle arrangement. Tables 2 and 3 in Appendix describe the flow, temperature, pressure, and different thermal properties like enthalpy and entropy at different locations of 210 MW power plants.

3 Mathematical Interpretation Used for Analysis

Intended to evaluate the performance of the plant in steady state and steady-flow process condition, the following basic equations are applicable:

Mass balance

$$\sum \dot{m}_{\text{in}} = \sum \dot{m}_{\text{out}} \tag{1}$$

Energy balance

$$\sum \dot{E}_{\text{in}} = \sum \dot{E}_{\text{out}} \tag{2}$$

$$\dot{Q} + \sum \dot{m}_{\text{in}} h_{\text{in}} = \dot{W} + \sum \dot{m}_{\text{out}} h_{\text{out}} \tag{3}$$

Exergy balance

$$\sum \dot{m}_{\text{in}} ex_{\text{in}} - \sum \dot{m}_{\text{out}} ex_{\text{out}} + \sum \dot{E}x^{Q} - \sum \dot{E}x^{W} + I = 0 \tag{4}$$

where: $I = T_0 \,.\, S_{\text{gen}}$

$\sum \dot{E}x^{Q} = \left(1 - \left(\frac{T_0}{T_t}\right)\right).Q_r$

Specific exergy

$$e_x = Q \times [(h_{\text{in}} - h_{\text{out}}) - T_0(s_{\text{in}} - s_{\text{out}})] \tag{5}$$

The equations used to analyze exergy in/out, exergy destruction and second-law efficiency condenser, regenerative feed heating, boiler, and turbine are as follows:

Condenser

$$\in_{\text{in}} = Q_{\text{sexh}} \times [(H_{\text{sexh}} - h_{\text{cond}}) - T_0(S_{\text{exh}} - s_{\text{cond}})] \tag{6}$$

$$\in_{\text{d}} = T_0 \times \left[Q_{\text{sexh}} \times (s_{\text{cond}} - S_{\text{exh}}) - Q_{\text{cw}} \times S \times \ln\left(\frac{t_2}{t_1}\right)\right] \tag{7}$$

$$\eta_{\text{II,cond}} = 1 - \frac{\in_{\text{d}}}{\in_{\text{in}}} \tag{8}$$

Regenerative Feed Heating

- Main ejector

$$\in_{\text{in}} = q_{\text{me}} \times [(H_{\text{me}} - h_{\text{dme}}) - T_0(S_{\text{me}} - s_{\text{dme}})] \tag{9}$$

$$\in_{\text{d}} = T_0 \times [q_{\text{me}} \times (s_{\text{dme}} - S_{me}) + Q_{\text{CEP}}(s_{\text{fwme}} - s_{\text{condo}})] \tag{10}$$

- Gland cooler-1

$$\in_{\text{in}} = q_{\text{GC1}} \times [(H_{\text{GC1}} - h_{\text{dGC1}}) - T_0(S_{\text{GC1}} - s_{\text{dGC1}})] \tag{11}$$

$$\in_{\text{d}} = T_0 \times [q_{\text{GC1}} \times (s_{\text{dGC1}} - S_{\text{GC1}}) + Q_{\text{CEP}}(s_{\text{fwGC1}} - s_{\text{fwme}})] \tag{12}$$

- Low pressure heater-1

$$\in_{\text{in}} = q_1 \times [(H_1 - h_{\text{d1}}) - T_0(S_1 - s_{\text{d1}})] \tag{13}$$

$$\in_{d} = T_0 \times [q_1 \times (s_{d1} - S_1) + Q_{CEP}(s_{fw1} - s_{fwGC1})] \tag{14}$$

- Gland cooler-2

$$\in_{in} = q_{sGC2} \times [(H_{GC2} - h_{dGC2}) - T_0(S_{GC2} - s_{dGC2})] \tag{15}$$

$$\in_{d} = T_0 \times [q_{sGC2} \times (s_{dGC2} - S_{GC2}) + Q_{CEP}(s_{fwGC2} - s_{fw1})] \tag{16}$$

- Low pressure heater-2

$$\begin{aligned}\in_{in} = (q_4 + q_3) \times [(h_{d3} - h_{d2}) - T_0(s_{d3} - s_{d2})] \\ + q_2 \times [(H_2 - h_{d2}) - T_0(S_2 - s_{d2})]\end{aligned} \tag{17}$$

$$\in_{d} = T_0 \times [q_2 \times (s_{d2} - S_2) + Q_{CEP}(s_{fw2} - s_{fwGC2}) + (q_4 + q_3) \times (s_{d2} - s_{d3})] \tag{18}$$

- Low pressure heater-3

$$\in_{in} = q_4 \times [(h_{d4} - h_{d3}) - T_0(s_{d4} - s_{d3})] + q_3 \times [(H_3 - h_{d3}) - T_0(S_3 - s_{d3})] \tag{19}$$

$$\in_{d} = T_0 \times [q_3 \times (s_{d3} - S_3) + Q_d(s_{fw3} - s_{fw2}) + q_4 \times (s_{d3} - s_{d4})] \tag{20}$$

- Low pressure heater-4

$$\in_{in} = q_4 \times [(H_4 - h_{d4}) - T_0(S_4 - s_{d4})] \tag{21}$$

$$\in_{d} = T_0 \times [q_4 \times (s_{d4} - S_4) + Q_d(s_{fw4} - s_{fw3})] \tag{22}$$

- De-areator

$$\begin{aligned}\in_{in} = (q_7 + q_6 + q_5) \times \left[\left(h_{d5} - h_{dD/A}\right) - T_0\left(s_{d5} - s_{dD/A}\right)\right] + q_d \\ \times \left[\left(H_{D/A} - h_{dD/A}\right) - T_0\left(S_{D/A} - s_{dD/A}\right)\right]\end{aligned} \tag{23}$$

$$\in_{\mathrm{d}} = T_0 \times [q_{\mathrm{d}} \times (s_{\mathrm{dD/A}} - S_{\mathrm{D/A}}) + Q_{\mathrm{fw}}(s_{\mathrm{fwD/A}} - s_{\mathrm{fw4}}) + (q_7 + q_6 + q_5) \times (s_{\mathrm{dD/A}} - s_{\mathrm{d5}})] \quad (24)$$

- High pressure heater-5

$$\in_{\mathrm{in}} = (q_7 + q_6) \times [(h_{\mathrm{d6}} - h_{\mathrm{d5}}) - T_0(s_{\mathrm{d6}} - s_{\mathrm{d5}})] + q_5 \times [(H_5 - h_{\mathrm{d5}}) - T_0(S_5 - s_{\mathrm{d5}})] \quad (25)$$

$$\in_{\mathrm{d}} = T_0 \times [q_5 \times (s_{\mathrm{d5}} - S_5) + Q_{\mathrm{fw}}(s_{\mathrm{fw5}} - s_{\mathrm{fwD/A}}) + (q_7 + q_6) \times (s_{\mathrm{d5}} - s_{\mathrm{d6}})] \quad (26)$$

- High pressure heater-6

$$\in_{in} = q_7 \times [(h_{d7} - h_{d6}) - T_0(s_{d7} - s_{d6})] + q_6 \times [(H_6 - h_{d6}) - T_0(S_6 - s_{d6})] \quad (27)$$

$$\in_{\mathrm{d}} = T_0 \times [q_6 \times (s_{\mathrm{d6}} - S_6) + Q_{\mathrm{fw}}(s_{\mathrm{fw6}} - s_{\mathrm{fw5}}) + q_7 \times (s_{\mathrm{d6}} - s_{\mathrm{d7}})] \quad (28)$$

- High pressure heater-7

$$\in_{\mathrm{in}} = q_7 \times [(H_7 - h_{\mathrm{d7}}) - T_0(S_7 - s_{\mathrm{d7}})] \quad (29)$$

$$\in_{\mathrm{d}} = T_0 \times [q_7 \times (s_{\mathrm{d7}} - S_7) + Q_{\mathrm{fw}}(s_{\mathrm{fw7}} - s_{\mathrm{fw6}})] \quad (30)$$

- Second-law efficiency (heater)

$$\eta_{\mathrm{II,AnyHeatter}} = 1 - \frac{\in_{\mathrm{d}}}{\in_{\mathrm{in}}} \quad (31)$$

Boiler

- Economizer

$$\in_{\mathrm{in,fg}} = Q_{\mathrm{fg}} \times C_{\mathrm{pfg}} \times \left[(t_{\mathrm{fgeci}} - t_{\mathrm{fgeco}}) - T_0 \times \ln\left(\frac{t_{\mathrm{fgeci}}}{t_{\mathrm{fgeco}}}\right) \right] \quad (32)$$

$$\in_{\text{rec,fw}} = Q_{\text{wec}} \times [(h_{\text{weco}} - h_{\text{weci}}) - T_0(s_{\text{weco}} - s_{\text{weci}})] \tag{33}$$

- Evaporator

$$\in_{\text{in,fg}} = Q_{\text{fg}} \times C_{\text{pfg}} \times \left[\left(t_{\text{fgevi}} - t_{\text{fgevo}}\right) - T_0 \times \ln \times \left(\frac{t_{\text{fgevi}}}{t_{\text{fgevo}}}\right)\right] \tag{34}$$

$$\in_{\text{rec,fw}} = Q_{\text{sg}} \times [(h_{\text{wevo}} - h_{\text{wevi}}) - T_0(s_{\text{wevo}} - s_{\text{wevi}})] \tag{35}$$

- Superheater

$$\in_{\text{in,fg}} = Q_{\text{fg}} \times C_{\text{pfg}} \times \left[\left(t_{\text{fgs}hi} - t_{\text{fgsh}ho}\right) - T_0 \times \ln \times \left(\frac{t_{\text{fgs}hi}}{t_{\text{fgs}ho}}\right)\right] \tag{36}$$

$$\in_{\text{rec,fw}} = Q_{\text{ms}} \times [(h_{\text{ws}ho} - h_{\text{ws}hi}) - T_0(s_{\text{ws}ho} - s_{\text{ws}hi})] \tag{37}$$

- Reheater

$$\in_{\text{in,fg}} = Q_{\text{fg}} \times C_{\text{pfg}}\left[\left(t_{\text{fgr}hi} - t_{\text{fgrho}}\right) - T_0 \times \ln \times \left(\frac{t_{\text{fgr}hi}}{t_{\text{fgr}ho}}\right)\right] \tag{38}$$

$$\in_{\text{rec,fw}} = Q_{\text{hrh}} \times [(h_{\text{hrh}} - h_{\text{crh}}) - T_0(s_{\text{hrh}} - s_{\text{crh}})] \tag{39}$$

- Exergy destruction and second-law efficiency (boiler component)

$$\in_{\text{d}} = \in_{\text{in,fg}} - \in_{\text{rec,fw}} \tag{40}$$

$$\eta_{\text{II, Boiler}} = 1 - \frac{\in_{\text{d}}}{\in_{\text{in,fg}}} \tag{41}$$

Turbine

- High pressure turbine

$$\begin{aligned}\in_{\text{rec,HPT}} = {} & Q_{\text{ms}} \times (H_{\text{ms}} - H_7) + (Q_{\text{ms}} - q_7) \times (H_7 - H_{\text{crh}}) \\ & - T_0 \times [Q_{\text{ms}} \times (S_s - S_7) + (Q_{\text{ms}} - q_7) \times (S_7 - S_{\text{crh}})]\end{aligned} \tag{42}$$

$$\in_{\text{rec}} = \in_{\text{in}} - \in_{\text{out}} \tag{43}$$

$$\in_{\text{d}} = \in_{\text{in}} - \in_{\text{out}} - W_{\text{HPT}} = \in_{\text{rec}} - W_{\text{HPT}} \tag{44}$$

- Medium pressure turbine

$$\begin{aligned}\in_{\text{rec,IPT}} = {} & Q_{\text{hrh}} \times (\text{H}_{\text{hrh}} - H_5) + (Q_{\text{hrh}} - q_5 - q_{\text{d}}) \\ & \times (H_5 - H_4) + (Q_{\text{hrh}} - q_5 - q_{\text{d}} - q_4) \\ & \times (H_4 - H_3) + (Q_{\text{hrh}} - q_5 - q_d - q_4 - q_3) \\ & \times (H_3 - H_2) - T_0 \times [Q_{\text{hrh}} \times (S_{\text{hrh}} - S_5) + (Q_{\text{hrh}} - q_5 - q_{\text{d}}) \\ & \times (S_5 - S_4) + (Q_{\text{hrh}} - q_5 - q_{\text{d}} - q_4) \\ & \times (S_4 - S_3) + (Q_{\text{hrh}} - q_5 - q_{\text{d}} - q_4 - q_3) \times (S_3 - S_2)]\end{aligned} \tag{45}$$

$$\in_{\text{rec}} = \in_{\text{in}} - \in_{\text{out}},\ \in_{\text{d}} = \in_{\text{in}} - \in_{\text{out}} - W_{\text{IPT}} = \in_{\text{rec}} - W_{\text{IPT}} \tag{46}$$

- Low pressure turbine

$$\begin{aligned}\in_{\text{rec,LPT}} = {} & Q_{\text{LPT}} \times (H_{\text{LPT}} - H_1) + (Q_{\text{LPT}} - q_1) \times (H_1 - H_{\text{bu}}) - T_0 \\ & \times [Q_{\text{LPT}} \times (S_{\text{LPT}} - S_1) + (Q_{\text{LPT}} - q_1) \times (S_1 - S_{\text{bu}})]\end{aligned} \tag{47}$$

$$\in_{\text{rec}} = \in_{\text{in}} - \in_{\text{out}},\ \in_{\text{d}} = \in_{\text{in}} - \in_{\text{out}} - W_{\text{LPT}} = \in_{\text{rec}} - W_{\text{LPY}} \tag{48}$$

- Second-law efficiency (turbine)

$$\eta_{\text{II,Turbine}} = 1 - {}^{\in_{\text{d}}}/_{\in_{\text{rec}}} \tag{49}$$

Other equations used for calculations will be:

$$\text{Efficiency Defect}(\delta) = \frac{\in_{\text{d}}}{\in_{\text{f}}} \tag{50}$$

where: $\in_{\text{f}}$ = Exergy Supplied by the fuel

$$\text{Efficiency defect of the plant}(\delta_{\text{pl}}) = \sum_{i=1}^{n} \delta_i \tag{51}$$

$$\text{Rational efficiency of the plant}(\varphi) = 1 - \delta_{\text{pl}} = 1 - \sum_{i=1}^{n} \delta_i \tag{52}$$

4 Results and Discussion

NTPC Badarpur plant was selected for analysis. Figure 1 shows the arrangement of components. As it is based on Russian design, it consists of impulse reaction turbines, seven nos. of heat exchangers (for regeneration), and reheater. A vertical, tangentially fired boiler uses low calorific value sub-bituminous coal on six elevations. On the basis of the above-mentioned equations and parameters, different values of efficiency defects, exergy efficiencies, irreversibility of different components, and the rational efficiency have been summarized and outcomes tabulated in Table 1.

The purpose for conduction of the exergy analysis in brief is to reflect the effect of different ambient temperatures on irreversibility as well as rational efficiency. The exergy analysis has been conducted for several ambient temperatures ranging from 288 to 313 K. After analysis, it was investigated that the boiler has much higher irreversibility rates in comparison with the other components. Irreversibility rates of boiler, turbine, RFH system, and the entire plant increase with increase in ambient temperature, while condenser reciprocates in the opposite direction, i.e., irreversibility rate decreases with increase in temperature. Figure 2 incorporates the variation of efficiency defects, exergetic efficiency, and irreversibility rates of components with respect to different ambient temperature.

5 Conclusion

The first portion of the study presents a blend of energy and exergy analysis with the data collected from real operational unit. Analysis has been conducted to illustrate the effect of end state temperature, i.e., T_0 on exergetic performance of the plant. Main emphasis of this investigation is to examine the rational efficiency with the help of energy and exergy calculation at 298 K, most common dead state temperature in north India. Another reason to conduct this elaborative exercise for exergy analysis is that it gives a better insight about exergy losses (irreversibility) and energy inefficiencies (efficiency defects) of the system. As per the study, the boiler is found to have a largest improvement potential as it is the major exergy-consuming unit in the plant. After boiler, performance of hcaters is not up to the mark as the exergetic efficiency lies between 71.41 and 81.03%, so it also has some improvement potential. This study reflects that 1 °C increase of the environmental temperature causes an enhancement of 0.29 MW in total irreversibility rate and reduction of 0.068% in rational efficiency of the plant.

Table 1 Overall analysis at different ambient temperatures for entire plant

Temp. (K)	288	293	298	303	308	313
Efficiency defects for different ambient temperatures						
Boiler	0.5229	0.5252	0.5275	0.5298	0.5320	0.5343
Turbine	0.0951	0.0962	0.0973	0.0984	0.0995	0.1006
Condenser	0.0215	0.0211	0.0208	0.0204	0.02	0.0196
Feed water Heater	0.0222	0.0225	0.0229	0.0233	0.0237	0.0241
Total eff. defect	0.6617	0.665	0.6685	0.6719	0.6752	0.6786
Rational efficiency	0.3383	0.335	0.3315	0.3281	0.3248	0.3214
Exergy efficiency for different ambient temperatures						
Boiler	0.4037	0.3973	0.3909	0.3843	0.3777	0.3709
Turbine	0.8351	0.8335	0.8319	0.8303	0.8287	0.8271
Condenser	0.6378	0.5573	0.4361	0.2331	0.0301	−0.1729
Feed water heater	0.7909	0.7802	0.7688	0.7565	0.7434	0.7293
Irreversibility (KW) for different ambient temperatures						
Boiler	222648.4	223614.4	224580.4	225546.5	226512.5	227478.6
Turbine	40470.3	40945.1	41419.9	41894.8	42369.6	42844.5
Condenser	9155.71	8999.47	8843.23	8686.99	8530.75	8374.51
Feed water heater	9434.42	9598.21	9762.01	9925.8	10089.6	10253.4
Total Irreversibility	281708.83	283157.18	284605.54	286054.09	287502.45	288951.01

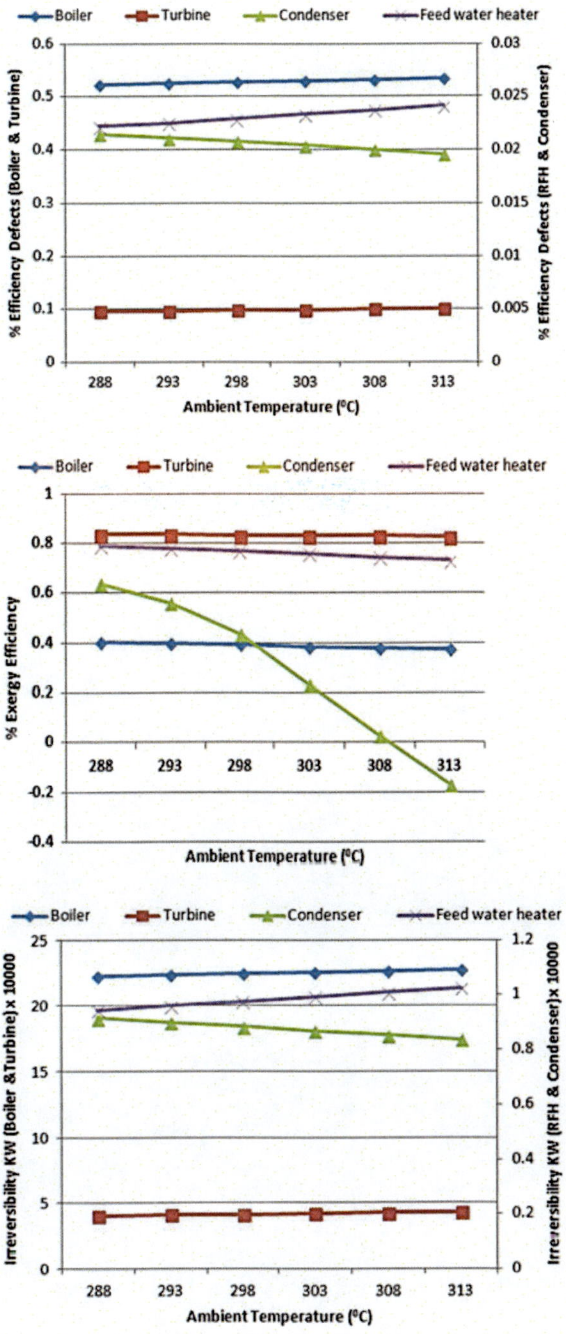

Fig. 2 Graph between different ambient temperatures and outcomes

Appendix

See Appendix Table 2 and 3.

Table 2 Parameters considered with value and abbreviations

Steam/condensate/feed water parameters

Pt.	Flow kg/s		Temp. °C	Press. kg/cm^2	Enthalpy kJ/kg		Entropy kJ/kg K	
	Abb.	Value	Value	Value	Abb.	Value	Abb.	Value
1.	Q_{ms}	171.57	540	127.88	H_{ms}	3447.43	S_{ms}	6.586
2.	q_7	7.7611	369	39.6	H_7	3137.09	S_7	6.6
3.	Q_{CRH}	150.59	333.65	24.1	H_{CRH}	3056.6	S_{CRH}	6.7
4.	Q_{HRH}	150.59	540	23.13	H_{HRH}	3531.6	S_{HRH}	7.45
5.	q_6	11.7	314	25.8	H_6	3041.04	S_6	6.686
6.	q_5	3.7542	440	12.1	H_5	3339.82	S_5	7.50
7.	q_4	7.4406	368	6.47	H_4	3183.2	S_4	7.6
8.	q_1	2.3503	68	0.28	H_1	2581.85	S_1	7.6835
9.	Q_{sexh}	126	46	0.1	h_{sexh}	2296	s_{sexh}	7.27
10.		126	45	0.1		191.81		0.649
11a.	Q_{CEP}	126	47.1	21	h_{condo}	188.3	s_{condo}	0.638
11b.		126	50	21	h_{fwme}	196.6	s_{fwme}	0.664
11c.		126	61	21	H_{fwGC1}	209.2	S_{fwGC1}	0.703
12.		19	103.9	1.28	h_{d2}	435.7	s_{d2}	1.35
13.	q_2	6.54	172	1.28	H_2	2813.73	S_2	7.8
14.		145	99.5	21	h_{fw2}	416.2	s_{fw2}	1.295
15.		126	68	21	h_{fw1}	255	s_{fw1}	0843
16.		7.44	157.8	6.47	h_{d4}	666.5	s_{d4}	1.922
17.		190	180	183	h_{fw5}	771.7	s_{fw5}	2.157
18.	q_d	3.9217	440	12.1	$H_{D/A}$	3346.91	$S_{D/A}$	7.50
19.	Q_{fw}	190	218.4	183	h_{fw6}	941.42	s_{fw6}	2.475
20.		19.46	198	25.8	h_{d6}	849.84	s_{d6}	2.310
21.		190	239	183	h_{fw7}	1035.15	s_{fw7}	2.66
22.		7.76	229	39.6	h_{d7}	985.76	s_{d7}	2.599
23.		126	68.2	21	h_{fwGC2}	284.4	s_{fwGC2}	0.929
24.		23.21	174	12.1	h_{d5}	736.93	s_{d5}	2.080
25.	q_3	5.4286	252	2.78	H_3	2976.16	S_3	7.65
26.		5.43	124.9	2.78	h_{d3}	524.7	s_{d3}	1.581
27.	Q_d	145	152.1	21	h_{fw4}	642.30	s_{fw4}	1.861
28.		6.54	172	1.28		2813.73		7.8
29.		2.04	67	0.28	h_{d1}	271.90	s_{d1}	0.893
30.		145	122.9	21	h_{fw3}	517.42	s_{fw3}	1.559
31.	Q_{LPT}	123.06	172	1.237	H_{LPT}	2818.75	S_{LPT}	7.61

(continued)

Table 2 (continued)

Steam/condensate/feed water parameters								
Pt.	Flow kg/s		Temp. °C	Press. kg/cm^2	Enthalpy kJ/kg		Entropy kJ/kg K	
	Abb.	Value	Value	Value	Abb.	Value	Abb.	Value
32.		190.5	165	7.5	$h_{fwD/A}$	697	$s_{fwD/A}$	1.991
33.	q_{sgGC2}	1.31	315.5	0.35	h_{GC2}	3107.2	s_{GC2}	8.88
34a.	Q_{wec}	184.288	239	183	h_{weci}	1035.15	s_{weci}	2.662
34b.		184.288	320	183	h_{weco}	1447.95	s_{weco}	3.408
35a.	Q_{sg}	184.288	329	143.11	h_{wevi}	1513.9	s_{wevi}	3.528
35b.		184.288	329	143.11	h_{wevo}	1990	s_{wevo}	4.354
36a.	Q_{ms}	171.57	329	127.88	h_{wshci}	2668	s_{wshci}	5.32
36b.		171.57	540	127.88	h_{wsho}	3447.33	s_{wsho}	6.586
37		0.42	137	0.25		2757.1		8.17
38.		0.42	53	0.25	h_{dme}	221.8	s_{dme}	0.74
39.	q_{GC1}	0.6	267	0.2	h_{GC1}	3010	s_{GC1}	8.864
40.		0.6	55	0.2	h_{dGC1}	230.1	s_{dGC1}	0.767
41.		1.31	72	0.35	h_{dGC2}	301.2	s_{dGC2}	0.978
42.			46	0.1 (DF = 0.88)	H_{bu}	2389.25	S_{bu}	7.7901
Cooling water parameter								
43.	Q_{cw}	7000	T_1	300				
44.	Q_{cw}	7000	T_2	309				

Table 3 Flue gas parameters

Point	Flow	Economizer		Evaporator		Superheater		Reheater	
Unit	kg/s	Inlet (°C)	Outlet (°C)	Inlet (°C)	Outlet (°C)	Inlet (°C)	Outlet (°C)	Inlet (°C)	Outlet (°C)
Abb.	Q_{fg}	t_{fgeci}	t_{fgeco}	t_{fgevi}	t_{fgevo}	t_{fgshi}	t_{fgsho}	t_{fgrhi}	t_{fgrho}
Value	315.2	360	540	950	1250	550	1080	750	980

References

1. Bhatt, M.S., Rajkumar, J.: Performance enhancement in coal fired thermal power plants. Part I: Boilers and Part II: Steam Turbines, Energy **23**(6), 489–515 (1999)
2. Aljundi, Islam H.: Energy and exergy analysis of a steam power plant in Jordan. Appl. Therm. Eng. **29**, 324–8 (2009)
3. Datta, A., Sengupta, S., Duttagupta, S.: Exergy analysis of a coal-based 210 MW thermal power plant. Int. J. Energy Res. **31**, 14–28 (2007)
4. Rosen, M.A.: Energy- and exergy-based comparison of coal-fired and nuclear steam power plants. Int. J. Exergy **3**, 180–92 (2001)
5. Hasan, H.E., Ali, V.A., Burhanettin, Ahmet D., Suleyman, H.S., Bahri, S., Ismail, T., Cengiz, G., Selcuk, A.: Comparative energetic and exergetic performance analyses for coal-fired thermal power plants in Turkey. Int. J. Therm. Sci. **48**, 2179–86 (2009)
6. Naterer, G.F., Regulagadda, P., Dincer, I.: Exergy analysis of a thermal power plant with measured boiler and turbine losses. Appl. Therm. Eng. **30**, 970–6 (2010)
7. Zubair, S.M., Habib, M.A.: Second-law-based thermodynamic analysis of regenerative-reheat Rankine-cycle power plants. Energy **17**, 295–301 (1992)
8. Ganapathy, T., Alagumurthi, N., Gakkhar, R.P., Murugesan, K.: Exergy analysis of operating lignite fired thermal power plant. J. Eng. Sci. Technol. Rev. **2**, 123–30 (2009)
9. Reddy, B.V., Butcher, C.J.: Second law analysis of a waste heat recovery based power generation system. Int. J. Heat Mass Transf. **50**, 2355–63 (2007)
10. Srinivas, T., Gupta, A.V.S.S.K.S., Reddy, B.V.: Performance simulation of 210MW natural gas fired combined cycle power plant. Int. J. Energy, Heat Mass Transf. **29**, 61–82 (2007)
11. Suresh MVJJ, Reddy KS, Ajit KK (2006) Energy and exergy analysis of thermal power plants based on advanced steam parameters. In: National conference on advances in energy research. IITB, India
12. Oktay, Z.: Investigation of coal-fired power plants in Turkey and a case study: can plant. Appl. Therm. Eng. **29**, 550–7 (2009)
13. Zhang, Chao, Wang, Yan, Zheng, Chuguang, Lou, Xinsheng: Exergy cost analysis of a coal fired power plant based on structural theory of thermoeconomics. Energy Convers. Manag. **47**(7–8), 817–843 (2006)
14. Jassim, R.K., Habeebullah, B.A., Habeebullah, A.S. Exergy analysis of carryover leakage irreversibilities of a power plant regenerative air heater. In: Proceedings of the Institution of Mechanical Engineers (Part A). Journal of Power and Energy **218**(1), 23–32 (2004)
15. Habib, M., Zubair, S.:. Second law based thermodynamics analysis of regenerative-reheat rankine cycle power plants. Energy **17**, 295–301 (1992)
16. Reddy, V.S., Kaushik, S.C., Tyagi, S.K., Panwar, N.L.: An approach to analyse energy and exergy analysis of thermal power plants: a review. Smart Grid Renew. Energy **1**, 143–152 (2010)
17. Kaushik, S.C., Reddy, V.S., Tyagi, S.K.: Energy and exergy analyses of thermal power plants: a review. Renew. Sustain. Energy Rev. **15**(4), 1857–1872 (2011)
18. Kopac, M., Hilalci, A.: Effect of ambient temperature on the efficiency of the regenerative LMZ power plant in Turkey Reheat Catalag **27**, 1377–1385 (2007)
19. Ahmadi, G.R., Toghraie, D.: Energy and exergy analysis of Montazeri Steam Power Plant in Iran. Renew. Sustain. Energy Rev. **56**, 454–463 (2016)
20. Rosen, M.A., Dincer, I., Kanoglu, M.: Role of exergy in increasing efficiency and sustainable and reducing environmental impact. Energy Policy **36**, 128–137 (2008)
21. Kotas, T.J.: The exergy method of thermal plant analysis. Butterworth and Co., London (1985)
22. Moran, M.J., Shapiro: Fundamental of Engineering Thermodynamics. New York: Wiley Publicaion (1995)

Microstructural Evolution and Enhanced Mechanical Properties of Atomization Cast Al–40% Si Alloys

Krishna Vijay, Subodh Kumar Sharma, Prashant Vashishtha, Ajay Kumar, Kumari Archana, and Pratibha Kumari

1 Introduction

Due to numbers of properties such as thermal conductivity, high strength-to-weight ratio, electrical conductivity and good corrosion resistance, aluminium alloys are widely used in day-to-day life. When wear is a main criterion of selection of these materials, aluminium alloys are extensively used. Aluminium has got a good capability to respond to various finishing processes. That is why, these alloys are widely used in automobile, electrical and aerospace applications.

Due to the silicon in aluminium silicon alloy, wear resistance improves [1, 4]. This is happened due to high hardness of Si in the alloy. From the several Al alloys, aluminium silicon alloys have observed good response for various applications such as in aerospace, automobile and electrical equipment industries due to mechanical properties. There are some of these applications, which include high-performance automobile engine parts. These parts have a possibility that they can be used in aviation industry. The atomization casting of Al-18% Si is performed previously [5].

2 Experimental Set-up

The schematic figure of atomization casting set-up is shown in Fig. 1. The photograph of atomized casting set-up is shown in Fig. 2. In the set-up, the atomization nozzle is placed above the cooling chamber. A heater is placed just above the nozzle, and a crucible is placed above the heater. The alloy was melt into a crucible in the furnace. The melt flow tube is connected to crucible passes through the heater to cooling chamber via nozzle. The molten metal is poured in the crucible after starting the flow

K. Vijay (✉) · S. K. Sharma · P. Vashishtha · A. Kumar · K. Archana · P. Kumari
KIET Group of Institutions, Ghaziabad, India
e-mail: krishna.ojha@kiet.eddu

M. Muzammil et al. (eds.), *Recent Advances in Mechanical Engineering*, Lecture Notes in Mechanical Engineering, https://doi.org/10.1007/978-981-15-8704-7_60

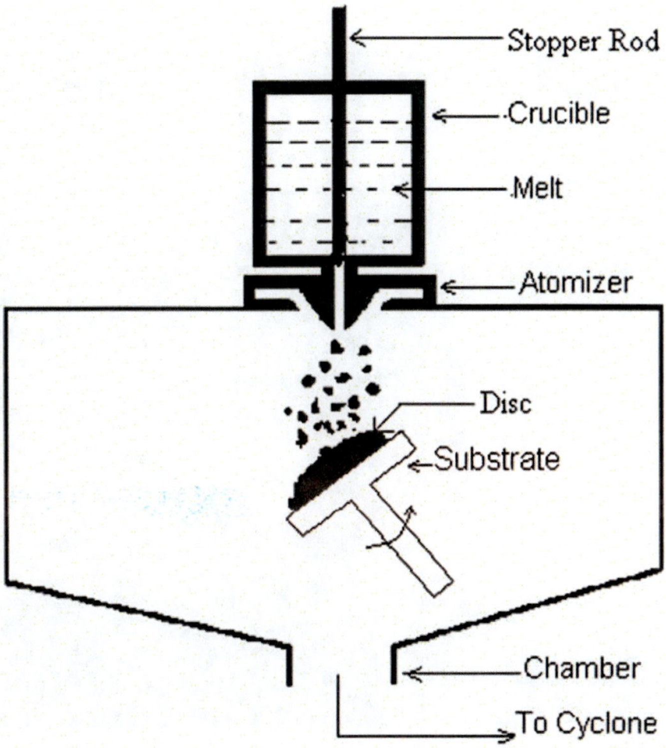

Fig. 1 Experimental set-up for atomization casting

of gas. The atomized melt stream was collected at 450 mm distance on to the copper substrate (of 6-mm thickness and 200-mm diameter).

About 1 kg Al–Si alloy of 40%Si was superheated to 200 K above the melting point in the crucible to carry out four runs.

Samples were taken from the substrate to see its microstructure using standard technique of viewing microstructure. The microstructures of sample were seen on Leitz optical microscope.

Brinell hardness is measured by hardness testing machine. The tensile strength of spray cast Al–40% Si alloy was determined using universal testing machine model on a specimen gauge length of 25- and 5-mm diameter.

3 Results and Discussion

Figure 3 illustrates ingot cast optical micrograph of the alloys. It shows the columnar grains of α-Al of 60–70 μm length. Figure 4 describes the optical micrograph of

Fig. 2 Photograph of atomization casting set-up

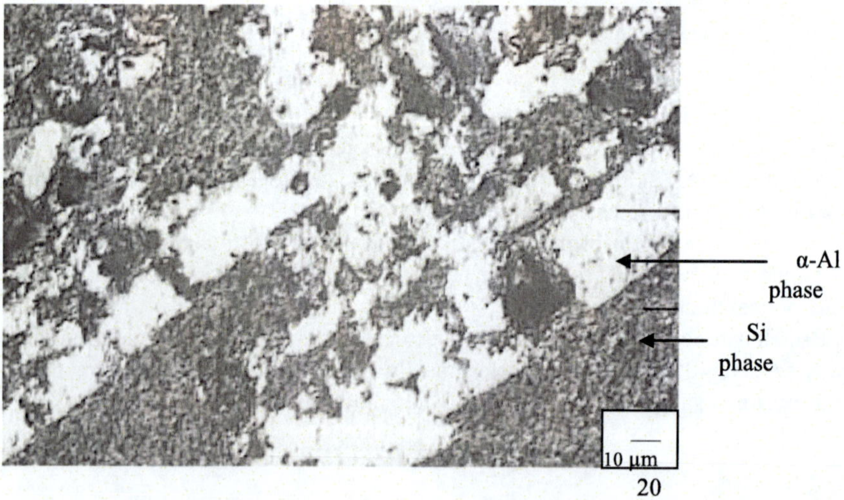

Fig. 3 Optical micrograph of as cast Al–40% Si showing long columnar grains of α-Al

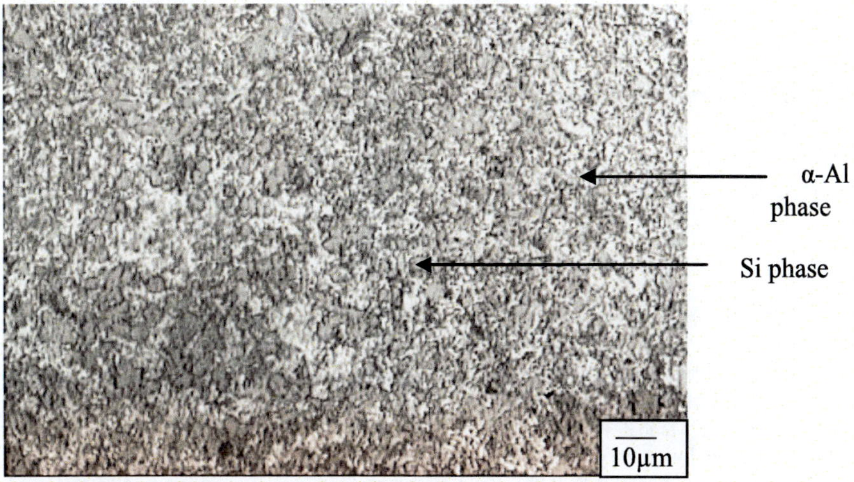

Fig. 4 Optical micrograph of atomization-deposited Al–40%Si uniform distribution of Si-rich grain

atomization cast alloy. It shows the distribution of primary Si is uniform throughout the figure having a size of 2–5 μm.

The Brinell hardness and tensile properties of ingot cast and atomization cast Al–40%Si are listed in the table given below.

The Brinell hardness of atomized cast is higher as compared to that of ingot cast alloy. The increase in hardness of Al–Si alloys with increase in Si content occurs because Si is a harder phase and it is uniformly distributed throughout the casting [2]. The yield strength and ultimate tensile strength data for ingot cast were not found due to difficulty in machining.

Fine grain formation in atomization casting and uniform distribution of second phase (due to rapid solidification) leads to high hardness (Table1) as compared to that of ingot cast alloys [3]. The yield strength and ultimate tensile strength of ingot cast Al–40%Si were not found due to the inability to machine the ingot cast alloy.

Table 1 Mechanical properties of ingot cast and atomized cast alloy

S. No.	Casting route	BHN	Yield strength (MPa)	UTS (MPa)
1	Ingot casting	90	–	–
2	Spray cast	125	95	108

4 Conclusions

The optical microstructure shows the grain refinement in the atomization cast alloy. The uniform distribution of secondary phase occurs in the spray casting. The Brinell hardness of ingot cast Al–Si alloys is lower than that of atomization cast alloys. The tensile strength for atomization cast alloys is found to be better than ingot casting.

References

1. Tenekdjiev, N., Gruzleski, J.E.: Hypereutectic aluminium-silicon casting alloys—a review. Cast Metals **3**(2), 96–105 (1990)
2. Srivastava, V.C., Mandal, R.K., Ojha, S.N.: Microstructure and mechanical properties of Al–Si alloys produced by spray forming process. Mat. Sci. Eng. A **304–306**, 555–558 (2001a)
3. Srivastava, V.C., Mandal, R.K., Ojha, S.N.: Microstructural evolution during spray forming of an Al-18Si alloy. J. Mater. Sci. Lett. **20**, 27–29 (2001b)
4. Singh, D., Tomar, A., Ojha, K.V.: Gas atomization of melts. Steel Res. Int. **78–3**, 241–247 (2007)
5. Raju, K., Ojha, S.N.: Effect of spray forming on the microstructure and wear properties of Al–Si alloys. Procedia Mater. Sci. **5**, 345–354 (2014)

Dynamic Analysis of Boring Bar for Advancement of Tool Life While Lessen the Machining Time Through FEA

Shweta Sonkar and Prabhat Kumar Sinha

1 Introduction

Boring is the process of enlarging the drilled hole or cast by means of single-point cutting tool, for achieving maximum shortest possible accuracy of hole diameter, or tapered hole can also be cut. Boring seems like the internal-diameter counterpart to turning, which cuts external diameters (Fig. 1) [1].

Micro-boring is acceptable for various diameters and chips can be easily removed due to "cooling through spindle." A small boring bar inserted into one of the hole accompanied by superior range through fastener. Chatter is sourced through difference in chip width. Vibration shows quick tool wearing and bad exterior appearance. The factor which affects chattering is tool edge kinematics [2]. An extended boring bar fixed at one end is used to unguarded because of their less inflexibility. To make it extremely effective, statically and dynamically inflexible improve static stiffness and optimize the bar geometry for dynamic stiffness, increase the damping of the structure. To avoid boring bar chattering, the different damping methods are used. The present work enquires an alternative method for straining of spontaneous amplitude or incrementing the dynamic flexibility. Representation of boring bar is inspected by finite element method. Between preceding and succeeding cutting stage shifting occurs, it energizes tool for chattering of frequency (ω_c) which is near about to the spontaneous frequency (ω_n). Superior vibration while boring is radial and therefore examines only one-dimensional vibration. The systemic shifting in radial direction can be found by:

$$\varphi(i) = \frac{\frac{\omega_n^2}{r}}{s^2 + 2X\omega_n + \omega_n^2} \tag{1}$$

S. Sonkar · P. K. Sinha (✉)
Department of Mechanical Engineering, SHUATS, Prayagraj, UP, India
e-mail: shwetasonkar478@gmail.com

M. Muzammil et al. (eds.), *Recent Advances in Mechanical Engineering*, Lecture Notes in Mechanical Engineering, https://doi.org/10.1007/978-981-15-8704-7_61

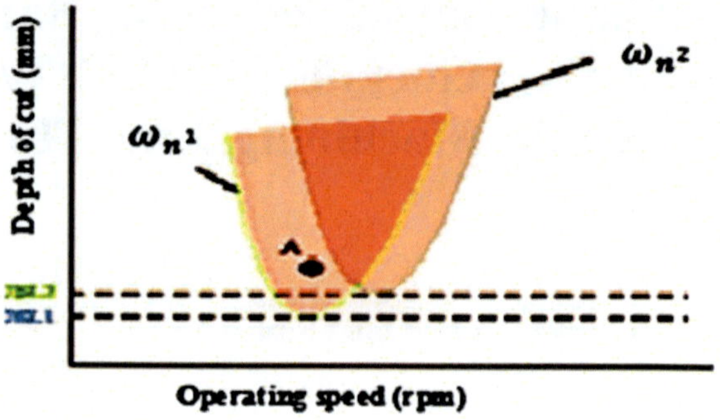

Fig. 1 Boring process

Here, ω_n = Spontaneous amplitude (Rad/s), r = Modal Stiffness (N/M), X = Damping Ratio. In the present work, only one means is investigated. For finding the condemn width use following:

$$n_{\text{lim}=\frac{-1}{2F_i * S(\omega_c)}} \tag{2}$$

Here F_i = Cutting force coefficient (feed direction), $S(\omega_c)$ = Chatter frequency transfer function, ω_m = Direction of Chip width. Chatter frequency is proportional to cutting speed by (Fig. 2):

$$\frac{\omega_m}{\omega_n} = 2\pi r + \varepsilon \tag{3}$$

According to the present work, spontaneous amplitude governs by the volume of the fluid. In order to change, the cutting operation tool moves in an unsteady region.

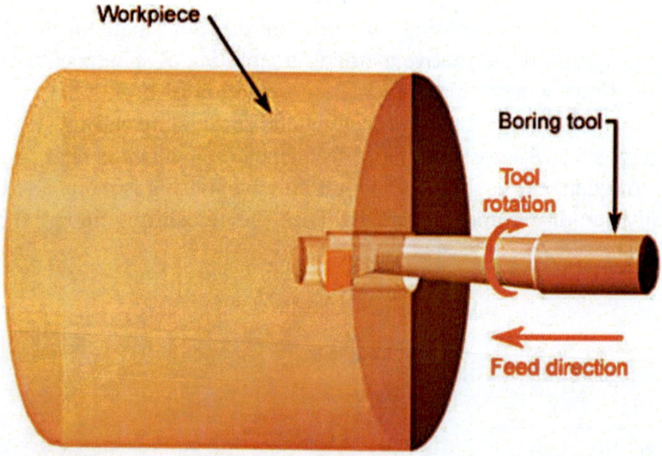

Fig. 2 Effect of changing ω_n and ζ on chatter stability

In order to continuous change in bar design, a governing system requires which sense variation. By changing the design of bar, amplitude variation (chatter) can be find [3].

2 Machining Parameters of Boring

The variables that include in the operation are machine rpm, amount of feed, and depth of cut, selected numerically [4]. Parameter that affects the boring operations is as follows:-

2.1 *Spindle RPM*

It refers the spindle speed and work. It is expressed by product of the spindle speed to time with perimeter meter and represented by **m/min**. By varying the diameter, cutting speed also varies while rotating speed remains same.

$$v = \pi DN/1000\,\text{m/min} \quad (4)$$

where v = Rotating/Cutting speed (m/min), D = Opening diameter of the work (mm), N = spindle speed (rpm).

2.2 *Rate of Feed*

It refers to the rate at which the tool moves along with the direction of cutting and is directly proportional to the spindle speed; it is represented by **mm/rev min**.

$$1m_f = FN\text{ mm} \quad (5)$$

where m_f = Feed (mm/min), F = Feed (mm/rev), N = Spindle speed (rpm).

2.3 *Cutting Depth*

It refers the width of chip remove by one time travel of tool, represented by **mm**.

$$d_{\text{cut}} = D - d/2\text{ mm} \quad (6)$$

3 Vibration Basics

Oscillation is a repetitive variation with respect to time. Mechanical systems are prone to vibrate if they can store energy in two different forms, usually potential and kinetic, in a way that energy can flow from one form to the other.

3.1 *Vibration in the Boring Operation and Its Influence*

In manufacturing industries, vibrations are introduced in various cutting operations. To eliminate vibration, proper parameters must be taken at the production planning stage like cutting velocity, feed rate, cutting depth with respect to the workpiece as an integration. It influences the tool life and surface integrity. Then the conclusion adopted from the following relation:

$$T = \frac{48.36 * 10}{v4f1.6d_w.28} \tag{7}$$

where v = Cutting velocity (m/min). f = Feed rate (mm/rev), d_w = Cutting depth(mm) [5].

3.2 *Cutting Force Modeling for Boring Operation*

In boring operation, the cutting forces are represented by friction force (F_f), which resolves into the feed direction to find feed force (F_f) and radial direction to find radial force (F_r) and tangential force (F_t).The distribution of chip thickness is different along the different cutting edge contact point and dependent on the tool dimension, rate of feed, and cutting depth in radial direction (Fig. 3) [6].

At any point, the differential functional cutting forces for the cutting edge contact modeled for local chip area (∂A) and contact length of the chip cutting edge (∂L) are:

$$\partial F_t = \partial F_{t1} + \partial F_{t2} = K_{t1} \cdot \partial A + K_{t2} \cdot \partial L \tag{8}$$

$$\partial F_f = \partial F_{f1} + \partial F_{f2} = K_{f1} \cdot \partial A + K_{f2} \cdot \partial L \tag{9}$$

where ∂F_{t1} and ∂F_{f1} = given by the chip removal, ∂F_{f1t2} and ∂F_{f2} = given by rubbed surface finish edge, K_{t1} and K_{f1} = cutting coefficients, which depends on the local rake, inclination, chip flow angles, cutting conditions and work–tool material properties, K_{t2} and K_{f2} = cutting coefficients, which depends on the cutting edge condition, It can be determined by orthogonal (2-D) to oblique (3D) alteration [7].

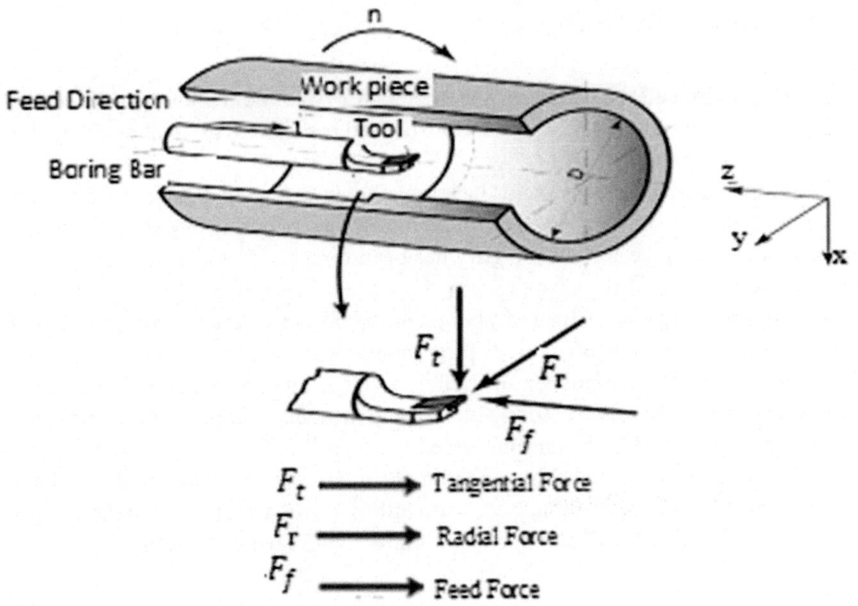

Fig. 3 Illustration of force directions in boring process

3.3 Vibration Control Techniques

It is a periodical motion, by means of suppression or elimination of an undesirable variation into desirable term, finding the maximum variations is the main aim of vibration engineering. Vibration can be minimized by active damping, passive damping or semi-active damping methods for the strengthening of boring bar [8].

Active Type Damping
This method is used to reduce variation in boring bar by means of system modification. In this vibration depletion, signal transfers during the machining operation, then inspecting the mode of amplitude. These systems need wire to transmit the energy.

Semi-Active Damping
Semi-active system is similar like active damping in which small amounts of vitality need without variation, energy cannot be released but it controls other functions.

Passive Damping
Passive damping is used to change the mechanical work into heat energy, which is responsible for the variation, exhausted by using visco-elastic materials.

4 Review of Literature

Lin et al. enquired all the necessary parameters for Aluminum alloy 6061T6 of CN boring optimization by GRA analysis. Rate of feed recognized as an effective factor of maximum roughness.

Balamuruga Mohan Raj et al. have conducted a test by taking the different values of cutting velocity, cutting depth, and rate of feed in boring. By using Gaussian regression analysis, the surface integrity has been forecasted, which is a function of tool vibration.

Alwarsamy et al. investigated about cutting forces, which is hypothetically resolves on the direction of indirect force component or tangent force (f_t), feed force (f_f), direct force component or radial force (f_r), whereas the cutting force is inversely proportional to the cutting velocity and the cutting depth or chip thickness is directly proportional to the cutting force.

Atabey et al. investigated about chip dimensions, cut from insert, used as a function of geometric design of tool, feed, and cutting depth. That results in the impact of cutting forces direct with chip area and indirect with feed and cutting depth.

5 Materials and Methods

Boring is a metal cutting operation that bores the accurate holes in the workpiece or enlarges the existing hole. Tool, which is used to execute and the boring operation is known as boring bar, is characterized by large *L*/*D* ratio. Vibration occurs because of using long slender kind of boring bar. During the boring operation, the motion of bar is considered as the force components in the direction of cutting velocity and depth of cut [9]. It is assumed that in an intrinsic variation the tool wear and *L*/*D* ratio is remained unchanged but the cutting velocity, cutting depth, and feed rate vary. This present work, study focused on various cutting parameters effect at the time of machining. Vibration is caused due to periodic oscillation of kinetic elements of cutting forces, which arises from the initial frequency of vibration.

6 Conclusion

In this research, an encyclopedic study of single-point cutting tool and micro-boring using CNC is presented. The collaborative function of cutting implemented and examined by machining at different feed rates, cutting depth and cutting speeds by inserting an asymmetrical design of the rake geometry which is approximately corresponded to the chip geometry and cutting forces, was examined through the cutting force equations and proved that the used material and their specification are verified.

The collaborative functions of cutting for even rake face inserts are assumed extraneous (2-D) to implicit or slang (3-D) alteration.

References

1. Lin, S.-S., Chuang, M.-T.: Optimization of 6061T6 CNC boring process using the Taguchi Method and Grey Relational Analysis. Open Ind. Manuf. Eng. J. **2**, 14–20 (2009)
2. Vohra, G., Singh, P., Sodhi, H.S.: Analysis and optimization of boring process parameters by using Taguchi method. Int. J. Comput. Sci. Commun. Eng. (IJCSCE) (2013). ISSN 2319-7080 (Special issue on Recent Advances in Engineering & Technology)
3. Alammari, Y., Sanati, M., Freiheit, T., Park, S.S.: Investigation of boring bar dynamics for chatter suppression. Procedia Manuf. **1**, 768–778
4. Borade, P. M., Deshmukh, M. J.: Optimization of process parameters in boring operation: a review. Int. J. Eng. Res. Technol. (IJERT) **3**(3) (2014). ISSN: 2278-0181
5. Vignesh, P., Prasath, S., Gowtham, C., Sasi Revathi, S.V., Sankar, N.: Mechanical vibration analysis of a boring bar in boring operation. Adv. Nat. Appl. Sci. **11**(7), 882–888 (2017). ISSN: 1995-0772
6. Alwarsamy, T., Vetrivel, S., Nadarajan, S.: Theoretical cutting force prediction and analysis of boring process using mathcad. World Appl. Sci. J. **12**(10), 1814–1818 (2011)
7. Atabey, F., Lazoglu, I., Altintas, Y.: Mechanics of boring processes. Int. J. Mach. Tools Manuf. **43**(5), 463–476 (2003)
8. Senthil Kumar, M., Mohana Sundaram, K., Sathish Kumar, B.: A case study on vibration control in a boring bar using particle damping. Int. J. Eng. Scie. Technol. **3**(8), 177–184 (2011)
9. Balamuruga Mohan Raj, G., Sugumaran, V.: Developing gaussian process model to predict the surface roughness in boring operation. Int. J. Eng. Trends Technol. **4**(3), 578–589 (2013)

Nanofuels: An Advancement in Biodiesel Applications

Apurv Yadav, Manish Kumar Shivhare, Vineet Kumar Vashishtha, and Harsha Yadav

1 Introduction

Energy demand is escalating exponentially per year as perpetual industrial expansion is happening. This demand is causing two of the world's major problem: pollution and depletion of conventional energy resources. To overcome the energy crisis, several steps have been taken by the researchers. Numerous policies are being implemented by various governments [1]. Efficient use of existing renewable energy resources is being considered [2–4]. Storage capabilities of solar energy systems are being enhanced [5, 6]. Neural network-based prediction technologies are being used for wind power systems [7]. Exploring various plants as biodiesels is also a major research topic [8]. Application of biodiesel is like the killing of two birds with one stone. India ranks third in the list of global energy consumers as shown in Fig. 1.

Due to this, there is a vast amount of diesel combustion, which increases pollution. The main source of diesel pollution is automobiles. Biodiesels are a solution to this pollution problem but their performance is not at par with diesel. Their efficiency needs to be enhanced to have a significant performance impact. Modification of fuel is adopted by several researchers as a promising method to enhance engine performance. The effect of additives in biodiesel is discussed by Riberio et al. [10]. Additives are required by biodiesel blends to stability, lubricity, and combustion

A. Yadav
Amity University Dubai, Dubai 345019, UAE

M. K. Shivhare
Gautam Buddha University, Greater Noida, UP 201308, India

V. K. Vashishtha
Krishna Institute of Engineering & Technology, Ghaziabad, UP 201206, India

H. Yadav (✉)
Indian Institute of Technology, Delhi, New Delhi 110016, India
e-mail: harsha.civil32@gmail.com

M. Muzammil et al. (eds.), *Recent Advances in Mechanical Engineering*, Lecture Notes in Mechanical Engineering, https://doi.org/10.1007/978-981-15-8704-7_62

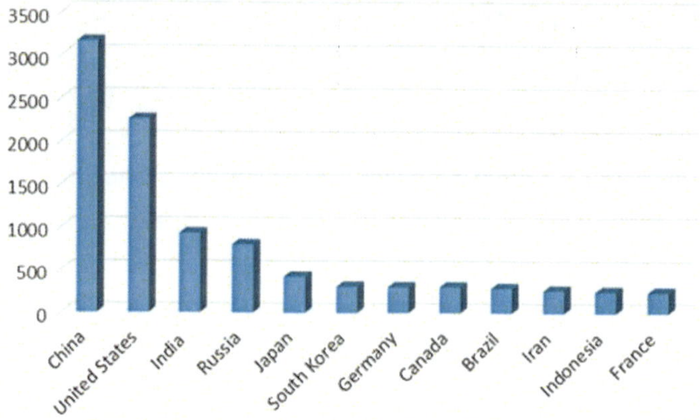

Fig. 1 Global energy consumption [9]

efficiency. There can be a classification of additives based on their purposes. Additives in biodiesel also affect their performance of combustion and their emissions.

2 Nanoparticle Additives in CI Engines

Nanoparticle additives have currently piqued the interest of the researchers worldwide. Jones et al. analyzed the outcome of adding nanoparticles of metal oxides of metal in biofuel and found that the quantity of heat released increases with additive concentration [11]. Kannan et al. found that adding of metal additives in blends of biodiesel greatly diminish its emissions and brake-specific fuel consumption (BSFC), while it increases its heat release rate, brake thermal efficiency, and cylinder pressure [12]. Figure 2 showcases the performance of BSFC against engine load for three different samples: a nano-biodiesel (B19MCE1), a diesel, and biodiesel (B20).

Sajith et al. stated that dispersing cerium oxide nanoparticles in blends of biodiesel reduces its NO_X and hydrocarbon emissions but it increases its viscosity [13]. Devarajan et al. found a reduction in emissions by including nanosilver oxide in biodiesel blends of neem oil [14]. These additives affect different parameters of an engine. Triratanasirichai observed that the on adding nano-sized additives, the kinematic viscosity of the biofuel is decreased [15]. Another researcher observed that the flash point of the biofuel reduces by the introduction of nano-additives [16]. Karthikeyan et al. reported that nano-sized additives in fuel increase its cetane number and calorific value [17]. BSFC of the test engine was reported to reduce by the addition of oxide nanoparticles as it oxidizes engine deposits leading to efficient and smooth operation [13]. Also, the brake thermal efficiency has been found to increase as higher calorific value reduces ignition delay, increases evaporation rate and increases flame temperature [18]. Figure 3 shows brake thermal efficiency characteristics for the same samples as described in Fig. 2.

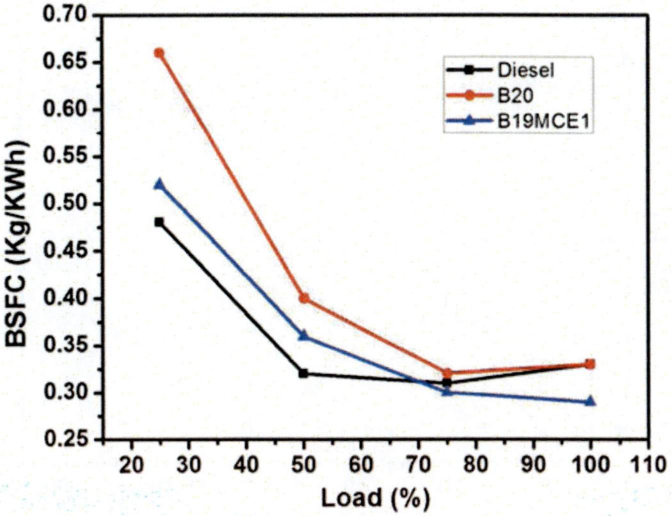

Fig. 2 BSFC characteristics for a nano-biodiesel, a diesel, and a biodiesel sample

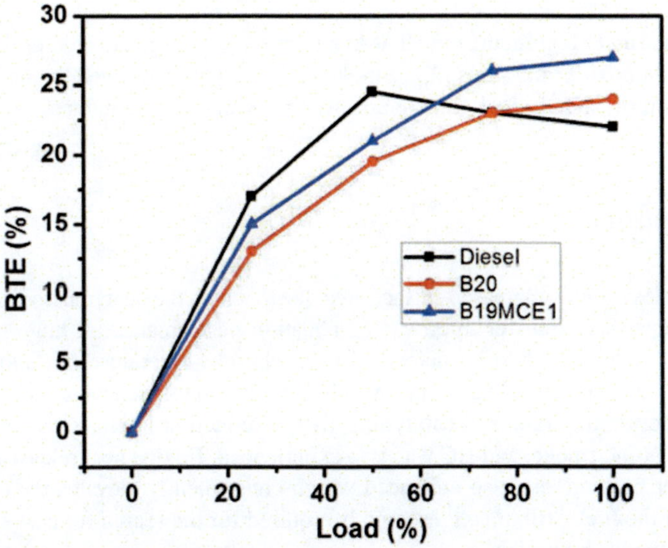

Fig. 3 BTE characteristics for a nano-biodiesel, a diesel, and a biodiesel sample

Emission from the diesel engine is also controlled by the use of biodiesel. Keskin et al. reported a 20% cutback in NO_X emissions by employing nano-cerium oxide [19]. Hydrocarbon (HC) emissions are also known to be reduced by addition of nanoparticles in biodiesel [20, 21]. Hydrocarbon emissions for samples described in Fig. 2 are presented in Fig. 4.

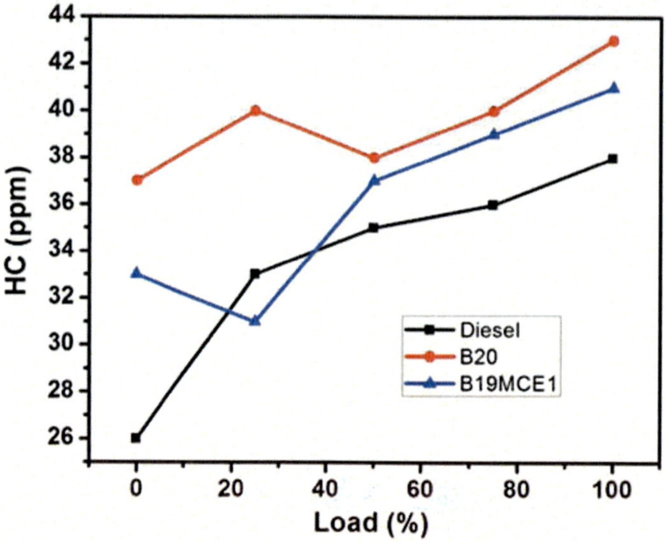

Fig. 4 HC characteristics for a nano-biodiesel, a diesel, and a biodiesel sample

Mehregan and Moghiman found that Al nanoparticle addition in biodiesel reduced its carbon monoxide emissions [22]. Smoke emissions also trimmed by the addition of nano-cerium oxide in fuel. Smoke opacity has been found to reduce.

3 Conclusion

Nanoparticles provide increased surface area due to which the heat transfer increases. The enhanced heat transfer improves combustion performance. A huge range of nano-sized additives could be dispersed in biodiesels. Literature has shown that nano-additives in biodiesel have improved brake thermal efficiency, proper combustion, air–fuel mixing rate, and catalytic activity. Reduction in emissions of carbon monoxide, NO_X, smoke, and HC was also experienced. Engine heat release rate and its chamber pressure are also enhanced which consequently elevates the evaporation rate of the fuel. Still, much research is required for the search for cost-effective nano-additive for biodiesels.

References

1. Stokes, L.C., Breetz, H.L.: Politics in the US energy transition: case studies of solar, wind, biofuels and electric vehicles policy. Energy Policy **113**, 76–86 (2018)

2. Yadav, A., Barman, B., Kumar, V., Kardam, A., Narayanan, S. S., Verma, A., Madhwal, D., Shukla, P., Jain, V. K.: A review on thermophysical properties of nanoparticle-enhanced phase change materials for thermal energy storage. In: Recent Trends in Materials and Devices, pp. 37–47. Springer, Cham (2017)
3. Yadav, A., Verma, A., Bhatnagar, P.K., Jain, V.K., Kumar, V.: Enhanced thermal characteristics of ng based acetamide composites. Int. J. Innov. Technol. Explor. Eng. (IJITEE) **8**(10), 4227–4231 (2019)
4. Yadav, A., Barman, B., Kardam, A., Narayanan, S.S., Verma, A., Jain, V.K.: Thermal properties of nano-graphite-embedded magnesium chloride hexahydrate phase change composites. Energy Environ. **28**(7), 651–660 (2017)
5. Yadav, A., Barman, B., Kumar, V., Kardam, A., Narayanan, S.S., Verma, A., Madhwal, D., Shukla, P., Jain, V.K.: Solar thermal charging properties of graphene oxide embedded myristic acid composites phase change material. In: AIP Conference Proceedings, vol. 1731, No. 1, p. 030030. AIP Publishing (2016)
6. Yadav, A., Verma, A., Narayanan, S. S., Jain, V.K., Bhatnagar, P.K.: Carbon based phase change nanocomposites for solar energy storage. In: AGU Fall Meeting Abstracts (2018)
7. Ramaswamy, S.: Forecasting of power in wind energy based on back propagation algorithm in neural networks. In: 2019 Amity International Conference on Artificial Intelligence (AICAI), pp. 461–464. IEEE (2019)
8. Kumar, S., Yadav, A.: Comparative experimental investigation of preheated thumba oil for its performance testing on a CI engine. Energy Environ. **29**(4), 533–542 (2018)
9. Global Energy Statistical Yearbook.: ENERDATA, https://yearbook.enerdata.net/total-energy/world-consumption-statistics.html (2019). Accessed August 2019
10. Ribeiro, N.M., Pinto, A.C., Quintella, C.M., da Rocha, G.O., Teixeira, L.S., Guarieiro, L.L., et al.: The role of additives for diesel and diesel blended (ethanol or biodiesel) fuels: a review. Energy Fuels **21**(4), 2433–2445 (2007)
11. Jones, M., Li, C.H., Afjeh, A., Peterson, G.P.: Experimental study of combustion characteristics of nanoscale metal and metal oxide additives in biofuel (ethanol). Nanoscale Res. Lett. **6**(1), 246 (2011)
12. Kannan, G.R., Karvembu, R., Anand, R.: Effect of metal based additive on performance emission and combustion characteristics of diesel engine fuelled with biodiesel. Appl. Energy **88**(11), 3694–3703 (2011)
13. Sajith, V., Sobhan, C.B., Peterson, G.P.: Experimental investigations on the effects of cerium oxide nanoparticle fuel additives on biodiesel. Adv. Mech. Eng. **2**, 581407 (2010)
14. Devarajan, Y., Munuswamy, D.B., Mahalingam, A.: Influence of nano-additive on performance and emission characteristics of a diesel engine running on neat neem oil biodiesel. Environ. Sci. Pollut. Res. **25**(26), 26167–26172 (2018)
15. Fangsuwannarak, K., Triratanasirichai, K.: Improvements of palm biodiesel properties by using nano-TiO_2 additive, exhaust emission and engine performance. Rom. Rev. Precis. Mech. Opt. Mechatron **43**, 111–118 (2013)
16. Arockiasamy, P., Anand, R.B.: Performance, combustion and emission characteristics of a DI diesel engine fuelled with nanoparticle blended jatropha biodiesel. Periodica Polytechnica Mech. Eng. **59**(2), 88–93 (2015)
17. Karthikeyan, S., Elango, A., Prathima, A.: Performance and emission study on zinc oxide nano particles addition with pomolion stearin wax biodiesel of CI engine. J. Sci. Ind. Res. **73**, 187–190 (2014)
18. Mehta, R.N., Chakraborty, M., Parikh, P.A.: Nanofuels: combustion, engine performance and emissions. Fuel **120**, 91–97 (2014)
19. Keskin, A., Gürü, M., Altıparmak, D.: Influence of metallic based fuel additives on performance and exhaust emissions of diesel engine. Energy Convers. Manage. **52**(1), 60–65 (2011)
20. Basha, J.S., Anand, R.B.: Performance, emission and combustion characteristics of a diesel engine using carbon nanotubes blended jatropha methyl ester emulsions. Alexandria Eng. J. **53**(2), 259–273 (2014)

21. Selvan, V.A.M., Anand, R.B., Udayakumar, M.: Effect of cerium oxide nanoparticles and carbon nanotubes as fuel-borne additives in diesterol blends on the performance, combustion and emission characteristics of a variable compression ratio engine. Fuel **130**, 160–167 (2014)
22. Mehregan, M., Moghiman, M.: Numerical investigation of effect of nano-aluminum addition on NO_x and CO pollutants emission in liquid fuels combustion. Int. J. Mater. Mech. Manuf. **2**(1), 60–63 (2014)

Prediction of Wind Power Curve Based on Wind Speed and Direction Utilizing Artificial Neural Network

Swaroop Ramaswamy Pillai, Apurv Yadav, and Vineet Kumar Vashishtha

1 Introduction

Electricity is a primary factor for economic advancement of any nation [1]. Production of electricity increases global energy demand. Conventional sources of energy are not sufficient to produce the current energy demand [2]. Additionally, these energy sources are one of the major cause of pollution. Lawmakers around the world are implementing several policies to reduce pollution [3]. There is a shift toward the exploration of new sources of energy and enhancing the efficiency of existing renewable energy resources. Various biodiesels are being explored to reduce the pollution from fossil fuel in automobiles [4]. Several materials have been investigated to enhance the performance of solar thermal systems [5–8]. Adoption of more and more solar or wind-based energy generation techniques are being considered [9]. Wind energy has grown popular due to its environment-friendly and modular nature [10]. Wind speed distribution of average 7 m/s is required for efficient functioning of a wind turbine. But, wind has an intermittent and stochastic nature. Prediction of the available wind energy is crucial for determining the feasibility of a wind power plant. Wind energy engineers are utilizing artificial intelligence-based techniques to assess the output of proposed wind farms. At different heights of the wind turbine, wind hitting force at the turbine tip is different [11]. Often analytical computer codes are used to gauge the performance and energy flow in wind power systems. These modeling codes uses complicated algorithms, which requires solving complex differential equation. Longer duration and a considerable amount of power are required to generate precise predictions. In lieu of confusing complex models, artificial neural

S. R. Pillai (✉) · A. Yadav
Amity University Dubai, Dubai 345019, UAE
e-mail: spillai@amityuniversity.ae

V. K. Vashishtha
Krishna Institute of Engineering & Technology, Ghaziabad, UP 201206, India

M. Muzammil et al. (eds.), *Recent Advances in Mechanical Engineering*, Lecture Notes in Mechanical Engineering, https://doi.org/10.1007/978-981-15-8704-7_63

networks (ANN) can grasp critical information patterns in the confines of multi-dimensional information field. As wind energy system has inherently noisy data, it provides suitable problems to be managed by neural networks due to their noise immunity and tolerance [12]. Earth receives about 1.74×10^4 KW of energy from the sun [13]. Equatorial region gets considerably more amount of heat than polar regions. Due to this, the density of air at the equator decreases, and it becomes lighter and rises. The pressure drop makes air at the polar region to flow there, and it causes wind. Wind power generation is greatly dependent on wind density and wind speed. Power generation from the wind turbine is obtained by

$$P = 0.5\rho AV^3 \tag{1}$$

where V is the wind velocity in ms^{-1}, A denotes the swept area of wind turbine blades in m^2, and ρ denotes air density in kg m^{-3} [14]. Other than these factors, many factors influence wind turbine power such as rated wind speed, cut-in speed, and surface roughness [15]. Interdependency on all these factors requires artificial intelligence for forecasting of performance. Power generation capability of a wind energy-based power park was predicted by an advanced ANN model developed by Kariniotakis et al. [16]. Catalao et al. did a forecast of short-term wind power by ANN in Portugal [17]. Wind speed profile is modeled using ANN in Nigeria by Fadare [18]. Kulkarni et al. used ANN with other techniques for wind speed prediction [19].

2 Prediction of Wind Power Curve

The hourly data for five days in August month is taken from a wind turbine [20]. Fig. 1 shows the wind rose diagram which gives the data on both wind speed and wind direction. From the figure, it is evident that the wind rose shows different wind speed and direction which is used for data training.

Figure 2a indicates the wind speed taken hourly. It shows that the wind speed was constantly varying between 4 m/s and 15 m/s. Figure 2b shows the wind direction measured in degrees. It is also evident that the wind speed varies from 120° and 300°. These two data mainly decides the energy per unit time produced by the wind turbine. Figure 2c indicates the power curve which varies between 1.5KWh and 14KWh. The energy theoretically depends only on the wind velocity, length of the blade, and the density which both are nearly constant. There are various other parameters which decides the energy curve. Hence, in this model, neural network is utilized for power curve prediction based on the past data.

Figure 3 indicates the learning curve from the neural networks. The error rate can be seen coming down with the number of iterations. The error is around 1.5 after thousand iterations. It can be also further reduced by adjusting the weights and by increasing the number of iterations. Figure 4 indicates that the actual and predicted power curve. It can be observed that the both actual and predicted follows the same

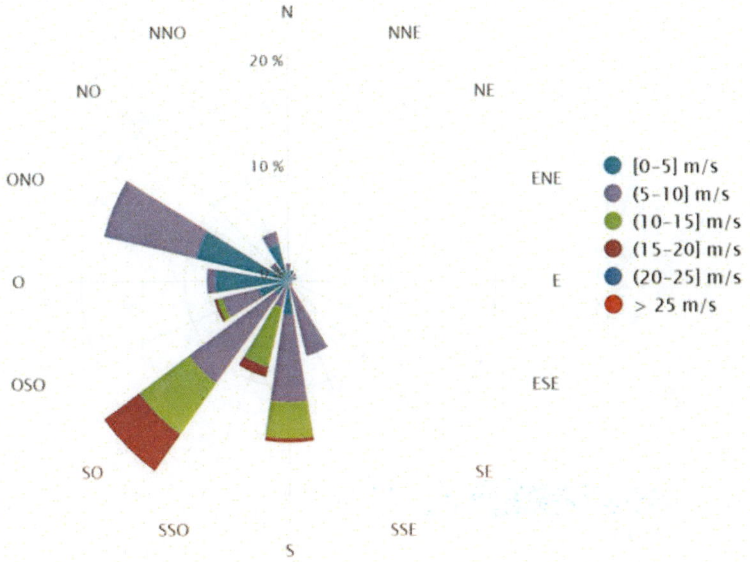

Fig. 1 Wind rose diagram [20]

path for little error. This can be also used for predicting the output for the new set of data.

Figure 5 also indicates the three-dimensional data for the power curve. The *x*- and *y*-axis showing the wind speed and direction. The *z*-axis indicates the power. It can be seen how the data are distributed. The blue dots indicates the predicted data, and the red dots indicates the actual data. It can be seen how both are clustered at one place which shows the accuracy of prediction. The wind direction here is represented in degrees after multiplying the scale by 10. The energy is expressed in Kwh with scale divided by 10. This scale is adjusted to obtain more accuracy when trained with the neural network.

Figure 6 indicates the neural network model which shows that inputs are wind speed and direction, and energy is the output. In this, it can be seen that two input layer, one output layer, and one hidden layer are used. The weights are updated based on the formula shown below. In this equation, $W_{\text{AB(New)}}$, $W_{\text{AB(Old)}}$ are the old and new weight matrix. η is the learning rate, E is the error, I is the input matrix, and O is the output matrix.

$$W_{\text{AB(new)}} = W_{\text{AB(old)}} - \eta \frac{\partial E^2}{\partial I_{\text{B}}} O_{\text{A}} \tag{2}$$

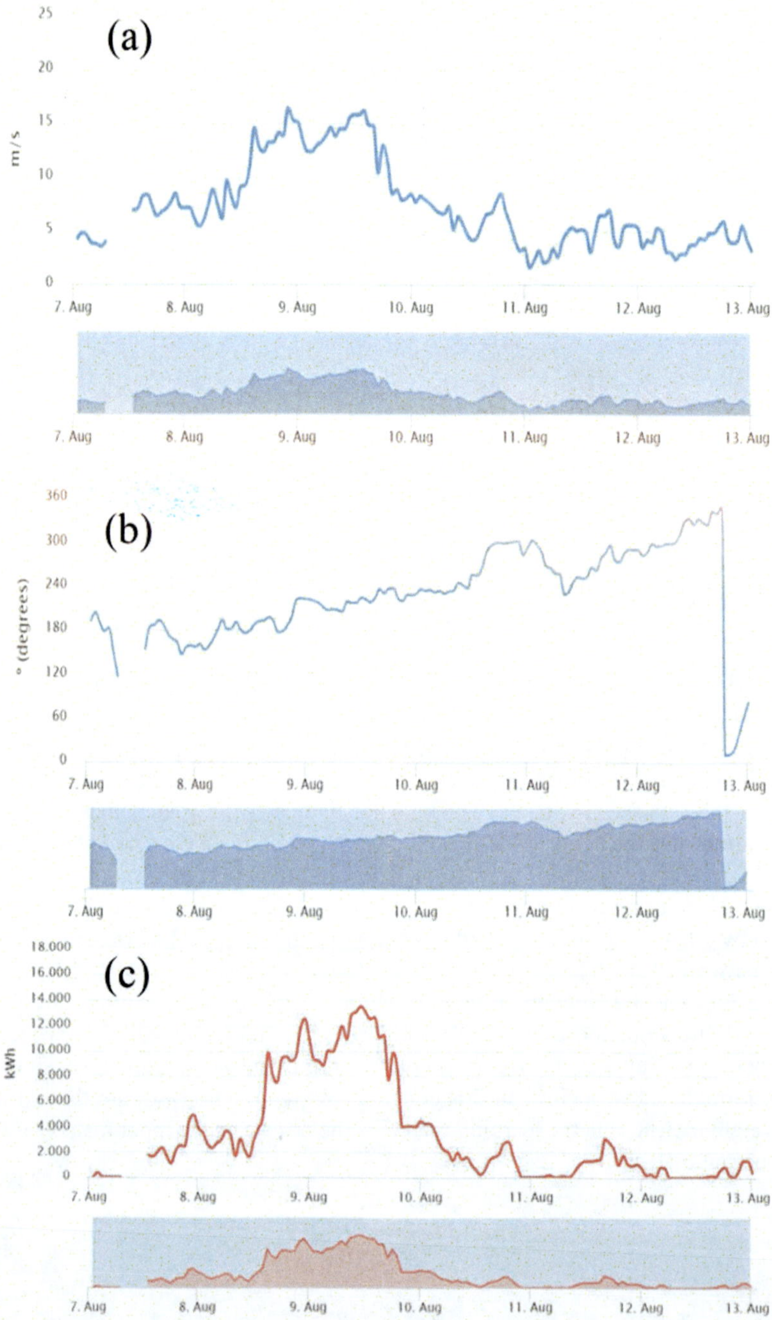

Fig. 2 **a** Wind speed curve, **b** wind direction, and **c** power curve based on wind speed and direction [20]

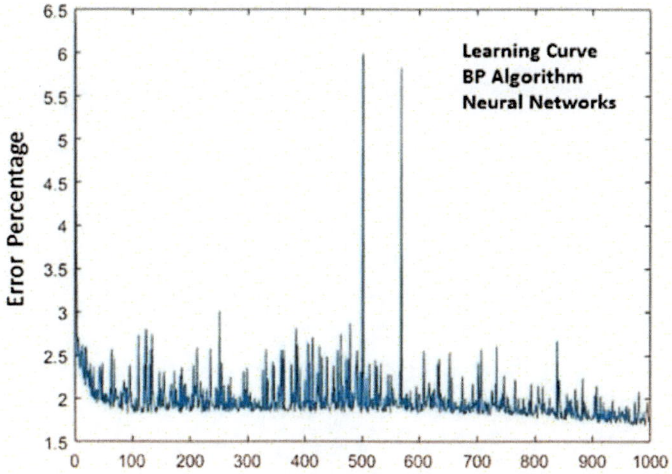

Fig. 3 Learning neural networks

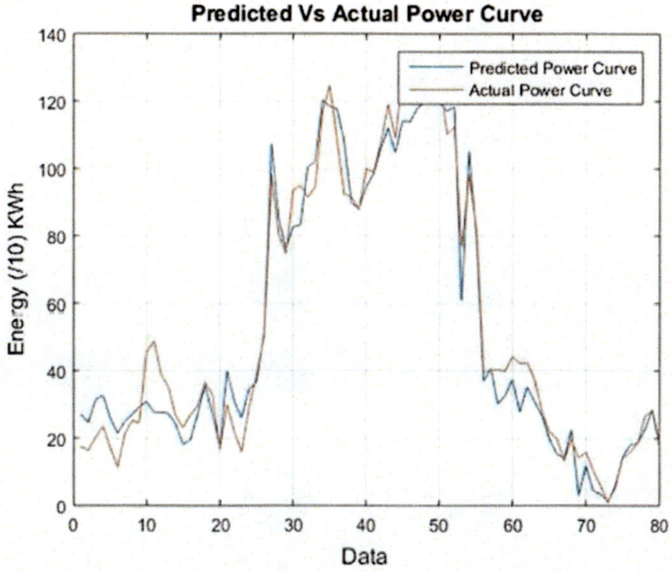

Fig. 4 Actual power curve versus predicted power curve

3 Conclusion

It can be observed in the simulation how that how the neural network is able to predict the energy. In this model, wind speed and direction data is taken as input, fed to the neural network back propagation algorithm after allotting the initial weights. After

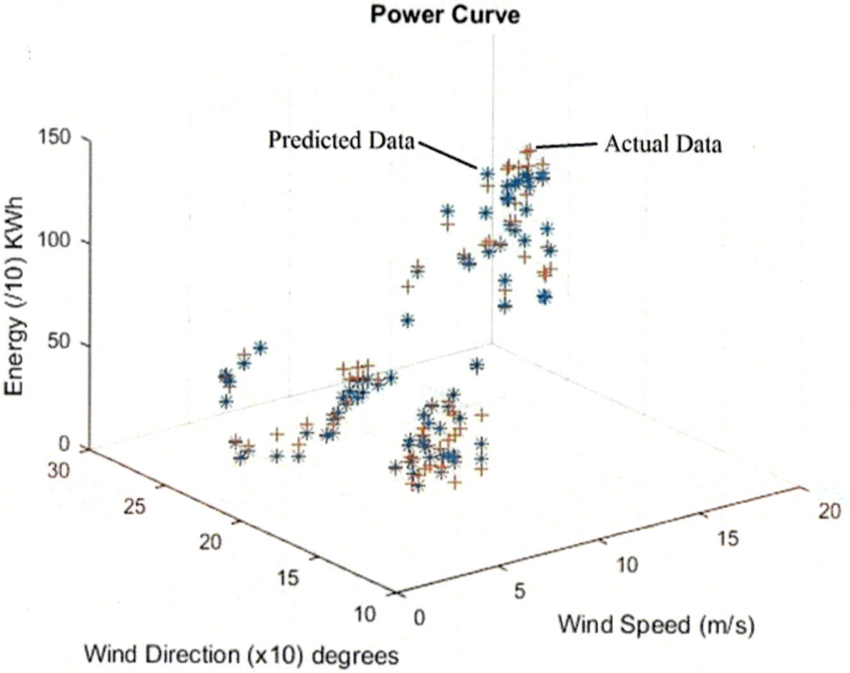

Fig. 5 Energy curve based on wind speed and direction

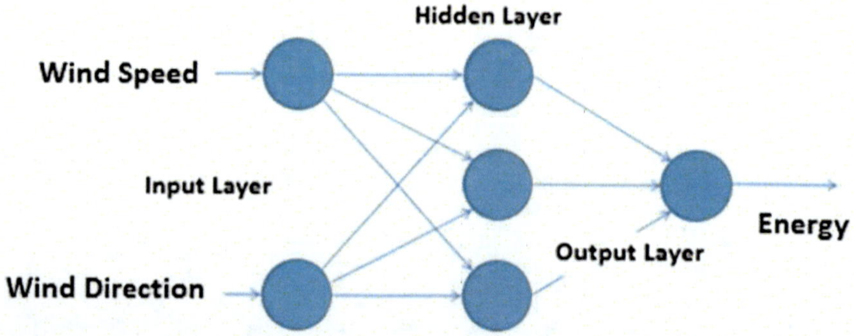

Fig. 6 Neural network model

many iterations, it can be found that the algorithm is able to predict the power based on the old and the new data. This model can be used forecasting the power curve and hence in the operation and management of wing power plants. ANN techniques could be applied in wide range of identification, field prediction, forecasting, evaluation, modeling of wind energy systems. ANN could be further extended to different functions like optimization of network configuration of large-scale wind energy conversion-interconnected systems, quick security evaluation, hybridization with other intelligent systems, stability enhancement, and preventive control measures. Suitable data selection is crucial for setting up of an ANN system. The training data set should be selected such that it is representing every operating mode of the system as the performance of the model depends on it. The application of ANN in wind energy systems are growing increasingly popular. Research on ANN application in wind power systems is expected to rise in the coming years.

References

1. Khare, V., Nema, S., Baredar, P.: Status of solar wind renewable energy in India. Renew. Sustain. Energy Rev. **27**, 1–10 (2013)
2. Kabir, E., Kumar, P., Kumar, S., Adelodun, A.A., Kim, K.H.: Solar energy: potential and future prospects. Renew. Sustain. Energy Rev. **82**, 894–900 (2018)
3. Stokes, L.C., Breetz, H.L.: Politics in the US energy transition: case studies of solar, wind, biofuels and electric vehicles policy. Energy Policy **113**, 76–86 (2018)
4. Kumar, S., Yadav, A.: Comparative experimental investigation of preheated thumba oil for its performance testing on a CI engine. Energy Environ. **29**(4), 533–542 (2018)
5. Yadav, A., Barman, B., Kumar, V., Kardam, A., Narayanan, S. S., Verma, A., Madhwal, D., Shukla, P., Jain, V. K.: A review on thermophysical properties of nanoparticle-enhanced phase change materials for thermal energy storage. In: Recent Trends in Materials and Devices, pp. 37–47. Springer, Cham (2017)
6. Yadav, A., Barman, B., Kardam, A., Narayanan, S.S., Verma, A., Jain, V.K.: Thermal properties of nano-graphite-embedded magnesium chloride hexahydrate phase change composites. Energy Environ. **28**(7), 651–660 (2017)
7. Yadav, A., Barman, B., Kumar, V., Kardam, A., Narayanan, S. S., Verma, A., ... & Jain, V. K. Solar thermal charging properties of graphene oxide embedded myristic acid composites phase change material. In: AIP Conference Proceedings, vol. 1731, No. 1, p. 030030. AIP Publishing (2016)
8. Yadav, A., Verma, A., Bhatnagar, P.K., Jain, V.K., Kumar, V.: Enhanced thermal characteristics of ng based acetamide composites. Int. J. Innov. Technol. Explor. Eng. (IJITEE) **8**(10), 4227–4231 (2019)
9. Khan, F.A., Pal, N., Saeed, S.H.: Review of solar photovoltaic and wind hybrid energy systems for sizing strategies optimization techniques and cost analysis methodologies. Renew. Sustain. Energy Rev. **92**, 937–947 (2018)
10. Elhadidy, M.A., Shaahid, S.M.: Parametric study of hybrid (wind + solar + diesel) power generating systems. Renew. Energy **21**(2), 129–139 (2000)
11. Junli, W., Xingjie, L., Jian, Q.: Wind speed and power forecasting based on RBF neural network. In: International Conference on Computer Application and System Modeling (ICCASM 2010), vol. 5, p. V5-298. IEEE (2010)
12. Rumelhart, D.E., Hinton, G.E., Williams, R.J.: Learning internal representations by error propagation (No. ICS-8506). California University, La Jolla Institute for Cognitive Science, San Diego (1985)
13. Ullah, K.R., Saidur, R., Ping, H.W., Akikur, R.K., Shuvo, N.H.: A review of solar thermal refrigeration and cooling methods. Renew. Sustain. Energy Rev. **24**, 499–513 (2013)
14. Azad, M.L., Singh, S.V., Khursheed, A.: Improving Voltage profile of a grid, connected to wind farm using static var compensator. Int. J. Adv. Eng. Technol. **7**(5), 1497 (2014)
15. Kalogirou, S.A.: Artificial neural networks in energy applications in buildings. Int. J. Low-Carbon Technol. **1**(3), 201–216 (2006)

16. Kariniotakis, G.N., Stavrakakis, G.S., Nogaret, E.F.: Wind power forecasting using advanced neural networks models. IEEE Trans. Energy Convers. **11**(4), 762–767 (1996)
17. Catalão, J.P.D.S., Pousinho, H.M.I., Mendes, V.M.F.: Short-term wind power forecasting in Portugal by neural networks and wavelet transform. Renew. Energy **36**(4), 1245–1251 (2011)
18. Fadare, D.A.: The application of artificial neural networks to mapping of wind speed profile for energy application in Nigeria. Appl. Energy **87**(3), 934–942 (2010)
19. Kulkarni, M.A., Patil, S., Rama, G.V., Sen, P.N.: Wind speed prediction using statistical regression and neural network. J. Earth Syst. Sci. **117**(4), 457–463 (2008)
20. Sotavento Plant Real Time Historical Data: http://www.sotaventogalicia.com/en/real-time-data/historical. Accessed 5 Sept 2019

A Review on Different Types of Hybrid Fiber Reinforced Composite

Ranjeet Kumar, Anurag Gupta, and Kunwar Laiq Ahmad Khan

1 Introduction

In composites, there are two main constituents, one is a matrix (primary phase), and the other is reinforcement (secondary phase). Sometimes, some other additives known as fillers are also used in composite materials to alter the properties of the same. In search of better and better materials, researchers start experimenting in the field of composites which lead to using multi-reinforcement in the single matrix for improvement of composite materials having single reinforcement. The composite materials having more than one type of reinforcing material are called hybrid composite materials.

Hybrid composites have large engineering applications where the tailored ratio of strength to weight at low cost is required. It provides a combination of mechanical and physical characteristics such as compressive strength, tensile strength, wear resistance, resistance toward chemical reactions, specific gravity, and many more. In the last two decades, hybrid composites are proved to be high-performance structural materials, consequently being used rapidly day by day. The use of hybrid composites is justified when a combined property of different types of fibers is to be achieved in a single FRP composite. In hybrid composite, it is possible to combine the advantages of individual properties and simultaneously justifying their less advantageous qualities. While selecting, the reinforcements special care should be taken toward the compatibility of two or more reinforcement in-between and with the matrix for developing HC. To study, the possibilities of different reinforcement that can be used to develop HC is of great interest to the researchers.

R. Kumar (✉) · A. Gupta · K. L. A. Khan
Department of Mechanical Engineering, KIET Group of Institutions, Delhi-NCR, Ghaziabad, India
e-mail: ranjeet.kumar@kiet.edu

M. Muzammil et al. (eds.), *Recent Advances in Mechanical Engineering*, Lecture Notes in Mechanical Engineering, https://doi.org/10.1007/978-981-15-8704-7_64

A review is carried out in this paper on the current status of available hybrid composite materials and their mechanical properties for various applications with some details of the hybrid composites.

Application of Hybrid Composite

- Aeronautical applications
- Wind power generation
- Marine applications
- Hybrid composites for civil construction and automotive industry
- Hybrid thermoplastic application
- Hybrid composites for telecom applications.

2 Literature Review

2.1 Glass and Carbon Fiber Hybrid Composite

Glass fiber reinforcement is one of the most popular reinforcement for FRP composite because of the ease of availability in a variety of forms for the manufacturing of FRP composite for different applications. The specific gravity of glass fiber is higher as compared to carbon fiber which restricts the use of glass fiber reinforced (GFRP) composite where a high strength-to-weight ratio is the prime requirement. At the same time, the cost of the carbon fiber is about 10 to 15 times more than glass fiber which restricts the use of carbon fiber. To take the advantages low cost of glass fiber and high strength of carbon fiber, researchers have started working over glass and carbon fibers FRP composites.

Mahdi et.al [1] investigated the hybridization effect on different types of composites that is, glass fiber epoxy, carbon fiber epoxy, glass–carbon–glass/epoxy, carbon glass–glass/epoxy, and glass–glass–carbon/epoxy. The structure with glass–carbon–glass/epoxy showed good energy absorption capability. Jesthi et.al [2] examined the effect of carbon and glass fiber stacking on mechanical characteristics of glass/carbon hybrid composite. The characteristics of the glass-carbon hybrid composite were compared under dry and seawater environments. The maximum strength of the composite was found 490 MPa with carbon-glass-glass-carbon- glass staking. It was observed that even a small amount of carbon fiber reinforcement in the GFRP composite can change the mechanical characteristics of FRP composite significantly. Wu et.al [3] reported that replacement of 10% glass fiber by carbon fiber improves the flexural modulus of GFRP composite by 35% and gives the best result at glass and carbon fiber ratio of 1:1. A layer of carbon was placed in a glass–epoxy composite by Wisnom et.al [4]. The magnitude of the effect of the hybridization depends on the ply thickness. There was a hybrid effect on the stress–strain diagram when the carbon layer was only 29 μm thick was of up to 20% only, and there was no significant effect reported for plies in excess of 80 μm thick. Zhang et.al [5] reduces the weight of the automotive vehicle body without excessive cost increase by using carbon fiber hybridization in GFRP composite along the main load path. The best flexural strength was found in carbon reinforcement which was 50%, and it was placed at the exterior, whereas compressive strength was highest by alternating carbon/glass lay-up. Bhagwat et.al [6] developed a glass/carbon hybrid FRP composite using 70% glass and 30% carbon fiber by hand layup process and found excellent improvement in the compressive and tensile strength and young's modulus. Chelliah et.al. [7] discussed the failure modes of GFRP, CFRP, and glass- carbon fiber hybrid epoxy resin laminates using acoustic emission (AE) monitoring for uni-axial tensile loading. It was concluded by the above study that the failure and accumulation of glass-carbon fiber hybrid/epoxy FRP composite were delayed along with better strain-to-failure rate as compared to GFRP and CFRP specimens. Dong and Davies [8] have hybridized S-2 glass and T700S CF, they found maximum hybrid effect at 0.125 hybrid ratio, and both fibers were reinforced in equal amount. The optimal hybrid effect was found 56.1% when glass and carbon were taken as 47.48% and 63.29% by volume, respectively.

2.2 *Glass and Natural Fiber Hybrid Composite*

Due to growing waste problems, environmental concerns, and sustainability, the researchers are focusing on the natural fibers having higher specific strength, stiffness, and lower weight and cost. Many natural fibers have been used by researchers such as jute, pineapple leaf, sisal, kenaf, bagasse, and banana. Mishra et.al [9] have reported that the mechanical properties of the hybrid composites were increased significantly by adding a relatively smaller amount of glass fiber into the hybrid of sisal and pineapple leaf reinforced into polyester resin. A composite by hybridization of jute sisal and glass fiber reinforcement in polyester resin was developed by Ramesh et.al [10, 11]. The mechanical properties were improved by the inclusion of sisal–jute fiber with GFRP. They further investigated the hybrid glass fiber-jute/sisal reinforced epoxy composites. The sisal fiber with GFRP displayed better tensile strength as compared to GFRP composites reinforced with jute fiber, whereas GFRP composites reinforced with jute fiber were having better flexural strength. An investigation was carried out by Deshpande and Rangaswamy [12] on E-glass and fiber-jute-fiber reinforced epoxy composite which was fabricated by mixing different amounts of coconut shell powder and bone. The flexural strength, shear strength, and hardness were improved when 15% coconut shell powder by volume was filled to the glass fiber-jute-fiber composite, whereas the impact strength was reported maximum by adding 15% (by volume) of bone powder. Atiqah et.al [13] manufactured kenaf-glass fibers reinforced composite with a matrix of unsaturated polyester resin in a ratio of 7:3(by vol.) with treated as well as untreated kenaf fiber. The treatment of kenaf fiber was done with a 6% sodium hydroxide diluted solution for 3 h. They found maximum tensile, bending, and impact strength as compared to treated kenaf with the ratio of 15:15 by volume of kenaf-glass fibers reinforced hybrid composite. Aslan et.al [14] analyzed mechanical properties and wear performance of polypropylene matrix reinforced with sisal-glass and sisal-carbon fiber hybrid composite. The tribological performances of the composite were decreased by the inclusion of sisal fibers on glass fiber and waste carbon reinforced PP composites. It was found that the hybrid ratio of 1:3 results in the best frictional and mechanical performances hybrid composite under consideration. Melo et.al [15] had manufactured the composite by hand layup process using glass fiber cross-ply fabric having nine layers of ply in an epoxy matrix reinforced with microparticles of Portland cement. Particle inclusions were done only up to the exterior four layers, with a curing time of 28 days and uniaxial compaction. A considerable increase was found in flexural strength and modulus. Gupta et.al [16, 17] developed pultruded glass fiber reinforced polymer (PGFRP) composite and pultruded jute fiber reinforced polymer (PJFRP) composite filled with hybrid filler containing calcium carbonate carbon black in bagasse fiber and found that hybrid filler significantly affects the mechanical properties of both the composites.

2.3 *Natural and Natural Fiber Hybrid Composite*

The researchers working in the field of material development are very vigilant about sustainable material which is environmentally friendly. The awareness about the environment leads the researchers to develop and analyze natural fiber FRP composite. It was also observed single natural fiber reinforcement do not produce the desired characteristics, but a combination of two or more type of reinforcement is helpful to tailor the characteristics as per requirement. Jawaid et.al [18] prepared hybrid composites by hybridizing oil palm fibers and jute in the epoxy matrix. It was reported that the addition of jute improves the elastic modulus and tensile strength of the above composite. Boopalan et.al [19] developed jute and banana fibers hybrid composites using different weight ratios of two fibers into the epoxy matrix. The mechanical and thermal properties of the hybrid composite were increased by reinforcing 50% (by weight) of banana fiber into jute-epoxy composites, whereas moisture absorption capabilities were found decreased. Ramlee et.al [20] fabricated hybrid composites made by sugarcane bagasse fiber and oil-palm-empty-fruit-bunch (OPEFB) reinforced with phenol–formaldehyde. Hybrid composites exhibited better performance than the other composites. They also found that OPEFB fiber-based composites had great mechanical quality because of higher cellulosic content, while SCB fiber decreased the water contents and void in fiber composites. Venkateshwaran et al. [21] prepared hybrid composites using sisal and banana fibers in the different ratios as 0:40, 10:30, 20:20, 30:10, and 40:0 with overall 40% fiber volume fraction. The author has examined the tensile strength and modulus of short, haphazardly oriented natural fiber (banana and sisal) reinforced hybrid composites. Sudhir Kumar and Chandan [22] examined the properties of jute fiber and bagasse hybrid composites for different bagasse stacking in the jute fiber reinforced composite. Furfuryl alcohol-grafted bagasse fiber varied from 20 to 50% in improvised jute fiber epoxy composite increased the strength of composite in tensile, flexural, and impact test by 52, 43, and 59% respectively. Filho et.al [23] analyzed the impact behavior of hybrid composites manufactured by hybridizing bagasse sugarcane fibers with disposed of rubber particles. By treating sustained chemicals of the bagasse fibers, composites stiffness enhanced, and energy absorption was decreased. Subsequently, higher energy absorption was reported in the composites made of untreated bagasse due to the enhanced pull-out mechanism of the fiber.

2.4 *Carbon and Natural Fiber Hybrid Composite*

Carbon fiber which is light and exceptionally strong is very expensive although it has high stiffness and strength-to-weight ratio. So, the researchers have started hybridizing carbon epoxy composites with natural fibers which are having comparative specific strength, stiffness, and lower weight and cost. Murty and De [24] investigated the effects of carbon black and silica on the bond between jute along with natural rubber and reported that the carbon black is better filler for reducing the requirement of least amount loading of jute fiber to accomplish reinforcement. Anuar et.al [25] have done hybridization of carbon fiber (natural rubber-based) with kenaf fiber. The hybrid composites with overall fiber contents of different % volume were investigated for flexural and impact testing, and up to 30% fiber content was considered in the samples. The strength and elastic modulus found increased up to 15% volume and then declined during flexural test, whereas during impact testing, improvement in the strength of treated and untreated composite was reported with the higher fiber contents. They also reported that the flexural properties of the composites having a single type of reinforcement better as compared to that of hybrid composite. Assarar et.al [26] have investigated damping characteristics of flax carbon epoxy composites due to staking sequences. The damping coefficients flax carbon epoxy composites were more than the carbon composite. The other mechanical and dynamic properties were also highly dependent on the stacking sequential arrangement of flax and carbon layers. Pathak et.al [27] investigated the flexural strength of polymer composites with graphene oxide which was used as filler for the manufacturing of carbon-fiber-graphene-oxide-epoxy hybrid composites. The flexural strength of the resulting hybrid composite was increased by 66%, whereas flexural modulus was increased by 72%.

3 Conclusion

It is concluded that various researches had been carried out to enhance the mechanical properties of composites by hybridization of composites. Further research may be carried out to improve the engineering applications of hybrid composites. Design, research, and product development efforts need to address the following issues.

- The manufacturing processes of FRP hybrid composites should be examined thoroughly, and more focus should be given to especially those factors that affect the microstructural integrity of the composites.
- A more detailed study should be done on the damaging tolerant properties such as fracture toughness, ductility, and creep in the hybrid composites.
- More focus should be given to manufacturing high quality at low-cost reinforcements from industrial/domestic wastes and by-products based on non-standard fibers & matrices to make the composite economical and nature-friendly.
- There is a great need to classify different types of hybrid composites based on their properties and manufacturing cost.

To quantify the undesirable defects in the hybrid composites, it is needed to develop simpler, more economical, and portable non-destructive methods.

References

1. Mahdi, E., Hamouda, A.M.S., Sahari, B.B., Khalid, Y.A.: Effect of hybridisation on crushing behaviour of carbon/glass fibre/epoxy circular–cylindrical shells. J. Mater. Process. Technol. **132**(1–3), 49–57 (2003)
2. Jesthi, D.K., Nayak, A., Mohanty, S.S., Rout, A.K., Nayak, R.K.: Evaluation of mechanical properties of hybrid composite laminates reinforced with glass/carbon woven fabrics. In: IOP Conference Series: Materials Science and Engineering, vol. 377(1), p. 012157. IOP Publishing (2018)
3. Wu, W., Wang, Q., Ichenihi, A., Shen, Y., Li, W.: The effects of hybridization on the flexural performances of carbon/glass interlayer and intralayer composites. Polymers **10**(5), 549 (2018)
4. Wisnom, M.R., Czél, G., Swolfs, Y., Jalalvand, M., Gorbatikh, L., Verpoest, I.: Hybrid effects in thin ply carbon/glass unidirectional laminates: accurate experimental determination and prediction. Compos. A Appl. Sci. Manuf. **88**, 131–139 (2016)
5. Zhang, J., Chaisombat, K., He, S., Wang, C.H.: Glass/carbon fibre hybrid composite laminates for structural applications in automotive vehicles. In: Sustainable Automotive Technologies, pp. 69–74. Springer, Berlin (2012)
6. Bhagwat, P.M., Ramachandran, M., Raichurkar, P.: Mechanical properties of hybrid glass/carbon fiber reinforced epoxy composites. Mater. Today: Proc. **4**(8), 7375–7380 (2017)
7. Chelliah, S.K., Kannivel, S.K., Vellayaraj, A.: Characterization of failure mechanism in glass, carbon and their hybrid composite laminates in epoxy resin by acoustic emission monitoring. Nondestructive Testing Eval. **34**(3), 254–266 (2019)
8. Dong, C., Davies, I.J.: Optimal design for the flexural behaviour of glass and carbon fibre reinforced polymer hybrid composites. Mater. Des. **37**, 450–457 (2012)
9. Mishra, S., Mohanty, A.K., Drzal, L.T., Misra, M., Parija, S., Nayak, S.K., Tripathy, S.S.: Studies on mechanical performance of biofibre/glass reinforced polyester hybrid composites. Compos. Sci. Technol. **63**(10), 1377–1385 (2003)
10. Ramesh, M., Palanikumar, K., Hemachandra Reddy, K.: Comparative evaluation on properties of hybrid glass fiber-sisal/jute reinforced epoxy composites. Procedia Eng. **51**, 745–750 (2013)
11. Ramesh, M., Palanikumar, K., Hemachandra Reddy, K.: Mechanical property evaluation of sisal–jute–glass fiber reinforced polyester composites. Compos. Part B Eng. **48**, 1–9 (2013)
12. Deshpande, S., Rangaswamy, T.: Effect of fillers on E-glass/jute fiber reinforced epoxy composites. Int. J. Eng. Res. Appl. 118–123 (2014)
13. Atiqah, A., Maleque, M.A., Jawaid, M., Iqbal, M.: Development of kenaf-glass reinforced unsaturated polyester hybrid composite for structural applications. Compos. Part B Eng. **56**, 68–73 (2014)

14. Aslan, M., Tufan, M., Küçükömeroğlu, T.: Tribological and mechanical performance of sisal-filled waste carbon and glass fibre hybrid composites. Compos. B Eng. **140**, 241–249 (2018)
15. Melo, A.B.L., Panzera, T.H., Freire, R.T.S., Scarpa, F.: The effect of Portland cement inclusions in hybrid glass fibre reinforced composites based on a full factorial design. Compos. Struct. **202**, 233–240 (2018)
16. Gupta, A., Singh, H., Walia, R.S.: Hybrid filler composition optimization for tensile strength of jute fibre-reinforced polymer composite. Bull. Mater. Sci. **39**(5), 1223–1231 (2016)
17. Gupta, A., Singh, H., Walia, R.S.: Effect of fillers on tensile strength of pultruded glass fiber reinforced polymer composite. IJEMS **22**(1), 62–70 (2015)
18. Jawaid, M., Abdul Khalil, H.P.S., Hassan, A., Dungani, R., Hadiyane, A.: Effect of jute fibre loading on tensile and dynamic mechanical properties of oil palm epoxy composites. Compos. Part B Eng. **45**(1), 619–624 (2013)
19. Boopalan, M., Niranjanaa, M., Umapathy, M.J.: Study on the mechanical properties and thermal properties of jute and banana fiber reinforced epoxy hybrid composites. Compos. B Eng. **51**, 54–57 (2013)
20. Ramlee, N.A., Jawaid, M., Zainudin, E.S., Yamani, S.A.K.: Tensile, physical and morphological properties of oil palm empty fruit bunch/sugarcane bagasse fibre reinforced phenolic hybrid composites. J. Mater. Res. Technol. **8**(4), 3466–3474 (2019)
21. Venkateshwaran, N., Elayaperumal, A., Sathiya, G.K.: Prediction of tensile properties of hybrid-natural fiber composites. Compos. B Eng. **43**(2), 793–796 (2012)
22. Sudhir Kumar, S., Chandan, D.: Thermomechanical properties of jute/bagasse hybrid fibre reinforced epoxy thermoset composite. Jute/bagasse composites. BioResources **4**(4), 1455–1476 (2019)
23. Filho, R., Moni, S.L., Oliveira, P.R., Panzera, T.H., Scarpa, F.: Impact of hybrid composites based on rubber tyres particles and sugarcane bagasse fibres. Compos. Part B Eng. **159**, 157–164 (2019)
24. Murty, V.M., De, S.K.: Effect of particulate fillers on short jute fiber-reinforced natural rubber composites. J. Appl. Polym. Sci. **27**(12), 4611–4622 (1982)
25. Anuar, H., Ahmad, S.H., Rasid, R., Ahmad, A., Wan Busu, W.N.: Mechanical properties and dynamic mechanical analysis of thermoplastic-natural-rubber-reinforced short carbon fiber and kenaf fiber hybrid composites. J. Appl. Polymer Sci. **107**(6), 4043–4052 (2008)
26. Assarar, M., Zouari, W., Sabhi, H., Ayad, R., Berthelot, J.M.: Evaluation of the damping of hybrid carbon–flax reinforced composites. Compos. Struct. **132**, 148–154 (2015)
27. Pathak, A.K., Borah, M., Gupta, A., Yokozeki, T., Dhakate, S.R.: Improved mechanical properties of carbon fiber/graphene oxide-epoxy hybrid composites. Compos. Sci. Technol. **135**, 28–38 (2016)

A Review Study on Solar Still: A Novel Approach of Solar Distillation

Javed Ahmed, Hardik Tyagi, Kartikey Joshi, Gaurav Bhardwaj, Ashok Kumar, Vineet Kumar Vashishtha, and Sandeep

1 Introduction

Solar energy is radiant heat from the sun that is an unlimited source of power, existing in exuberance and causes no pollution. Our Earth obtains huge power from the sun. Solar distillation is one of the ways to obtain clean water from brackish water with the help of solar energy. It uses natural rain water cycle in its operation and provides the purest form of water.

As in remote areas, distillation processes are not feasible as they are uneconomical; therefore, to do so, we have this process of desalination which works on the natural evaporation and condensation phenomenon which is instrumental in natural rain water cycle [1].

The first solar-still plant was constructed around 1872 by a Swedish Engineer Charles Wilson. In this plant, the basin water was heated up by the solar radiation. The evaporated vapour condensed on inner glazing surface; the distillate condensed output was collected in the collecting tank/channel. A solar still has many advantages over other devices such as low cost, simplicity, ease of maintenance, and low environmental impact. The only drawback is its low performance. The representation of solar still is shown in Fig. 1.

2 Classifications of Solar-Still

Solar-stills are classified in two parts: active and passive solar-still (Fig. 2).

J. Ahmed · H. Tyagi · K. Joshi · G. Bhardwaj · A. Kumar · V. K. Vashishtha · Sandeep (✉)
KIET Group of Institutions, Delhi-NCR, Ghaziabad, UP 201206, India
e-mail: Sandeep.chhabra@kiet.edu

M. Muzammil et al. (eds.), *Recent Advances in Mechanical Engineering*, Lecture Notes in Mechanical Engineering, https://doi.org/10.1007/978-981-15-8704-7_65

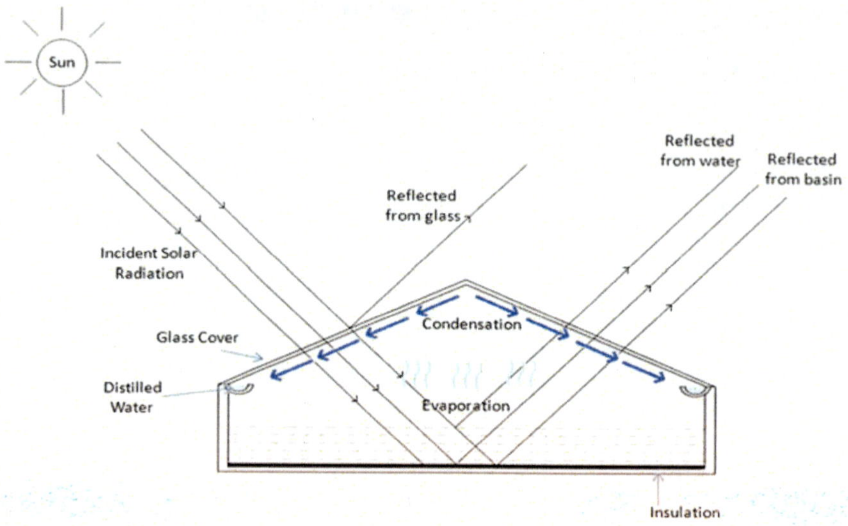

Fig. 1 Conventional double slope solar still

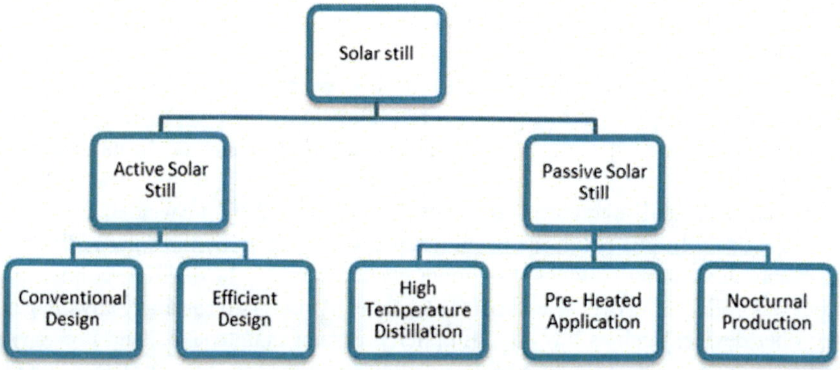

Fig. 2 Categorizations of solar-still

2.1 Passive Solar Still

It is the basic form of solar-stills which operates in day time only with no other added source of heat. It is available with a single basin coated with some absorber usually black paint. There is a glass cover which entraps the solar heat inside the still and helps in the evaporation of water to be desalinated. Here, efficiency is dependent on the strength of sun rays and the clean water distillate daily output. The water depth is kept small here and the distance between the inner surface of the glazing surface and water surface is kept low. Generally, the basic design of a passive solar-still is as shown in Fig. 3

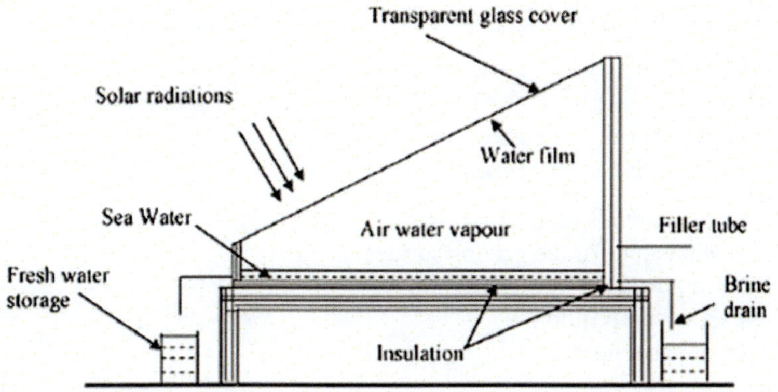

Fig. 3 Passive solar still

2.2 *Active Solar Still*

It is the modified form of passive still. Due to low efficiency, the passive still is attached with added heat supplying accessories transforming it to active solar still. A better productivity is obtained with high basin water temperature after preheating and refining. The design of active type solar-still is depicted in Fig. 4.

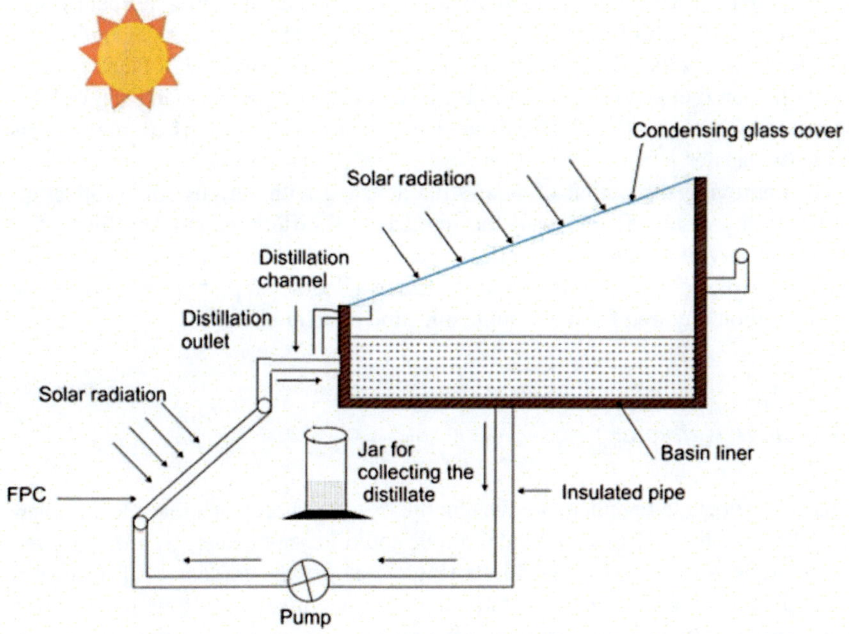

Fig. 4 Active solar-still

3 Design Alterations in Solar-Stills to Enhance Their Performance

To enhance performance of passive-stills, various changes are made.

Some of the variations possible are:

(i) Types of materials used
(ii) Inclination angle of the glazing surface
(iii) Absorbing materials inside the stills
(iv) feed water composition
(v) Types of basin lines.

Panchal and Shah [2] made experimental study and showed using solar stills made up of aluminium and galvanized iron types that the choice of material has an impact on the performance of a still. He found that more daily distillate output was obtained (around 3.8 L/m^2 day) when using aluminium solar stills than when using galvanized iron solar still (around 2.6 L/m^2 day). Different researchers obtained different results when it comes to the inclination angle of the glass cover.

Tiwari et al. [3] determined that the optimal angle is 15° in Delhi (with latitude angle approximately 28.6°). Muhammed et al. found the inclination angle of the glass surface to be optimal at 33.3° in southern-west arid area of Pakistan (with latitude angle 33.7°) [4]. Vinoth Kumar and Kasturi Bai reached a conclusion that the feed water composition also impacts the performance of the solar still as they found sea and tap water caused better productivity of the still than the industrial effluent which was treated by the still [5]. Badran through his experimental studies concluded that with the output of the still was more when he used asphalt instead of black paint as liner [6]. Tabrizi et al. [7] worked on design on cascade solar-stills and showed that the mass flow rate was found to be dependent on various factors such as internal heat and mass transfer.

For improving the output of solar-stills attached with various solar collectors, Sathyamurthy et al. [8] reviewed the influence of various parameters like water depth, intensity of radiation, water flow rate, etc. El-Maghlany et. al. [9] studied the prime angle of inclination of the glass surface of a double-slope-type solar-stills and the orientation required for maximum collection of solar energy.

3.1 *Solar Reflector*

Solar reflectors help in enhancing the amount of solar-energy that reaches the solar-still. Hiroshi obtained that distillate output could be augmented by inclining the exterior reflectors at various positions (Fig. 1) as per the seasons [10]. Boubekri and Chaker performed numerical study on the output of solar-stills, with the adding of external and internal reflectors to have an augmentation in the performance of 72.8%. They found that the angle of inclination for internal and external reflectors must be under 25° and the optimal angle of the glass surface should be between 10° and 50° [11]. Madhlopa et al. [12] completed an investigation on calculation of solar-radiation distribution in single slope type solar-stills with exterior and interior reflectors. Quality of surface finish & optical view factor is considered as an elementary parameter.

3.2 Wicked and Stepped Basin Solar Still

These help to improve mass transfer and heat transfer inside stills to increase its productivity. Halimeh et al. performed a relative study between exergy efficiency and energy in a cascade stepped-type solar-stills and recorded maximum exergy efficiency and energy of 10.5 and 83.3%, respectively (Fig. 2) [13]. Sadineni et al. achieved theoretical and experimental researches in stepped-type solar-stills and recorded that the productivity came out to be 20% more than a conventional-type solar-still [14]. Agouz performed some modification in the stepped-type solar-stills along with continuous water flow with the help of a cotton wick and found an increment of about 48% in the productivity.

Samuel performed tests with various wick materials such as coir mate, wool, cotton, jute cloth, charcoal cloth, nylon sponge, and water coral fleece and found that the water coral fleece was the best material [15]. Mahdi et al. thru tilted wicked solar still and could accomplish an increment of around 53% daily productivity with a charcoal wick [16]. Thus, it is determined that stepped and wicked-evaporators could augment the productivity by 20–53%.

3.3 Thermal Heat Storage

This fundamental principle helps the solar still functions even after dusk. Scholars have exasperated various assignments of the heat-absorbing materials. Storage of heat can be of two types: the sensible heat storage and the latent heat storage.

Sensible Heat Storage

These are the materials that can store heat of the solar irradiation when there is abundant of solar radiations and release heat when the irradiation is much less. Examples include sponge, pebbles, red brick pieces, etc. Rahim used a black-painted aluminium sheet just below the water surface and could obtain an output during nocturnal period along with an efficiency of about 47.2% [17]. Velmurugen et al. studied the performance of a stepped-type solar-stills having sensible-type heat storage under the basin and found an increment in the same of about 65 and 68% along with pebble and sponge material, respectively [18].

Latent Heat Storage
Materials like phase changing materials (PCMs) like paraffin wax are used as a PCMs. Omar et al. performed tests on a stills having PCM under the basin and recorded enhanced productivity and efficiency both [19].

3.4 *Basin with Fins*

Fins enhance the surface area of the base and thus increases the heat transfer rate to water. Velmurugan et al. conducted a trial in a single basin-type stills by modifying it and recorded that the productivity increased by 45.5%, 29.6%, and 15.3% for fins, wick, and sponge, respectively [20]. Velmurugan et al. designed a solar-still having fins to heal industrial effluents. Even though productivity enhanced using fins but greater improvements were obtained when pebbles and sponges were used. It is obtained from different researches that too many fins can reduce the output [21].

4 Conclusions

From this review work, it is being observed that solar still is one of the best alternatives for the purification of water. It works on the principle of natural rain water cycle and provides drinkable water for human consumption. Some of the conclusions which comes out though this review is that

- Via this technique, one can replace the use of existing fossil fuel and secure the environments from various harmful by-products like CO_2, CO, NO_x, etc.
- The productivity of the solar-stills can be improved by improving the evaporation rate, condensation rate, and by minimizing the losses in still. The evaporation rate can be upgraded successfully by the use of various absorbing materials like PCMs.
- By the use of computational study like CFD analysis, one can consume the experimental time and also save the expenditure cost in the setup of the project.

References

1. Sampath Kumar, K., Senthil Kumar, P.: Utilization of solar water heater in a single basin solar still—an experimental study. Desalination **297**, 8–19 (2012)
2. Panchal, H.N., Shah, P.K.: Char performance analysis of different energy absorbing plates on solar stills. Iran J. Energy Environ. **4**, 297–301 (2011)
3. Tiwari Anil, Kr., Tiwari, G.N.: Annual performance analysis and thermal modelling of passive solar still for different inclinations of condensing cover. Int. J. Energy Res. **31**, 1358–1382 (2007)

4. Muhammad Ali, S., Mirza Umar, K., Tariq, M., Nasir, A.: Design and performance of a simple single basin solar still. Renew. Sustain. Energy Rev. **11**, 543–549 (2007)
5. Vinoth Kumar, K., Kasturi, B.R.: Performance study on solar still with enhanced condensation. Desalination **230**, 51–61 (2008)
6. Badran, O.O.: Experimental study of the enhancement parameters on a single slope solar still productivity. Desalination **209**, 136–143 (2007)
7. Tabrizi, F.F., Dashtban, M., Moghaddam, H., Razzaghi, K.: Effect of water flow rate on internal heat and mass transfer and daily productivity of a weir-type cascade solar still. Desalination **260**(1–3), 239–247 (2010)
8. Sathyamurthy, R., El-Agouz, S.A., Nagarajan, P.K., Subramani, J., Arunkumar, T., Mageshbabu, D., Madhu, B., Bharathwaaj, R., Prakash, N.: A review of integrating solar collectors to solar still. Renew. Sustain. Energy Rev. **77**, 1069–1097 (2017)
9. El-Maghlany, W.M.: An approach to optimization of double slope solar still geometry for maximum collected solar energy. Alexandria Eng. J. **54**, 823–828 (2015)
10. Hiroshi, T.: Monthly optimum inclination of glass cover and external reflector of a basin type solar still with internal and external reflector. Sol Energy **84**, 1959–1966 (2010)
11. Boubekri, M., Chaker, A.: Yield of an improved solar still: numerical approach. Energy Procedia **6**, 610–617 (2011)
12. Madhlopa, A., Johnstone, C.M.: Computation of solar radiation distribution in a solar stillwith internal and external reflectors. Sol Energy **85**(2), 217–233 (2011)
13. Aghaei, Z.H., Farshchi, T.F., Faramarz, S., Fazlollah, H.: Comparison between energy and exergy efficiencies in a weir type cascade solar still. Desalination **325**, 113–121 (2013)
14. Sadineni, S.B., Hurt, R., Halford, C.K., Boehm, R.F.: Theory and experimental investigation of a weir-type inclined solar still. Energy **33**, 71–80 (2008)
15. Hansen, R.S., Narayanan, C.S., Murugavel, K.K.: Performance analysis on inclined solar still with different new wick materials and wire mesh. Desalination **358**, 1–8 (2015)
16. Mahdi, J.T., Smith, B.E., Sharif, A.O.: An experimental wick-type solar still system: design and construction. Desalination **267**, 233–238 (2011)
17. Rahim, N.H.A.: New method to store heat energy in horizontal solar desalination still. Renew. Energy **28**, 419–433 (2003)
18. Velmurugan, V., Naveen Kumar, K.J., Noorul Haq, T., Srithar, K.: Performance analysis in stepped solar still for effluent desalination. Energy **34**, 1179–1186 (2009)
19. Omar, A., Mohamed, A., Abdallah, B., Abdelaziz, A., Ahmed, K.: Desalination of the brackish water using a passive solar still with a heat energy storage system. Desalination **324**, 10–20 (2013)
20. Velmurugan, V., Gopala Krishnan, M., Raghu, R., Srithar, K.: Single basin solar still with fin for enhancing productivity. Energy Convers. Manag. **49**, 2602–2608 (2008)
21. Velmurugan, V., Deenadayalan, C.K., Vinod, H., Srithar, K.: Desalination of effluent using fin type solar still. Energy **33**, 1719–1727 (2008)

A Comparative Analysis of Single Basin Solar Still with and Without Nanofluids

Kartikay Kumar, Ashish Kumar Dubey, Ankit Singh, Bhavya Singhal, Ashok Kumar, Vineet Kumar Vashishtha, and Sandeep

1 Introduction

The population of India is increasing year by year and the number of industries is also increasing, which give rise to the need of drinking water in a tremendous manner. The only possible way of getting drinking water is from lake, ponds, rivers, well, etc., which have to be purified as they may contain harmful micro-organisms, bacteria, virus, and mineral contents. And the general purification processes are chlorination, boiling, and sand filtration. India is almost having 18% of the total world population and the water sources available for Indian community is only 4% of the total water sources present across the globe. The annual per capita of water available is from 6042 m^3 during 1947, 1545 m^3 in 2011, whereas during 2001, the total annual per capita of water available is only 1816 m^3. The latest survey concluded that the water availability will reduce to 1340 m^3 by 2025. The major utilization of ground water source is for drinking and domestic purpose and the other major ground water consuming sector is in irrigation.

Indian ministry of water resources and United Nations, UNICEF stated that approximately 90% of the water discharged from the rivers does not meet the environmental norms. Roughly, 60% of the rain water goes into oceans as a waste [1]. Except domestic and industrial sector, the next major use of water is in agriculture sector [2]. Several techniques are available for getting the fresh and purified water. These techniques include membrane and thermal processes. As the name suggested, thermal processes utilize the energy in the form of heat; this heat energy is used to evaporate the water to convert it into the vapour and after which these vapours are condensed by giving its latent heat to the surrounding. The condensed water is collected in the separate tank. Desalination of water through membrane includes the use of perforated membrane having very tiny pores. The saline water is pumped to the

K. Kumar · A. K. Dubey · A. Singh · B. Singhal · A. Kumar · V. K. Vashishtha · Sandeep (✉)
KIET Group of Institutions, Delhi-NCR, Ghaziabad, UP 201206, India
e-mail: sandeep.chhabra@kiet.edu

M. Muzammil et al. (eds.), *Recent Advances in Mechanical Engineering*, Lecture Notes in Mechanical Engineering, https://doi.org/10.1007/978-981-15-8704-7_66

membrane through the pump. This operation includes the usage of electrical energy to pump water. The main problem in using the membrane is the deposition of salt on the membrane, blockage of pores of membrane, and fouling effect. But the use of fossil fuels, wood, and the other fuels to operate the system make the system non-eco-friendly and uneconomical. Solar distillation is a better solution to this problem. Solar energy is renewable, green, and available free of cost and in abundance.

Solar desalination systems are divided into two major categories, namely:

- Direct desalination system
- Indirect desalination system.

Direct desalination system uses the solar energy and converts impure water into distillate water. While in indirect desalination system, a solar thermal collector is used to collect the solar energy and after this, stored energy is used in the running of the desalination system. During the nineteenth century, the use of renewable energy came into existence and after which the solar desalination units are designed and fabricated to get rid from the shortage of potable water.

Solar still is a device which works on the solar energy. The solar radiation incident on the glass of the still and goes to the base of the solar still where it helps to increase the temperature of the water and water get evaporated to form vapour.

The vapour gets condensed on the inner surface of the glass due to the temperature difference and the condensed vapour is collected in the fresh water tank. The principle of solar still is based on the continuous evaporation and condensation phenomenon.

The constant water level inside the basin is maintained with the help of flow regulating valve, which regulates and maintains the constant supply of water from the reservoir (Fig. 1).

Solar stills are classified as shown in Fig. 2.

2 Literature Review

Abhinav and Hari Singh [3] performed experiments with and without nanofluids on single basin passive type solar still. By using the Al_2O_3 as nanofluid, it increases the thermal conductivity of water by 20% and the distillation process become faster by 20% as compared to without using the nanofluid. By providing insulation of thermocol and aluminium sheet and painting inner portion with black colour increase the efficiency by 5%. It was noted that the overall efficiency of solar still increased by 5% with usage of nanofluids.

Gan and Qiao [4] proved experimentally that the radiation absorption capacity of Al_2O_3 nanofluids is less than the capacity of aluminium nanofluids. Nanofluids with Al_2O_3 particles, the condition is different because of the different optical properties.

Filho et al. [5] conducted an experiment on silver nanoparticle as the direct solar energy absorber. They concluded that the maximum stored thermal energy was increased by 52%, 93%, and 144% for the concentration of silver particle of 1.62 ppm, 3.25 ppm, and 6.5 ppm, respectively.

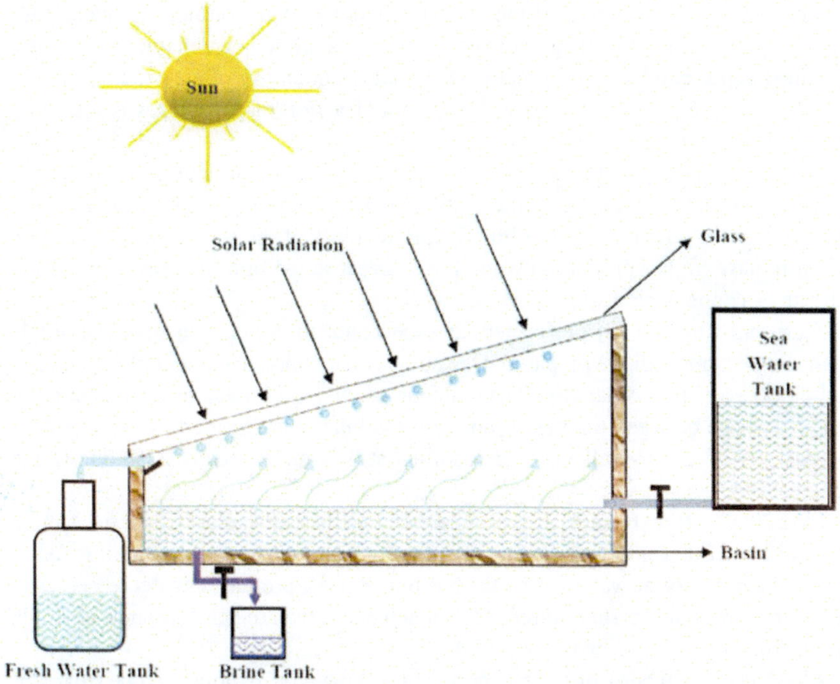

Fig. 1 Schematic representation of solar still

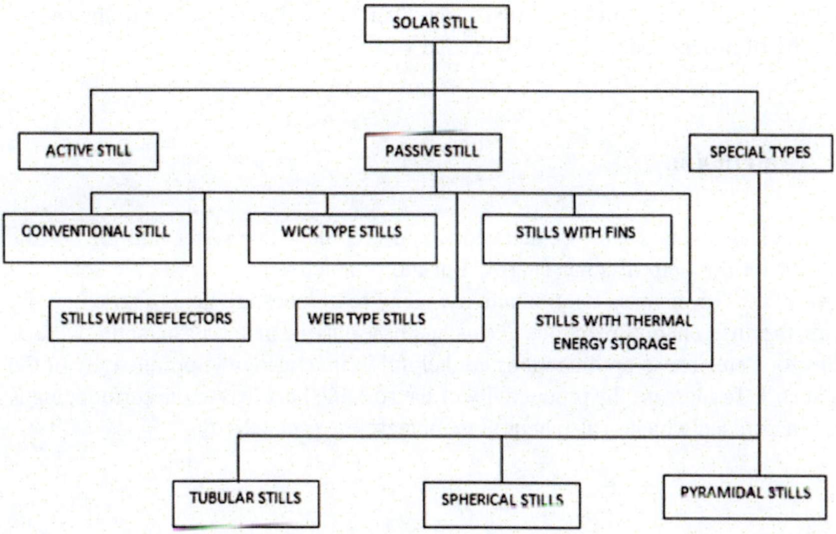

Fig. 2 Classification of solar still

Hong et al. [6] investigated thermal conductivity of nanofluids. Fe nanofluids were prepared with ethylene glycol and Fe nanocrystalline powder synthesized by a chemical vapor condensation process. Thermal conductivity of a Fe nanofluid was increased nonlinearly up to 18% as the volume fraction of particles was increased up to 0.55 vol. %.

Subhash et al. [7] performed experiment on single basin double slope solar still in two types; in type one, top of the glass is kept dry and in type two, the top of the glass is covered with thin water film. The productivity of the solar still covered with thin film have higher productivity and this is because the thin film which increases the rate of condensation.

Jyoti and Ruchika [8] performed an experiment in which two identical single basin single slope still are prepared. One still has the concrete cement block in basin and other is a simple solar still. The outcome of the experiments is as the temperature difference between the basin and the glass inner surface increase productivity increases. And the overall yield of modified still is increased by 25% due to the concrete block.

Pavan et al. [9] performed an experiment on two different solar still; one is simple solar still and other is external water source single slope still and the result is found that as the quantity of water inside the basin increases, productivity decreases. And the performance of external source still is much better than the performance of single slope still.

Akash et al. [10] performed an experiment on single basin double slope solar still with and without phase change materials and effect of reflectors and fins. He uses paraffin wax as a phase change material and found improvement in the efficiency up to 40.53%. But the cost of per litre of water is increased as compared to the cost of per litre of water produced by simple solar still.

3 Conclusion

Solar still is a very simple device which converts the saline water into the potable water with the help of solar energy. But the productivity of the simple solar still is very low. To increase its productivity, many researches are being carried out by using the different optimizations like using phase change materials, nanofluids, solar reflectors, etc. These optimizations are helpful in increasing the productivity of the solar still. To increase the productivity of the still, the heat loss to the surrounding is also minimized which is also helpful to increase the productivity.

References

1. WHO/UNICEF: 2015 update and MDG assessment, p. 90. World Health Organisation (2015)

2. Roca, L., Sanchez, J.A., Rodríguez, F., Bonilla, J., de la Calle, A., Berenguel, M.: Predictive control applied to a solar desalination plant connected to a greenhouse with daily variation of irrigation water demand. Unsustain. Energies **9**, 194 (2016)
3. Abhinav, K.S., Hari Singh, K.: Performance evaluation of solar still with and without nanofluid. Int. J. Sci. Environ. Technol. **3**(4) (2015). ISSN (O): 2348-4098
4. Gan, Y., Qiao, L.: Radiation enhanced evaporation of ethanol fuel containing suspended metal nanoparticles. Int. J. Heat Mass Transfer **55**, 5777–5782 (2012)
5. Filho, E.P.B., Mendoza, O.S.H., Beicker, C.L.L., Menezes, A., Wen, D.: "Experimental Investigation of a silver nanoparticle based direct absorption solar thermal system. Energy Convers. Manag. **84**, 261–267 (2014)
6. Hong, T.K., Yang, H.S., Choi, C.J.: Study of the enhanced thermal conductivity of Fe nanofluids. J. Appl. Phys. **6**(97), 1–4 (2005)
7. Chandra, S., Singh, D., Sharma, A.K.: Experimental investigation on a single basin double slope solar still. **05**(06) (2016). e-ISSN: 2395-0056
8. Bhalavi, J., Saini, R.: Performance enhancement of solar still using energy storage material. Int. J. Curr. Eng. Technol. **8**(4) (2018). e-ISSN 2277-4106
9. Pathak, P.K., Sharma, H.: Design improvement of solar still device using external water source. Int. J. Adv. Res. Develop. **3**(1) (2018). ISSN: 2455-4030
10. Lokhande, A.A., Shaikh, S. M.: Performance investigation of single basin double slope solar still with and without phase change material and effect of reflector and fins. Int. Res. J. Eng. Technol. **05**(08) (2018). e-ISSN: 2395-0056

Expedited Heat Transfer Rate of Mesoporous Carbon-Enhanced PCM

Apurv Yadav, Vivek Kumar, Abhishek Verma, P. K. Bhatnagar, and V. K. Jain

1 Introduction

Nowadays, energy storage is a major issue for researchers worldwide. As the technological advancement of the world demands more energy, these demands can be met by conventional or non-conventional energy-generating techniques. Resources of conventional energy sources are limited, hence there is a push for utilization of renewable energy generation methods [1]. Solar thermal energy has direct applications such as swimming pool heating, crop drying, domestic water heating, freshwater production, space heating, etc. Therefore, solar thermal energy storage is crucial for both commercial and non-commercial applications. Phase change or phase transition materials (PCM) are immensely popular among various techniques of latent heat-based energy storage. These storage density of these materials are almost ten-fold than that of sensible energy storage materials [2].

PCMs have found application in a variety of systems, such as energy-efficient structures, solar energy systems [1–3], waste heat recovery systems, and storage systems for central air-conditioning [4]. Increased efficiency of PCMs also means a considerable reduction in storage system size and consequently cost reduction. Inorganic salt hydrates are most promising PCMs. Their low cost, broad melting temperature range, and high energy storage density are advantageous in many applications [3]. However, the bulk of inorganic PCMs have an immensely low value of thermal

A. Yadav (✉)
Amity University Dubai, Dubai 345019, UAE
e-mail: apurv0210@gmail.com

AIRAE, Amity University, Noida 201313, India

V. Kumar · A. Verma · V. K. Jain
AIARS (M&D) & AIRAE, Amity University, Noida 201313, India

P. K. Bhatnagar
Department of Electronic Science, Delhi University, New Delhi 110021, India

M. Muzammil et al. (eds.), *Recent Advances in Mechanical Engineering*, Lecture Notes in Mechanical Engineering, https://doi.org/10.1007/978-981-15-8704-7_67

conductivity, rendering them unsuitable for energy storage [5–9]. Amidst various inorganic salt hydrate PCMs, hexahydrate magnesium nitrate ($Mg(NO_3)_2{\cdot}6H_2O$), due to its many favorable characteristics, has risen as an appropriate PCM for employability in energy storage applications with latent heat as their basis [10–13].

To solve the issue of poor thermal conduction of PCM and to amplify their heat transfer performance, numerous approaches like the use of fins, addition of metallic or non-metallic additives have been employed [14–17]. Out of different techniques, the addition of highly thermally conductive nanostructures to PCMs to prepare nanostructure enhanced composite PCMs have been proposed. Since a long time ago, it has been investigated and reported that the incorporation of nanoparticles in a poorly conductive PCM appreciably amplifies its thermal conductivity [18].

Among the various nanomaterials investigated, use of carbon-based nanoparticles has shown great potential in the thermal conduction amplification of PCM because they possess better heat transfer characteristics than other additives and also their density comparatively less [19–24]. For example, Warzoha et al. [25], found thermal conduction amplification for pure paraffin by dispersing in it, herringbone style graphite nanofibres. Wang et al. [26] observed 35–40% thermal conductivity increment in paraffin wax by adding graphite. Zhong et al. [27] reported 28–180 times increase in compressed expanded natural graphite (CENG)/Paraffin nano-composite. Kalaiselvam et al. [28] inferred that the number of nanoparticles incorporated in the PCMs should also have an optimal value for better heat transfer and thermos-physical characteristics. Experiments on myristic acid-graphene oxide composites showed an increase of 48 and 70% in melting and solidification rates, respectively, for 0.5 wt.% of graphene oxide added to myristic acid as compared to myristic acid alone [32]. Investigation on $Mg(NO_3)_2{\cdot}6H_2O$—nanographite composite showed 48% elevation in melting rate and 77% elevation in solidification rates, by adding a small amount of nanographite (0.6 wt.%) in $Mg(NO_3)_2{\cdot}6H_2O$ [33]. Similarly, melting and solidification times of $MgCl_2{\cdot}6H_2O$ was found to be decreased by 22% and 75%, respectively, by the inclusion of (0.5 wt.%) nanographite in $MgCl_2{\cdot}6H_2O$, as reported in our recently published paper [34]. Authors are also involved in the development of a eutectic gel with nanographite fillers. The prepared PCM nanocomposites were employed in (i) solar-based rechargeable glove and (ii) solar-based water heating system [35].

Present work investigated the thermal and heat transfer characteristics of PCM nanocomposites, consisting of magnesium nitrate hexahydrate as pristine PCM with minimal amounts of mesoporous carbon (MC) as nanofillers, in order to explore the further possibility of amplifying the thermal conduction and augmenting the rate of melting/freezing. Therefore, low concentrations of mesoporous carbon have been employed for the current study.

2 Experimental Section

2.1 Materials and Methods

$Mg(NO_3)_2{\cdot}6H_2O$ (MNH) with a purity of 99% and a molecular weight of 256.41 g/mol were purchased from Merck, India. Mesoporous carbon (MC) was obtained from Aldrich with a purity of 99.95% and dynamic light scattering particle size of less than 500 nm, an average pore diameter of 64 Å, and pore volume of 0.342 cm^3/g.

2.2 Sample Preparation

PCM nanocomposite samples were formed by the dispersion of MC nanopowder of desired amounts into the molten pristine PCM ($Mg(NO_3)_2{\cdot}6H_2O$). The composite was then positioned on a hot plate magneto-stirrer. The set temperature was at 100 °C for a duration of 30 min. For the next 30 min, aggressive sonication was done. Samples of composites of various weight percentage of MC (i.e., 0.1, 0.2, 0.3, 0.4 and 0.5 wt.%) were formed with a pristine PCM reference sample. Higher loadings were avoided due to the dispersion problem, which causes non-uniformity in the heating and cooling rates, throughout the sample. The optical images of pristine PCM ($Mg(NO_3)_2{\cdot}6H_2O$) and the prepared MC-PCM composite samples with a varying weight percentage of MC are shown in Fig. 1, which depicts the growth in the blackness of MC-PCM nanocomposites as the weight percentage of MC is increased.

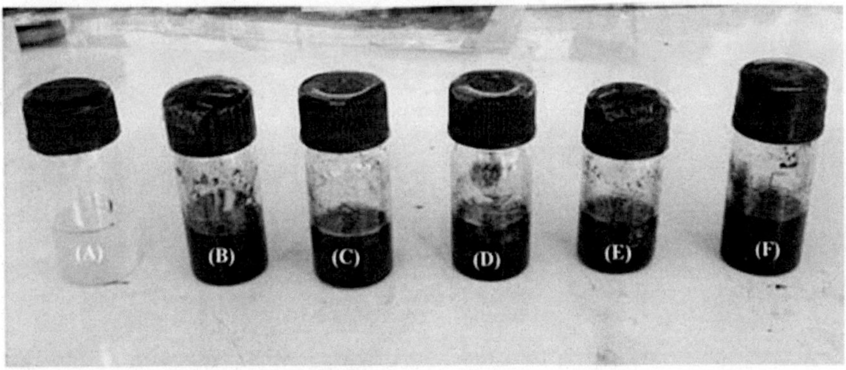

Fig. 1 Samples of pristine ($Mg(NO_3)_2{\cdot}6H_2O$ and nanocomposite having 0.1–0.5 wt.% of MC (labeled as A, B, C, D, E, and F, respectively)

2.3 Characterization Techniques and Experimental Setup

Scanning electron microscope apparatus (SEM, Zeiss EVO 18) was utilized for morphological study of MC and MC-PCM composites. Transient Hot Bridge (THB6N43) Thermal Conductivity Meter from Linseis was employed for measuring the thermal conduction values for all nanocomposite samples. A conventional heating setup was used for the examination of heating and cooling pattern behavior of pristine PCM (MNH) and MC-MNH nanocomposites. The description of the conventional heating setup can be seen in our previous publication [33]. For obtaining melting characteristics, the increase in temperature of each sample was set at 5 °C/min. For solidification characteristics, the vial was made to undergo a natural cooling process.

3 Results and Discussion

3.1 Characterization of Pristine PCM and MC-PCM Composites

The size of as-received MC and their distribution after dispersion in composite samples was investigated using SEM images. The SEM image of as-received mesoporous carbon (Fig. 2a) revealed that they are agglomerated together. The measured size of the MC dispersed in the PCM agrees with the specification of the supplier. The SEM images of pristine PCM (MNH) and MC-PCM nanocomposites (Fig. 2b, c, respectively) clearly indicate that the MCs are well dispersed in the composite samples.

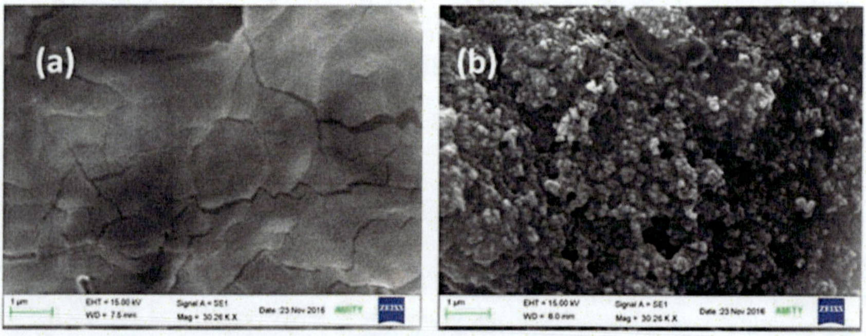

Fig. 2 SEM images of **a** pure PCM and **b** MC-PCM nanocomposites

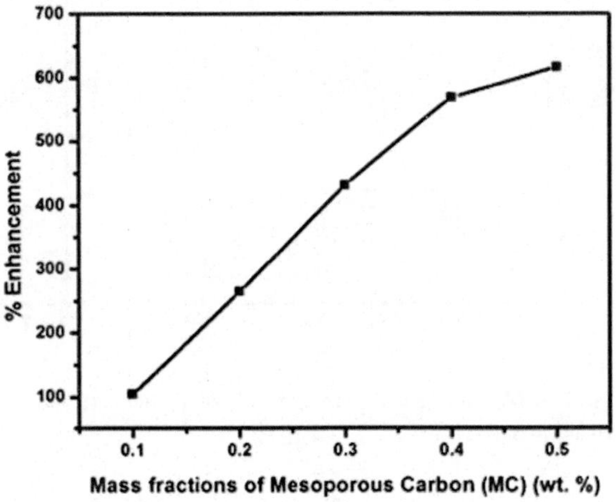

Fig. 3 Thermal conductivity enhancement of various concentration of additives

3.2 *Thermal Conductivity*

To study the impact on MNH's thermal conductivity by dispersion of MC, measurements were conducted on the various concentration of MC-MNH samples. The obtained thermal conductivity of pristine MNH 0.48 W/m K at an ambient temperature of 25 °C. With an increase in MC concentration, MC-PCM nanocomposites showed amplification in their thermal conduction. The calculated percentage augmentation in thermal conduction based on measured data at room temperature and its dependence upon the weigh percentage MC powder along with the values are presented in Fig. 3.

A maximum percentage enhancement of more than 616% was obtained for the 0.5 wt.% MC-MNH composite sample. The augmented thermal conduction in this study can be caused by the significantly enlarged surface area of mesoporous carbon which elevates heat transfer.

3.3 *Heat Transfer*

Heat transfer experiments on pure MNH were conducted on a conventional heating setup. A similar procedure was then repeated for the nanocomposites. Graphs illustrating the comparison of the melting behavior of pure PCM and its nanocomposites are shown in Fig. 4a. The experiments were started at ambient temperature. Initially, the temperature trend of the base MNH and its nanocomposites showed gradual increase until their respective temperature of the phase change (89 °C) was achieved. Approximately, after 52 min from the start, the melting of PCM started. As the concentration of MC varies from the value 0.1 wt.% up to the value of 0.5 wt.%, the time took by the composite to reach the melting temperature decreased sharply.

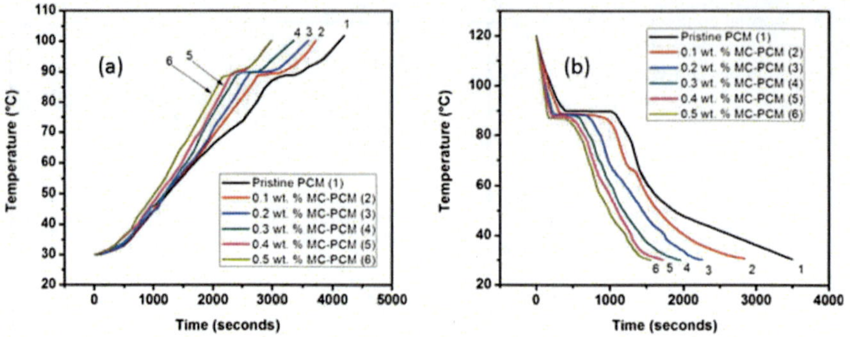

Fig. 4 **a** Melting and **b** characteristics of pure PCM and its nanocomposites at various MC concentrations

For 0.5 wt.% MC concentration, the time took by the composite to attain the melting temperature was only 36 min, which corresponds to an overall melting time reduction of 30%.

4 Conclusion

This research was done to prepare a novel PCM-based nanocomposite for solar thermal energy storage applications with improved thermal conduction. MC-MNH nanocomposite samples have been prepared. The concentration of MC powder was varied from 0.1 to 0.5 wt.%. The prepared MC-MNH composite samples exhibit enhanced thermal conductivity. The addition of MC powder leads to significantly enhanced heat conduction of the prepared PCM nanocomposite samples with the relative thermal conductivity enhancement increasing with the increase in MC concentration. A maximum percentage increment of more than 616% was obtained for the maximum loading of 0.5 wt.% MC-MNH composite sample. The melting and solidification experiments on PCM nanocomposites indicate that the rates of charging/discharging of composite samples were increased by about 30% and 55%, respectively, for 0.5 wt.% MC. This significant increase with low concentration of additives is desirable as there will be no noticeable decrease in the latent heat of the PCM. The obtained results are a confirmation of the suitability of MC-MNH composite as a prospective candidate for employability in latent heat storage.

Acknowledgements Authors are in gratitude of Dr. Ashok K. Chauhan, President, and founder of Amity University, for his perpetual encouragement, motivation, and guidance. The SEM characterization was carried out with the kind support of Mr. Shiv Kumar.

References

1. Kumar, S., Yadav, A.: Comparative experimental investigation of preheated thumba oil for its performance testing on a CI engine. Energy Environ. **29**(4), 533–542 (2018)
2. Kuznik, F., David, D., Johannes, K., Roux, J.J.: A review on phase change materials integrated in building walls. Renew. Sustain. Energy Rev. **15**(1), 379–391 (2011)
3. Yadav, A., Barman, B., Kumar, V., Kardam, A., Narayanan, S.S., Verma, A., Madhwal, D., Shukla, P., Jain, V.K.: A review on thermophysical properties of nanoparticle-enhanced phase change materials for thermal energy storage. In: Recent Trends in Materials and Devices, pp. 37–47. Springer International Publishing, Berlin (2017)
4. Sharma, A., Tyagi, V.V., Chen, C.R., Buddhi, D.: Review on thermal energy storage with phase change materials and applications. Renew. Sustain. Energy Rev. **13**(2), 318–345 (2009)
5. Sharma, S.D., Sagara, K.: Latent heat storage materials and systems: a review. Int. J. Green Energy **2**(1), 1–56 (2005)
6. Kenisarin, M.M.: Thermophysical properties of some organic phase change materials for latent heat storage. A review. Solar Energy **107**, 553–575 (2014)
7. Pielichowska, K., Pielichowski, K.: Phase change materials for thermal energy storage. Prog. Mater. Sci. **65**, 67–123 (2014)
8. Sari, A., Alkan, C.: Preparation and thermal energy storage properties of poly (n-butyl methacrylate)/fatty acids composites as form-stable phase change materials. Polym. Compos. **33**(1), 92–98 (2012)
9. Tamme, R., Bauer, T., Buschle, J., Laing, D., Müller, H., Steinhagen-Steinhagen, W.D., Steinmann, W.D.: Latent heat storage above 120 °C for applications in the industrial process heat sector and solar power generation. Int. J. Energy Res. **32**(3), 264–271 (2008)
10. Iverson, B.D., Broome, S.T., Kruizenga, A.M., Cordaro, J.G.: Thermal and mechanical properties of nitrate thermal storage salts in the solid-phase. Sol. Energy **86**(10), 2897–2911 (2012)
11. Lane, G.A.: Phase change materials for energy storage nucleation to prevent super cooling. Solar Energy Mater. Solar Cells **27**(2), 135–160 (1992)
12. Naumann, R., Emons, H.H., Köhnke, K., Paulik, J., Paulik, F.: Investigation on the thermal behaviour of $Mg(NO_3)_2{\cdot}6H_2O$. J. Therm. Anal. **34**(5–6), 1327–1333 (1988)
13. Naumann, R., Emons, H.H.: Results of thermal analysis for investigation of salt hydrates as latent heat-storage materials. J. Thermal Anal. **35**(3), 1009–1031 (1989)
14. Zalba, B., Marın, J.M., Cabeza, L.F., Mehling, H.: Review on thermal energy storage with phase change materials, heat transfer analysis and applications. Appl. Therm. Eng. **23**(3), 251–283 (2003)
15. Lacroix, M.: Study of the heat transfer behavior of a latent heat thermal energystorage unit with a finned tube. Int. J. Heat Mass Transf. **36**(8), 2083–2092 (1993)
16. Agyenim, F., Hewitt, N., Eames, P., Smyth, M.: A review of materials, heat transfer and phase change problem formulation for latent heat thermal energy storage systems (LHTESS). Renew. Sustain. Energy Rev. **14**(2), 615–628 (2010)
17. Fan, L., Khodadadi, J.M.: Thermal conductivity enhancement of phase change materials for thermal energy storage: a review. Renew. Sustain. Energy Rev. **15**(1), 24–46 (2011)
18. Choi, S.U., Eastman, J.A.: Enhancing thermal conductivity of fluids with nanoparticles (No. ANL/MSD/CP--84938; CONF-951135--29). Argonne National Laboratory, IL (United States) (1995)
19. Wang, J., Xie, H., Xin, Z., Li, Y., Chen, L.: Enhancing thermal conductivity of palmitic acid based phase change materials with carbon nanotubes as fillers. Sol. Energy **84**(2), 339–344 (2010)
20. Wang, J., Xie, H., Xin, Z., Li, Y.: Increasing the thermal conductivity of palmitic acid by the addition of carbon nanotubes. Carbon **48**(14), 3979–3986 (2010)
21. Cui, Y., Liu, C., Hu, S., Yu, X.: The experimental exploration of carbon nanofiber and carbon nanotube additives on thermal behavior of phase change materials. Sol. Energy Mater. Sol. Cells **95**(4), 1208–1212 (2011)

22. Harish, S., Ishikawa, K., Chiashi, S., Shiomi, J., Maruyama, S.: Anomalous thermal conduction characteristics of phase change composites with single-walled carbon nanotube inclusions. J. Phys. Chem. C **117**(29), 15409–15413 (2013)
23. Yavari, F., Fard, H.R., Pashayi, K., Rafiee, M.A., Zamiri, A., Yu, Z., Ozisik, R., Borca-Tasciuc, T., Koratkar, N.: Enhanced thermal conductivity in a nanostructured phase change composite due to low concentration graphene additives. J. Phys. Chem. C **115**(17), 8753–8758 (2011)
24. Kim, S., Drzal, L.T.: High latent heat storage and high thermal conductive phase change materials using exfoliated graphite nanoplatelets. Sol. Energy Mater. Sol. Cells **93**(1), 136–142 (2009)
25. Warzoha, R.J., Weigand, R.M., Fleischer, A.S.: Temperature-dependent thermal properties of a paraffin phase change material embedded with herringbone style graphite nanofibers. Appl. Energy **137**, 716–725 (2015)
26. Wang, J., Xie, H., Xin, Z.: Thermal properties of paraffin based composites containing multi-walled carbon nanotubes. Thermochim. Acta **488**(1–2), 39–42 (2009)
27. Zhong, Y., Li, S., Wei, X., Liu, Z., Guo, Q., Shi, J., Liu, L.: Heat transfer enhancement of paraffin wax using compressed expanded natural graphite for thermal energy storage. Carbon **48**(1), 300–304 (2010)
28. Kalaiselvam, S., Parameshwaran, R., Harikrishnan, S.: Analytical and experimental investigations of nanoparticles embedded phase change materials for cooling application in modern buildings. Renew. Energy **39**(1), 375–387 (2012)
29. Yadav, A., Barman, B., Kumar, V., Kardam, A., Narayanan, S.S., Verma, A., Madhwal, D., Shukla, P., Jain, V.K.: Solar thermal charging properties of graphene oxide embedded myristic acid composites phase change material. AIP Conf. Proc. **1731**(1), 030030 (2016)

Biomedical Applications of Additive Manufacturing

Ankita Jaisingh Sheoran, Arunesh Chandra, and Harish Kumar

Abbreviations

PEEK	Polyether ether ketone
SLA	Stereolithography
FDM	Fused deposition modeling
DMLS	Direct metal laser sintering
SLS	Selective laser sintering
API	Active pharmaceutical ingredients
EBM	Electron beam melting
SLM	Selective laser melting

1 Introduction

The intrinsic principle of AM for building parts in a layer-wise manner has allowed extensive customization in fabricating patient-specific medical constructs. These medical constructs include implants, surgery tools, orthoses, prosthesis, dental casts, tissue, or foreign body scaffold, etc., which are being manufactured with advancements in AM techniques such as FDM, SLA, DMLS, EBM, SLM, etc. The variety of biomedical and health-care applications of AM can be classified into five broad sub-groups as depicted in Table 1 [1].

A. J. Sheoran (✉) · H. Kumar
Department of Mechanical Engineering, NIT Delhi, New Delhi 110040, India
e-mail: ankitasheoran0@gmail.com

A. Chandra
Department of Mechanical Engineering, KIET Group of Institutions, Ghaziabad 201206, India

M. Muzammil et al. (eds.), *Recent Advances in Mechanical Engineering*, Lecture Notes in Mechanical Engineering, https://doi.org/10.1007/978-981-15-8704-7_68

Table 1 Five sub-groups of biomedical and health-care applications of AM [1]

S. No	Major biomedical and health-care applications sub-group of AM	Types	Purpose
1	Biomedical model	• Porotypes of anatomical parts • Exact physical replica of anomalies, fractures, deformities, etc.	For educational, training, and research purpose, to better represent, visualize, and understand the medical problem and plan the surgical process
2	Surgical tools	• Patient-specific dental crowns, bridges, casts, retractors, etc. • Hearing-aids, glass frame and lenses • Drilling, cutting tools • Other surgical devices, instruments, etc.	AM builds patient-specific surgical tools in lesser time and cost, and higher quality and accuracy. Such patient-specific customizations are required to address to anatomical variability from patient-to-patient
3	Implants	• Insert implants for replacement or aid in the recovery of fractured, non-functional or missing bone, tissue or other anatomical parts, etc.	AM prints complex-structured implants with a variety of bio-materials
4	Bio-printing	• Living cells are seeded or selectively placed on an implant or in-vivo scaffold after it has been printed • Biocompatible materials containing living cells called as bio-ink are directly utilized as the printing material to build the implant or in-vivo scaffold	For tissue engineering, to replace or promote tissue regeneration or tissue growth
5	External medical aid	• Prosthesis, casts, surgical guide instruments, etc. • Unlike implants, these are non-invasive and wearable	For using along with some standard medical devices to ensure patient-specific fitting, longer-term support, etc.

The various research fields and their contribution in utilizing AM for biomedical and health-care applications are as shown in Fig. 1.

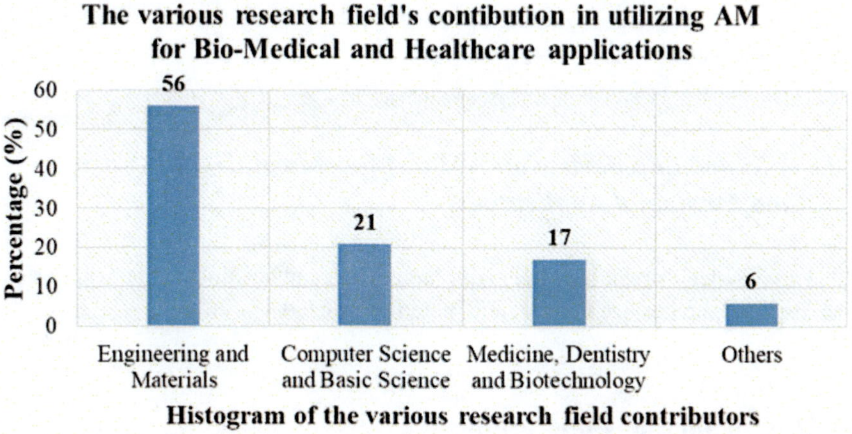

Fig. 1 Histogram of various research field contributors in utilizing AM for biomedical and health-care applications [2]

2 Process-Steps Involved in AM for Biomedical and Health-Care Applications

The customer-specific medical products can be personalized, designed, and fabricated using AM techniques. AM has made possible the fabrication of ideal, perfectly fitting medical parts and implants in lesser time and cost, and that too at higher accuracy and quality [3]. The process for this begins by diagnosis and then through precision imaging techniques like CT scan or MRI scan, collecting 2D data. These 2D images or data are then used by a stacking software like Mimics, wherein these 2D images are stacked and the contours in each of the 2D images are interpolated and connected to obtain a 3D virtual CAD model. This CAD model is then converted to the Standard Tessellation Language (STL) file formatted. The STL file is then sliced into multiple-stacked layers depending upon layer thickness. This sliced CAD data is given as the input to the AM machine. It then prints the implants or medical parts in a layer-wise manner, and the flow chart of the process is shown in Fig. 2.

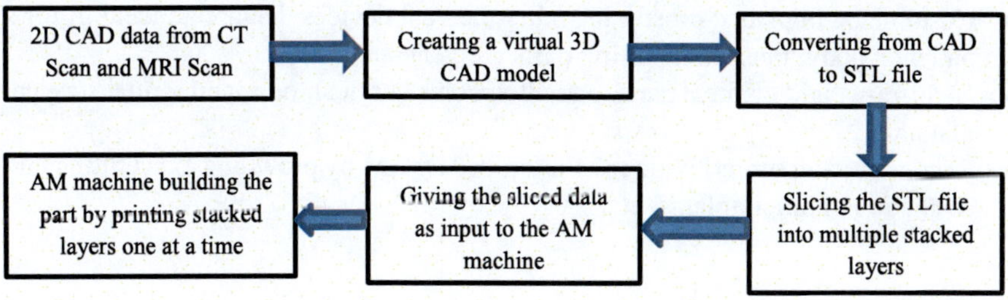

Fig. 2 Process for AM patient-specific medical parts and implants

Figure 3 shows details regarding the stages or the process-steps involved in AM for biomedical and health-care applications.

2.1 A Systematic Review on AM in Biomedical, Health-Care, and Medicinal Applications

Table 2 gives details regarding the major biomedical and health-care applications of AM, their advantages, techniques, and materials utilized.

3 Discussion and Conclusion

AM has revolutionized the medical field with its flexibility of manufacturing extremely complex-structured implants and other biomedical applications. Also with the available variety of bio-materials, many pharmaceutical and biomedical parts, anatomical part printing, scaffold printing have been successful. However, certain research gaps still need to be addressed, such as discovering new bio-materials, bio-inks, bio-printing must overcome its limitations like micro-vascularization and unique bio-inks with longevity, printing fully functional organs, etc. The major advantages and potential as demonstrated by AM are owing to the principle of creating parts in a layer-wise fashion, no requirement for tooling, minimal waste generation, flexibility in fabricating extremely complex designs and structures, etc. Some significant advantages of AM in biomedical, health-care, and medicinal applications is:

- Development of physical 3D models to aid in understanding of surgeons and hence better performance of the surgical operations
- Designing, developing, and fabricating even the complex-structured patient-specific implants and medical devices
- Reduced time and cost in manufacturing these medical instruments, parts, and implants
- Better quality increased strength of parts, better surface finishing, etc.
- Customized implants, prosthesis, other medical devices that are an ideal fit for a patient's body; thus, exact fitting parts are generated.
- Light-in-weight medical parts, patient-specific surgical tools, and virtual surgical planning.
- Reconstruction surgeries are revolutionized by AM by providing exact fitting and aesthetic looking implants, etc.

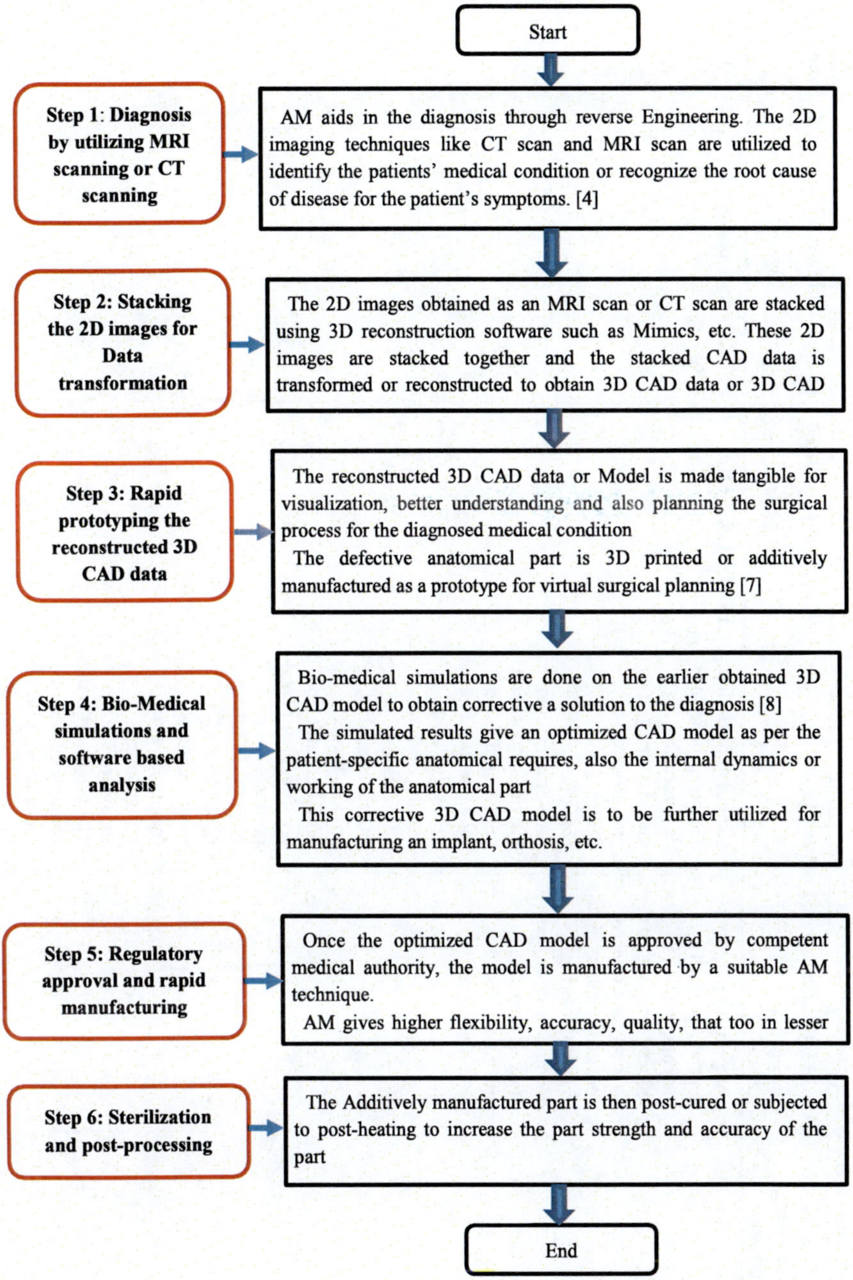

Fig. 3 The stages or the process-steps involved in AM for biomedical and health-care applications [3–10]

Table 2 Major biomedical and health-care applications of AM

S. No	Biomedical or health-care application	Aim of implementing AM	Advantages	AM technique utilized	Materials utilized	References
1	Dentistry	• The primary purpose is to print a perfectly matching replica in reduced time and cost • To meet the extensive needs of customization in manufacturing dentures, guards, crown, bridges, dental prostheses, dental orthodontics, etc.	• Extensive flexibility and potential in the design and manufacturing of patient-specific dentures in a cost-effective and time-efficient manner • Provides innovation, radical shifts, and fundamental changes in designing and making of dentistry applications • Improves quality, reduces waste, provides safer, and eco-friendlier dental products • Both the additively manufactured prosthodontics (dentures) and orthodontics (retainers) are • The metal and ceramic dental implants are replaced with lighter, biocompatible, and effective materials like PEEK, etc.	FDM, SLA, DMLS	PEEK, high-noble metal alloys, etc.	[11–13]

(continued)

Table 2 (continued)

S. No	Biomedical or health-care application	Aim of implementing AM	Advantages	AM technique utilized	Materials utilized	References
2	Tissue engineering and bio-printing of implants	• The primary purpose is to fabricate cell-seeded biocompatible implants, organoids, in vivo, and in vitro scaffolds that accelerate tissue growth for tissue development/regeneration • AM has revolutionized and replaced the traditional manufacturing techniques of these cell-seeded biocompatible implants, organoids, in-vivo and in-vitro scaffolds	• Easier fabrication of porous implants and scaffolds, and this facilitates tissue regeneration or development • Enables controlled cell-seeding for the implants, higher flexibility and accuracy to print biocompatible complex-structured implants and support scaffolding • These bio-compatible implants, organoids, in vivo and in vitro scaffolds effectively aid in bone or tissue recovery, or even for replacing a damaged bone or tissue	SLA, FDM is generally utilized for bio-printing; SLS, FDM, SLA, and material jetting are utilized for tissue engineering	Bio-materials like gelatin, alginate, fibrin (fibrinitrogen), gellan gum, agarose, collagen, hyaluronic acid (HA), etc. also, certain ceramics and thermoplastics for support scaffolding	[2, 14–16]

(continued)

Table 2 (continued)

S. No	Biomedical or health-care application	Aim of implementing AM	Advantages	AM technique utilized	Materials utilized	References
3	Virtual surgical planning	• The main aim is to create perfectly accurate physical models or replicas of critical bone deformities or certain organs, which are to be operated, to provide superior visualization and understanding, which otherwise cannot be attained by precision imaging techniques like MRI and CT scan images • This allows surgeons to test out the surgical procedure before the actual operation, this increases safety and ease during the actual surgery	• AM has allowed for extensive help in understanding the surgical procedure and operation planning required for severe organ abnormalities, fractures, bone deformities, etc. • This has tremendously increased safety and reduced risk of complex surgeries like spine-related deformities, craniofacial and heart-related surgeries, by enabling better pre-planning of complex surgery, better understanding of defects, deformities or anomalies and also early cause diagnostics	Generally SLA, polyjet	Polymer, rubber, thermoplastics	[17–20]

(continued)

Table 2 (continued)

S. No	Biomedical or health-care application	Aim of implementing AM	Advantages	AM technique utilized	Materials utilized	References
4	Medical devices, instruments, and surgical tools	• The main purpose is to design and fabricate patient-specific medical devices, instruments, and tools required for surgeries involving anatomical anomalies and deformities, wherein standard sizes of tools cannot be used, rather only custom-made tools are to be utilized for the surgical process	• AM is the solution to fabricating patient-specific medical devices and surgical tools in lesser time and cost • AM is the best solution to account for anatomical variability from patient-to-patient while manufacturing patient-specific medical devices and tools like cutting and drilling tools, hearing aids, glass frames, lenses, dental casts, surgical guide tools, custom, and inexpensive stethoscope, etc. • These surgical devices and tools are manufactured faster and hence have lesser cost and higher quality, precision, and reliability	DMLS for spinal cord surgical tools, jigs, retractors, etc., and generally FDM, SLA for other lightweight medical devices like eyewear frames, casts, hearing aids, surgical guide tools, etc.	Metal alloys, PLA	[21–23]

(continued)

Table 2 (continued)

S. No	Biomedical or health-care application	Aim of implementing AM	Advantages	AM technique utilized	Materials utilized	References
5	Design and fabrication of medicine delivery techniques	• The main purpose is to design and fabricate new drug delivery techniques: for both invasive and non-invasive methods • Utilizing AM to manufacture invasive drug delivery methods including: drug-releasing implants and ingestible tables in solid-form • Also, utilizing AM to manufacture non-invasive drug delivery methods (transdermal delivery techniques) including: patches and microneedle array	• AM is capable of manufacturing both in-vivo and in-vitro drug-releasing implants. AM not only facilitates manufacturing complex-shaped implants but also allows for incorporating medicine or remedial drugs within the implant, which release the drugs at a controlled rate within the body • AM allows for fabricating solid-tablets having varying proportions of API, customized as per varying prescriptions specified from patient-to-patient • AM has made fabrication of different transdermal delivery techniques possible	FDM utilized for drug-releasing implants and FDM, Inkjet printers with UV photo-initialization and SLA for ingestible tables in solid-form. SLA for micro-needle array as well as patches. Inkjet printer for coating the micro-needle array with medicine or API's	Antibiotic-infused-PLA for drug delivery implants, UV photopolymer resin for micro-needles and patches, which are then coated with insulin, API's, medicines, etc	[24–26]

(continued)

Table 2 (continued)

S. No	Biomedical or health-care application	Aim of implementing AM	Advantages	AM technique utilized	Materials utilized	References
6	Orthopedics and prosthetics	• The main purpose of utilizing AM for orthopedic implants is to cater to anatomical variability and thereby manufacture patient-specific orthopedic implants that will restore the motion and structural integrity • Also, utilizing AM to manufacture prosthetic implants as replacements for amputated, missing, or non-functional anatomical parts	• The patient-specific prosthesis and orthosis are manufactured by AM in lesser time and cost, higher reliability, accuracy, and quality • AM enables controlled porous	DMLS for orthopedic implants like knee-joint implant, etc. SLA for prosthesis	Generally, metal-alloys like cobalt–chromium alloy, titanium alloy, etc., for orthopedic implants. Thermoplastics, biomaterials, etc., for prostheses	[27, 28]

(continued)

Table 2 (continued)

S. No	Biomedical or health-care application	Aim of implementing AM	Advantages	AM technique utilized	Materials utilized	References
7	For medicine practice, research, education, learning, and training purpose	• The main purpose is to enable better visualization and hence understanding of all internal and external geometrical features of an anatomical part • Variety of colors and printing materials can be utilized in AM to help practitioners, educators, researchers, learners, and trainees learn and recognize how the actual organ looks	• AM has enabled creating perfect replica prototypes or scaled prototypes of anatomical models, to allow new medical practitioners, researchers, learners, etc., to learn from and practice on these prototype models before the actual organ	FDM, SLA	Different thermoplastics, polymers, resins, etc.	[29, 30]

References

1. Tuomi, J., Paloheimo, K.S., Vehviläinen, J., et al.: A novel classification and online platform for planning and documentation of medical applications of additive manufacturing. Surg. Innov. **21**, 553–559 (2014)
2. Javaid, M., Haleem, A.: Additive manufacturing applications in medical cases: a literature based review. Alexandria J. Med. 1–12 (2017). https://doi.org/10.1016/j.ajme.2017.09.003
3. Bibb, R., Eggbeer, D., Evans, P., Bocca, A., Sugar, A.: Rapid manufacture of custom fitting surgical guides. Rapid Prototyp. J. **15**, 346–354 (2009)
4. Marchelli, G., Prabhakar, R., Storti, D., Ganter, M.: The guide to glass 3D printing: developments, methods, diagnostics and results. Rapid Prototyp. J. **17**, 187–194 (2011)
5. Sedghi, S., Sanderson, M., Clough, P.: Medical image resources used by health care professionals. Aslib Proc. **63**, 570–585 (2011)
6. Vestal, K., Massey, R.: Work transformation in health care. Health Manpower Manage. **20**, 9–13 (1994)
7. Eyers, D., Dotchev, K.: Technology review for mass customisation using rapid manufacturing. Assembly Autom. **30**, 39–46 (2010)
8. Sun, L.W., Lee, R.Y.W., Lu, W., Luk, L.D.K.: Modelling and simulation of the intervertebral movements of the lumbar spine using an inverse kinematic algorithm. Med Biol. Eng. Comput. **42**, 740–746 (2004)
9. Arrieta, C., Uribe, S., Ramos, G., et al.: Quantitative assessments of geometric errors for rapid prototyping in medical applications. Rapid Prototyp. J. **18**, 431–442 (2012)
10. AhnSung, H., Lee, C.S., Jeong, W.: Development of translucent FDM parts by post processing. Rapid Prototyp. J. **10**, 218–224 (2004)
11. Haleem, A., Javaid, M.: Polyether ether ketone (PEEK) and its manufacturing of customized 3D printed dentistry parts using additive manufacturing. Clin. Epidemiol. Global Health **7**(4), 571–577 (2019)
12. Tahayeri, A., Morgan, M., Fugolin, A.P., Bompolaki, D., Athirasala, A., Pfeifer, C.S., Ferracane, J.L., Bertassoni, L.E.: 3D printed versus conventionally cured provisional crown and bridge dental materials. Dent. Mater. **34**, 192–200 (2018)
13. Gan, N., Ruan, Y., Sun, J., Xiong, Y., Jiao, T.: Comparison of adaptation between the major connectors fabricated from intraoral digital impressions and extraoral digital impressions. Sci. Rep. **8**, 529 (2018)
14. Chocholata, P., Kulda, V., Babuska, V.: Fabrication of scaffolds for bone-tissue regeneration. Materials **12**, 568 (2019)
15. Peltola, S.M., Melchels, F.P.W., Grijpma, D.W., Kellomäki, M.: A review of rapid prototyping techniques for tissue engineering purposes. Ann Med. **40**, 268–280 (2008)
16. Yeong, W.Y., Chua, C.K., Leong, K.F., Chandrasekaran, M.: Rapid prototyping in tissue engineering: challenges and potential. Trends Biotechnol. **22**, 643–652 (2004)
17. Singhal, A.J., Shetty, V., Bhagavan, K.R., Ragothaman, A., Shetty, V., Koneru, G., Agarwala, M.: Improved Surgery planning using 3-D printing: a case study. Indian J. Surg. **78**, 100–104 (2016)
18. Sodian, R., Weber, S., Markert, M., et al.: Stereolithographic models for surgical planning in congenital heart surgery. Ann. Thorac. Surg. **83**, 1854–1857 (2007)
19. Paiva, W.S., Amorim, R., Bezerra, D.A.F., Masini, M.: Application of the stereolithography technique in complex spine surgery. Arq Neuropsiquiatr. **65**, 443–445 (2007)
20. Zenha, H., Azevedo, L., Rios, L., et al.: The application of 3-D bio-modelling technology in complex mandibular reconstruction experience of 47 clinical cases. Eur. J. Plast. Surg. **34**, 257–265 (2010)
21. Van Noort, R.: The future of dental devices is digital. Dent. Mater. **28**, 3–12 (2012)
22. George, M., Aroom, K.R., Hawes, H.G., Gill, B.S., Love, J.: 3D Printed surgical instruments: the design and fabrication process. World J. Surg. **41**, 314–319 (2017)
23. Rankin, T.M., Giovinco, N.A., Cucher, D.J., Watts, G., Hurwitz, B., Armstrong, D.G.: Three-dimensional printing surgical instruments: are we there yet? J. Surg. Res. **189**, 193–197 (2014)

24. Martinez, P.R., Goyanes, A., Basit, A.W., Gaisford, S.: Influence of geometry on the drug release profiles of stereolithographic (SLA) 3D-printed tablets. AAPS PharmSciTech. **19**, 3355–3361 (2018)
25. Clark, E.A., Alexander, M.R., Irvine, D.J., Roberts, C.J., Wallace, M.J., Sharpe, S., Yoo, J., Hague, R.J.M., Tuck, C.J., Wildman, R.D.: 3D printing of tablets using inkjet with UV photo initiation. Int. J. Pharm. **529**, 523–530 (2017)
26. Pere, C.P.P., Economidou, S.N., Lall, G., Ziraud, C., Boateng, J.S., Alexander, B.D., Lamprou, D.A., Douroumis, D.: 3D printed microneedles for insulin skin delivery. Int. J. Pharm. **544**, 425–432 (2018)
27. Noorani, R.: Rapid prototyping: principles and applications. Wiley, Hoboken (2006). ISBN: 978-0-471-73001-9
28. Haglin, J.M., Eltorai, A.E., Gil, J.A., Marcaccio, S.E., Botero-Hincapie, J., Daniels, A.H.: Patient-specific orthopaedic implants. Orthop. Surg. **8**, 417–424 (2016)
29. Mori, K., Yamamoto, T., Oyama, K., Nakao, Y.: Modification of three-dimensional prototype temporal bone model for training in skull-base surgery. Neurosurg. Rev. **32**, 233–239 (2008)
30. Liu, Q., Leu, M.C., Schmitt, S.M.: Rapid prototyping in dentistry: technology and application. Int. J. Adv. Manuf. Technol. **293**, 317–335 (2006)

Dissimilar Friction Stir Spot Welding of AA2014 and AA7075 Aluminum Alloys

Shubham Jaiswal, Vijay Verma, and Chaitanya Sharma

1 Introduction

Automobile industries and aerospace industries are under great pressure to reduce weight and energy cost by reduction in consumption of fuel and hence greenhouse emissions [1]. The demand for welding dissimilar aluminum alloy is continuously increasing because it gives better mechanical properties [2]. There are so much challenges in joining the aluminum alloys by resistance spot welding (RSW) method. RSW is most commonly used welding technique for high strength, low carbon and coated steels only. On the other hand, solidification shrinkage, oxide formation at surface and high solubility of hydrogen restrict the use of RSW technique for joining aluminum alloys [3]. The above-mentioned problems can be effectively eliminated by friction stir spot welding (FSSW). FSSW was invented by Mazda Motor Corporation and Kawasaki Heavy Industries, Japan, in 2003 for making lap spot joints. FSSW is basically an offshoot of conventional friction stir welding (FSW) and is solid-state, energy-efficient, environment-friendly technique. The process is capable in producing high-quality, stronger similar and dissimilar spot welds in 7xxx and 2xxx grade aluminum alloys so as is increasingly used in aerospace and automotive industries. The basic principle of FSSW is fairly simple and is illustrated in Fig. 1 [4]. In this process, the two sheets to be joined are arranged in lap configuration and held in position by the use of a suitable robust fixture. The lower plate is placed on a baking plate, while upper sheet comes in contact with rotary tool. Then, non-consumable profiled rotating tool plunges into the securely positioned sheets to a pre-defined depth of the total thickness of plates (Fig. 1a). During plunging, first pin comes in contact with upper sheets and further plunging results in intimate contact

S. Jaiswal · V. Verma
Mechanical Engineering Department, BIET, Jhansi 284128, India

C. Sharma (✉)
Mechanical Engineering Department, RJIT, BSF Academy, Tekanpur 475005, India
e-mail: chaitanya.sharmaji@gmail.com

M. Muzammil et al. (eds.), *Recent Advances in Mechanical Engineering*, Lecture Notes in Mechanical Engineering, https://doi.org/10.1007/978-981-15-8704-7_69

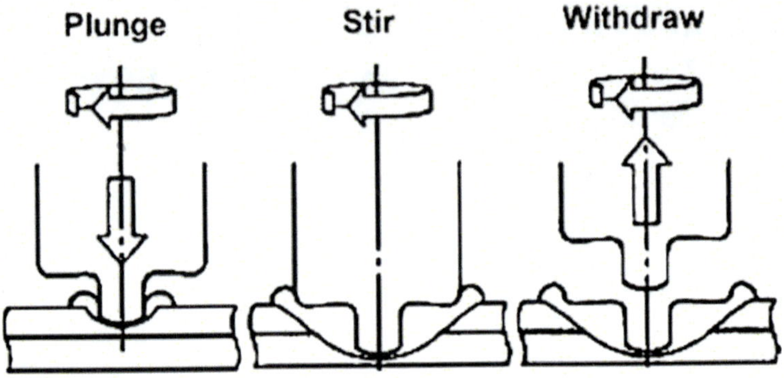

Fig. 1 A schematic illustration of FSSW process [4]

of shoulder with top surface of upper sheet. The friction at the interface of rotating tool pin and shoulder with sheets produces heat, and softens and plasticizes the sheet materials. Plastic deformation of (sheets) material by rotating and plunging tool also contributes to the heat generation. The extended stirring during dwelling period adds additional heat to the deformed sheets and further softens and plasticizes them (Fig. 1b). The rotating tool stirs this softened and plasticized material to complete solid-state joining under the action of extreme axial force. Finally, the tool is drawn out from the sheets leaving behind an exit hole in spot welded sheets (Fig. 1c).

For a particular combination of materials and sheet thickness, the development of sound friction stir spot welded joints (FSSWed) as well as evolution of microstructure, mechanical and other properties of the joints is greatly influenced mainly by process parameters such as tool geometry, welding parameters and joint configuration. Dwell time plays a very important role in deciding the strength of FSSWed joint [4]. The tool rotational speed, dwell time, plunge rate, tool penetration depth, downward force, etc., are main welding parameters, while pin and shoulder diameter and their geometry are main tool parameters [5–7]. Rotational speed, downward force, dwell time and tool geometry and dimension are selected to generate sufficient heat by frictional heating and plastic deformation for proper softening and plasticization of sheets and formation of defect-free friction stir spot welds with optimum weld properties.

Tozaki et al. [8] have reported that with increase in pin length bond width increases but beyond optimum value increase in pin length led to significant increase in hook height also. With increase in shoulder diameters, bond width increases. Beyond optimum plunge depth, hook height increases. Very short stir time might lead to insufficient mixing of plate materials. The hardness decreases on the HAZ to a minimum at the interface between heat-affected zone (HAZ) and thermo-mechanically affected zone (TMAZ) and becomes maximum in stir zone [9]. Hooking works as crack nucleation site. At sharp hooking site, crack nucleates and under small loading it can grow on upper sheet material. Hooking becomes sharp due to high plasticity, i.e., at longer dwell time [10]. Babu et al. investigated that there is no major effect on joint strength and joint formation by FSSW of base metal tempered condition and presence of alclad layers in aluminum alloy AA2014 [11]. In this study, dissimilar friction stir spot welding of precipitation hardening aluminum alloys AA2014 and AA7075 was performed in order to develop defect-free spot welds and to investigate the evolution of microstructure and mechanical properties of developed spot welds.

2 Experimental Procedure

In this study, AA2014 and AA7075 aluminum alloy sheets of 1.7-mm and 2.5-mm thickness, respectively, were used as base material to produce dissimilar FSSW lap joints. The chemical compositions of base material alloy AA7075-T6 and of alloy AA2014-T6 are Zn—5.299%, Cu—1.703%, Mg—2.496%, Cr—0.221% and Cu—4.046%, Zn—0.13%, Mg—0.772%, Cr—0.028%, respectively.

The pieces of width 24 mm and length 120 mm were machined from the received plates. The sheets were held in lap joint configuration using indigenously fabricated fixture. The sheets of AA2014 were placed over the sheets of AA7075. Friction stir spot welding was performed on computer numerical controlled (CNC) milling machine MCV-350 (maximum spindle speed 6000 rpm and maximum axial load 400 kgf) at Jai Bhagwati Enterprises, Kanpur, using rotary speed of 1000 rpm, plunge rate of 10 mm/min, plunge depth of 3.8 mm and dwell time of 8 s. Tool was machined from die steel and had flat shoulder with tapered cylindrical pin. The tool shoulder diameter was 10 mm, while 3 mm long pin had diameter of 5 mm at root and diameter of 4 mm at tip.

After FSSW, sound weld joint was sectioned from the center of exit hole of tool pin along the plane perpendicular to thickness of sheets to prepare specimen for optical microscopy. The sectioned specimen was polished and etched in Keller's reagent (95 ml H_2O, 1.5 ml HCl, 1 ml HF, 2.5 ml HNO_3) for 40 s for microstructural examination. The optical microscopy was performed on Leica microscope (DM750, Leica, Germany) at Rustamji Institute of Technology, BSF Academy, Tekanpur, Gwalior (M.P.). After FSSW, the developed welded lap joints were subjected to tensile testing on servo hydraulic computerized universal testing machine (25 kN, BISS, India) at MNNIT Allahabad. Tensile testing was conducted at room temperature condition, at a crosshead speed of 1.0 mm/min to obtain tensile shear strength of friction stir spot weld joints. Three tests were conducted, and average value was used for discussion. After tensile testing, fractured surfaces were investigated using scanning electron microscope (JXA-8230, JEOL, Japan) at IIT Kanpur to identify the mode of failure.

3 Results and Discussion

3.1 *Microstructure*

Figure 2 shows the photograph of friction stir spot weld joints. The upper plate is AA2014-T6, and lower plate is AA7075-T6. The plates were welded due to the formation of FSSW joint. The FSSW joint showed well-defined outer ring and keyhole at the center of welded zone. The keyhole is created due to withdrawal of rotating pin, and diameter of this hole was equal to the base diameter of pin.

Figure 3a represents the macroscopic cross-sectional view of FSSW joint obtained by cutting the weld from center of weld along a plane parallel to the thickness of plate. The careful observation of macrograph revealed complete joining of both the plates and formation of stir zone, TMAZ and HAZ. The tool geometry is also seen in dark region. The FSSW joints can be divided into two regions where metallurgical bonding between the plates is either partial or complete and between them hook formation takes place. Micrographs in Fig. 3b–d show the evolution of all zones in FSSWed joints and microstructure in stir zone, and HAZ of FSSW joints. The shiny region adjacent to keyhole represents the stir zone. Stir zone had dynamically recrystallized finer grains (Fig. 3b), significantly smaller than base metals and other zones. The TMAZ is a region between stir zone and HAZ where thermo-mechanical action takes place. TMAZ exhibited distorted and coarse grains whose average size was smaller than base metals and HAZ but greater than WNZ. HAZ exhibited coarsened grains of size larger than base metals. Secondary precipitates were prefrentially

Fig. 2 Photograph of friction stir spot weld joints

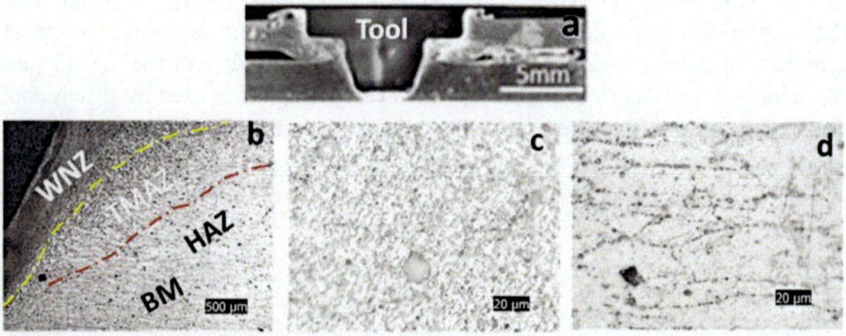

Fig. 3 Evolution of microstructure in FSSW joints

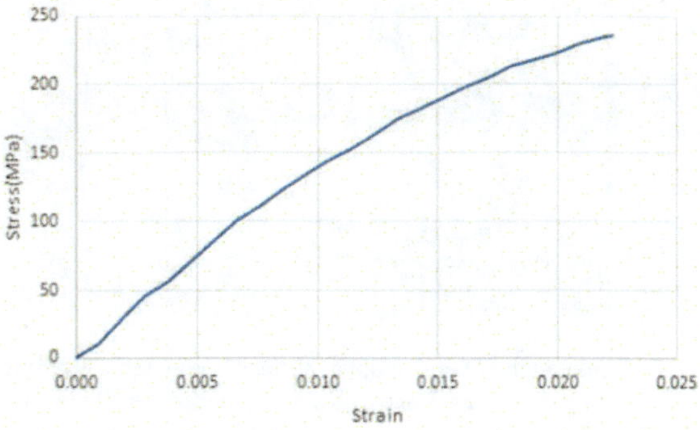

Fig. 4 Stress–strain diagram of FSSW joints

aligned along the grain boundaries however, few were found inside the grain boundry (Fig. 3d). The thermo-mechanical deformation is responsible for the evolution of different kinds of microstructure in different zones of weld joints [12].

3.2 Mechanical Properties

Stress–strain curve obtained from tensile testing is shown in Fig. 4. The average tensile shear load was 4.15 kN. The shear tensile strength and % strain of the FSSW joint were 228.1 MPa and 2.22%, while toughness was 3.13J/mm^3 which is defined as the area under the stress–strain curve.

The tensile shear strength of the FSSW joints is dependent on bond length. Larger the stir zone, better is the tensile shear load/strength of joints due to longer bond length. Up to yielding, graph was linear, and after yielding it lost its linearity. Toughness depends on both stress and strain, so for high toughness good combination of stress and strain is essential. High strain combined with low stress will result in better ductility but low toughness because the curve will be flattened and small area under the graph will cause low toughness.

3.3 Fractography

The photographs of surfaces of weld joints before and after tensile tests are shown in Fig. 5. Figure 5a shows appearance of joint of upper plate before shear load failure, and Fig. 5b, c represents upper and lower plates, respectively, after the shear load failure.

There is a circular hole in the center of joint which is made by pullout of tool in the joint named as keyhole. Around the keyhole, there is a ring of stirred material. The appearance of upper surface of lower plate after tensile testing suggested that failure is because of nugget pullout. Figure 6 presents the SEM micrographs of lower plate

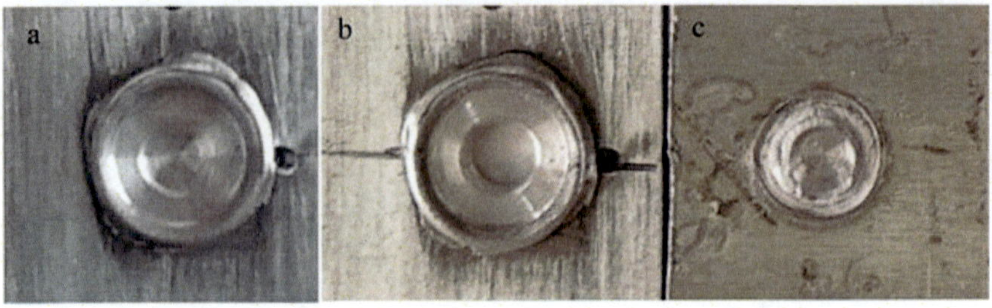

Fig. 5 Photograph of weld: **a** upper surface before tensile testing, **b** upper surface of top plate after tensile testing, **c** upper surface of lower plate after tensile testing

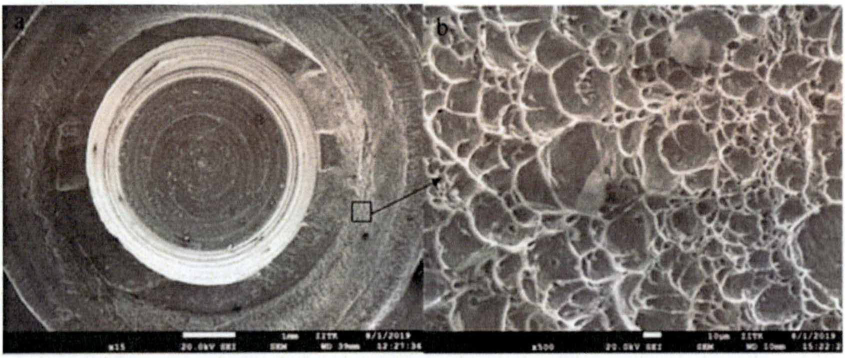

Fig. 6 SEM micrographs showing the features of fracture surface of FSSW joint

shown in Fig. 5c. After tensile testing, fracture surfaces were studied using SEM for getting the mode of failure. Figure 6a represents whole image of lower plate, while Fig. 6b shows the enlarged view of square region which is indicated by an arrow in Fig. 6a. From Fig. 6b, it is clear that there are large numbers of dimples and there is a variation in the size of these dimples. The dimples were elongated during tensile shear testing. The presence of dimples of varying sizes in large number confirmed that welded joints underwent ductile failure. If mode of failure will be ductile in nature, then strain will be more and in that case area under the stress–strain will be more; i.e., toughness of this joint will be more, so there will be requirement of more energy for breaking the welded joint.

4 Conclusion

AA7075-T6 of thickness 2.5 mm and AA2014-T6 of thickness 1.7 mm were welded by friction stir spot welding in a lap configuration. Proper metallurgical bonding of these two plates was observed in stir zone. Besides this WNZ, TMAZ and HAZ were observed in the stir zone of welded joint with varying microstructure than base metals. The lap shear load of welded specimen at 1000 rpm, 10 mm/min and 8 s is found to be 4.15 kN. SEM analyses confirmed that the mode of failure was ductile in nature as elongated and deep dimples ere present on fractured surface.

References

1. Doshi, S.J., Gohil, A.V., Mehta, N.D., Vaghasiya, S.R.: Challenges in fusion welding of Al alloy for body in white. Mater. Today Proc. **5**(2), 6370–6375 (2018)
2. Anawa, E.M., Olabi, A.G.: Using Taguchi method to optimize welding pool of dissimilar laser-welded components. Optics Laser Technol. **40**, 379–388 (2008)
3. Praveen, P., Yarlagadda, P.K.D.V.: Meeting challenges in welding of aluminum alloys through pulse gas metal arc welding. J. Mater. Process. Technol. **164–165**, 1106–1112 (2005)
4. Lathabai, S., Painter, M.J., Cantin, G.M.D., Tyagi, V.K.: Friction spot joining of an extruded Al–Mg–Si alloy. Scripta Mater. **55**(10), 899–902 (2006)
5. Akinlabi, E.T., Sanusi, K.O., Muzenda, E., Akinlabi, S.A.: Material behaviour characterization of friction stir spot welding of copper. Mater. Today: Proc. **4**(2), 166–177 (2017)
6. Shen, Z., Yang, X., Yang, S., Zhang, Z., Yin, Y.: Microstructure and mechanical properties of friction spot welded 6061–T4 aluminum alloy. Mater. Des. **1980–2015**(54), 766–778 (2014)
7. Hirasawa, S., Badarinarayan, H., Okamoto, K., Tomimura, T., Kawanami, T.: Analysis of effect of tool geometry on plastic flow during friction stir spot welding using particle method. J. Mater. Process. Technol. **210**(11), 1455–1463 (2010)
8. Tozaki, Y., Uematsu, Y., Tokaji, K.: Effect of tool geometry on microstructure and static strength in friction stir spot welded aluminium alloys. Int. J. Mach. Tools Manuf. **47**(15), 2230–2236 (2007)
9. Rosendo, T., Parra, B., Tier, M.A.D., Da Silva, A.A.M., Dos Santos, J.F., Strohaecker, T.R., Alcântara, N.G.: Mechanical and microstructural investigation of friction spot welded AA6181-T4 aluminium alloy. Mater. Des. **32**(3), 1094–1100 (2011)
10. Liu, H.J., Fujii, H., Maeda, M., Nogi, K.: Tensile properties and fracture locations of friction-stir-welded joints of 2017–T351 aluminum alloy. J. Mater. Process. Technol. **142**(3), 692–696 (2003)
11. Babu, S., Sankar, V.S., Janaki Ram, G.D., Venkitakrishnan, P.V., Madhusudhan Reddy, G., Prasad Rao, K.: Microstructures and mechanical properties of friction stir spot welded aluminum alloy AA2014 JMEPEG **22**, 71–84 (2013)
12. Sharma, C., Dwivedi, D.K., Kumar, P.: Effect of post weld heat treatments on microstructure and tensile properties of friction stir welded joints of Al-Zn-Mg alloy AA7039. Mater. Design **43**(1), 134–143 (2013)

Analysing Crane Hook of Different Cross Sections and Different Materials

Prajwal Singh, Oshi Jain, Prashant Yadav, Nikhil, and Subodh Kumar Sharma

1 Introduction

A crane hook or simply a hook is used for lifting the load with the help of a device such as a hoist or chain or wire rope. They help lift heavy loads at construction sites as well as in workshops and industries. It is subjected to bending stresses which makes it highly prone to failure. It is equipped with a latch which is a safety mechanism put in place to avoid detachment from the lifting chain or rope to which the freight is attached. Crack development in crane hooks, especially at the areas of high stress concentrations, is the leading cause of fracture. It leads to severe accidents in the workplace, most of which are fatal. There are two types of fractures: ductile fracture—the crack spreads continuously which can be detected easily, and brittle fracture, which has sudden propagation of the crack sudden failure. This is very difficult to detect. So, ductile is preferred over brittle fracture. We do the stress analysis of these hooks to determine the areas with high stress (i.e. failure points) and those with low stresses—these are potential points for material removal, weight reduction and reduced costs.

2 Literature Reviews

Shaban et al. [1] modelled a crane hook on ABAQUS software to study the stress concentration of the hook under full load conditions. This allows us to determine the real-time pattern of stress pattern in a 3D model. To verify its accuracy, an acrylic model of the hook, making use of the shadow optical method (also known as caustic method), is set up. Then, the shape of the hook is modified, after determining the

P. Singh · O. Jain · P. Yadav · Nikhil · S. K. Sharma (✉)
Department of Mechanical Engineering, KIET Group of Institutions, Ghaziabad, India
e-mail: subodhmeerut@gmail.com

M. Muzammil et al. (eds.), *Recent Advances in Mechanical Engineering*, Lecture Notes in Mechanical Engineering, https://doi.org/10.1007/978-981-15-8704-7_70

stress concentration, to prolong the work-life along with reducing the rate of failure. To reduce failure, it is essential to identify the extent and site of stress. The FEA method is used to validate results. The authors draw the conclusion that if the area of the inner side of the crane hook, the site of maximum stress, is broadened, then the stress concentration and thereby failure can be reduced. In addition to this, Narvydas et al. [2] used FEA to determine circumferential stress concentration factors in the shallow notches of a trapezoidal cross-sectional hook. These factors are used to evaluate the structural strength as well as the durability. The FEA results were then used to form equations that can be used to derive formulas for quick calculations to determine stress concentration of structures and machine elements. Most hooks use ductile materials because ductile failure is easier to investigate then brittle failure. Keeping this in mind, the authors studied the strain-based criteria of failure, taking into account the stress triaxiality.

Uddanwadiker [3] used finite element method to study stress analysis of a crane hook. She used photo-elasticity to validate the results obtained. Photo-elasticity test works on the birefringence property of light. The study of stress concentration in the hook under loading conditions was carried out in two steps—first, as mentioned above, FEM stress analysis followed by validation via photo-elastic experiment, and secondly, by assuming hook to be a curved beam and verifying by using FEM. The last step was comparing the analytical calculations with the ANSYS result. The results were found to have a meagre percentage error (=8.26%). Shape was modified to increase the strength of the crane hook based on the stress pattern obtained. Further, Trifković et al. [4] analysed the state of stress in the hook by making use of both exact and the approximate methods. They calculated the stresses in various fragments of the hook, first, by assuming it to be a straight beam and after that a curved beam.

Singh et al. [5] made a solid model and then used finite element analysis of a crane beam. They used PRO/E WILDFIRE 2.0 and ALTAIR HYPER MESH using OPTISTRUCT 8.0 SOLVER Software to evaluate the difference of stress and displacement in various sections of the beam. Weight reduction, limiting stresses and cost cuts were some of the objectives achieved by using finite element analysis. One of the chief conclusions was that maximum stress is found around the fixing positions. Bhagyaraj et al. [6] described their work on crane hook design and did analysis using finite element method. Considering the weight reduction and heavy load dealing of hooks in this paper, the lead alloy is used to build a resistant block hook, which has high strength as compared to presently used alloy materials. SolidWorks has been used for part design and assembly. Further, an analysis of the hook was done on SolidWorks simulation. Results of von Mises stress, factor of safety, strain and deformation are discussed based on their maximum and minimum values. According to Bhagyaraj et al., von Mises stress are very less as compared to yield strength of the material used. The hooks made up of these type of high-strength alloys can reduce the size of hooks withstanding the heavy loading conditions.

Reddy et al. [7] have discussed load-carrying capacity of different cross sections of hooks in their study. They have studied I-section and T-section with combination of different materials, so that the material used to manufacture a crane hook should have ductile fracture property instead of brittle fracture to avoid sudden and serious accidents. The crane hooks are subjected to bending and axial stresses, so the design of the hook has done considering the loads, cross-sectional area and curvatures used in application of hooks at industries. The finite element method results using analysis software were compared with theoretically calculated values of stresses induced and deformation in the hook model. The studies showed that stresses induced in T-section are less than that of I-section with same materials. The method for optimization of material and the various cross-sectional areas of crane hooks for better results were carried to fulfil the industrial requirements. Meanwhile, Krishna and Suresh [8] studied the design and analysis of crane hook using various materials. They predicted stress concentration areas on the crane hook to prolong the work-life and reduce failure causing stresses. They have also stated about the crane hook safety by avoiding improper and unskilled manual handling. Continuous working can also result in an increase of stresses in crane hook. The basic model is designed which was imported for finite element structural analysis. The results obtained for stress were compared for different materials. Further, the total deformation was compared for selection of material.

Bergaley and Purohit [9] performed a structural analysis of the hook using FEM. They purchased a crane hook to perform finite element analysis. To locate yield point and the area with maximum stresses, the hook was tested on UTM machine. The model with that dimensions and material property was made using CAE software and analysed for stresses induced. The modifications were done using FEM at cross section with minimum stress induced. Adding on to this, Gopichand [10] discussed Taguchi optimization method for design parameters. The radius of curvature, cross section and material selection for crane hook are used as design parameters. The need of design optimization is very necessary to avail the better results of load lifting without casing failure. Singh and Rohilla [11] have presented their work on optimization and fatigue analysis of a crane hook using FEM. In their study, they have used SolidWorks for modelling of a hook which had a trapezoidal cross section. After that, they have analysed the model on ANSYS workbench. The results are obtained for different cross sections of crane hook as per loading capacity on the basis of mass, von Mises stress and total displacement. In addition to this, they also performed fatigue analysis for best response to load-bearing capability of the hook.

Devaraj [12], has discussed design criteria of crane hook and analyse stress distribution using ANSYS workbench. The design of crane hook using different materials has been done to calculate von Mises stress distribution and total deformation under loading conditions at ambient temperature. This paper also shows the factors for material selection on the basis of continuous loading and unloading which causes fatigue failure and tells that the crack propagation in ductile fracture is slow and easily detectable. Based on the results of analysis, maximum value of von Mises stress and maximum total deformation were individually compared and modifications were done where the stress concentration was found to be maximum in the model, to get best suited results for material of hook which can withstand higher loading capacity. Further, Ratnakumar et al. [13] discussed design and stress analysis of various cross section of hooks. Also, they used different radii of curvature. Stresses induced against loads were calculated using curved beam theory and experiment done on universal testing machine (UTM). By using curved beam theory, solutions were obtained for circumferential stress and radial stress in curved beams. Further tests were made on UTM so that experimental values can be calculated. Theoretical stresses are calculated using finite element method, for that modelling of crane hook with various cross sections and curvature radii. FEA is done for different models to obtain different stress distribution results. The experimental stresses obtained were compared with theoretical stresses obtained by finite element method, and graphs were studied. The experimental stresses were found greater than theoretical stresses obtained through FEA because of the certain assumptions used in numeric solution.

Benkar and Wankhade [14] studied the stress pattern of the hook under loading conditions. ANSYS 14 workbench was used to prepare a solid model of the hook. The real-time stress distribution pattern in the 3D model of the crane hook is obtained. Then, they performed finite element analyses on various models of a hook—circular, triangular, trapezoidal and rectangular cross sections. Meanwhile, Sahu and Yadav [15] worked on bigger crane hooks carrying a comparatively larger load. Existing drawings were used to prepare the CAD model followed by implementing the modified cross section of the hook using the static structural analysis workbench of CATIA V5. These results helped in determining stresses and deflections in that model. To evaluate the optimum dimensions, a wide range of models of varying dimensions of the hook were tested and then the best of them was picked. The selection was based on a strict criterion and only those that satisfied all the factors such as load-carrying capacity, stress-induced and deflection. In another paper, Sahu et al. [16] worked on crane hook of the trapezoidal section. Modelling was done on in CATIA V5R20. Finite element method was used to identify the area of maximum stress concentration. After this, they applied design of experiment (DOE) by changing the length of the parallel sides of the trapezoidal cross section and studying its effect on energy stored in the hook, displacement, stress and mass.

Mehendale et al. [17] wrote their research paper on design and stress analysis of an EOT crane hook of different cross sections. They used three different cross sections—rectangular, circular and trapezoidal, for analysis of a varying load (in tones). Failure of a crane hook mainly depends on three major factors i.e. dimension, material, overload. They used curved beam formula for calculating stresses produced in different areas. The results obtained were validated on computer-aided software. Apart from this, Kumhar et al. [18] worked on design and analysis of crank hook of trapezoidal cross section using FEA. Analysis of trapezoidal cross section is done using Winkler-Bach formula for curved beams. This formula is helpful in evaluating and setting standards of safety of crane. They also used software like FEM, Creo Parametric, etc., to verify the results. The theoretical results obtained were checked using von Mises stress theory. Post this, they were compared with ANSYS.

Kurt et al. [19] did finite element analysis of a tower crane. These cranes are mostly used in the construction sites and in the manufacturing industries for carrying of the heavy loads. It is a fixed crane which has to be mounted on site. The tower crane can carry up to thousand kilograms with varying different internal and external loads due to dead weight, wind load and the dynamic loads. After the calculation of these loads, the result obtained was analysed with the help of FEM. In this research paper, the load combinations (such as horizontal load) resulting from vertical as well as inertia loads applied on the main beam can lead to sag which is why their values are compared with permissible stress values too using both analytical and FEM methods. Rajurkar et al. [20] established magnitude of stresses involved in the crane hook by several methods including analytical, numerical and experimental methods. Curved beam formula was used for analytical analysis, and CAD models were used for numerical analysis. Meanwhile, FEM and photo-elasticity methods were used for analysis of stress concentration on different cross sections of the hooks. Photo-elasticity allows us to see changes in the optical properties of a material under mechanical deformation, and this helps in determining the stress distribution in any given material.

3 Conclusion

Identifying the structural strength of the crane hook is of paramount importance in determining its load-bearing capacity. After reviewing the previous papers, the authors of this paper have come to the conclusion that establishing the best material, geometry and dimensions for particular loading conditions needs a lot more research and investigation. Far too few papers have been published to state anything definitively. We can tentatively conclude that a lead and iron alloy hook of trapezoidal cross section is the most optimum. However, it is safe to say that the finite element method has come up as an indispensable tool for design and analysis of these components.

References

1. Shaban, M., Mohamed, M.I., Abuelezz, A.E., Khalifa, T.: Determination of stress distribution in Crane Hook by caustic. Int. J. Innov. Res. Sci. Eng. Technol. **2**(5), 1834–1840 (2013)
2. Narvydas, E., Puodžiūnienė, N.: Circumferential stress concentration factors at the asymmetric shallow notches of the lifting hooks of trapezoidal cross-section. MECHANIKA **18**(2), 152–157 (2012)
3. Uddanwadiker, R.: Stress analysis of crane hook and validation by photo-elasticity. Engineering **4**(3), 935–994 (2011)
4. Trifković, S., Radić, N., et. al, "Stress analysis of crane hook using FEM. INFOTEH-JAHORINA **10**(Ref. C-2), 244–248 (2011)
5. Singh, B., Nagar, B., Kadam, B.S., Kumar, A.: Modelling and finite element analysis of crane boom. Int. J. Adv. Eng. Res. Stud. **1**(1), 51–52 (2011)
6. Bhagyaraj, G., Suryaprakash, K., Subba Rao, K.: Crane hook design and analysis. Int. Res. J. Eng. Technol. **4**(9), 61–65 (2017)
7. Amareswari Reddy, M., Krishnaveni, M.N.V., Nagaraju, B., Raja Roy, M.: Static analysis of crane hook with I-section and T-section using ANSYS. Int. J. Eng. Trends Technol. **26**, 72–79 (2015)
8. Sai Krishna, E., Dr. Suresh Kumar, S.: Design and analysis of crane hook with different materials. Int. J. Mech. Eng. Technol. **9**(4), 786–791 (2018)
9. Bergaley, A., Purohit, A.: Structural analysis of crane hook using finite element method. Int. J. Sci. Mod. Eng. **1**(10) (2013)

10. Gopichand, A.: Crane hook using taguchi method. Int. J. Sci. Mod. Eng. **2**(12), 7780–7784 (2013)
11. Singh, A., Rohilla, V.: optimization and fatigue analysis of a crane hook using finite element method. Int. J. Recent Adv. Mech. Eng. **4**(4), 31–43 (2015)
12. Devaraj, A.: Design of a crane hook of different materials and stress analysis using ANSYS workbench. Int. J. Res. Appl. Sci. Eng. Technol. **3**(VII), 310–314 (2015)
13. Ratnakumar, G.E.V., Jitendra Kumar, B., Prasad, K.: Design and stress analysis of various cross section of hook. Int. J. Innov. Eng. Technol. **4**(4), 90–99 (2014)
14. Benkar, C.N., Dr. Wankhade, N.A.: Finite element stress analysis of crane hook with different cross sections. Int. J. Technol. Res. Eng. **1**(9), 868–872 (2014)
15. Sahu, G.N., Yadav, N.: Design and stress analysis of various cross-section of hook. Int. J. Mod. Eng. Res. **34**, 2187–2189 (2013)
16. Sahu, S., Dewangan, R., Patnaik, M., Yadav, N.: Study of crane hook having trapezoidal section by finite element method & design of experiments. Int. J. Mod. Eng. Res. **2**(4), 2779–2781 (2012)
17. Mehendale, S.A., Wankhade, S.R.: Design and analysis of EOT crane hook for various cross sections. Int. J. Curr. Eng. Sci. Res. **3**(12), 53–58 (2016)
18. Kumhar, V., Talapatra, P.K., Sahu, N.K.: Design and analysis of crank hook trapezoidal cross section using finite element analysis. VSRD Int. J. Mech. Civil Automob. Prod. Eng. **5**(05), 89–94 (2015)
19. Kurt, S., Gerdemeli, I., Deliktaş, O.: Finite element analysis of the tower crane. In: Trends in the Development of Machinery and Associated Technology, Mediterranean Cruise, pp. 561–564 (2010)
20. Rajurkar, G.U., Dr. Bhope, D.V., Khamankar, S.D.: Investigation of stresses in crane hook by FEM. Int. J. Eng. Res. Technol. **2**(8), 117–122 (2013)

Exploration on Wear Characteristics: Performance of Gears of Polyamide66

Shashank Singh, Yogesh Kumar Yadav, Siddhartha, and Arunesh Chandra

1 Introduction

Since 1950s, polymer gears are preferred in many low load applications. These gears have some inherent properties such as noiseless operations, lightweight, low cost, self-lubrication and so on. After proliferation in manufacturing techniques, these gears are being used in many essential practices such as food processing machines, ATM machines, wiper devices in automobiles, copier machines and watches [1]. Previously, plastic gears were used only for below ¼-hp drives due to changes in their properties and uncertainties in their behaviour under varied environmental conditions such as temperature and moisture [2]. Nowadays, research practices of better moulding controls encompassing environmental factors have boosted the plastic gear drive power to ¾ hp [3]. Plastic gears might be failed under different modes. These failure modes are wearing, pitting and cracking at root circle as well as at pitch circle. Wearing is defined as material removal from a surface and its deformation after its interaction with any mechanical act in counter directions. Many of the researchers have found the occurrences of thermal failure in the plastic gears generally after the generation of heat at the surface, which is due to the friction and materials' hysteresis. This has been seen that wear rate grows with rise in temperature [3–9].

S. Singh · Y. K. Yadav (✉) · Siddhartha
Department of Mechanical Engineering, National Institute of Technology, Hamirpur 177005, India
e-mail: yadav.yogeshkumar@gmail.com

S. Singh
e-mail: er.shashank994@gmail.com

Siddhartha
e-mail: sidmech@nith.ac.in

A. Chandra
Department of Mechanical Engineering, KIET Group of Institutions, Ghaziabad 201206, India
e-mail: arunesh.chandra@kiet.edu

M. Muzammil et al. (eds.), *Recent Advances in Mechanical Engineering*, Lecture Notes in Mechanical Engineering, https://doi.org/10.1007/978-981-15-8704-7_71

2 Experimental Details

Most studies have exposed the usage of polymer-metal gear pairs in various experiments, while polymer–polymer gear pairs are still to be utilized and explored [6, 10–16]. Many researchers have used alike and unlike gear pairs [3, 15]. Polyamide66 gears were selected just as the polymer gears. Polyamide66 which is the amalgamation of two monomers is a modification and alteration of vitally used polyamide. The two monomers of Polyamide66 named as hexa-methylenediamine and adipic acid have six carbon atoms. Polyamide66, a thermoplastic polymer has advantages such as excellent impact properties and very low notch sensitivity in comparison with that of other thermoplastic materials. On the other hand, Polyamide66 has some disadvantages such as low flexural modulus and which can be counterbalanced by affixing any filler material for improving its properties [17]. Injection moulding process was employed for manufacturing polymer gears designated as NPA66G, H15PA66G and H30PA66G which have glass fibre contents by weight as 0%, 15% and 30%, respectively. Table 1 shows the various parameters for injection moulding process of polymer gears. Gear pairs of Polyamide66 were investigated and tested on the polymer gear test rig, manufactured by DUCOM Instruments, India (CM-9108). The specifications of gears to be manufactured are shown in Table 2. The governing factors for gear testing are driver material (A), driven material (B) and torque (C). The possible combinations of polymer gear pairs are examined at constant speed of 1200 rpm and at different torques such as 1, 1.2 and 1.4 Nm one by one run up to 1 $\times 10^5$ cycles. Specific wear rates (S.W.R.) of polymer gears are calculated using the formula [18]:

Table 1 Polymer gears' parameters for injection moulding process

Parameters	NPA66G	H15PA66G	H30PA66G
Drying temperature (°C)	90	90	90
Injection pressure (MPa)	7	7	7
Press temperature (°C)	190	210	220
Waiting time in moulds (s)	120	120	120

Table 2 Specifications

Specifications of polymer gears	
Module	2 mm
Number of teeth	20
Pressure angle	20°
Face width	8 mm
Pitch circle diameter	40 mm

$$W_S = W_V/2zmbN_T$$

where W_S is specific wear rate in mm^3/mm-mm-revolution, W_V is volumetric wear in mm^3 (which is estimated by the ratio of gear mass change to the density of its material), z is number of teeth on gear, m is module in mm, b is tooth face width in mm and N_T is total number of revolution (rev.).

Table 3 exhibits the outcomes of possible combination of mated polymer gears in the form of specific wear rates. S.W.R. increases with the increase in torque.

Table 3 Experimental pattern/design: outcomes of mated polymer gears (A-B)

Passes	Gear material (driver) (A)	Gear material (driven) (B)	Torque (C) (Unit: Nm)	S.W.R. (A) (Unit: mm^3/mm-mm-rev.)	S.W.R. (B) (Unit: mm^3/mm-mm-rev.)
1st	NPA66G	NPA66G	1	9.48465×10^{-7}	2.62856×10^{-6}
2nd	NPA66G	NPA66G	1.2	1.17133×10^{-6}	2.79454×10^{-6}
3rd	NPA66G	NPA66G	1.4	1.24836×10^{-6}	4.95381×10^{-6}
4th	NPA66G	H15PA66G	1	1.89241×10^{-6}	1.13425×10^{-6}
5th	NPA66G	H15PA66G	1.2	4.37185×10^{-6}	2.2949×10^{-6}
6th	NPA66G	H15PA66G	1.4	6.71135×10^{-6}	2.94445×10^{-6}
7th	NPA66G	H30PA66G	1	1.09333×10^{-5}	*8.71399×10^{-7}*
8th	NPA66G	H30PA66G	1.2	1.19226×10^{-5}	1.07492×10^{-6}
9th	NPA66G	H30PA66G	1.4	*1.2476×10^{-5}*	2.20258×10^{-6}
10th	H15PA66G	NPA66G	1	1.05365×10^{-7}	6.04934×10^{-6}
11th	H15PA66G	NPA66G	1.2	1.17687×10^{-7}	7.18051×10^{-6}
12th	H15PA66G	NPA66G	1.4	1.70621×10^{-7}	8.65447×10^{-6}
13th	H15PA66G	H15PA66G	1	6.19744×10^{-7}	1.61606×10^{-6}
14th	H15PA66G	H15PA66G	1.2	7.87474×10^{-7}	3.17869×10^{-6}
15th	H15PA66G	H15PA66G	1.4	1.12759×10^{-6}	4.3025×10^{-6}
16th	H15PA66G	H30PA66G	1	2.11623×10^{-6}	2.44465×10^{-6}
17th	H15PA66G	H30PA66G	1.2	3.41167×10^{-6}	4.20335×10^{-6}
18th	H15PA66G	H30PA66G	1.4	5.97602×10^{-6}	5.69904×10^{-6}
19th	H30PA66G	NPA66G	1	*5.85737×10^{-8}*	8.19805×10^{-6}
20th	H30PA66G	NPA66G	1.2	7.43612×10^{-8}	8.60622×10^{-6}
21st	H30PA66G	NPA66G	1.4	9.91864×10^{-8}	*1.00391×10^{-5}*
22nd	H30PA66G	H15PA66G	1	2.24456×10^{-7}	4.94588×10^{-6}
23rd	H30PA66G	H15PA66G	1.2	3.07169×10^{-7}	5.97892×10^{-6}
24th	H30PA66G	H15PA66G	1.4	5.39176×10^{-7}	7.05734×10^{-6}
25th	H30PA66G	H30PA66G	1	9.34434×10^{-7}	6.46244×10^{-6}
26th	H30PA66G	H30PA66G	1.2	1.04746×10^{-6}	7.93091×10^{-6}
27th	H30PA66G	H30PA66G	1.4	2.21814×10^{-6}	9.00434×10^{-6}

3 Wear Performance of Driver Materials

At distinct torques, for different combinations of driver and driven gear materials, wear performances of driver gear materials are plotted in Figs. 1–4.

Figure 1 exhibits the wear performance of NPA66G (driver) having combinations one by one with NPA66G, H15PA66G and H30PA66G. On mating NPA66G with same gear material, it does not change the S.W.R. with torque considerably. When NPA66G is mated with other two materials, the S.W.R. of NPA66G grows with rise in torque. Figure 2 exhibits the wear performance of H15PA66G (driver). H15PA66G on mating with itself and NPA66G does not show considerable deviation of S.W.R. on advancing torque, and on mating with H30PA66G exhibits an increasing tendency for S.W.R. up to 1.2 Nm torque. Beyond 1.2 Nm torque, its S.W.R. grows rapidly.

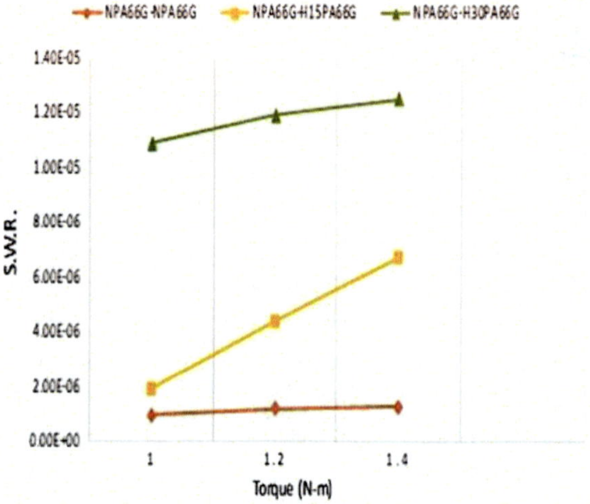

Fig. 1 Wear performance of NPA66G

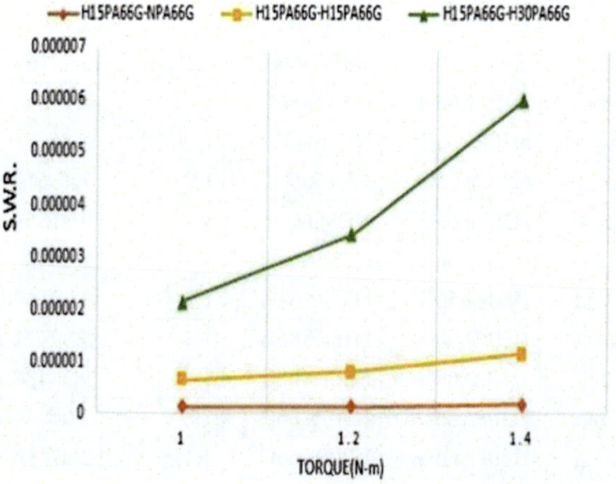

Fig. 2 Wear performance of H15PA66G

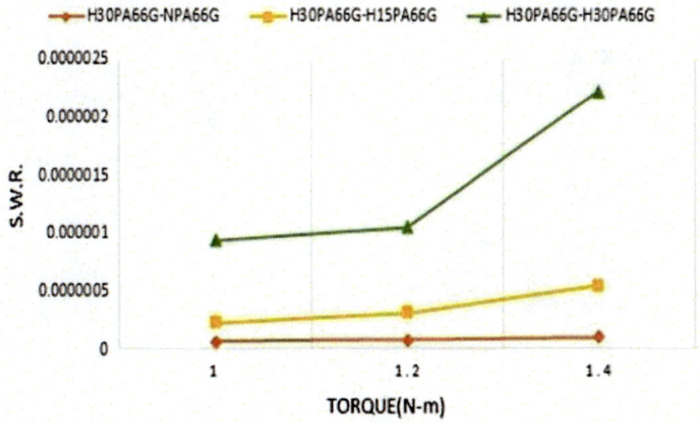

Fig. 3 Wear performance of H30PA66G

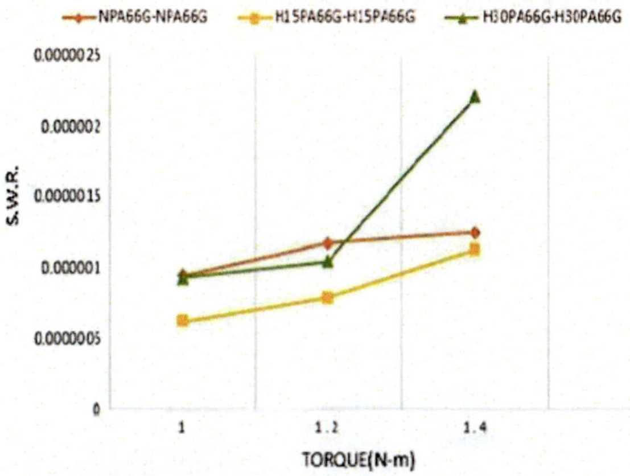

Fig. 4 Driver mated with same material

Figure 3 exhibits the wear performance of H30PA66G (driver). On its combination with NPA66G (driven), its S.W.R. rate does not vary much on growing torque. On its mating accordingly with the rest two combinations, its S.W.R increases. After 1.2 Nm torque level, its S.W.R. increases rapidly for H30PA66G-H30PA66G combination. Figure 4 exhibits the wear performances of driver gears mated with driven gears of same materials. The S.W.R. (driver) increases with increasing torque in each of three possible mating combinations. H15PA66G shows the least S.W.R.

4 Wear Performance of Driven Materials

Wear performances of driven gear materials are plotted in Figs. 5–8 for different driver-driven gear materials combinations at different torque levels.

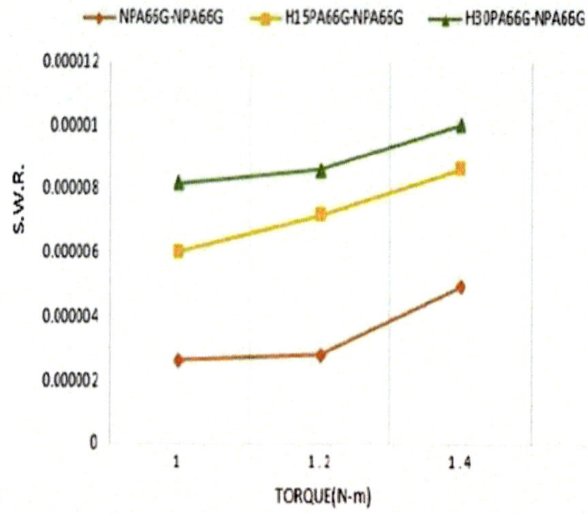

Fig. 5 Wear performance of NPA66G

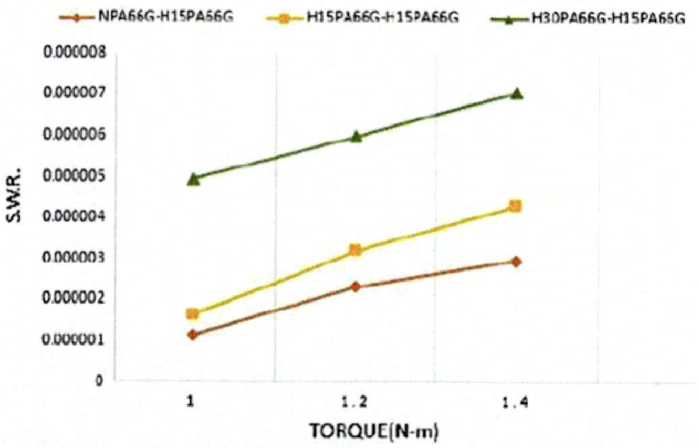

Fig. 6 Wear performance of H15PA66G

Figure 5 exhibits the wear performance of NPA66G (driven) having combinations one by one with NPA66G, H15PA66G and H30PA66G. In all combinations, S.W.R. (driven) is increased succeeding 1.2 Nm torque. After this torque level, S.W.R. raises rapidly. H30PA66G has the extreme considerable outcome on S.W.R. of NPA66G. Rest of the two materials has an insignificant effect on NPA66G. Figure 6 exhibits the wear performance of H15PA66G (driven) having combinations with NPA66G, H15PA66G and H30PA66G. S.W.R. (driven) has increasing tendency in all combinations. H30PA66G (driver) has the extreme considerable effect.

Figure 7 exhibits the wear performance of H30PA66G (driven) mated with itself and rest other two materials. NPA66G (driver) has the lowest outcome on wear rate

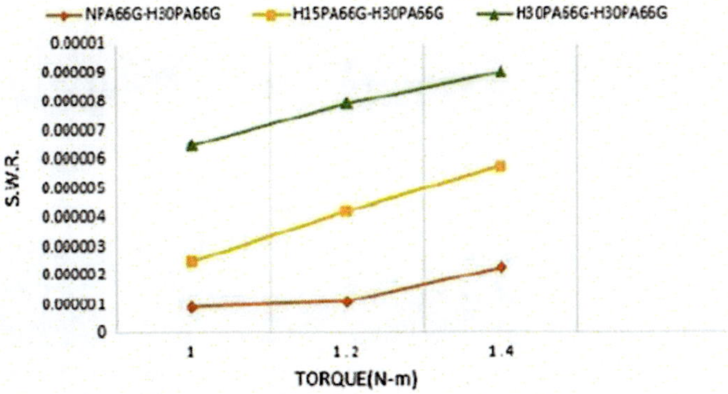

Fig. 7 Wear performance of H30PA66G

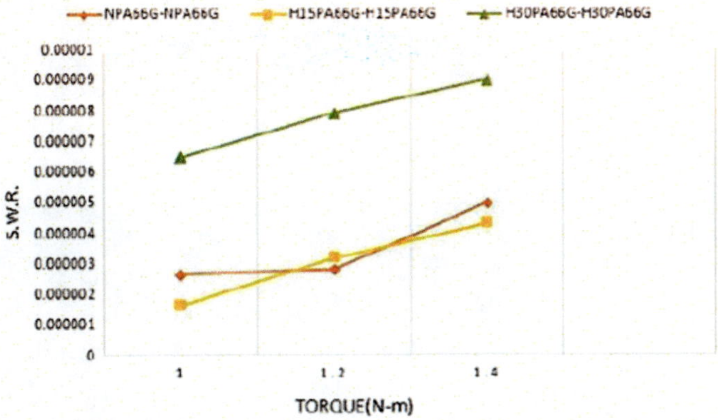

Fig. 8 Driven mated with same gear material

of H30PA66G, whereas H30PA66G-H30PA66G plot shows the most considerable effect on wear rate. Figure 8 exhibits the wear performances of driven gears mated with driver gears of same material. They show increasing wear tendency on increasing torque levels. H30PA66G-H30PA66G combination shows the highest wear rate.

5 Morphology: Wear Surfaces of Polymer Gears

Polymer gears show very low heat transfer rate and so increased specific wear rate. Figure 9 shows the accumulation of some portion of NPA66G (driven) on H30PA66G (driver) which is on account of adherence of NPA66G gear teeth particles on H30PA66G gear teeth. So, 19th row (Table 3) represents the least S.W.R. for driver. This might be on account of restoration of H30PA66G wear with sticking of NPA66G particles. For mating of H30PA66G (driven) and NPA66G (driver), least

(a) Before experiment (b) After experiment

Fig. 9 H30PA66G driver gear

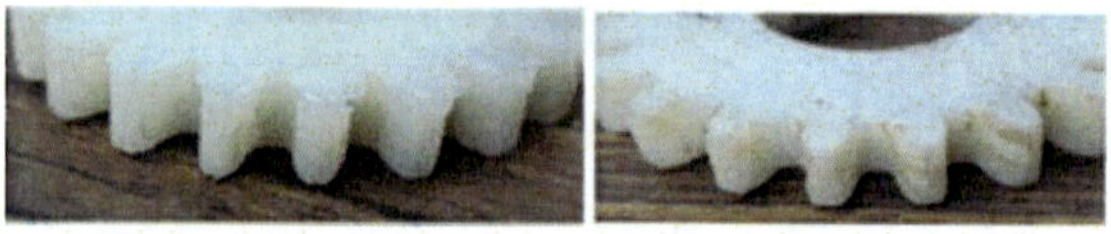

(a) Before experiment (b) After experiment

Fig. 10 H30PA66G driven gear

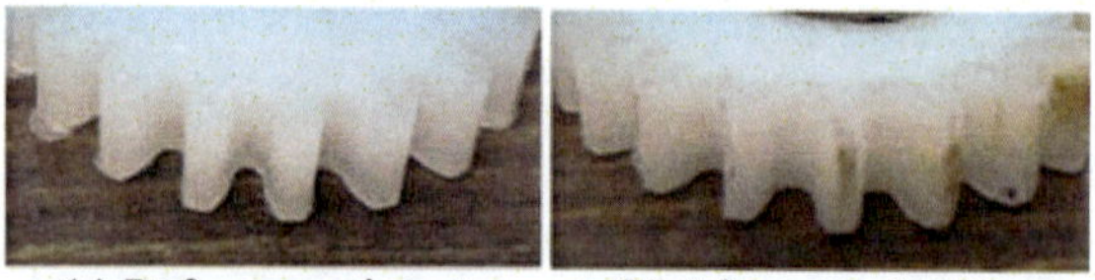

(a) Before experiment (b) After experiment

Fig. 11 NPA66G gear (driver)

(a) NPA66G (b) H15PA66G (c) H30PA66G

Fig. 12 Immense wear of driven gears at 1.6 Nm torque

S.W.R. for driven gear (Fig. 10) is shown in 7th row (Table 3). H30PA66G has faced the low fatigue wear. The basis might be the hardness of H30PA66G (30% glass fibre content) and the lowest torque (1 Nm).

For mating of NPA66G (driver) and H30PA66G (driven), the maximum S.W.R. for driver (Fig. 11) is shown in the 9th row (Table 3). Figure 12 exhibits the failure of NPA66G gear on mating with H30PA66G (driver) due to rapid and excessive wear of teeth at torques equal/above 1.6 Nm. NPA66G is remarked as showing maximum wear at 1.4 Nm torque in 21st row (Table 3). For the maximum wear, high torque (1.4 Nm) and 30% glass fibre content of H30PA66G gear might be responsible.

6 Results and Conclusions

The polymer gears of NPA66G, H15PA66G and H30PA66G materials have been successfully manufactured. Applying governing factors, Polyamide66 gears have been successfully examined on the basis of their performance characteristics using factorial method of design of experiment. Wear performances and behaviour of different pairs of polymer gears have been explored under distinct torque levels. It is obvious and can be concluded that the polymer gear of superior glass fibre content shows lesser wear to that of inferior one. The wearing rate of polymer gears grows on increasing torque levels. Some extra wear can be watched at higher torques.

References

1. Adams, C.E.: Plastic gearing: selection and application, pp. 49–50. Marcel Dekker (1986)
2. Yelle, H., Gauvin, R.: Ultra high molecular weight high density polyethylene as a gear material. Polym. Eng. Sci. **15**, 90–96 (1975)
3. Li, W., Wood, A., Weidig, R., Mao, K.: An investigation on the wear behaviour of dissimilar polymer gear engagements. Wear **271**, 2176–2183 (2011)
4. Mao, K., Greenwood, D., Ramakrishnan, R., Goodship, V.: The wear resistance improvement of fibre reinforced polymer composite gears. Wear **426–427**, 1033–1039 (2019)
5. Mao, K.: A numerical method for polymer composite gear flash temperature prediction. Wear **262**, 1321–1329 (2007)
6. Mao, K., Li, W., Hooke, C.J., Walton, D.: Friction and wear behaviour of acetal and nylon gears. Wear **267**, 639–645 (2009)
7. Kansal, G., Rao, P.N., Atreya, S.K.: Study: temperature and residual stress in an injection molded gear. J. Mater. Process. Technol. **108**, 328–337 (2001)
8. Yians, J., Quinonez, P.D.: Gear surface temperature monitoring. J. Eng. Tribol. **219**, 99–105 (2005)
9. Mertens, J., Senthilvelan, S.: Effect of mating metal gear surface texture on the polymer gear surface temperature. In: 4th International Conference on Materials Processing and Characterization, Material Today: Proceedings, vol. 2, pp. 1763–1769 (2015)
10. Hoskins, T.J., Dearn, K.D., Kukureka, S., Walton, N.D.: Acoustic noise from polymer gears—a tribological investigation. Mater. Des. **32**, 3509–3515 (2011)
11. Breeds, A.R., Kukureka, S.N., Mao, K., Walton, D., Hooke, C.J.: Wear behaviour of acetal gear pairs. Wear **166**, 85–91 (1993)
12. Keresztes, R., Zsidai, L., Kalacska, G., Baets, P.D.: Friction of polymer/steel gear pairs. Mech. Eng. Lett. **2**, 1–11 (2009)
13. Mao, K., Langlois, P., Hu, Z., Alharbi, K., Chetwynd, D.: The wear and thermal mechanical contact behaviour of machine cut polymer gears. Wear **332–333**, 822–826 (2015)
14. Mao, K., Hooke, C.J., Walton, D.: Acetal gear wear and performance prediction under unlubricated running condition. J. Synthetic Lubric. **23**, 137–152 (2006)
15. Mao, K., Hooke, C.J., Walton, D.: The wear behaviour of polymer composite gears. J. Synthetic Lubric. **12**, 337–345 (1996)
16. Singh, P.K., Siddhartha, Singh, A.K.: An investigation on the thermal and wear behavior of polymer based spur gears. Tribol. Int. **118**, 264–272 (2018)
17. Gyoo, P., Min, V., Sriram, K., Sung, C.: Morphology, thermal, and mechanical properties of polyamide 66/clay nanocomposites with epoxy-modified organoclay. J. Appl. Polym. Sci. **101**, 1711–1722 (2006)
18. Yakut, R., Demirci, M.T.: The load capacity of PC/ABS spur gears and investigation of gear damage. Arch. Mater. Sci. Eng. **40**, 41–46 (2009)

Assessing the Thermal Performance of the Walls in a Building Using the Phase Change Material Methodology

Shaheen Hasan, J. A. Usmani, Mohd. Islam, and Saif Uddin

1 Introduction

After the realization of the PCMs energy storage concept, many researchers work on this concept and apply PCMs in the building by using different methods in the wall and roof [1–3]. Those results were calculated by different authors in their study when using different quantities of PCMs material used in building walls [4, 5].

Several studies has been done by the authors and conclude that the main factor which influences the property of PCMs is melting point, mass and airflow rate. For the optimum performance of PCMs, the melting point must be lying within the range of thermal comfort temperature. Several researchers had presented their study on the assessment of PCM's property in different kinds of concrete block [6]. Concrete has also played a significant role in determining the thermal performance of PCM wall. Concrete property changed due to immersion time, alkalinity, immersion temperature and dilution [7].

2 Methodology

To calculate a favorable temperature for a particular commercial building, a field survey was conducted, as per ASHRAE 2007 guidelines. The Basic building property has been taken from ECBC 2017, which is Indian standard building code and mandatory to applied in the building. After picked up HVAC system one by one from ASHRAE guiodeline book, then simulate the building for the whole year and find

S. Hasan (✉) · J. A. Usmani · Mohd. Islam
Mechanical Engineering, Jamia Millia Islamia, New Delhi, Delhi 110025, India
e-mail: shasan39@gmail.com

S. Uddin
Global Evolutionary Energy Design, New Delhi 110025, India

M. Muzammil et al. (eds.), *Recent Advances in Mechanical Engineering*, Lecture Notes in Mechanical Engineering, https://doi.org/10.1007/978-981-15-8704-7_72

Table 1 Baseline building minimum performance specifications

Subject component	Material	Base case
Roof	U factor with film (W/m^2 K)	0.393
Wall	U factor with film (W/m^2 K)	0.677
Floor	U factor with film (W/m^2 K)	1.995
Glazing	W/m^2 K	1.199
Glass visible transmittance		0.301
Glass SHGC		0.252
Window-to-wall ratio		28

out the results. In this research work, EnergyPlus software was used. The EnergyPlus is a collection of many program modules that works together to calculate the energy required for heating and cooling a building using a variety of systems and energy sources.

This basic property is defined in Table 1, now using HVAC system one by one and found the results.

3 Description of HVAC System

ASHRAE defined 8 type of HVAC system, in this research work picked one by one correspondingly simulated and then find out the effect of these HVAC system on load means heating load and cooling load, it is basically a comparative study and tried to find out which one is better for a particular building, location and occupancy level.

3.1 System 1 (Package Terminal Air Condition)

See Table 2.

Table 2 Energy consumption and gas emission

	Electricity (kWh)	SO_2 (kg)	CO_2 (kg)	CO (kg)	NOx (kg)
Heating	4250	36.93	2499.94	14.4	10.28
Cooling	360,806.47	3135.41	212,237.2	1223.13	873

Table 3 Energy consumption and gas emission

	Electricity (kWh)	SO_2 (kg)	CO_2 (kg)	CO (kg)	NOx (kg)
Heating	760.35	6.6	447.085	2.578	1.804
Cooling	626,840.1	5447.24	368,582	2124.98	1516.7

Table 4 Energy consumption and gas emission

	Electricity (kWh)	SO_2 (kg)	CO_2 (kg)	CO (kg)	NOx (kg)
Heating	2924.01	25.41	1719.3	9.912	7.07
Cooling	172,335.47	1497.6	101,333.25	584.2	416.98

Table 5 Energy consumption and gas emission

	Electricity (kWh)	SO_2 (kg)	CO_2 (kg)	CO (kg)	NOx (kg)
Heating	2868.72	23.35	1686.8	9.72	6.94
Cooling	145,613.83	1265.4	85,620.9	493.6	4720.7

3.2 *System 2 (Package Terminal Heat Pump)*

See Table 3.

3.3 *System 3 (Packaged Rooftop Air Conditioner)*

See Table 4.

3.4 *System 4 (Packaged Rooftop Heat Pump)*

See Table 5.

3.5 *System 5 (Packaged Rooftop VAV with Reheat)*

See Table 6.

Table 6 Energy consumption and gas emission

	Electricity (kWh)	SO_2 (kg)	CO_2 (kg)	CO (kg)	NOx (kg)
Heating	445,132.55	3868.2	261,737.94	1508.99	1077
Cooling	535,565.01	4654.05	314,912.22	1815.56	1254.27

Table 7 Consumption of energy in different areas

	Electricity (kWh)	SO_2 (kg)	CO_2 (kg)	CO (kg)	NOx (kg)
Heating	1,082,731.90	9408.9	636,646.35	3670.46	2619.8
Cooling	263,113.35	2286.45	154,710.65	891.95	631.5

Table 8 Energy consumption and gas emission

	Electricity (kWh)	SO_2 (kg)	CO_2 (kg)	CO (kg)	NOx (kg)	Electricity (kWh)
Heating		1,462,051.12				
Cooling	229,879.58		1997.65	135,169.2	779.3	551.74

3.6 *System 6 (Packaged VAV with PFP Box)*

See Table 7.

3.7 *System 7 (Packaged Rooftop VAV with Reheat)*

See Table 8.

3.8 *System 8 (VAV with Reheat)*

See Table 9.

Table 9 Energy consumption and gas emission

	Electricity (kWh)	SO_2 (kg)	CO_2 (kg)	CO (kg)	NOx (kg)
Heating	1,161,142.16	10,090.28	682,751.59	3936.3	2809.5
Cooling	229,878.84	1997.62	135,168.758	776.9	556.2

4 Modeling of PCM-Based Building Wall System Assumptions

(1) Heat transfer is one-dimensional (x-direction only). The sunbeam incident on the wall has been uniformly distributed throughout the wall with changes in its intensity. In this paper, for numerical calculation it is assumed that the radiation fall over the wall is constant. EnergyPlus software is used for the calculation of PCM effect. For the numerical calculation, a constant temperature is assumed for ambient atmosphere. But in an actual ambient atmosphere, it is not constant. So, any two points on the surface of the wall are to be considered in thermodynamic equilibrium condition (no heat exchange with each other). So, the heat transfer takes place along with the thickness of the wall.

(2) The ambient temperature does not maintain constant, it is low in the morning than increase upto 5–6 pm and reached higher temp and then reduced till morning 6 am, so basically if we observe ambient temp. don't get a constant temp and this ambient temp directly affect building inside room temp , it also change (if don't use any air conditioning). So if we applied PCM material into the building wall section, it tried to reduce this fluctuation inside the building and provide thermal comfort constant temperature.

(3) In this research work firstly to calculate energy consumption load by selecting an HVAC system one by one, then high energy consumption HVAC system was picked up. In my research works, PTHP was a high energy consumption HVAC system. So, the PCM material was used in the same building with PTHP HVAC system. Basically, the HVAC system maintained a human comfort temperature and PCM material was tried to minimize the variation in temperature inside the building walls. So due to this, natural convective heat transfer between the inner wall and room was reduced. The PCM material cannot eliminate the need of HVAC system totally, but cooperates with the system and minimizes the load on the HVAC system. PCM helps HVAC system in terms of reducing heat gain through the wall which reduces the load on the system (Tables 10 and 11).

Table 10 Property used in PCM

PCM	T_m (°C)	Heat of fusion (J/kg)	Latent heat of fusion (Lf), (J/g)	Density of solid phase g mL^{-1}	T_f (°C)	Flash point (°C)
Butyl stearate	23.7	121	128	0.861 g mL^{-1} at 20 °C	24.4	160

Table 11 Thermal conductivity at different temperatures

T (°C)	40	50	60	70	80	90	100	110
Butyl stearate	155	154	132	150	149	147	146	144

5 Results and Conclusion

From the study, several conclusions were drawn as described below.

1. With the help of this study, we conclude that system 4 is more efficiently working. For the same building area and environment, system 4 consumes less amount of energy and provides same human comfort as provided by other systems. It is not necessary that system 4 is effectively working for all kinds of area.
2. As per study, we concluded that system 2 (PTHP) consumes more energy as compared to other 7 systems. So for the research work, we picked up the worst case among them and applied PCM in the building walls. Through this, researcher tried to improve the worst case (system 2), i.e., to reduce building load at maximum level.
3. The effect of the PCM on the system 2 is shown in Fig. 1. The graph of system 2 with PCM declined to second last number. Figures 1 and 2 concludes that PCM helps to reduce temperature in the space and saves energy to 55.18%.
4. Table 12 shows the negative sign in heating row; it means that in winter PCM does not effectively work in this building.

The most important results obtained in this research work are that the PCM emerged as having a low melting point temperature in the building wall receiving the least amount of sun radiation. The PCM would not work effectively if PCM kept at a high latent heat of fusion in the building walls in combination with huge quantity of sun radiation. If PCM does not reached at their melting phase, PCM does not work effectively. Authors explain if orientation, shading or any PCM applied in a cold region cause PCM not melt properly correspondingly does not work effectively. This

Table 12 Consumption of energy in different areas and savings

	Electricity (kWh) system 2 with PCM	Electricity (kWh) system 2 without PCM	Savings (kWh)
Heating	6067.45	760.35	−5307.1*
Cooling	393,397.27	626,840.1	233,442.83

*Negative sign shows that system 2 with PCM is consuming more energy as compared to system 2 without PCM

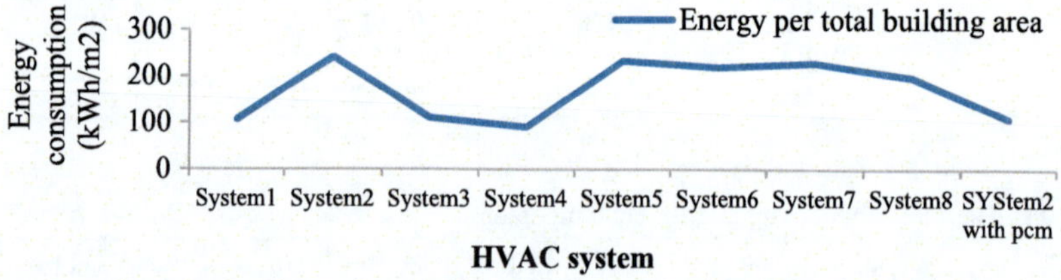

Fig. 1 Comparisons of HVAC systems with HVAC system 2 with PCM

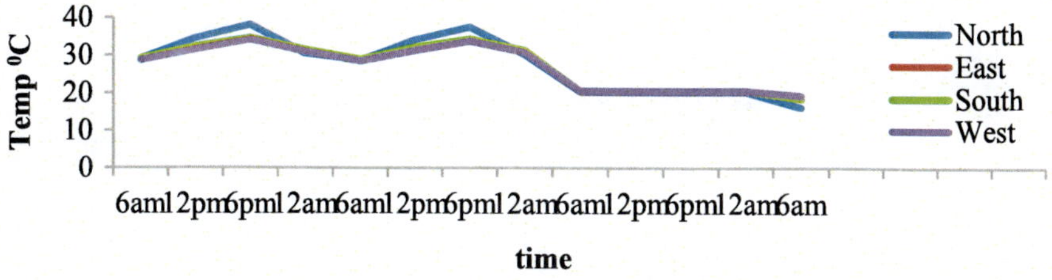

Fig. 2 Temperature variation inside building space with PCM

research work concluded that with the help of PCM worst case could be changed into good cases. The results also concluded that PCM properties are the function of building wall geometry. So for best results of PCM in a different wall orientation, then a different PCM in a different mass concentration should be used rather than applying same PCM in the same concentration. It is not concluded that north wall is best choice for the using PCM but it depends on building orientation. A particular type of PCM is used for a particular environment condition.

References

1. Peippo, K., Kauranen, P., Lund, P.D.: A multicomponent PCM wall optimized for passive solar heating. Energy Build **17**, 259–270 (1991)
2. Feustel, H.E., Stetiu, C.: Thermal performance of phase change wallboard for residential cooling application, Lawrence Berkeley National Laboratory, Report LBL: 38320 (1997)
3. Shapiro, M.: Development of the enthalpy storage materials, mixture of methyl stearate and methyl palmitate sub-contract report to Florida Solar Energy Center (1989)
4. Zhang, Y., Zhou, G., Lin, K., Zhang, Q., Di, H.: Application of latent heat thermal energy storage in buildings. Build Environ. **42**, 2197–2209 (2007)
5. Khudhair, A., Farid, M.: A review on energy conservation in building applications with thermal storage by latent heat using phase change materials. Energy Convers. Manag. **45**, 263–275 (2004)
6. Stritih. U., Novak, P.: Thermal storage of solar energy in the wall for building ventilation, Second workshop, IEA, ECES IA Annex 17. In: Advanced thermal energy storage techniques, feasibility studies and demonstration projects, pp. 3–5. Ljubljana, Slovenia, (2002)
7. Hawes, D.W., Feldman, B.D.: Latent heat storage in bottom concrete slab. Sol. Energy Mater. **21**, 61–80 (1990)

Structural Analysis of Spur Gear with Composite Material Under Different Loading Conditions

Roshni Khanna and Prabhat Kumar Sinha

1 Introduction

A material which is made from more than two material with different chemical and physical properties when combined they produce a material with characteristics totally different from the individual components. The composite materials are preferred over traditional material for many reasons such as these materials are stronger, less expensive or lighter. An attempt is being made for replacing the gear material of conventional steel with different composite materials in the required application. In this work of the spur gear pair pinion as a driver and the gear as the driven part made in the solid works software. Ansys 16.0 is that analysis software which is used as the analysis tool for the static structural analysis and for the replacement of traditional material with different composites under variable load conditions. Simulated result gives the total deformation, equivalent von Mises stress.

1.1 Review of Literature

Yakut et al. [1] presented examination of PC/ABS material spur gears, and they investigated damage. They also examined the use of PC/Abs as a material in spur gear. The result shows that PC/ABS material is advantageous in many industrial areas as they are durable against flame, holds low moister.

Siva et al. [2] investigated and described the design of metallic gears and proposed the replacement of the metallic gears in sugarcane juice machine with the polymer gears for the reduction in weight and the noise. They concluded and proved that the composite gears give the useful properties of conventional gear.

R. Khanna (✉) · P. K. Sinha
Department of Mechanical Engineering, SHUATS, Prayagraj, Uttar Pradesh, India
e-mail: rosh20khanna@gmail.com

M. Muzammil et al. (eds.), *Recent Advances in Mechanical Engineering*, Lecture Notes in Mechanical Engineering, https://doi.org/10.1007/978-981-15-8704-7_73

Patil et al. [3] presented free vibration behavior of spur gear with composite material through FEM methodology. They concluded and found that natural frequency of gear increases with the increase in the fiber orientation.

Singh et al. [4] investigated using ANSYS workbench software, bending stress, contact stress, and static load on the tooth of spur gear drive is found. The Hertz theory and Lewis formula also were used for theoretical calculation of contact stress and bending stress of spur gear. They observed and found that theoretically results obtained by Lewis formula and Hertz equation were comparable with finite element analysis of spur gear, keeping in mind the comparison they concluded that the finite element analytical result can be better as a problem solving software and used for other analyzing purpose.

Kapoor et al. [5] They studied the parametric model of differential gear box which is developed using some parameters. The case study shows that the composite material can be used effectively in place of metallic material because the weight of glass-filled polyamide composite material of differential is reduced by 60% comparing with the traditional materials (auminum alloy, alloy steel, cast iron). Finally, they concluded that glass-filled polyamide composite material is selected as a best material for differential gear box.

Olguner et al. [6] studied about the design of spur gears with asymmetric teeth in external pump applications. The process of generating the gear model calculations is automated,, and using the data from the model, tooth contact and root bending stresses were evaluated during a mesh cycle by finite element analyses. They investigated seventeen different spur gear pair designs for the different drive side pressure angles coast side pressure angle and profile shift coefficients. They concluded that the effects of these parameters on tooth contact stress, bending stress, flow rate and flow rate fluctuations for specific gear pump.

Jabbour et al. [7] determined the root and contact stresses for metal, spur and helical gears. The results of this method were verified by finite element calculations. This method enables the modeling of a pair of meshed gears during an interval of rotation for helical gears, non-uniform load distribution along the lines of contact. They concluded that the parameters as well as the gear configurations were maximum for the bending and contact stresses.

Sjoberg et al. [8] presented that the high-performing gear transmission was needed to reduce CO_2 emissions. The gear surface roughness was investigated in parallel with testing. Higher efficiency was observed for test using a high a running-in load, and for low lubricant temperatures. The running-in load had larger effect on the mesh efficiency than temperature. They concluded that high running-in load also yielded higher gear mesh efficiency and had a larger impact on surface parameters.

Ghosh et al. [9] studied how to reduce the vibration level with the modification of the tooth profile of a of a spur gear pair. The severity of vibration and noise at different operating conditions had been measured by different metrices, namely dynamic mesh force, dynamic tooth force, bearing force along line of action and offline action directions. The optimal modification of the tooth profile has been obtained by solving an optimization problem. They concluded that the vibration caused by geometrical errors like profile and pitch error could be reduced.

Keerthi et al. [10] studies and analyzed the stresses for safety conditions, and weight reduction of gear was studied. The gear was modeled in Tworks and was analyzed in ANYSYS 16. Static structural analysis was performed to determine the deformation, and modal analysis was performed to determine the mode shapes and natural frequencies, and the results were validated. They found that stress induced was same in steel and composite material and natural frequencies vary.

2 Finite Element Analysis of Gear

The FEM is a methodology for finding the solutions of engineering and physics. FEM simulates results to algebraic equations. This method approximates the unknown function over the domain [4]. In this chapter, FEM tool is used to analyze and carried by using the ANSYS 16.0 FEM tool to determine the total deformation, equivalent von Mises stress (Table 1).

2.1 Static Structural Analysis of Gear Assembly with Different Composite Materials at Different Torques

At torque 350 NM after designing gear and meshing (equivalent stress, total deformation) (Fig. 1 and Table 2)

Table 1 Table showing the parameters of gear design

	GEAR	PINION
Number of teeths	50	20
Module	5	5
Pitch circle diameter	260	104
Face width (in mm)	16	16
Dedendum circle diameter(mm)	247.5	91.5
Pressure angle (in degrees)	20°	20°
Face width (b)	20	20

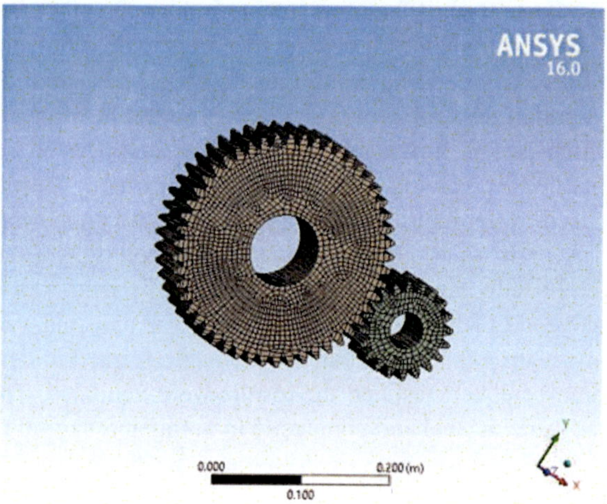

Fig. 1 Meshing in Ansys analysis software

Table 2 Table showing material properties

Properties	Unit	Structural steel	Carbon epoxy	Carbon fiber high modulus
Young's modulus	GPA	210	450	320
Tensile strength	MPA	500	52	274
Density	KG/m^3	7850	1800	7800
Poison's ratio	–	0.3	0.3	0.2

1. For structural steel

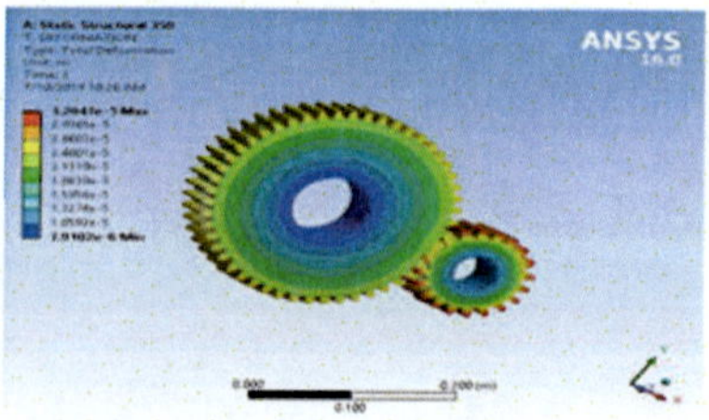

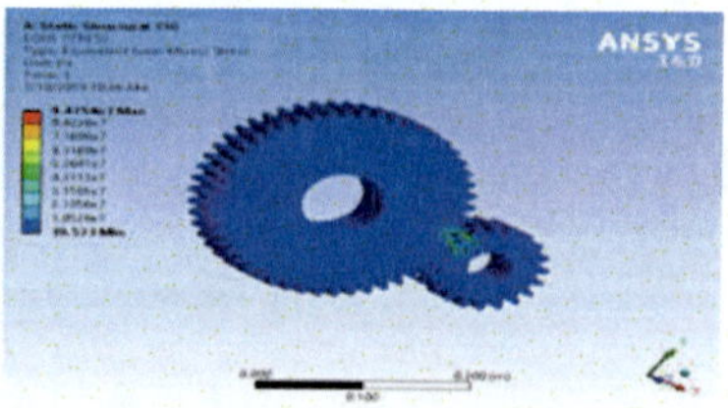

2. For carbon epoxy

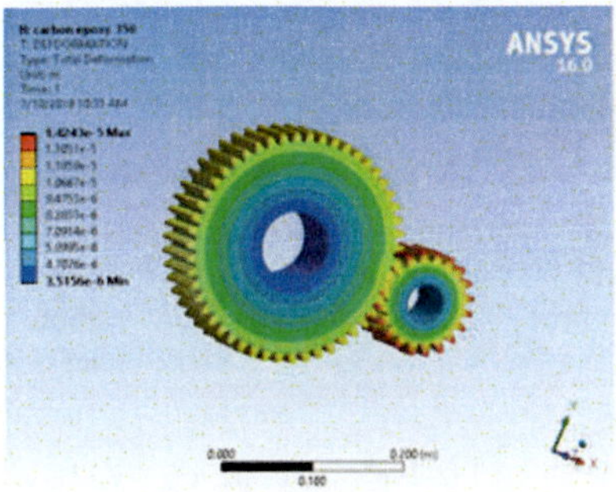

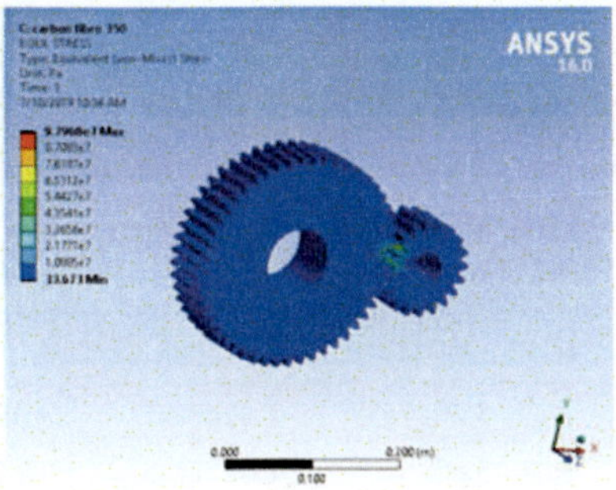

3. For carbon fiber high modulus

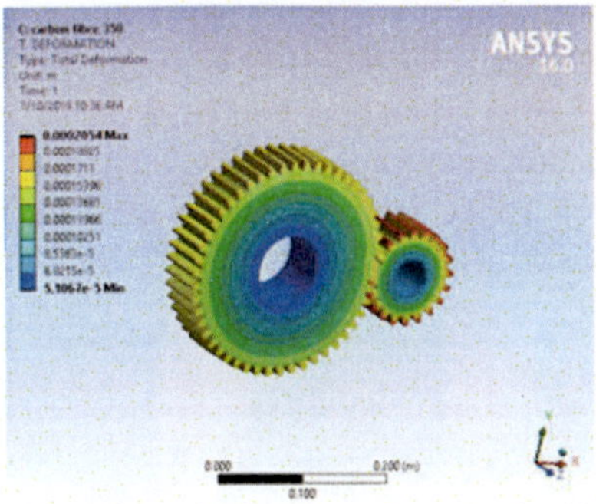

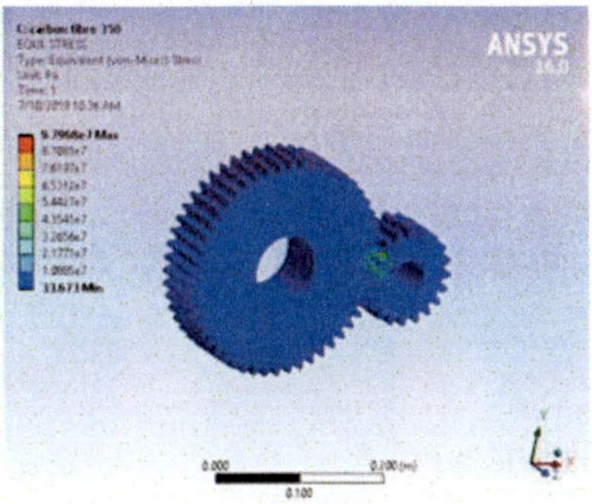

3 Conclusion

The main purpose of this paper is the reduction of stress and the gear deformation with the use of other materials for the application of transmission system. For that purpose, the solid modeling of the gear was modeled in solid works using the parameters. Theoretical analysis and software analysis were done for finding the generation of stress at different load conditions and were compared with one other. The FEM tool ANYSYS 16.0 was used to analyse the composite material gear and steel gear. In the images of total deformation and equivalent stress, the color range shows the different values of result. The value was increasing from downwards to upwards from blue to red. The red portion indicates that the area is most affected. The evaluation was done under three conditions of torque. The total deformation and equivalent von Mises stress for both materials conclude with the results.

1. The comparison between steel gear and composite material gear was carried out for the required applications successfully under 350, 400 and 450 N M, and Table 3 shows the corresponding results for total deformation and equivalent stress.

Table 3 Table showing the comparative results

Torque conditions (NM)	Total deformation (m)			Equivalent von Mises stress		
	Structural steel	Carbon epoxy	High modulus carbon fiber	Structural steel	Carbon epoxy	High modulus carbon fiber
350	$3.2047e^{-5}$	$1.4243e^{-5}$	$2.054e^{-4}$	$9.4754e^{7}$	$9.4754e^{7}$	$9.7968e^{7}$
400	$3.6625e^{-5}$	1.6278e-5	$2.3474e^{-4}$	$1.0829e^{9}$	$1.0029e^{9}$	$1.1196e^{8}$
420	$3.5238e^{-5}$	0.000017092	$2.4648e^{-4}$	$1.137e^{8}$	$1.137e^{8}$	$1.1756e^{8}$

2. The inner portion of the gear pair was less stress; hence, the deformation in that area was less and contact area of the meshed gear was most affected; in each case, the stress induced is maximum and the deformation was also more.
3. While comparing the data from Table 3, it was found that deformation and von mises stress induced were less for the carbon epoxy composite as compared to other composite materials. Hence, carbon epoxy material can be used for its useful properties to replace the conventional material gear.

References

1. Yakut, R., Duzcukoglu, H., Demirci, M.T.: The load capacity of PC/ABS spur gears and investigation of gear damage. Arch. Mater. Sci. Eng. **40/1**, 41–46 (2009)
2. Altaf Hussain, S.V., PalaniKumar, P.: Modeling and analysis of spur gear for sugarcane juice machine under static load condition by using FEA. Int. J. Modern Eng. Res. **2/4**, 2862–2866 (2012)
3. Herakal, P.M., Kerur, B.: Dynamic analysis of composite spur gear. In: Proceedings' of 3rd IRF International Conference, pp. 562–567 (2014)
4. Mohit, S., Waris, K., Kumar, W.: Structural analysis of composite material helical gear under different loading conditions. Int. J. Eng. Sci. Res. Technol. 31–34. ISSN: 2277-9655
5. Kapoor, N., Pradeep, K., Rahul, G., Bhool, B.: Parametric modeling and weight analysis of glass filled polyamide composite differential gearbox. Int. J. Sci. Eng. Technol. Res. **3/6**, 203–208 (2014)
6. Filiz, O.S.: A study on the design of assymetric spur gear in gear pump application. In: Yon, I. (ed.), International Gear Conference, 26th–28th, pp. 406–417 (2014)
7. Toni, J., Anmar, G.: Tooth stress calculation of metal spur and helical gears. In: Mechanism and Machines Theory, vol. 92, pp. 375–390 (2015)
8. Sjoberg, S., Andersson, M., et al.: Analysis of efficiency of spur gear ground gears and the influence of running. Tribiol. Int. **93**, part A, 145–163 (2016)
9. Chakraborty, G.G.: An optimum tooth profile modification for the reduction of vibration and noise in spur gear pairs. In: Mechanism and Machine Theory, vol. 105, pp. 145–163 (2016)
10. Keerthi, M., Sandya, K.: Static & dynamic analysis of spur gear using different materials. Int. Res. J. Eng. Technol. (IRJET) **03**(01), 40–41 (2018). e-ISSN: 2395-0056, p-ISSN: 2395-0072

Mitigation of Harmonic Current in Grid-Connected Solar Power System

Uma Yadav, Anju Gupta, H. K. Rai, and D. K. Bhalla

1 Introduction

As far as the solar power is concerned, it is first concentrated by solar cells. Assembly of these cells is called solar module and the combination of modules, forms a Panel. Multiple panels connected together form an array, which is designed to supply significant power to the grid. When solar power is directly connected with the grid, certain problems are faced, like change in voltage, change in frequency, spikes, and harmonics of current. The grid-connected loads are nonlinear and complex in nature. Moreover the power obtained by solar PV is not uniform. To solve these problems, we use multiple approaches. One technique is MPPT. Simulator is used to simulate these faults. One example of simulator is SPI-Sun Simulator 350i [1]. Use of inverter can control the power requirement of the changing load [2]. Distribution board (DB) interfacing the inverters and solar panels is used to meet the ever-increasing demand of electricity and to cope up with the interrupted supply. It reduces the need for installing new transmission lines which are very costly and difficult to install. Mitigation of harmonics improves quality of power supply, commercially viable, and sustainable. It is environmental friendly too. Of course, the safety while installing these large panels must be kept on priority [3].

MPPT technique is used to track maximum power at optimum temperature. Interestingly, maximum solar power output is not at highest temperature during noon time in summers.

In this paper, a model is made to simulate and analyze the harmonics and method found to compensate the reactive power, when PV generated power is connected to

U. Yadav (✉) · A. Gupta
Electrical Engineering Department, J.C. Bose University of Science & Technology, YMCA, Faridabad, India
e-mail: er.yadavuma11@gmail.com

H. K. Rai · D. K. Bhalla
Mechanical Engineering Department, IPEC, Ghaziabad, India

M. Muzammil et al. (eds.), *Recent Advances in Mechanical Engineering*, Lecture Notes in Mechanical Engineering, https://doi.org/10.1007/978-981-15-8704-7_74

the main supply. PI type controllers are used. The tailor-made current controller (i.e., combining a PR controller and an RC controller) can achieve relatively high control accuracy as well as less complexity [4]. Synchronous reference frame method is used to obtain reference current against compensation of complex load. This synchronous reference frame method, separates the reactive and harmonic components of the load current. This control method reduces the calculations and gives more accurate results.

This shows the working of grid-connected DG system driven by boost converter and a three-phase VSI. There is considerable improvement in quality of supply current using this method. This is seen that solar electricity supplied to the grid by this system has the same frequency, amplitude, voltage, and phase with no spikes and harmonics, as available on the grid.

2 Representation of Power from PV to Grid

A typical assembly of solar power generation and connections to grid i shown below in Fig. 1. Here, a model has been selected for simulation, and schematic diagram of the same is shown.

The basic elements of grid-connected PV system are solar panel, renewable power system inverter (typically used for 600 KW solar plant manufactured by Bonfiglioli Vectron), DC distribution board, MCB box, and auxiliary power supply with battery bank. These all are important and play vital role. To convert DC into AC, an inverter is must which is connected in series; this ensures the correctness of frequency, phase, and amplitude of the current. Also, the inverter must be of good quality, and it should be reliable to bear loads of power grid. With mixed and varying loads, the harmonic sources are more which hamper the quality of current. Our effort is to bring power factor to unity (one), at appropriate voltage. This ensures efficient energy transfer with reduced reactive losses.

To detect harmonics and reactive current and to mitigate them, synchronous reference frame method has been used.

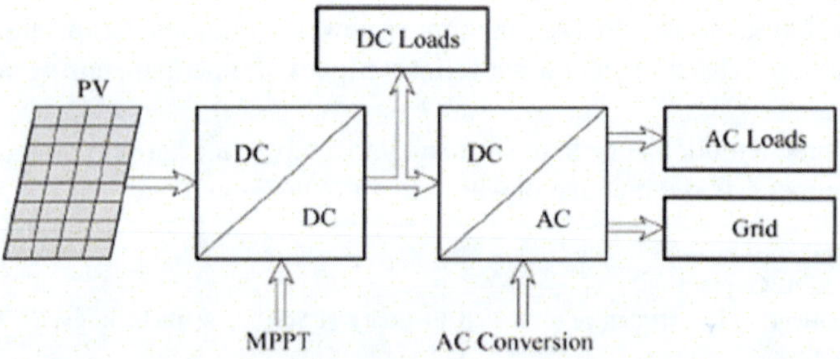

Fig. 1 Schematic diagram of power from PV to grid

3 Analyses of Reference Current and Output Current

In fact, there are three components of total current: Active component, reactive component, and the harmonic. The operating principle of PV based grid-connected system is discussed in following steps.

(i) At first, detection of grid voltage and current of total load is observed.
(ii) Calculation for harmonic component, active and reactive components of load current is done, and subsequently command signal is generated.
(iii) Then, the controller generates the current reference. Actual current needed is calculated to compensate, harmonic, active, and reactive components of total load.
(iv) The coordinates are generated to differentiate the reactive and harmonic current components of the grid.
(v) Then, the inverter supplies steady current and voltage and absorbs harmonics. It reduces reactive power also.

The purpose of using synchronous reference frame (SRF) method is to separate out the reactive and harmonic components of the load current. When reference current is generated, the design of controller is made. When value of reference current is obtained, the controller is designed. Once value of reference current is known, this controller supplies the output current of the inverter according to this reference current. The transformations of Park and Clarke are used for simulation. Clarke transformation makes use of two equations, whereas Park transformation uses the three equations. Both have their own advantages. Clarke transformation is simple, whereas Park transformation tells the voltage direction as well as it calculates the load current components. A high pass filter, which is used in d axes, will not permit the DC component of current, which has less frequency range of around 10–15 Hz.

But, we see from the other angle Park transformation which has other added advantage that it does not make use of compensator.

4 Results and Discussion

In Figs. 2 and 3, characteristics of the PV module (P–V and I–V curve) at constant irradiance and constant temperature have been shown, which indicates an open circuit voltage of 20 V and short-circuit current of 2.5 A. For simulation of this model grid-connected PV system, an equivalent source of current is designed and used, with a diode connected in series. This is obvious that the current generated by the solar array will be proportional to the solar power attained from the Sun. We obtained 52.5 W of power at 2.5 A circuit current and at 20 V. To step up the DC voltage, before inverter, a boost converter is used. Finally, it is the inverter, which supplies AC current, to the grid, although it extracts the power from Sun, in DC form. This inverter has a controller to ensure high-quality output of power and compensating

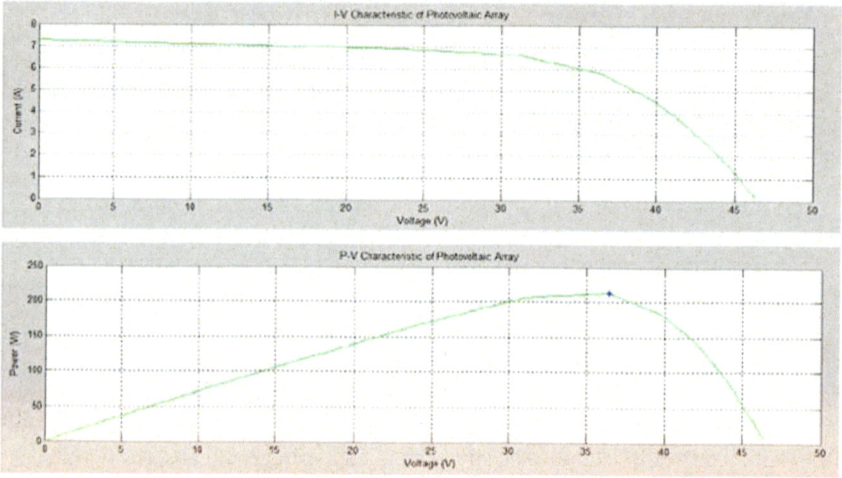

Fig. 2 I–V characteristics of photovoltaic array

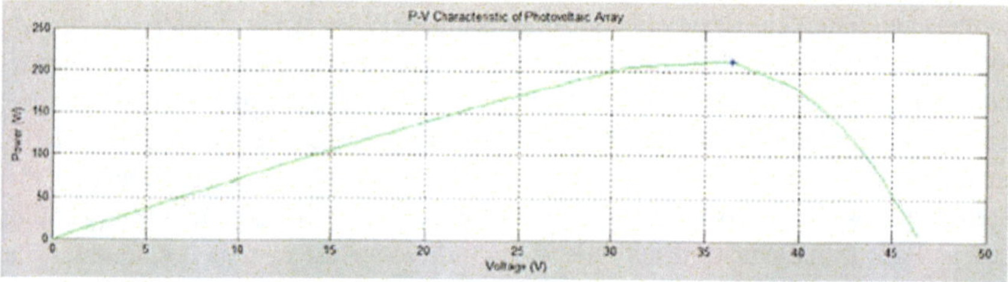

Fig. 3 P–V characteristics of photovoltaic array

reactive power and harmonics as well. The point where inverter is connected with the grid is called PCC or point of common connection. A DC link capacitor is used, in between the PV array and inverter, which acts as a current reservoir.

Figures 2 and 3 shows the I–V characteristics versus current and P–V characteristics versus power.

The model as discussed above is simulated at MATLAB/Simulink software. A typical reading of DC boost converter is taken at 0.05 s, giving output voltage of 80 V, the full-wave diode rectifier, which is used, generates RL load on the circuit. The harmonic currents on the main load are continuously generated due to complex load. The control circuit tracks the reference current as shown in Fig. 4. This is with respect to time. Figure 5 is showing the output tracking current with respect to time.

We see from the obtained results that the reactive power of grid is zero after 0.05 s. This shows that DC voltage is not varying and reactive power compensation is taking place. Both grid voltage and current are in same phase. The overall current is non-harmonic with non-reactive.

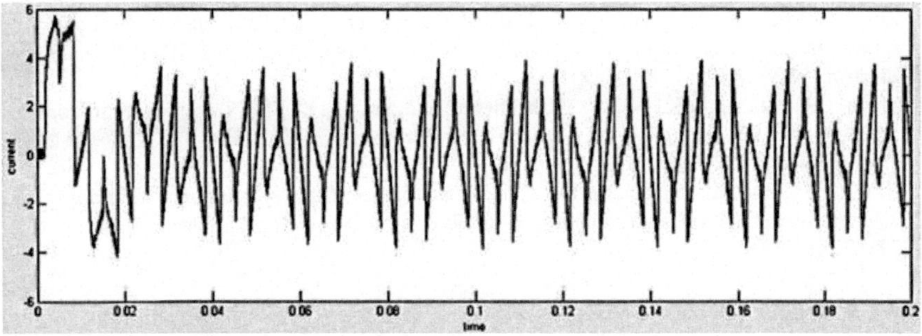

Fig. 4 Time versus reference current

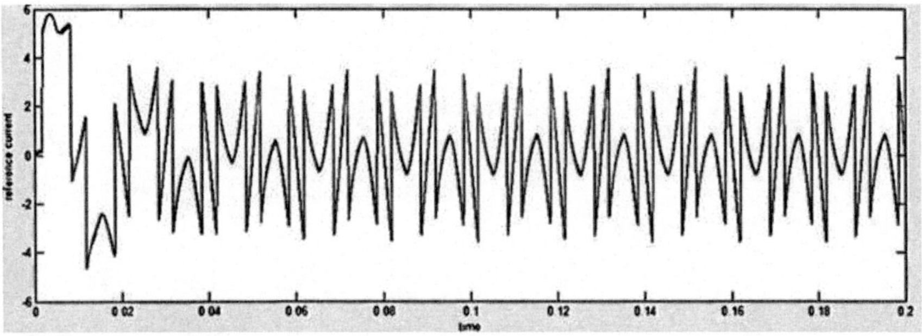

Fig. 5 Time versus output current

5 Conclusion

The results obtained from SRF controller, for grid-connected solar PV system, were stimulated and studied. These show that the system gives extremely good results with varying and complex loads of the grid. It gives very favorable results on reactive and harmonic components of the load current. Phase is also same. Calculations are reduced. Supply current quality is considerably improved. Although complete harmonic compensation is not there, due to limitation of inverter and varying load. But, it is need of time to connect solar PV system with the grid, so that solar power is not only fully utilized, but also it can be supplied to the main grid, and if solar power availability is less, it can be compensated by the grid current.

References

1. Singh, B., Kumar, S.: Harmonics mitigation and phase compensation technique for 3P4W SPV system, pp. 1–6 (2016). IEEE, 978-1-4799-5141-3/14

2. Zhang, L., Sun, K., Li, J.W., Lu, X., Zhao, J., Zhang, et al.: A distributed power control of series-connected module-integrated inverters for PV grid-tied applications. IEEE Trans. Power Electron. **33**(9), 7698–7707 (2018)
3. Singh, O.P., Bhalla, D.K., Yadav, U.: Modeling and analysis of MPTT control techniques for grid connected PV generation. Adv. Res. Electric. Electron. Eng. **3**(4), 324–330 (2016). p-ISSN: 2349-5804, e-ISSN: 2349-5812
4. Yang, Y., Zhou, K., Blaabjerg, F.: Current harmonics from single-phase grid-connected inverters—examination and suppression. IEEE J. Emer. Sel. Topics Power Electron. **4**(1), 221–233 (2016)

Investigation of Optimum Process Parameter on CNC Turning for Aluminium Alloy AA6262 Using Grey Relational Analysis

Rahul Sharma, Kamal Sharma, Devendra Singh, Kuwar Mausam, and Ravindra Pratap Singh

1 Introduction

Manufacturing industry has challenge to achieve high productivity and high quality with minimum input like machining parameters, time, manpower, fuel and electricity consumption. To achieve this industrial goal, it is important to optimize the machining parameters such as speed, feed and depth of cut [1]. This chapter gives an overview of the optimization method which is used to optimize the machining parameters for CNC turning on AA6262 T6 aluminium alloy [2, 3]. AA6262 T6 is used for screw, machine product, nuts, coupling, marine fitting, Hinge pin, Oil line fitting and Valves. So we need to optimize the machining parameters for the faster production.

The experiment has been performed for minimum surface roughness (SR) and maximum material removal rate (MRR) because these two factors are important for any product and machining process. Surface roughness is the main considerable factor for the product [4]. Material removal rate is an important factor for production; each industry wants to remove maximum material with minimum time [4]. It reduces the cost and production time and increases productivity.

Saravanakumar is used a grey relation analysis in Taguchi method for the optimization of machining parameters of turning of aluminium alloy AA6063 T6 and determines that feed rate is most influencing factor for grey relation grade followed by depth of cut and speed [5, 6]. Jayaraman is used a grey relation analysis in Taguchi method for the optimization of machining parameters of turning of aluminium alloy AA6063 T6 and determines that feed rate is most influencing factor for grey relation grade followed by depth of cut and speed [7, 8]. The experiment has been performed

R. Sharma (✉) · D. Singh
Department of Mechanical Engineering, Sachdeva Institute of Technology NH 2, Farah, Mathura, Uttar Pradesh 281122, India
e-mail: rahulsharma15394@gmail.com

R. Sharma · K. Sharma · K. Mausam · R. P. Singh
Department of Mechanical Engineering, GLA University, Mathura, Uttar Pradesh 281406, India

M. Muzammil et al. (eds.), *Recent Advances in Mechanical Engineering*, Lecture Notes in Mechanical Engineering, https://doi.org/10.1007/978-981-15-8704-7_75

with multiple responses based on orthogonal L9 array with grey relation analysis (GRA) [9, 10].

2 Experimental Setup

2.1 Machining Process

The machining process is the method to obtain to machining part with good surface finish. Machining process involves many factors like feed, speed and depth of cut to achieve desired surface quality and MRR which depends on many factors like geometry to be generated, shape [11]. Grey relation grade in Taguchi is used for the optimization process. There is a lot of research work is carried out to obtain the optimization of machining parameters. For the optimization condition, we study machining parameters with power consumption and lubrication which has been used in the CNC machine [5, 6].

2.2 Machine Tool

The machine tool which has been used for performing the experimental work is Speed Turn make CNC turning Centre Model ST-400 G X750 (Siemens 808D) CNC system shown in Fig. 1a and machine specification has mentioned in Table 2. The factor which is considerable for the experiment is feed, speed and depth of cut (DOC). Experiment work has performed on three levels which is shown in Table 1 ADDIN EN.CITE ADDIN EN.CITE.DATA [12, 13].

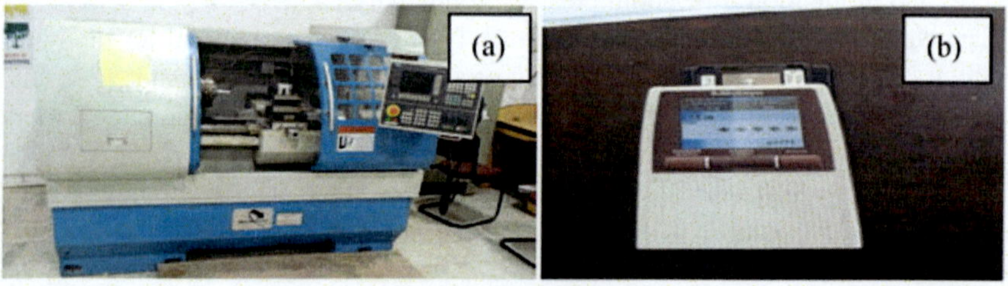

Fig. 1 **a** CNC turning machine. **b** Surface roughness tester

Table 1 Levels and parameters

Description	Measurement
Max. swing over bed	Ø 400 mm
Max. swing over carriage	Ø 240 mm
Max. length of work piece	1000 mm
Range of spindle speed	200–2000 rpm
Spindle bore	Ø 60 mm

Table 2 CNC machine specification

Machining parameters	Level 1	Level 2	Level 3
Feed (mm/rev)	0.05	0.08	0.11
Speed (RPM)	700	900	1100
Depth of cut (mm)	0.8	1.2	1.6

2.3 Experimental Plan

In this research work, we use the L9 standard array in Table 3. It is a suitable array for three levels and three parameters. It provides data with a change in a variable. Machining levels and parameters are shown in Table 1.

Material removal rate (MRR) is calculated by

$$\text{MRR} = f * v * d \tag{1}$$

$$v = (\pi * D * s)/60 \tag{2}$$

where

f Feed in mm/rev.

Table 3 L9 standard array

Exp. No.	A	B	C
	Speed	Feed	DOC
1	1	1	1
2	1	2	2
3	1	3	3
4	2	1	2
5	2	2	3
6	2	3	1
7	3	1	3
8	3	2	1
9	3	3	2

v	Velocity in mm/s.
d	Depth of cut in mm.
D	Original diameter in mm.
s	Speed in RPM.
MRR	Material removal rate (mm^3/s).
SR	Surface roughness in μm.

2.4 Surface Roughness Tester

The values of surface roughness (SR) of experimental work piece were measured by using surface Roughness tester of Mitutoyo SJ-210 4 mN shown in Fig. 1b. SR testers get in contact with the surface of work piece for a few seconds to get roughness value. This process is repeated for performing many operations and get value. Aim is to be achieving minimum surface roughness. SR texture is a component of surface texture [12].

2.5 Work Piece Material

Aluminium alloy AA6262 T6 has been used as a work material in this experiment because of their properties shown in Table 4. It is a heat treatable alloy with high strength and good corrosion resistance. It also has an excellent machinability and surface finish. AA6262 T6 shown in Fig. 2 is used for screw, machine product, nuts, coupling, marine fitting, Hinge pin, Oil line fitting and Valves. Experimental values are shown in Table 5.

Table 4 Chemical composition of aluminium alloy AA6262 T6 by weight%

Chemical elements	Fe	Cu	Si	Cr	Mg	Mn	Ti	Bi	Pb	Zn	Other	Al
Per. (%)	0.50	0.25	0.60	0.10	0.10	0.15	0.10	0.60	0.50	0.15	0.15	98

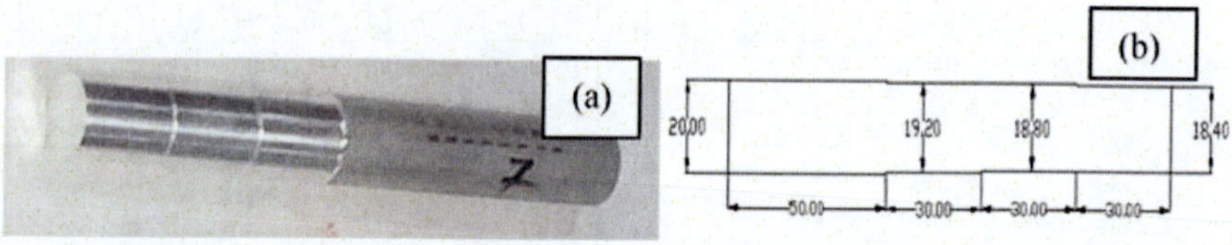

Fig. 2 **a** Specimen after machining. **b** Drawing of specimen

Table 5 Experimental values for L9 array

Exp. No.	Speed	Feed	DOC	SR	MRR
1	700	0.05	0.8	0.516	29.74627
2	700	0.08	1.2	0.515	71.39104
3	700	0.11	1.6	1.515	130.8836
4	900	0.05	1.2	1.017	57.3678
5	900	0.08	1.6	1.711	122.3846
6	900	0.11	0.8	0.71	84.13944
7	1100	0.05	1.6	0.554	93.48827
8	1100	0.08	0.8	5.102	74.79061
9	1100	0.11	1.2	2.123	154.2556

3 Grey Relational Analysis (GRA) Method for Optimization

GRA is applied to convert multi-variables into a single variable and find out the optimum condition in which MRR in mm^3/s is maximum and surface roughness in μm is minimum [13]. The following steps are used for grey relation grade (GRG) [7, 8].

Step 1: Grey relational generation

In this step, we find out the normalization of the experiment performed.

$$x_i^*(k) = \frac{x_i^k(k) - \min\left(x_i^0(k)\right)}{\max\left(x_i^0(k)\right) - \min\left(x_i^0(k)\right)} \tag{3}$$

In the present work, the surface roughness has to be minimized and the normalization of experimental data can be calculated as follows.

$$x_i^*(k) = \frac{\max\left(x_i^0(k)\right) - x_i^k(k)}{\max(x_i^0(k)) - \min\left(x_i^0(k)\right)} \tag{4}$$

Step 2: Calculated deviation sequence is for each experiment and shown in Table 6.

Step 3: Calculated grey relation coefficient for surface roughness and MRR.

$$\zeta_i\,(K) = \frac{\Delta\min + \mu * \Delta\max}{\Delta_{0i}\,(k) + \mu * \Delta\max} \tag{5}$$

Here, $\Delta_{0i}(y)$ is deviation sequence

Table 6 Grey relational normalization and Δ_{0i} is deviation sequence calculated with reference sequence 1

Exp. No.	Normalized data		Deviation sequence Δ_{0i}	
	MRR	SR	MRR	SR
1	0.00000	0.99978	1.00000	0.00022
2	0.33447	1.00000	0.66553	0.00000
3	0.81229	0.78199	0.18771	0.21801
4	0.22184	0.89056	0.77816	0.10944
5	0.74403	0.73926	0.25597	0.26074
6	0.43686	0.95749	0.56314	0.04251
7	0.51195	0.99150	0.48805	0.00850
8	0.36177	0.00000	0.63823	1.00000
9	1.00000	0.64944	0.00000	0.35056

$$\Delta_{0i}(y) = \left|x_0^*(y) - x_i^*(y)\right| \tag{6}$$

Minimum value for $\Delta_{0i}(y)$ is $\Delta_{\min}$, and maximum for Δ_{0i} is $\Delta_{\max}$ [21].

$$\Delta_{0i}(y) = \Delta_{\min} = \min\left|x_0^*(y) - x_i^*(y)\right| \tag{7}$$

$$\Delta_{0i}(y) = \Delta_{\max} = \max\left|x_0^*(y) - x_i^*(y)\right| \tag{8}$$

0.5 is the value of (M) I.E. identification coefficient its range is $0 \le M \le 1$

$$\Delta_{\max} = 1(\text{MRR})$$

$$\Delta_{\min} = 0(\text{SurfaceRoughness})$$

Table 7 shows the grey relational coefficient of experimental values.

Step 4: Calculate overall GRG. Surface roughness and MRR are two factors; Table 7 shows GRG for machining parameters [14].
Table 7 The maximum value of GRG is shown in experiment No. 9.

Step 5: Calculate the mean of GRG.
The mean of GRG is the average sum of grey relational coefficient.

$$\gamma\left(x_0^*, x_i^*\right) = \gamma_i = \frac{1}{n}\sum_{k=1}^{n}\zeta_i(k) \tag{9}$$

Here, n is the number of process parameters.

Step 6: Grey relational ordering.

Table 7 Grey relation coefficient for each experimental value and overall GRG

Exp. No.	Grey relational coefficient		Grey relational grade
	MRR	SR	
1	0.33333	0.99956	0.66645
2	0.42899	1.00000	0.71449
3	0.72705	0.69637	0.71171
4	0.39119	0.82043	0.60581
5	0.66140	0.65726	0.65933
6	0.47030	0.92164	0.69597
7	0.50604	0.98328	0.74466
8	0.43928	0.33333	0.38631
9	1.00000	0.58785	0.81974

Greatest grey relational grade gives an order 1. Equation 7 is used to calculate grey relational grades. From Table 7, we find out that experiment 9 has the greatest grey relational grade. The larger value in the S/N curve is considered as an optimum parameter. So the optimum turning parameters have $s_1f_3d_2$ [7, 15].

4 Result and Discussions

Mean of GRG shown in Table 8. It represents the significant levels of parameters. Significant level is the difference between maximum and minimum values of process parameters. Greater significant is the greatest difference. From Tables 8 and 9 it is

Table 8 Means for GRG response table

Level	Speed	Feed	Depth of cut
1	0.6976	0.6723	0.5829
2	0.6537	0.5867	0.7133
3	0.6502	0.7425	0.7052
Delta	0.0473	0.1558	0.1304
Rank	3	1	2

Table 9 S/N ratios response table

Level	Speed	Feed	Depth of cut
1	−3.133	−3.480	−1.978
2	−3.706	−4.933	−3.000
3	−4.183	−2.610	−3.044
Delta	1.050	2.324	1.978
Rank	3	1	2

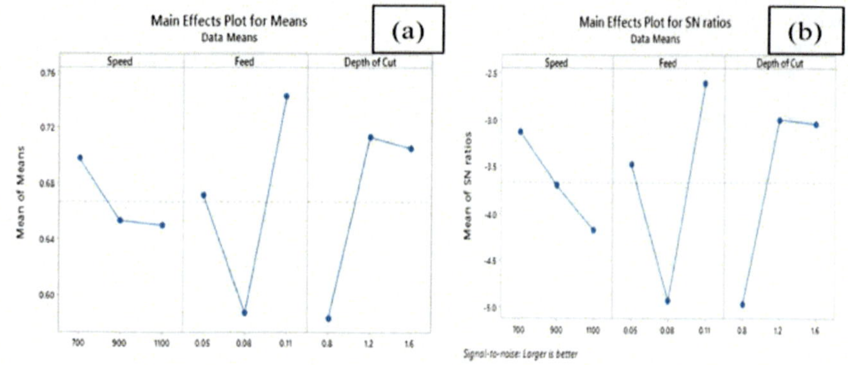

Fig. 3 **a** Main effect plots for mean for grey relation grade. **b** Main effect plots for S/N ratio for grey relation grade

Table 10 Results

Experiment	Parameters	SR	MRR
Initial prediction	$s_3f_3d_2$	2.123	154.25564
Experimental	$s_1f_3d_2$	1.045	163.88357

clear that feed has more contribution to optimization followed by depth of cut and speed.

It gives a maximum MRR and minimum surface roughness. Figure 3a shows the mean of GRG and Fig. 3b shows the main effect plots of S/N ration. Initial prediction and experimental results are shown in Table 10. Figure 3a shown that,

- With an increase speed, GRG decrease up to level 3.
- With an increase of feed, GRG decreases up to level 2, then increases up to level 3.
- With an increase of depth of cut, GRG increases up to level 2 and then decreases up to level 3.

5 Conclusions

The seed, feed and depth of cut are measured under the different cutting condition for find out the optimum combination. In this present work, we found out the optimum combination of process parameters, naming speed, feed and depth of cut for maximum MRR and minimum surface roughness through GRG; it is the best method to find out the optimum value with least time. Following experiment results are concluding.

- Optimum process parameters are $s_1f_3d_2$, i.e. speed 700 RPM, feed 0.11 mm/rev and DOC 1.2 mm.

- Optimum surface roughness is 1.045 μm.
- Optimum MRR is 163.884 mm^3/s.
- For the optimization, value feed has the main factor, it has first rank.
- Depth of cut (DOC) has the second rank.
- Speed has a third Rank.

References

1. Tzeng, C.-J., et al.: Optimization of turning operations with multiple performance characteristics using the Taguchi method and Grey relational analysis. J. Mater. Process. Technol. **209**(6), 2753–2759 (2009)
2. Lumley, R.N.: Introduction: aluminium, the strategic material. In: Lumley, R.N. (ed.), Fundamentals of Aluminium Metallurgy, pp. xvii–xxx. Woodhead Publishing (2018)
3. Rathod, P., Aravindan, S., Venkateswara Rao, P.: Performance evaluation of novel micro-textured tools in improving the machinability of aluminum alloy (Al 6063). Proc. Technol. **23**, 296–303 (2016)
4. Mausam, K., et al.: Investigation of process parameter of EDM using genetic algorithm (GA) approach for carbon fiber based two phase epoxy composites. Mater. Today Proc. **3**(10, Part B), 4102–4108 (2016)
5. Silva, G.B., et al.: Designing a novel feeding system for CNC turning machines. Proc. Manuf. **17**, 1144–1153 (2018)
6. Otto, A., Radons, G.: Application of spindle speed variation for chatter suppression in turning. CIRP J. Manuf. Sci. Technol. **6**(2), 102–109 (2013)
7. Jayaraman, P., Kumar, L.M.: Multi-response optimization of machining parameters of turning AA6063 T6 aluminium alloy using grey relational analysis in Taguchi method. Proc. Eng. **97**, 197–204 (2014)
8. Hussain, M.Z., Khan, S., Sarmah, P.: Optimization of powder metallurgy processing parameters of Al_2O_3/Cu composite through Taguchi method with Grey relational analysis. J. King Saud Univ. Eng. Sci. (2019)
9. Eapen, J., Murugappan, S., Arul, S.: A study on chip morphology of aluminum alloy 6063 during turning under pre cooled cryogenic and dry environments. Mater. Today Proc. **4**(8), 7686–7693 (2017)
10. Sirichaivetkul, R., et al.: In-situ study of microstructural evolution during thermal treatment of 6063 aluminum alloy. Mater. Lett. **250**, 42–45 (2019)
11. Yusup, N., Zain, A.M., Hashim, S.Z.M.: Evolutionary techniques in optimizing machining parameters: review and recent applications (2007–2011). Expert Syst. Appl. **39**(10), 9909–9927 (2012)
12. Saravanakumar, A., et al.: Optimization of CNC turning parameters on aluminum alloy 6063 using Taguchi robust design. Mater. Today Proc. **5**(2, Part 2), 8290–8298 (2018)
13. Uzun, G.: Analysis of grey relational method of the effects on machinability performance on austempered vermicular graphite cast irons. Measurement **142**, 122–130 (2019)
14. Mausam, K., et al.: Multi-objective optimization design of die-sinking electric discharge machine (EDM) machining parameter for CNT-reinforced carbon fibre nanocomposite using grey relational analysis. J. Braz. Soc. Mech. Sci. Eng. **41**(8), 348 (2019)
15. Ajith Arul Daniel, S., et al.: Multi objective prediction and optimization of control parameters in the milling of aluminium hybrid metal matrix composites using ANN and Taguchi-grey relational analysis. Def. Technol. (2019)

Role of Performance Measures in the Configuration Selection of Reconfigurable Manufacturing System

Gaurav Kumar, Kapil Kumar Goyal, and Neel Kamal Batra

1 Introduction

Manufacturing plays a crucial role in economic growth of a nation. Customization and flexibility in demands are creating challenges for modern industries. To cope up these challenges, industries required to reduce production time and cost along with improvement in quality to fulfill the customer requirements. To take these challenges as an opportunity, industries are using such manufacturing system which provide product with changing capabilities and functionalities [1]. Nowadays, there is a great advancement in automation, soft computing, and IT sectors which are playing a great role in manufacturing industries to cover the above challenges easily. Finally, a reconfigurable manufacturing system (RMS) have the capabilities which permit it to be replanned and redesigned as and when required so as to reconfigure it within a short time to meet the changed production requirements. In RMS, for designing a desired product, both software and hardware can be re-oriented quickly and trustworthily [3]. Koren et al. [3] first described how system configurations can have a significant effect on RMS in terms of performance like throughput and quality. The RMT which are located at the heart of RMS is configured to accommodate capacity and functionality by selecting the required auxiliary and basic modules, including kinetic modules and structural modules, from the module catalog. The RMT is a modular machine consisting of a variety of modules and required controllers [4, 5]. The RMT may be used to relocate or adjust modules without varying its topological features, RMT may be used [6, 7]. This paper summarizes the reconfigurable manufacturing system managed within the framework of this study. In the present investigation, authors intend to propose a design of flow line configuration for the RMS based upon the

G. Kumar (✉) · N. K. Batra
M. M. Deemed to be University, Mullana, Ambala, India
e-mail: gaurav.me86@gmail.com

K. K. Goyal
Dr. B R Ambedkar National Institute of Technology, Jalandhar, India

M. Muzammil et al. (eds.), *Recent Advances in Mechanical Engineering*, Lecture Notes in Mechanical Engineering, https://doi.org/10.1007/978-981-15-8704-7_76

performance metrics discussed for RMS and various other manufacturing layouts [1, 6, 8, 9].

2 Need of the Performance Issues Measure

The performance indices play a very crucial role in the design and development of these new classes of manufacturing systems. The availability of various configurations and the ability of theses configurations to perform multiple operations with different capabilities generate lot many alternatives which need to be graded based upon the economic and responsiveness capabilities [8, 9].

The present study considers a reconfigurable flow line layout with paralleling of similar machines to fulfill the required capacity. In order to perform the desired operations on the part having operation sequence 1-3-8-11, four stages are proposed (see Fig. 1). Each stage is able to perform the desired operation through allotted RMT. The possible number of configurations may vary depending upon the number of feasible RMTs available to perform the desired operation at each stage. Number of machines required on a particular production stage depends upon the production capacity of the allotted RMT to perform a desired operation in its existing configuration. In the considered illustration, following are the feasible configuration designs:

Appropriate performance parameters are required for finding the best alternative among the vast set of alternatives generated mostly due to the multiple configuration of RMTs being capable of performing the same operation on varied rate of outputs with different costs involved. Various performance parameters considered in this investigation are cost, reliability, machine utilization, and operational capability.

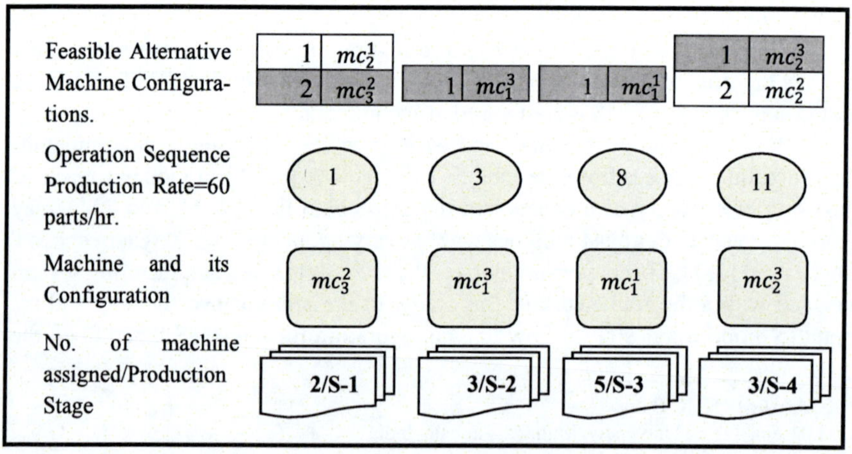

Fig. 1 RMS flow line feasible machine configurations mapping

3 Performance Measure

Cost (C), reliability (R), machine utilization (MU), and operational capability (OC) are the basis of choosing a feasible machine configuration to perform the operation. Despite the low level of comprehensiveness of manufacturing systems, the cost is the most important concern in analyzing the performance of manufacturing systems while the other parameters like reliability, machine utilization, and operational capability show responsiveness [10]. An illustration is presented to access the impact of performance metrics on a flow line configuration based on the random data generated for various parameters taking into the relevant literature into consideration.

3.1 *Cost (C)*

The cost of a combination is calculated by,

$$C = n \times CM, \text{ Where } n = D/P$$

n is the number of machines needed to meet the demand (D), *P* is the rate of production of the machine, and CM is the cost of the machine. The cost is an important and non-beneficial performance parameter and is to be minimized. For illustration, consider a layout configuration $mc_3^2 \rightarrow mc_1^3 \rightarrow mc_1^1 \rightarrow mc_2^3$, for which the demand is 60 units. In Table 1 [1], the production rates of RMTs for the selected machine configurations $mc_3^2 \rightarrow mc_1^3 \rightarrow mc_1^1 \rightarrow mc_2^3$ are 30, 20, 12, and 25 units respectively. The cost of considered RMT configurations as per Table 1 are 1825, 1025, 750, and 1140 units, respectively. The numbers of machine required for the combination are 2, 3, 5, and 3, respectively. Then, the cost of this configuration is 1825 × 2 + 1025 × 3 + 750 × 5 + 1140 × 3 = 13,895 US$. Similarly, the cost of other machine configurations is given in Table 2.

3.2 *Reliability (R)*

Reliability of the system is defined as the probability that the system will run without failure for a specified period of time. Reliability is useful attribute that is to be maximized. Configuration having single machine at each stage, R is determined by:

$$R_{\text{series}} = \sum_{i=1}^{N} R_i$$

For each configuration having multiple machines, the reliability is determined by:

Table 1 Operational capabilities of RMTs

Operation (k)	RMT production rate in parts/hour for performing various operations												Cost (10^3 of US$)	Reliability
mc_i^j	1	2	3	4	5	6	7	8	9	10	11	12		
mc_1^1	–	–	–	14	–	–	–	12	–	–	–	8	750	0.5
mc_1^2	–	–	–	–	15	–	–	–	20	–	–	–	955	0.5
mc_1^3	–	–	20	–	–	–	15	–	–	–	–	–	1025	0.6
mc_1^4	–	–	–	–	–	–	–	–	–	15	–	–	840	0.6
mc_2^1	14	–	–	–	–	15	–	–	–	–	–	12	1215	0.7
mc_2^2	–	15	–	–	–	–	–	–	–	–	15	–	910	0.4
mc_2^3	–	–	25	–	–	–	–	18	–	–	25	–	1140	0.6
mc_2^4	–	20	–	–	20	–	18	–	–	–	–	–	1350	0.6
mc_2^5	–	–	–	18	-	–	–	–	–	–	–	–	1050	0.5
mc_3^1	–	12	–	–	–	–	–	–	15	–	–	10	780	0.7
mc_3^2	30	–	–	26	–	–	–	24	–	–	24	–	1825	0.6

Table 2 Overall performance measures

Configuration design s. no.	Cost in US$	Reliability	Machine utilization	Operational capability
1	13,895	0.108	0.95	58
2	14,115	0.072	1	58
3	16,320	0.126	0.92	31
4	16,540	0.084	0.97	31

$$R_{\text{parallel}} = 1 - \sum_{i=1}^{N} (1 - R_i)$$

where R_i is the reliability of the ith configuration, e.g., if the combination $mc_3^2 \rightarrow mc_1^3 \rightarrow mc_1^1 \rightarrow mc_2^3$ is taken into account. The reliability of each machine configuration is 0.6, 0.6, 0.5, and 0.6 from Table 1. Thus, the reliability of the combination is determined by $R = 0.6 \times 0.6 \times 0.5 \times 0.6 = 0.108$. Similarly, the reliability of other machine configurations is given in Table 2.

3.3 *Machine Utilization (MU)*

The decrease in utilization of machine capacity can affect economic performance, which can cause serious threat to the existence of the industry in machines. Therefore, machine utilization is calculated by:

$$MU = \frac{D}{P \times n}$$

where D is the demand rate, P is the rate of production, and n is the no. of machine. Machine utilization is a useful property to maximize. For example, the required production rate is 60 parts/h at production stage 1 for the selected machine configuration mc_3^2. The production rate of selected machine configuration is 30 parts/h; therefore, machine utilization according to above equation is 1.0. The machine utilization of the combination $mc_3^2 \rightarrow mc_1^3 \rightarrow mc_1^1 \rightarrow mc_2^3$ would be the average of the all stages which is 0.95 in this case. Similarly, the machine utilization of another machine configuration is given in Table2.

3.4 *Operational Capability (OC)*

Depending on the various operations that the machine can perform in the current configuration, the operational capability is calculated for performing a specific operation k ($1 < k < K$) of the machine configuration in a feasible alternative. Our goal is to maximize operational capability. A power index Y is used to represent that consideration. The operational capability is calculated using:

$$OC = \left[\left(\sum_{k=1}^{K} \delta_{p,k}^{q}\right) - 1\right]^{Y}$$

δ represents 1 if operation q can be performed with machine p having its kth configuration, otherwise 0. The machine mc_3^2 is capable of performing operations 1, 4, 8, 11, 15, 17, and 19. Thus, the operation capability for mc_3^2 is $[(7) - 1]^2 = 36$. The operational capability of the combination $mc_3^2 \rightarrow mc_1^3 \rightarrow mc_1^1 \rightarrow mc_2^3$ would be the sum of all production stages which is 58. Similarly, the operational capability of other machine configurations is given in Table 2.

4 Discussion

Performance measures considered in the present study for the sample configurations have been given in Table 2. It can be observed that the configuration 1 is the most economical solution but lacks in the reliability and machine utilization point of view. Similarly, the machine 2 has highest machine utilization but is the lowest in the reliability from which it can be envisaged that the desired level of production output may not be achieved. The operational capability is important to gage the level of future adaptability to adjust the variations in the mix and volume. The configurations 1 and 2 are certainly better than 3 and 4 on the basis of adaptability. Thus, it is evident from the illustrative configurations considered that the no single configuration can serve as best configuration on all the performance metrics. The weightage assigned to various configurations by experts would certainly affect the decision on the configuration selection. In the present era, such kind of configurations is preferred which are cost effective and reliable enough with maximum machine utilization and operational capability to enhance the throughput and the overall economy of the reconfigurable manufacturing system.

5 Conclusions and Future Scope

In the modern industries, a process engineer always keep planning to improve the cost effectiveness, reliability, and the ability to respond faster to the sudden unpredictable changes through selection of appropriate flow line configurations with different combination of reconfigurable machine tools. No single configuration can fulfill all the expectations on various metrics; thus, role of selection of performance metrics ensures the survival and growth of an industry. In the present study, four performance measures viz cost, machine utilization, operational capability, and reliability have been presented through sample flow line configurations. Further, it has been concluded that cost and machine utilization contribute to the cost effectiveness, the reliability leads to on time deliveries, and the operation capability helps to cater the needs of upcoming new products and helps in handling the product volume and mix variations. It has been observed that no single configuration gives the best values among all the performance measures, and hence based on the diverse needs, suitable flow line configurations can be designed implementing the presented performance metrics. It will enable the system for adjust capabilities and functionality based on the demand and response to market fluctuations. In future, authors plan to develop a multicriteria decision-making approach for the sensitivity analysis of various configurations.

References

1. Groover, M.P.: Automation Production Systems and Computer Integrated Manufacturing, 1st Indian edition. Pearson Education Asia (2001)
2. Bortolini, M., Galizia, F.G., Mora, C.: Reconfigurable manufacturing systems: literature review and research trend. J. Manuf. Syst. **49**, 93–106 (2018)
3. Koren, Y., Ulsoy, A.G.: Reconfigurable manufacturing systems. Engineering research center for reconfigurable machining systems (ERC/RMS) report # 1. The University of Michigan, Ann Arbor (1997)
4. Moon, Y.M., Kota, S.: Design of reconfigurable machine tools. J. Manuf. Sci. Eng. Trans. ASME **124**(2), 480–483 (2002)
5. Koren, Y., Gu, X., Guo, W.: Reconfigurable manufacturing systems: principles, design, and future trends. Front. Mech. Eng. **13**(2), 21–136 (2018)
6. Goyal, K.K., Jain, P.K., Jain, M.: A novel methodology to measure the responsiveness of RMTs in reconfigurable manufacturing system. J. Manuf. Syst. **32**(4), 724–730 (2013)
7. Kumar, G., Goyal, K.K., Batra, N.K.: Evolution, principles and recent trends in reconfigurable manufacturing system. J. Phys. Conf. Ser. **1240**, 012161 (2019)
8. Gumasta, K.: Developing a reconfigurability index using multi-attribute utility theory. Int. J. Prod. Res. **49**(6), 1669–1683 (2011)
9. Yigit, A.S., Allahverdi, A.: Optimal selection of module instances for modular products in reconfigurable manufacturing systems. Int. J. Prod. Res. **41**(17), 4063–4074 (2003)
10. Mittal, K.K., Jain, P.K.: An overview of performance measures in reconfigurable manufacturing system. Proc. Eng. **69**, 1125–1129 (2014)

Investigation of Variation in Stress Concentration Factor with the Change in Orientation of Central Hole on a Rectangular Plate

Manish Pandey, Aprajita Patel, Kshitiz Jaiswal, Lalit Kirola, and Subodh Kumar Sharma

1 Introduction

A stress concentration, also known as stress risers, is a location in an object where stress is concentrated. When an object has uniform geometry, stress is distributed evenly across its cross section. Any reduction in the area results in a localized increase in the stress. An abrupt change in the cross section area due to irregularities causes a variation in the distribution of stresses; this happens for various types of stress-axial, bending or shear due to the presence of fillets, holes, grooves, or irregularities on the surface. The usage of structural steel plates is inevitable in mechanical structures, civil engineering structures, marine structures, aviation structures, and various other places. In order to reduce the stress concentration factor, deciding the shape of the holes on the structural steel plates has been an important engineering problem. These holes are mostly created in plates to cut the weight of the structure or to create points of entry and exit. These holes change the plate geometry and lead to severe local stresses called stress concentration around the holes. Stress concentration reduces strength and causes premature failures in the structures and plastic deformations at the point of stress concentration. We have come to know from our experience that most failures occur in aerial structures due to the presence of stress concentration at joints. Therefore, it becomes necessary to keep a check over the stress concentration factor before the manufacturing of structures.

Stress concentration in a plate under uniaxial loading and weakened by a hole was identified in early periods of development, and some fundamental analytical methods to eradicate them were presented by Muskhelishvilli [1]. Materials applied in this study were assumed homogeneous, isotropic, and linearly elastic. By using methods such as complex variables method and conformal mapping, Savin [2] solved the infinite plate with a rounded triangular hole. Inglis [3] by determining the stresses

M. Pandey · A. Patel · K. Jaiswal · L. Kirola · S. K. Sharma (✉)
Department of Mechanical Engineering, KIET Group of Institutions, Ghaziabad, India
e-mail: subodhmeerut@gmail.com

M. Muzammil et al. (eds.), *Recent Advances in Mechanical Engineering*, Lecture Notes in Mechanical Engineering, https://doi.org/10.1007/978-981-15-8704-7_77

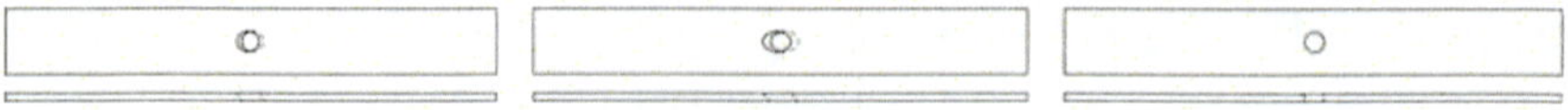

(a) (b) (c)

Fig. 1 Hole inclination **a** 30°, **b** 45°, **c** 90° from vertical axis

around the elliptical holes presented that stresses developed near the crack tip is much more than other places in any structure under loading. Fracture theory based on energy balance in an ideal brittle material was presented by Griffith et al. [4]. The stress analysis by Inglis was used by him to obtain a method to reduce elastic energy at a plate; boundary element alternating method was developed by Chen et al. [5] to study the stress concentration of two-dimensional plate with perforation. Pilkey et al. [6] presented a paper on stress concentration factor for different possible specimen configurations. Abuelfoutouh [7] brought about the circumferential stress around different holes such as triangular, square, circular, and circular holes for a special conformal mapping using composite shells. Also, Rezaeepazhand and Jafari [8] dispensed the solution for the study of anisotropic plates with holes. They also derived stress concentration factor for various shapes of the hole in metallic plates of infinite sizes and followed the effect of the loading directions and bluntness on the stress concentration. Stress distribution was studied by Ardalani [9] for the different holes in a metallic plate of finite length, the complex variable method, and the expansion of research conducted by various authors [10–19] used finite element method to find the augmentation of the composite plates with multiple holes. They considered the effect of hole spacing relative to each other and the number of holes. Based on the finite element method and particle swarm optimization, an optimization method for designing the reliability of composite structures was presented.

Study of variation in stress concentration factor is done in this paper, taking into consideration the cases where the orientation of holes has differed, and it is inclined at different angles as shown in Fig. 1.

2 Finite Element Approach

FEA is a computational tool used to perform engineering analysis. It involves the generation of mesh, for dividing the whole geometry into thousands of finite elements to perform the required analysis, which is done by solving the equations over these elements. The smaller the mesh size and larger the number of elements, the more accurate are results. The FEA solver algorithm firstly calculates the deformation at nodes and based on these results, it calculates strain in the element using the shape function. Hence, the problems of thermal stresses, elasticity, solid mechanics, or computational fluid dynamics are solved by the help of simultaneous differential equations and boundary equations.

3 Methodology

The objective here is to obtain various parameters such as deformation, equivalent stress, normal stress, equivalent elastic strain, and stress concentration factor by simulating the model of a flat plate using ANSYS software which undergoes application of different uniaxial loads. The object under consideration is a flat plate of structural steel having a circular slot at its geometric center. Reference geometry.

A flat plate having the following dimensions with no hole-thickness—0.00635 m, width—0.0508 m, length—0.4 m.

- Model 1—Plate with a hole of diameter 0.0254 m, inclined at 30° from vertical.
- Model 2—Plate with a hole of diameter 0.0254 m, inclined at 45° from vertical.
- Model 3—Plate with a hole of diameter 0.0254 m, in the perpendicular direction.

Structural steel is the chosen material which has following properties—Elastic modulus (E = 205 GPa), yield strength in tension (Sy = 250 MPa), ultimate strength (Su = 460 MPa), dimensions of the simulated model were—thickness—0.00635 m, width—0.0508 m, length—0.4 m, hole diameter—0.0254 m. The rectangular sheet of above-specified material having a hole in its centroid was then analyzed under different loading conditions—5000, 10,000, 15,000, and 20,000 N.

4 Results and Discussion

- A flat plate without hole loaded with an axial tensile force is ranging from 5000 to 20,000 N, since no cut-out is present in this case, we get the maximum value of normal stress (15.5–62 MPa), when compared to all other cases. Also, stress is not concentrated at any specific point; it is evenly distributed across the plate.
- Figure 2 shows a flat plate having hole inclined at 30°, loaded with an axial tensile force ranging from 5000 to 20,000 N. When the plate having hole inclination at 30° is loaded, it gives the median value of stress concentration factor and total deformation.
- Figure 3 shows a flat plate having hole inclined at 45°, loaded with an axial tensile force ranging from 5000 to 20,000 N. In the case where the hole is inclined at 45° to the longitudinal axis, minimum value of normal stress and stress concentration factor is found. It shows that the inclination angle of the hole plays an important role, and we can minimize the stress concentration for plates with uniaxial tensile loading by changing the value of this parameter.
- Figure 4 shows a flat plate having hole inclined at 90° (straight), loaded axially ranging from 5000 to 20,000 N. The maximum value of stress concentration factor comes in this case. But, the value of maximum total deformation is the lowest in this case. This change is due to the inclination of the slot.

The results show that maximum stress concentration occurs at the boundary of the hole. Hence, this area is most vulnerable to fracture or fatigue and often crack

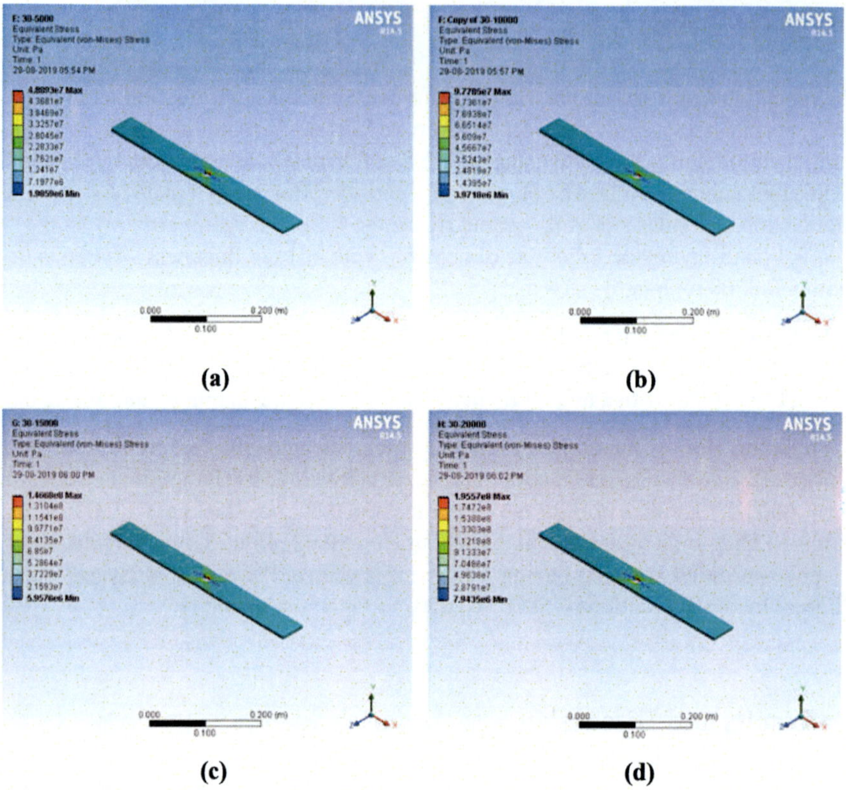

Fig. 2 Stress distribution in plate with hole inclined at 30° at applied load of **a** 5000 N, **b** 10,000 N, **c** 15,000 N, **d** 20,000 N

generation and propagation starts from here only. This area is depicted by red and orange colors in the contour diagram. In addition to that, the following inferences are drawn to see Fig. 5, i.e.,

1. The minimum value of stress concentration factor is found in the case of the plate having hole inclined at 45°.
2. The maximum value of the stress concentration factor is found in the case of the plate having a vertical hole (in longitudinal axis).
3. The minimum value of total deformation comes in the case of the plate having a vertical hole (90°).
4. Value of stress concentration factor is independent of the value of forces applied and depends on the subject and its geometry.

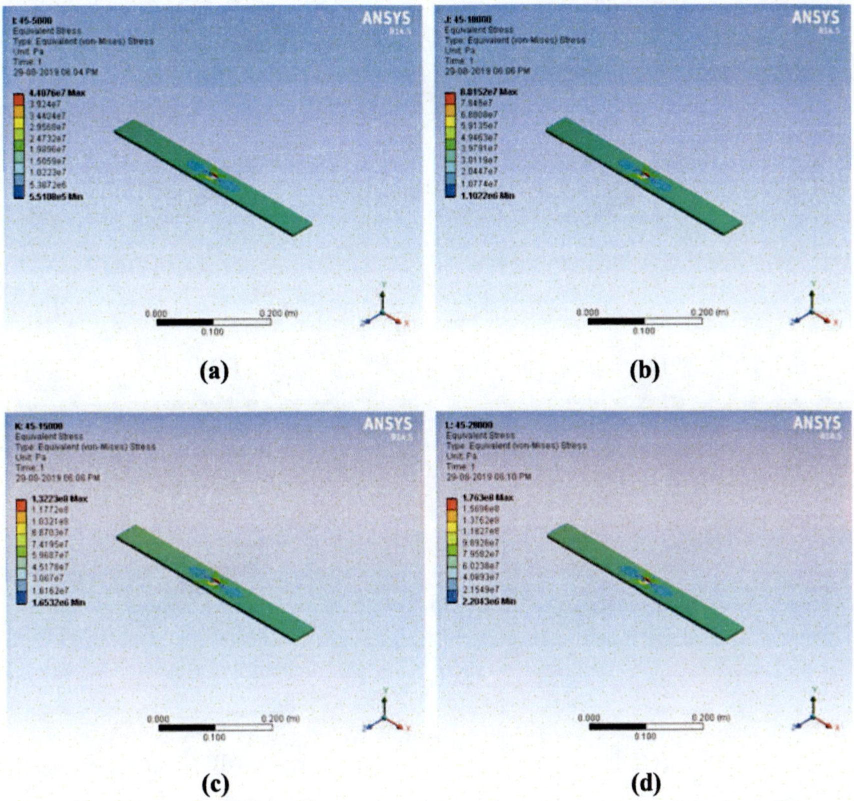

Fig. 3 Stress distribution in plate with hole inclined at 30° at applied load of **a** 5000 N, **b** 10,000 N, **c** 15,000 N, **d** 20,000 N

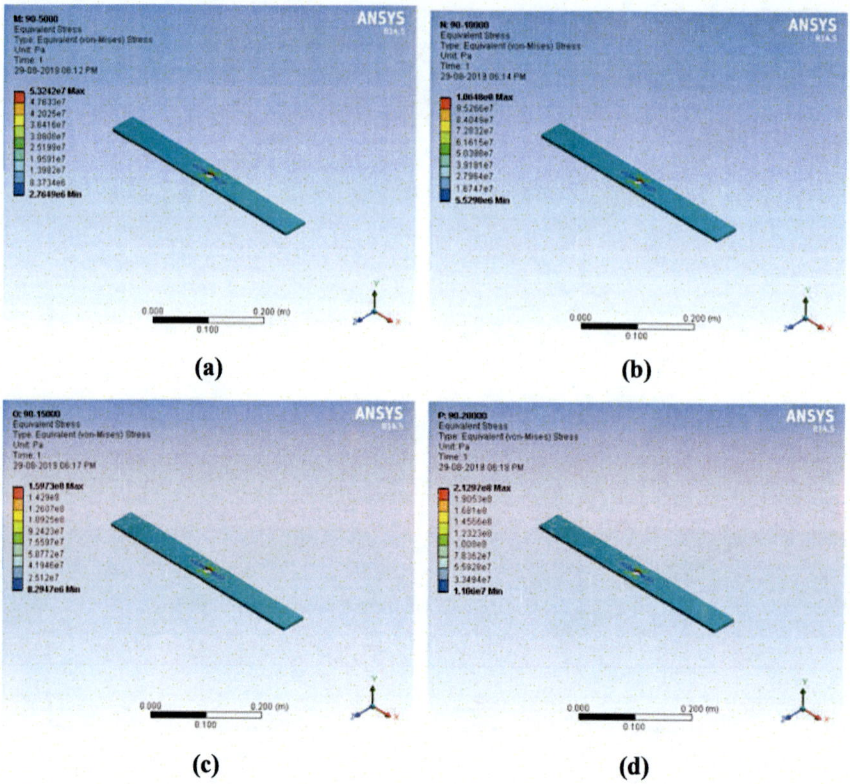

Fig. 4 Stress distribution in plate with hole inclined at 30° at applied load of **a** 5000 N, b 10,000 N, **c** 15,000 N, **d** 20,000 N

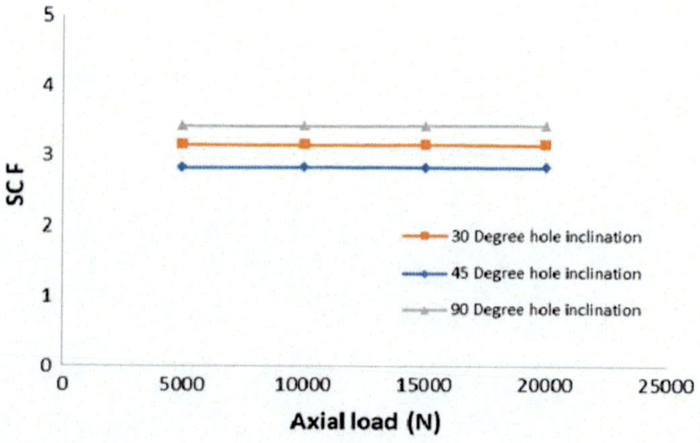

Fig. 5 Load versus stress concentration factor

References

1. Muskhelishvili, N.I.: Some basic problems of the mathematical theory of elasticity. P. Noordhoff Ltd., Groningen (1953)
2. Savin, G.: Stress concentration around the holes. Pergamon Press, Oxford (1961)
3. Inglis, C.E.: Stresses in a plate due to the presence of cracks and sharp corners. Proc. Inst. Naval. Arch. **55**, 219–230 (1913)
4. Griffith, A.A.: The phenomena of rupture and flow in solids. Philos. Trans. R. Soc. Lond. Ser. A **221**, 163–198 (1921)
5. Chen, K.T., Ting, K., Yang, W.S.: Stress analysis of two-dimensional perforated plates using boundary element alternating method. Comput. Struct. **75**, 515–527 (2000)
6. Pilkey, W.D.: Peterson's stress concentration factors, 2nd edn. Wily, United States, America (1997)
7. Abuelfoutouh, N.M.: Preliminary design of unstiffened composite shells. In: Symposium of 7th Technical Conference of ASC (1993)
8. Rezaeepazhand, J., Jafari, M.: Stress analysis of perforated composite plates. Compos. Struct. **71**, 463–468 (2005)
9. Jafari, M., Ardalani, E.: Stress concentration in finite metallic plates with regular holes. Int. J. Mech. Sci. **106**, 220–230 (2016)
10. Pan, Z., Cheng, Y., Liu, J.: Stress analysis of a finite plate with a rectangular hole subjected to uniaxial tension using modified stress functions. Int. J. Mech. Sci. **75**, 265–277 (2013)
11. Santos, A.: Determination of stress concentration factors on flat plates of structural steels. J. Phys: Conf. Ser. **466**, 012–035 (2013)
12. Chen, H.C.: Special finite element including stress concentration effects of a hole. J. Finite Element Anal. Design **13**, 249–258 (1993)
13. Rezaei Pour Almasi, A., Fariba, F., Rasoli, S.: Modifying stress-strain curves using optimization and finite element simulation methods. J. Solid Mech. **7**(1) (2015)
14. Batista, M.: The stress concentration around a hole in an infinite plate subject to a uniform load at infinity. Int. J. Mech. Sci. **53**, 254–261 (2011)
15. Nagpal, S., Jain, N., Sanyal, S.: Stress concentration and its mitigation techniques in flat plate with singularities—a critical review. Eng. J. **16**(1) (2012)
16. Watsar, S.D., Bharule, A.: Stress analysis of finite plate with special shaped cutout. Int. J. Sci. Eng. Res. (IJSER) **3**(4) (2015)
17. Nilugal, R., Hebbal, M.S.: A closed-form solution for stress concentration factor around a circular hole in a linearly varying stress field. Int. J. Mech. Eng. Technol. (IJMET) **4**(5), 37–48 (2013)
18. Momcilovic, N., Motok, M., Maneski, T.: Stress concentration on the contour of a plate opening: analytical, numerical and experimental approach. J. Theor. Appl. Mech. **51**(4), 1003–1012 (2013)
19. Handa, D., Dondapati, R.S., Usurumarti, P.R.: Investigation on the stress concentration in metallic flat plates due to holes with different configuration. Int. J. Mech. Eng. Technol. (IJMET) **8**(7), 1718–1725 (2017)

Recognition of Mechanical Tools Using Artificial Neural Network

Pradeep Jain and Meenu

1 Introduction

Object recognition is considered as a vast research field with numerous applications in a wide range of disciplines, like computer vision, robotics, military, industry, medical, automation, and many more. Object recognition is a very complex task, whose goal is to recognize and classify the objects into their respective categories. In vision applications, these objects are acquired in the form of images to the system. The problem of object recognition is effortlessly solved by humans. But, computers find it very complicated to deal with such problems. With the increasing use of technology, several algorithms have developed to overcome these problems. Many achievements have been made during the last few years; the existing object classification techniques have problems like inaccuracy, common adaptability, incapability to recognize an object without a given model.

The difficulties of object recognition are due to the complexity and variability of objects in the images, occlusion, deformations, and various types of noise. Human can easily identify objects in the surroundings, regardless of their circumstances, whether they are upside down, different of color or texture, partly occluded, etc. Even objects that appear in many forms, like vases, or objects that are subject to much shape deviation, such as trees, can easily be generalized for our brain to one kind of object.

In this research work, a general purpose real-time fast and reliable object recognition system is developed for recognizing mechanical tools irrespective of their position and orientation.

P. Jain (✉) · Meenu
NIT, Kurukshetra, Kurukshetra, India
e-mail: pradeepjain05@yahoo.co.in

M. Muzammil et al. (eds.), *Recent Advances in Mechanical Engineering*, Lecture Notes in Mechanical Engineering, https://doi.org/10.1007/978-981-15-8704-7_78

2 Literature Survey

Zhuang et al. [1] developed an efficient method for controlling the amount of image enhancement. In addition, the final good quality enhanced image is obtained each sub-histogram separately.

Hayfron et al. [2] developed the algorithm for performance evaluation of the image enhancement schemes, and compared different fuzzy logic-based contrast enhancement techniques.

Zeng et al. [3] developed algorithm by using gray level and its distance, and this algorithm also improved the gray level difference and increased the contrast of metals.

Rahman et al. [4] developed a image enhancement algorithm to decrease the contrast of the image. The golden search method is used to find a suitable division into sub-images to conserve the original brightness.

Lal et al. [5] developed an technique in which the given image quality is enhanced by a modified sigmoid transfer function.

Kumar et al. [6] discussed the fundamental methods used in image segmentation. Segmentation of images divided in to fully and semi-automatic approach, and the developed algorithm used these methods. Segmentation of given image objects are very crucial step in the image processing, as it is responsible for the success of feature detection and object recognition.

Chaturvedi et al. [7] developed the region-based image segmentation method, and the success of good quality segmentation depends on the seed selection method.

Wang [8] developed an modified segmentation algorithm based on 3D region growing algorithm. This algorithm successfully used for the liver segmentation.

Fan et al. [9] developed an modified segmentation technique by combining the edge detection approach and seeded region growing. The final output of both edge extraction and region growing were incorporated to get the better segmented images.

Cheng et al. [10] presented an new method derived from threshold and region merging for better image segmentation. The homogram analysis was done by considering the fuzzy entropy as a measure to obtain all major uniform regions. The homogram reviewed by the incident of gray levels and the uniformity among the neighboring pixels. On the basis of color similarity among these regions, region merging was employed to obtain the segmented regions.

Kale [11] presented the methods for extraction of features using rotation invariant Zernike moment.

Marouf1 et al. [12] proposed new efficient facial-based identical twins recognition that is proposed according to geometric moment. The utilized geometric moment is Zernike moment (ZM) as a feature extractor inside the facial area of identical twins images. The developed method is evaluated on two datasets, Twins Days Festival and Iranian Twin Society which contain scaled and rotated facial images of identical twins in different illuminations. The results prove the ability of the proposed method to recognize a pair of identical twins.

Mukundan et al. [13] introduced a new set of orthogonal moments, these moments not involve any approximation, as the basis functions were orthogonal in discrete domain.

Haddadnia et al. [14] presented a feature extraction technique using the invariant Pseudo-Zernike moments for face recognition. The lower order moments represented the information of the facial expressions, while the higher-order moments contained information useful for the face recognition process. These moments provided invariance to change in size, position, orientation, and reflection.

Yu et al. [15] developed a novel low-level feature method based on color texture moments. Eight orthogonal templates or characteristic maps were derived using the local Fourier transform.

3 Methodology

The developed object recognition and classification system is an automatic system which can work in real-time applications. The major parts of the proposed system are: image enhancement, image segmentation, feature extraction, and classification.

Image enhancement is an important step for object recognition, as it enhances the quality of the input image and makes the image more accurate and clear for recognition. In this work, a fuzzy logic-based image enhancement algorithm is proposed.

Another major step in the process of object recognition is the segmentation of images to obtain meaningful objects from an image and remove the undesired parts of the image. The traditional image processing techniques of seeded region growing are used for the image segmentation.

In feature extraction, a set of feature is extracted from the image, which completely represents the information of the image, and thus reduces the processing of the system. These extracted features should be able to differentiate the patterns individually and also should be invariant to any kind of distortions. After having extracted the shapes, the descriptors are created. This is another central issue that will be researched in this thesis. These descriptors are made specifically to be invariant representations entailing as much important information of the shape as possible. They are created in such a way that they can be fed into a neural network and allow for easy categorization. In this work, rotation-invariant Zernike moments are used as a feature in the object recognition system.

Classification refers to recognizing of objects on the basis of the features extracted from the images. Artificial neural networks perform better for classification purpose, as they can easily generalize among patterns and can mimic the working of the human brain. Back propagation used in this work is the most efficient algorithm for classification in object recognition applications.

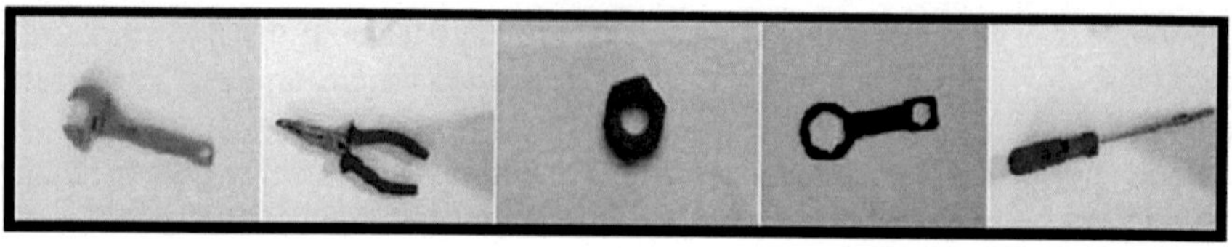

Fig. 1 Five categories of mechanical tools

4 Results and Discussions

In this paper, objects are classified using artificial neural network. For object classification, two steps are required (i) to train the classifier and (ii) to test the classifier to unknown data. For classification of objects, images are acquired using industrial camera by presenting the object at different scale, position, and orientation. Images are then resized to 100 × 100 pixels. Quality of the image is enhanced using fuzzy logic image enhancement algorithm. Object is segmented by region growing algorithm. Zernike moments are used to extract useful features from the given image, and objects are classified using artificial neural network.

4.1 Dataset Used for Classification

For classification of objects, five mechanical tools are taken as shown in Fig. 1. For each tool, 40 images are captured at different scale, rotation, and translation. The data is divided into two sets: (i) training set consists of 100 images (20 for each object) chosen randomly and (ii) testing set consists of 100 images (20 for each object).

4.2 Conversion of RGB to GRAY Image

The captured images are resized to 100 × 100 pixels with 256 shades of black and white, i.e., gray to reduce the computation time.

4.3 Image Enhancement

The gray scale image is improved by minimizing the noise and enhancing the contrast of the image. For image enhancement, the proposed fuzzy-based automatic enhancement algorithm is implemented in MATLAB.

Table 1 Different partitions used for experimentation

Experiments	1	2	3	4
Training/validation/testing samples ratio	50:25:25	60:25:15	70:15:15	80:10:10
No. of neurons in hidden layer	5	5	5	5

4.4 Image Segmentation

For image segmentation, seeded region growing algorithm is implemented. The enhanced image then undergoes segmentation, which separates the object from its background.

4.5 Feature Extraction

After segmentation, the image features are extracted using the Zernike moments. The Zernike moments are computed from 0 to 12th order which results in 49 moment feature vectors.

4.6 Recognition/Classification

For recognition of objects, neural pattern recognition (NPR) tool of MATLAB is used. A two-layer feed-forward network with sigmoid activation function in hidden layer and linear activation function in output layer is chosen. For recognition, 49 Zernike moments are taken as input and objects are classified into five classes using five neurons in output layer. Cross-validation approach is used for classification of objects. For this, data is divided into three sets:

1. Training set, 2. Validation set, 3. Testing set.

4.6.1 Effect of Training/Validation/Testing Sample Ratio

Table 1 shows the different cross-validation partitions of data used for experimentation. Adaptive learning is used for experimentation.

4.6.2 Summary of Results

Table 2 shows the summary of results for all the four partitions.

Table 2 Summary of results for all the four partitions

Experiments	1	2	3	4
Training/validation/testing samples ratio	50:25:25	60:25:15	70:15:15	80:10:10
Neurons in hidden layer	5	5	5	5
Training accuracy	66	68.3	61.4	83.8
Validation accuracy	44	32	26.7	80
Test accuracy	44	60	60	70
Over all accuracy	55	58	56	82
Gradient	0.19	0.0275	0.026	0.10062
No. of epoch	134	124	123	171
Learning rate	1.8	4.241	4.039	5.70
Cross-entropy performance	0.80	0.82	0.83	0.69

4.6.3 Effect of Hidden Nodes

80:10:10 partitions show the best performance for training, validation, and test data set. So for further analysis, 80:10:10 partitioning of data set is used. Hidden nodes are varied from 5 to 30 in step of 5.

This research work refers to the recognition/classification of objects from the images. In this work, an automatic image contrast enhancement algorithm is proposed. The technique is based on employing an automatic selection of the suitable value of exponential fuzzifier to enhance the images. The presented algorithm can be used over a wide variety of images with good quality results without user intervention. In addition, the results of the proposed image enhancement method provide more consistency in data distribution with smooth images than other classical image enhancement techniques.

The proposed segmentation algorithm provides a novel method for image segmentation. This method can be applied to any number of images, as the calculation of seed points is reliant only on the image information. The algorithm continues carried out segmentation process until all the pixels of the image are not labeled. This prevents under-segmentation.

The proposed feature extraction algorithm using Zernike moments provides the flexibility to neural network for testing with different number of features, which gives the more accurate results as compared to other moment features.

The proposed object recognition algorithm was successfully implemented and tested for various images subject to several transformations, like scaling, translation, and rotation. The developed object recognition system has the capability to produce good results, and work efficiently in identification and recognition of objects, irrespective of translation, scaling, and rotation, in real time.

The dataset with different partitions for training, validation, and testing is used. 80:10:10 dataset with ten hidden nodes performance is better than other partitions at the same number of hidden nodes. This is due to the fact that with less number of

Table 3 Summary of results with different hidden nodes

Experiments	1	2	3	4	5	6
Training/validation/testing samples ratio	80:10:10	80:10:10	80:10:10	80:10:10	80:10:10	80:10:10
Neurons in hidden layer	5	10	15	20	25	30
Training accuracy	83.8	96.3	88.8	87.5	82.5	33.8
Validation accuracy	80	90.0	90	90	30	0.0
Test accuracy	70	80	100	70	60	20.0
Overall accuracy	82	94	90	86	75	29
Gradient	0.1006	0.123	0.109	0.133	0.275	0.105
No. of epoch	171	0.176	162	162	119	53
Learning rate	5.70	3.99	3.17	2.016	1.27	0.1325
Cross-entropy performance	0.69	0.60	0.59	0.63	0.83	0.91

samples in training in other partitions, network is not able to learn the characteristics of data. When the dataset for training is increased from 50 to 80, performance increases. Further, it is found that with increase in hidden nodes from 5 to 10, overall accuracy is increased from 82 to 94%, but with further increase in hidden nodes, there is decrease in test accuracy because with the less number of hidden nodes, network is not able to learn the characteristics of data, but with a large number of hidden nodes its started over-fitting, so accuracy decreased. There are an optimum number of hidden nodes at which accuracy is maximum. Ten numbers of hidden nodes gave better accuracy. So, 80:10:10 partitions with ten numbers of hidden nodes is selected for testing of data. Table 3 shows the summary of results.

Result Summary: Overall accuracy increased from 82 to 94% when number of hidden nodes is increased from 5 to 10, with further increase in number of hidden nodes, overall accuracy decreased because when the number of hidden nodes were less, than network could not learn the characteristics of network, but with increase in number of hidden nodes, it starts crumbling, and it could not generalize on new data. So, ten hidden nodes were optimum for our data that gave 94% accuracy, and optimum learning rate was 3.99.

5 Conclusion

The proposed object recognition algorithm was successfully implemented and tested for various images subject to several transformations, like scaling, translation, and rotation. The developed object recognition system has the capability to produce good results, and work efficiently in identification and recognition of objects, irrespective of translation, scaling, and rotation, in real time. The dataset with different partitions for training, validation, and testing is used. 80:10:10 dataset with ten hidden nodes performance is better than other partitions at the same number of hidden nodes. This is due to the fact that with less number of samples in training in other partitions, network is not able to learn the characteristics of data. When the dataset for training is increased from 50 to 80, performance increases. Further, it is found that with increase in hidden nodes from 5 to 10, overall accuracy is increased from 82 to 94%, but with further increase in hidden nodes there is decrease in test accuracy because with the less number of hidden nodes, network is not able to learn the characteristics of data, but with a large number of hidden nodes its started over-fitting, so accuracy decreased. There is an optimum number of hidden nodes at which accuracy is maximum. Ten numbers of hidden nodes gave better accuracy. So, 80:10:10 partitions with ten numbers of hidden nodes is selected for testing of data.

References

1. Zhuang, L., Guan, Y.: Adaptive image enhancement using entropy-based sub histogram equalization. In: Computational Intelligence and Neuroscience. Hindawi Publishing Corporation (2018)
2. Hayfron, J.B., Panford, J.K., Poakwah Gyimah, Y.: Enhancing quality of images using fuzzy logic and singleton parameters. Int. J. Comput. Appl. **165**(13) (2017)
3. Zeng, L., Yan, B., Wang, W.: Contrast enhancement method based on gray and its distance double-weighting histogram equalization for 3D CT images of PCBs. In: Mathematical Problems in Engineering. Hindawi Publishing Corporation (2016)
4. Rahman, M.A., Liu, S., Lin, S.C.F., Wong, C.Y., Jiang, G., Kwok, N.: Image contrast enhancement for brightness preservation based on dynamic stretching. Int. J. Image Process. **9**(4) (2015)
5. Lal, S., Chandra, M.: Efficient algorithm for contrast enhancement of natural images. Int. Arab J. Inf. Technol. **11**(1) (2014)
6. Kumar, S., Singh, B.M., Yadav, K.: A review on image segmentation techniques. Int. J. Current Eng. Sci. Res. **4**(10) (2017)
7. Chaturvedi, A., Khanna, R., Kumar, V.: An analysis of region growing image segmentation schemes. Int. J. Comput. Trends Technol. (IJCTT) **34**(1) (2016)
8. Wang, Q., Song, X., Jiang, Z.: An improved image segmentation method using three-dimensional region growing algorithm. The authors—Published by Atlantis Press (2013)
9. Fan, J., Yau, E., Aref: Automatic image segmentation by integrating color-edge extraction and seeded region growing. IEEE Trans. Image Process. **10**(10), 1454–1466 (2001)
10. Cheng, H.D., Jiang, X.H., Wang, J.: Color image segmentation based on homogram thresholding and region merging. Pattern Recogn. **35**, 373–393 (2002)
11. Kale, K.V., Deshmukh, P.D., Chavan, S.V., Majharoddin, M., Rode, Y.S.: Zernike moment feature extraction for handwritten Devanagari (Marathi) compound character recognition. Int. J. Adv. Res. Artif. Intell. **3**(1) (2014)
12. Marouf, H., Faez, K.: Zernike moment-based feature extraction for facial recognition of identical twins. Int. J. Comput. Sci. Eng. Inf. Technol. (IJCSEIT) **3**(6) (2013)
13. Mukundan, R., Ong, S.H., Lee, P.A.: Image analysis by Tchebichef moments. IEEE Trans. Image Process. **10**(9) (2001)
14. Haddadnia, J., Faez, K., Moallem, P.: Neural network based face recognition with moment invariants. In: IEEE International Conference on Image Processing (2001)
15. Yu, H., Li, M., Zhang, H., Feng, J.: Color texture moments for content-based image retrieval. In: IEEE International Conference on Image Processing (2002)

Design and Development of Automated Manipulator in TIG Welding

Pushp Kumar Baghel

1 Introduction

In comparison to other well-versed nations, we are still in the developing stage of modernization of equipment's and machineries. Instead of manufacturing and make use of automated machines, we are quite conventional in producing the welds. Whereas, well-versed nations are more concentrated to bring automatic technology in every area of manufacturing. This has not only helped them in reducing the variation in quality production of welds but also reduces wastage and manufacturing time which leads to more economical welds. Despite of all the intricacies and deficiencies in old methods of developing welds, small-scale industries rely heavily on conventional machines or specially designed machines for particular work. Most of the work in producing the weldment is accomplished by application of manual arc welding machines. Use of non-automatic welding process machines instead of the using automatic machine is normal practice in small-scale industry as automation is unaffordable due to large cost of setup, power required for running, repair and replacement, etc. To compete at global level and in order to reap more benefits in terms of cost in export of manufacturing items, we need to increase the robotization and automization of conventional machineries [1].

The state of affair is somehow totally on different page in significant and large manufacturing industries. They typically purchase the special purpose machine for TIG welding that is dear and not economical. However, within the long-standing time, it might once more turnout to be costly affair due to periodic breakdown and servicing of any of the tool component.

This generates the necessity for a newly created, design and develop automatic setup for TIG welding to produce the welds of thickness of 1–3 mm. So automatic manipulator is developed which prevents close proximity of welder to dangerous

P. K. Baghel (✉)
Noida International University, Plot 1, Yamuna Expy, Sector 17A, Noida 203201, India
e-mail: pushpkumar.baghel@niu.edu.in

M. Muzammil et al. (eds.), *Recent Advances in Mechanical Engineering*, Lecture Notes in Mechanical Engineering, https://doi.org/10.1007/978-981-15-8704-7_79

fumes and arc of high intensity and brightness [2], conjointly make the weld manufacturing economical and speedy process, so has edge as compared to present manual welding process.

Srinivasa and Gurudutta [3] developed automated TIG welding for cylindrical parts in which cylindrical job is rotated and torch is held stationary. Takalkar and Kakarparthy [4] indigenously designed special purpose machine integrated with TIG welding to manufacture bowl feeder; here, again torch is held stationary and workpiece is rotated. Al-min Li [5] used PLC control to increase the efficiency for circular weld. Haque [6] developed the manipulator which can move in linear and vertical motion within the specified velocity. The motion control of the manipulator is established by controlling the motors using the H-bridge circuit through parallel port of the computer. Patel [7] discussed the application of parallel manipulators in industry. He observed that parallel mechanisms like Stewart platform, Hexaglides, hexa-pods and delta robots more suitable for high speed operation. Richard [8] developed the six-axis manipulator which is suitable for digital programmed control movement to achieve universal work processing. Apart from this, various TIG welding variants have been developed like which has achieved significant advancements in this field of automation.

In conventional TIG fabrication process, the manufacturing time is more due to the inconsistency and non-synchronization of various linear motions in the machine links of setup. The pin point accuracy and standard of welding get adversely affected. The new mechanism designed and fabricated introduced in this chapter solves this problem. The motion in mechanism used is more uniform which helps in avoiding various defects typically include pits and bulges which are formed due to an-uniformity within the axis movement in mechanism of weld design equipment.

2 TIG Welding

Tungsten inert gas (TIG) welding is additionally known as gas tungsten arc welding (GTAW). It is a fusion fabrication process that uses a non-utilizable tungsten electrode with inert atmosphere provided by shielding gas to fabricate the weld.[9, 10]. It can be autogeneous, homogeneous and heterogeneous. Some welds are autogeneous welds, i.e., they can be welded without filler. Thickness of 1–3 mm can be welded without filler. Homogeneous welds are often takes place where the thickness of metal is greater than 5 mm and when welding dissimilar welds heterogeneous welding takes place. Conventional GTAW is usually considered highly difficult of all the fabrication process commonly utilized in manufacturing industry today. Because the fabricator must keep a minimum arc length, high keeping and maneuvering skill are essentially desired to limit contact between the tool and the workpiece. The method permits the fabricator more control over the weldment than any other competitor arc welding processes letting stronger, higher standard welds. GTAW continued with its apex position in complicability and difficulty in mastering as it needs bigger welder decisiveness than MIG and other processes, and moreover, it is considerably more gradual than most alternative fabrication technique.

3 Automation in Pulse TIG Welding Machines

The TIG fabrication process is performed for joining mainly aluminum and magnesium alloys. Subsequently, its use spread to almost all metals. The main metal nowadays used in industrial application is aluminum. Generally, material having higher thickness than 5 mm requires the automated welding equipment [11]. Earlier this task was done with conventional machinery. The certain reasons why we require automation are discussed as follows,

- Negative and poor effect on body of fabricator due to close proximity to dangerous poisonous fumes generated in fabrication process.
- Zig-zag motion of the weld because of agitation in hand while torch is in motion.
- Protection of the fabricator.
- The obtainable SPM within the merchandise is extremely expensive, miniature scale industries unable to bear cost of this SPM.

3.1 Automatic Manipulator Movement

In this specially designed mechanism using various links, components used in manipulator:

1. Lead screw.
2. Stepper motor which transmits the rotary motion through crown wheel or pulley belt to lead screw:

   Specifications of stepper motor.

- Size: 42.3 mm square × 48 mm, not including the shaft (NEMA 17), weight: 350 g (13 oz), shaft diameter: 5 mm "D", steps per revolution: 200, current rating: 1.2 A per coil, voltage rating: 4 V, resistance: 3.3 Ω per coil, holding torque: 3.2 kg cm (44 oz-in.), inductance: 2.8 mH per coil, lead length: 30 in., output shaft supported by two ball bearings, bipolar
- Belt
- Pulley
- Fixture for clamping workpiece
- Supporting rods
- Mounting Table.

  Components used for controlling stepper motor:

- Step-down transformer (1 A 12 V)

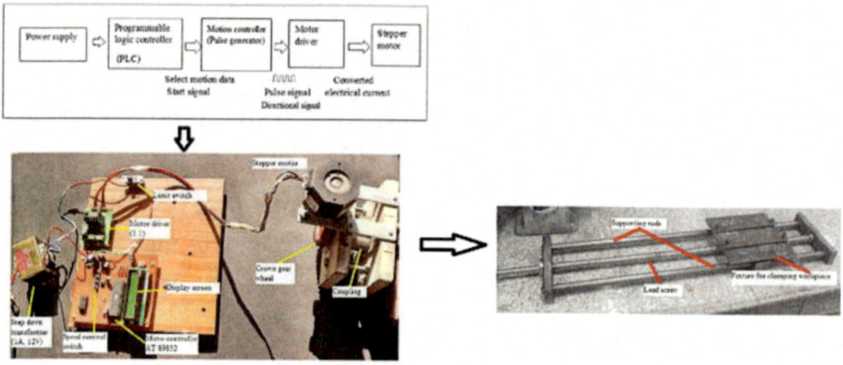

Fig. 1 Principle of working and control system for single axis-driven mechanism

- PLC which consist of micro-controller (AT 89,852)
- Motion controller (DRV877)
- Motor driver (1.1).

In this setup, the motion of fixture on which the workpiece is kept is given by lead screw attached and the welding torch is held at rest. The pulley belt is used to transfer the rotatory motion of stepper motor to the lead screw shaft on which the workpiece fixture is connected. The speed of stepper motor is controlled by sending the input signals in the form of pulses. First, the AC 220 V 50 Hz power supply is provided to the programmable logic control which sends motion data signal to pulse generator for regulating the movement. Then pulse creator pushes the binary pulse signal into motor speed controller which transforms the binary pulse signal into current signal and sends it to input positive terminal of the motor. Movement of fixture is provided by the single axis-driven mechanism as shown in Fig. 1. The design of single axis-driven mechanism consists of pulley belt, lead screw and connecting rods. The design of both components is vital for uniform movement of manipulator as follows.

Design of Component:

(i) Design of pulley

Size of larger pulley = 8 in.

Size of smaller pulley = 2 in. [9]

$$\frac{D_1}{D_2} = \frac{N_2}{N_1} \tag{1}$$

$$\frac{2}{D_2} = \frac{300}{1200}, \text{ or } D_2 = 8 \text{ in.}$$

(ii) Design of belt

$$\begin{aligned}\text{Centre distance} &= 3 \times \text{Diameter of larger pulley} \\ &= 3 \times 8 \\ &= 24\ \text{in.}\end{aligned} \tag{2}$$

$$\text{Length of belt, } L = 2C\frac{\pi(D_2 + D_1)}{2} + (D_2 - D_1)^2/4C \tag{3}$$
$$L = 64\ \text{in.}$$

(iii) Design of lead screw

Length of lead screw: 30 in., Length of threaded portion in lead screw: 24 in.
Power, $P = 150$ W

$$T = \frac{P}{\omega} = \frac{P \times 60}{2\pi N} = \frac{150 \times 60}{2\pi \times 1200} = 1.19\ \text{N m} \tag{4}$$

$$T = 1/16 \times \tau \times d^3 \text{ where } \tau = 42\ \text{Mpa (given)}$$
$$d = 7.681\ \text{mm} \tag{5}$$

- By considering bending moment

  When torch is at center position then,

$$M = \frac{WL}{4} = \frac{2 \times 9.81 \times 600}{4} = 2943\ \text{N m} \tag{6}$$

- Considering twisting moment

$$T_e = \sqrt{M^2 + T^2} \text{ or } T_e = 3174.4\ \text{N mm} \tag{7}$$

$$T_e = \frac{1}{16} \times \tau \times \text{d}^3 \text{ or d} = 10.65\ \text{mm}$$

- Selection of stepper motor

$T = 3.17$ N m

Stepper motor with 5 N m is selected

FOS = 5/3.17 = 1.57

Required factor of safety is 1.57.

4 Conclusion

Automation in TIG fabrication process is utilized to fabricate not only similar welds but also dissimilar metals. An automated manipulator in TIG welding setup is designed and developed whose development cost is three times less in contrast to automation accessible in merchandise. This makes automation cheaper which leads to economical fabrication of welds.

References

1. Cary, H.B.: Modern welding technology. Prentice-Hall, (NJ) (1989)
2. Trade India catalogue (2010)
3. Shrinivasa, D., Gurudatta, N.J.H., Mandath, B.S., Anil, K.H.S., Prajwal, B.J.: Int. J. Ac R De 222 (2018)
4. Takalkar, M., Kakarparthy, V., Khan, I.R.: Int. J. Res. App. Sci. Eng. Tech. 1344 (2017)
5. Li, A.M., Zhang, C.H., Li, H.L., Xu, Z.Y., Chen, X.H., Ye, S.W.: Design of automatic welding machine based on PLC. In: Fourth International Conference on Intelligent Computation Technology and Automation (2011)
6. Md. Anisul Haque, T.T.: Design and Development of a Robotic Manipulator for Automatic Arc Welding, vol. 27 (2013)
7. Yasenchak, R.W.F.: Six-axis Manipulator. United States Patent (1972)
8. Patel, Y.D.: Parallel manipulators applications—a survey. Modern Mech. Eng. (2012)
9. Hussain, A.K., Lateef, A., Javed, M., Pramesh, T.: Int. J. Ap Eng. Re Dindigul **70**, (2010)
10. Tseng, K.H., Chuang, K.J.: Pow. Tech. **228** (2012)
11. Minnick, W.H.: Gas tungsten arc welding. Goodheart Willcox co. (1995)

Design and Optimization of Suspension Geometry

Mrinal Deep, Pranav Upadhyay, and Pikesh Bansal

1 Introduction

The objective of designing the suspension system at the front is to maintain the traction with the road at different driving conditions and at the same time optimize the ride comfort of the vehicle so as to maintain the driving conditions for the driver when the vehicle encounters various bumps and droops while running. To achieve the above targets, we need to control the camber and toe variation while the vehicle is in running condition and this is done by designing the geometry on LOTUS software while considering all the dimensional restrictions of the vehicle so as to confirm that vehicle has the appropriate dimension as decided for the vehicle. To design the geometry, we followed the procedure showed in the chapter.

2 Literature Review

Cherian et al. [1] have worked on nonlinear model of a double wishbone suspension and found a procedure in which we can find the effect on transmission, tire distribution forces acting on wheel spindle and chassis by variation in suspension parameters. The equations of motion are derived with the help of Lagrange multiplier method and solved numerically using MATLAB. Lamers [2] has worked on suspension kinematics by using numerical optimization techniques. Suspension kinematic objectives have to be examined as close as possible by relocating the suspension coordinates. The most important improvements of the suspension kinematics are harmonization of camber dependent kinematics which result in the optimal camber angles of the tires during driving. The suspension is designed to have a steady ride height during

M. Deep (✉) · P. Upadhyay · P. Bansal
ABES Engineering College, Ghaziabad, Uttar Pradesh, India
e-mail: mrinal.deep98@gmail.com

M. Muzammil et al. (eds.), *Recent Advances in Mechanical Engineering*, Lecture Notes in Mechanical Engineering, https://doi.org/10.1007/978-981-15-8704-7_80

cornering which causes the suspension to work in the intended region. Esfahani et al. [3] explained the effect of camber angle on, rubber abrasion, acceleration and braking. A suspension mechanism is acquainted with adjustable camber angle in the vehicle. Popovic et al. [4] have applied the system engineering methods and system approach in the initial phase of vehicle active suspension development. A study has been done to find the interrelation between computer-aided simulation and other elements of the development process. Application of active suspension simulation has many benefits: Mechatronic components/devices have new and improved functions and increase the system reliability. Yarmohamadi and Berbyuk [5] investigated on both dynamic and kinematic properties of the vehicle with the help of simulations and analytical expressions. Result of their study was an evaluation between the trucks with rigid front axle and IFS with respect to handling and comfort. Analysis was done by seeing the responses of the vehicles on different steering inputs. Dhamodharan et al. [6] have presented an optimal dimensional synthesis of a five-link suspension mechanism in terms of length of the five links and positions of the ten spherical joints, which are installed on the vehicle chassis and wheel carrier. The influence of the complaint joints on the dynamic behavior of the suspension system is approached. MATLAB optimization tool box is used for this optimization. The aim was to take the motion of the body to the wheel carrier with respect to the vehicle chassis as close as possible to a vertical translation. Liu et al. [7] have researched on the relation of the hardpoints of the suspension. The correlation of the hardpoints is used to analyze and design optimal suspension. Vehicle design was referred from the hard points of the rotational design. Shijil et al. [8] have done a study on dynamics of the vehicle when driven on an off-road racetrack conditions is analyzed on dynamic and static suspension system of an ATV. The performance of the ATV is affected by many parameters. The chapter work focuses on design and analysis, determination, optimization of suspension systems and their assembly best performance. The aim is to identify and optimize the parameters which affects the dynamic performance of suspension systems.

3 Procedure

- Finalize the type of suspension to be used.
- Finalize track width and wheelbase of the vehicle.
- Calculation of load transfer.
- Design the geometry on LOTUS software.

3.1 Type of Suspension

The type of suspension to be used at the front is double wishbone damper to lower suspension system. The reason for choosing the same is that it gives better camber control in case of bump and droop. Using damper to lower suspension system at front of the vehicle helps in absorbing more load at the front tire which results in better ride frequency of the vehicle.

3.2 *Trackwidth and Wheelbase*

Now, the trackwidth and wheelbase of the vehicle is decided by keeping in consideration all the needs of different department as well. Steering and suspension has the major role in deciding the trackwidth and wheelbase of the vehicle.

We decided.

TRACKWIDTH: 52.3 in.

WHEELBASE: 62 in.

This trackwidth and wheelbase was chosen by using different iterations where we get the maximum turning angle and the Ackerman angle was according to our need. The main basis was to maintain the trackwidth to wheelbase ratio below 85%.

3.3 *Load Transfer Calculations*

The vehicle is a single seater tadpole configuration vehicle, which runs on electric as well as solar drive, and forces are calculated while calculated while considering the same.

To find the stiffness of shocker, we calculated the weight transferred during different driving condition, like braking, acceleration and turning, etc. The formulae used are the equations of weight transfer from the book Race Car Vehicle Dynamics by Millikan [9].

Formula: $F = (\text{weight} \times cg \times \text{acceleration}) \div \text{wheelbase}$.

- Weight transfer during acceleration = 14.79 kg toward rear
- Weight transfer during deacceleration = 178 kg toward front
- Weight transfer during turning = 26 kg at the side.

According to the weight transfer in the body stiffness of the shocker is calculated by the equation of stiffness, i.e., $F = Kx$, the deflection at various condition is assumed, i.e., at acceleration, deacceleration and turning and then the stiffness is calculated.

The stiffness calculated for the shocker was 20 N\mm.

3.4 Lotus Geometry

We are designing a double wishbone damper to lower suspension system for tadpole configuration three-wheeler vehicle at the front of the vehicle.

Equal and Parallel A-arms:

There are various other factors which are taken into consideration while entering the data for geometry of the suspension.

- The type of suspension was selected as double wishbone suspension system damper to lower arm with unequal and non-parallel A-arms on the frontal axis.
- Here, we are not using equal and parallel A-arms because it was providing negligible amount of camber gain at dynamic conditions which resulted in lesser area of contact patch on road.
- In Fig. 1, a geometry is drawn with equal and parallel combination and analysis also is given in Figs. 1 and 2.
- Here, Figure 3 shows wireframe of the simulation of roll over in LOTUS with Full + Half + Static Screen Display, which explains max condition with roll angle 2°.

Geometry of suspension. The finalized geometry of suspension that we are installing in our vehicle is given in Fig. 4; this geometry of suspension is of unequal and unparallel A-arms with damper to lower wishbone.

Analysis of geometry (finalized). There are various factors which are taken in consideration while designing:

- The roll center of the car must be below the center of gravity of the car during different driving conditions.

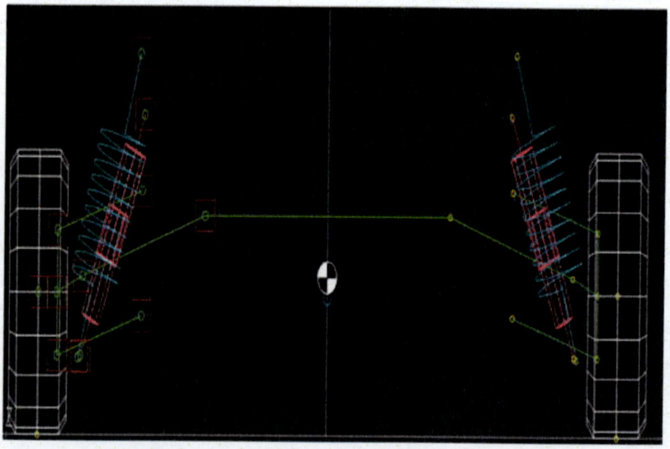

Fig. 1 Suspension geometry for equal and parallel A-arm

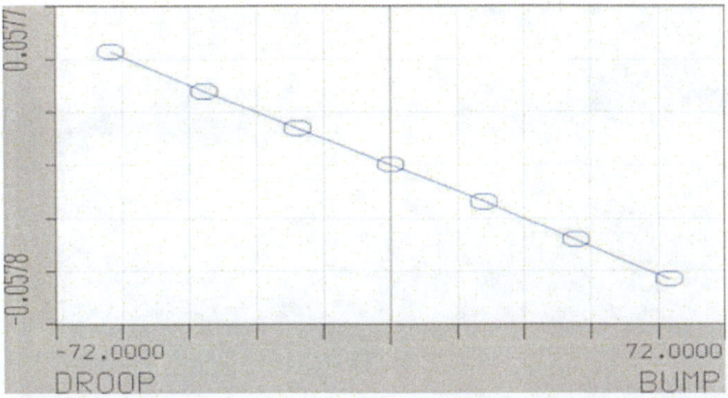

Fig. 2 Degree versus displacement for bump and droop for equal and parallel A-arm configuration (Camber Graph)

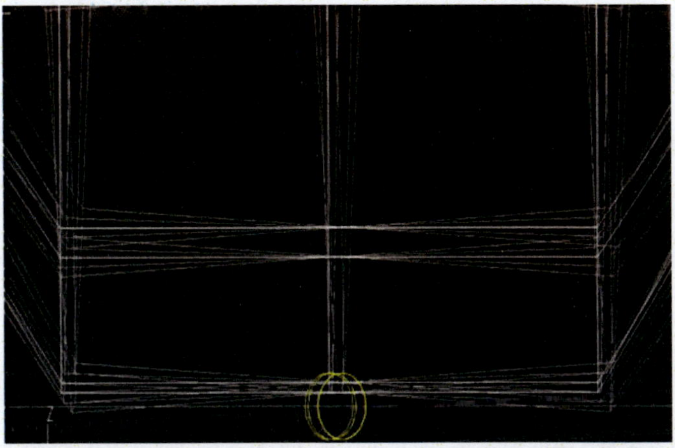

Fig. 3 Equal and parallel A-arm (full + half + static screen display) (Roll angle 2°)

- The camber must be controlled and there must be as less camber gain as possible because we are using radial tire; here, the geometry designed is with zero camber at static condition and the camber graph for dynamic condition due to bump and droops while driving is explained in Fig. 5.
- The maximum camber gain is 5.8° at 60 mm (Fig. 5) droop; therefore, the motion of the A-arms is such that the roll center is below the center of gravity.
- Appropriate size and width of tire was inserted according to the tire chosen.
- Value of bumps and droops is input according to the dynamic condition.
- Figure 6 shows wireframe of tires contact patch with roll angle 5° which has lesser contact patch than before at max roll condition.
- Figure 7 shows wireframe of tires contact patch with roll angle 10° which brings lesser contact patch than at 5° at max roll condition.

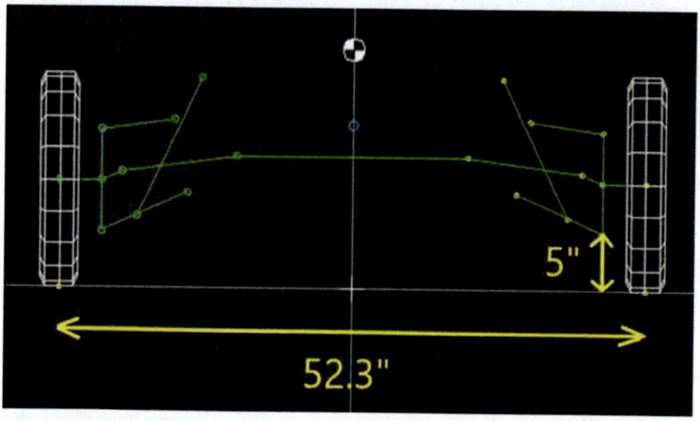

Fig. 4 Suspension geometry for unequal and unparalleled A-arm (Finalized)

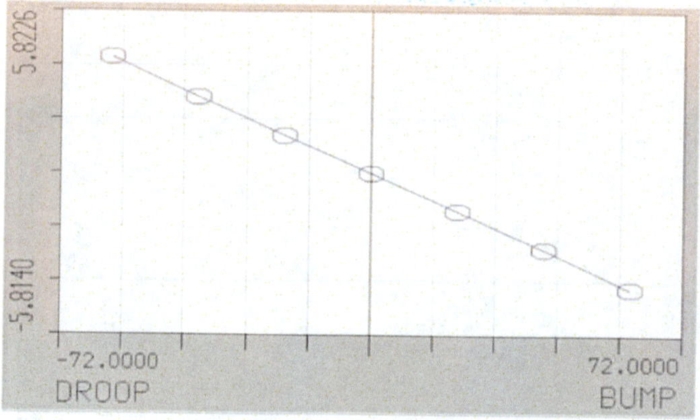

Fig. 5 Degree versus displacement for bump and droop for unequal and unparallel A-arm configuration (Camber Graph)

Hence, we try to prevent using equal and parallel geometry of A-arms.

Castor and Toe Optimization.

- Castor is given for generating self-centering force on steering wheel whenever steering wheel is turned.
- Generally, castor angle ranges from 0° to 5°–6° for the four-wheel automobiles specially which have manual steering.
- When we give castor angle more than 7° then generally, we use power steering because by increasing castor steering effort increases.
- So, for vehicle without power steering may chose castor near 3–4.
- The toe for the given geometry can be adjusted by adjusting the height of the rack.
- The toe optimization graph is shown in Fig. 8.

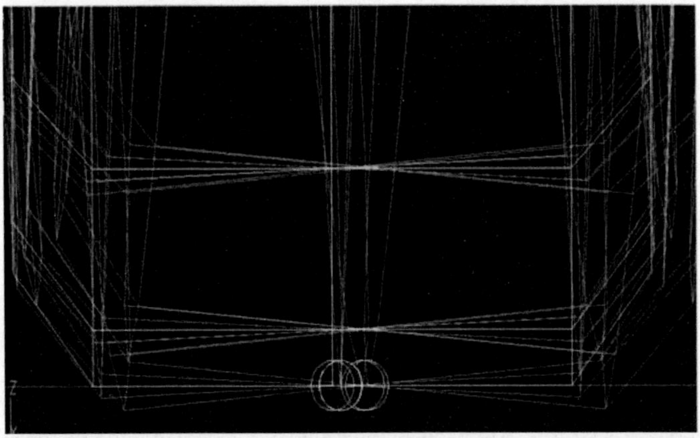

Fig. 6 Unequal and unparalleled A-arm (contact patch) (roll angle 5°)

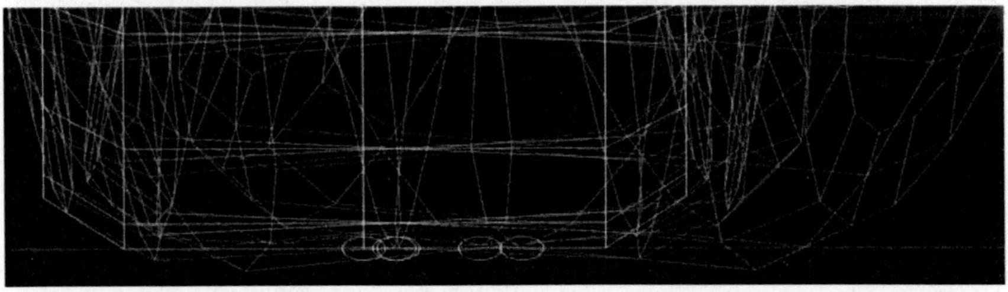

Fig. 7 Unequal and unparalleled A-arm (contact patch) (roll angle 10°)

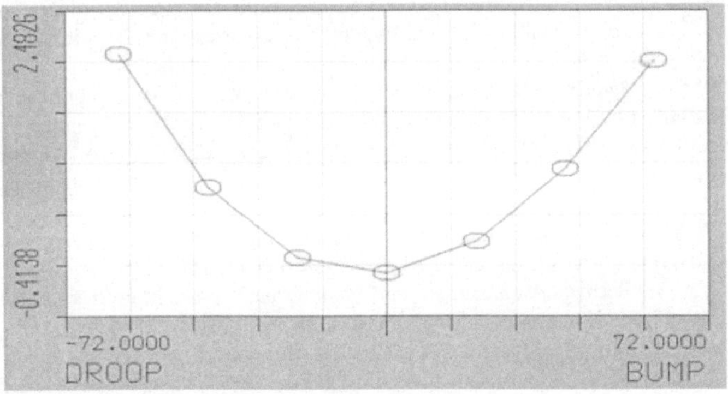

Fig. 8 Degree versus displacement for bump and droop for unequal and unparalleled (TOE)

- The basic criteria to optimize the castor and toe in the wheel are by adjusting the steering arm and the ball joint position. We opted for zero-degree camber as well as toe while considering all the variations that will be caused by the variation in the camber and toe angles.

4 Conclusion

The geometry was made with different trackwidth and wheelbase of the vehicle and the most optimized and feasible value of trackwidth and wheelbase is selected as 52.3 in. and 62 in., respectively. Calculations were performed according to the dimension of the vehicle and the assumed vehicle running condition. The stiffness of shocker is calculated as 20 N/mm. After that the geometry was designed considering all the factors, it gave the optimized camber variation at various bumps and droop which the vehicle will encounter when it is in the running state. Parallel geometry gives zero camber variation according to the parallelogram law which is not good for our vehicle at the time of turning. The contact area will change as zero camber gain vehicle will lose traction. The optimized suspension geometry design is made for the vehicle stability.

References

1. Cherian, V., Jalili, N., Haque, I.: Development of a Non-linear model of a double wishbone suspension for the characterization of force transmission to the steering column and chassis. In: ASME 2004 International Mechanical Engineering Congress and Exposition, pp. 775–780 (2004)
2. Lamers, W.: Development and analysis of a multi-link suspension for racing applications. Master Thesis, Eindhoven University of Technology, Department of Mechanical Engineering (2008)
3. Mohammad Esfahani, I.M., Mosayebi, M., Mohammad P., Keshavarz, A.: Optimization of double wishbone suspension system with variable camber angle by hydraulic mechanism. Int. J. Mech. Mechatron. Eng. **4**(1), 60–67 (2016)
4. Popovic, V., Vasic, B., Petrovic, M., Mitic, S.: System approach to vehicle suspension system control in CAE environment. J. Mech. Eng. **57**(2), 100–109 (2011)
5. Yarmohamadi, H., Berbyuk, V.: Kinematic and dynamic analysis of a heavy truck with individual front suspension. Int. J. Veh. Mech. Mobil. **51**(6), 877–905 (2013)
6. Dhamodharan, P., Mohamed Raafiq, N., Madhu, S.: A review paper on multilink suspension mechanism with compliant joints. Int. J. Appl. Eng. Res. **10**(33), 25491–25493 (2015)
7. Liu, X., Wang, M., Wang, X., Li, C., et al.: Hardpoint correlation analysis and optimal design for front suspension of a formula SAE car. Aust. J. Mech. Eng. **13**(2), 67–76 (2015)
8. Shijil, P., Vargheese, A., Devasia, A., Joseph, C., Jacob, J.: Design and analysis of suspension system for all-terrain vehicle. Int. J. Sci. Eng. Res. **7**(3), 164–190 (2016)
9. Milliken, W.F., Milliken, D.L.: Race car vehicle dynamics. society of automotive engineers, Inc.,400, Commonwealth Drive, Warrendale, PAI5096–0001 U.S.A.

Perspective on Effect of Metallic Fillers on Electrical Conductivity of FRP Composites

Aditya Pratap Singh, Avinash Yadav, Srashti Mishra, Kunwar Laiq Ahmad Khan, and Anurag Gupta

1 Introduction

Over the past few years, FRPs have emerged as the principal material for various industries such as aeronautics, electronics, structure, communication, etc. This has led to the exploration of various properties of FRP in context with its mechanical, thermal and electrical performances. The area of mechanical and thermal properties has been well explored, and there are numerous applications in this era based on the results and discussion of various experiments performed in the field. Though the domain of their electrical properties is being studied, it has not reached its full potential and much more is yet to be done. Various applications are using the inherent electrical insulation of most of the FRPs, but less work has been done to improve the conductivity to be able to use them more vividly. Industries like aeronautics and automobiles require their structures to be strong as well as a good electrical conductor to deal with situations like a lightning strike, electromagnetic interference, etc., and this has led to the need of improvement in electrical conductivity of the FRPs along with a minor change in their strength. The excessive weight addition by metals to fulfill this requirement plays an important role in the dependence on FRPs. Properties for various reinforcement materials are mentioned in Table 1.

As inferred from the above table, the tensile strengths of carbon and glass fibers are almost similar, i.e., in the range of 2200–5600 MPa and 3400–4800 MPa, respectively. However, aramid and other fibers show a lower strength compared to these. Modulus of elasticity shows a larger deviation for different reinforcement materials. Modulus of elasticity to density ratio (or specific modulus) is an essential factor of material selection for stiffness-driven structures such as wings of an airplane. For structural weight to be minimum along with high stiffness, the high specific modulus is required.

A. P. Singh (✉) · A. Yadav · S. Mishra · K. L. A. Khan · A. Gupta
Department of Mechanical Engineering, KIET Group of Institutions, Ghaziabad, India
e-mail: aditya.1640011@kiet.edu

M. Muzammil et al. (eds.), *Recent Advances in Mechanical Engineering*, Lecture Notes in Mechanical Engineering, https://doi.org/10.1007/978-981-15-8704-7_81

Table 1 Comparison of properties of various materials [1]

Material	Tensile strength (MPa)	Modulus of elasticity (GPa)	Density (kg/m^3)	Modulus of elasticity-to-density ratio (Mm^2/s^2)
Carbon	2200–5600	240–830	1800–2200	130–380
Aramid	2400–3600	130–160	1400–1500	90–110
Glass	3400–4800	70–90	2200–2500	31–33
Epoxy	60	2.5	1100–1400	1.8–2.3
Carbon fiber-reinforced polymer (CFRP)	1500–3700	160–540	1400–1700	110–320
Steel	280–1900	190–210	7900	24–27

In a study performed by Ha et al. [2], used aluminum mesh in FRP composite, to be used in B787 aircraft, has electrical conductivity 73,000 S/m (Siemens/meter), whereas for maiden FRP composites could reach up to 15,000 S/m. When silver nanoparticles were used as filler material it achieved up to 26,000 and 28,000 S/m at 5% colloid for spray time of 3 and 5 s respectively while 56,000 and 63,000 S/m at 10% colloid for spray time of 3 and 5 s, respectively, which reached about 86% of the aluminum-filled composites along with reduced weight. These results are shown in Fig. 1.

Electrical conductivities of commonly used metals are within the range of 10^6 S/m which is still much greater than that of FRPs [3].

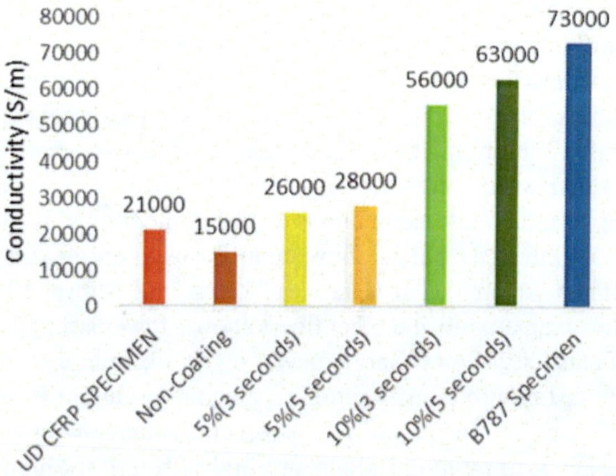

Fig. 1 Comparison of conductivity from B787 aircraft spaceman and the hybrid CFRP by conductive nanoparticle colloid coating [2]

2 Basic Process of FRP Fabrication

The initialization of the process begins with the selection of appropriate matrix material, reinforcement, additives and choosing the technique as per the application. The matrix material is liquefied and molded in required shape with reinforcement being done either by hand, spray or with any other technique. The impregnated resin is then left to dry or is oven-dried. The extraction of a semi-finished product is done, and then, the final finishing is done by various machining methods. Every FRP goes through this process with some little changes in the environment or addition of one or more processes which depends on the future use of that fabricated FRP [4].

3 Fillers in FRP

Fillers are the substances that are used in composites to improve their inherent properties along with the lesser effect on the cost which may affect some other properties. They also are used to reduce shrinkage, control viscosity, and improve part stiffness. Commonly used fillers include kaolin, calcium carbonate, silica, feldspar, talc and glass microspheres. Fillers are not as common in high-performance composites because they may adversely affect the fiber-resin load transfer and decrease the toughness of the resin at high filler content. Some of the previously used metallic fillers are copper, zinc, silver, tin, etc. [4].

4 Effect of Fillers on Characteristics of FRP

An investigation conducted by Ha et al. [2] to improve the electrical conductivity of composites was done with the assistance of conductive fillers for the better performance of FRP in the aircraft lightning protection. Silver nanoparticles were used as fillers for this purpose. A colloid of silver nanoparticles was sprayed on the carbon fibers, which were then to be impregnated in the epoxy resins. The experiment was conducted for various amounts of the colloid, and the spray time was also varied to examine the effect of the period for which colloid is sprayed. The electrical resistance was first measured with the contact-type resistance meter which used the principle of AC 4-terminal method. The resistance was then converted into resistivity, and then, resistivity was finally converted to electrical conductivity. Scanning electron microscopy (SEM) and energy-dispersive X-ray spectroscopy (EDS) verification of the silver nanoparticles was also done. The results showed that electrical conductivity was increased as much as to four times the ordinary CFRP [2].

Another experiment was carried out on CFRP by Zhu et al. [5] for their application as the anode material in the impressed current cathode protection (ICCP) for the protection of the steel-reinforced concrete structure. This was done to find out the performance and service life of CFRP under various configurations when used as an anode in the ICCP system. The current densities applied are maintained to be constant for the idealization of conditions. The results of the experiment showed that there was no significant degradation in the electrical and mechanical properties of CFRP when subjected to anodic polarization with various current densities. The recordings were helpful to judge the property change of CFRP, and the performance was determined based on the study of the practical reinforced concrete structure layout [5].

FRPs are also used for reinforcement in concrete structures for their strengthening purpose. This is done for obtaining various features in the concrete structures such as seismic retrofitting, additional live or dead load. This reinforcement is done by various methods such as steel jacketing, externally bonded steel plates, external retrofitting, concrete jacketing and post-tensioning. The technique is selected based on interference material or vice versa. According to JSCE, 1999 [6], the steps followed are the identification of requirements, an inspection of an existing structure and its evaluation of performance, design of retrofitting structure, selection of material and technique, evaluation of the retrofitted structure and implementation of the technique [1].

Srinivas et al. [7] performed a study on the electrical conductivity of room temperature (RT)-cured epoxy resin (LY556 + HY951) containing three different particulate fillers. The particles taken for the study were classified as soft material graphite, a hard material silicon carbide (SiC), and a hybrid of both of them. The weight fractions of the fillers in the specimen were also varied from 10 to 40% in the steps of 10% for all to study the impact of the weight of the filler on the electrical conductivity. It was observed that the change in electrical conductivity of the epoxy resin was directly proportional to the weight fraction of the filler. Another observation was that the resins with 40% filler fraction showed a significant change in electrical conductivity especially the one with 40% hybrid fillers (20% graphite, 20% SiC, and 60% epoxy resin) had electrical conductivity about 2.5 times the original epoxy. The results of experiment were verified with the Maxwell–Wagnor and Wiedeman–Franz law which relates the thermal and electrical conductivity. The experimental results showed a good correlation with analytical equations [7].

According to Tarafder and Swain [8], the stress vs strain curve for the carbon fiber is much more linear than any other fiber including various types of steel as shown in Fig. 2. This results in the higher strength of CFRP and thus gives it priority over others in the group for more use in the strengthening purpose. Tarafder and Swain also suggested different types of reinforcement fibers, according to the purpose they are required for, by a case study on the durability of FRP.

Another study by Caradonna et al. [9] to study thermal and electrical conductivity of composites with carbon-based fillers which makes the use of graphene-like nanoplatelets and graphite fillers has been done. The epoxy resin and the fillers were first cast into a mold and then cured to obtain the experimental specimens. Multiwall carbon nanotubes were also used as fillers along with nanoplatelets and graphite. Electrical and thermal conductivity was investigated for various filler composition and weight fractions. The study was accompanied by the percolation theory, and it was inferred that the percolation paths enhance the electrical conductivity but were

Fig. 2 Schematic diagram of stress vs strain with respect to behaviour of reinforcing fibers in comparison with steel [8]

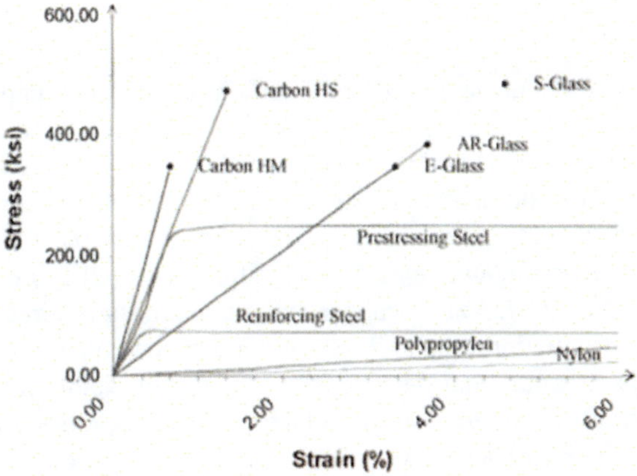

not so helpful for the increment of the thermal conductivity. Also, the results of single filler composites were compared to the hybrid fillers composites which were occupied with the combination of any two fillers out of the three. Results showed that fillers having different aspect ratios were observed to have a synergetic effect in relation to electrical conductivity showing noticeable improvement in it but showed a smaller effect if compared in case of thermal conductivity [9].

Wang et al. [10] performed an experiment to study the mechanical properties of FRP reinforcement bars when used as internal reinforcement in concrete structures. The experiment was conducted at different temperatures up to the failure of FRP bars. FRP bars of different geometry were procured from an external source, and then specimens were prepared. It is observed in the experiment that stress–strain relationship of FRP bars remained linear to the range of 300–400 °C. After this temperature, there was a sharp drop in the elastic modulus up to the point of failure at 500 °C [10].

A study conducted by Tsangaris and Kazilas of National Technical University of Athens [11] used the metallic fillers that resulted in the increment of electrical conductivity with their concentration. These materials were complex and considered to be a chaotic mixture of conductive particles which are randomly distributed in an insulating matrix. The conductivity of the materials in DC electrical field is studied with the assistance of percolation theory where there is a rapid increase in the electrical conductivity of the material at a particular concentration of conductive phase, referred as the critical phase. In this experiment, the various specimens with multiple conductive fillers were used. The fillers used were powders of copper, aluminium, or zinc. The temperature range at which study conducted was also varying from 20° to 140 °C. The results obtained from this experiment were also analyzed using percolation theory, and semi-empirical algorithm was used for the determination of some new parameters [11].

5 Conclusion

The literature reviewed by the authors in the present paper shows that by using various fillers, the conductivity of FRP composites can be increased without degrading its mechanical and thermal properties significantly. Metallic and non-metallic both type of filler materials can be used to improve the conductivity of FRP composite. The metallic fillers increase the conductivity four times, while non-metallic fillers increase conductivity 2.5 times. Effect of metallic fillers such as silver nanoparticles and powdered zinc, copper and aluminum have already been studied for different concentrations and at different temperature conditions. It was also concluded from the literature that the viscosity of the resin rapidly increases due to the addition of metallic filler to increase conductivity. We can infer that the electrical conductivity of FRP composites may further increase by introducing some other metallic fillers at the various concentration in different forms. Some other aspects such as variation in electrical conductivity with temperature should also be studied. The systematic alteration of electrical conductivity of FRP composites can widen the scope for the usage of FRP in applications of aeronautics where fly-by wire systems are used, automobiles, communication to prevent the noise due to electromagnetic interference, etc.

Acknowledgements Authors are grateful to have been supported by the Royal Academy of Engineering under the auspices of the "UK-India Industry-academia partnership program" between Cranfield University, UK, and KIET Group of Institutions, India (Grant No. IAPP18-19\295).

References

1. Günaslan, S.E., Karaşin, A., Öncü, M.E.: Properties of FRP materials for strengthening. Int. J. Innov. Sci. Eng. Technol. **1**(9), 656–660 (2014)
2. Ha, M.S., Kwon, O.Y., Choi, H.S.: Improved electrical conductivity of CFRP by conductive nano-particles coating for lightning strike protection. Compos. Res. **23**(1), 31–36 (2010)
3. Jean Tiberghien, G.: TIBTECH innovations: metal properties comparison: electric conductivity, thermal conductivity, density, melting temperature. [Online] Tibtech.com. Available at: https://www.tibtech.com/conductivite.php?lang=en_US. Accessed 25 Sep. 2019
4. Agarwal, B., Broutman, L., Chandrashekhara, K.: Analysis and Performance of Fibre Composite, 3rd edn., p. 562. Wiley, New Delhi (2015)
5. Zhu, J.H., Zhu, M., Han, N., Liu, W., Xing, F.: Electrical and mechanical performance of carbon fibre-reinforced polymer used as the impressed current anode material. Materials **7**(8), 5438–5453 (2014)
6. Japan Society of Civil Engineers (JSCE): Guidelines for retrofit of concrete structures—draft. Concr. Library Int. **6**, 61–112 (2000)
7. Srinivas, K., Bhagyashekar, M.S., Darshan, B.G.: Effect of fillers on electrical conductivity of epoxy composites. J. Polym. Compos. **6**(3), 25–30 (2018)
8. Tarafder, N., Swain, R.: Durability and case study of fibre reinforced polymer (FRP) IOSR J. Mech. Civil Eng. IOSR-JMCE **6**(3), 53–62 (2016)
9. Caradonna, A., Badini, C., Padovano, E., Pietroluongo, M.: Electrical and thermal conductivity of epoxy-carbon filler composites processed by calendaring. Materials **12**(9), 1522 (2019)

10. Wang, Y.C., Wong, P.M.H., Kodur, V.K.R.: Mechanical properties of fibre reinforced polymer reinforcing bars at elevated temperatures. In: SFPE/ASCE Specialty Conference: Designing Structures for Fire, Baltimore, Maryland, September, pp. 183–192
11. Tsangaris, G.M., Kazilas, M.C.: Conductivity and percolation in epoxy resin/conductive filler composites. Mater. Sci. Technol. **18**(2), 226–230 (2002)

PID Design-Utilized Controller for DC Engines Employed in Wheeled Movable Golem

Shahida Khatoon, Huma Khan, and Prerna Gaur

1 Introduction

The unmanned ground vehicle may be a conveyor that works with the skin of earth and without the closeness of a person. UGVs may be used for a few applications wherever it would be woefully designed, perilous, or arduous to own somebody's administrator gift. Wheeled many-sided robots may be used in sundry applications like operating in industrial facility, terminal or emergency clinic condition. At the purpose, once condition is totally kenned, free ways that square measure engendered through street map or cell decay.

A run of the mill development of the two wheel engine gets together with haggle in wheeled movable golem appeared in Fig. 1. It consists of three parts: motor, drive assembly, and controller. The engine and drive assembly comprised of an engine associated with wheel, encoder, and driver. The driver is the drive circuit for driving the motor based on the signal given by the controller. During this orchestration, robot kineticism is the result of the adjustment of close observation.

2 Literature Survey

Wheeled portable golems are regularly utilized for support of round channels that may normally be spoken to as a barrel-shaped workspace. Functioning of pipes, dynamic models of wheeled movable golems are not the same as those applying on

S. Khatoon · H. Khan (✉)
Department of Electrical Engineering, JMI, New Delhi, India
e-mail: humazakhan@yahoo.com

P. Gaur
Division of ICE, NSUT, New Delhi, India

M. Muzammil et al. (eds.), *Recent Advances in Mechanical Engineering*, Lecture Notes in Mechanical Engineering, https://doi.org/10.1007/978-981-15-8704-7_82

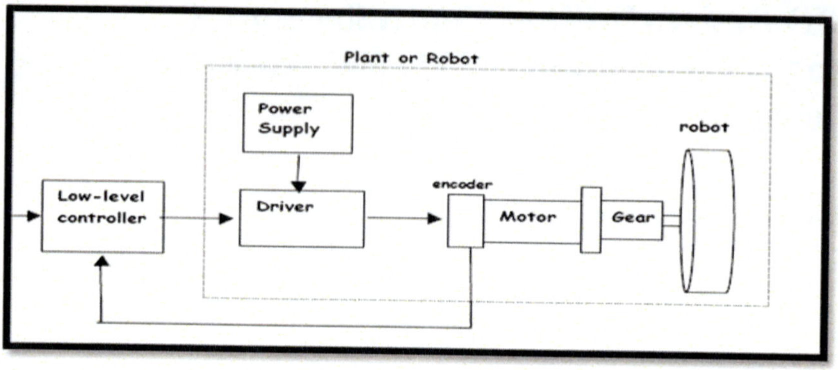

Fig. 1 Two-wheel motor and controller assembly in wheeled movable golem

a plane and consequently present critical difficulties regarding kinematic examination and movement control. In view of these examinations, dynamic properties of wheeled movable golem in round and hollow workspaces are talked about with screw hypothesis. Analyses utilizing a vehicle like portable golem rotating in conduits are completed to demonstrate the attainability of the proposed calculation in [1].

The execution of two controllers is accomplished for the main connection of the two-wheeled portable golem for the presentation examination between regular controller and clever controller by using PID controller in [2, 12–14].

Another following controller used in portable wheeled golem dependent upon back venturing way is planned. Lyapunov work technique is built up for checking the controller. A divider behind control calculation for the portable wheeled golem is created utilizing the Lyapunov strategy. The direction mistake between the direction of the portable golem and the divider is clarified by state factors in [3–7].

A control plot for certain issues on following an impediment shirking for the wheeled portable golem with non-holonomic requirement is clarified. An all-encompassing state eyewitness is exhibited to assess the speed data and obscure unsettling influences of the wheeled portable golem. To achieve most extreme following objective and snag shirking in complex situations, one indirect controller has been planned in [8, 11].

Such kind of a controller set down, without precedent for composed task, the elements associated three subsystems which establish a wheeled versatile golem in [9, 10].

3 Mathematical Modeling

A DC engine can be designed through speaking to its equivalent circuit in MATLAB simulation.

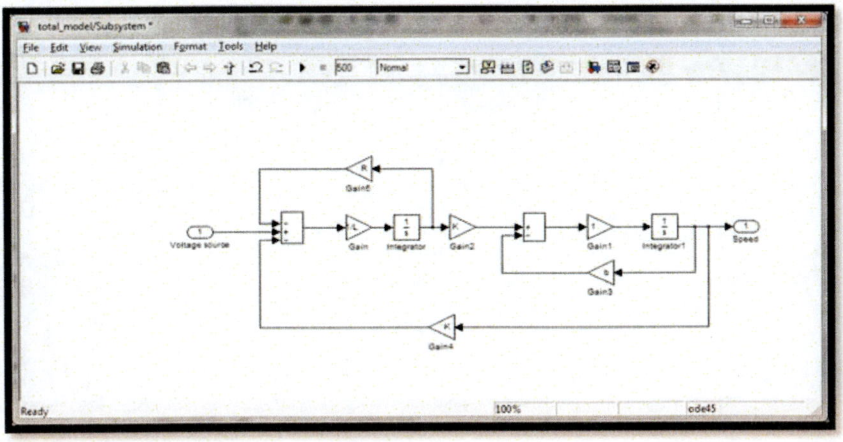

Fig. 2 Simulink model of DC motor in MATLAB

$$G_v(s) = \frac{w(s)}{V(s)} = \frac{K}{\{[(R + Ls)(Js + b) + K^2]\}}$$

The simulation of dynamic model for the DC engine is appeared underneath (Fig. 2).

The drive reaction of the engine is appeared in Fig. 3, and bode plots of the point and speed of the engine are appeared in Fig. 4 by utilizing the general exchange capacity of DC engine as far as edge and speed.

4 Controller Design

The control yield, u(t), is acquired by accompanying condition,

$$u(t) = k_p e(t) + K_i \int_0^T e(\tau)\mathrm{d}\tau + K_d \frac{\mathrm{d}e(t)}{\mathrm{d}t}$$

$$C(s) = \frac{U(s)}{E(s)} = k_p + \frac{K_i}{s} + K_d s$$

$$C(s) = \frac{k_d s^2 + K_p s + K_i}{s}$$

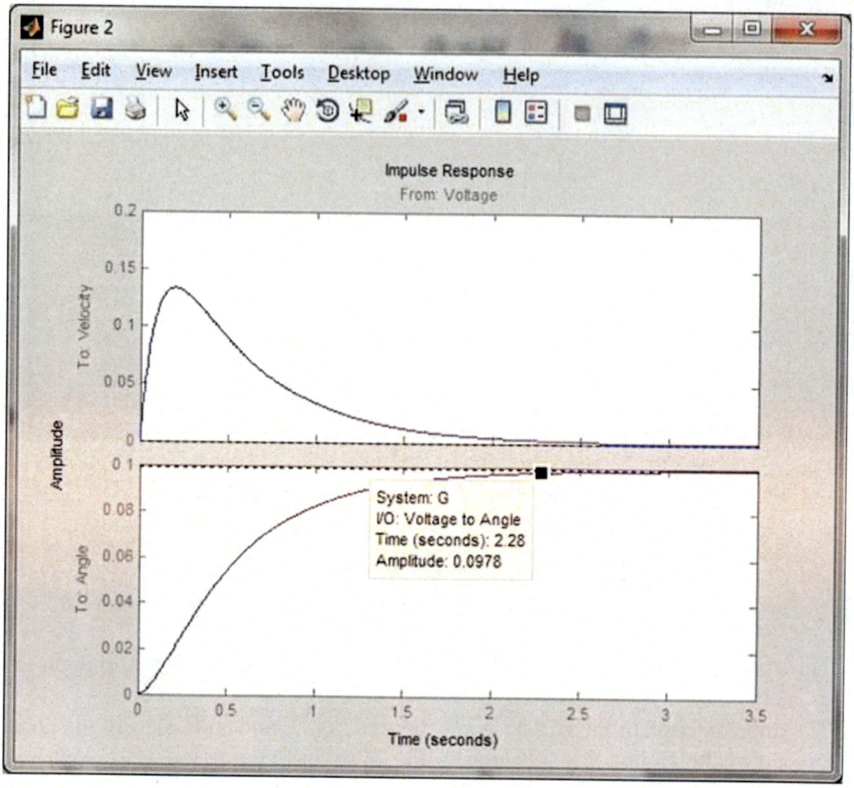

Fig. 3 The impulse response of DC motor

5 Results

See Figs. 5, 6, 7 and 8.

6 Conclusion

The distinguished drive golem essentially comprised of wheels driven through motors. On the basis of variations of speed of the motors controlled by PID controller, the golem will push ahead. The golem can flip within other way to the facet wheel pivoting gradually. In this way, the speed of DC motor employed in wheeled movable golem has been controlled. Simulink model of DC engine inferred and reproduced, and the PID is intended for improving its performance. A distinguished drive wheeled golem application has been presented with simulated model in Simulink.

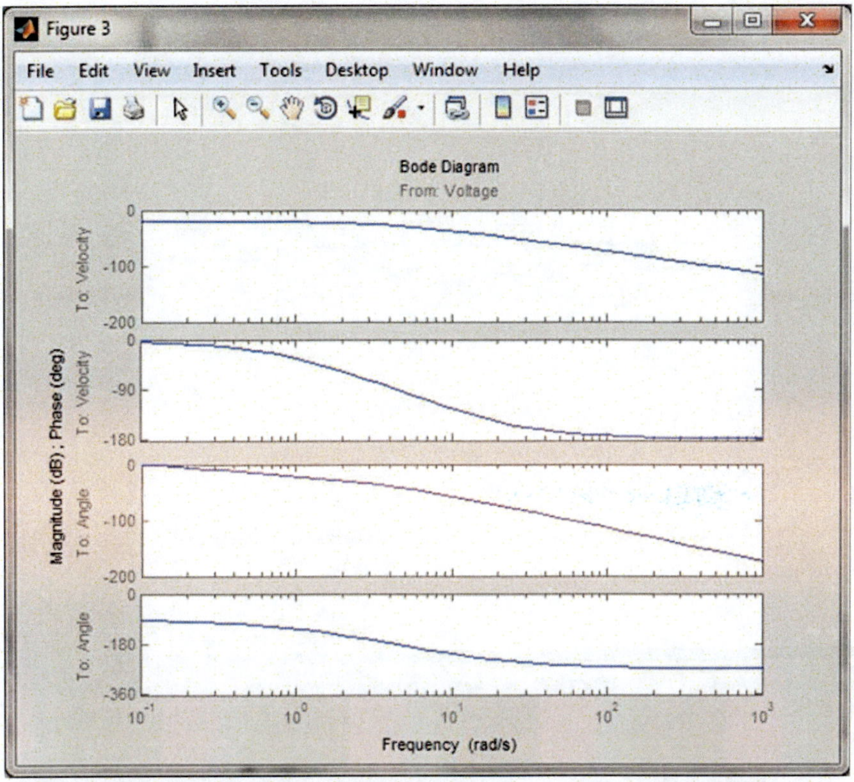

Fig. 4 Angle and velocity bode plot of DC engine

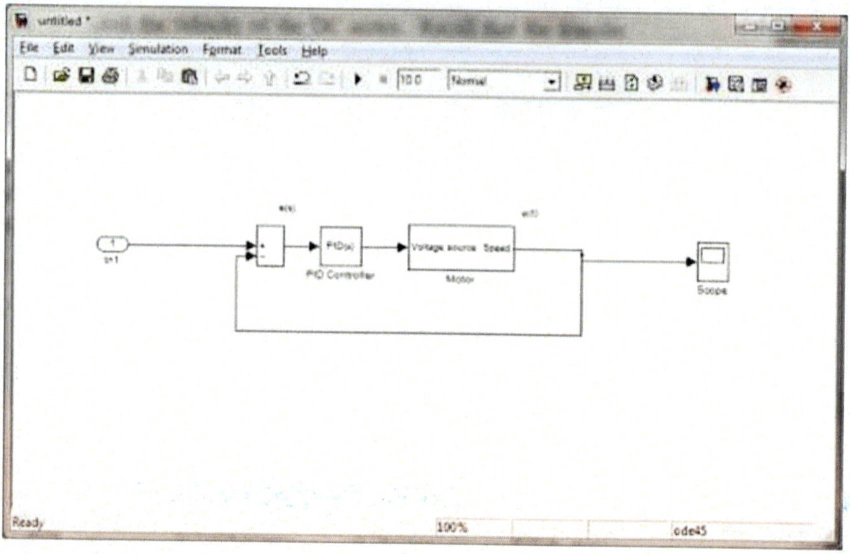

Fig. 5 Motor controlled by PID controller in Simulink

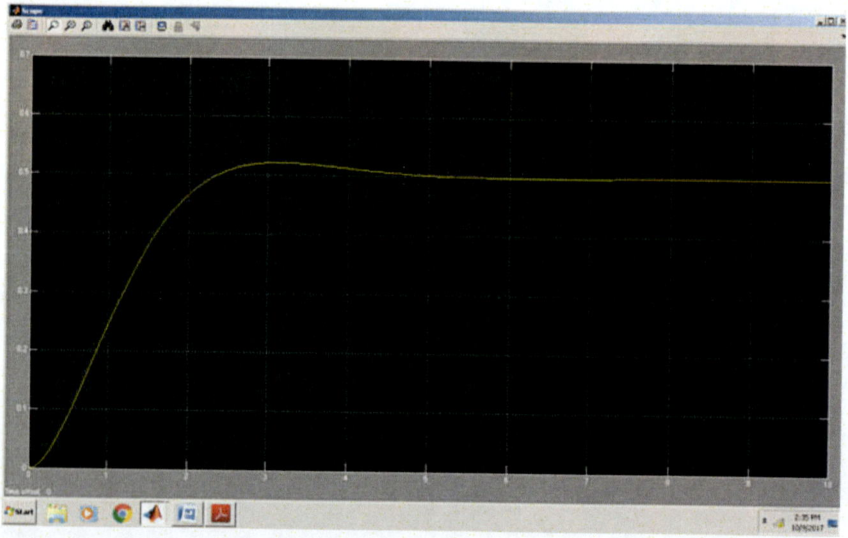

Fig. 6 Motor speed response controller by PID controller

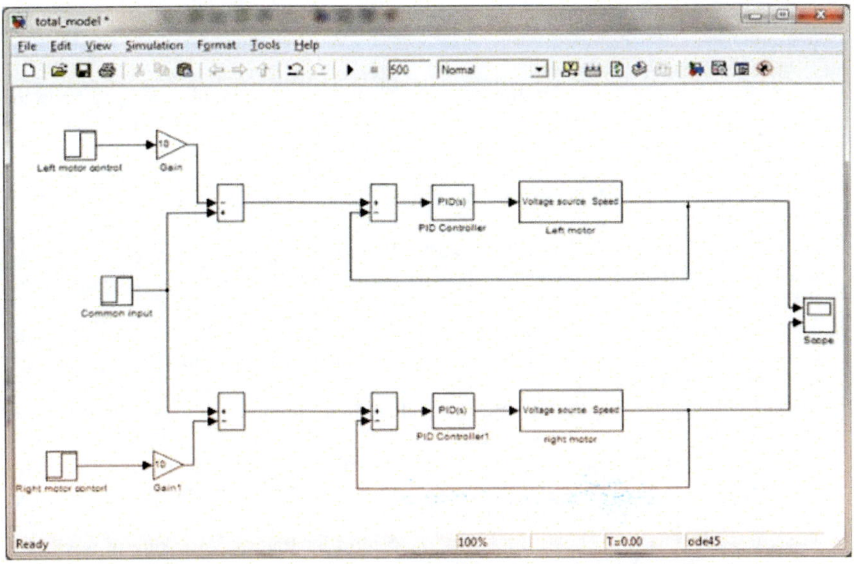

Fig. 7 PID controller applied on both the motors of the movable golem in Simulink

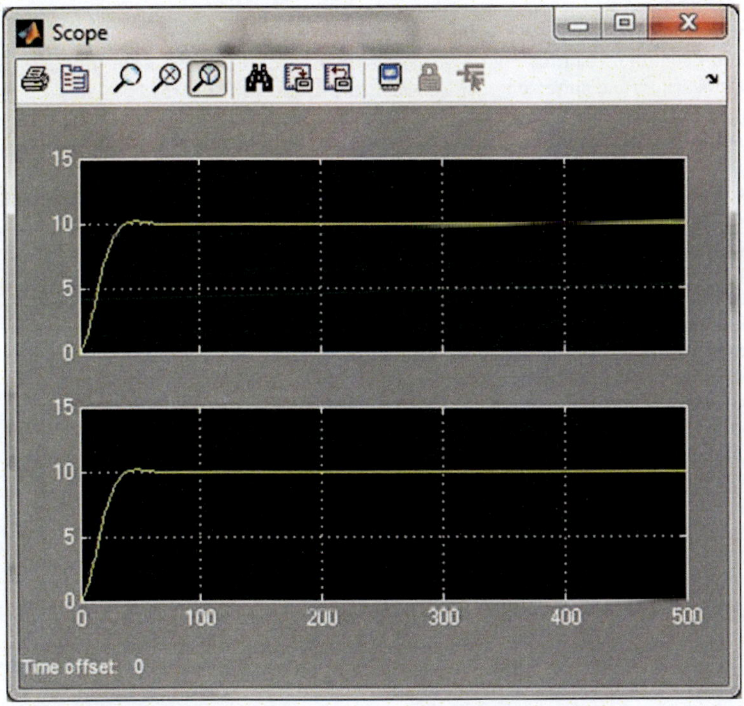

Fig. 8 Speed response of both the motors of the movable golem in Simulink

Reference

1. Koopaee, M.A.J., Majd, V.J.: An H(infinity) Switching T-S fuzzy controller design for position control of two-wheeled mobile robots. In: Proceedings of the 3rd RSI International Conference on Robotics and Mechatronics October 7–9, 2015, Tehran, Iran
2. Abdul Rahman, M.T., Ahmad, S.: Performance comparison between PD-fuzzy and PID controller towards the stability of the extendable double-link two-wheeled mobile robot. IEEE 978-1-4799-7862-5/2015
3. Xiao, H., Li, Z., Yang, C., Zhang, L., Yuan, P., Ding, L., Wang, T.: Robust stabilization of a wheeled mobile robot using model predictive control based on neurodynamics optimization. IEEE Trans Ind Electron **64**(1) (2017)
4. Mahgoub, W.M.E., Sanhoury, I.M.H.: Back stepping tracking controller for wheeled mobile robot. In: IEEE, International Conference on Communication, Control, Computing and Electronics Engineering (2017)
5. Krick, L., Broucke, M.E., Francis, B.A.: Stabilisation of infinitesimally rigid formations of multi-robot networks. Int. J. Control **82**(3), 423–439 (2009)
6. Prayudhi, L.H., Widyotriatmo, A., Hong, K.-S.: Wall following control algorithm for a car-like wheeled-mobile robot with differential-wheels drive. In: IEEE, International Conference on Control, Automation and Systems (ICCAS 2015)
7. Aissa, B.C., Fatima, C.: Adaptive neuro-fuzzy control for trajectory tracking of a wheeled mobile robot. IEEE. 978-1-4673-6673-1/2015
8. Yang, H., Fan, X., Shi, P., Hua, C.: Nonlinear control for tracking and obstacle avoidance of a wheeled mobile robot with NonholonomicConstraint. IEEE Trans. Control Syst. Technol. **24**(2) (2016)
9. Ortigoza, R.S., Sánchez, J.R.G., Guzmán, V.M.H., Sánchez, C.M., Aranda, M.M.: Trajectory tracking control for a differential drive wheeled mobile robot considering the dynamics related to the actuators and power stage. IEEE Latin AM. Trans. **14**(2) (2016)
10. Bayoume, M.O., Abd El-Geliel, M., Rezeka, S.F.: Supervisory position control for wheeled mobile robot. In: IEEE, International Conference on System Theory, Control and Computing (ICSTCC), October 13–15, 2016
11. Sorour, M., Cherubini, A., Fraisse, P., Passama, R.: Motion discontinuity-robust controller for steerable mobile robots. IEEE Robot. Autom. Lett. **2**(2) (2017)
12. Chan, R.P.M., Stol, K.A., Roger Halkyard, C.: Review of modelling and control of two-wheeled robots. Ann. Rev. Contrssol **37**, 89–103 (2013)
13. Yao, J., Jiao, Z., Ma, D.: Adaptive robust control of DC motors with extended state observer. IEEE Trans. Ind. Electron. **61**(7) (2014)
14. Dinesh Kumar, U., Nisha, M., Mathivanan, N.: Tracking of a PID driven differential drive mobile robot. Int. J. Mechatron. Electr. Comput. Technol. **8**(27), 3690–3704 (2018)

Multiperformance Optimization of Parameters in Deep Drilling of SS-321 by Taguchi-Based GRA

Mukesh Kumar, Gaurav Kumar, Om Prakash Singh, and Ankit Tomer

1 Introduction

The improved surface quality, i.e. decreased surface roughness, is the matter of concern for industries. With the improvement of equipments, machining process technology has progressed to generate high surface quality with decreased surface roughness, precise dimensions, high MRR as main features and to reduce cost of machining, it is also required to increase the MRR [1–4].

Deep drilling is mainly used machining method in automobile, surgical equipments, power plant and aircrafts industries, etc. SS-321 is often used in manufacturing for various equipments, and its machining is very complex. Earlier, an experimental study conducted by Sun and Guo [5] on Ti–6Al–4V for many milling conditions to completely characterize the surface integrity and found that the surface quality depends on cutting speed and feed. Singh et al. [6] investigated the effects of turning parameters on tool wear and surface roughness in eco-friendly turning of SS-304. Tosun [7] conducted an experiment and find optimized drilling parameter for surface roughness and burr height using Taguchi-based GRA. Shivapragash et al. [8] find optimized parameters of radial drilling using GRA for MRR and surface roughness as performance parameters.

The literature review reveals that the machining of stainless steels like SS-321 is complex due to excessive tool wear, high surface roughness and low MRR. Deep-drilling machining parameters such as cutting fluid, depth of hole, speed and feed have crucial effect on performance variables. So, the experiment was design to optimize deep-drilling parameters for multiperformance that is roughness and MRR in machining of SS-321 using optimization tools such as Taguchi-based GRA.

M. Kumar (✉) · G. Kumar · O. P. Singh
Vidya College of Engineering, Meerut, India
e-mail: mukesh.ankwan@gmail.com

A. Tomer
J.P. Institute of Technology, Meerut, India

M. Muzammil et al. (eds.), *Recent Advances in Mechanical Engineering*, Lecture Notes in Mechanical Engineering, https://doi.org/10.1007/978-981-15-8704-7_83

Table 1 Experimental design values of deep-drilling process parameters

Symbols	Process parameters	Value 1	Value 2	Value 3
A	Cutting fluid	off	on	–
B	Cutting speed (rpm)	300	400	500
C	Feed rate (mm/s)	0.03	0.04	0.05
D	Hole-depth(mm)	25	30	35

2 Experimental Design

According to the experimental design of L_{18} orthogonal array on CNC lathe machine (Manufacturer—Jyoti India Pvt. Ltd., 5500 W/4500 rpm), the experimental research was carried out by using solid carbide tool (K—series of 10 mm diameter) and cutting fluid Hocut 795-H. The workpiece in geometry (25 mm diameter and 60 mm length) was selected for deep-drilling process. Optimized combination of deep-drilling process parameters having multiperformance feature was determined by ANOVA. The description of deep-drilling process parameters with their different values is given below in Table 1.

Surface roughness of all eighteen workpieces was measured by Mitutoyo roughness tester. MRR is calculated using formulas.

$$\text{Material Removal Rate} = \frac{\pi}{4} d^2 f \tag{1}$$

where d = diameter of bore in mm and f = feed in mm/s;

The experimental values of measured performance variables are given in Table 2.

3 Taguchi Technique

3.1 S/N Ratio

The Taguchi technique is a systematic way of optimization. The ratio of mean to the standard deviation is known as S/N ratio. Depending on the conditions, S/N ratios (signal-to-noise ratios) are of three types; first one is smaller-the-better, where second larger-the-better, and the third is nominal-the-better. S/N ratios computational details are explained and well defined in [9, 10]. The S/N ratios to minimize the surface roughness were determined by smaller-the-better criteria and to maximize the MRR by larger-the-better criteria as shown in Table 2.

Table 2 Experimental values of surface roughness and MRR with their S/N ratios

Experiments	L_{18} orthogonal array				Measured values		S/N ratios	
	A	B	C	D	SR (μm)	MRR (mm^3/s)	SR	MRR
1	1	1	1	1	4.08	2.36	−12.213	7.448
2	1	1	2	2	3.98	3.14	−11.998	9.946
3	1	1	3	3	4.12	3.93	−12.298	11.885
4	1	2	1	1	3.75	2.36	−11.481	7.448
5	1	2	2	2	4.11	3.14	−12.277	9.946
6	1	2	3	3	4.19	3.93	−12.444	11.885
7	1	1	1	2	3.93	2.36	−11.888	7.448
8	1	1	2	3	3.8	3.14	−11.596	9.946
9	1	1	3	1	3.04	3.93	−9.657	11.885
10	2	2	1	3	4.07	2.36	−12.192	7.448
11	2	2	2	1	3.24	3.14	−10.211	9.946
12	2	2	3	2	3.41	3.93	−10.655	11.885
13	2	1	1	2	4.15	2.36	−12.361	7.448
14	2	1	2	3	3.68	3.14	−11.317	9.946
15	2	1	3	1	3.64	3.93	−11.222	11.885
16	2	2	1	3	3.25	2.36	−10.238	7.448
17	2	2	2	1	3.141	3.14	−9.939	9.946
18	2	2	3	2	3.09	3.93	−9.799	11.885

3.2 Pre-processing and Grey Relational Analysis

In this process, an analogous series is generated by the original series. This is obtained by normalizing the results of experiments in a range from 0 to 1. The details of normalization are available in [11].

Following relations are used for grey relational analysis.

$$\xi_i(m) = \frac{\Delta_{\min} + \xi.\Delta_{\max}}{\Delta_{0i}(m) + \xi.\Delta_{\max}} \tag{2}$$

where $\xi_i(m)$ is grey relational coefficient and $\Delta_{0i}(m)$, $z_0^*(m)$ & $z_i^*(m)$ are deviation sequence, reference sequence, and comparability sequence, respectively.

$$\Delta_{0i}(m) = \left\| z_0^*(m) - z_i^*(m) \right\| \tag{3}$$

$$\Delta_{\min} = \overset{\max}{\forall \Delta} \in i \overset{\min}{\forall m} \left\| z_0^*(m) - z_i^*(m) \right\| \tag{4}$$

$$\Delta_{\max} = \overset{\max}{\forall \Delta \in i} \ \overset{\min}{\forall m} \left\| z_0^*(m) - z_i^*(m) \right\| \tag{5}$$

Usually, the identification coefficient (ζ) is taken as 0.5. Grey relational grades (γ_i) are obtained by equations given below:

$$\gamma_i = \frac{1}{n} \sum_{k=1}^{n} \xi_i(m) \tag{6}$$

As impact of each characteristic is not similar in practical applications, so Eq. (6) is changed as:

$$\gamma_i = \frac{1}{n} \sum_{m=1}^{n} w_m.\xi_i(m) \sum_{m=1}^{n} w_m \tag{7}$$

Equations (6) and (7) are equal if the normalized weight (w_m) similar and i varies from 1 to 18.

4 Outcomes

4.1 *Optimized Parameters*

The combination of optimized process parameters of deep drilling is determined by GRA. The maximum grey relational levels of each deep-drilling parameter are A_1, B_2, C_3, and D_3 as shown in Fig. 1. Therefore, deep-drilling parameters with values $A_1B_2C_3D_3$ are an optimized combination for multiperformance.

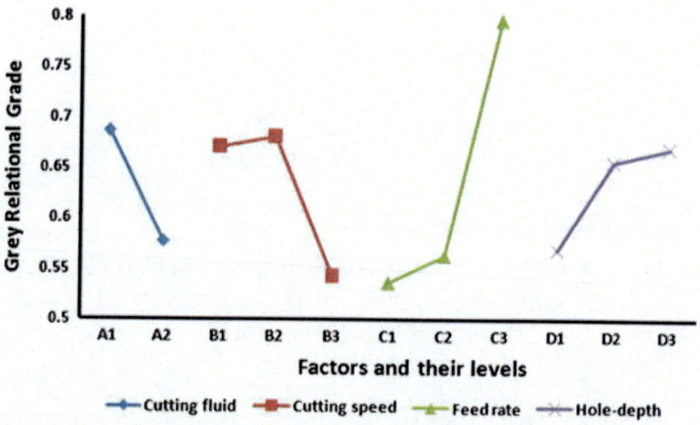

Fig. 1 Grey relational grades of drilling process parameters

Table 3 Grey relational grade responses

Symbol	Drilling parameter	Value 1	Value 2	Value 3	Max–Min
A	Cutting fluid	0.6856	0.5776		0.1080
B	Cutting speed	0.6768	0.6759	0.5422	0.1346
C	Feed rate	0.5356	0.5633	0.7959	0.2602
D	Hole-depth	0.5691	0.6563	0.6694	0.1003

Difference in maximum and minimum grey relational grade values of feed rate is more than any other process parameters, so feed rate is most influencing parameter. Similarly, remaining process parameters B, A and D are in descending order of their influences on multiperformance as shown in Table 3.

4.2 Results of ANOVA

ANOVA is used to determine the contribution of deep-drilling parameters on multi-performance factors. The variances of the grey relational grades were separated to determine the contribution of deep-drilling parameters. To find the impact of drilling parameters on multiperformance, F test was carried out (see Table 4). If F-value for a process parameter is less than four, then changes in that process parameter have insignificant effect on performance variables. Results of F-test describe that feed rate, cutting fluid and cutting speed are effective parameters with *F*-values 23.8768, 9.3007 and 6.8314, respectively, and role of hole-depth is insignificant with *F*-value (3.5268) which is less than 4.

The ANOVA results state that the most significant deep-drilling parameter feed rate is with 54.41% contribution, followed by cutting speed (15.57%), cutting fluid (10.60%) and hole-depth (8.04%) (see Fig. 2).

Table 4 Result of ANOVA

Symbol	Deep-drilling parameters	DOF	Sum of squares	Mean square	*F*-value	Contribution (%)
A	Cutting fluid	1	0.00494	0.0494	9.3007	10.60
B	Cutting speed	2	0.0726	0.0363	6.8314	15.57
C	Feed rate	2	0.2537	0.1268	23.8768	54.41
D	Hole-depth	2	0.0375	0.0187	3.5268	8.04
Error		10	0.0531	0.0053		11.39
Total		17	0.4663			100

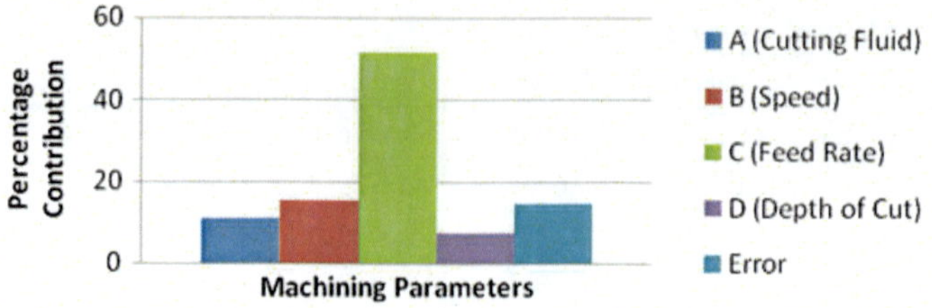

Fig. 2 Role of machining parameters

Table 5 Optimized deep-drilling parameters

	Predicted values	Experimental values	% improvement
Values	$A_1B_2C_3D_3$	$A_1B_2C_3D_3$	
Surface roughness		1.0000	
Material removal rate		1.0000	
Grey relational grade	0.9473	1.0000	5.56

4.3 Confirmation Experiments

There is an increment of 5.56% in experimental grade relative to predicted grades of grey relation, and they are satisfactory as shown in Table 5.

5 Conclusions

The finest combination of deep-drilling process parameters is $A_1B_2C_3D_3$, that is, with cutting fluid off, cutting speed 400 r.p.m., feed rate 0.05 mm/s and hole-depth 35 mm optimum multiperformance characteristics are obtained. The contribution percentages of feed rate, cutting speed, cutting fluid and hole-depth determined by ANOVA are 54.41, 15.57, 10.60 and 8.04%, respectively. Using this method, an improvement of 5.56 is obtained. As a result, Taguchi is a very useful technique for optimizing process parameters to minimize costs and time.

Acknowledgements The authors of this paper are deeply grateful to Vidya College of Engineering, Meerut for the inspiration and support during this research.

References

1. Munoz De Escalona, P., Maropoulos, P.G.: A geometrical model for surface roughness prediction when faces milling Al 7075-T7351 with square insert tools. J. Manuf. Syst. **36**, 216–223 (2015)
2. Azuddin, M., Abdullah, W.: A study on surface roughness and burr formation of Al6061 with different spindle speed and federate for small end milling cutter. Int. J. Integr. Eng. **1**(1), (2009)
3. Atul, S.C.: Investigation on chromizing of C45 steel using response surface methodology. Int. J. Chem. Tech. Res. **9**(1), 82–91 (2016)
4. Singh, O.P., Kumar, G., Kumar, M.: Multi Performance optimization of shoulder milling process parameters of AA6063 T6 aluminium alloy by Taguchi based GRA. Int. J. Innov. Technol. Explor. Eng. **8**(10S), 420–425 (2019a)
5. Sun, J., Guo, Y.B.: A comprehensive experimental study on surface integrity by end milling Ti–6Al–4V. J. Mater. Process. Technol. **41**(12), 1705–1719 (2009)
6. Singh, T., Dureja, J.S., Dogra, M., Bhatti, M.S.: Multi-response optimization in environment friendly turning of AISI 304 austenitic stainless steel. Multidiscipline Model. Mater. Struct. **15**(3), 538–558 (2019)
7. Tosun, N.: Determination of optimum parameters for multi-performance characteristics in drilling by using grey relational analysis. Int. J. Adv. Manuf. Technol. **28**, 450–455 (2006)
8. Shivapragash, B., Chandrasekaran, K., Parthasarathy, C., Samuel, M.: Multiple response optimizations in drilling using Taguchi and grey relational analysis. Int. J. Mod. Eng. Res. **3**(2), 765–768 (2013)
9. Siddiquee, A.N., Khan, Z.A., Goel, P., Kumar, M., Agarwal, G., Khan, N.Z.: Optimization of deep drilling process parameters of AISI 321 steel using Taguchi method. Procedia Mater. Sci. **6**, 1217–1225 (2014)
10. Singh, O.P., Kumar, G., Kumar, M.: Role of Taguchi and grey relational method in optimization of machining parameters of different materials: a review. Acta Electronica Malaysia (AEM) **3**(1), 19–22 (2019b)
11. Khan, Z.A., Kamaruddin, S., Siddiquee, A.N.: Feasibility study of use of recycled High-Density Polyethylene and multi response optimization of injection moulding parameters using combined grey relational and principal component analyses. Mater. Des. **31**(6), 2925–2931 (2010)

Optimization of End Milling Machining Parameters of SS 304 by Taguchi Technique

Gaurav Kumar, Mukesh Kumar, and Ankit Tomer

1 Introduction

Milling is mainly used machining method in automobile, surgical equipments, power plant, mining and aircrafts industries, etc. A large number of the components are manufactured from machined SS-304. The main concerns for industries are to decrease surface roughness to improve surface quality and also increase in material removal rate to reduce the cost to compete with flexible requirement of market [1–4]. By the regular improvement of tools and equipments, a huge number of processing techniques have developed for generating high surface quality, surface roughness, precise size, MRR as the main features and so on [5].Earlier, Hayajneh et al. [6] performed a set of end milled experiments on aluminum pieces for finding the effects depth of cut, feed and speed on surface roughness from which feed is the most important factor. Maiyar et al. [7] performed end milling experiments on Inconel 718 alloy using L9 orthogonal array for multi-objective optimization.

After reviewing the literature, it was found that the poor surface quality, wearing of tools and material removal rate are the main issues in manufacturing which creates complexity in machining of SS-304, and main process parameters affecting surface quality and material removal rate are depth of cut, spindle speed and feed. So, keeping in mind above discussed complexity in milling of SS-304 problem was designed for improving surface roughness. The experiments were performed for optimization of end milling machining parameters of SS-304 by Taguchi technique.

G. Kumar (✉)
M. M. Deemed to be University, Mullana, Ambala, India
e-mail: gaurav.me86@gmail.com

M. Kumar
Vidya College of Engineering, Meerut, India

A. Tomer
J.P. Institute of Technology, Meerut, India

M. Muzammil et al. (eds.), *Recent Advances in Mechanical Engineering*, Lecture Notes in Mechanical Engineering, https://doi.org/10.1007/978-981-15-8704-7_84

Fig. 1 SS-304 pieces

2 Detail of Work Material and Equipments

2.1 Work Material

Work material for experiment was SS-304, and it was divided into eighteen parts of geometry 80 mm × 50 mm × 10 mm as shown in Fig. 1.

2.2 Milling Machine

The end milling operations were performed on the work material by solid carbide end milling tool(CR 2.5 Ø12, M—series) fitted on vertical milling center machine (BMV 60 TC 24) (see Fig. 2) with coolant (Blastocut-500), and surface roughness was measured by Mitotoyo roughness tester as shown in Fig. 3.

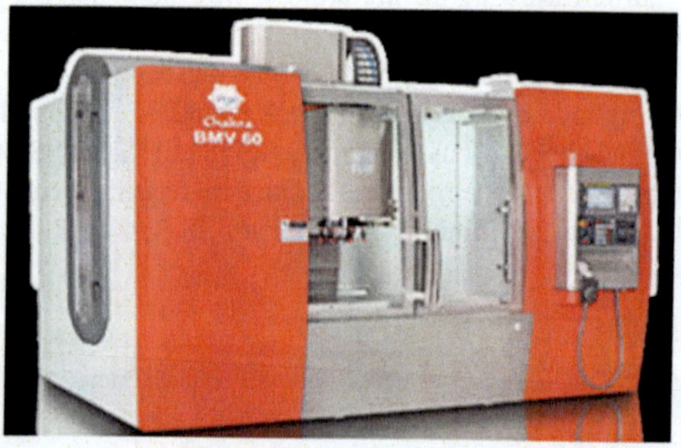

Fig. 2 Vertical milling center machine

Fig. 3 Tester for measuring roughness

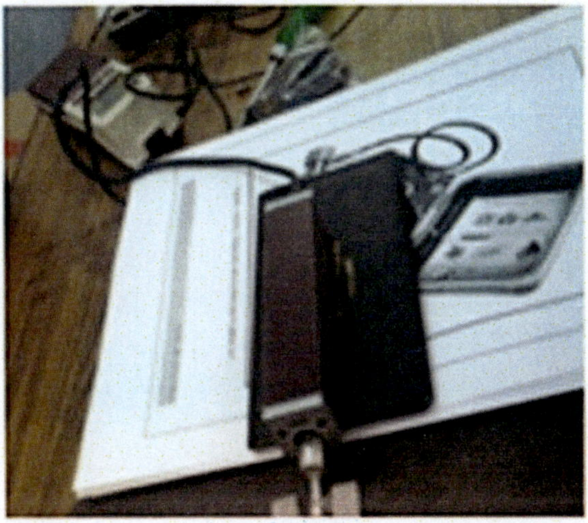

Table 1 Selection of end milling process parameters

S. No.	Parameters	Value 1	Value 2	Value 3
1	Coolant	OFF	ON	–
2	Feed (mm/min)	1500	2000	2500
3	DOC (mm)	0.2	0.3	0.4
4	Speed (rpm)	1500	2500	3500

2.3 *Selection of End Milling Process Parameters*

Coolant, feed, depth of cut and speed were used as end milling parameters for finding the effect on surface roughness. L_{18} orthogonal array was designed to conduct the experiments with following end milling parameters given in Table 1.

3 Outcomes

3.1 *Analysis of S/N (Signal-to-Noise) Ratio*

In this experiment, smallest-is-better criterion was used for minimizing the surface roughness.

Smallest-is-better:

$$S/N_{\mathrm{dB}} = -10\log_{10}\left[\frac{1}{n}\sum_{x=1}^{n} R_x^2\right] \tag{1}$$

where R_x shows the surface roughness value of xth experiment, and N is the total experiments. Formulae for calculation of S/N ratio are also represented in references [8, 9]. Signal-to-noise ratio values of 18 experiments are shown in Table 2. Equation 2 was used for computing the mean value (Y) of signal-to-noise ratios (Table 3).

$$\Upsilon = \frac{1}{N}\sum_{j=1}^{N}\Upsilon_j \tag{2}$$

Table 2 S/N ratio values

Exp. No.	Coolant	Feed rate	Doc	Cutting speed	Ra (μm)	S/N ratio Y_j (dB)
1	Off	1500	0.2	1500	1.91	−5.62
2	Off	1500	0.3	2500	1.45	−3.23
3	Off	1500	0.4	3500	1.75	−4.86
4	Off	2000	0.2	1500	2.35	−7.42
5	Off	2000	0.3	2500	2.3	−7.23
6	Off	2000	0.4	3500	2.18	−6.77
7	Off	2500	0.2	2500	1.96	−5.85
8	Off	2500	0.3	3500	1.79	−5.06
9	Off	2500	0.4	1500	2.2	−6.85
10	On	1500	0.2	3500	1.35	−2.61
11	On	1500	0.3	1500	1.29	−2.21
12	On	1500	0.4	2500	1.14	−1.14
13	On	2000	0.2	2500	1.41	−2.98
14	On	2000	0.3	3500	1.29	−2.21
15	On	2000	0.4	1500	1.99	−5.98
16	On	2500	0.2	3500	1.7	−4.61
17	On	2500	0.3	1500	1.98	−5.93
18	On	2500	0.4	2500	1.55	−3.81
				Avg.	1.76	−84.36

Table 3 Response of average S/N ratio

Level	Coolant	Feed	Doc	Speed
1	−5.88	−3.28	−4.85	−5.67
2	−3.50	−5.43	−4.31	−4.04
3		−5.35	−4.90	−4.35
Delta	2.38	2.16	0.59	1.63
Rank	1	2	4	3

Fig. 4 Response graph

where $j = 1, 2, \ldots, N$ (and $N = 18$) and Y_j is signal-to-noise ratio [10].

The maximum value of signal-to-noise ratio of each end milling parameter was $A_2B_1C_2D_2$with mean value—4.69 as shown in Fig. 4 that means coolant is on; feed of 1500 mm/min, doc 0.3 mm and cutting speed of 2500 rpm.

3.2 *Analysis of Variance (ANOVA)*

ANOVA is used to find out the contribution of end milling process parameters on surface roughness. To find the impact of end milling parameters on multi-performance, F test was carried out. If F- value is greater than four, then variations in parameters play an effective role on surface roughness [11].

The results of ANOVA are shown in Table 4 that coolant is a parameter with the highest percentage contribution (40.05%); whereas, feed, speed and depth of cut are with their percentage contribution of 28.16%, 14.12% and 2%, respectively.

Table 4 Results of the ANOVA

Source	Dof	SS	MS	F-value	Contribution (%)
Coolant	1	25.46	25.46	25.54	40.05
Feed	2	17.90	8.95	8.98	28.16
Doc	2	1.27	0.63	0.64	2.00
Speed	2	8.97	4.49	4.50	14.12
Error	10	9.97	1.00		15.68
Total	17	63.56			100.00

Table 5 Optimum conditions of parameter

	Prediction	Experiment	% change
	$A_2B_1C_2D_2$	$A_2B_1C_2D_2$	
Surface roughness	1.131	1.14	0.79
S/N ratio (db)	−1.07	−1.14	6.28

3.3 *Result Confirmation*

Confirmation test shows the average surface roughness (prediction and testing) values.

From Table 5 shows that S/N ratio decreased 6.28% due to increase in surface roughness of 0.79% which justifies optimal combinations ($A_2B_1C_2D_2$) are better for improving the surface roughness of end milled products.

4 Conclusions and Future Scope

This work presents an experimental analysis in which end milling operations were performed on SS-304. Taguchi is a very useful technique for optimizing industry process parameters to minimize costs and time. The effect of four parameters, namely coolant, feed, depth of cut and speed on the integrity of the machined surface were found. $A_2B_1C_2D_2$ (i.e., coolant—On, feed—1500 mm/min, doc—0.3 mm and cutting speed—2500 rpm) is the optimal combination of the current end milling process. The determination of the optimum cutting parameters is verified by the validation experiment. Coolant is the most important input parameter affecting the performance characteristics due to its highest percentage contribution (40.05%) among the process parameters followed by feed (28.16%), speed (14.12%) and depth of cut (2%). Speed (rpm) has little effect but depth of cut has no statistically effect on the performance characteristics. Therefore, the surface roughness can be improved by using Taguchi technique. In future, other techniques such as principal component analysis, fuzzy logic, etc., can be used to determine the grades/index of multi-performance characteristics.

References

1. Satish Kumar, P., Dharmalingam, S., Raja, K., Lingadurai, K., Padmanaban, G.: Investigation on electrochemical micro machining of Al 6061–6% wt Gr based on Taguchi design of experiments. Int. J. Chem. Tech. Res. **7**(1), 203–211 (2015)
2. Munoz De Escalona, P., Maropoulos, P.G.: A geometrical model for surface roughness prediction when faces milling Al 7075-T7351 with square insert tools. J. Manuf. Syst. **36**, 216–223 (2015)

3. Atul, S.C.: Investigation on chromizing of C45 steel using response surface methodology. Int. J. Chem. Tech. Res. **9**(1), 82–91 (2016)
4. Singh, O.P., Kumar, G., Kumar, M.: Role of Taguchi and grey relational method in optimization of machining parameters of different materials: a review. Acta Electronica Malaysia (AEM) **3**(1), 19–22 (2019a)
5. Singh, O.P., Kumar, G., Kumar, M.: Multi Performance optimization of shoulder milling process parameters of AA6063 T6 aluminium alloy by Taguchi Based GRA. Int. J. Innov. Technol. Explor. Eng. **8**(10S), 420–425 (2019b)
6. Hayajneh, M.T., Tahat, M.S., Bluhm, J.: A study of the effects of machining parameters on the surface roughness in the end-milling process. Jordan J. Mech. Ind. Eng. **1**(1), 1–5 (2007)
7. Maiyar, L., Ramanujan, R., Venkatesan, K., Jerald, J.: Optimization of Machining parameters for end milling of Inconel 718 super alloy using Taguchi based grey relational analysis. Int. Conf. Des. Manuf. **64**, 1276–1282 (2013)
8. Kuram, E., Ozcelik, B.: Optimization of machining parameters during micro-milling of Ti6Al4V titanium alloy and Inconel 718 materials using Taguchi method. Proc. Inst. Mech. Eng. Part B J. Eng. Manuf. **231**(2), 228–242 (2017)
9. Sosa, P.B., Makwana, R.D., Acharya, G.D.: Optimization of machining parameters on end milling of EN 8 back shaft for power press. Trends Mech. Eng. Technol. **8**(3), 49–58 (2018)
10. Chang, C.L., Tsai, C.H., Chen, L.: Applying grey relational analysis to the decathlon evaluation model. Int. J. Comput. Internet Manage. **11**(3), 54–62 (2003)
11. Siddiquee, A.N., Khan, Z.A., Goel, P., Kumar, M., Agarwal, G., Khan, N.Z.: Optimization of deep drilling process parameters of AISI 321 steel using Taguchi method. Procedia Mater. Sci. **6**, 1217–1225 (2014)

Evaluation of Mechanical Properties and Tribological Behaviour of Self-lubricating Aluminium Matrix Hybrid Composites

B. K. Badhan, H. Nautiyal, R. Moharana, U. S. Rao, R. K. Gautam, and R. Tyagi

1 Introduction

Aluminium matrix composites (AMCs) offer low weight, high specific strength/stiffness, corrosion and creep resistance with superior mechanical properties and are attractive materials for military, aerospace, electronics and automobile industries [1–3]. Al_2O_3 as reinforcement offers advantage of easy availability, low cost and better wear resistance properties [2]. Size of reinforcing particles influences the final properties of the composite. Researchers have found that with the addition of micron-sized reinforcement powders into a metal matrix, although mechanical properties like ultimate tensile strength, yield strength, hardness and wear performance are increased, but there is a considerable decrease in ductility of composites with high particle concentrations [4]. Metal matrix nano-composites (MMNC's) overcome these shortcomings and offer good mechanical properties, high fracture toughness and machinability while maintaining the ductility [5]. Properties are significantly enhanced with the addition of small quantity of nano-particles.

Addition of hard particles in metals leads to higher coefficient of friction, and after certain load, their abrasiveness is also high at contact [6]. Efforts were also put to overcome friction and wear in automobiles due to engine transmission, tyre and brakes [7]. Most of the conventional lubricants are petroleum-based which do not work under extreme temperatures and pressures and are usually toxic, non-biodegradable

B. K. Badhan
Saint Gobain Research India, Chennai 600113, India

H. Nautiyal · U. S. Rao (✉) · R. K. Gautam · R. Tyagi
Department of Mechanical Engineering, Indian Institute of Technology (BHU), Varanasi, UP 221005, India
e-mail: sruppu.mec@iitbhu.ac.in

R. Moharana
Department of Mechanical Engineering, Indian Institute of Technology Delhi, New Delhi 110016, India

M. Muzammil et al. (eds.), *Recent Advances in Mechanical Engineering*, Lecture Notes in Mechanical Engineering, https://doi.org/10.1007/978-981-15-8704-7_85

and cause damage to the environment. Therefore, researchers have put great attention to reduce friction and wear with the help of development of new materials and self-lubricating materials. Materials having very low shearing strength such as WS_2, MoS_2 and graphite are added to reduce the friction [6]. In self-lubricating metal matrix composites, one of the reinforcements is a solid lubricating material such as WS_2, MoS_2, graphite or hexagonal boron nitride. Quality of these composites depends upon properties of matrix and lubricant, the interaction of matrix and lubricant, distribution and concentration of lubricant. Anitha et al. [8] investigated dry sliding wear behaviour of Al7075/Al_2O_3/Gr hybrid composite and found that even a small amount of Al_2O_3 and Gr have significant effect on wear and coefficient of friction (CoF) where wear is mainly affected by Al_2O_3 and Gr content mainly affecting CoF.

Stir casting route is one of the important and cheapest routes to synthesize the aluminium matrix composites. Melt processing technique offers the advantages of low cost, simplicity, wide selection of materials and easy control of matrix structure [9].

In the present study, aluminium matrix is reinforced with nano-alumina and MoS_2/WS_2 which are added as a solid lubricant to synthesize self-lubricating metal matrix hybrid composites. Four different composites are made to study the effect of addition of solid lubricant on mechanical and tribological properties of the developed material.

2 Experimental Details

2.1 *Materials and Synthesis of Composites*

In the present study, different materials used for synthesis of composites and the designations assigned to them are as mentioned in Table 1.

Composites have been synthesized using stir casting route in the present investigation. Calculated amount of reinforcement powders was preheated in a microwave

Table 1 Material used for synthesis of composites

S. No.	Material	Designation	Pure Al (wt%)	Al_2O_3 (wt%)	MoS_2 (wt%)	WS_2 (wt%)
1	Pure Al	C1	100	–	–	–
2	Al/Al_2O_3	C2	98	2	–	–
3	Al/Al_2O_3/MoS_2	C3	96	2	2	–
4	Al/Al_2O_3/WS_2	C4	96	2	–	2
5	Al/Al_2O_3/MoS_2/WS_2	C5	96	2	1	1

oven up to 110 °C for 1 h for moisture removal. For synthesizing the composites by stir casting, preheated reinforcement powders were injected into molten aluminium and stirred continuously for proper mixing.

2.2 Mechanical and Wear Testing

The samples fabricated with different compositions as mentioned in Table 1 were investigated by different techniques. Micrographs were taken under scanning electron microscope at 500 × and 1500 × , respectively. Theoretical and actual density measurements were carried out by rule of mixture and using Archimedes principle, respectively. Hardness of the composites was measured by using Brinell hardness testing machine (make: Mechatronics India Pvt. Ltd). Tensile specimens were made from all the composites with ASTM E8 standards having gauge diameter of 4.5 mm and gauge length of 15.5 mm, and tensile tests were performed on Instron tensile testing machine having a maximum capacity of 25 kN with an integrated computer system which plots load vs extension curves. XRD of the samples were carried out with CuK_{α} radiation ($\lambda = 1.5402$ Å). Dry sliding wear tests were performed on pin-on-disc rotary tribometer (make: DUCOM). Pin samples of 8 mm diameter and 35 mm length were made from all the composites and were tested against EN31 bearing steel as counterface material. Tests were carried out with different loads of 10, 20, 30 and 40 N at constant sliding speed of 0.5 m/s for total 6000 revolutions and average values of COF are reported. Worn surfaces were cut from the pin samples and examined under scanning electron microscope for possible wear mechanism.

3 Results and Discussion

3.1 Metallographic Analysis

The distribution of particles in the matrix has been studied through scanning electron microscope (SEM) at 500 × and 1500x. It has been observed that composite C2 and C3 show the uniformly distributed particles. However, small clusters were observed in composites C4 and C5 (Fig. 1), respectively, which may be due to intermixing of more number of particles into the matrix.

3.2 XRD Analysis

Figure 2 shows the X-ray diffraction pattern of composites. From the XRD results, it can be seen that major peaks observed are related to aluminium and very few minor

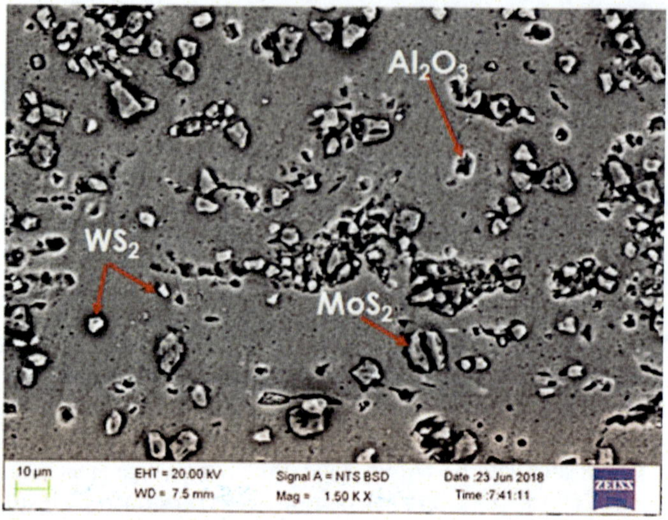

Fig. 1 SEM micrographs of Al/Al2O3/MoS2/WS2 (C5) composites

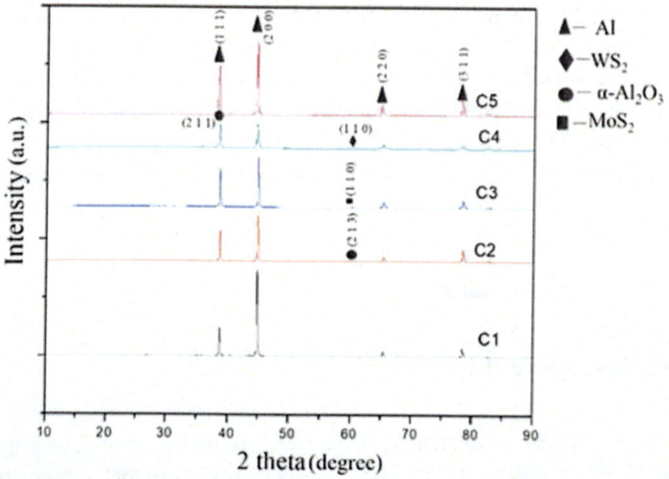

Fig. 2 XRD pattern of all the composites

peaks are observed for MoS_2 and WS_2 since the weight percentage of these is below 5%. However, it can be confirmed that all elements are present in the composites.

Table 2 Density and hardness of the composites

Material	Theoretical density ρ_t (g/cm^3)	Actual density (ρ_t) (g/cm^3)	% porosity	BHN
Pure Al (C1)	2.7000	2.6660	1.2592	16
Al/Al_2O_3 (C2)	2.7172	2.6831	1.2549	24
Al/Al_2O_3/MoS_2 (C3)	2.7430	2.7241	0.6890	22
Al/Al_2O_3/WS_2 (C4)	2.7527	2.7105	1.0389	21
Al/Al_2O_3/MoS_2/WS_2 (C5)	2.7478	2.7019	1.6704	19

3.3 Density and Hardness

Increase in density is observed with addition of reinforcement which can be attributed to filling of pores inside the aluminium matrix (Table 2). C3 composite shows maximum density among all the composites due to uniform distribution of particles but it further decreased in C4 and C5 composites due to particle clustering. Brinell hardness has shown the same trend as that of density. Hardness of the matrix phase increases with the addition of external hard and brittle phase of ceramic particles. Further, with the addition of solid lubricants, the hardness of hybrid composites decreases due to the softer nature of MoS_2 and WS_2 (Table 2).

3.4 Tensile Testing

The maximum value of tensile strength was obtained in Al/Al_2O_3/MoS_2 hybrid composite which may be attributed to the uniform dispersion of particles and less porosity (Table 3). Al/Al_2O_3 shows a lower value of UTS because of agglomeration of hard phase particles which act as notches and thus brittleness increases, and hence, percentage elongation decreases.

Table 3 Tensile properties of composites

S. No.	Material	UTS (MPa)	YS (MPa)	% elongation
1	Pure Al	57.98	27.10	41.70
2	Al/Al_2O_3	58.36	24.40	29.47
3	Al/Al_2O_3/MoS_2	64.35	32.82	37.57
4	Al/Al_2O_3/WS_2	60.31	29.83	40.52
5	Al/Al_2O_3/MoS_2/WS_2	59.22	23.86	42.99

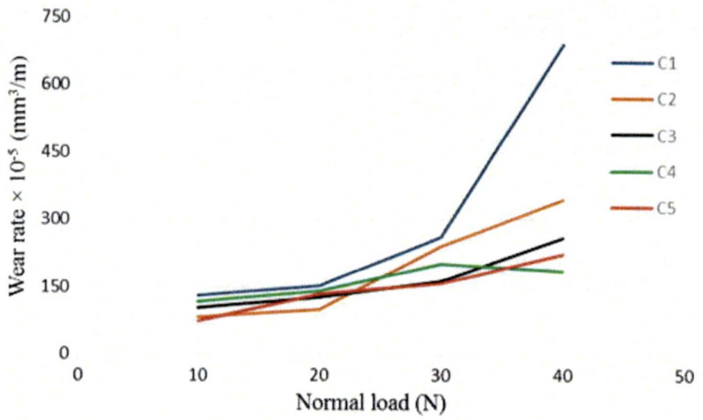

Fig. 3 Variation of wear rate with normal load for all composites

3.5 Tribological Behaviour of Composites

Wear rate of all the composites was calculated at different loads. Variation of wear rate with normal load for all the composites is shown in Fig. 3.

It has been observed that wear rate increases with increasing load for most of the composites following Archard's law of wear. At higher load particularly at 40 N load, wear rate of all materials significantly increases except C4 composite, due to strong adhesion with counterface material, resulting in a large amount of material loss. However, wear rates of composites with solid lubricants are lower as compared to pure Al and Al/Al_2O_3 composites which is due to the presence of MoS_2/WS_2 which reduces friction at contacting surfaces and thus allowing less material removal.

The fluctuation of coefficient of friction in the present study is observed in the range 0.4–0.6. Further addition of hard phase particles and solid lubricants significantly reduced the COF values. This may be attributed to the presence of soft phase solid lubricant particles which are present in the composite materials and give lubricating action, thereby reducing the friction at the interface. Among all the composites, Al/Al_2O_3/MoS_2 (C3) shows the lowest values of the coefficient of friction.

SEM micrographs of worn surfaces have been studied for all the composites fabricated at 10 and 40 N. It is observed that at 10 N load abrasion accompanied by delamination occurs with severe oxidation and at 40 N load severe delamination has been observed (Fig. 4).

4 Conclusions

From the above results and discussion, the following conclusions may be drawn:

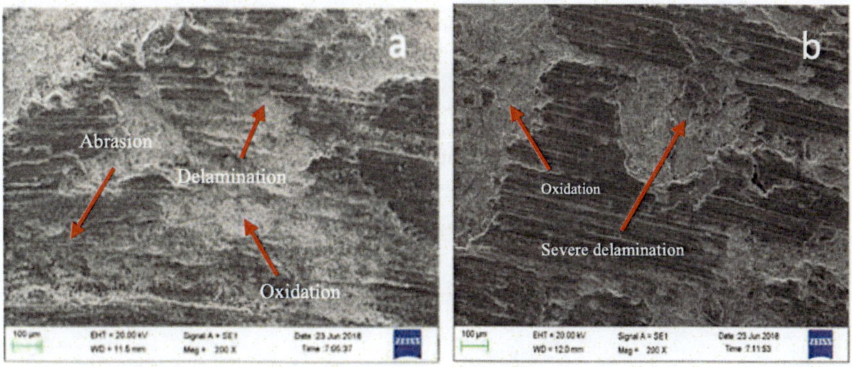

Fig. 4 SEM micrographs of worn surfaces of composite C5 at **a** 10 N, **b** 40 N

- Al/Al_2O_3, Al/Al_2O_3/MoS_2, Al/Al_2O_3/WS_2, and Al/Al_2O_3/MoS_2/WS_2 hybrid composites were successfully prepared by stir casting.
- All the composites exhibit higher density than pure Al, and Al/Al_2O_3/MoS_2 (C3) composite has the highest density among all composites.
- All the composites exhibit higher hardness than pure Al. However, decrease of hardness is observed with the addition of soft phase particles of MoS_2 and WS_2.
- Al/Al_2O_3/MoS_2 (C3) composite has higher tensile strength than other developed composites.
- Wear resistance of the composites is higher as compared to pure aluminium.
- Hybrid composites have lower coefficient of friction due to the presence of MoS_2/WS_2 particles.

References

1. Singh, M., Rana, R.S., Purohit, R., Sahu, K.: Development and analysis of Al-Matrix nano composites fabricated by ultrasonic assisted squeeze casting process. Mater. Today Proc. **2**, 3697–3703 (2015)
2. Murthy, N.V., Prasad Reddy, A., Selvaraj, N., Rao, C.S.P.: Dispersion of alumina nano particles in Al 2219 alloy by ultrasonic assisted stir casting technique. Mater. Today Proc. **4**, 10113–10117 (2017)
3. Hai, Su., Gao, W., Feng, Z., Zheng, Lu.: Processing, microstructure and tensile properties of nano-sized Al2O3 particle reinforced aluminum matrix composites. Mater. Des. **36**, 590–596 (2012)
4. Mazahery, H., Abdizadeh, H.R.: Baharvandi, "Development of high-performance A356/nano-Al2O3 composites." Mater. Sci. Eng. A **518**, 61–64 (2009)
5. Raturi, A., Mer, K.K.S., Pant, P.K.: Synthesis and characterization of mechanical, tribological and micro structural behavior of Al 7075 matrix reinforced with nano Al_2O_3 particles. Mater. Today Proc. **4**, 2645–2658 (2017)
6. Niste, V.B., Ratoi, M., Tanaka, H., Xu, F., Zhu, Y., Sugimura, J.: Self-lubricating Al-WS2 composites for efficient and greener tribological parts. Sci. Rep. **7**, 14665

7. Rengifo, S., Zhang, C., Harimkar, S., Boesl, B., Agarwal, A.: Effect of WS_2 addition on tribological behavior of aluminum at room and elevated temperatures. Tribol. Lett. **65**, 76 (2017)
8. Anitha, P., Balraj, U.S.: Dry sliding wear performance of Al/7075/Al2O3p/Grp hybrid metal matrix composites. Mater. Today Proc. **4**, 3033–3042 (2017)
9. Karbalaei Akbari, M., Baharvandi, H.R., Mirzaee, O.: Fabrication of nano-sized Al_2O_3 reinforced casting aluminum composite focusing on preparation process of reinforcement powders and evaluation of its properties. Composites Part B **55**, 426–432 (2013)

Geometry Optimization of Magneto-Rheological Damper Based on Magnetic Saturation

Bhupesh Ojha

1 Introduction

Magneto-rheological fluids are materials that exhibit a change in rheological properties with the application of a magnetic field. The MR effect can be applied to a hydraulic damper for better control of vibration [1, 2]. In technical and laboratory context, the most common configuration used with magneto-rheological fluids is based on a modification of the typical hydraulic actuator structure (Fig. 1). An outer cylinder is filled up with MR fluid, and an inner piston moves inside. Between the two main components, a constant annular gap is ensured, in which MR fluid is free to flow during the piston displacements in the device chamber. The piston, which is one of the two paramagnetic elements, has a central groove along its lateral surface, where one or more electric coils of copper are housed. The cylinder corresponds to the second pole of the field. When current is supplied to the coil, some areas of the fluid in the gap piston–cylinder are "activated" (i.e., material is polarized), and the formation of particle chains starts. Increasing the intensity of the magnetic field, the number and strength of the chains of particles between the two poles of the circuit increase. The flow resistance can be adjusted, starting from the total absence of magnetic field to the state of magnetic saturation of the fluid, where a complete alignment of ferromagnetic particles takes place and a further increase in current does not produce, at constant speed, greater damping force. In this solution, an external oil reservoir must to be used in order to collect the fluid volume moved by the entrance of the piston rod in the cylinder chamber [3].

B. Ojha (✉)
Mechanical Engineering Department, I.T.S Engineering College, Gautam Budh Nagar, Greater Noida 201308, India
e-mail: bhupeshojha@its.edu.in

M. Muzammil et al. (eds.), *Recent Advances in Mechanical Engineering*, Lecture Notes in Mechanical Engineering, https://doi.org/10.1007/978-981-15-8704-7_86

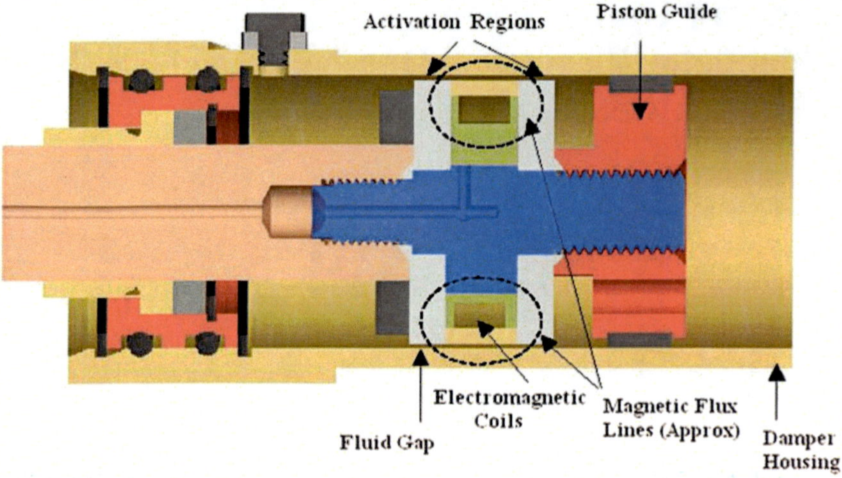

Fig. 1 MR damper: main components [3]

2 MR Damper Design

2.1 Input Data with Design Conditions

The specifications are used for input data, and the limiting conditions are used for design solutions.

The specifications are as follows:

a. controllable level of mechanical power: $W_m = 80$ W
b. maximum velocity of MR damper: $v_0 = 0.03$ m/s.

The design conditions are as follows:

a. internal diameter of damper cylinder: $d_{\text{in}} = 0.04$ m
b. piston rod diameter: $d_{\text{rod}} = 0.02$ m
c. length of the piston: $l = 0.036$ m.

2.2 Selection Criterion of MR Fluid

To choose the most effective MR fluid, the normal characteristics of different MR fluid must be studied. The normal characteristics of MR fluid indicate by the figure of merit. The MR fluid provides the maximum force when it generates the maximum yield stress [4]. This condition is achieved when the MR fluid is magnetically saturated. In this study, the Lord Corporation 132 DG MR fluid is chosen. For shear strain rate, $\gamma = 140\ \text{s}^{-1}$ and magnetic field, $H = 250$ kA/m, the following values can be obtained:

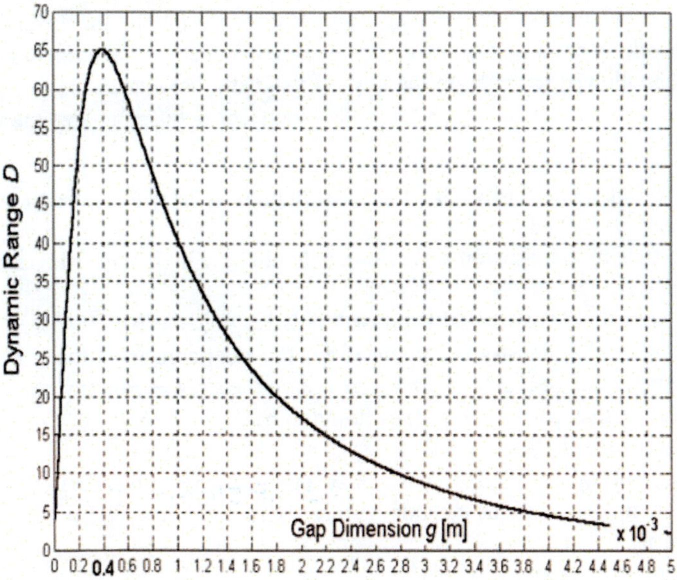

Fig. 2 Variation of dynamic range (D) with gap dimension (g) [5]

- The magnetic flux $B = 0.83$ T.
- The yield stress $\tau_y = 44.1$ kPa.
- The off state plastic viscosity $\eta = 0.25$ Pa s.

2.3 *Hydraulic Design*

The MR damper will be more effective when the controllable range of force should be as high as possible. The controllable force will increase when the gap is small but for small gap the viscous force increases and dynamic range decreases. For large gap size, both the viscous force and controllable force decrease [5]. Hence, there exists an optimal gap size which maximizes the dynamic range D. From Fig. 2, $g = 0.4 \times 10^{-3}$ m for $D_{\max} = 65$. Now the hydraulic circuit parameters are calculated on the basis of above gap size.

2.4 *Magnetic Circuit Design*

Parallel to the working mode of MR fluid, the realization of a suitable magnetic circuit in the device is fundamental, because the final load damping performances depend on fluid-activated shear stress τ_y that depends on magnetic field produced B.

The main objective of circuit design, once chosen the maximum current intensity i for the working condition of the device, was to calculate the necessary number of coils N that allowed to generate the desired magnetic field, according to the limits on the geometry and on safety [6, 7]. Following steps are used to size the magnetic circuit:

(1) Operating point of MR fluid: magnetic field, $H = 250$ kA/m, the magnetic flux density, $B = 0.83$ T.
(2) The steel used for piston and cylinder should have high magnetic permeability (C% < 0.15).
(3) Two equal poles are considered with pole length; $l_p = 6 \times 10^{-3}$ m.
(4) Magnetic flux density is calculated for the steel core from the continuity of magnetic flux (ϕ), ϕ in the MR gap = ϕ in the steel = constant.
(5) The effective area of the pole is found to be $A'_f = 7.38 \times 10^{-4}$ m^2 and $B_{steel} = \phi/A_0 = B_f A'_f/A_0 = 1.5$ T.
(6) $H_{steel} = 1000$ A/m (on the basis of B versus H magnetic curves).
(7) Magnetic coil contains the ampere turns (Kirchoff's law): $NI = \Sigma H_i L_i = 300$ A-turns.
(8) Taking $I = 2$A yields $N = 150$.
(9) Calculated inner coil diameter for a coil: $d_c = 2.28 \times 10^{-2}$ m.
(10) Calculated coil length: $l_c = 2.4 \times 10^{-2}$ m.

Iterative calculation is used to design the hydraulic circuit and magnetic circuit.

3 Finite Element Analysis of MR Damper

Through hydraulic and magnetic design, the main design parameters for MR damper have been selected. Now the piston head geometry is analyzed in detail because the design parameters directly affect the behavior magnetic saturation. For better vibration control, it is always desired that all of the components are below their saturation fields [8]. The main design parameters have been analyzed through a series of numerical magnetic analysis using the finite element method "ANSYS" (Fig. 3).

4 Results and Discussions

4.1 *Effect of Core Diameter (DC) (d_c)*

The flux density should reach its maximum value across the fluid gap for optimized damper during its operation. The core diameter is a critical design parameter to affect the limiting magnetic flux density. In Fig. 4, the magnetic flux density is plotted in

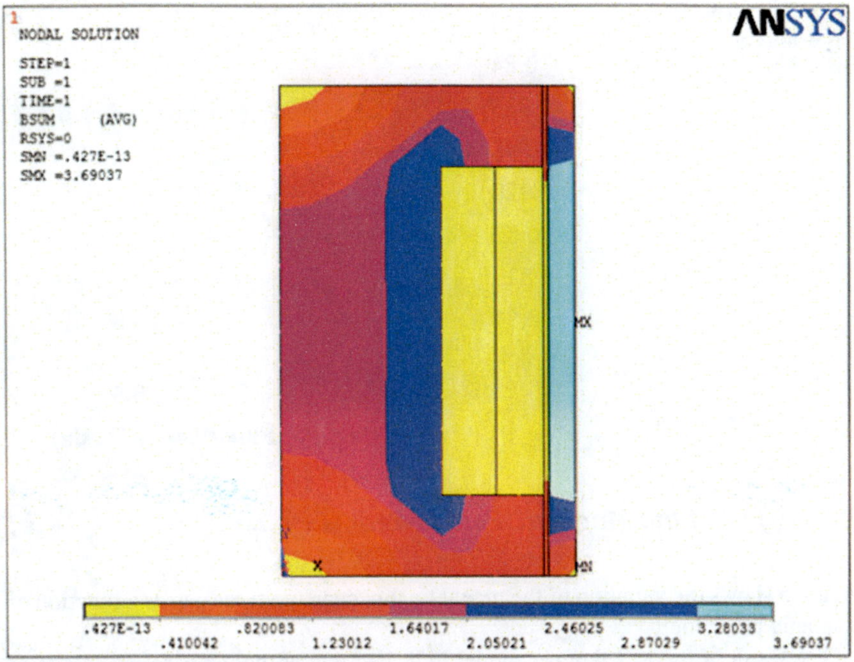

Fig. 3 Two-dimensional magnetic flux density (B_{SUM})

Fig. 4 Variation of magnetic flux density with core diameter

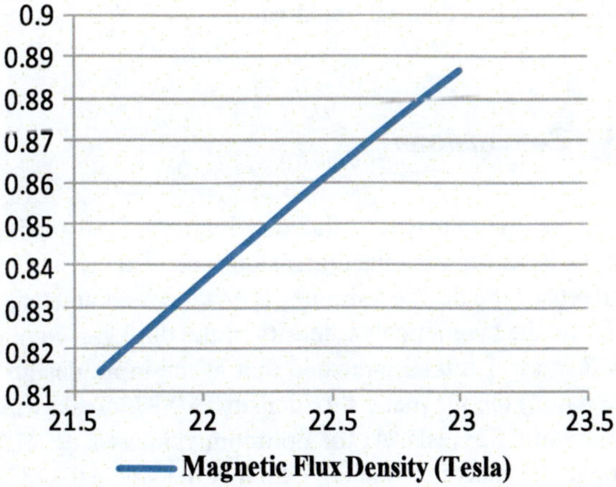

the gap versus the core diameter.

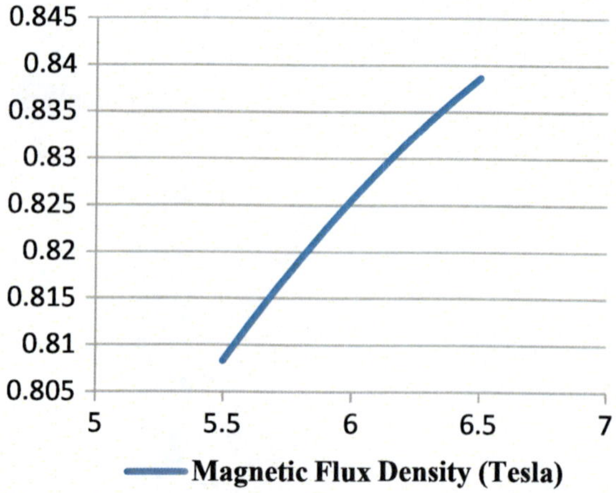

Fig. 5 Variation of magnetic flux density with magnetic pole length

4.2 *Effect of the Magnetic Pole Length (LP) (l_p)*

Figure 5 shows the variation of the magnetic flux density versus gap as a function of magnetic pole length.

The result of the MR damper model indicated that at the input design variables ($d_c = 22.8$ mm, $l_p = 6$ mm), the magnetic flux density is 0.877 Tesla which is more than the saturated value of 0.83 Tesla. At the optimum values ($d_c = 21.8$ mm, $l_p =$ 6.1 mm) through magnetic analysis, the magnetic flux density is 0.828 T, which is close and less than saturated value.

5 Conclusions

The maximum magnetic flux density in the fluid gap is generated below the saturated value by optimizing the dimensions and also verified by simulation. In this study, the effects of the design variables (DV's), core diameter (d_c) and magnetic pole length (l_p) on the magnetic flux density in the fluid gap were investigated. The result of the MR damper model indicated that at the input design variables ($d_c = 22.8$ mm, l_p $= 6$ mm) the magnetic flux density is 0.877 Tesla which is more than the saturated value of 0.83 Tesla. At the optimum values ($d_c = 21.8$ mm, $l_p = 6.1$ mm) through magnetic analysis, the magnetic flux density is 0.828 T, which is close and less than the saturated value. So this study will help to design the MR damper for better control of vibrations.

References

1. Sung, K.G., Choi, S.B., Park, M.K.Y.U.: Geometry optimization of magneto-rheological damper for vehicle suspension via finite element method. Adv. Sci. Lett. **4**, 805–809 (2011)
2. Sallon, M.Y., Samad, Z.: Finite element modeling and simulation of proposed design magneto-rheological valve. Int. J. Adv. Manuf. Technol. **54**, 421–429 (2011)
3. Samad, Z., Salloom, M.Y., Hawary, A.F.: Simulation and design optimization of magneto rheological control valve. Int. J. Mech. Mater. Eng. IJMME **6**(2), 231–239 (2011)
4. Cesmeci, S., Engin, T.: Modeling and testing of a field controllable magnetor heological fluid damper. Int. J. Mech. Sci. **52**, 1036–1046 (2010)
5. Kciuk, S., Turczyn, R., Kciuk, M.: Experimental and numerical studies of MR damper with prototype magnetorheological fluid. J. Achiv. Mater. Manuf. Eng. **39**(1), 52–59 (2010)
6. Nam, Y.J., Park, M.K.: Electromagnetic design of a magnetorheological damper. J. Intell. Mater. Syst. Struct. **20**, 181–191 (2009)
7. Grunwald, A., Olabi, A.-G.: Design of magneto-rheological (MR) valve. Sens. Act. A Phys. **148**(1), 211–223 (2008)
8. Case, D., Taheri, B., Richer, E.: Finite element modeling and analysis of magnetorheological dampers. In: Proceedings of the ASME, International Mechanical Engineering Congress and Exposition (2011)

Comparative Investigation of Process Capability of Surface Finish in Milling of EN19 Steel Using VMC

Rupesh Chalisgaonkar

1 Introduction

Nowadays, the process capability investigation has an important role to enhance manufactured product's quality in industries. The process capability represents and assures that the dimension or surface quality of part/product would fall between engineering specifications/customer specification.

Process capability index (C_{pk}) shows the process centering with respect to the specified specification limits of the product. C_{pkl} represents that the process capability index of manufacturing process is neater to lower limit while C_{pku} indicates the proximity of process nearer to upper limit. The minimum value of process capability index is selected considering precautionary safety [1]. Process capability investigation is implemented in all stages of product development from design to quality inspection to ensure part's specification lies between tolerance limit [2]. The two parts in nature could not be similar which yields the possible variation of product dimensions in manufacturing process. The assignable cause of variation could be controlled by process parametric setting during the process, while non-assignable causes are inherent and difficult to control. Motorola emphasized to produce defect-free product using six sigma limits during manufacturing process. If the specified limits of product dimension are greater than three sigma limits, then the number of rejected parts would be minimized. Apart from it, process center could be changed without resulting in any loss in rejection. Process capability index (C_{pk}) exhibits the centering of the process while tails of the curve represent the proximity to design specifications. If the process is not centered during manufacturing process then there will be more chance to produce defective parts [3]. In today's competitive market to satisfy customer demand (high material removal rate, surface finish and tolerance), the other factors such as cutting tool, machining conditions and machine tool are to

R. Chalisgaonkar (✉)
KIET Group of Institutions, Ghaziabad, U.P., India
e-mail: rupesh.chalisgaonker@kiet.edu

M. Muzammil et al. (eds.), *Recent Advances in Mechanical Engineering*, Lecture Notes in Mechanical Engineering, https://doi.org/10.1007/978-981-15-8704-7_87

Table 1 Cutting parameters with their level

Process parameters (unit)	Parameter designation	Level 1	Level 2	Level 3
Cutting speed (rpm)	A	420	710	1000
Feed rate (mm/rev)	B	0.02	0.0765	0.133
Depth of cut (mm)	C	0.1	0.2	0.3

be selected accordingly [4]. EN 19 alloy steel possesses higher range of ductility, impact resistance and wear resistance properties. The EN 19 alloy steel is has a wide applications for making components in automobiles, aircraft, oil and gas industries and transportation industries [5]. Several research works was reviewed [6–8]. It was evident from available literature review that a very little work has been reported containing comparative evaluation of process capability index of surface roughness in milling process of EN 19 steel. The proposed work would also focus to explore the best tool among coated and uncoated carbide tool considering process capability of surface finish. The research work carried out could be beneficial for manufacturing industries.

2 Experimentation (Material, Method and Measurement)

The rectangular workpiece of EN 19 steel was chosen for experimental study.

The chemical composition of EN 19 steel was C: 0.43%, Mn: 0.79%, Si: 0.23%, S: 0.017%, Cr: 0.93%, Ni: 0.40, Mo:0.21% and balance Fe.

CNC vertical machining center of Jyoti PX 40 make was utilized to perform the experiments. End mill cutters of two different types (each having 6 mm diameter) titanium nitride coated carbide and uncoated carbide were taken for machining. Machining slot was cut of dimension 30 mm × 6 mm on workpiece. The part programming was done using DELCAM software of the CNC VMC. Three cutting parameters mainly cutting speed (RPM), feed rate (mm/rev) and depth of cut (mm) were selected for study (Table 1). Taguchi's L9 orthogonal array was implemented for design of experiments. The experiments were repeated three times to reduce the experimental error. Surface roughness (R_a) of the milled workpiece was computed using surface roughness tester (Mitutoyo make).

3 Results and Discussion

3.1 Computation of Process Capability

Computational steps to calculate process capability of surface roughness are as following.

Mean ($\overline{X}$):

Process mean was computed for each experimental trial by using the following equation.

$$\text{Mean } (\overline{X}) = \sum_{i=1}^{N} y_i/N \tag{1}$$

where

y_i value of surface roughness for each experimental trial.

N Number of trials.

Standard Deviation (σ)
Standard deviation for each experiment trial was computed by using the following equation.

$$\text{Standard deviation } (\sigma) = \sqrt{\frac{\sum_{i=1}^{N}(xi - \overline{X})^2}{N}} \tag{2}$$

Process Capability Index (Cpi)
The Eq. 3 was used to calculate Cpi of all experimental trial.

$$\text{Cpi} = \left\{\frac{\overline{X} - \text{LSL}}{3\sigma}, \frac{\text{USL} - \overline{X}}{3\sigma}\right\} \tag{3}$$

The specifications limits set for setting limits are as follows.
Upper specification limit (USL) for SR (coated tool) = 2 μm.
Lower specification limit (LSL) for SR (coated tool) = 0 μm.
Upper specification limit (USL) for SR (uncoated tool) = 4 μm.
Lower specification limit (LSL) for (uncoated tool) SR = 1 μm.
The value of mean, standard deviation and process capability index of surface roughness for both tools were computed and reported in Table 2.

3.2 *Single Response Optimization*

In order to minimize the deviation of target of output parameters (SR_{coated} and $SR_{uncoated}$) from defined specification limit, the S/N ratio of process capability index (Cpi) was maximized. Following formulae were used for calculating S/N ratio.

Table 2 Computing of process capability index (Cpi)

Exp. No.	SR_{coated} $(\overline{X})$	SR_{coated} (σ)	SR_{coated} (Cpi)	$SR_{uncoated}$ $(\overline{X})$	$SR_{uncoated}$ (σ)	$SR_{uncoated}$ (Cpi)
1	0.700	0.0816	2.857	1.813	0.223	1.215
2	1.160	0.1947	1.438	2.090	0.448	0.811
3	1.515	0.0206	7.841	2.540	0.639	0.803
4	0.900	0.1083	2.769	3.100	0.588	1.077
5	1.375	0.1592	1.308	3.713	0.180	2.386
6	1.550	0.0656	2.287	3.190	0.394	1.532
7	1.160	0.0252	11.126	3.940	0.057	6.182
8	1.260	0.0346	7.120	4.330	0.070	3.201
9	1.375	0.0126	16.556	4.253	0.223	0.733

$$S/N_{Cpi} = -10\log_{10}\left[\frac{1}{n}\sum_{i=1}^{n}\frac{1}{y_{ij}^{2}}\right] \tag{4}$$

where

y_{ij} surface roughness in *i*th replicate of the *j*th trial.

n total number of trials.

Figures 1 and 2 illustrate the S/N ratio plots for SR_{coated} and $SR_{uncoated}$, respectively. The single response optimal parametric setting for surface roughness using coated tool was found at cutting speed (A) 1000 rpm, feed rate (B) 0.1333 mm/min and depth of cut (C) 0.3 mm, while using uncoated tool it was obtained cutting speed (A) 1000 rpm, feed rate (B) 0.02 mm/min and depth of cut (C) 0.3 mm.

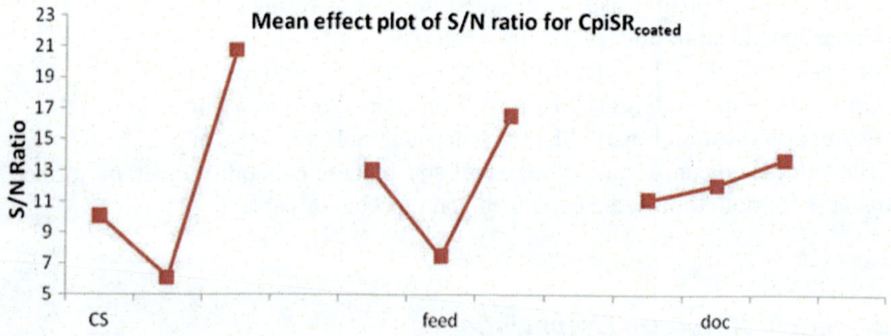

Fig. 1 Mean effect plot of S/N ratio for Cpi (SR_{coated})

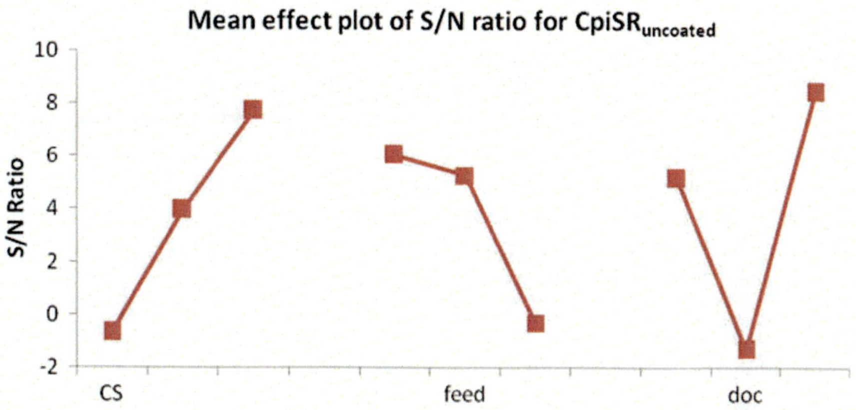

Fig. 2 Mean effect plot of S/N ratio for Cpi ($SR_{uncoated}$)

Table 3 Confirmatory results

Process parametric condition	Cpi SR_{coated} process parametric setting	Cpi $SR_{uncoated}$ process parametric setting
Initial level	**2.857** (A1, B1, C1)	**1.215** (A1, B1, C1)
Predicted value	**15.578** (A3, B3, C3)	**6.182** (A3, B1, C3)
Experimental value	**15.698** (A3, B3, C3)	**6.163** (A3, B1, C3)

4 Confirmation Experiments

The confirmatory experimental results are tabulated in Table 3. The improvement of the process capability index (Cpi) from the initial parametric setting was remarkably higher (Table 3). Hence, the machining performance was found to be improved significantly.

5 Conclusions

The process capability analysis regarding surface finish was investigated using titanium nitride coated carbide tool and uncoated carbide tool for machining of EN 19 steel in this study. The following points were concluded.

1. The optimized machining conditions for process capability index of surface roughness (Cpi SR_{coated}) using titanium nitride coated tool was found to be: cutting speed-1000 rpm, feed rate-0.1333 mm/min and depth of cut-0.3 mm. For process capability index of surface roughness (Cpi $SR_{uncoated}$) using uncoated

carbide tool, the optimal setting is different (cutting speed-1000 rpm, feed rate-0.02 mm/min and depth of cut-0.3 mm).

2. The optimized process capability index (Cpi SR_{coated} and Cpi $SR_{uncoated}$) was found to be more than one (15.698 and 6.163, respectively), which exhibits that the higher process capability in milling process of EN-19 steel could be achieved at evaluated optimized parametric condition for surface roughness.
3. The comparative evaluation of surface roughness (Table 2) shows that the coated tool should be preferred as it gives lower surface roughness in comparison with uncoated tool for milling of EN 19 steel.

References

1. Sharma, G.V.S.S., Srinivasa Rao, P.: Process capability improvement of an engine connecting rod machining process. J. Ind. Eng. Int. **9**, 37 (2013)
2. Rajvanshi, P.K., Belokar, R.M.: Improving the process capability of a boring operation by the application of statistical techniques. Int. J. Sci. Eng. Res. **3**, 5 (2012). ISSN 2229-5518
3. A technical note on process capability and statistical quality control, https://www.ateneonline.it/chase2e/studenti/tn/6184-7_tn07.pdf
4. Reddy, S.M., Reddy, A.C.: Studies on tool wear, cutting forces and chip morphology during high-speed milling of Al-Si-Mg-Fe Alloys. Int. J. Eng. Sci. Res. Technol. **2**(8), 1–9 (2013)
5. Chandan, B.R., Ramesha, C.M.: Evaluation of mechanical properties of medium carbon low alloy forged steels quenched in water, oil and polymer. J. Mater. Sci. Eng. **6**(2), 1–5. https://doi.org/10.4172/2169-0022.1000320
6. Uddin, M.S., Phamh, B., Ahmed, S., Basak, A., Pramanik, A.: Comparative study between wear of uncoated and TiAlN-coated carbide tools in milling of Ti6Al4V. Adv. Manuf. **5**(1), 83–91 (2017)
7. Sonawane, G., Sargade, V.G.: Comparative performance evaluation of uncoated and coated carbide inserts in dry end milling of stainless steel (SS 316L). Proc. Int. J. Comput. Appl. IJCA. 1–9
8. Sukumar, M.S., Ramaiah, P.V., Nagarjuna, A.: Optimization and prediction of parameters in face milling of Al-6061 using Taguchi and ANN approach. Procedia Eng. **97**, 365–371 (2014)

Ergonomic Evaluation of Foundry Operations Using Energy Expenditure Prediction Program Software

Sohail Akhtar and Imtiaz Ali Khan

1 Introduction

Ergonomic evaluation of a work place is very significant aspect which increases the productivity and quality of work. Energy expenditure prediction is one of the most widely used methods to evaluate the working environment such as foundry environment and lathe environment. Skeletal muscles may be assumed as biochemical machines in which the chemical energy is stored in the form of adenosine triphosphate (ATP), which ultimately goes into muscles and converted into mechanical work and heat energy, Umberger et al. [1]. In the case where the loading is static, and the work done by the muscles is zero, the muscle energy is mainly dissipated as heat. Mechanical power can be expressed as the product of joint actuator torque and velocity of joint. We can find the total mechanical power of the whole system as the summation of mechanical power of all joints. Ergonomist, applied physiologist, sports scientist, nutritionist and epidemiologist require the estimates of activity patterns and energy expenditures. Methods generally used for measurement of energy expenditure are accelerometer, motion sensors, combined heart rate and motion sensors, integrated electromyography, etc.

In activities such as repeated loading and unloading large number muscles perform submaximal, dynamic contractions as described by Astrando [2], when a person is doing this type of work, his endurance is mainly limited by the amount of oxygen, transported and being utilized by the system (maximum aerobic power). The process of determination of a person's maximum limit of aerobic power will not be discussed in this research, but other excellent observations will be presented in this paper. By relating the energy spent in a job to the aerobic power of the person for endurance efforts, an objective evaluation can be made for a worker for carrying a particular task without getting fatigue.

S. Akhtar (✉) · I. A. Khan
Department of Mechanical Engineering, AMU, Aligarh, U.P., India
e-mail: sohail3422@gmail.com

M. Muzammil et al. (eds.), *Recent Advances in Mechanical Engineering*, Lecture Notes in Mechanical Engineering, https://doi.org/10.1007/978-981-15-8704-7_88

For predicting thermal as well as mechanical energy released during simulated muscle contractions, a model of muscle energy expenditure was established EEPP [3]. The muscle energy model was assessed at varying levels of complication, which ranges from simulated contractions of isolated muscles, to whole body locomotion simulation. In all cases, the satisfactory agreement was found among the simulated and experimental energy liberation. Gunn [4] investigated four household tasks such as sweeping, window cleaning, vacuuming and mowing, conducted in the subjects' homes and a standardized laboratory environment. Energy expenditure was predicted via indirect methods. The findings propose that the given household tasks can contribute to the 30 min per day of moderate-intensity activity required to consult health benefits. The important metabolic equivalent (MET) changes between the home and laboratory highlight the effects of 'environment and terrain' and the 'mental approach to a task' on self-paced energy outflow. Levine [5] reviewed and assessed metabolic requirements, fuel consumption and the relative effect of different drink, food, drugs and emotional factors for evaluation of energy expenditure in individuals. He recommended where high accuracy is required and sufficient resources are available, an open-circuit indirect calorimeter can be used, whereas at a point, where facilities are constrained as well as we can sacrifice the ideal accuracy, flexible systems and non-calorimetric techniques are possibly helpful. For detailed information on free-living subject, factorial method is used. Przybyszewski [6] utilizes physical activity (PA), heart rate (HR) and anthropometry strategies in female Indian tea pluckers to calculate the energy expenditure. An energy expenditure (EE) prediction equation was created utilizing an expanded method that initially separates time during typical workday movement such as resting, plucking and walking, using accelerometer counting. Resting EE was anticipated from weight and age, while minute-by-minute non-resting EE was forecasted from HR and body mass index BMI. He set up that the energy expenditure can be accurately estimated with a bifurcated equation based on PA, HR, height, weight and age for a specific population taking part in a known set of activities. Limited data is available about the longitudinal changes in vitality necessities in late life Cooper [7]. The reasons for this investigation were to (i) assess the vitality necessities in late life and how they change during time span of 6 years, (ii) decide whether the variations in fat-free mass (FFM) were linked with changes in resting metabolic rate (RMR) and (iii) decide the accuracy off anticipated total energy expenditure (TEE) to know TEE. In this study, it was established that TEE, RMR and activity energy expenditure (AEE) diminished in men, yet in not women, from the 8th to 9th decade of life. Ergonomic evaluation of CNC lathe environment was performed by Mohsin [8] in 2018, ZHCET, AMU in which energy expenditure of different sub-tasks were evaluated using EEPP software and then it was compared with the standards.

2 Experimentation

2.1 Experimental Setup

In this study, all the foundry operations were performed in the foundry shop of Mechanical Engineering Department in Aligarh Muslim University. The time taken in each task was recorded using the stopwatch and the distances and heights involved were measured using the measuring tape. The weights were measured using the digital weighing machining. The task started with the pattern making. The weight of each part of flask is 8.6 kg. The height of table on which the pattern was prepared is 84 cm. Time taken in making the one pattern is 15 min. The height and weight of the subject are 178 cm and 80 kg. The distance walked by the subject holding the ladle and crucible is 10 m and the total weight is 7.8 kg. The crucible and ladle were lifted from the height of 0 to 89 cm. The time taken in pouring the molten metal into the mold is 3 min. Before the start of the experiment, the subject was given the sufficient time to be relaxed. Since this job consists od five different sub-tasks, every task was monitored and the time was recorded using the stopwatch.

2.2 Experimental Procedure

There are various operations in foundry given as follows.

A. **Making furnace ready**: For melting metal, furnace is used. The unadulterated metal, interior waste, outer waste and alloying components are utilized to make ready the furnace. In this process, the charge is melted, the melt is refined, adjusted the melt chemistry and tapped into a vessel. Harmful gases and elements are removed by refining from the molten metal to get rid of all types of casting defects. To bring the final chemistry into the specified limits, some elements are added during the melting process.

B. **Making equipment ready**: Next task is to make equipment such as flasks and rammer ready on the working table. Each past of flasks has the weight of 8.6 kg which is lifted from the ground to the table of the height 84 cm and other tools as well.

C. **Mold making**: In the casting process, a pattern is prepared in the shape which is required, as shown in Fig. 1. Some uncomplicated designs can be prepared in a solid pattern or single piece. The complex designs are used to be prepared in two parts, given the name split patterns. An upper section of the split pattern is known as a cope, and a bottom one is drag. To create the hollow areas in the mold, the cores are used.

D. **Lifting ladle and walking**: The subject lifted and carried the ladle and crucible to the working table as shown in Fig. 1(b) and 1(c).

Fig. 1 **a** Stoop and hand work; **b** semi-squat and lift; **c** carry and walk; **d** hold and arm work

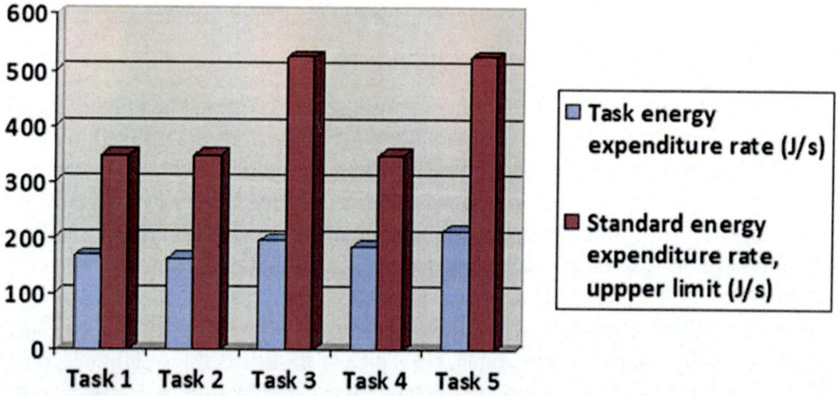

Fig. 2 Comparison of standard and measured values of energy expenditure rates

E. **Poring metal**: In a foundry, the metal in the molten state is poured into the mold as shown in the figure. Nowadays, numerous advance foundries are taking the help of robots or programmed poring machines to help the pouring of liquid metal. Molds are poured by simply hands with the help of ladle.

2.3 Results and Discussion

An outline of the research work studies on human performance using energy expenditure related to foundry operations shows that very little or no research work has been done on this topic. Whereas the foundry business is being enormously used all over the world. In today's time, the very large number of the workers are related to the foundry works. Tables 1 and 2 are showing the energy expenditure rates and task descriptions for various tasks performed during the foundry operations, respectively. Table 3 is showing the standards based on energy expenditure level, adapted from textbook and human factors in engineering [9]. Grades of physical work of the operations that are formed in the foundry shops can be derived from Table 3, such as task 1 falls under the category of light work, task 2 falls under the category of moderate work, task 3 is also moderate work and so on. Comparing the energy expenditure rates obtained in Table 1, using the energy expenditure prediction program to the standard values from Table 3, it can be observed that energy expenditure rates of all the tasks are in the range of standard values.

Table 1 Energy expenditure measurement by EEPP software

Task title	Cycle duration (min)	Task duration (h)	Total posture energy (kJ)	Total element energy (kJ)	Cycle energy (kJ/cycle)	Task energy rate (J/s or W)	Total task energy (kJ)
Task 1	15.00	0.25	138.82	12.55	151.38	168.05	151.37
Task 2	3.00	0.05	27.61	2.05	29.58	164.56	29.66
Task 3	20.00	1.00	182.42	50.20	232.63	193.84	697.89
Task 4	1.00	0.0167	9.12	1.96	11.08	184.78	11.08
Task 5	1.00	0.05	9.12	3.68	12.80	213.37	38.40

Table 2 Description of task

Tasks	Description
Task 1	Making furnace ready
Task 2	Making equipment ready
Task 3	Mold making
Task 4	Lifting ladle and walking
Task 5	Pouring the molten metal into mold

Table 3 Standards based on energy expenditure level [9]

Grade of work	Energy expenditure (J/s or W)
Rest	104.60
Very light work	111.56–174.32
Light work	174.32–348.65
Moderate work	358.65–522.97
Heavy work	522.97–697.30
Very heavy work	697.30–871.62
Unduly heavy work	>871.62

3 Conclusion

The World Health Organization (WHO) and the Occupational Safety and Health Administration (OSHA) consider work related energy expenditure to have multi-factorial causes. The managers and workers in the present scenario are significantly worried about the workplace, ergonomics, the nature of work and work related safety and health. This study presented an effective approach for evaluation of foundry operations using energy expenditure prediction program. For various tasks, a detailed report through software was generated for the analysis. On the basis of present study for discrete and mass production environment, the following concluding remarks are drawn.

The energy expenditure rate while making the furnace ready (Task 1) for the use determined by the EEPP software is 168.05 J/. Similarly, in the second task which is making equipment ready (Task 2) at the table, the energy expenditure rate is 164.56 J/s. The rate during mold making (Task 3) is determined as 193.84 J/s. During lifting the ladle and walking (Task 4) task, the energy rate is 184.78 J/. The value for pouring the molten metal (Task 5) is 213.37 J/s. All these values are in the range of permissible values. Although these values are within the standard limits but some ergonomic changes still can be suggested for the redesigning of the posture, process and system to make the values of the rate of energy expenditure further reduced. For example, from this study, it is observed that the maximum total task energy is for the task 3, viz., mold making, in which the subject was in the standing stoop posture. So in some work place, alterations can be introduced which gives the subject a sitting posture. Also, for the task 5, viz., pouring the metal, the energy expenditure rate is maximum, here, some modifications can be done in design of ladle so as to reduce its weight because it is significantly heavy in this particular case. These changes will further help in increasing the efficiency of the workers and reduce the risks of the health hazards.

References

1. Umberger, B.R., Gerritsen, K.G.M., Martin, P.E.: A model of human muscle energy expenditure, computer methods in biomechanics and biomedical. Engineering **6**(2), 99–111 (2003)
2. Astrando, P.O., Rodahi, K.: Textbook of Work Physiology. McGraw-Hills, New York (1970)
3. University of Michigan Center for Ergonomics Energy Expenditure Prediction Program Version 2.0. https://c4e.engin.umich.edu/tools-services/eepp-home/eepp-download/
4. Gunn, S.M., van der Ploeg, G.E., Withers, R.T., Gore, C.J., Owen, N., Bauman, A.E., Cormack, J.: Measurement and prediction of energy expenditure in males during household and garden tasks. Euro. J. Appl. Physiol. **91**(1), 61–70
5. Levine, J.A.: J. Measure. Energy Expenditure Public Health Nutrition **8**(7A), 1123–1132
6. Przybyszewski, E.M.: Predicting energy expenditure from physical activity, heart rate and anthropometry in female Indian tea pluckers. Ph.D. Thesis, College of Agriculture and Life Sciences, Division of Nutritional Sciences of Cornell University (2011)
7. Cooper, J.A., Manini, T.M., Paton, C.M., Harris, T., Schoeller, D.A.: Longitudinal change in energy expenditure and effects on energy requirements of the elderly. J. Nutr. **12**, PMC3679966. https://doi.org/10.1186/14752891-12-73
8. Mohsin, S., Khan, I.A.: Ergonomic Evaluation of a CNC lathe environment using EEPP. In: International Conference on HWWE, Mechanical Engineering Department, AMU, Aligarh (2017)
9. Sanders, M.S., McCormick, E.J.: Human Factors in Engineering and Design. McGraw-Hill, New York (1992)

Dynamic Analysis of Isotropic Annular Plate with Eccentric Hole Using Finite Element Method

Manoj Narwariya, Premanand Singh Chauhan, Avadesh Kumar Sharma, and Yogendra Singh Yadav

1 Introduction

Sometimes, in annular plate structure, eccentric holes are provided for functional requirements such as assembling of parts or units, for making passage required for cables and controlling mechanisms. Presence of holes as well as position of holes in these components commonly responsible for variation of natural frequency and also reduction in respective potential. The natural frequency of component with eccentric holes is predicted to vary significantly from a component with central one. Such types of plates are mostly subjected to dynamic force which create vibration. These unwanted vibrations can be minimized to make the careful design. Present analysis is done to fulfil these requirements. Various studies have been worked out to analyse the vibration characteristics of annular plates using various methods till date. Lee et al. [1] analysed free vibration of circular plates containing several holes of circular geometry. Zhou et al. [2] presented the 3D analysis of vibration behaviour for thick sector plates under different constraints with the help of Chebyshev–Ritz technique. Malekzadeh et al. [3] developed a 3D elasticity approach for analysing modal characteristics of laminated composite plate with annular and circular geometry putted on flexible support using FEM technique. Hosseini et al. [4] studied the modal analysis for annular plates coupled piezoelectrically of various thickness using Mindlin plate theory. Mirtalaie et al. [5] investigated the modal behaviour of thick FG annular sector plates using DQM technique. Saeedi et al. [6] analysed the free vibration of a circular plate with multiple perforations by using the Rayleigh–Ritz method. Thakare and Damale [7] did the theoretical and experimental study for

M. Narwariya (✉) · P. S. Chauhan · Y. S. Yadav
Mechanical Engineering Department, IPS College of Technology and Management, Gwalior 474001, India
e-mail: manoj.narwariya@gmail.com

A. K. Sharma
Mechanical Engineering Department, Rajkiya Engineering College, Mainpuri, India

M. Muzammil et al. (eds.), *Recent Advances in Mechanical Engineering*, Lecture Notes in Mechanical Engineering, https://doi.org/10.1007/978-981-15-8704-7_89

free vibration of a plate with multiple slots and holes. Boutahar et al. [8] examined the vibrational behaviour of thin isotropic axisymmetric annular plates with both edges clamped boundary condition to determine the behaviour of large vibration amplitudes. Powmya and Narasimhan [9] presented the solutions of free vibration analysis for axisymmetric laminated, polar orthotropic, annular and circular plates employing FSDT technique. Khare and Mittal [10] studied the numerical results for modal characteristics for thick laminated circular plates for different constraints, i.e., simply supported, free and clamped. Mercan et al. [11] examined the free vibration analysis of orthotropic composite annular plate using the FSDT technique for two different kernels. Senjanovic et al. [12] worked out for analysing vibrational behaviour of laminated plates having circular geometry with number of openings in the plate area. Rao et al. [13] analysed the vibrational characteristics of annular and circular plate under different boundary conditions. Behera and Kumari [14] developed a solution for analysing modal response of isotropic, laminated and sandwich rectangular plate using zig-zag and third-order theory.

In present work, dynamic response of isotropic annular plates with eccentric holes are studied using FEM.

2 Modelling

2.1 Geometrical Modelling

Geometry of the isotropic annular plate with eccentric holes is presented in Fig. 1.

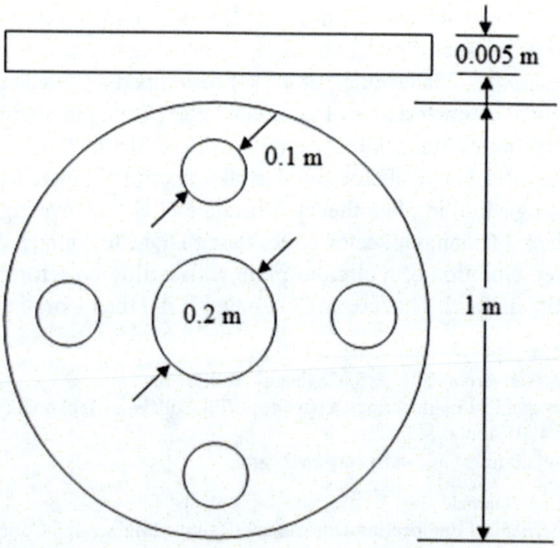

Fig. 1 Geometry of annular plate with eccentric hole

The dimensions of the plate are:

Outer diameter (D) = 1 m, inner diameter = 0.2 m, eccentric hole diameter = 0.1 m and thickness of plate = 0.005 m.

2.2 *Material Properties*

Material = steel, mass density per unit area of the plate, $\rho = 7800$ kg/m^3.

Young's modulus, $E = 210$ GPa, Poisson's ratio, $\mu = 0.3$.

2.3 *Loading*

For modal analysis, an isotropic annular plate with eccentric holes are chosen in this study. All edges of inner rings of centre as well as eccentric holes and outer ring of circular plate, have been fixed to all degrees of freedom, i.e., clamped–clamped boundary condition is used to analyse the plate. For harmonic analysis, the same plate which is subjected to a pressure 1 MPa is used at the whole area of the plate. Figure 2 shows an annular plate with clamped–clamped boundary condition.

3 Verification of Present Analysis

In order to accuracy the results, current investigation is primarily validated via performing the comparison study with available literature of Rao et al. [13] as listed in Table 1. Natural frequency is compared for isotropic annular plate for different diameter ratios taking thickness as unity and outer radius is chosen to be 0.1 m. From this comparison, it has been obtained that results observed close propinquity.

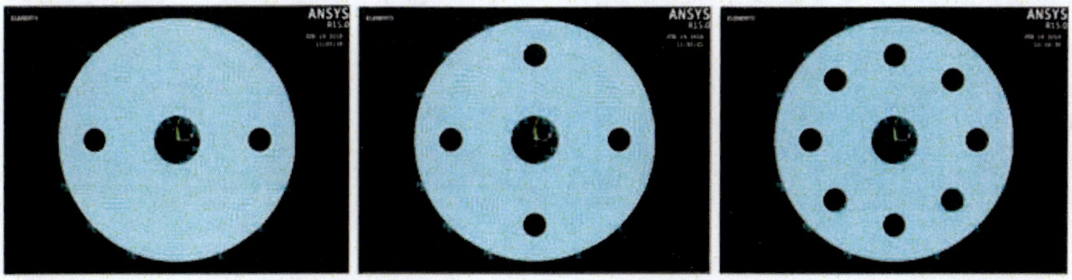

Fig. 2 Annular plate with eccentric holes with outer edge and inner edge clamped boundary

Table 1 Comparison for natural frequencies of isotropic annular plate for different diameter ratios with clamped–clamped boundary conditions

References	Inner diameter to outer diameter (*d*/*D*) ratios			
	0.01	0.25	0.5	0.8
Present	19.303	22.778	32.744	80.908
Rao et al. [13]	19.625	22.802	32.742	80.653
% error	1.64076	0.10525	−0.0061	−0.3161

4 Numerical Results and Discussion

4.1 Modal Analysis

Modal analysis is done to find the natural mode shapes and frequencies of any component during free vibration. In this study, the first six modes of frequencies have been obtained for isotropic annular plate with eccentric holes. The results have been found for increasing number of eccentric holes. The 0.02 element length is used to mesh the geometry. Figure 3 presents the first six mode shapes of frequency of annular plate with outer and inner edges clamped.

The response of increasing number of eccentric holes is shown in Table 2. From this table, it has been observed that the frequency of the plate increases while increasing number of eccentric holes.

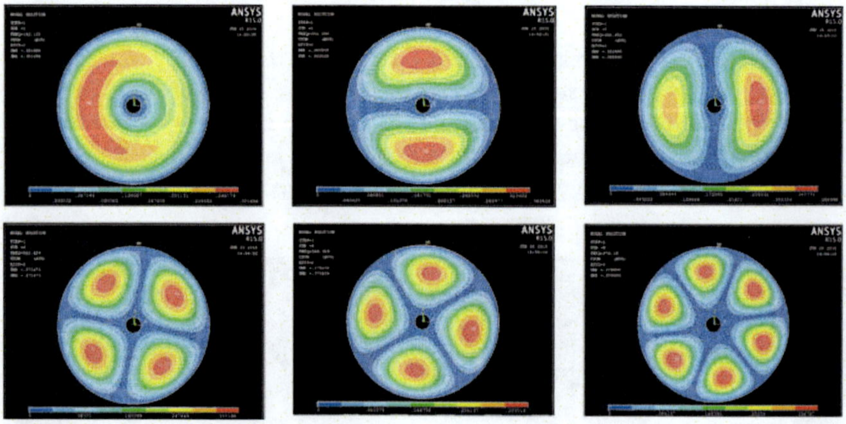

Fig. 3 First six modes of the annular plate

Table 2 Response of increasing number of eccentric holes on fundamental frequencies

No. of eccentric holes	Mode sequence number of fundamental frequency					
	1	2	3	4	5	6
No hole	102.25	264.2	264.53	513.58	513.67	566.79
One centre hole	342.12	354.01	356.45	542.43	546.06	879.15
One centre hole and two eccentric holes	405.7	421.14	484.53	632.19	661.9	961.51
One centre hole and four eccentric holes	448.88	478.05	482.83	613.65	673.51	948.58
One centre hole and Eight eccentric holes	478.89	546.15	669.61	857.2	865.49	1089.5

4.2 Harmonic Analysis

The resonant amplitudes have been determined at natural frequency to know the harmonic behaviour of the plate having number of eccentric holes under clamped boundary conditions taking the following parameters:

Inner diameter to outer diameter ratio of the plate $(d/D) = 0.2$ m, eccentric hole diameter $= 0.1$ m, thickness to outer diameter $= 0.005$ m and clamped–clamped boundary condition is used.

The effect of increasing number of eccentric holes on the resonance amplitude at its natural frequency is shown in Table 3.

From the study, it has been found that resonance amplitude decreases with increasing natural frequency as the number of holes is increased. Decrement of resonance amplitude occurs because of clamping the inner edges of centre as well as eccentric holes.

Frequency Response Function (FRF)

An FRF is used to present the harmonic behaviour of a component. It identifies the maximum amplitude of a component. Harmonic response of isotropic annular plate having eccentric holes for clamped–clamped boundary condition with increasing number of eccentric holes are demonstrated on FRF.

Table 3 Effect of increasing number of eccentric holes on resonance amplitude and frequency

No. of holes	Frequency (Hz)	Amplitude (mm)
No hole	102.25	9493.06
One centre hole	342.12	655.417
One centre hole and two eccentric holes	405.7	585.976
One centre hole and four eccentric holes	448.88	495.534
One centre hole and eight eccentric holes	478.89	483.51

Figure 4a–e shows frequency response function for isotropic circular plate, isotropic annular plate, isotropic annular plate with two eccentric holes, isotropic annular plate with four eccentric holes, isotropic annular plate with eight eccentric holes, respectively. Maximum amplitudes have been observed at natural frequencies for each type of plates which are listed in Table 3.

5 Conclusions

In this paper, FEM technique is used for dynamic analysis of isotropic annular plates with eccentric holes for clamped boundary conditions. The effect of increasing eccentric holes was fully investigated. By matching the obtained results with published literature, the present work is examined and a very good agreement is observed. From the study, it has been observed that as the number of eccentric holes increase, natural frequency also increases. A very high amplitude has been found for circular plate under clamped boundary condition on outer edge only. Resonance amplitude decreases as number of eccentric holes increase. Maximum resonance occurs at first mode of frequency.

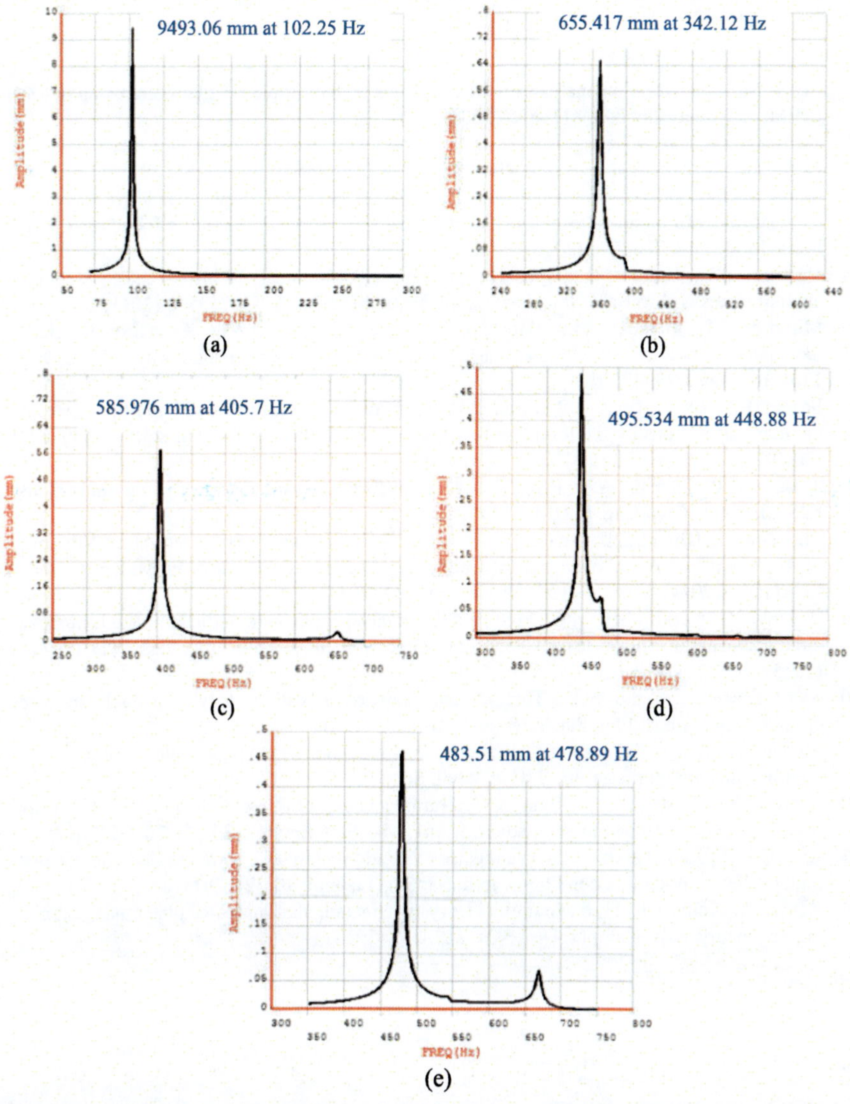

Fig. 4 Harmonic response on isotropic circular plate for **a** No hole, **b** one centre hole, **c** one centre hole and two eccentric holes, **d** one centre hole and four eccentric holes, **e** one centre hole and eight eccentric holes with thickness ratio (*t*/*D*) as 0.005, diameter ratio (*d*/*D*) as 0.2 and eccentric hole diameter as 0.1

References

1. Lee W.M.: Free vibration analysis of circular plates with multiple circular holes using indirect BIEMs. J. Sound Vib. **304**, 811–830 (2007)
2. Zhou, D., Lo, S.H., Cheung, Y.K.: 3-D vibration analysis of annular sector plates using the Chebyshev-Ritz method. J. Sound Vib. **320**, 421–437 (2009)
3. Malekzadeh, P., Afsari, A., Zahedinejad, P., Bahadori, R.: Three-dimensional layerwise-finite element free vibration analysis of thick laminated annular plates on elastic foundation. Appl. Math. Model. **34**, 776–790 (2010)
4. Hashemi, S.H., Kalbasi, H., Taher, H.R.D.: Free vibration analysis of piezoelectric coupled annular plates with variable thickness. Appl. Math. Modell. **35**, 3527–3540 (2011)
5. Mirtalaie, S.H., Hajabasi, M.A., Hejripour, F.: Free vibration analysis of functionally graded moderately thick annular sector plates using differential quadrature method. Appl. Mech. Mater. **110**(116), 2990–2998 (2012)
6. Saeedi, K., Leo, A., Bhat, R.B., Stiharu, I.: Vibration of circular plate with multiple eccentric circular perforations by the Rayleigh-Ritz method. J. Mech. Sci. Technol. **26**(5), 1439–1448 (2012)
7. Thakare, S.B., Damale, A.V.: Free vibration analysis of circular plates with holes and cutouts. IOSR J. Mech. Civil Eng. **8**(2), 46–54 (2013)
8. Lhoucine, B., Khalid, E.B., Rhali, B.: Geometrically Non-linear Axisymmetric Free Vibrations of Thin Isotropic Annular Plates. In: Cunha, A., Caetano, E., Ribeiro, P., Müller, G. (eds.) EURODYN 2014, pp. 1859–1866. Porto, Portugal (2014)
9. Powmya, A., Narasimhan, M.C.: Free vibration analysis of axisymmetric laminated composite circular and annular plates using Chebyshev collocation. Int. J. Adv. Struct. Eng. **7**(2), 129–141 (2015)
10. Sumit Khare, S., Mittal, N.D.: Three-dimensional free vibration analysis of thick laminated circular plates. Int. J. Eng. Sci. Technol. **8**(2), 11–29 (2016)
11. Mercan, K., Akgöz, B., Demir, C., Civalek, O.: Frequencies values of orthotropic composite circular and annular plates. Int. J. Eng. Appl. Sci. **9**(2), 55–65 (2017)
12. Senjanovic I., Hadzic N., Vladimir, N.: Vibration analysis of thin circular plates with multiple openings by the assumed mode method. J. Eng. Maritime Environ. **231**(1), 70–85 (2017)
13. Rao, G.V.R., Ramakanth, U., Suryaprakash: Vibration analysis of circular and annular plate under different boundary conditions. Indian J. Sci. Technol. **10**(32) (2017)
14. Behera, S., Kumari, P.: Free vibration of Levy-type rectangular laminated plates using efficient zig-zag theory. Adv. Comput. Des. **3**(3), 213–232 (2018)

Effect of Low Weight Fraction of Nano-reinforcement on Tensile Properties of Polymer Nanocomposites

Deepak Soni, Avadesh Kumar Sharma, Manoj Narwariya, and Premanand Singh Chauhan

1 Introduction

The reinforcing of plastic material by the organic and inorganic fillers is the common phenomena from last long period. Initially, the fillers are added to the polymeric materials to cut down the production cost and to lighten the final weight of fabricated objects. The addition of fillers will have some positive as well as some ill effects also, i.e., some fillers will have positive effect on some properties while have some adverse effect on other properties.

To overcome the problem of orthotropic characteristics of the filled plastic and to have somewhat homogeneous properties, the use of particles of regular and irregular shape are the area of research of many research groups from last many years. The filler particles of micrometer size range give very promising results in enhancing the mechanical properties of composites. The main disadvantage of reinforcing microsize particles is there higher volume fraction requirement for having marginal enhancement in mechanical properties [1]. Whereas in case of nanometer size range fillers, the same level of enhancement in the mechanical properties can be achieved by relatively low volume fraction of filler content [1].

Fu et al. [1] have a detail review of the effect of particle size, particle content and particle–matrix interfacial bonding on the tensile and other mechanical properties of the polymer matrix nanocomposites. The author says that as the size of filler material reduces to nanometer size range there is increase in yield strength of the composite. The increment in the strength and modulus of the composites also depend upon the stress transfer between matrix and reinforcement phase. The volume fraction

D. Soni · M. Narwariya (✉) · P. S. Chauhan
Department of Mechanical Engineering, IPS College of Technology and Management, Gwalior 474001, India
e-mail: manoj.narwariya@gmail.com

A. K. Sharma
Department of Mechanical Engineering, Rajkiya Engineering College, Mainpuri, India

M. Muzammil et al. (eds.), *Recent Advances in Mechanical Engineering*, Lecture Notes in Mechanical Engineering, https://doi.org/10.1007/978-981-15-8704-7_90

content of the filler particles will also have a very significant role in the improvement of tensile and other mechanical properties of the polymer nanocomposites.

Loos et al. [2] observed a great improvement on properties of polyurethane in addition of MWCNT in the polyurethane/MWCNT composite. The author showed that by reinforcing low amount of CNT (0.3% weight fraction), the tensile energy to break was increased by 38%.

The reinforcement of high stiffness ceramics material also produces promising results in developing various properties of the composites. Shukla et al. [3, 4] reinforced 200 nm thick platelets of alumina in highly cross-linked brittle epoxy thermoset processed through in situ polymerization. Though the modulus was improved by 110% with the reinforcement of as received platelets, but the failure and tensile strength were reduced.

Clay is also one of the most extensively used filler materials for reinforcing the polymer matrix. Clay dispersion varies from the microsize tractoids to the nanometer size range fully exfoliated dispersion. The clay is basically the layered silica having the dimension of the stacked platelets in the range 1–2 nm (thickness) and length in the range of few microns [5–10].

Mallick et al. [5] investigated effects of reinforcing the clay on polypropylene and polyamide-6 matrix. The composites are fabricated through melt blending technique and two-percent weight fraction (3 and 5%) reinforcement of the clay was investigated.

Khan et al. [10] prepared the hybrid CFRP reinforced nanoclay epoxy composite through hand lay-up technique. The nanoclay was dispersed in epoxy with high speed shear mixing followed by hot ultra-sonication. A mixture of intercalation to fully exfoliation dispersion was confirmed by TEM micrographs. The author reported on enhancement of more than 10% in both ultimate tensile strength and modulus at 5% weight content of clay in hybrid composites. The increased fiber–matrix interaction was main reported cause of increased tensile properties.

The reinforcement of some novel metals was also seen in the literature to form the polymer matrix composites. Evora et al. [11] have reinforced unsaturated polyester resin with spherical TiO_2 of 36 nm average size in varying volume fraction. The optimum enhancement in the tensile properties was reported with 1% volume fraction reinforcement of the TiO_2 in polymer matrix.

Lee et al. [12] did the excellent review on use of various nanocellulose as reinforced material in polymer composites and discussed the tensile properties of nanocellulose theoretically as well as experimentally. Bhattacharya [13] compared for various materials like graphene, clay and carbon nanotubes as nanofillers in polymer nanocomposites to demonstrate thermal, electrical and mechanical properties. Anil Kumar et al. [14] demonstrated the various combinations of two filler materials, graphene and silica in polymer nanocomposite by changing the weight percentage to analyze the mechanical properties. Ashraf et al. [15] investigated the role of stiffening efficiency, specific surface area and density of nanoparticles in polymer nanocomposites. Fu et al. [16] reviewed critically for basic aspects of polymer nanocomposites.

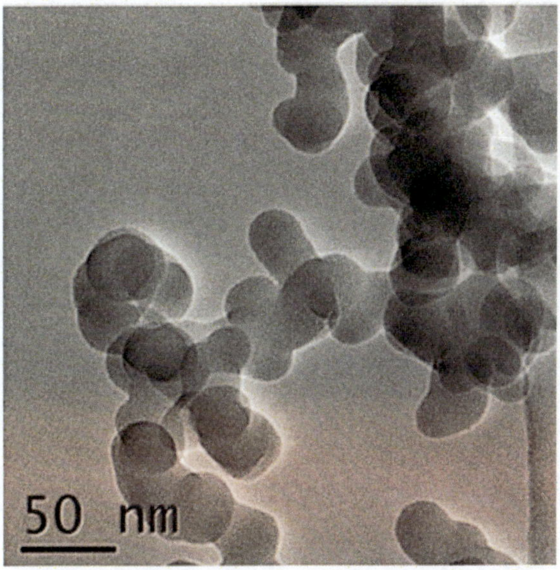

Fig. 1 Transmission electron microscope image of SiO_2

2 Material Used

2.1 *Particles (Reinforcement)*

Nanosilicon dioxide (SiO_2) particles were used in this work. Figure 1 shows the TEM image of the SiO_2 nanoparticles provided by the supplier.

Technical details for silicon dioxide particles:

Nanoparticles: SiO_2
Purity: 99.8%
Specific surface: 150 m^2/g
Density: 0.05 g/cm^3
Particle size: Average diameter in the range of 7–14 nm.

2.2 *Weight Fraction of Particles*

110 g matrix material is taken in this work and 0.553 g nanoparticles of the SiO_2 are added in the matrix material to be 0.5 wt% by the formula as shown below.

$$w_t\% = \frac{w_p}{w_p + w_m}$$

where

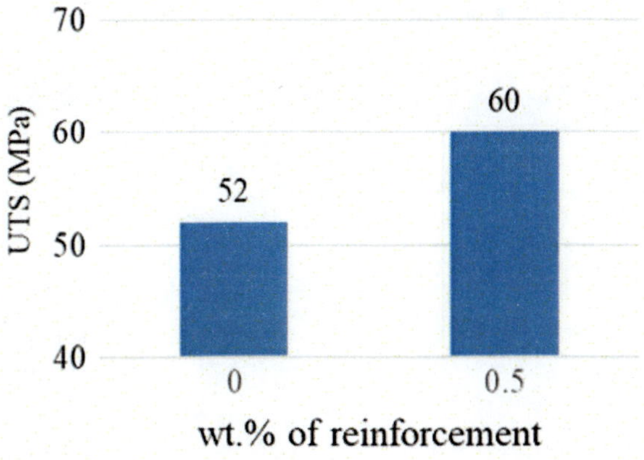

Fig. 3 Variation of the ultimate tensile strength of composites with the weight fraction of SiO_2 content

3.3 *Failure Strain at Fracture*

The main deciding parameter for classifying the material as brittle or ductile is the strain at failure, i.e., the failure strain. The variation in the failure strain in the composites with the variation of weight fraction of the SiO_2 nanoparticles was shown in Fig. 4. From Fig. 4, it can be clearly be observed that the failure strain increases with increase in the weight fraction of the reinforcement. The increase in failure strain at 0.5 wt% of SiO_2 was 33%. The increase in failure strain was due to increase interaction of the nanoparticles with the polymer matrix and supporting the motion of polymer chain during the stretching against the tensile load. The increase in the failure strain also represents the better filler–matrix interface interaction. Thus, the reinforcement of SiO_2 enhances the properties of the epoxy by causing the brittle to tough transition. As if the material has higher deformation before fracture, it has higher ductility and less prone to the fracture before showing any sign of plastic deformation.

The picture of the tensile test fracture specimen is shown in Fig. 5. From the

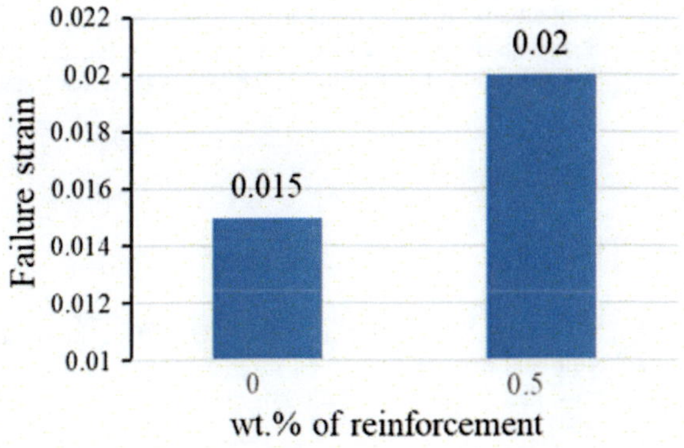

Fig. 4 Variation of failure strain at fracture with percent weight fraction content of nano-SiO_2 particles

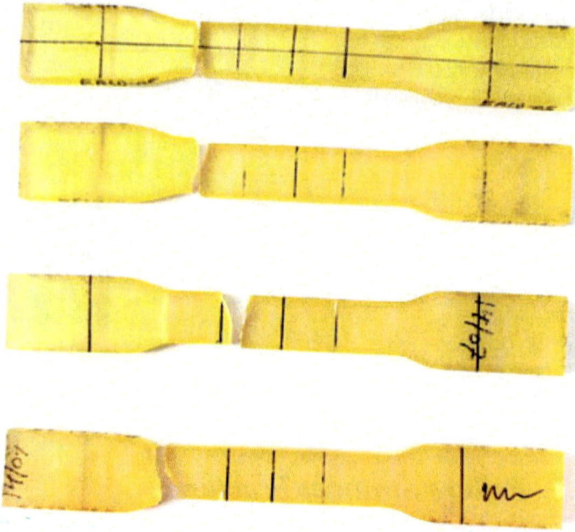

Fig. 5 Broken specimen of nano-SiO_2 particles

figure, it is clear that the fracture in all cases is brittle in nature as expected in case of thermoset brittle epoxy.

4 Conclusion

The tensile properties of the epoxy nanocomposites and neat epoxy were evaluated in accordance with ASTM standard D-638 by molding the dog-bone specimens directly by using vertical acrylic molds. The results show that the addition of SiO_2 nanoparticles contributes significant improvement in the elastic modulus, ultimate tensile strength and failure strain of the SiO_2 epoxy nanocomposites. The increment in elastic modulus, ultimate tensile strength and failure strain is 18%, 15% and 33%, respectively.

Acknowledgements Authors thank to Dr. Vijay Verma, Assistant Professor, BIET, Jhansi (U.P) for his suggestions during the entire work of research. We also thankful to the MNNIT, Allahabad, Prayagraj (U.P.) which allows us to test the results on various laboratory equipment.

References

1. Fu, S.Y., Feng, X.Q., Lauke, B., Mai, Y.W.: Effects of particle size, particle/matrix interface adhesion and particle loading on mechanical properties of particulate–polymer composites. Compos. Part B **39**, 933–961 (2008)

2. Loos, M.R., Yang, J., Feke, D.L., Zloczower, I.M., Unal, S., Younes, U.: Enhancement of fatigue life of polyurethane composites containing carbon nanotubes. Compos. B **44**, 740–744 (2013)
3. Shukla, D.K., Parameswaran, V.: Epoxy composites with 200 nm thick alumina platelets as reinforcements. J Mater Sci. **42**, 5964–5972 (2007)
4. Shukla, D.K., Kasisomayajula, S.V., Parameswaran, V.: Epoxy composites using functionalized alumina platelets as reinforcements. Compos. Sci. Technol. **68**, 3055–3063 (2008)
5. Mallick, P.K., Zhou, Y.: Yield and fatigue behavior of polypropylene and polyamide-6 nanocmposites. J. Mater. Sci. **38**, 3183–3190 (2013)
6. Qian, Z., Hu, G., Zhang, S., Yang, M.: Preparation and characterization of Montmorillonite—silica nanocomposites; a Sol–gel approach to modifying clay surface. Phys. B **403**, 3231–3238 (2008)
7. Fu, X., Qutubuddin, S.: Synthesis of Polystyrene—clay nanocomposites. Mater. Lett. **42**, 12–15 (2000)
8. Tien, Y.I., Wei, K.H.: Hydrogen bonding and mechanical properties in segmented montmorillanite/polyurethane nanocomposites of different hard segment ratio **42**, 3213–3221 (2001)
9. Chin, J., Albrecht, T.T., Kim, H.C., Russell, T.P., Wang, J.: On exfoliation of montmorillonite in epoxy. Polymer **42**, 5947–5952 (2001)
10. Khan, S.U., Munir, A., Hussain, R., Kim, J.K.: Fatigue damage behaviors of carbon fiber-reinforced epoxy composites containing nanoclay. Compos. Sci. Technol. **70**, 2077–2085 (2010)
11. Evora, M.F., Shukla, A.: Fabrication, characterization, and dynamic behavior of Polyester/TiO_2 nanocomposites. Mater. Sci. Eng. A **361**, 358–366 (2003)
12. Lee, K.Y., Aitomäki, Y., Berglund, L.A., Oksman, K., Bismarck, A.: On the use of nanocellulose as reinforcement in polymer matrix Composites. Compos. Sci. Technol. **105**, 15–27 (2014)
13. Bhattacharya, M.: Polymer Nanocomposites—a comparison between carbon nanotubes, graphene, and clay as nanofillers. Materials **9**, 262 (2016)
14. Anil Kumar, P.R., Venkategowda, T., Manjunatha, L.H.: Effect of graphene and silica fillers on mechanical properties of polymer nano composites. IJASRE **3**(9), 1–6 (2017)
15. Ashraf, M.A., Peng, W., Zare, Y., Rhee, K.Y.: Effects of size and aggregation/agglomeration of nanoparticles on the interfacial/interphase properties and tensile strength of polymer nanocomposites. Nanoscale Res. Lett. **13**, 214 (2018)
16. Fu, S., Sun, Z., Huang, P., Li, Y., Hu, N.: Some basic aspects of polymer nanocomposites: a critical review. Nano Mater. Sci. **1**, 2–30 (2019)